An Introduction to

INTERFACES & COLLOIDS

The Bridge to Nanoscience

An Introduction to

INTERFACES & COLLOIDS
The Bridge to Nanoscience

John C. Berg
University of Washington, USA

World Scientific

NEW JERSEY · LONDON · SINGAPORE · BEIJING · SHANGHAI · HONG KONG · TAIPEI · CHENNAI

Published by

World Scientific Publishing Co. Pte. Ltd.

5 Toh Tuck Link, Singapore 596224

USA office: 27 Warren Street, Suite 401-402, Hackensack, NJ 07601

UK office: 57 Shelton Street, Covent Garden, London WC2H 9HE

British Library Cataloguing-in-Publication Data
A catalogue record for this book is available from the British Library.

First published 2010
Reprinted 2012, 2014, 2015

AN INTRODUCTION TO INTERFACES AND COLLOIDS
The Bridge to Nanoscience

ISBN 978-981-4293-07-5
ISBN 978-981-4299-82-4 (pbk)

Printed in Singapore by B & Jo Enterprise Pte Ltd

To my students,
past and present,
with gratitude and affection

PREFACE

The importance of Interfacial and Colloid Science across the spectrum from industrial manufacturing to energy development to biomedical research to everyday activities from cooking to cleaning is beyond dispute. Still, it is to be found in relatively few courses, particularly required courses, in science and engineering curricula in our colleges and universities. More often, it is that chapter in one's physics or chemistry text that is never assigned. The early stirrings of a shift in curricula to include this material is underway, however, particularly as new tools and insights are rapidly emerging and it is recognized as the bridge to the new era of nanoscience and nanotechnology. This text is addressed to both undergraduate and graduate students in science and engineering programs as well as to practitioners, although even high school students should enjoy parts of it. Its evolving versions have been used, I believe successfully, in both undergraduate and graduate elective courses in Chemical Engineering at the University of Washington, as well as in a variety of industrial short courses since the mid 1980's. It is now used as the text for a course in Interfacial and Colloid Science, with a significant laboratory component, that has just become required for undergraduate Chemical Engineering students at Washington as the Department embraces a shift toward molecular engineering and nanoscience.

The text is an Introduction in that it assumes the reader to have no more than the background common to third-year university students in science and engineering and to have no prior experience with the subject. An over-arching goal in preparing the book has been to keep it User-Friendly, but in the end, it seeks to bring the reader to a level permitting comfortable entry into the current scientific literature.

Part of the joy of learning science, I believe, comes from its tangibility and its relationship to everyday experience. Therefore, at the end of each chapter, after the Introduction, are described "some fun things to do," *i.e.*, simple experiments to illustrate some of the concepts of the chapter. They require essentially no instrumentation or expensive, hard-to-get materials, and most are suitable for "Mr. Science" type classroom demonstrations.

Learning does not take place as it did even a decade ago. Now, as soon as students become familiar with the basic concepts and terminology of a topic, their next step is a visit to the Internet. A Google request, using the correct key words, for "images" or "videos" for example, instantly opens a world of information. It also, unfortunately, often produces a world of extraneous material, which only a degree of prior knowledge can sort out.

I must include here a pre-emptive apology for three things: first, the number of topics there wasn't space to cover; second, for the enormous amount of important work that has not been cited, and lastly, for the inevitable number of typographical and other errors the manuscript is sure to contain despite all attempts to minimize them.

The subject of Interfacial and Colloid Science is still one of awe and wonder to me, and it is hoped that this text will help at least in some way to convey this feeling to the reader.

John C. Berg
Seattle, WA
August, 2009

CONTENTS

Chapter 1

INTRODUCTION

The focus of this text is interfaces and colloids, and it makes the argument that this is the "bridge to nanoscience." It often appears that finite systems, even at equilibrium, cannot be described neatly in terms the usual set of intensive and extensive variables, *i.e.*, size matters. The most basic reason for this is that all tangible material systems have boundaries, or "interfaces," that have properties of their own. Physical, as opposed to mathematical, interfaces are nano-scale strata of material whose structure is markedly different from that of "bulk phase material," and as the objects of interest are made smaller and smaller, more and more of their total mass is part of "the interface," with major consequences for their properties. Also, as the "kinetic units" are made smaller, their thermal energy assignments begin to exert influence, and ultimately one must acknowledge the corpuscular nature of matter: molecules, atoms, sub-atomic particles, *etc*. How do these regions blend into one another? It is the "in-between" region that is of concern in this text, and it too finds itself divided into parts. In the first, continuum concepts are retained, but a range of "new" properties and phenomena emerge, such as the influence of droplet size on vapor pressure, spontaneous clumping together of particles, Brownian motion, etc. This is the traditional domain of "colloid science," and extrapolations from it into the nano region (down to as small as 1-10 nm) often provide valid descriptions of properties and behavior. Such extrapolations, dealt with in this text, constitute "generic nanoscience." But one may also encounter other behavior, if continuum descriptions start to break down. Most notable are the properties attributable to quantum confinement, and it is sometimes to these kinds of systems that the term "nanoscience" is restricted. Despite their more exotic attributes, however, the objects in question are still subject to most, if not all, of the rulebook of generic nanoscience, the bridge to which is knowledge of interfaces and colloids. The following brief overview seeks to set the stage.

A. Interfaces

"Interfaces" (or "surfaces") are the thin boundary regions separating macroscopic chunks of matter from their surroundings or from one another. "Interface" is the more general term for any phase boundary, while the word "surface" generally refers to the boundary between a condensed phase and a gas. Use of the term "interface" for all situations would be desirable because it reminds us that the properties of all boundaries depend on *both* phases.

The "surface" of a silica micro-particle in contact with air, for example, is different from its "surface" in contact with water, or in contact with an apolar solvent or *in vacuo*. While interfaces appear abrupt to the naked eye, they all have in fact a finite thickness, typically ranging from a few Ångströms to a few nanometers, as suggested in Fig. 1-1. The material in

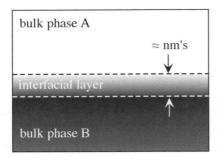

Fig. 1-1: Schematic of material interface, showing zone of inhomogeneity and adjacent bulk phases.

these thin regions is first of all inhomogeneous, and its properties differ profoundly from those of material in a bulk phase state. No system is free of the influence of its interface(s), and in many applications or situations, it is essential that one know not just the bulk properties, but the *interfacial properties* as well. For example:

- *Surface tension* determines the shape of small fluid masses or menisci.

- *Surfaces* of adjacent systems determine whether they stick to, slide over, or repel one another.

- It is often only the interaction of light with the *surfaces* of systems that we see.

- It is the dependence of *surface tension* on system composition that makes possible the formation of such unlikely liquid structures as foams or froths.

- It is often the accumulation of trace components at *surfaces*, i.e., "adsorption," that governs the chemistry of their interaction with adjacent phases.

Consider some of the most fundamental ways in which material in interfaces differs from bulk materials. First, the state of *internal mechanical stress* in the interfacial stratum is different from that in the bulk phases. In the case of fluid interfaces, this difference is manifest as a measurable contractile tendency, *viz.*, the surface or interfacial tension. A fluid interface acts as an elastic membrane seeking a configuration of minimum area. The origin of the boundary tension can be stated in terms of local internal mechanical stresses as follows. In the interior of unstrained bulk fluid at rest, the stress at any point is given by a single scalar quantity, *viz.*, the pressure p. But in the fluid interfacial layer under the same conditions, the state of stress requires a tensor for its specification; the components pertaining to stresses parallel to the layer are equal to one another but *less* than the component normal to the layer. It is this tangential pressure deficit in the interfacial layer that leads to boundary tension. The study of fluid interfaces is referred to as *capillarity*. Solid–fluid and solid-solid interfaces are also generally in a state of tension, although it is not readily measurable. These boundary strata have more complicated stress fields than fluid interfaces, and often are not in states of internal equilibrium. Thus they depend not just on the thermodynamic state but also on how the interface was formed, whether from the fracture of a bulk phase, from the stretching of a pre-existing surface, from precipitation out of solution, or from some other process.

The *chemical composition* of the interfacial stratum is also different from that of the bulk phases it separates. Consider, for example, the apparent interface between a piece of metal, such as iron or aluminum, and air, as shown in Fig. 1-2. Unless it is a noble metal, it will be covered with an oxide layer, which is at least partially hydroxylated from contact with water vapor in the air. The -OH groups have an acid or a base character depending on the metal. On top of this will probably be a layer of tightly bound water and possibly an additional layer of loosely bound water, depending on the relative humidity. The very top layer is likely to be an adsorbed scum of grease or other organic contaminant. The ability of interfaces of all kinds to contain components that may or may not be present in the bulk phases at the instant of interest makes their composition more complex than might be expected.

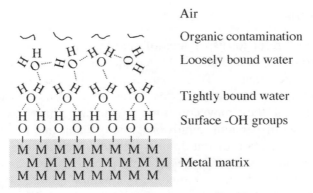

Fig. 1-2: Schematic representation of the typical chemistry at a nominal metal-air interface.

The interfacial stratum also often exhibits *electrical charge separation*. Interfaces dividing electrically neutral bulk phases may appear to bear a charge when viewed from one side or the other because positive and negative charges separate in the direction normal to the interface, leading to the formation of an "electrical double layer." There are a number of mechanisms by which the charge separation arises, and its existence has many consequences. Particles dispersed in water, for example as shown in

Fig. 1-3, may repel one another upon close approach, due to overlap of the like-charged outer portions of the double layers at their surfaces, keeping the particles from clumping together, and when placed in an external electric field, migrate by the process of electrophoresis. Fluid interfaces with electrical charge separation in an external electric field may become unstable to wavy disturbances, such that they are distorted or even torn apart in the processes of electro-spraying or electro-emulsification.

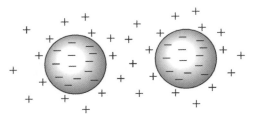

Fig. 1-3: Electrical charge separation at the interfaces of particles against their dispersion medium, water.

The unique mechanical, chemical and electrical properties of interfaces exert great and highly varied effects upon the behavior of material systems that cannot be described or explained in terms of bulk phase behavior alone. Interfacial effects reserve their greatest impact for systems with large area-to-volume ratios, such as thin films, fine fibers and small particles, or pushed to the limit: "nanofilms," "nanowires" or "nanorods" and "nanoparticles."

B. Colloids

"Colloids" refer to dispersions of small particles, usually with linear dimensions from 1 nanometer to 10 micrometers, thus spanning the "nano" to "micro" size range. The "particles" may be either dissolved macromolecules or macromolecular structures formed from smaller structural units, or they may constitute a separate phase, as in aerosols, powders, pigment dispersions, emulsions, micro-foams and finely pigmented plastics. The description of multiphase colloids, such as those just named, must take account of the properties of both phases as well as the interface between them, so that their investigation is a natural adjunct to the study of interfaces. Reaching down to the size of colloid particles, there are six over-arching aspects that distinguish their behavior from that of their larger counterparts:

- Mobility due to thermal kinetic energy
- Absence of inertial effects
- Negligibility of gravitational effects
- Inter-system molecular interactions: adhesion
- Size effects on thermodynamic properties, and
- Interaction with electromagnetic radiation

Simple kinetic theory assigns to all objects ("kinetic units"), whether they be molecules, bowling balls or planets, a fixed amount of translational kinetic energy, viz. $3/2\ kT$ on the average, where k is Boltzmann's constant, and T is the absolute temperature. This amounts at room temperature to about 10^{-20} Joules, completely negligible for objects of macroscopic size. For molecules in a gas phase, however, it is sufficient to cause them to fly about in straight lines at velocities of several hundred m/s, colliding with each other and their confining boundaries. In an ideal gas at room temperature and atmospheric pressure, the average distance between collisions, the "mean free path," ranges from 10-100 nm, depending on molecular size. In liquids and solids, the mean free path is of the order of one Ångström (0.1nm), and the motion of molecules resulting from their intrinsic kinetic energy resembles more closely vibration within a cage of nearest neighbors, with an occasional "escape." For colloid particles, usually consisting of hundreds up to 10^{10} molecules or so, kinetic effects are still important. They are sufficiently small that when they are dispersed in a gas or liquid medium, any unevenness in the bombardment they receive from the surrounding molecules causes them to move about in a process known as "Brownian motion." Thus in contrast to macroscopic objects, colloid particles do not stay put, and the smaller the particles, the more pronounced is their *Wanderlust*. The effect of kinetic energy on other objects of colloidal dimensions, such as ultra-thin deformable films (*e.g.*, bilayer lipid membranes) or long nanowires, is also evident in the wave-like undulations these objects exhibit. The effect of intrinsic kinetic energy on fluid interfaces is manifest as microscopic waves called "riplons."

The second consequence of smallness is the absence or near absence of inertial effects. The response of colloids to the presence of external fields (gravitational, electric, magnetic, *etc.*) that act on the particles to orient them, move them about and generally to concentrate them in some region of the system is effectively instantaneous. This derives from the law that Impulse = Momentum, *i.e.*, $Ft = mv$, where F is the average force acting on the object during the time interval t, m is the mass of the object, and v is velocity it attains. Since the mass of a colloid particle is so small, the time required for it to reach its steady state velocity upon application of a force is essentially nil. For example, a one-micrometer diameter sphere of density 2.0 g/cm^3 sedimenting in water reaches its terminal velocity in about 50 microseconds. The movements of small particles resulting from of an external field are termed *phoretic* processes, and may be exemplified by electrophoresis of particles with electrical double layers in an electric field, or magnetophoresis of magnetic particles in a magnetic field. The ability to "herd" colloid particles around through the control of external fields provides one of the important strategies for controlling colloidal systems. As the particles get down into the nano size range, the randomizing effects of Brownian motion may start to overtake the phoretic processes, and one must be able to deal with or perhaps even exploit the balance between the two.

The third difference between ultra-small objects and their larger counterparts is the relative unimportance of gravity compared with other forces acting on them. Gravitational forces scale with the third power of the object's linear dimensions, while many of the "colloid forces" scale with the second power of the linear dimensions, *i.e.*, they are "surface forces."

The fourth difference is the effect that intermolecular interactions have upon them. Just as molecules interact across a distance with one another through a variety of forces (dispersion, dipolar, electrostatic, *etc.*), objects consisting of many molecules interact with one another by virtue of the *collective* effect of these molecular interactions. Particles "feel" each other across distances up to tens of nm's or greater. Since this is much greater than the range of interaction between molecules, they are termed "long range" forces. The interaction between like particles in a given medium is always attractive, so that they are drawn together and stick, and a significant part of colloid science is the design and use of strategies to prevent such sticking. These rely on repulsive interactions, which are often the result of electrical charges or the presence of adsorbed macromolecules at the surfaces. The intervening fluid between the particle surfaces may be viewed as a film which, upon reaching a certain degree of thinness, may spontaneously thin further, re-thicken or break up into drops or bubbles. Such behavior is critical to the formation of ultra-thin coatings as well as to the drawing together (aggregation and coalescence) of colloid particles in fluid media.

The fifth distinguishing feature is that many intensive thermodynamic properties begin to change with particle size as size is reduced. For example, the vapor pressure of a tiny droplet of liquid depends on its diameter. The vapor pressure of a 1-μm droplet of water at room temperature is approximately 1% higher than the handbook value. For 100-nm droplets, it is 12% higher, while for 10-nm droplets it is higher by a factor of three. The vapor pressure enhancement is one manifestation of the effect of curvature on thermodynamic properties, known as the "Kelvin effect." Other consequences of the Kelvin effect include the condensation of vapors into small pores, crevices and capillaries at partial pressures below their vapor pressure, the enhanced solubility of small particles over larger ones, and the mechanism of phase change by nucleation. The Kelvin effect governs the process of cloud formation and ultimately their evolution into raindrops.

The sixth distinction concerns the interaction of electromagnetic radiation, in particular visible light, with small objects, an interaction dependent in part on the ratio of the particle size to the wavelength. For objects large relative to the wavelength, light is refracted or absorbed by the system, or reflected from its surface and diffracted at its edges, leading to a blurring of the edges by an amount approximating the wavelength. Thus ordinary optical microscopy produces images with a resolution no better than that wavelength (400–700 nm). Particles in the colloidal domain *scatter* visible light, giving a colloid a turbid appearance. Turbidity is a maximum when the scattering centers (particles) are roughly comparable in size to the wavelength of visible light, while for smaller particles, falls off as the sixth power of the particle diameter. For diameters less than approximately 50 nm, they may become effectively invisible. Most sunscreens use particles that absorb ultraviolet light but are in a size range that only negligibly scatters visible light and are thus clear on the skin. In another example, nanoparticles may be incorporated into optical coatings to improve mechanical properties (scratch resistance, *etc.*) without compromising transparency.

It is fun to think of questions regarding the behavior of every-day systems or processes that cannot be answered without knowledge of interfaces and colloids. A few examples are listed in Table 1-1. They are posed here without answers, but all are dealt with in the text. The first question concerns the fact that one may easily "float" a needle or other dense object, such as a paper clip, on water if it gently placed on the water surface.

Table 1-1: Examples of questions that can be answered only with knowledge of interfacial and colloid science.

1. How can a metal needle (7-8 times as dense as water) be made to "float" on water?

2. Why will a teaspoon of certain materials spread spontaneously over several acres of water surface, and then suppress both waves and evaporation?

3. Why do liquids stick to some surfaces but not to others? How does an adhesive work?

4. How do soaps and detergents help us to wash things?

5. How can water remain as a liquid at temperatures more than 20°C below its freezing point?

6. How can we make water into a froth or a foam?

7. How can we dissolve large amounts of oil in water using just a trace of a third component?

8. How can particles much denser than water be suspended in water almost indefinitely?

9. How can pumping gasoline through a hose lead to a spark, and possible disaster?

10. How does an absorbent paper towel soak up spills?

11. Why does a liquid jet break up into droplets?

12. How does the addition of salt to turbid water cause it to clarify?

It is evident that factors other than gravity are involved. The second question is illustrated in the frontispiece of the pioneering monograph by Davies and Rideal: **Interfacial Phenomena**, and is reproduced in Fig. 1-4. It shows the mirror-like surface of Loch Laggan, in Scotland, resulting from the application of a small amount of hexadecanol, which spreads out into a monolayer and damps all small wavelets. The monolayer also significantly suppresses evaporation. Question 3 addresses the everyday observation that liquids appear to stick to some surfaces, but not others. Teflon-coated cookware is designed to avoid the sticking of food; carpet fibers are surface treated with anti-soil coatings; Gore-Tex™ outer garments are designed to shed water. On the other hand, liquids may be formulated to be pressure-sensitive adhesives. Question 4 notes the everyday need for soaps and detergents to facilitate the cleaning of clothes, dishes and our hands, and asks: What is it that makes the dirt come loose and disappear? Question 5 reminds us that the freezing point of water, or any liquid, found in the Handbook, does not necessarily tell us the temperature at which freezing

Fig. 1-4: A monolayer of hexadecanol spread on Loch Laggan, Scotland. From: [Davies, J.T., and Rideal, E.K., **Interfacial Phenomena**, Academic Press, New York, 1961.]

will actually be observed. If one is careful, for example, water may be chilled to nearly -40°C without freezing, but if subjected to even a small disturbance, sudden phase change occurs. Question 6 asks why water can form itself into the delicate polyhedral structure known as a froth or foam, but apparently only if the water is *dirty*. Question 7 addresses the experiment in which water containing dissolved detergent is then capable of dissolving significant amounts of an otherwise-insoluble oil, such as gasoline. Question 8 is exemplified by an unsettled dispersion of gold particles [sp. gr. 13.7] in water prepared in Michael Faraday's lab in the mid 1800's, as shown in Fig. 1-5, and which can be viewed in the British Museum in London today. Question 9 recalls a problem associated with the pumping of volatile fuels, such as during the gassing of an automobile. If the hose is not grounded, or the gasoline does not have additives giving it a certain degree of electrical conductivity, a potential may develop along the hose great enough to produce a spark, which in the fuel-air mixture can lead to an explosion. Question 10 notes the development of absorbent products capable of imbibing many times their weight in water, and asks how this can be achieved. Question 11 addresses the fact that cylinders of liquid are unstable

Fig. 1-5: Gold sol prepared by Michael Faraday, on view in the Faraday Museum, London.

and will break up into droplets, which in jets occurs too rapidly to be seen by the naked eye. Figure 1-6 shows a flash photograph of the phenomenon. The clarification of a turbid dispersion of clay or silt in water by adding salt, as suggested in Question 12, is observed on a large scale in the formation of river deltas bordering on saltwater bodies. Looking at an atlas of maps will confirm that similar deltas do not form where rivers empty into freshwater bodies.

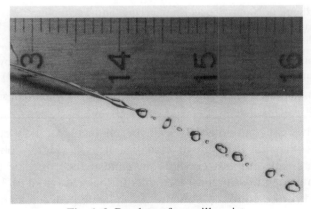

Fig. 1-6: Breakup of a capillary jet.

C. The bridge to nanoscience

1. *What is "nanoscience?"*

A search of amazon.com for books, using the keyword "nanoscience," reveals 2,061 (!) titles currently (Summer, 2009) in print. The first dozen are listed in Table 1-2. Even a cursory perusal of the literature reveals the

---

Table 1-2: The first 12 titles listed by amazon.com with a search request of "nanoscience."

- **Introduction to Nanoscience** by Gabor L. Hornyak, H.F. Tibbals, Joydeep Dutta, and Anil Rao (2008)
- **Nanophysics and Nanotechnology: An Introduction to Modern Concepts in Nanoscience** by Edward L. Wolf (2006)
- **Nanoscience: Nanotechnologies and Nanophysics** by Claire Dupas, Philippe Houdy, and Marcel Lahmani (2006)
- **An Introduction to Nanosciences and Nanotechnology** by Alain Nouailhat (2008)
- **Introduction to Nanoscience and Nanotechnology** by Gabor L. Hornyak, H.F. Tibbals, and Joydeep Dutta (2008)
- **Introduction to Nanoscale Science and Technology (Nanostructure Science and Technology)** by Massimiliano Di Ventra, Stephane Evoy, and James R. Heflin (2004)
- **Nanotechnology For Dummies** by Richard D. Booker and Mr. Earl Boysen (2005)
- **Nanotechnology: A Gentle Introduction to the Next Big Idea** by Mark A. Ratner and Daniel Ratner (2002)
- **Understanding Nanotechnology by Scientific American and editors at Scientific American** (2002)
- **Handbook of Nanoscience, Engineering, and Technology, Second Edition** by William A. Goddard III, Donald W. Brenner, Sergey Edward Lyshevski, and Gerald J. Iafrate (2007)
- **Nanotechnology: Basic Science and Emerging Technologies** by Mick Wilson, Kamali Kannangara, Geoff Smith, and Michelle Simmons (2002)
- **Nanosciences: The Invisible Revolution** by Christian Joachim and Laurence Plevert (2009)

---

amazing diversity and inter-disciplinary character of this emerging field, making it somewhat difficult to visualize it as a single, coherent body of knowledge. In its simplest conception, nanoscience may be defined as the study of objects with structural elements in the size range of 1–100 nm, *i.e.*, the low end of the colloidal domain. As indicated earlier, many of the size-dependent aspects of the behavior of entities in the colloidal size range may be successfully extrapolated into the nano range. The point at which continuum concepts start to break down is different for different materials and conditions, and different properties. The bulk-phase equations of state may become inapplicable, and the phenomenological equations of transport (Fourier's Law of heat conduction, Fick's Law of species transport, Newton's Law of viscosity, Ohm's Law, *etc*.) begin to fail. Thus what may be required is not just new size-dependent properties, but "new physics."

Even though systems requiring "new physics" are not the focus of this text, a few examples are recounted here. One of the best known of these is the observation that the color of nanoparticles of various semiconductors (called "quantum dots") is a function of their size. The electrons in a quantum dot are confined to distinguishably discrete energy levels (quantum confinement), dependent on dot size, so that when the dot is illuminated, it emits only light of a given wavelength, dependent on the particle size and composition. Figure 1-7 shows a set of dispersions of CdSe/ZnS core-shell nanocrystals varying in size from ≈ 1.7 nm to ≈ 5 nm from left to right.

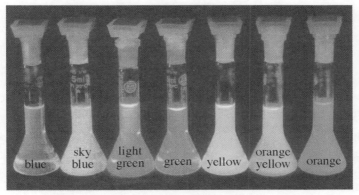

Fig. 1-7: Size-dependent change of the emission color of colloidal dispersions of of CdSe/ZnS core-shell nanocrystals, varying in size from left to right: ≈ 1.7 to ≈ 5 nm. For full color, see image at: [http://www.nanopicoftheday.org/2003Pics/QDRainbow.htm]

This property gives rise to many potential applications, including use as biological markers. Quantum dots of a known range of size and concentration might be incorporated into a single colloidal latex particle, which in turn may be attached to a particular type of tissue using an appropriate biological targeting agent. Upon illumination, such a particle will produce a unique spectrum (a spectral bar code) characteristic of the particular collection of quantum dots, thus identifying the existence and location of the tissue in question. As another example, the transport of electricity along a nanowire or a carbon nanotube (virtually the poster child for nanotechnology) is a function of diameter and may be governed by quantum effects. This means it may be possible to construct transistors, diodes, switches, gates, as well as conductors and other components of microcircuits from such objects, leading to ultra-small electronic computer logic systems. Recent discussion has moved to the possibility of using properly designed *single molecules* for this purpose. One of the problems of using these tiny elements for controlling charge flow is the fact that whenever they are bumped by neighboring nano objects (kinetic energy!), they are easily discharged.

Another example of the specific importance of size in the nano range concerns magnetic properties. A particle of magnetite (Fe_3O_4) in this range, for example, will retain its magnetism more effectively the larger it is, making it a better magnetic data storage medium than smaller particles, which are subject to loss of their magnetic information by being bumped by their neighbors. When it becomes too large, however, it will split into two magnetic domains of opposite polarity, and much reduced total magnetism. Thus there is an optimum single-domain magnetic nano-crystal size (called the super-paramagnetic limit, and often about 50 nm) for data storage, well known to the manufacturers of magnetic data storage media. Other examples of the special photonic, electronic and magnetic properties of nano-objects could be listed.

2. Nanostructures and assemblies

In the nano range, one often cannot refer to the objects of interest as "phases" or "particles," but must think of them as "structures." Sometimes nanoparticles are single molecules, or consist of a countable number of atoms or molecules, arranged in particular ways. It is useful to describe just a few of these "nano-structures" to give an idea of what is possible.

Buckyballs: As pictured in Fig. 1-8, Buckyballs are hollow spherical molecules of carbon atoms, more formally known as "Buckminster-fullerene," in view of their resemblance to structures created by the architect and inventor, Buckminster Fuller (1895-1983). The structure first discovered in 1985 upon examining the debris formed by vaporizing graphite with a laser was C_{60}. It consists of 5-member rings isolated by 6-member rings and is approximately one nm in diameter. Since then, many other fullerenes of different numbers of carbon atom and different structures have been discovered and characterized. Furthermore, fullerene-like structures of other

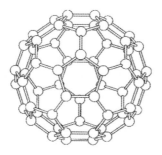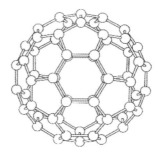

Fig. 1-8: Two views of the structure of a C_{60} "Buckyball"
From: [www.bfi.org/?q=node/351]

materials, notably tungsten and molybdenum sulfides, WS_2 and MoS_2, have recently been made and characterized. Atoms or small molecules can be put inside these structures or attached to their exteriors. These structures have some remarkable properties. For example, when certain metal atoms are trapped inside the cages, they exhibit superconductivity.

Carbon nanotubes: As shown in Fig. 1-9, carbon nanotubes are cylinders whose walls are made from monomolecular sheets of graphite, termed *graphene*, resembling chicken wire. Graphene layers themselves, including their exfoliation from graphite and their bottom-up synthesis, are objects of current interest. The tubes may be single-walled (SWNT's), about 1.5 nm in diameter, or multi-walled nanotubes (MWNT's) of somewhat larger diameter. Different types of SWNT's are formed when the graphene

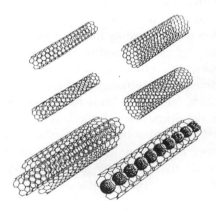

Fig. 1-9: Carbon-wall nanotubes, showing single-walled (SWNT) and multi-walled (MWNT) configurations an a tube filled with nanoparticles. From: [www.wtec.org/loyola/nano/04_03 .htm]

sheet is rolled in different ways, and the different structures have different electrical properties. One type is conductive, one is semi-metallic and another is a semiconductor, with quantum confinement of electrons. They may be filled with metal or other atoms and chemically functionalized externally. Carbon nanotubes are usually formed in a plasma arcing process, although other methods of synthesis are available, and still others are under investigation. They are incredibly strong for their size and weight, and may be the basis for super reinforced nanocomposites. It has even been speculated that they may be the basis for a cable supporting a "space elevator" on which one might ride to the stratosphere. Nanotubes can also be made from other inorganic materials producing structures different from those of carbon and holding the promise of a range of new applications.

Vesicles: As shown in Fig. 1-10, vesicles are spherical bilayer structures of lipids or other surfactants, and may be single-walled or multi-walled, with an aqueous interior. They form spontaneously in water upon gentle agitation when the right monomeric species (usually di-tail surfactants or lipids) are present. Vesicles may be as small as 30 nm in diameter, or as large as one micrometer or more. The smaller ones in particular are showing promise as drug delivery vehicles, with an appropriate drug in the interior

Fig. 1-10: Single-walled
vesicle formed from di-tail
lipid monomers.

and targeting agents attached to their outer surfaces, or nano-reactors of
various sorts.

Dendrimers: As shown in Fig. 1-11, dendrimers are highly ordered,
regularly branched single globular macromolecules, sometimes called
starburst polymers. They are formed by the successive addition of layers (or
"generations") of chemical branches, and have reached sizes up to 100 nm.
The sponginess of their structure has led to their use as impact modifiers in
composite materials. These remarkable molecules can also be designed to
allow the incorporation of desired guest molecules in their interiors (for
applications like drug delivery) or chemical functionality of their exteriors.
Their controllable chemistry and structure has led to their use as synthetic
catalysts themselves, and recently a dendrimer-based method has been
reported for controllably synthesizing 1-nm and subnanometer-sized metal
catalyst particles containing a well-defined number of atoms. It turns out that
3-nm diameter particles show higher catalytic activity for the oxygen-
reduction reaction (ORR) than either smaller or larger particles.

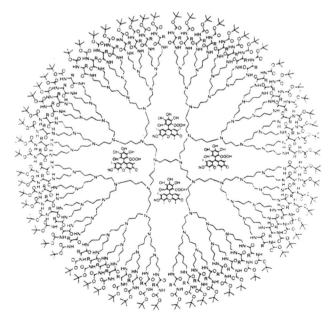

Fig. 1-11: Bengal rose encapsulation in a dendrimer center. From: [Zeng, F. and
Zimmerman, S. C., *Chem. Rev.*, **97**, 1681 (1997).]

Many more nanostructures are displayed in the delightful website: "Nanopicture of the Day:" www.nanopicoftheday.org, from which three more examples are drawn.

Nanoforests: Figure 1-12 shows a scanning electron micrograph (SEM) of a "forest" of uniform gold nanowires produced by plasma etching a polycarbonate film containing the nanowires. The loaded film had been

Fig. 1-12: A forest of gold nanowires produced by plasma etching a polycarbonate film containing the nanowires. From: [Yu, S., Li, N., Wharton, J., and Martin, C. R., Nano Letters, 3, 815 (2003).]

prepared by electrochemically filling holes that had been etched in the polycarbonate by electron beam lithography. When the nanowires-containing membranes are exposed to an O_2 plasma, the polymer at the membrane surface is selectively removed, exposing the ends of the gold nanowires. The length of the nanowires is controlled by varying the etch time. Other types of nanoforests can be constructed in other ways. Tree-like nanostructures ('nanotrees') can be formed through the self-assembled growth of semiconductor nanowires via a vapor–liquid–solid growth mode. This bottom-up method uses initial "seeding" by catalytic nanoparticles to form the trunk, followed by the sequential seeding of branching structures. This controlled seeding method has potential as a generic means to form complex branching structures, and may also offer opportunities for applications. One of the methods for synthesizing carbon nanotubes is by chemical vapor deposition using catalyst seeds that produce carpets of the nanotube fibers.

Nanohelices: Figure 1-13 shows a scanning electron micrograph of nanohelices formed from single-crystal zinc piezo-electric nano-belts produced by chemical vapor deposition. They are formed in response to the electrostatic energy introduced by the spontaneous polarization across the thickness of the nanobelt, owing to the presence of large polar surfaces. The shape of the nanohelix is determined by minimizing the energy contributed by electrostatic interaction and elasticity. The piezoelectric and possibly ferroelectric nanobelts have potential as surface selective catalysts, sensors, transducers, and micro-electromechanical devices.

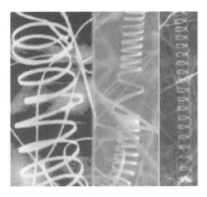

Fig. 1-13: Nanohelices.
From: [Kong, X. Y., and
Wang, Z.L., *Nano Letters*,
3, 1265 (2003).]

Nanoarrays: Figure 1-14 shows a regular array of platinum catalyst
nanoparticles, 20 nm in diameter, 15 nm high and spaced 100 nm apart, as
viewed by an atomic force microscope. It is found that catalytic activity
often depends critically on the particle size, aspect ratio and spacing, and
these features can now be controlled in the synthesis process.   In the case
shown, the structure was created using a template formed by a thin polymer
film coated onto a smooth silicon oxide surface subsequently etched into the
desired pattern by an electron beam.   A platinum film was then evaporated
onto the polymer, filling in the holes.   Finally, the polymer film was
removed, leaving the pattern shown.

Fig. 1-14: An array of
platinum catalyst
nanoparticles. From: [Gabor
Somorjai: Nano Picture of
the Day, Jan. 24, 2004]

        Structures such as those described above and others may in turn be
dispersed in a medium or assembled (or *self*-assembled) into mesoscopic or
macroscopic   superstructures   (monoliths)   with   unique   and   important
properties. Micro- or nano-composites may be constructed of units of these
dimensions to produce materials of almost any desired properties; for
example, materials with the heat resistance of ceramics and the ductility of
metals, or materials with the strength of steel, but the weight of a polymer,
etc. Sintered arrays may produce nano-filters or membranes, super thermal
insulation materials, or storage media for liquefied gases at modest
pressures. One exciting possibility is that of constructing nano-circuits from
quantum dots, nanowires and other elements, put into a proper array onto
some surface. Another possibility is the creation of a device for bringing
chemicals together in miniscule quantities so that they may react to produce
some recordable result. One may imagine an assembly of nano- or micro-
sized tubes, channels, reservoirs, *etc*., assembled on some surface for this

purpose. These are referred to as LOC ("*lab-on-a-chip*") configurations and may be the basis for chemical sensors more accurate than any known today and small enough to be injected into the blood stream. The motion of fluids through such a system is governed by "*micro-* or *nano-fluidics*," which in the ultimate may not be able to treat the fluids as continua. Still other configurations made of nano building blocks might be motors or mechanical actuators, converting chemical energy into mechanical work or electricity with nearly 100% efficiency. These structures, known as MEMS or NEMS ("*micro-* or *nano-electromechanical systems*") may someday be the molecular assembly devices for making structured nano-objects. One concept for the realization of such devices is that of harnessing the random kinetic energy of all kinetic units by devising ratchets that allow net movement to occur only in one direction, and another speculated that they could be the basis for nano-manufacturing ("*nanobots*"), perhaps even to replicate themselves.

As their potential has come into clearer view over the past decade, "nanoscience" and its application, "nanotechnology," have become among the hottest topics for academic inquiry, government research funding and fanciful speculation. George Whitesides (Professor of Chemistry, Harvard University), in a recent (March, 2009) plenary lecture before the American Chemical Society, described nanoscience and nanotechnology as now entering late adolescence, *i.e.*, past the exaggerated expectations of infancy (self-replicating "nanobots," space elevators, single-molecule computing, *etc.*) as well as the over-reaction to the inevitable disappointments, and moving into a more mature status in which the focus has returned to fundamentals, "discovery," and "science."

3. *Generic nanoscience*

In addition to the dazzling array of "special effects" associated with many of the objects of nanoscience, perhaps one of the most amazing observations is how many of them may be described by just properly taking into account the generic effects of smallness while extrapolating the world of interfaces and colloids into the nano domain. Despite any special properties they might have, all of these micro or nano entities must contend with the six universal aspects of smallness listed earlier, and it is to these that principal attention is given in this text. The practical issues concerning these entities are that they must be:

- *Produced*, either by somehow subdividing larger systems in so-called "top-down" procedures, or grown out of molecular or atomic media by so-called "bottom-up" procedures;
- *Characterized*, *i.e.*, one must be able to "see" or image the particles or structures and determine their mechanical, chemical, electrical, *etc.*, properties

- *Assembled*, often into two-dimensional or three-dimensional arrays by means of flow, external fields, or by various self-assembly processes;

- *Manipulated*, *i.e.*, placed where they are wanted, possibly one at a time, and usually on some surface, in accord with some blueprint; and

- *Protected*, against moving out of position, becoming attached to unwanted particles, vaporizing, *etc.*

4. *New tools of generic nanoscience*

The recent emergence of nanoscience and nanotechnology owes as much to the development and refinement of tools for the accomplishment of the above tasks as it does to the birth of quantum dots, carbon nanotubes, and the like. Some of these new tools are listed in Table 1-3. One of the most important developments is the family of scanning probe techniques, described briefly here but in more detail in Chap. 4.M. The first to emerge was Scanning Tunneling Microscopy (STM), pictured in Fig. 4-92. When a sharp conducting tip is poised a few nanometers above a conducting surface,

---

Table 1-3: Some new tools for the characterization and manipulation of nanosystems.

SPM (scanning probe microscopy
- STM (Scanning Tunneling Microscopy)
- AFM (Atomic Force Microscopy)
- NSOM (Nearfield Scanning Optical Microscopy)

Colloid force-distance measurement

- SFA (Surface Forces Apparatus)
- MASIF (Measurement and Analysis of Surface Interactions and Forces)
- TIRM (Total Internal Reflection Microscopy)

Particle manipulation

- Optical tweezers (optical trapping)
- Traveling-wave dielectrophoresis

---

and a slight bias is imposed across the gap, electrons may tunnel directly across. It occurs when the highest occupied molecular orbital (HOMO) of the material on one side of the gap overlaps with the lowest unoccupied orbital (LUMO) of the material on the other side of the gap. The tunneling current is exponentially sensitive to the distance between the tip and the nearest atom on the surface. As the tip is rastered across the surface, it is moved up and down by a piezoelectric controller to maintain the current constant. The up-and-down motion of the controller is tracked and recorded by a computer to produce a topographical map of the surface with atomic resolution. A related device, the Atomic Force Microscope (AFM), does not require that the tip and surface be conductive. In the simplest mode of operation, shown in Fig. 4-93, the sharp tip is held with constant force in direct contact with the surface by means of an ultra-thin cantilever beam. As the sample is rastered below the tip, the cantilever moves up and down to follow the topography. The motion of the cantilever is tracked by means of a laser reflected from its back to a photodiode system, revealing the topography of the scanned sample. The topographical resolution of STM or AFM is of the order of Ångströms, far better than that of the wavelength-limited optical microscopy. The techniques are thus ideally suited to the examination of nano features and nano structures. AFM may also be configured to map adhesion, stiffness, lubricity, and many other properties of the scanned surface, or may be operated in an intermittent contact mode to probe soft samples or in a non-contact mode (in which the tip vibrates above the surface) to probe the force fields emanating from the surface.

Another scanning probe method for overcoming the wavelength limitations of optical microscopy is that of Nearfield Scanning Optical Microscopy (NSOM), shown in Fig. 4-91, in which the object of interest is illuminated by the evanescent light which emerges a few tens of nm from the back side of a medium in which there is total internal reflection of light. Information pertaining to a spot a few nm^2 is captured by a sharp transparent probe tip a few nm above the spot and transmitted to a computer. Images are again built up by rastering the tip over the surface, or the sample beneath the probe.

A variety of methods have been developed for the *direct* measurement of forces of the type existing between proximate colloid particles, as described in Chap. 7 in detail. One of the most important of these devices is the Surface Forces Apparatus (SFA), pictured in Fig. 7-1, in which the force acting between a pair of approaching crossed cylinders (of radii of a few mm) attached to sensitive leaf springs. Positioning and force measurement are effected using piezo-electric positioners, and the distance of separation (with Ångström resolution) is determined interferometrically. An important breakthrough in the development of the SFA was the discovery that freshly-cleaved Muscovite mica is atomically smooth, and is thus used for the surfaces of the approaching cylinders. Other materials of interest may be coated on to these substrates. The MASIF device is similar to the SFA accept that a bimorph strip (two slices of piezo-electric material sandwiched together) is used to measure the force between the approaching surfaces. TIRM (Total Internal Reflection Microscopy) monitors the level of particles suspended against gravity by repulsive forces above a surface from which an evanescent beam is emerging.

A number of methods are being developed for the manipulation of micro- or nano-particles one at a time. The AFM probe may be used to push or pull nano structures (and even single atoms!) around on a surface. In a famous image now more than a decade old, IBM researchers in Zürich produced the company logo in this way by arranging an array of 35 individual xenon atoms on a Ni(110) surface into the letters IBM. Figure 1-15 shows AFM images of strands of DNA adsorbed onto a modified graphite surface. Manipulation was effected by bringing the AFM tip into contact with the surface and moving it using homemade hardware and software: (a) shows molecules as deposited, (b) shows two the molecules

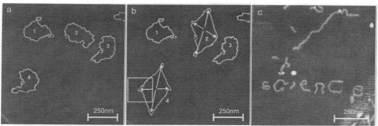

Fig. 1-15: Manipulation and overstretching of genes on a solid substrate. From: [Severin, N., Barner, J., Kalachev, A. A., and Rabe, J. P., *Nano Letters*, **4**, 577 (2004).]

stretched into diamond shapes, and (c) shows fragments drawn into the word "science." Nano objects can also be picked up using a pair of carbon nanotubes like chopsticks attached to a cantilever tip. The tweezer-like device is closed or opened with an applied electrical bias. In a simpler way, particles can be plucked out of an array, or put down in a desired position on a surface by charging and discharging a cantilever tip.

Another technique born out of the technology of atomic force microscopy is Dip-Pen Nanolithography (DPN), a scanning probe nano-patterning technique pictured in Fig. 1-16 in which an AFM tip is used to

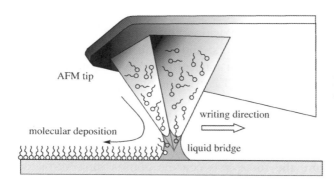

Fig. 1-16: Schematic of dip-pen nanolithography.

deliver molecules to a surface via a solvent meniscus, which naturally forms in the ambient atmosphere. This direct-write technique offers high-resolution patterning capabilities for a number of molecular and bio-molecular "inks" on a variety of substrate types such as metals, semiconductors, and monolayer functionalized surfaces. Figure 1-17 shows a paragraph from Richard Feynman's 1959 speech: "There's Plenty of Room at the Bottom," written using DPN on a 7-μm square surface. Feynman's speech[1] is regarded as the birth document of nanoscience, and the portion of is shown in Fig. 1-17 reads:

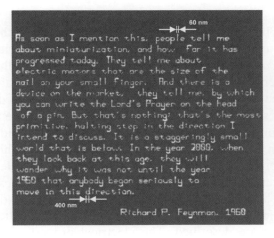

Fig. 1-17: DPN-written portion of Feynman's speech: "There's Plenty of Room at the Bottom." From: Mirkin Group, Northwestern University [http://www.nanotech-now.com/basics.htm]

"As soon as I mention this, people tell me about miniaturization, and how far it has progressed today. They tell me about electric motors that are the size of the nail on your small finger. And there is a device on the market, they tell me, by which you can write the Lord's Prayer on the head of a pin. But that's nothing: that's the most primitive, halting step in the direction I intend to discuss. It is a staggeringly small world that is below. In the year 2000, when they look back at this age, they will wonder why it was not until the year 1960 that anybody began seriously to move in this direction."

A powerful technique for positioning and holding a micro-particle at a desired position in three-dimensional space is that of "optical trapping," pictured in Fig. 6-44. A particle whose refractive index differs significantly from that of the suspending medium, and whose size is greater than the wavelength of the light, is held by photonic forces in the focused beam of a laser. The device is also referred as "laser tweezers," and may be used to assemble structures one particle at a time or several at a time into aggregates or arrays of desired structure using independently-controlled, time-shared optical traps or "holographic optical tweezers," in which a propagating laser wave front is modified by passing through a pattern of interference fringes. Laser tweezers may also be used as *in situ* force sensors as well as micromanipulators. They thus provide, in principle, an additional method for probing colloid force-distance relationships.

Another method for moving particles is that of "traveling wave dielectrophoresis," in which particles are acted upon by a traveling,

---

[1] Feynman, R. P., *Engineering and Science (Caltech)*, Vol. XXIII, No. 5, pp. 22-36 (1960).

sinusoidal electric field, which induces dipoles in the particles. When there is a time lag between the induced dipole and the field, a particle experiences a force, which induces motion.

5. *The plan*

The journey in this text begins with a study of fluid interfaces and the many consequences of the contractile tendency (surface or interfacial tension) they exhibit, *i.e.*, *capillarity*. Next is an explicit look at the thermodynamics of interfacial systems, particularly those that are multicomponent. Adsorption at both fluid and solid interfaces is examined, and the wondrous diversity of surfactants and their solutions is investigated. It moves on to the examination of the physical interaction between liquids and solids, and the interaction of fluid interfaces with solid surfaces. This includes wetting, coating, spreading, wicking and adhesion. It includes examination and characterization of the properties of solid surfaces: topography, surface energy, and surface chemistry. Next, the world of colloids, including nano-colloids, is explored. Their morphological, kinetic, phoretic and optical properties are described. The electrical properties of interfaces, and their many consequences, are examined next, and the following chapter deals with the interaction between colloid particles. Following a consideration of colloid system rheology, fluid phase colloids, *i.e.*, emulsions and foams, are examined, and the journey concludes with a consideration of interfacial hydrodynamics, symbolized by the image of "wine tears" on the cover.

Chapter 2

FLUID INTERFACES AND CAPILLARITY

A. Fluid interfaces: Young's Membrane Model

1. *The thinness of interfaces*

Fluid-fluid interfaces are a good place to start because they are far simpler to describe than fluid-solid or solid-solid interfaces. The molecular mobility in fluids makes it reasonable to assume that they will be in internal mechanical and diffusional equilibrium. Thus when the composition and the required number of thermodynamic state variables are set, the system is uniquely defined. In solids, non-equilibrium structures are frozen in place over time scales of practical interest. Fluid interfaces are smooth (as opposed to generally rough), morphologically and energetically homogeneous (as opposed to heterogeneous) and free of all internal shear stresses when at rest (as opposed to supporting un-relaxed internal stresses). Consider first the simplest case of all, *viz.*, the interface between a pure liquid (water) and its equilibrium vapor at 20°C. The pressure is then the vapor pressure of water at 20°C, *i.e.*, 2.33 kPa.

As noted earlier, the interface is not a mathematical discontinuity, but rather a thin stratum of material whose intensive properties vary across it from those of the liquid phase to those of the gas phase, as suggested in Fig. 1-1. In going from the liquid phase to the gas phase in the present case, the density decreases by a factor of approximately 58,000!

It is known that, except when one is very near to the critical point, the stratum of inhomogeneity at a liquid surface is very very thin, usually of the order of a few Ångströms. The abruptness is verifiable from experimental observations of the nature of light reflected from a surface. In accord with Fresnel's Laws of reflection,[1] if the transition between a gas and a medium of refractive index n (> 1) is abrupt (*i.e.*, thickness $\ll$ wavelength of light), the reflected light will be completely plane polarized when the angle of incidence is equal to $\tan^{-1}n$ (called the *polarizing angle* or Brewster's angle).[2] (An important technique for studying the structure of interfaces

---

[1] Jenkins, F. A., and White, H. E., **Fundamentals of Optics**, 3rd Ed., pp. 509ff, McGraw-Hill Co., NY, 1957.

[2] Hennon, S., and Meunier, J., *Rev. Sci. Inst.*, **62**, 936 (1991);
Hennon, S., and Meunier, *Thin Solid Films*, **234**(1-2), 471 (1993);
Hönig, D., and Möbius, D., *J. Phys. Chem.*, **95**, 4590 (1991);
Hönig, D., and Möbius, D., *Thin Solid Films*, **210, 211**, 64 (1993).

examines them when illuminated by laser light at precisely Brewster's angle ("Brewster angle microscopy" or BAM). It is the polarization of reflected light that makes it possible for sunglasses or Polaroid filters, polarized vertically, to block glare from horizontal surfaces. If the density transition through the interface is more gradual, the reflected light is *elliptically polarized*. Light reflected from most smooth solid surfaces and unclean liquid surfaces show at least some "ellipticity." Lord Rayleigh showed,[3] however, that when liquid surfaces are swept clean (by a technique to be described later), light reflected from them at Brewster's angle indeed shows virtually no ellipticity whatsoever. One monolayer of "foreign" molecules at the surface can measurably change this, and the technique of "ellipsometry"[4] (in which the extent of ellipticity is measured) is used to study the thickness and optical properties of material at surfaces.

The abruptness of a clean interface is also supported by statistical mechanical calculations, which have provided quasi-theoretical pictures of the density profile across the interfacial layer in simple systems. An example is shown in Fig. 2-1, showing density profiles for noble gases computed for various reduced temperatures. It reveals that gas-liquid interfacial layers for

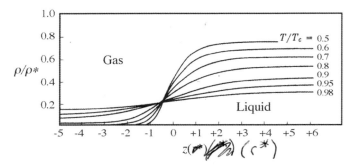

Fig. 2-1: Density profiles across the interfacial layer computed for noble gases at various reduced temperatures, T/T_c, where T_c = critical temperature. r^* is the molecular radius, and ρ^* is the maximum packing density. The point of $z = 0$ is arbitrary. After [Hill, T. L., **Introduction to Statistical Thermodynamics**, Addison-Wesley Publ., Reading, MA, 1960, p. 318.]

such systems (and by inference, for other gas-liquid systems) have thicknesses of the order of molecular dimensions, except very near to their critical points (where $T \rightarrow T_c$, and the distinction between the phases vanishes), where the interfacial layer becomes thicker and eventually envelops the entire system. Partially miscible liquid-liquid systems may also exhibit critical points. A critical solution point occurs at the temperature just below (or just above) which two liquid phases coexist. Liquid-liquid systems may have an upper critical solution temperature (UCST), a lower critical

[3] Lord Rayleigh (J. W. Strutt), *Phil. Mag.*, **33**, 1 (1892).
[4] Tompkins, Harlan G., **A User's Guide to Ellipsometry**, Academic Press, Boston, 1993.

solution temperature (LCST) or both. The thinness of either the gas-liquid or liquid-liquid interfacial layer (at least that associated with clean interfaces removed from their critical points) allows it to be treated, for purposes of macroscopic mechanical modeling, as a *membrane of zero thickness.*

2. Definition of surface tension

Everyday experience reveals that a fluid interface wants to contract in order to assume a minimum area, subject to whatever external forces or constraints are put upon it. For example, a mass of liquid undistorted by gravity, such as an oil drop of density equal to water suspended in water, or an air-filled soap bubble in air, assumes the shape of a sphere to produce the minimum area/volume. The contractile tendency of fluid interfaces can be quantified with reference to the zero-thickness-membrane model in terms of a "surface tension" or "interfacial tension" (σ) of the membrane, defined with reference to Fig. 2-2. Consider a point P on a small patch of surface (membrane). We can consider the state of tension at point P by imagining the patch to be divided into two parts by line MM′ passing through P and regarding each as a "free body." One such body exerts a pull on the other

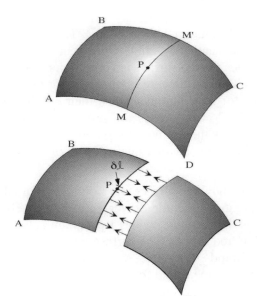

Fig. 2-2: Mechanical definition of surface or interfacial tension.

across the line MM′. The |force|/length along this line at P is the surface tension at P, *i.e.*, $dF = \sigma\, d\ell$ or $\sigma = dF/d\ell$, a scalar quantity with units of force/length. "σ" is the notation used for surface tension in this text, but it is also common to see the symbol "γ" used for it. For all fluid systems this force/length at P has the same value (at P) regardless of how MM′ is drawn, *i.e.* the system is *isotropic* with respect to its surface tension. Also, for uniform composition, isothermal surfaces, the surface tension is uniform. The macroscopic mechanical model of a fluid interface is thus a zero-thickness membrane in uniform, isotropic tension, σ. This is "Young's membrane model" (after Thomas Young, who first described fluid interfaces in this way in 1805).[5] The units of surface tension, force/length, are the same as those of energy (or work) per unit area, so that surface tension can also be interpreted as the mechanical energy required to create unit new area of a *liquid* surface. (The surfaces of solids require additional considerations.)

Consideration of the *thermodynamics* of capillary systems (examined later in Chap. 3) leads to another definition of surface tension, *viz.*, $\sigma = (\partial F/\partial A)_{T,V,eq}$, where F is the Helmholtz free energy of the system, A is the surface area, and the subscript "eq" refers to full internal equilibrium. If the *mechanical* definition of σ restricts itself to conditions of constant T,V and internal equilibrium, the two definitions are equivalent. It is useful for present purposes to think in terms of the mechanical model of the fluid interface and the mechanical definition of the surface or interfacial tension.

B. The surface tension of liquids

1. *Pure liquids*

Surface or interfacial tension values are usually expressed in either cgs units (dynes/cm or erg/cm^2) or SI units (mN/m or mJ/m^2). The numerical values are the same in either system, and they range from near zero to as high as nearly 2000. Literature values for the surface tension of pure liquids[6] are plentiful and usually reliable, although they are often given for only one temperature. Some specific values for pure liquids against their equilibrium vapor are shown in Table 2-1. The lowest values are those for liquefied gases. Most organic liquids (at or below their atmospheric boiling points) are in the range of 20-40 mN/m, while water has a value at 20°C of about 73 mN/m.

Essentially the only liquids having surface tensions substantially below 20 mN/m at room temperature are the lower molecular weight silicone oils and the fluorocarbons. Highest are values for molten salts and metals, being generally several hundred mN/m. The surface tensions of pure liquids are assumed to apply to liquids in contact with their equilibrium vapor when in fact, they are more often measured for the liquid against air at atmospheric pressure. The difference in surface tension between the two cases is generally negligible,[7] however, and even though it is not strictly correct to do so, we assume surface tensions of pure liquids against air to be functions of temperature only.

---

[5] Young, T., *Phil. Trans. Roy. Soc. (London)*, **95**, 55 (1805).
[6] a large database is given by: J. J. Jasper, *J. Phys. Chem. Ref. Data*, **1**, 841-1008 (1972).
[7] Defay, R., Prigogine, I., Bellemans, A., and Everett, D. H., **Surface Tension and Adsorption**, pp. 88-89, Longmans, London, 1966.

Table 2-1: Surface tension values for various liquids		
Liquid	$T\,(°C)$	Surface Tension (mN/m)
Helium	-272	0.16
Hydrogen	-254	2.4
Perfluoropentane	20	9.9
Oxygen	-183	13.2
Silicone (HMDS)	25	15.9
n-Heptane	20	20.3
Ethanol	20	22.0
Benzene	20	28.9
Olive oil	18	33.1
Ammonia	-33	34.1
Nitric acid	21	41.1
Glycerol	20	63.4
Methylene iodide	20	67.0
Water	20	72.7
Sodium chloride	801	114.
Lithium	181	394.
Zinc	360	877.
Iron	1530	1700.

2. Temperature dependence of surface tension

The surface tension of all pure liquids decreases with temperature and goes to zero as their respective critical points are approached. Over modest ranges of temperature, the decrease is nearly linear for most liquids, as suggested by the data of Fig. 2-3, and the coefficient, $d\sigma/dT$, is approximately -0.1 mN/m-°K for most cases. This rule of thumb may be used for rough extrapolation of surface tension values in the absence of any further data. Jasper's extensive data collection provides linear expressions of the form: $\sigma = a - bT$ for many liquids. Many semi-empirical relationships have been proposed for the dependence of surface tension on temperature, one of the oldest of which is the Eötvös Law,[8]

$$\sigma v^{2/3} = k_E(T_c - T),\tag{2.1}$$

where v is the molar volume of the liquid at the temperature of interest, T; T_c is the critical temperature, and k_E is the "Eötvös constant," equal approximately to 2.5 erg/°K (0.25 mJ/°K) for apolar, non-associating liquids (although there are many exceptions). The term $\sigma v^{2/3}$ corresponds to a molar surface free energy of the liquid, seen to decrease linearly with the approach to the critical temperature. A second semi-empirical law, based on the

---

[8] Eötvös, R., *Wied. Ann.*, **27**, 456 (1886).

principle of corresponding states, and valid for apolar, non-associating liquids, is that due to van der Waals (1894)[9] and Guggenheim (1945)[10] *viz.*

$$\sigma = \sigma^*\left(1 - \frac{T}{T_c}\right)^{11/9}.$$ (2.2)

σ^* is a "characteristic surface tension," given initially in terms of the critical properties of the liquid as $\sigma^* = 4.4(T_c/v_c^{2/3})$ [=] mN/m, with T_c [=] °K and v_c the critical molar volume [=] cm³/mol. The exponent of 11/9 in Eq. (2.2) reproduces the slight upward concavity of the σ-T curves for apolar

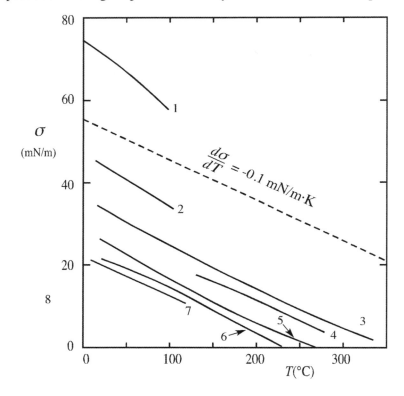

Fig. 2-3: Surface tension dependence on temperature for a variety of liquids: (1) water, (2) furfural, (3) chlorobenzene, (4) acetic acid, (5) carbon tetrachloride, (6) ethanol, (7) *n*-octane. Dashed line has slope: -0.1 mN/m·K, in reasonable agreement with that for most liquids.

liquids, and its format is excellent for generally good for interpolating $\sigma(T)$ data for such systems. The equation has been much refined yielding more elaborate expressions for σ^*, usually involving a third parameter obtained

---

[9] van der Waals, J. D., *Z. Phys. Chem.,* **13,** 716 (1894).
[10] Guggenheim, E. A., *J. Chem. Phys.,* **13,** 253 (1945).

from vapor pressure data.[11] It has also been extended to include polar or self-associating liquids, but this requires more general expressions for the exponent as well.

3. *Surface tension of solutions*

The surface tension of solutions depends on both temperature and composition. Some representative data for binary systems at 20°C are shown in Fig. 2-4, and an extensive bibliography for binary solutions has been compiled by McClure *et al.*[12] A more nearly complete discussion of the surface tension dependence on composition must await the discussion of capillary thermodynamics, but the figure suggests a few generalizations:

i) The surface tension of binary solutions is usually intermediate to those of the pure components, but less than the mole-fraction-average value. Some systems show extrema (minima or maxima) at intermediate values of the composition. The pronounced maximum in the water-sulfuric system is thought to be associated with the formation of the hydration complex: $H_2SO_4 \cdot 4H_2O$.

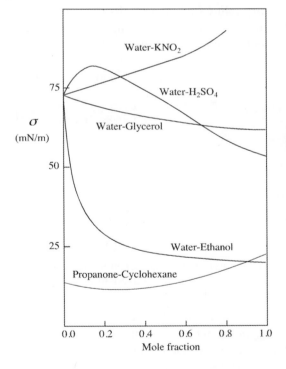

Fig. 2-4: Surface tension dependence on composition for several binary solutions at 20°C.

[11] Reid, R.C., Prausnitz, J.M., and Sherwood, T.K., **The Properties of Gases and Liquids**, 3rd Ed., Chap. 12, McGraw-Hill, New York, 1977.

[12] McClure, I. A., Pegg, I. A., and Soares, V. A. M., "A Bibliography of Gas-Liquid Surface Tensions for Binary Liquid Mixtures," in **Colloid Science (A Specialist Periodical Report)**, Vol. 4, D. H. Everett (Ed.), The Royal Society of Chemistry, London, 1983.

   ii) The surface tension of water is increased approximately linearly by dissolved salts, although the increases are small for low concentrations.

   iii) The surface tension of water is usually decreased sharply by organic solutes.

   The surface tension of water may be reduced *very* sharply at low concentrations of certain solutes (termed "surface active agents" or "surfactants") as shown in Fig. 2-5. These will be discussed in more detail later. To classify as a surface active agent, a solute must generally reduce the surface tension of water by 30 mN/m or more at a concentration of 0.01M or less. Although "surface activity" may also be identified in non-aqueous media, the reductions in surface tension involved are generally much less.

   Certain surfactants are but vanishingly soluble in water, as well as being nonvolatile, and monomolecular films of these compounds at the water-air interface represent an important class of systems. Their surface activity may be represented by plots of surface tension against *surface concentration*, Γ, [=] moles/cm^2, or more conveniently μmole/m^2, as shown in Fig. 2-5(b). Surfactant solutions have a monomolecular layer at the surface that is highly enriched in the solute (called an "adsorbed" or "Gibbs" monolayer), whereas insoluble surfactants form such a monolayer (called a "spread" or "Langmuir" monolayer) by direct spreading of the surfactant at

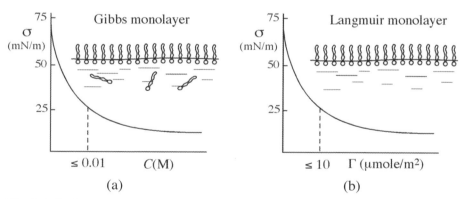

Fig. 2-5: Surface tension dependence on composition for surfactant monolayers (a) adsorbed or "Gibbs monolayer," or (b) spread or insoluble "Langmuir monolayer." Insert shows schematic of surfactant monolayer for either case.

the surface. In either case, the surfactant molecules consist of segregated hydrophilic and hydrophobic portions which orient themselves at the interface with their hydrophilic portions dissolved in the water and their hydrophobic portions directed outward, as shown. One of the most powerful of all insoluble surfactants is that which lines the moist inner lining of the *alveoli* of the mammalian lung. This surfactant mixture is primarily dipalmitoyl lecithin (DPL), pictured in Fig. 2-6. It consists of a glycerol molecule with an adjacent pair of its hydroxyl groups esterified with

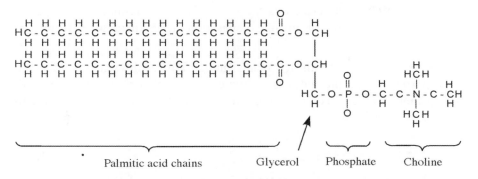

Fig. 2-6: Stylized diagram of a lung surfactant molecule: L-a, dipalmitoyl lecithin.

palmitic (C_{15}) acid, and the remaining group with phosphoric acid. The opposite hydrogen atom of the phosphoric acid is substituted with a choline group. A monolayer of these molecules is capable, upon compression, of reducing the tension of water to less than one mN/m.

C. Intermolecular forces and the origin of surface tension

1. *Van der Waals forces*

The existence of surface tension, and all of its manifestations, derives ultimately from the forces that exist between molecules.[13] These may be purely physical in nature or they may involve chemical complexation (association), such as that due to hydrogen bonding. If liquid metals are involved, metallic bonding exists, in which a cationic matrix of metal atoms is held together in part by a "sea" of free (conductance) electrons. Media containing ions introduce net electrostatic (Coulombic) interactions.

The purely physical interactions between neutral molecules are referred to as "van der Waals interactions." One type is that which exists between permanent molecular charge distributions, such as dipoles or quadrupoles. The interaction is obtained through vectorial summation of the Coulombic interactions between the various charge centers of the molecules, and is mutual-orientation dependent. Boltzmann-averaging over all possible mutual orientations of a pair of permanent dipoles yields the result (due to Keesom) for the potential energy of interaction Φ as a function of their distance of separation, r:

$$\Phi_{\text{dip-dip}} = -\frac{B_{\text{polar}}}{r^6}, \tag{2.3}$$

where B_{polar} varies as the square of the dipole moments of the molecules and inversely with the dielectric constant, ε, and with absolute temperature. The

---

[13] A comprehensive account is given by: Israelachvili, J. N., **Intermolecular and Surface Forces**, 2nd Ed., Academic Press, London, 1991.

potential function Φ is defined as the reversible work required to bring the two molecules, initially at infinite separation, to a distance r from one another.

A second type of van der Waals interaction results when a molecule with a permanent dipole induces a dipole in a neighboring molecule, with which it then interacts. The resulting pair interaction energy function (due to Debye) takes the form:

$$\Phi_{ind} = -\frac{B_{ind}}{r^6},\tag{2.4}$$

where B_{ind} depends on the permanent dipole moment of the first molecule and the molecular polarizability of the second molecule. Of course both molecules may possess permanent dipoles so that two such terms may be involved.

A final type of van der Waals force results from the oscillations of the electron clouds of all molecules, which produce strong *temporary* dipole moments. These induce strong temporary dipole moments in neighboring molecules with which they then interact in accord with the relationship (due to London):

$$\Phi_{disp} = -\frac{B_{disp}}{r^6},\tag{2.5}$$

where B_{disp} depends on the ground-state energies of the molecular oscillations, which in turn are closely proportional to the first ionization potentials of the molecules and their molecular polarizabilities.[14] (The close analogy of the effects described by London to that of light impinging on a medium has led to their being termed "dispersion" interactions, and hence the notation above.) For the interaction between a molecule i and a molecule j, to good approximation:

$$B_{disp(ij)} \approx \frac{3}{2}\alpha_i\alpha_j\left(\frac{I_iI_j}{I_i+I_j}\right) \approx \frac{3}{4}\alpha_i\alpha_jI = \sqrt{B_{ii}B_{jj}},\tag{2.6}$$

where I_i and I_j, and α_i and α_j are the first ionization potentials and the molecular polarizabilities of molecules i and j, respectively. The second approximate equality derives from the fact that the first ionization potentials seldom differ by more than a factor of two between molecules, so that it is

---

[14] It should be appreciated that this is a simplified picture. Attractive interactions may also arise from other than the ground state oscillation frequencies, as well as from fluctuating quadrupole and higher multipole interactions. Also, as the distance between molecules increases, the induced molecular oscillations become increasingly out of phase with the inducing oscillations, due to the finite speed of electromagnetic radiation. This effect, called *retardation*, generally begins at separations of about 50 nm or so, and by 100 nm the interaction energy approaches a $-1/r^7$ dependence. Retardation will be addressed again in the discussion of interactions between colloid particles in Chap. 7.

generally reasonable to put $I_i \approx I_j = I$. This gives the important result, used later, that *when dispersion forces predominate*, the interaction between unlike molecules is given by the geometric mean of the interactions between the like molecules (Berthelot's Principle).

An important observation related to the Φ-functions for all the principal van der Waals interactions is that they vary as $1/r^6$, so that they may be combined to give:

$$\Phi_{vdW} = -\frac{B_{vdW}}{r^6}.$$

(2.7)

The content of B_{vdW} is discussed with somewhat more sophistication in Chap. 7. The above relationship holds only so long as the electron clouds of the interacting molecules do not overlap. Under such conditions, strong repulsive forces arise. While the exact functional form of the r-dependence of these repulsive interactions has not been established, they are known to be very steep and can reasonably be represented by:

$$\Phi_{rep} = \frac{B_{rep}}{r^{12}}.$$

(2.8)

The net (or total) physical interaction between a pair of molecules is given by the sum of the van der Waals and repulsive interactions, and takes the form shown in Fig. 2-7.

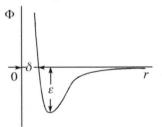

Fig. 2-7: Schematic of Lennard-Jones potential function.

Defining δ as the distance of separation for which $\Phi = 0$ (representative of the molecular diameter) and ε as the depth of the "potential well," Lennard-Jones derived expressions for B_{vdW} and B_{rep} leading to what is now termed the "Lennard-Jones potential:"

$$\Phi = 4\varepsilon\left[\left(\frac{\delta}{r}\right)^{12} - \left(\frac{\delta}{r}\right)^6\right].$$

(2.9)

Its application is restricted to approximately spherical, apolar or weakly polar molecules.

The *attractive force* of interaction between two molecules is the slope of the potential energy function, *viz.*

$$F_{att} = \frac{d\Phi}{dr} = -\frac{24\varepsilon}{r}\left[2\left(\frac{\delta}{r}\right)^{12} - \left(\frac{\delta}{r}\right)^{6}\right].\tag{2.10}$$

Both Φ and F_{att} are shown in Fig. 2-8, computed for carbon tetrachloride (for which $\delta = 5.881$ Å, and $\varepsilon = 4.514 \times 10^{-14}$ erg).[15] It is to be noted that at the average distance of molecular separation in the gas at standard conditions, both Φ and the intermolecular force are effectively zero. In the liquid, due to thermal motion, the average molecular separation lies not at the potential minimum, but just to the right of it, and the attractive intermolecular force at this point has a finite positive value.

The intermolecular forces and energies associated with chemical complexation, such as hydrogen bonding, require essentially direct molecular contact, and are thus shorter-ranged than attractive van der Waals forces. They are not generally represented in terms of an r-dependence, but it is clear that such dependence would be very steep and would produce a deep potential energy minimum. Metallic "bonding" is very strong at close range, and ionic interaction energies are both very strong and long-ranged, varying as $1/r$, as compared with $1/r^6$ for van der Waals energies.

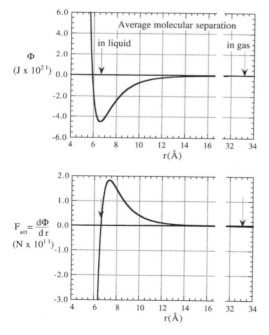

Fig. 2-8: Pair interaction curves for CCl_4.

---

[15] Bird, R. B., Stewart, W. E., and Lightfoot, E. N., **Transport Phenomena**, 2nd Ed., p. 865, Wiley, New York, 2007.

2. Surface tension as "unbalanced" intermolecular forces; the Hamaker constant

Surface tension may be interpreted directly in terms of intermolecular forces. The simplest picture, suggested by Fig. 2-9, contrasts the intermolecular forces acting on a molecule at the surface of a liquid with those acting on a molecule in the interior. The latter is acted upon equally in all

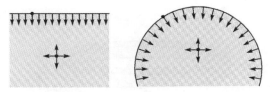

Fig. 2-9: Unbalanced intermolecular forces on molecules at a liquid surface.

directions, while a molecule at the surface experiences intermolecular forces directed only inward towards the interior. The net inward attraction tends to draw surface molecules toward the interior, causing the surface to seek minimum area (subject to whatever additional forces and constraints act on the system). The stronger the intermolecular attractions are, the greater the expected surface tension. This may be made somewhat more quantitative as follows.

Consider a body of liquid conceptually divided into an upper and a lower half extending to infinity away from the imaginary interface dividing them. We may compute the total energy of interaction between the molecules above the interface with those beneath it, on a per unit area basis. If the two liquid half-spaces were conceptually separated (to infinity), the work required to do so can be identified with (the negative of) this energy, and may be equated to twice the energy/area of the new surface created, *i.e.*, 2σ, because two new interfaces would be produced by the process. Begin by considering the interaction of a single molecule in the upper layer a distance D above the interface with the lower layer, as shown in Fig. 2-10. First we compute the energy of the molecule's interaction with a ring of diameter x of molecules in the lower half-space a vertical distance z away from it. The volume of the ring is $2\pi x dx dz$; and the number of molecules in it is $\rho_m 2\pi x dx dz$, all a distance $r = \sqrt{x^2 + z^2}$ from the subject molecule in the upper half-space. ρ_m here is the *molecular* density. The energy of interaction between the subject molecule and the entire ring, assuming a van der Waals fluid, Eq. (2.7), and pairwise additivity of the molecular interactions, is

$-\dfrac{B_{vdW}}{r^6} \cdot \rho_m 2\pi x dx dz = -2\pi B_{vdW}\rho_m \dfrac{xdxdz}{\left(x^2 + z^2\right)^3}$. Integration then gives the

interaction between that molecule and the entire lower half-space:

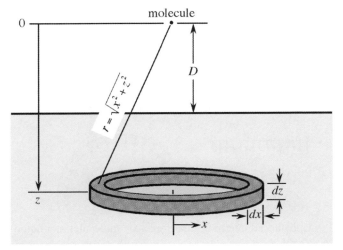

Fig. 2-10: Computation of the interaction of a single molecule with a semi-infinite half-space of the same molecules.

$$\Phi_{\text{molec-half-space}} = -2\pi B_{\text{vdW}} \rho_{\text{m}} \int_{D}^{\infty} dz \int_{0}^{\infty} \overset{1/4z^4}{\overbrace{\frac{x}{(x^2+z^2)^3}}} dx$$

$$= -\frac{\pi}{2} B_{\text{vdW}} \rho_{\text{m}} \int_{D}^{\infty} \overset{1/3D^3}{\overbrace{\frac{1}{z^4}}} dz = -\frac{\pi B_{\text{vdW}} \rho_{\text{m}}}{6D^3}. \qquad (2.11)$$

Next we compute the interaction of a thin sheet (of thickness dD and unit area) of molecules in the upper half space at a distance D from the interface with the lower half-space. The number of molecules in this thin sheet of unit area is $\rho_{\text{m}}dD$, so the interaction of the sheet with the lower half-space is $-\dfrac{\pi B_{\text{vdW}} \rho_{\text{m}}^2 dD}{6D^3}$, and the total interaction energy of a unit area (signified by the superscript $^{\sigma}$) of infinite depth[16] with the lower half-space is

$$\Phi^{\sigma} = -\frac{\pi B_{\text{vdW}} \rho_{\text{m}}^2}{6} \int_{D_0}^{\infty} \frac{dD}{D^3} = -\frac{\pi B_{\text{vdW}} \rho_{\text{m}}^2}{12 D_0^2}, \qquad (2.12)$$

where D_0 is the closest distance of molecular approach. It is common to put

$$(\pi \rho_{\text{m}})^2 B_{\text{vdW}} = A, \qquad (2.13)$$

---

[16] A simplification is made here in that the effects of "retardation" are neglected.

where A is called the Hamaker constant,[17] embodying the integrated molecular interactions. This is discussed in more detail in Chap. 7. Then

$$\sigma = -\frac{1}{2}\Phi^{\sigma} = \frac{A}{24\pi D_0^2}. \tag{2.14}$$

D_0 may be estimated from the molecular size and packing density in the liquid and is usually of the order of 1-2 Å. The accepted value of D_0 is 1.65 Å.[18] The important point that has been demonstrated here, however, is the direct relationship between the surface tension and the strength of the intermolecular interactions in the liquid.

3. Pressure deficit in the interfacial layer; Bakker's Equation

A somewhat different picture focuses upon intermolecular forces within the zone of inhomogeneity that constitutes the interfacial layer. Consider a pure liquid at rest in the absence of external force fields and facing its own vapor across the interfacial layer, and divided into two parts by a plane AB drawn normal to the interface,[19] as shown in Fig. 2-11. One may consider the forces acting across the plane AB by one part on the other. Out in the bulk of each phase, the forces acting on the plane are those of the hydrostatic pressure, which, in the absence of flow in the bulk material and external force fields such as gravity, is uniform and isotropic. In general, the net pressure force exerted on a plane drawn in the fluid depends on the following two factors:

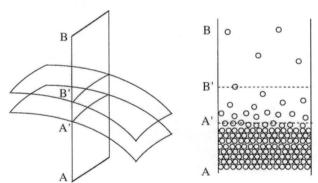

Fig. 2-11: Molecular interpretation of surface tension.

1) The kinetic energy (thermal agitation) of the molecules: Pressure is manifest at the (real or conceptual) confining boundaries of the phase through the change in momentum of the molecules colliding with them, per

---

[17] Hamaker, H. C., *Physica*, **4**, 1058 (1937), who was one of the first to perform such integrations.

[18] Israelachvili, J. N., **Intermolecular & Surface Forces**, 2nd Ed., p. 203, Academic Press, London, 1991.

[19] "Normal to the interface" is defined as the direction of the density gradient, $\nabla\rho$.

unit area, and this pressure is transmitted throughout the bulk of the phase. This is the only factor at play in ideal gases.

2) Configurational potential energy, due to intermolecular forces: This is expressed as the sum of interaction energies between pairs of molecules, per unit area of the conceptual interface between them, and is a function of the distance of their separation, as shown in the preceding paragraphs. The *forces* between the molecules are generally attractive and thus *subtract* from the pressure effect that would be attributable to thermal agitation alone.

Out in the bulk phases on either side of the interfacial layer, the net effect of these two contributions is a pressure that is uniform (neglecting external field effects) and isotropic. As detailed below, this condition does *not* exist within the interfacial layer, between A′ and B′ in Fig. 2-11.

Since density is presumed to vary continuously from that of the liquid to that of the gas as one moves upward through the interfacial layer, intermediate molecular separations are forced to exist, and for these, the lateral attractive forces will be much greater than they are in either the bulk liquid or gas. This is suggested by Fig. 2-8, in which it is seen that the forces between molecules at the average spacing in either bulk liquids or gases are quite small relative to those at intermediate spacings. Thus the net local pressure forces in the lateral direction, p_T, *i.e.*, the difference between the pressure forces due to thermal motion and those due to the attractive intermolecular forces, are substantially reduced in the interfacial region relative to the bulk regions, as shown schematically in Fig. 2-12. They may even become negative, producing a net local tension. The lateral pressure component becoming negative, however, is not a requirement for the occurrence of surface tension; merely the reduction in the pressure component is sufficient.

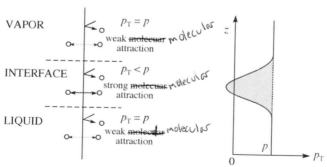

Fig. 2-12: Deficit in the tangential component of pressure in the interfacial layer.

One can be more specific about the "pressure" quantities being discussed. In general the state of (compressive) stress in a fluid is given by the tensor $\underline{\underline{p}}$. In a bulk fluid at rest this tensor reduces to

$$\underline{\underline{p}} = \begin{bmatrix} p & 0 & 0 \\ 0 & p & 0 \\ 0 & 0 & p \end{bmatrix}. \tag{2.15}$$

The off-diagonal elements must be zero because they represent shear stresses, which a fluid cannot support in equilibrium at rest. The three normal components of stress must furthermore be equal since a bulk fluid phase (in the absence of external force fields) is uniform in all intensive variables. One can still balance the forces even when the pressure components are different, *i.e.*, if the requirement of isotropy is relaxed, so that

$$\underline{\underline{p}} = \begin{bmatrix} p_{xx} & 0 & 0 \\ 0 & p_{yy} & 0 \\ 0 & 0 & p_{zz} \end{bmatrix}, \tag{2.16}$$

with $p_{xx} \neq p_{yy} \neq p_{zz}$. In the interfacial layer, these pressure components *will* be different, as full isotropy with respect to the pressure can no longer exist. In particular, one must have two different pressure components, one normal and one tangential to the interface, $p_N(z)$ and $p_T(z)$, respectively. Mechanical stability requires that the gradient of the pressure tensor be everywhere equal to zero[20], and symmetry requires isotropy in the plane tangent to the surface ("lateral" or "transverse" isotropy) so that $p_{xx}(z) = p_{yy}(z) = p_T(z)$, where z is the coordinate normal to the surface. The pressure component normal to the surface is $p_{zz}(z) = p_N(z) \neq p_T(z)$. The pressure component being discussed in terms of intermolecular forces is p_T.

Knowledge of the pressure tensor across the interfacial layer (which can be expressed only to the extent that one has detailed knowledge of the intermolecular potential functions and the molecular distribution functions) and application of Young's membrane model thus permits, in principle, a computation of the surface tension. The result is:

$$\sigma = \int_{z_a}^{z_b} (p_N - p_T)dz, \tag{2.17}$$

where z is the coordinate normal to the interface, and the gap $z_a \Leftrightarrow z_b$ defines the region of molecular inhomogeneity constituting the interfacial layer. If the interface is flat, $p_N(z) = p = $ constant, and one may replace the finite limits with $\pm\infty$, since the regions outside the zone $z_a \Leftrightarrow z_b$ contribute nothing to the integral:

$$\sigma^{\text{flat surf.}} = \int_{-\infty}^{+\infty} (p - p_T)dz. \tag{2.18}$$

---

[20] Ono, S., and Kondo, S., "Molecular Theory of Surface Tension in Liquids," pp. 134-304 in **Handbuch der Physik, Vol. 10**, E. Flügge (Ed.), Springer-Verlag, Berlin, 1960.

The above result is due to Bakker.[21] Using an assumed *local* equation of state similar to that describing the bulk phase, together with an expression for free energy minimization in the interfacial layer, van der Waals made computations of the interfacial density profiles and the resulting surface or interfacial tension. Others[22] have since refined these arguments and have applied statistical mechanics to the problem.[23,24]

What has been done using Bakker's Equation is to project the integrated excess lateral stress (or alternatively, the compressive stress deficit), as shown in Fig. 2-12, onto the mathematical surface defined as the interface. The mathematical surface of the model is termed the "surface of tension," and for a flat interfacial layer, its exact location is immaterial, *i.e.*, its location in no way impacts the unambiguous and physically measurable surface tension. For an interface that is not flat, it is useful to distinguish between weak curvature and strong curvature. (A more detailed discussion of surface curvature is given later.) Weakly curved surfaces are those whose mean radius of curvature is large relative to the thickness of the zone of inhomogeneity, whereas strongly curved surfaces are those whose radius of curvature is comparable to that of the interfacial layer thickness. In the latter case, since interfacial layers are of the order of only a few Å in thickness, one might expect the continuum concept of surface tension to break down, or alternatively, to require that the surface tension be regarded as a *function* of the curvature. For the moment, we shall consider only weakly curved surfaces. As will be proved later, curved fluid surfaces (whether strongly or weakly curved) require a difference in the equilibrium bulk pressures, with the pressure on the concave side larger than the pressure on the convex side.

The molecular picture of the interfacial layer, particularly with respect to the interpretation of the states of stress that exist within it, gives one the idea of why fluid interfacial layers should exhibit a tension. It also explains qualitatively why there should be a difference in the surface tension from one pure liquid to the next in terms of the type and strength of the intermolecular forces that prevail. Intermolecular forces such as those of ionic bonds, metallic bonds, or hydrogen bonds, which yield very strong attractions, lead to much higher boundary tensions than those for liquids with only van der Waals interactions. Thus it is that molten salts and liquid metals have very high surface tensions and that water's surface tension is high relative to that of organic liquids. One may also explain the nature of the temperature dependence of surface tension. The portion of the pressure component due to kinetic energy increases linearly with temperature, while

---

[21] Bakker, G., **Kapillarität u. Oberflächenspannung**, Vol. 6 of **Handb. d. Experimentalphysik**, W. Wien, F. Harms and H. Lenz (Eds.), Akad. Verlags., Leipzig, 1928.

[22] Cahn, J. W., and Hilliard, J. E., *J. Chem. Phys.*, **28**, 258 (1958).

[23] Rowlinson, J. S., and Widom, B., **Molecular Theory of Capillarity**, Clarendon Press, Oxford, 1982.

[24] Davis, H. T., **Statistical Mechanics of Phases, Interfaces and Thin Films**, VCH, New York, 1996.

that due to intermolecular attractive forces remains essentially constant. Thus, as temperature increases, the difference between p and p_T diminishes, and σ decreases in approximately linear fashion. The reason for the effectiveness of surface active agents in reducing surface tension can also be understood. These molecules orient themselves in the interface so that in both the upper and the lower portions of the layer, *portions* of molecules are present which interact favorably with the predominant component of the respective bulk phases. This reduces the impact of the lateral intermolecular forces, *i.e.*, the magnitude of p - p_N , in passing from one phase to the other. The high-energy clean water surface is effectively replaced by the lower energy hydrocarbon moieties of the surfactant. Finally, the molecular picture of the interfacial layer makes clear the reason for its thinness. The zone of inhomogeneity constituting the interfacial layer is necessarily limited by the range of the intermolecular forces. Van der Waals forces, for example, seldom are significant beyond the second- or third- nearest neighbors. Intermolecular forces leading to hydrogen bonding are even shorter-ranged. Ionic interactions are longer ranged, and interfacial layers involving these types of forces may be somewhat thicker.

4. *Components of the surface tension*

The direct dependence of surface tension on the intermolecular forces in the fluid has led Fowkes and others to divide the contributions to surface tension into the various contributions to the intermolecular forces that may exist. Specifically, Fowkes[25] first wrote

$$\sigma = \sigma^d + \sigma^p + \sigma^i + \sigma^H + \sigma^m + ...,   \tag{2.19}$$

where "d" refers to dispersion forces, "p" to forces between permanent dipoles, "i" to induced dipoles, "H" to hydrogen bonds, "m" to metallic bonds, *etc*. It is now known from the theory of intermolecular forces in condensed-phase media[26] that the contributions of dipole-dipole (Keesom) and dipole-induced dipole (Debye) interactions to the surface energy are essentially negligible, as a result of the self-cancellation that occurs when multiple dipoles interact. This is in contrast to the situation in gases, where dipoles interact predominantly in pairwise fashion. The portion of the 72 mN/m surface energy of water at room temperature that is attributable to such polar effects, for example, has been computed to be only 1.4 mN/m.[27] In addition to dispersion force interactions, the major contributor to σ is that due to donor-acceptor interactions, *i.e.* Lewis acid-base association. A donor (base) donates a pair of electrons in an adduct-forming complexation with an acceptor (acid). This picture can be made to include hydrogen bonding.[28]

[25] Fowkes, F. M., *A.C.S. Advances in Chemistry Series,* **43**, 99-111 (1964).

[26] Israelachvili, J. N., **Intermolecular and Surfaces Forces,** 2nd Ed., Academic Press, London, 1992.

[27] van Oss, C. J., Chaudhury, M. K., and Good, R. J., *J. Colloid Interface Sci.*, **111**, 378 (1986).

[28] Fowkes, F. M., *J. Adhesion Sci. Tech.*, **1**, 7 (1987).

Thus σ^H is replaced with the more general σ^{ab}. Many liquids (most notably water) may act as both acids *and* bases, and thus *self*-associate. For most liquids then, Eq. (2.19) reduces to: $\sigma = \sigma^d + \sigma^{ab}$, where σ^{ab} refers to the contribution of acid-base self-association. In molten metals, metallic bonding is important, so in those cases: $\sigma = \sigma^d + \sigma^m$.

   As will be seen later, the components of surface tension for a given liquid may be determined experimentally from interfacial tension measurements between that liquid and an immiscible, non-associating reference liquid, or from the measurement of contact angles against reference solids (see Chap. 4). Of particular importance, for water at 20°C: $\sigma = 72.8$ mN/m, with $\sigma^d = 21.2\pm0.7$ mN/m. For mercury, $\sigma = 485$ mN/m and $\sigma^d = 200$ mN/m. A list of values for various liquids is given in Table 2-2.

Table 2-2: Components of surface tension (in mN/m at 23.±0.5°C). From [Fowkes, F. M., Riddle, F. L., Pastore, W. E., and Webber, A. A., *Colloids Surfaces*, **43**, 367 (1990)].

Liquid	σ	σ^d	σ^{ab}	Type
Water	72.4	21.1	51.3	both
Glycerol	63.4	37.0	26.4	both
Formamide	57.3	28.0	29.3	both
Methyl iodide	50.8	50.8	0	Neither
a-Bromonaphthalene	44.5	44.5	0	Neither
Nitrobenzene	43.8	38.7	5.1	Both
Dimethylsulfoxide	43.5	29.0	14.5	Both
Aniline	42.5	37.3	5.1	Both
Benzaldehyde	38.3	37.0	1.3	Both
Pyridine	38.0	38.0	0	Basic
Formic acid	37.4	18.0	19.4	Both
Pyrrole	37.4	32.6	4.8	Both
Dimethylformamide	36.8	30.2	6.6	Both
1,4-Dioxane	33.5	33.5	0	Basic
cis-Decaline	32.2	32.3	0	Neither
Squalane	29.2	29.2	0	Neither
Acetic acid	27.6	22.8	4.8	Both
Chloroform	27.1	27.1	0	Acidic
Methylene chloride	26.6	26.6	0	Acidic
Tetrahydrofuran	26.5	26.5	0	Basic
Ethyl acetate	25.2	25.2	0	Basic
Acetone	23.7	22.7	1.0	Both
Ethanol	22.2	20.3	1.9	Both
Triethylamine	20.7	20.7	0	Basic
Ethyl ether	17.0	17.0	0	Basic

D. Interfacial tension

1. *Experimental interfacial tension*

As stated earlier, the terminology "surface tension" is usually reserved for the tension observed at a liquid-vapor interface, whereas "interfacial" tension is used in reference to fluid interfaces of all kinds, but in the present context to liquid-liquid interfaces. The same molecular picture developed earlier explains the existence of interfacial tension between liquids. If the liquids are dissimilar enough to form an interface, then the molecules of each bulk phase prefer to stay together rather than mix. They resist the enforced molecular separation between like species that must exist throughout the interfacial layer, where intermediate compositions prevail, and manifest this resistance as interfacial tension. For example, at the water-oil interface, the hydrogen bonds between the water molecules are disrupted.

Values for liquid-liquid interfacial tensions are less plentiful in the literature than those of surface tension and are generally less reliable, due to uncertainty as to the extent of mutual saturation of the liquid phases during the measurement. Some representative experimental values for interfacial tensions between water and various liquids are shown in Table 2-3.

Table 2-3: Interfacial tension values		
Liquids	T(°C)	Interfacial tension (mN/m)
Water/Butanol	20	1.8
Water/Ethyl Acetate	20	6.8
Water/Benzene	20	35.0
Water/HMDS (Silicone)	20	44.3
Water/Perfluorokerosene	25	57.0
Water/Mercury	20	415
Water/Oil (with surfactant)	20	as low as < 0.001

The effect of temperature on interfacial tension is somewhat more complex than that for surface tension, because changes in temperature may strongly change the extent of mutual solubility of the liquids. For systems having an upper critical solution temperature (UCST), σ decreases with temperature, but for those with a lower critical solution temperature (LCST), σ increases with T. For systems with both a UCST and an LCST, σ passes through a maximum at an intermediate temperature.

2. *Combining rules for interfacial tension*

Effort has been put into developing semi-empirical equations allowing interfacial tension to be calculated in terms of known values for the surface tensions of the two liquids forming the interface. Some are based on the simple picture of the molecular origin of surface tension displayed in Fig. 2-9. If surface tension represents the "unbalanced" inward-pulling intermolecular forces, then interfacial tension should represent the *net* inward force, directed toward the liquid of greater surface tension, as suggested in Fig. 2-13. This is the basis for Antanow's Law,[29] which states that the interfacial tension between two liquids is the absolute value of the difference between their surface tensions:

$$\sigma_{AB} = \left| \sigma_{A(B)} - \sigma_{B(A)} \right|. \tag{2.20}$$

It often does well if the surface tension values used correspond to mutually saturated liquids, as suggested by the subscripts in the equation. Table 2-4 shows data for several mutually saturated, water–organic systems in comparison with calculations based on Antanow's Law.

Table 2-4: Interfacial tensions of mutually saturated water-organic liquid systems. From [Voyutsky, S., **Colloid Chemistry**, p. 129, Mir. Pub., Moscow, 1978.]

Liquid T (°C)	Surface tension, against air (mN/m)		Interfacial tension (mN/m)	
	Water layer	Organic layer	Antanow's Law	Experimental
Benzene (19°)	63.2	28.8	34.4	34.4
Aniline (26°)	46.4	42.2	4.2	4.8
Chloroform (18°)	59.8	26.4	33.4	33.8
Carbon tetrachloride (17°)	70.2	26.7	43.5	43.8
Amyl alcohol (18°)	26.3	21.5	4.8	4.8
Cresol (18°)	37.8	34.3	3.5	3.9

Another equivalent approach to estimating interfacial tensions also derives from their direct computation in terms of intermolecular forces. Consider two immiscible liquids A and B as semi-infinite half-spaces

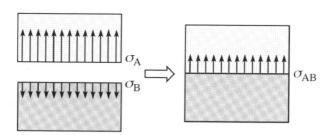

Fig. 2-13: Model for evaluating interfacial tension in terms of surface tensions.

meeting at their common interface. The work required to separate the two phases (to infinity, *in vacuo*) is the energy required to form the surfaces of A and B, minus the energy recovered by the destruction of the AB interface, *viz.*

$$W_{\text{separation}} = \sigma_A + \sigma_B - \sigma_{AB}. \tag{2.21}$$

This work of separation can be computed as the sum of the intermolecular forces between the molecules of phase A with those of phase B, by analogy with Eqs. (2.12)-(2.14):

$$W_{\text{separation}} = -\Phi^{\sigma}_{AB} = \frac{\pi B_{AB}\rho_A\rho_B}{12D_0^2} = \frac{A_{AB}}{12\pi D_0^2}, \tag{2.22}$$

where B_{AB} is the cross van der Waals molecular interaction constant between molecules A and B, ρ_A and ρ_B are molecular densities, and A_{AB} is the cross Hamaker constant. Applying the geometric mean rule, Eq. (2.6), for B_{AB} (assuming the dominance of dispersion forces), we see that the cross Hamaker constant is given by

$$A_{AB} = \pi^2 \rho_A\rho_B\sqrt{B_{AA}B_{BB}} = \sqrt{A_{AA}A_{BB}}. \tag{2.23}$$

Then relating the Hamaker constant to surface tension, Eq. (2.14):

$$W_{\text{separation}} = \Phi^{\sigma}_{AB} = 2\sqrt{\sigma_A\sigma_B}, \tag{2.24}$$

so that finally, substituting in Eq. (2.21):

$$\sigma_{AB} = \sigma_A + \sigma_B - 2\sqrt{\sigma_A\sigma_B}. \tag{2.25}$$

Girifalco and Good[30] wrote Eq. (2.25) in the form:

$$\sigma_{AB} = \sigma_A + \sigma_B - 2\Phi\sqrt{\sigma_A\sigma_B}, \tag{2.26}$$

with the factor Φ (presumably ≤ 1) accounting for the fact that not all of the molecular interactions across the interface may be of the dispersion force type.

An alternative formulation was given by Fowkes,[31] who argued that, in the absence of acid-base interactions (or metallic bonding), only dispersion forces were operative *across* the interface. The result was thus:

$$\sigma_{AB} = \sigma_A + \sigma_B - 2\sqrt{\sigma_A^d\sigma_B^d}, \tag{2.27}$$

To use Eq. (2.27) one needs to know the dispersion force contributions to the surface tension values, such as given in Table 2-2. The interpretation of the Girifalco-Good Φ-factor becomes rather awkward (and quite different from its original interpretation, which involved presumed polar interactions). The Fowkes equation is easily extended, at least in a formal way, to include the possibility of acid-base interactions *across* the interface, I^{ab}:

$$\sigma_{AB} = \sigma_A + \sigma_B - 2\sqrt{\sigma_A^d\sigma_B^d} - I^{ab}. \tag{2.28}$$

[30] Girifalco, L.A., and Good, R.J., *J. Phys. Chem.,* **67**, 904 (1957).
[31] Fowkes, F. M., *A.C.S. Advances in Chemistry Series,* **43**, 99-111 (1964).

Situations in which I^{ab} is significant may lead to $\sigma_{AB} \leq 0$, suggesting miscibility between liquids A and B. While strictly applicable only in the case of total immiscibility, the Fowkes Equation may be applicable to partially miscible systems if the values of surface tension correspond to those of mutual saturation, *i.e.* $\sigma_{A(B)}$, $\sigma^d_{A(B)}$, *etc*. This idea, however, seems not to have been tested.

A word of caution must be raised concerning the use of equations for interfacial tension employing Berthelot's principle. The mixing rule applies to energy quantities (such as internal energy or enthalpy), whereas surface and interfacial tensions are *free* energies.[32] Thus equations such as those of Girifalco and Good or Fowkes ignores the entropy effect associated with bringing together or disjoining the phases.

E. Dynamic surface tension

The surface and interfacial tensions referred to in the foregoing are assumed to be *equilibrium* values. Before exploring further the mechanical consequences of capillarity for equilibrium systems, a word should be said about systems which may *not* be in equilibrium, and which exhibit a time-dependent, or *dynamic* surface or interfacial tension $\sigma(t)$. Examples would include "fresh" surfaces created in coating operations, for liquids emerging from orifices or spray nozzles, or when bubbles are formed within liquids. Practical situations exist, such as in ink-jet printing, in which surface ages as low as fractions of a millisecond are important. The time required for molecular re-orientation at fresh interfaces of non-macromolecular *pure* liquids is less than one *micro*-second, so that dynamic surface tension behavior of pure liquids is of little practical significance, but for solutions, particularly dilute solutions of surface active agents, surface tension may be found to vary from its value at $t \rightarrow 0$ (when σ is presumably close to that corresponding to the pure solvent) to its equilibrium value over times from less than one millisecond to several hours. As has been noted, such solutes reduce surface tension as they accumulate at the interface, and to do so requires at least the time for diffusion. Additional time may be required for the solute molecules to enter the surface and possibly to re-orient themselves. Further discussion of dynamic surface tension is deferred to Chap. 3, following discussion of surfactant adsorption.

F. Capillary hydrostatics: the Young-Laplace Equation

1. *Capillary pressure: pressure jump across a curved fluid interface*

Consider next the problem of determining the shape and location of fluid interfaces, the fundamental problem of capillary hydrostatics. The solution to this problem is the basis for most of the methods of measuring

---

[32] Lyklema, J., *Colloids Surfaces A*, **156**, 413 (1999).

surface or interfacial tension and has important consequences for the formation of adhesive bonds, for the motion of liquids in porous media, for the thermodynamic properties of small drops or bubbles, and for the process of phase change by nucleation. For the fluid interface, nothing more complicated than Young's membrane model is needed.

It is a matter of experience that when an elastic membrane is deformed, as when air is blowing on a soap film suspended on a frame as shown in Fig. 2-14, the pressure on the concave side (p'') must be greater than the pressure on the convex side (p'). The pressure difference is found

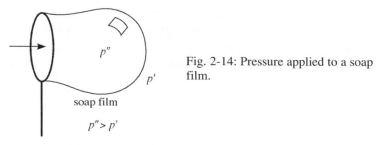

Fig. 2-14: Pressure applied to a soap film.

to be directly proportional to the *curvature* of the soap film, as demonstrated by the example shown in Fig. 2-15. When two bubbles of different sizes are connected by a tube, the larger one will grow at the expense of the smaller one since the curvature of the smaller bubble is greater (has a higher pressure inside) than the larger one. Flow continues until the curvatures are equal, as shown, with the smaller bubble eventually becoming a spherical cap with the same radius as the larger bubble.

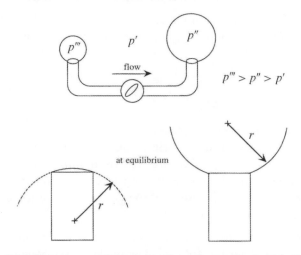

Fig. 2-15: Spontaneous flow occurs from the smaller bubble (higher curvature) to the larger bubble until the spherical cap at the location where the smaller bubble started has the same radius of curvature as the final larger bubble.

Thomas Young (in 1805, *loc. cit.*) and P. S. Laplace (in 1806)[33] derived the exact relationship which must hold between the pressure jump across a fluid interface, $\Delta p = p'' - p'$, and its local curvature, κ, *viz.*,

$$\Delta p = \sigma \kappa, \tag{2.29}$$

with pressure on the concave side higher. This is the *Young-Laplace Equation*, and is derived below. In order to understand and use Eq. (2.29), the curvature κ, of a surface in space (at a point) must be defined.

2. The curvature of a surface

One may first recall the definition of the curvature of a *plane* curve, with reference to Fig. 2-16. The curvature of a plane curve C at P is its rate of change of direction with arc length S, at P, measured along the curve, *i.e.*, $\kappa = d\phi/dS$, where ϕ is the angle made between the tangent to the curve at the

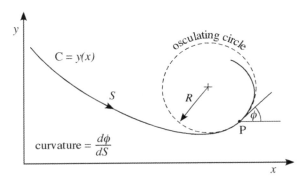

Fig. 2-16: Construction for defining plane curvature.

point of interest and some arbitrary direction (say, the x-direction, as shown in the figure). Its units are length[-1], and its sign is ambiguous. In terms of the equation of the curve $y(x)$:

$$\kappa = \pm \frac{d^2y}{dx^2}\left[1 + \left(\frac{dy}{dx}\right)^2\right]^{-3/2}. \tag{2.30}$$

The curvature of a circle is computed from its equation: $x^2 + y^2 = R^2$ (for a circle of radius R centered at the origin) and is seen to be $\pm 1/R$. It is thus possible to define a *circle of curvature* (or "osculating circle") for any point P along any curve C, as the circle passing through point P and having the same curvature as C at P. The radius of the circle of curvature at P is referred to as the *radius of curvature* of C at P.

It is next possible to define the curvature of a *surface* in space, with reference to Fig. 2-17, in the following way. We first erect a normal, $\underline{n}$, at the point of interest and pass a pair of orthogonal planes through it. These

[33] De Laplace, P. S., **Traité méchanique céleste, supplement au Livre X**, 1806.

cut the surface in two plane curves. The curvature of the surface is the *sum* of the curvatures of these two plane curves:

$$\kappa = \pm\left(\frac{1}{R_1} + \frac{1}{R_2}\right) = \pm\frac{2}{R_m},$$

(2.31)

where R_m is the mean radius of curvature of the surface. It may be identified as the radius of the osculating *sphere*. The sum is invariant as one rotates the planes about the normal. The R_1-value that is maximum, and the corresponding R_2-value which is minimum, are referred to as the *principal*

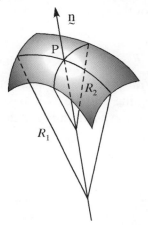

Fig. 2-17: General surface in Cartesian space.

radii of curvature. The sign of the curvature of the surface is ambiguous until a physical context is specified.

Recalling the expression for the curvature of a plane curve, it is easy to appreciate that the general expression for the curvature of a surface is quite complex. For the general case shown in Fig. 2-18, where the surface is

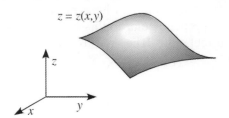

Fig. 2-18: General surface in Cartesian space.

given as the elevation z as a function of the planform variables x and y, *viz.* $z = z(x,y)$, the general expression for the curvature becomes:

$$\kappa = \pm\frac{\left(\frac{\partial^2 z}{\partial x^2}\right)\left[1+\left(\frac{\partial z}{\partial y}\right)^2\right] - 2\left(\frac{\partial z}{\partial x}\right)\left(\frac{\partial z}{\partial y}\right)\left(\frac{\partial^2 z}{\partial x \partial y}\right) + \left(\frac{\partial^2 z}{\partial y^2}\right)\left[1+\left(\frac{\partial z}{\partial x}\right)^2\right]}{\left[1+\left(\frac{\partial z}{\partial x}\right)^2+\left(\frac{\partial z}{\partial y}\right)^2\right]^{3/2}}.$$

(2.32)

Many of the cases of special interest, however, possess certain symmetries that simplify the expressions considerably. Some examples of special cases are discussed below.

For spheres or segments of spherical surfaces, as might be created by soap films, it is evident, as shown in Fig. 2-19, that any normal to the surface will pass through the center of the sphere, and any plane containing this line will cut the sphere to yield a great circle. The radius of this circle, R_1, is the radius of the sphere, R. The plane containing the normal and orthogonal to the first plane will also cut the surface of the sphere in a great circle, so we see that the curvature of the spherical surface is

$$K = \frac{1}{R_1} + \frac{1}{R_2} = \frac{1}{R} + \frac{1}{R} \equiv \frac{2}{R}. \qquad (2.33)$$

Right cylindrical surfaces (or portions of such surfaces) may be similarly analyzed. Any normal to the surface will pass through and be orthogonal to the axis of the cylinder, as shown in Fig. 2-20. One convenient

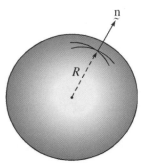

Fig. 2-19: Curvature of a spherical surface.

plane passing through the normal will be perpendicular to the axis of the cylinder, and cut the cylindrical surface in a circle whose radius, R_1, is the radius of the cylinder. Then the plane perpendicular to this circle and cutting the surface of the cylinder will be a rectangle. The "radius" of this curve = ∞. Thus for the circular cylinder,

$$K = \frac{1}{R_1} + \frac{1}{\infty} = \frac{1}{R} . \qquad (2.34)$$

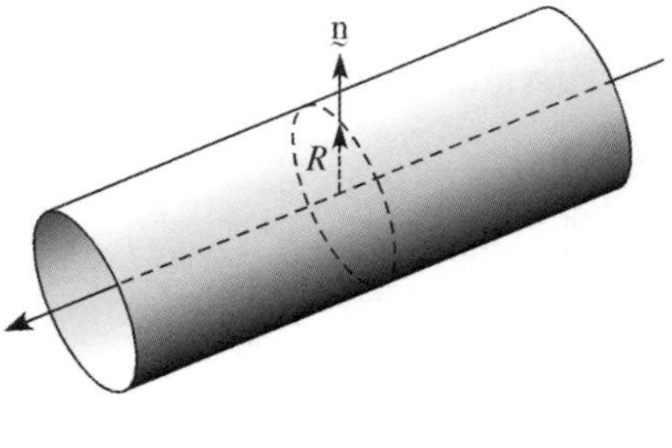

Fig. 2-20: Curvature of a right circular cylinder

Cylindrical surfaces in general are those swept out by moving a straight line (the *generatrix*) normal to itself, as shown in Fig. 2-21. It is evident that the curvature of any general cylindrical surface will be the

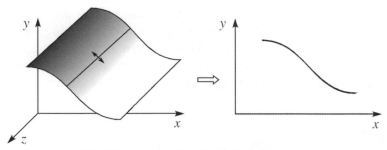

Fig. 2-21: A general cylindrical surface.

curvature of a plane curve in the plane perpendicular to the generatrix, *i.e.*,

$$\kappa = \frac{1}{R_1} + \frac{1}{\infty} = \pm \frac{y''}{\left[1 + (y')^2\right]^{3/2}}. \tag{2.35}$$

Some practical situations yielding this type of surface are shown in Fig. 2-22 and include menisci against flat walls contacting a liquid, and menisci between flat plates or between cylinders and plates.

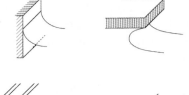

Fig. 2-22: Examples of cylindrical liquid surfaces.

Finally, there are surfaces of axial symmetry, some examples of which are shown in Fig. 2-23. The first three are closed surfaces, *i.e.*, cut by the axis of symmetry. Case (d) is a soap film suspended between opposing open

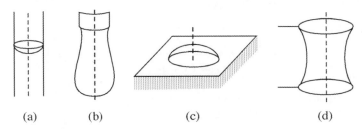

(a)   (b)   (c)   (d)

Fig. 2-23: Examples of surfaces of axial symmetry. The meniscus in a round tube (a), the pendant drop (b) and the sessile drop (c) are cut by the axis of symmetry, whereas the soap film suspended between circular rings (c) is not.

wire loops, and is an example of an axisymmetric surface that is not closed. One may derive the expression for the curvature of closed surfaces of axial symmetry by considering that any normal to such a surface, when extended, will intersect the axis of symmetry, and the plane established by the normal and the axis of symmetry will cut surface yielding its profile, as shown in Fig. 2-24. The profile may be given the equation $y(x)$, defining the origin as the point where the surface is cut by its axis of symmetry, y as the coordinate along the axis of symmetry, and x the distance measured away from it. One of the two principal radii of curvature, R_1, at the point of interest, will be the

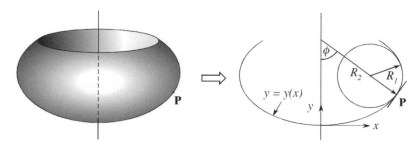

Fig. 2-24: Axisymmetric interface; profile view in plane passing through axis of symmetry.

plane curvature of $y(x)$ at that point. A little examination of the figure reveals that the second principal radius of curvature, R_2, must be the distance measured from the point of interest on the surface back to the axis of symmetry along a line perpendicular to the tangent of the curve $y(x)$. It is evident that as such a radius swings around the axis of symmetry, it will trace out a circle on the surface. R_1 thus swings in the plane of the figure, while R_2 swings around the axis as shown. R_1 is given by the usual expression for plane curvature. R_2 is x divided by $\sin\phi$, where ϕ is the angle whose tangent is dy/dx, $i.e.$ y'. This works out to be:

$$R_2 = \pm \frac{x\left[1+(y')^2\right]^{1/2}}{y'} .$$
(2.36)

Thus for an axisymmetric surface (~~which is cut by the axis of symmetry~~) the expression for the curvature becomes:

$$\kappa = \frac{1}{R_1} + \frac{1}{R_2} = \pm\left\{ \frac{y''}{\left[1+(y')^2\right]^{3/2}} + \frac{y'}{x\left[1+(y')^2\right]^{1/2}} \right\} .$$
(2.37)

Fluid interfaces that are cylindrical or axially symmetric represent perhaps the majority of cases of practical interest.

3. *Derivation of the Young-Laplace Equation*

The Young-Laplace Equation, Eq. (2.29), which takes the general form:

$$\Delta p = \sigma\left(\frac{1}{R_1} + \frac{1}{R_2}\right),$$

(2.38)

may be derived with reference to Fig. 2-25. Consider a small patch of surface centered at P and enclosed by a curve drawn in the surface everywhere a distance ρ (measured along the surface) from P. ρ is taken to be very small. Phase (") is on the lower, concave side of the patch, while phase (') is above it. Construct orthogonal lines AB and CD as shown, as lines made by a pair of orthogonal planes passing through a normal to the surface at P, *viz.*, $\underline{n}$. A normal force balance, *i.e.*, in the direction of $\underline{n}$, on the patch, requires:

$$\begin{bmatrix} \text{net pressure force} \\ \text{on patch} \end{bmatrix} = \begin{bmatrix} \text{normal component of surface tension} \\ \text{force acting on patch perimeter} \end{bmatrix}.$$

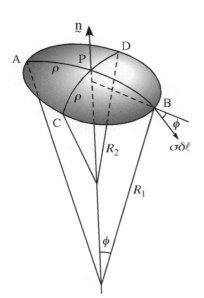

Fig. 2-25: Figure for derivation of the Young-Laplace Equation.

The net upward (+$\underline{n}$ direction) force on the patch is

$$F\uparrow = (p'' - p')\pi\rho^2$$

(2.39)

to any desired degree of accuracy by making ρ sufficiently small. To compute the downward force, consider first the force on an element of perimeter, $\delta\ell$, at point B, as shown. The force pulling downward (-$\underline{n}$ direction) on the element of perimeter is $\delta F\downarrow = -\sigma\delta\ell\sin\phi$. Since ϕ is very small (because ρ is small), $\sin\phi \approx \tan\phi \approx \rho/R_1$, and this component of force is $-\sigma(\rho/R_1)d\ell$. At point A, the component of force is the same. At points C and D, the force is $-\sigma(\rho/R_2)d\ell$. Adding these gives $-2\sigma\rho(1/R_1+1/R_2)d\ell$. To obtain the total downward force acting on the perimeter, the above expression is integrated around one-fourth the perimeter, *i.e.*, a distance of $1/4(2\pi\rho) = 1/2(\pi\rho)$, to get:

$$F\downarrow = -\int_0^{\pi\rho/2} 2\sigma\rho\left(\frac{1}{R_1}+\frac{1}{R_2}\right)\delta\ell = -\sigma\pi\rho^2\left(\frac{1}{R_1}+\frac{1}{R_2}\right). \qquad (2.40)$$

Equating $F\uparrow = F\downarrow$, and canceling $\pi\rho^2$ from both sides yields Eq. (2.38).

The general expression for curvature renders the result a second-order, non-linear partial differential equation, whose solution will give the shape of a fluid interface under given conditions.

Next the Young-Laplace Equation requires the appropriate expression for Δp. For a static system, such as that of the meniscus against a vertical plate shown in Fig. 2-26, in the absence of force fields other than gravity, the local pressure *on each side* of the interface at a point P is given by the hydrostatic equation written in the appropriate phase at the elevation h of point P above the datum plane[34] *i.e.*,

$$p'' = p_0'' - \rho''gh, \text{ and } p' = p_0' - \rho'gh. \qquad (2.41)$$

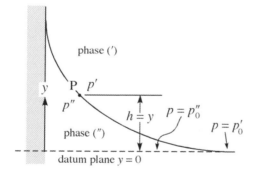

Fig. 2-26: Hydrostatic pressure difference across a curved fluid interface.

Note that both pressures must be referenced to the location of the datum plane *in the appropriate phase*, whether or not either or both the actual phases exists at the datum plane. Δp is then the difference between the hydrostatic pressures of Eq. (2.41):

$$\Delta p = \Delta p_0 - (\rho'' - \rho')gh. \qquad (2.42)$$

If $y = 0$ is located at the datum plane, then $h = y$, and

$$\Delta p = \Delta p_0 - (\rho'' - \rho')gy. \qquad (2.43)$$

If, as in Fig. 2-26, the datum plane is located at an elevation where the surface happens to be flat (one cannot always do this, as for example, in any

----

[34] Under more general conditions, stagnation flows and/or rigid body rotation may also contribute to the local value of p at a given point on either side of the interface.

of the cases of Fig. 2-23), $p_0' = p_0''$, and $\Delta p = -(\Delta\rho)gy$. Using expressions of the above type for Δp, the Young-Laplace Equation for the special cases of curvature discussed earlier may be written as follows:

1. Sphere: $\Delta p = \Delta p_0 = \dfrac{2\sigma}{R}$. $\qquad\qquad\qquad\qquad\qquad\qquad\qquad$ (2.44)

2. Circular cylinder: $\Delta p = \Delta p_0 = \dfrac{\sigma}{R}$. $\qquad\qquad\qquad\qquad\qquad$ (2.45)

3. General cylindrical surface:

$$\Delta p = \Delta p_0 - (\Delta\rho)gy = \pm\sigma\frac{y''}{\left[1+(y')^2\right]^{3/2}} \qquad\qquad (2.46)$$

4. Axisymmetric (closed) surface:

$$\Delta p = \Delta p_0 - (\Delta\rho)gy = \pm\sigma\left\{\frac{y''}{\left[1+(y')^2\right]^{3/2}} + \frac{y'}{x\left[1+(y')^2\right]^{1/2}}\right\} \qquad (2.47)$$

A quick calculation of the pressure jump Δp between the inside and the outside of a soap bubble of radius 5 mm, using Eq. (2.44), reveals that it is quite small for curvatures of this magnitude. Taking the surface tension of the soap solution as 35 mN/m, and noting that the soap film has both an inside and an outside surface, gives $\Delta p \approx 2.8\cdot10^{-4}$ atm.

4. *Boundary conditions for the Young-Laplace Equation*

In order to obtain solutions to the Young-Laplace Equation in general, one must provide information equivalent to two boundary conditions. If the surface has an edge, there are two types of conditions that may prevail there (as pictured in Fig. 2-27): 1) the "fixed edge location" condition, and 2) the "fixed contact angle" condition. The latter condition states that a given fluid interface must meet a given solid surface at some specified angle. When a fluid interface terminates on a solid surface, the angle drawn in one of the fluid phases (which must be specified) is termed the "contact angle."

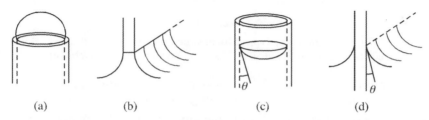

| (a) | (b) | (c) | (d) |

Fig. 2-27: Boundary conditions for Young-Laplace Equation: (a) and (b) are examples of "fixed-edge" conditions, and (c) and (d) are "fixed-angle" conditions.

When a fluid interface terminates at another fluid interface, as when a liquid drop rests upon another immiscible liquid (as shown in Fig. 2-28), the angles between the three interfaces meeting at the interline must be such as to satisfy the vectorial equation

$$\underline{\sigma}_{AB} + \underline{\sigma}_{BC} + \underline{\sigma}_{AC} = 0. \tag{2.48}$$

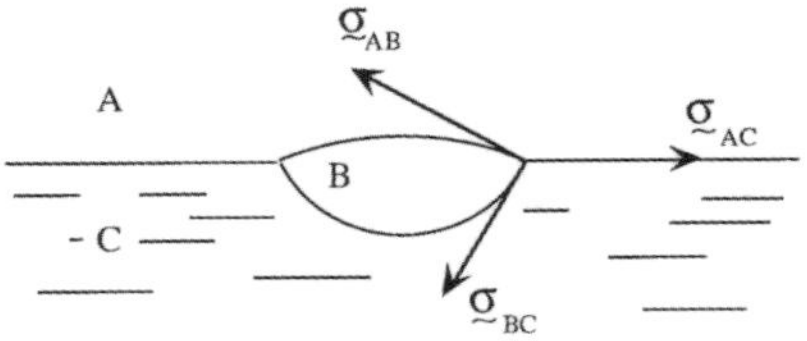

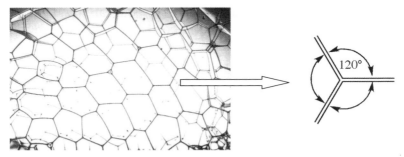

Fig. 2-28: Boundary tension forces at a tri-fluid interline.

Equation (2.48) is referred to as "Neumann's triangle of forces,"[35] and it fixes the angles between the surfaces. For the intersection of soap film lamellae in a foam (as shown in Fig. 2-29), the σ's are all equal, and the films must thus intersect at 120° angles. One might inquire about an intrinsic tension associated with the fluid interline itself, $i.e.$ a "line tension" τ_{ℓ} (which would presumably tend to contract the interline). Such a property would have units of force, or energy/length, and estimates for its magnitude range from -10^{-9} to $+10^{-9}$ N[36] (from which it is seen that it may take on negative values).[37] If the value of the line tension is positive, it will contribute a radially inward force on the interline of magnitude τ_{ℓ}/R, where R is the radius of the lens. For a lens of radius 1 mm, and a line tension of 10^{-9} N, this would contribute a tension of only 10^{-3} mN/m, quite negligible in comparison with typical surface or interfacial tensions. Line tension may play an important role, however, for micro or nano lenses.

Fig. 2-29: Photograph of a typical foam structure. After [Everett, D. H., **Basic Principles of Colloid Science**, p. 178, Roy. Soc. of Chem., Letchworth (1988).]

With reference to Fig. 2-28, consider the possibility that $\sigma_{AC} > \sigma_{AB} + \sigma_{BC}$, $i.e.$, the force pulling the interline to the right is larger than the maximum possible force pulling it to the left. Under such circumstances it

---

[35] Neumann, F., **Vorlesungen über die Theorie de Capillarität**, B. G. Teubner, Leipzig, 1894.

[36] Toshev, B. V., Platinakov, D., and Sheludko, A., *Langmuir*, **4**, 489 (1988).

[37] Kerins, J., and Widom, B., *J. Chem. Phys.*, **77**, 2061 (1982).

would be impossible to satisfy Neumann's equilibrium condition. The droplet of liquid B would have no recourse but to spread indefinitely as a thin liquid film, possibly all the way to becoming a monolayer. The driving force for the spreading of liquid B at the A-C interface is the *spreading coefficient*, defined as

$$S_{B/AC} = \sigma_{AC} - (\sigma_{AB} + \sigma_{BC}). \tag{2.49}$$

If the spreading coefficient is positive, one may expect to see the spontaneous spreading of the liquid at the interface. Spreading of this type may also be observed at a solid-fluid interface, discussed further in Chap. 4.

G. Some solutions to the Young-Laplace Equation

1. *Cylindrical surfaces; meniscus against a flat plate*

One solution to the Young-Laplace Equation that can be obtained analytically is that for the shape of the meniscus formed by a flat plate dipping into a liquid pool, as shown in Fig. 2-30. This is an example of a general cylindrical surface and satisfies the differential equation given for this case for interfaces of liquids at rest in a gravitational field, acting in the -y direction, *viz.*

$$\sigma \frac{y''}{\left[1 + (y')^2\right]^{3/2}} - \rho g y = 0, \tag{2.50}$$

where the datum plane of $y = 0$ has been chosen as the elevation where the surface is flat. The derivatives y' and y'' are taken with respect x, the horizontal coordinate measured away from the location where the interface is (or would be) vertical. The analytical solution in dimensionless form[38] is explicit in x, specifically:

$$(x/a) = \frac{1}{\sqrt{2}} \ln \left[\frac{\sqrt{2} + \left[2 - (y/a)^2\right]^{1/2}}{(y/a)} \right] - \left[2 - (y/a)^2\right]^{1/2} + C, \tag{2.51}$$

where "a" is the "capillary length," defined as: $a = \sqrt{2\sigma/\rho g}$.[39] It is a useful yardstick characterizing the size of a meniscus. The constant of integration, C, is determined by the value of x corresponding to the location of the solid-liquid-gas interline. The curve (c) in Fig. 2-30 includes all possible situations, with the relevant piece of that curve being determined by the angle made by the meniscus at the interline with the vertical axis. This angle is the difference η btween the contact angle θ and the tilt angle of the plate α, i.e. $\eta = \theta - \alpha$. C is given by:

---

[38] Princen, H.M., in **Surface and Colloid Science, Vol. 2**, E. Matijevic (Ed.), pp. 1-84, Wiley-Interscience, New York,1969.

[39] It must be noted that some authors define the capillary length as: $\sqrt{\sigma/\rho g}$.

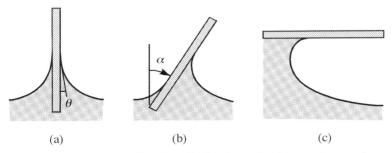

(a)   (b)   (c)

Fig. 2-30: Meniscus against a flat plate dipping into a liquid at various angles.

$$C = \left(1 + \sin\eta\right)^{1/2} + \frac{1}{\sqrt{2}} \ln\left[\frac{\left(1 - \sin\eta\right)^{1/2}}{\sqrt{2} + \left(1 + \sin\eta\right)^{1/2}}\right]. \tag{2.52}$$

Figure 2-31 shows the complete solution. The value of C corresponding to η = 0° is 0.3768... Under these conditions, the meniscus against a vertical wall

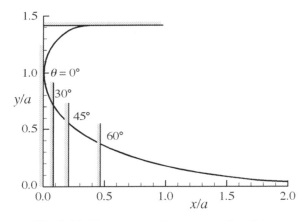

Fig. 2-31: Meniscus profile against a flat plate.

with $\theta = 0°$ rises to a height of precisely the capillary length, a. Other values of the contact angle θ yield other values of C, shifting the location of the wall, *i.e.*, where $(x/a) = 0$, but not altering the shape of the curve. The locations of the wall corresponding to contact angles of 30°, 45° and 60° are shown. A useful result that can readily be derived is that for the maximum height, h_m, of a meniscus against a vertical flat wall, *viz.*

$$\left(\frac{h_m}{a}\right)^2 = 1 - \sin\theta. \tag{2.53}$$

2. *Axisymmetric and other surfaces*

Figure 2-32 shows the solution for the axisymmetric surface of sessile drops (or captive bubbles, when inverted.) Problems of this type, *i.e.*,

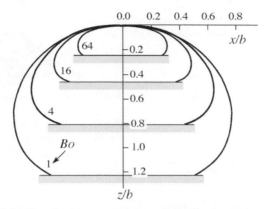

Fig. 2-32: Sessile drop (or if inverted, captive bubble) profiles.

axisymmetric interfaces cut by the axis of symmetry and described by Eq. (2.47), were solved numerically by Bashforth and Adams.[40] Over a twenty year period, these authors compiled solutions for closely-spaced values of the dimensionless parameter $\beta = (\Delta\rho)gb^2/\sigma$ using seven-place tables of logarithms. (The parameter β is now known as the Bond Number,[41] *Bo*.) This achievement was of great importance because it yielded solutions for most of the cases encountered in measuring surface tension.

Computer solutions have been obtained for meniscus shapes in which simplifying symmetries do not exist, and examples are shown in Figs. 2-33 and 2-34. The first shows the profile of a drop on an inclined surface, and the second shows the meniscus about a rectangular object immersed at an angle into a liquid surface.

3. *Non-dimensionalization of the Young-Laplace Equation; the Bond Number*

It is useful to consider limiting cases where the Young-Laplace Equation and its solution take on especially simplified forms. The most important way of delineating these is in terms of the relative importance of surface tension and gravity forces in determining the interface shape. This can be done in a systematic way by nondimensionalization, requiring only the specification of an appropriate characteristic length L for the system. Some examples are shown in Fig. 2-35. If one is interested in the shape of

---

[40] Bashforth, F., and Adams, J.C., **An Attempt to Test the Theories of Capillary Action**, Univ. Press, Cambridge, UK, 1883.

[41] Following: Bond, W. N., and Newton, D. A., *Phil. Mag.*, **5**, 794 (1928), in which the group was used in describing the rise of bubbles in liquids.

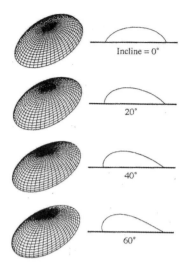

Fig. 2-33: Drop shape on an inclined
plane. From [Brown, R.A., Orr, F.M.,
and Scriven, L.E., *J. Colloid Interface
Sci.*, **73**, 76 (1980).]

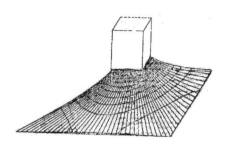

Fig. 2-34: Meniscus around a square pin.
From [Orr, F.M., Scriven, L.E., and Chu,
Y.T., *J. Colloid Interface Sci.*, **60**, 402
(1977).]

the liquid surface in a large container, such as that in a laboratory beaker
shown at the right, the characteristic length depends on what aspects of the
shape are sought. If the entire surface is of interest, the diameter of the
beaker, D, is appropriate, but for only the meniscus near the wall, the
characteristic length is usually chosen as the capillary length, *i.e.* $L = a$.

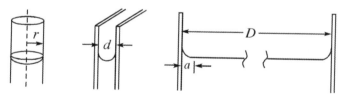

Fig. 2-35: Characteristic lengths for various systems.

Non-dimensionalization of the Young-Laplace Equation in the form:

$$\Delta p_0 + (\Delta \rho) g y = \sigma \left(\frac{1}{R_1} + \frac{1}{R_2} \right) \tag{2.54}$$

proceeds from the definition of the dimensionless length variables:
$\hat{y} = y/L$; $\hat{x} = x/L$, *etc.*, and gives

$$\left[\frac{(\Delta p_0)L}{\sigma} \right] + \left[\frac{(\Delta \rho) g L^2}{\sigma} \right] \hat{y} = \left(\frac{1}{\hat{R}_1} + \frac{1}{\hat{R}_2} \right). \tag{2.55}$$

It is seen to yield two dimensionless groups. The first, $(\Delta p_0)L/\sigma$, is a
dimensionless reference curvature. The second is the Bond Number, *Bo*:

$$Bo = \frac{(\Delta\rho)gL^2}{\sigma} = \frac{\text{(gravity forces)}}{\text{(surface tension forces)}}, \qquad (2.56)$$

which is seen to be the ratio of the gravity forces to the surface tension forces that are responsible for determining the shape of a fluid interface.

When the Bond Number is sufficiently small (< 0.01), gravity is unimportant in determining the shape of the interface, and under such circumstances, the second term of the equation drops out, giving *surfaces of constant curvature*. All confined cylindrical surfaces become portions of right circular cylinders, and all closed surfaces of revolution become portions of spherical surfaces (spherical caps, *etc.*). For example, for a spherical cap of radius R, $L = R$, and $\hat{R}_1 = \hat{R}_2 = R/R = 1$, so that:

$$\left[\frac{(\Delta p_0)R}{\sigma}\right] + 0 = \left(\frac{1}{\hat{R}_1} + \frac{1}{\hat{R}_2}\right) = 2, \text{ or } \Delta p_0 = \frac{2\sigma}{R}. \qquad (2.57)$$

On the other hand, if Bo is very large (> 100), surface tension forces will be unimportant relative to gravity in determining the interface shape. The interface (at rest) will be just a flat surface perpendicular to the g-vector. This would be the case considering the *entire* surface of liquid in a large beaker (large $L = D$), as opposed to the shape of the meniscus near the wall. Low-to-moderate Bo cases are thus of importance in capillary hydrostatics. Various ways in which very low Bond Numbers can be achieved might be:

- Characteristic length (L) is small.
- Density difference between phases ($\Delta\rho$) is small.
- Gravitational acceleration (g) is low.
- Surface tension (σ) is large.

When fluid systems are small, low Bo conditions often exist; for liquids of ordinary surface tension, this is usually the case when $L \leq 1$ mm. Thus menisci in small tubes, small liquid bridges between solid particles, *etc.* will have surfaces strongly affected (and sometimes totally determined) by capillary forces. Surface tension forces are also dominant in determining the shape of interfaces across which the density difference is small. An oil drop suspended in a liquid of nearly the same density will assume the shape of a sphere, undistorted by gravity, which would flatten the drop if it were denser than the medium, and distend it, if it were lighter. A soap bubble, with air inside and outside at nearly the same pressure, is also a sphere. A soap bubble deposited on a flat surface pre-moistened with the soap solution, will be a perfect hemisphere.

Another situation leading to very low Bond Numbers is that of low-to-zero g, as realized on board spacecraft, and many capillary hydrodynamics experiments have been performed on space flights (as well as in zero-g maneuvers in ordinary aircraft). When gravity is no longer operative to contain or transport liquids, they behave in ways that are often counter-intuitive to one's experience on Earth. Wetting liquids, for example, when let loose in the capsule, do not fall to the "floor," but may contact and spread out over the entire solid inner surface of the capsule and all the equipment, *etc.*, contained in it.[42]

The condition of high σ may yield a small Bond Number. Surface tension values are not large enough to render very large fluid interfaces free of gravitational influence, but it is a matter of experience that a droplet of mercury (with a very high surface tension) will be more nearly spherical than a drop of water or organic liquid of comparable size.

4. *Saddle-shaped surfaces*

A soap film open to the same pressure on both sides, as in the case of Fig. 2-23(d) is interesting. The surface must be one of zero mean curvature everywhere because Δp is zero, yet the film is clearly "curved." This does not mean that the surface must be flat, but may be saddle-shaped, as shown in Fig. 2-36. The two plane curvatures must be equal in magnitude and opposite in sign so that the *sum*: $(1/R_1 + 1/R_2)$ is zero. The problem of determining the surface satisfying the condition of zero mean curvature and passing through a given closed (non-planar) curve (or set of curves) in space is known as "Plateau's problem," after the blind Belgian physicist, J. Plateau, who published work on capillary hydrostatics in the late 1800's[43].

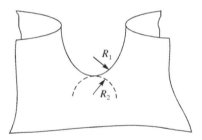

Fig. 2-36: Saddled-shaped surface (surface of zero mean curvature).

It can be proven that this is also the surface of minimum area passing through the given closed curve(s). A number of fascinating experiments with Plateau's problem can be done with a soap solution and wire frames of various shapes, as suggested by Fig. 2-37. A delightful account of experiments that can be done with soap films has been written by C.V. Boys[44]. It is the substance of a series of lectures delivered to juvenile and

---

[42] A delightful 47-minute suite of zero-*g* experiments conducted by NASA on board the Space Station can be viewed at: http://www.youtube.com/watch?v=jXYlrw2JQwo

[43] Plateau, J. A. F., **Statique expérimentale et théorique des liquides soumis aux seules forces moléculaires,** Gauthier-Villars, Trubner et cie, F. Clemm, 2 Vols., 1873.

[44] Boys, C.V., **Soap Bubbles: Their Colors and the Forces Which Mold Them**, Dover, New York, 1959.

popular audiences in 1889-1890. A more detailed description of much of its contents is given by Isenberg.[45]

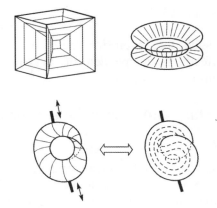

Fig. 2-37: Some soap-films suspended on wire frames. Many interesting saddle-shaped surfaces may be created by selectively puncturing different panels (using an alcohol-dipped pencil point) on the cubical frame. Puncturing the center panel on the structure between circular loops leads to a film as shown in Fig. 2-23(d). The film on the spiral wire may jump between two shapes as the spiral is squeezed or distended.

The apparent mechanical equivalence between a flat surface and a saddle-shaped surface of zero mean curvature belies the assumption that an interface shape may be specified completely in terms of a single variable, such as κ or R_m. The interface shape in fact requires *two* variables for its specification. These may be chosen as the two principal radii of curvature, R_1 and R_2, defined earlier, but more commonly one uses the curvature κ, defined by Eq. (2.31), as the first variable, and

$$\lambda = \pm \frac{1}{R_1 R_2},\qquad(2.58)$$

termed the *Gaussian* curvature, as the second. Different nomenclature and notation are sometimes used. The *mean* curvature, defined as $H = 1/2\ \kappa$, is termed the "Hermitian curvature." Gaussian curvature is generally important in those cases when the interfacial layer has a highly organized structure, as might be the case for close-packed surfactant monolayers or bilayers. Such interfaces may resist bending deformations in accord with the relationship given by Helfrich:[46]

$$F^\sigma \text{(bending)} = \frac{1}{2}k_1(\kappa - \kappa_o)^2 + k_2\lambda,\qquad(2.59)$$

where F^σ(bending) is the Helmholtz free energy/area of an interface attributable to its state of bending, and k_1 and k_2 are the "mean bending modulus" and the "saddle splay modulus," respectively. These constants have units of energy and magnitudes of order kT. κ_0 is the "spontaneous curvature," taken as zero for interfacial layers of symmetrical structure (Critical Packing Parameter ≈ 1, *cf.* Chap. 3.I), but non-zero otherwise. The spontaneous Gaussian curvature is taken as zero in all cases. A second

[45] Isenberg, C., **The Science of Soap Films and Soap Bubbles**, Dover Publ., New York, 1992.
[46] Helfrich, W., *Z. Naturforsch.*, **28c**, 693 (1973); **33a**, 305 (1978).

Helfrich Equation expresses the interfacial tension in such systems as a function of curvature:

$$\sigma_{\text{curved}} = \sigma_{\text{flat}} + \frac{1}{2}k_1\kappa^2 - k_1\kappa_o\kappa + k_2\lambda. \tag{2.60}$$

The magnitudes of k_1 and k_2 are such that even for these structured interfaces, the interfacial tension is effectively independent of curvature for radii of curvature in excess of a few tens of nanometers.

H. The measurement of surface and interfacial tension

1. *Geometric vs. force methods*

A large number of methods and devices for measuring surface or interfacial tension in the laboratory have been proposed, and many are now represented by commercial instrumentation. A few examples are listed in Table 2-5. Solutions to the Young-Laplace Equation, one way or another, provide the basis for their use. Some, termed geometric methods, are based

Table 2-5: Some methods for measuring surface or interfacial tension.

Geometric methods

- Capillary rise
- Sessile drop (captive bubble)
- Pendant drop (pendant bubble)
- Spinning drop
- Oscillating jet
- Contracting circular jet

Force methods

- Du Nüoy ring detachment
- Wilhelmy slide (or rod)
- Langmuir barrier
- Drop weight (volume)
- Maximum bubble pressure

on a direct determination of an interface shape or position. In these cases, the boundary tension is determined by finding the value for it that produces the best match between a measured interfacial profile or location and the appropriate solution to the Young-Laplace Equation. Force methods, on the other hand, are based on the measurement of a force or its equivalent, such as a mass, volume or pressure, and its comparison with the value computed using the Young-Laplace Equation. In the latter case, one most often deals with a solid object suspended in or detached from a fluid interface or a liquid drop detached from an orifice. Since the geometry of the experimental situation can be designed to be convenient, one is essentially always dealing with interfaces of a high degree of symmetry. The interfaces are usually closed-axisymmetric or cylindrical (in the general sense). A few of the methods that are important historically or are commonly used in present-day laboratories are described briefly below.

2. Capillary rise

One of the oldest methods for measuring surface tension is based on determining the position of the meniscus of the liquid in a capillary tube. If the liquid wets the tube wall, which is generally glass, its surface is constrained to meet the wall at a contact angle less than 90°. The meniscus is thus concave upward, requiring that at equilibrium, the pressure above it be greater than the pressure beneath. To achieve equilibrium, the meniscus rises in the tube, as shown in Fig. 2-38, until the hydrostatic pressure beneath the surface ($p_0 - \rho gh$) is sufficiently below atmospheric (p_0) to support the

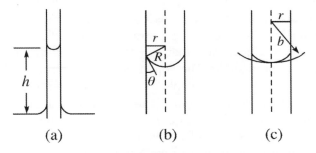

(a)   (b)   (c)

Fig. 2-38: The capillary rise method for measuring surface tension. (a) h is the equilibrium rise height, (b) If the meniscus is spherical, but the contact angle is > 0°, the radius of the meniscus is $R = r/\cos\theta$, where r is the radius of the capillary, and θ is the contact angle, (c) if the meniscus is flattened by gravity, the radius of the curvature at its apex is $b > r$.

curvature of the meniscus. The simplest situation arises when $Bo = (\Delta\rho)gr^2/\sigma << 1$ (where r = capillary radius) so that gravity will not flatten the meniscus, and it will be a segment of a sphere. If the contact angle is 0°, the meniscus will be a hemisphere of radius r, so that:

$$\Delta p = \rho gh = \sigma\left(\frac{1}{R_1} + \frac{1}{R_2}\right) = \frac{2\sigma}{r} , \qquad (2.61)$$

where ρ is the density of the liquid (the density of the overlying gas has been neglected) and h is the height of rise from the flat surface in the vessel to the bottom of the meniscus. (In order to assure a flat surface in the vessel, it should be at least five cm wide, so that for the vessel, $Bo > 100$.) The above formula simplifies to:

$$\sigma = \frac{1}{2}\rho grh . \qquad (2.62)$$

Parenthetically, it was noted by early investigators (including Leonardo da Vinci) that for a given capillary tube and liquid, the product rh was a constant. It is evident now that

$$rh = \frac{2\sigma}{\rho g} = \text{constant},$$
(2.63)

from which one obtains the original (da Vinci) definition of the capillary constant:

$$a^2 = \frac{2\sigma}{\rho g} = \text{the capillary constant, } [=] \text{ length}^2.$$
(2.64)

The capillary length, a, has been defined earlier and is seen to be the square root of the capillary constant. If the contact angle is different from 0°, (but $Bo \ll 1$) one would still have a spherical segment as the meniscus, but the radius of curvature would be $R = r/\cos\theta$, as shown in Fig. 2-38(b), and the amount of capillary rise would be correspondingly less. As a practical matter, however, it is necessary that $\theta = 0°$, as this is the only condition which is reliably reproducible. The condition is generally satisfied for most liquids against glass (with the notable exception of mercury, for which $\theta > 90°$, resulting in capillary *depression*) if the glass is scrupulously clean and has been put in contact with the liquid for a sufficient period of time (*i.e.*, "seasoned"). The liquid is generally brought to a level above the equilibrium rise height and allowed to recede to the equilibrium position. It is thus the receded angle that must be 0°. Also, one usually uses a device with two tubes of different radius with a precision cathetometer to measure the difference in capillary rise between them.

The radius of the capillary required to achieve $Bo < 0.01$ is generally less than 0.2 mm, often impracticably small. For larger tubes, the meniscus must be treated as a general surface of revolution, flattened to some extent by gravity. The surface tension in this case can be expressed in terms of the rise height and the radius of curvature, b, of the meniscus at its apex, as shown in Fig. 2-38. But determining b amounts to solving for the entire meniscus configuration using the Bashforth-Adams tables. This is extremely inconvenient since the unknown (σ) is buried in β and b, and tedious trial and error is required. Thus Sugden[47] derived tables from those of Bashforth and Adams for use with the capillary rise method. They give values of (r/b) *vs.* (r/a) (where a = the capillary length) for the case of $\theta = 0°$, as shown in Table 2-6. The procedure for using the tables is one of successive approximation.

[47] Sudgden, S., *J. Chem. Soc.*, **1921**, 1483.

Table 2-6: Sugden's tables for capillary-rise corrections.

Values of r/b for values of r/a

r/a	0.00	0.01	0.02	0.03	0.04	0.05	0.06	0.07	0.08	0.09
0.00	1.0000	9999	9998	9997	9995	9992	9988	9983	9979	9974
0.10	0.9968	9960	9952	9944	9935	9925	9915	9904	9893	9881
0.20	9869	9856	9842	9827	9812	9796	9780	9763	9746	9728
0.30	9710	9691	9672	9652	9631	9610	9589	9567	9545	9522
0.40	9498	9474	9449	9424	9398	9372	9346	9320	9293	9265
0.50	9236	9208	9179	9150	9120	9090	9060	9030	8999	8968
0.60	8936	8905	8873	8840	8807	8774	8741	8708	8674	8640
0.70	8606	8571	8536	8501	8466	8430	8394	8358	8322	8286
0.80	8249	8212	8175	8138	8101	8064	8026	7988	7950	7913
0.90	7875	7837	7798	7759	7721	7683	7644	7606	7568	7529
1.00	7490	7451	7412	7373	7334	7295	7255	7216	7177	7137
1.10	7098	7059	7020	6980	6941	6901	6862	6823	6783	6744
1.20	6704	6655	6625	6586	6547	6508	6469	6431	6393	6354
1.30	6315	6276	6237	6198	6160	6122	6083	6045	6006	5968
1.40	5929	5890	5851	5812	5774	5736	5697	5659	5621	5583
1.50	5545	5508	5471	5435	5398	5362	5326	5289	5252	5216
1.60	5179	5142	5106	5070	5034	4998	4963	4927	4892	4857
1.70	4822	4787	4753	4719	4686	4652	4618	4584	4549	4514
1.80	4480	4446	4413	4380	4347	4315	4283	4250	4217	4184
1.90	4152	4120	4089	4058	4027	3996	3965	3934	3903	3873
2.00	3843	3813	3783	3753	3723	3683	3663	3633	3603	3574
2.10	3546	3517	3489	3461	3432	3403	3375	3348	3321	3294
2.20	3267	3240	3213	3186	3160	3134	3108	3082	3056	3030

The first estimate is $a \approx \sqrt{rh}$. Then r/a is computed and r/b is obtained from the table, giving an estimate of b. The next estimate of a is: $a \approx \sqrt{bh}$, from which r/a is computed, *etc.*, to convergence. It seldom requires more than three rounds.

Lord Rayleigh[48] proposed a convenient approximate solution, valid for $Bo \leq 0.04$, in the form:

$$a^2 = r\left(h + \frac{r}{3} - \frac{0.1288r^2}{h} + \frac{0.1312r^3}{h^2} \cdots\right). \tag{2.65}$$

For wide tubes ($Bo > 10$), he proposed the approximate formula:

$$\ln\left(\frac{h}{a}\right) = 0.6648 + 0.1978\left(\frac{a}{r}\right) - \sqrt{2}\left(\frac{r}{a}\right) + \frac{1}{2}\ln\left(\frac{r}{a}\right). \tag{2.66}$$

---

[48] Lord Rayleigh (J. W. Strutt), *Proc. Roy. Soc.*, **A92**, 184 (1915).

3. *Sessile drop and pendant drop*

Among the most commonly used methods for measuring boundary tension are those in which the shape profiles of drops or bubbles are determined and compared with solutions of Bashforth and Adams with the value of σ chosen that produces the best fit. The most common methods of this type are those of the sessile drop (or captive bubble) and the pendant drop (or bubble) as pictured in Fig. 2-39. An expedited method obtains the fit in terms of a pair of descriptive parameters, commonly the maximum diameter, and the height above it (or below it). Tables derived from the Bashforth-Adams calculations can then be used to estimate surface tension.

At present it is more common to use a commercially available axisymmetric drop shape analysis (ADSA) system, as shown schematically in Fig. 4-22 (where it is discussed in the context of determining contact angle). An accurate drop or bubble profile is obtained using a precision CCD camera, and the full profile match is effected using a computer.

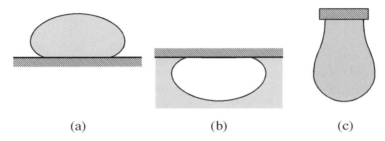

(a)   (b)   (c)

Fig. 2-39: Axisymmetric drop profiles: (a) sessile drop, (b) captive bubble, (c) pendant drop.

4. *Du Noüy ring detachment*

In this method, pictured in Fig. 2-40, one measures the force required to detach a ring (usually of platinum) from a surface. A schematic plot is shown of the measured downward force as a function of the height of the ring above the undisturbed surface of the liquid. It is the maximum in the measured force that is used to determine the boundary tension. After the maximum is reached, the meniscus beneath the ring begins to contract before final detachment occurs. Since the maximum force is what is needed, most current instrumentation does not actually detach the ring. The maximum downward force on the ring is given by

$$(\text{Force}) \downarrow \; = \frac{4\pi R\sigma}{F_c}, \qquad\qquad (2.67)$$

where R is the radius of the ring and F_c is a correction factor for which there are tables[49] or empirical fitting formulas.[50-51] It represents the weight of the

---

[49] Harkins, W. D., and Jordan, H. F., *J. Amer. Chem. Soc.*, **52**, 1751 (1930).

subtended liquid at the point of maximum force. The needed correction factor is generally automatically implemented in current commercial instrumentation. The method is generally suitable for interracial as well as surface tension measurements. Other objects of known wetted perimeter (*e.g.*, plates, cylinders, *etc.*) can also be detached from surfaces and the appropriate forces measured and analyzed to give the boundary tension.

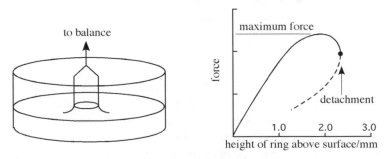

Fig. 2-40: du Nüoy ring detachment method.

5. *Wilhelmy slide*

In a simple but powerful method one measures the downward force on an object partially immersed in the liquid. The usual configuration is a dipping slide, as shown in Fig. 2-41 (left), known as a Wilhelmy slide. In some cases a rod is convenient to use rather than a slide, as shown at the right. It is assumed that the contact angle of the liquid against the solid surface is 0°, in which case the downward force on it consists of its weight

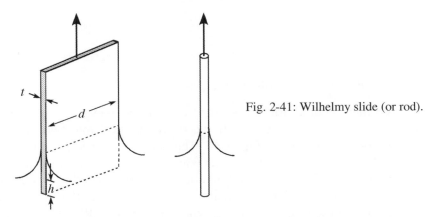

Fig. 2-41: Wilhelmy slide (or rod).

in air, plus the downward-pulling interline force of surface tension acting around its wetted perimeter, minus the buoyancy force due any protrusion of the plate (or rod) beneath the level of the undisturbed liquid surface. The weight in air is usually zeroed out, so the net downward force is given by:

[50] Freud, B. B., and Freud, H. Z., *J. Amer. Chem. Soc.*, **52**, 1772 (1930).
[51] Zuidema, H. H., and Waters, G. W., *Ind. Eng. Chem.*, **13**, 312 (1941).

$$F_{net} = \sigma \cdot (\text{wetted perimeter}) - (\text{buoyancy}). \tag{2.68}$$

For a rectangular plate of width d and thickness t, extending a distance h beneath the flat, undisturbed surface of the liquid:

$$F_{net} = \sigma \cdot (2d + 2t) - \rho ghtd. \tag{2.69}$$

Since the plates used are usually thin ($t \ll d$), and measurements are made at the point where $h = 0$, σ is evaluated from:

$$\sigma = \frac{F_{net}}{2d}. \tag{2.70}$$

The Wilhelmy technique is often used for measurement of the surface tension, σ, of water in a rectangular trough, covered with an insoluble (Langmuir) surfactant monolayer. In a method pioneered by Agnes Pockels,[52] and shown schematically in Fig. 2-42, one compresses or expands the surfactant film by means of a movable barrier that seals the liquid surface on one side from that on the other. Surface tension is monitored during compression or expansion by measuring the force on the plate held in null position.

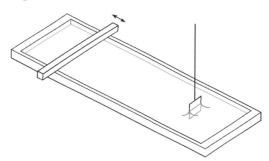

Fig. 2-42: Schematic of Pockels' trough, with Wilhelmy slide.

6. *Langmuir film balance*

Another method for measuring the surface tension of monolayer-covered liquid surfaces is to divide the trough surface into two parts by a movable boom, as shown in Fig. 2-43. The boom is connected to the trough edges by flexible hydrophobic threads that seal the parts of the surface from one another. The film is deposited on one side, while the other side presents a clean water surface. A force equal to σW, where W is the boom length, pulls on the film side of the boom, while the force $\sigma_0 W$, where σ_0 is the surface tension of pure water, pulls in the opposite direction. The net force on the boom, $F_{net} = (\sigma_0 - \sigma)W$, is measurable by means of rigid connection to a calibrated torsion wire. When the monolayer is soluble to some extent, the subphase portions between the two parts of the surface may be kept apart by

---

[52] Pockels, A., *Nature*, **43**, 437 (1891).

a flexible membrane connecting the boom and its tethers to the bottom of the trough.

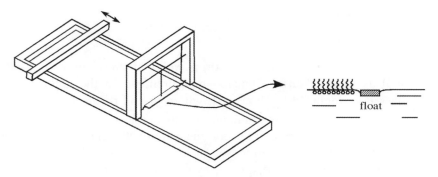

Fig. 2-43: Langmuir film balance.

7. Drop weight (or volume)

In this method, drops are formed, as shown in Fig. 2-39(c), and made to grow until they break away by gravity. The collective volume (or weight) of several drops is measured. Despite the complexity of the break-off event, as shown in the rapid sequence photographs of Fig. 2-44, the size at break-off is a reproducible function of σ for a given nozzle radius, and fluid density difference.[53] To a rough approximation, the weight of the drop is given by $2\pi r\sigma$, where r is wetted tube radius, assuming the surface is vertical around the perimeter at the time of rupture, and that no liquid is retained on tip when detachment occurs. Actual results may be conveniently expressed as

$$ mg = r\sigma \frac{1}{F_c}, \qquad (2.71) $$

Fig. 2-44: Break-off of water drops in air. From [Pierson, F. W., and Whitaker, S., *J. Colloid Interface Sci.*, **54**, 219 (1976).]

(a) early stages   (b) late stages

where m is the drop mass and the correction factor F_c is a universal function of (V/r^3), where V is the drop volume. F_c accounts primarily for liquid

[53] Harkins, W. D., and Brown, F. E., *J. Amer. Chem. Soc.*, **41** 499 (1919).

retained on the tip after detachment and has been given in tabular form,[54] and fit analytically by[55]:

$$F_c = 0.14782 + 0.27896\left(\frac{r}{V^{1/3}}\right) - 0.1662\left(\frac{r}{V^{1/3}}\right)^2, \qquad (2.72)$$

valid for $0.3 < (r/V^{1/3}) < 1.2$. The drop weight method may be used for interfacial as well as surface tension measurement. In the latter case, the drop mass m is replaced by $V|\rho'' - \rho'|$, where ρ' and ρ'' are the densities of the two liquids involved. Instruments are available commercially,[56] but a home-built setup for interfacial tension measurement is shown schematically in Fig. 2-45. Oil drops are formed in an inverted water-filled vessel with a side arm. The mass of water displaced by the formation of a given number of oil drops is used to determine the oil drop size.

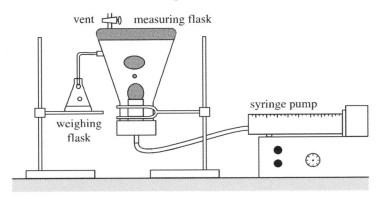

Fig. 2-45: Drop weight method for interfacial tension measurement.

8. *Maximum bubble pressure and dynamic surface tension*

In the formation of a bubble from a nozzle tip, as shown in Fig. 2-46, maximum pressure is required when the radius of curvature of the bubble is minimum (for $Bo \ll 1$), and under these conditions the surface is a hemisphere, with $p_{max} - p_{liq} = 2\sigma/r$, where r is the radius of the capillary. For larger capillaries, the appropriate corrections can be worked out for the bubble flattening using the Bashforth and Adams tabulations, as in the capillary rise method. A useful approximate formula (with reference to Fig. 2-46) is:

based on an approximation for small tubes given in Johnson, C. H. J., and Lane, J. E. (2.73) Colloid Interface Sci, 47, 117 (1974)

$$\sigma = \frac{r}{2}p_{max} - \frac{1}{3}\rho g r^2 - \frac{1}{2}\rho g r h - \frac{(\rho g)^2 r^3}{12(p_{max} - \rho g h)}, \qquad (2.73)$$

[54] Lando, J. L., and Oakley, H. T., *J. Colloid Interface Sci.*, **25**, 526 (1967).

[55] Heertjes, P.M., De Nie, L.H., and De Vrie, H.J., *Chem. Eng. Sci.*, **26**, 441 (1961).

[56] Gilman, L. B.,(Krüss USA) "A Review of Instruments for Static and Dynamic Surface and Interfacial Tension Measurement," presented at 84th AOCS Ann. Mtg. and Expo, Anaheim, CA Apr. 27, 1993.

where h is the depth of the capillary tip beneath the surface of the liquid. For very small tubes (giving $Bo < 0.01$), the final three terms of Eq. (2.73) are negligible. The maximum bubble pressure method has the advantage of being very rapid, and the surface formed is fresh. The method is good for difficult-to-access liquids, such as molten metals, polymer melts, *etc.*, and for rapid "on-line" determinations of surface tension in general.

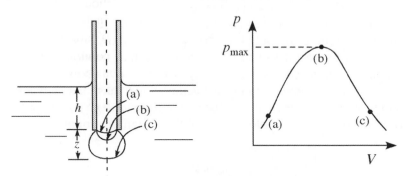

Fig. 2-46: Maximum bubble pressure method.

One of the most important advantages of the maximum bubble pressure method is that it is amenable to dynamic surface tension measurement, usually down to surface ages of a few milliseconds or less. Surface tension is often time dependent as a result of slow solute diffusion to the interface, kinetic barriers to adsorption or desorption, surface chemical reaction rates, including denaturation, *etc.*, as mentioned earlier and discussed further in Chap. 3. Many of these processes, such as spray coating, pesticide or herbicide spray applications and ink-jet printing, produce changes over time scales of practical interest. Structural changes accompanying the formation of fresh surfaces of pure liquids occur over time scales of *micro*seconds or less, and are not of practical interest. While the methods of drop (or bubble) shape analysis, Wilhelmy tensiometry and drop weight determination are useful for surface ages of the order of one second or greater, they are not applicable for the much shorter times that are often of practical interest. In commercial maximum bubble pressure devices, the bubbling rate may be varied so that the surface tension corresponding to a range of surface ages may be obtained. A bubble tensiometer currently available from Lauda Instruments (Model MPT-2)[57] is capable of measurements for surface ages from about three seconds down into the sub-millisecond time range.

9. *The pulsating bubble "surfactometer"*

A useful variation on the maximum bubble pressure method employs a single small bubble suspended at the end of a capillary and made to pulsate and therefore produce sinusoidal time variations in the bubble surface area.

---

[57] Munsinger, R. A., *Amer. Lab. News*, Jan. (1997).

The device is termed commercially a "surfactometer,"[58] and is shown schematically in Fig. 2-47. The bubble radius is made to oscillate between 0.40 and 0.55 mm at frequencies from 1 to 100 cycles/min. The sample chamber contains 25 μL, usually of a surfactant solution, that communicates with ambient air through a capillary, which also serves as the airway for bubble formation. The liquid sample is pulsed by means of a volume displacement piston to produce the desired radius variation, while the pressure inside the chamber is monitored using a sensitive pressure transducer. Assuming the bubble to be spherical, knowledge of its size and the chamber pressure suffices to calculate the surface tension.[59] The method is especially applicable to the study of the dynamics of surfactant monolayers, in particular lung surfactant, and has the advantage of producing wider ranges of surface compression/expansion rates and requiring much smaller sample sizes than the Langmuir trough.

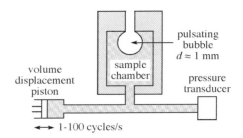

Fig. 2-47: Schematic of the pulsating bubble "surfactometer."

10. *Elliptical (vibrating) jet*

One of the earliest devices for measurements of dynamic surface tension in the millisecond range was the elliptical (oscillating) jet. In this method, now primarily of historical interest, a liquid is forced through an elliptical orifice at a sufficient rate to form a jet. The jet attempts to "correct" its noncircular cross-section, and in so doing overshoots and oscillates about a circular shape, as shown in Fig. 2-48. The value of the surface tension can be computed from knowledge of the jet parameters, fluid density and the measured wavelength of the oscillations, as originally shown by Nils Bohr, but in simplified form for liquids of low viscosity by Sutherland:[60]

$$\sigma \approx \frac{2\rho Q^2 \left(1 + 37a^2/24b^2\right)}{3a\lambda^2 \left(1 + 5\pi^2 a^2/3\lambda^2\right)},$$ (2.74)

where Q = volumetric flow rate; $a = r_{max} + r_{min}$, and $b = r_{max} - r_{min}$. When σ is changing with time (surface age), as by the adsorption of a solute that must first diffuse to the surface, λ will vary with distance along the jet (*i.e.*, time).

[58] Enhorning, G., *J. Appl. Physiol.*, **43**, 198 (1977).
[59] This assumption has been relaxed: Seurynck, S.L., *et al.*, *J. Appl. Physiol.*, **99**, 624 (2005).
[60] Sutherland, K. L., *Aust. J. Chem.*, **7**, 319 (1954).

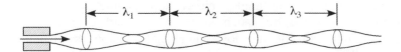

Fig. 2-48: Elliptical oscillating jet.

Since the flow rate is known, one can measure the wavelength λ as a function of position and, knowing the rate of jet flow, extract the time variation of the surface tension. Usually up to seven or eight wavelengths can be realized before the jet disintegrates. It is generally not a good method for interfacial tension, because the jet breaks up too quickly. A number of practical difficulties have precluded the method from being realized commercially, but its use in earlier studies was indispensable in the study of dynamic processes at interfaces.

11. *Contracting circular jet*

For surface or interface lifetimes in the hundredths of a second range, the elliptical jet is not practical, but the contracting circular jet method may be used, in particular for oil-water interfacial tensions.[61] The extent of contraction in a given distance for a given set of jet parameters and fluid properties can be related directly to the interfacial tension.

12. *Problems with interfacial tension measurement*

There are sometimes difficulties in the measurement of interfacial tension that do not arise in the measurement of surface tension. While the drop shape analysis, the ring detachment method and drop weight methods are all in principle adaptable to interfacial tension measurements, their use is limited to systems with an adequate density difference between the liquids and/or a sufficiently high value of the interfacial tension itself. From a practical point of view, a density difference of at least 0.10 specific gravity units is required. Otherwise, the drop shape will be too close to spherical for accurate matching with the Bashforth-Adams computed profiles, and for the drop weight method, drops of impractically large size are obtained. In the ring detachment method, very large displacements of the ring above the surface are required for the maximum force to be attained. The recommended method for handling the problem is to carefully equilibrate the two liquids (with respect to any mutual solubility), as in a separatory funnel, remove samples of each liquid and measure their respective surface tensions. The needed interfacial tension can then be computed using Antanow's Law, Eq. (2-20).

Another problem arises if the interfacial tension between the liquids is extremely small (< 0.1 mN/m). It is then difficult to adequately pin a drop for profile determination, *i.e.*, either one liquid or the other will wet out the

---

[61] England, D. C., and Berg, J. C., *AIChE J.*, **17**, 313 (1970).

surface, producing contact angles either approaching 0° or 180°. The detachment method requires such small forces and displacements that accurate determination of the low interfacial tension is difficult. The problem is addressed by the spinning drop method described below.

13. *Spinning drop method*

The spinning drop method[62], shown in Fig. 2-49, is especially useful for determining ultra-low interfacial tensions of the type encountered in the polymer-surfactant flooding strategies for tertiary oil recovery. A drop of the less dense liquid is injected into a capillary tube containing the denser fluid. The tube is spun on its axis until the suspended drop is elongated into a cylindrical shape with hemispherical caps. The lower the interfacial tension, the greater will be the elongation. When the drop length is much greater than the radius, r_m, the result is:

$$\sigma = \frac{(\Delta\rho)\omega^2 r_m^3}{4}, \tag{2.75}$$

where $\Delta\rho$ is the density difference between the liquids, and ω is the angular velocity of rotation about the tube axis.

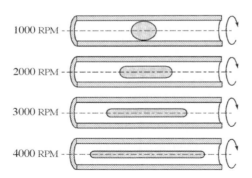

Fig. 2-49: Spinning drop method. Drawing taken from photographs showing a heptane drop (0.156 cm^3) in glycerol rotating at various speeds about a horizontal tube axis. Redrawn from [Princen, H. M., Zia, I., and Mason, S. G., *J. Colloid Interface Sci.*, **23**, 99 (1967).]

I. Forces on solids in contact with liquids: capillary interactions

1. *Liquid bridges*

Recall that liquids exert forces on solids in contact with them through the action of boundary tension. When two or more solid objects are in contact with the same liquid mass, such that they share a meniscus there are effective forces (referred to as capillary forces) acting between the solid bodies, tending either to draw them together or to push them apart. It is convenient to distinguish between two categories of systems, *viz.*, liquid bridges and shared menisci, as shown in Fig. 2-50.

---

[62] Vonnegut, B., *Rev. Sci. Inst.*, **13**, 6 (1942).

When a finite liquid mass separates two solids, it is referred to as a "liquid bridge." A similar configuration (a "vapor bridge") is created when a bubble joins two particles immersed in a liquid medium. Consider as an example, a drop of liquid between a pair of horizontal flat plates wet by the

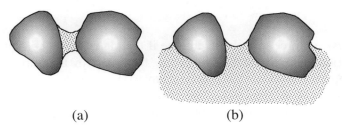

(a)   (b)

Fig. 2-50: (a) Liquid bridge, (b) shared meniscus.

liquid as shown in side view in Fig. 2-51. The volume of the drop is small, but since the plate spacing, h, is also considered very small, the diameter of the drop D (as would be observed in a top view) may be large. Focusing on the upper plate, one may note that there are two types of forces (apart from gravity) drawing it toward the lower plate. The first of these is the downward component of the *interline force* acting around the perimeter of the drop, *viz.* $\pi D \sigma \sin\theta$, where θ is the contact angle of the liquid against the

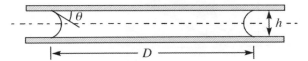

Fig. 2-51: Liquid bridge between horizontal flat plates.
Not in perspective, since $h \ll D$.

glass plate. The second contribution is the *capillary pressure force*, i.e., the force pulling down on the upper plate because the liquid is on the convex side of the fluid interface, rendering the pressure inside the liquid lower than that outside by an amount Δp, the Young-Laplace pressure jump. This produces a pressure force acting to hold the plates together equal to Δp times the area which is wetted ($\pi D^2/4$). The capillary pressure force, which can be considerable if θ is low and the plate spacing is small, is often much larger than the interline force, which may then be neglected.

Δp is given by the Young-Laplace Equation: $\Delta p = \sigma(1/R_1 + 1/R_2)$. The meniscus is a saddle with $R_2 = -D/2$ (the minus sign being used to account for the fact that this curvature is opposite in sign to that of the profile of the figure). Since $h \ll D$, one may neglect $1/R_2$, and treat the meniscus as effectively a cylindrical surface, and if it is assumed further that h is sufficiently small that $Bo \ll 1$, it is a right circular cylindrical surface, whose radius is $R_1 = (h/2)/\cos\theta$. Thus:

$$F \downarrow = \Delta p \frac{\pi D^2}{4} = \sigma \frac{\cos\theta}{(h/2)} \frac{\pi D^2}{4} \equiv \frac{\sigma \pi D^2 \cos\theta}{2h}. \tag{2.76}$$

For the case of perfect wetting, $\theta = 0°$, and

$$F \downarrow = \frac{\sigma \pi D^2}{2h}. \tag{2.77}$$

This "adhesive" force must be overcome if the plates are to be separated, and the formation of a liquid bridge is the first step in the formation of an adhesive bond. It is clear that it becomes very large as $h \to 0$. It is also seen to go to zero as θ approaches 90°, and becomes negative (*i.e.*, a force pushing the plates apart) if $\theta > 90°$, *i.e.*, the plates are unwet.

Liquid bridges, as described above, can be formed between solids of various shapes. For example, Fig. 2-52 shows the establishment of a liquid bridge between a cylinder and a plane. The total force acting to hold the objects together consists of three terms. The first two are capillary forces of

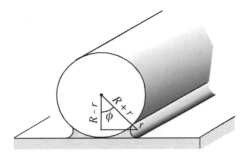

Fig. 2-52: Liquid bridge between a cylinder and a flat plate.

the same type as those described above for the flat plates, *viz.*, the downward force of the liquid surface tension acting along the cylinder-liquid interline and the force due to the Young-Laplace pressure deficit within the liquid. A third term may arise because the solids in this system are in direct contact, leading to a solid-solid adhesion term, as described later in Chap. 4. This last effect can often be neglected because the presence of even small degrees of roughness precludes significant direct solid-solid contact (unless the solids are soft and subject to flattening). If the liquid masses are sufficiently small that $Bo \ll 1$, and that the contact angle θ is 0°, the downward capillary forces per unit length acting on the cylinder are given by:

$$(F/L) \downarrow = 2\sigma \sin\phi + \Delta p (2R \sin\phi). \tag{2.78}$$

Thus with a circular meniscus, $\Delta p = \sigma/r$, and with reference to Fig. 2-52, it can be seen that the cylinder radius R, the meniscus radius r and the filling angle ϕ are related to one another in accord with

$$\frac{(R-r)}{(R+r)} = \cos\phi, \quad \text{or} \quad r = R \frac{(1-\cos\phi)}{(1+\cos\phi)}, \tag{2.79}$$

and $\Delta p = \dfrac{\sigma\,(1+\cos\phi)}{R\,(1-\cos\phi)}$. Substituting into Eq. (2.78):

$$\frac{F}{L} = 2\sigma\sin\phi + \frac{\sigma}{R}\left[\frac{1+\cos\phi}{1-\cos\phi}\right](2R\sin\phi) = 4\sigma\left[\frac{\sin\phi}{1-\cos\phi}\right]. \qquad (2.80)$$

It is to be noted that as $\phi \to 0$, $F/L \to \infty$. Thus the adhesive force is maximized when the amount of liquid is *very* small. Anyone who has glued together the parts of model airplanes knows this. Equation (2.80) is not valid, of course, when $\phi \approx 0$, since the continuum concept of surface tension breaks down as the meniscus approaches molecular dimensions, although there is evidence that this does not occur until r is less than a few nm.[63]

The adhesive force of a liquid bridge between two cylinders also approaches infinity as the bridge size approaches zero. For equal-sized cylinders,

$$\frac{F}{L} = \frac{2\sigma\sin\phi}{1-\cos\phi} \qquad \text{(half the cylinder-plane value).} \qquad (2.81)$$

The expression for the adhesive force between a sphere and a plane is:

$$F = \pi R\sigma(1+\cos\phi)(3-\cos\phi), \qquad (2.82)$$

which approaches a *finite* constant as $\phi \to 0$, *viz.*,

$$F_{\text{lim}} \to 4\pi R\sigma. \qquad (2.83)$$

a result confirmed by experiment.[64] The corresponding results for equal-sized spheres are:

$$F = \frac{1}{2}\pi R\sigma(1+\cos\phi)(3-\cos\phi), \text{ with} \qquad (2.84)$$

$$F_{\text{lim}} \to 2\pi R\sigma. \qquad (2.85)$$

The physically important case of a liquid bridge between *crossed* circular cylinders of geometric mean radius R, shown in Fig. 2-53, yields a

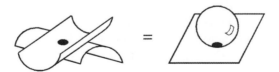

Fig. 2-53: Equivalence of small liquid bridges between crossed cylinders and a sphere and plane.

[63] Fisher, L. R., and Israelachvili, J. N., *Colloids Surfaces*, **3**, 303 (1981).
[64] McFarlane, J. S., and Tabor, D., *Proc. Roy. Soc. A*, **202**, 224 (1950).

force the same as that given for the bridge between a sphere of radius R and a flat plate, as long the condition of $r \ll R$ holds.[67] It should be recalled that all of the formulae developed above refer to the case in which the contact angle is 0°. They may all be modified to accommodate the case in which θ is finite, or is different against the two solid surfaces.[65] It should also be noted that the results above assume either a perfect line or point of direct contact between the solid surfaces subtending the liquid bridge. If these solid surfaces are held even a very small distance (a few nm) apart, as by intervening particles or asperities, the relationship between the bridge strength and bridge size is very different. Instead of rising monotonically as ϕ decreases, the inter-particle force rises to a maximum at some finite ϕ and then falls abruptly to zero as $\phi \to 0$.

Liquid bridge formation underlies the important process for the size-enlargement of fine powders or particulates known as *spherical agglomeration*. The bridging liquid is immiscible with the dispersion medium and must preferentially wet the particles. The process is used in granulation, balling, pelletization, tabletting and sintering to produce, *e.g.*, ceramic powders, carbon blacks, catalysts, commercial fertilizers, pesticides and pharmaceutical products.[66] It is especially useful for the *selective* collection of one dispersed phase from among many, such as may be desired in mineral beneficiation. The process agglomerates the more hydrophobic particles to a size that can be easily separated from an aqueous dispersion by screening or other mechanical means.

2. Shared menisci

Shared menisci refer to fluid interfaces between neighboring solids partially immersed in a common liquid pool. This configuration also leads to apparent forces acting between the solid objects in the direction parallel to the undisturbed interface, tending either to draw them together or push them apart. Solid objects, such as those shown in Fig. 2-50(b), find themselves located in fluid interfaces in the first place by virtue of forces on them *normal* to the undisturbed fluid interface. Aside from the interline forces that act around their wetted perimeters to hold them in the interface, the objects may be floating, *i.e.*, trapped at the interface by gravity. This occurs whenever the object is intermediate in density between the lower and upper fluids. If the objects are denser than the lower fluid, they may be supported by a rigid surface from below (or conversely, if they are lighter than the upper phase, they may be supported from above). In some cases, the interline forces are sufficient to retain the objects in the interface under such conditions (*e.g.*, the "floating" needle or paperclip mentioned in Chap. 1).

The concern here is with forces acting parallel to the interface. To fix ideas, consider the parallel dipping plates held or supported in the interface

[65] Orr, F. M., Scriven, L. E., and Rivas, A. P., *J. Fluid Mech.*, **67**, 723, 1975.
[66] Pietsch, W., **Size Enlargement by Agglomeration**, Wiley, New York, 1991.

as shown in Fig. 2-54. It is assumed that liquid may flow between the region between the plates and the outer pool. If the contact angle θ is less than 90°, liquid will rise to some level h required to satisfy the Young-Laplace pressure jump across the curved meniscus. Focusing on the left plate, it is seen to be acted upon by interline forces and hydrostatic (Δp) forces. The interline forces have equal and opposite horizontal components in the amount of $L\sigma\sin\theta$.[67] The capillary rise between the plates (to the level h)

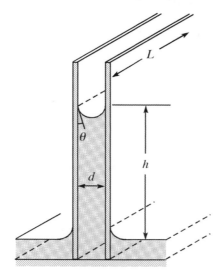

Fig. 2-54: Capillary rise between vertical flat plates. Example of a shared meniscus, with the plates held in vertical position by a supporting surface from below.

creates a pressure deficit on the right side of the free body, yielding a force

$$F = L\int_0^h \rho g y \, dy = \frac{1}{2}\rho g h^2 L,$$ (2.86)

tending to pull the plates together. Assuming the plate spacing d is very small, so that $Bo = (\rho g d^2/\sigma) \ll 1$, the meniscus will be a portion of a right circular cylindrical surface of radius $R_1 = d/2\cos\theta$. R_2, in the plane perpendicular to the figure, is infinite. Therefore:

$$\rho g h = \sigma\left(\frac{1}{R_1} + 0\right) = \frac{2\sigma\cos\theta}{d},$$ (2.87)

so that $h = \dfrac{2\sigma\cos\theta}{\rho g d}$, and

$$\frac{F}{L} = \frac{1}{2}\rho g h^2 = \frac{2\sigma^2\cos^2\theta}{\rho g d^2}.$$ (2.88)

---

[67] These forces are applied at different elevations, imparting a clockwise torque to the plate, causing it to tip toward the opposite plate at the top.

If neither plate is wet by the liquid, there will be a *depression* of the liquid between them, and examination of the diagram shows that there will be a net force tending to *push* them together. In this case h is *negative*, but h^2 of course remains positive, and Eq. (2.88) is still valid.

The case in which one of the plates is wet by the liquid, while the other is not, is shown in Fig. 2-55. In this event, the meniscus has a point of inflection, which must occur at an elevation of $h = 0$ (since the surface has no curvature there). The equation of this inflected meniscus cannot be obtained in closed form, but it can be expressed in terms of elliptic integrals. The solution[68] shows that the curvature of the meniscus at the interline on

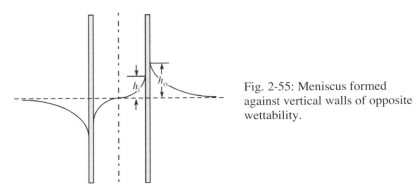

Fig. 2-55: Meniscus formed against vertical walls of opposite wettability.

the *inside* is always less than that on the outside. Thus the interline location will be further away from $h = 0$ on the outside than the inside, *i.e.*, $h_0 > h_i$, as shown. The pressure will be greater on the inside than on the outside, and the plates will be seen to *repel* each other.

The qualitative conclusions concerning the forces acting between neighboring flat plates in contact with a liquid pool also apply of course to objects of other shapes. Thus if two solid particles of the same material are floating on a liquid and happen to approach one another, a meniscus forms between them, and since the wetting is the same on both (whether it be wetting *or* non-wetting) will attract them together, leading to surface flocculation. On the other hand, if particles of two *different* materials are floating on the surface, they will be attracted if $\theta > 90°$ for both or $\theta < 90°$ for both, but if they have contact angles on opposite sides of 90°, they will be mutually repelled. If particles that are wet by the liquid come near to the edge of the containing vessel whose walls are also wet by the liquid, they will be drawn into the meniscus at the wall and accumulate there, as shown in Fig. 2-56(a). Particles un-wet by the liquid ($\theta > 90°$) will shun the edge meniscus. When the meniscus at the edge is reversed, it is the unwet particles that will be drawn to it, as shown in (b). Capillary forces are thus seen to be responsible for the clumping together of particulate materials at

[68] Princen, H.M., "The Equilibrium Shape of Interfaces, ...," in **Surface and Colloid Science, Vol. 2**, E. Matijevic (Ed.), pp. 1-84, Wiley-Interscience, New York, 1969.

fluid interfaces, as in the formation of "rafts" or "rags" of solid contaminants at liquid-liquid interfaces in extraction equipment, and they also explain the collection of particulates at the meniscus around the containing vessel, if the vessel walls and the particles are of like wettability.

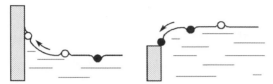

Fig. 2-56: (a) Wetted particles (○) are drawn to a wetting meniscus at the container wall while non-wetted particles (●) shun the meniscus. (b) Non-wetted particles are drawn to a non-wetting meniscus at the edge.

The action of shared menisci is also responsible for the coherence of the fibers in a paintbrush, as shown in Fig. 2-57. Immersed in either the paint or in air, the bristles are separated from one another, but when a liquid meniscus is formed between them, they are drawn together in a coherent bundle.

Fig. 2-57: Paint brush in air, in water and in air after being dipped in water. From [Boys, C.C., **Soap Bubbles and the Forces which Mould Them**, Doubleday Anchor Books, Garden City, NY, p. 22, 1959.]

The consequences of the forces of the type described above are widespread. The formation of liquid bridges or shared menisci is not only an important first step in the action of liquid adhesives, but also plays a vital role in the consolidation of wet-formed non-woven fibrous materials (such as paper). The enormous capillary forces developed between adjacent fibers during dry-down are believed to be sufficient to produce inter-fiber hydrogen bonding. These large capillary forces also pose a serious problem for the drying of porous media whose structural integrity one seeks to preserve. Waterlogged specimens of archeological interest, as an example, will implode if they are simply dried in air. Strategies for addressing this problem include the successive exchange of the water with volatile liquids of lower surface tension, or freezing the specimen followed by freeze- drying. One of the most successful methods, however, is the exchange of the water with supercritical carbon dioxide, followed by its drying without the presence of a liquid interface at all. Capillary interactions between particles bound to interfaces are discussed further in Chap. 4, and more detailed descriptions, with additional references, are given by Kralchevsky and Nagayama.[69]

J. Effect of curvature on the equilibrium properties of bulk liquids: the Kelvin Effect

1. *The vapor pressure of small droplets and liquids in pores*

When a liquid is bounded at least in part by a strongly curved interface against another fluid, the phase equilibrium properties of the system are not the same as in the case when the phases are divided by a flat interface. This is a direct consequence of the required *pressure* difference that must exist across a strongly curved fluid interface. For example, the vapor pressure of a tiny droplet of radius r, p_r^s is higher than that associated with a flat surface of the same liquid, p_∞^s, the "handbook" value. Similarly, the vapor pressures of liquids in finely porous solids are different from those over flat surfaces, leading to the phenomenon of capillary condensation described below. Analogously, the solubility of tiny droplets, bubbles or solid particulates in a liquid will be different from the solubility of their larger counterparts.

Consider here a small droplet of radius r of a pure liquid. The dependence of its vapor pressure on its radius is derived as follows. At equilibrium, the fugacities of the vapor and liquid are equal, *i.e.*, $f^V = f^L$. If it is assumed that the vapor phase behaves as an ideal gas, its fugacity is equal to its vapor pressure, *i.e.*, $f^V = p_r^s$. The fugacity of the liquid, $(f^L)_{drop}$, is to be evaluated at the pressure: $p_r^s + 2\sigma/r$. It is computed by referencing it to the fugacity of the pure liquid beneath a flat interface. Recall from thermodynamics that, at constant T:

$$RT d\ln f^L = v^L dp. \tag{2.89}$$

Thus the change in fugacity of a liquid in going from its pressure when the surface is flat (p_∞^s) to the pressure inside a droplet of radius r ($p_r^s + 2\sigma/r$) is

$$\ln\frac{(f^L)_{drop}}{(f^L)_{ref}} = \int_{p_\infty^s}^{p_r^s + 2\sigma/r} \frac{v^L}{RT} dp. \tag{2.90}$$

Assuming 1) the vapor above the flat surface is also an ideal gas, so that $(f^L)_{ref} = p_\infty^s$; 2) that the liquid is incompressible, *i.e.*, $v^L = $ constant, and 3) that $2\sigma/r >> p_\infty^s$, Eq. (2.90) becomes

_____

[69] Kralchevsky, P. A., and Nagayama, K., *Adv. Colloid Interface Sci.*, **85**, 145 (2000).

$$(f^L)_{drop} = p_\infty^s \exp\left(\frac{2\sigma v^L}{rRT}\right).$$
(2.91)

Finally, equating $(f^L)_{drop}$ to the fugacity in the vapor phase around the droplet:

$$p_r^s = p_\infty^s \exp\left(\frac{2\sigma v^L}{rRT}\right),$$
(2.92)

known as the *Kelvin Equation*. The vapor pressure is seen to increase as droplet size decreases, as shown in Fig. 2-58, in which values for water and mercury at room temperature (20°C) are plotted. One-micron radius water

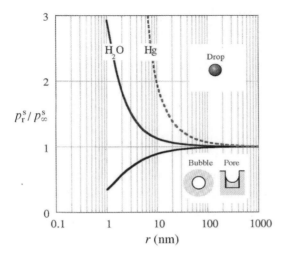

Fig. 2-58: Vapor pressure dependence for water drops or inside bubbles as a function of drop or bubble size, vapor pressure of mercury as a function of drop size.

droplets show a vapor pressure enhancement of approximately 0.1%, while those of radius one nm are increased by about a factor of three. One would not expect smaller droplets to be described by the Kelvin Equation due to the continuum assumption inherent in it. Mercury droplets of a given size show much larger vapor pressure enhancements than water since the surface tension of mercury is much larger, with a one-nm droplet of mercury showing a vapor pressure nearly 300 times its handbook value. Figure 2-58 also shows the vapor pressure of water surrounding small bubbles or inside small wetted capillaries or pores. In this case the liquid is on the con*vex* rather than the concave side of the surface, so that the vapor pressure should be *reduced* rather than increased. The Kelvin Equation is the same, but the sign of the argument of the exponential is negative. The vapor pressure of water inside a one-nm radius bubble is only about 1/3 its handbook value.

The Kelvin effect has many important consequences. In a mixture of droplets at the conditions (T, p_r^s), droplets smaller than radius r evaporate, while larger ones grow. As the small drops shrink, the driving force for their evaporation increases until they disappear. Thus raindrops "condense" out of fog, or fog evaporates, *i.e.*, "lifts." The equilibrium of a droplet of radius r to which the conditions (T, p_r^s) refer is thus an *unstable* equilibrium. The slightest addition or subtraction of material from such a drop will lead to further condensation or evaporation, respectively, until the drop has either grown to a very large size or disappeared.

2. *The effect of curvature on boiling point*

One might also consider the effect of curvature on the equilibrium temperature (boiling point) of a droplet at constant pressure, *i.e.*, constant vapor pressure p_r^s. This requires evaluation and integration of the coefficient: $\left(\dfrac{\partial T_r^s}{\partial r}\right)_{p_r^s}$. Using ordinary partial derivative reductions, together with the Kelvin and the Clausius-Clapeyron Equations, one obtains:

$$\left(\frac{\partial T_r^s}{\partial r}\right)_{p_r^s} = -\left(\frac{\partial p_r^s}{\partial r}\right)_{T_r^s} \bigg/ \left(\frac{\partial p_r^s}{\partial T_r^s}\right)_r = \frac{2\sigma v^L T_r^s}{r^2 \lambda^{vap}}, \qquad (2.93)$$

in which, in the last step, it has been assumed that the heat of vaporization λ^{vap} and the liquid molar volume v^L are constant and that $v^L \ll RT/p_r^s$. Integration of the above equation leads to:

$$T_r^s = T_\infty^s \exp\left(-\frac{2\sigma v^L}{r\lambda^{vap}}\right), \qquad (2.94)$$

which is known as the *Thomson Equation* (after J. J. Thomson, elder brother of Lord Kelvin). It shows that the boiling point is lower the smaller the droplet. Thus in order to condense a vapor to droplets, the latter must be sub-cooled below the handbook value of the boiling point. A more detailed treatment of the effect of curvature on the thermodynamic properties of both pure and multicomponent systems is given by Defay *et al.*[70]

3. *Capillary condensation*

As illustrated in Fig. 2-58, the Kelvin effect leads to the condensation of vapor into finely porous solids wet by the condensate at partial pressures below the equilibrium vapor pressure. Such "capillary condensation" is often observed in the study of adsorption of vapors onto porous solids, as pictured in Fig. 2-59 for the case of nearly uniform sized pores. Assuming that the contact angle $\theta = 0°$ and that $Bo \ll 1$, so that the surface of the liquid is

---

[70] Defay, R., Prigogine, I., Bellemans, A., and Everett, D. H., **Surface Tension and Adsorption**, pp. 217-285, Longmans, London, 1966.

hemispherical with radius equal to that of the pores, r, vapor will begin to condense when its partial pressure reaches

$$p = p_\infty^s \exp\left(-\frac{2\sigma v^L}{rRT}\right).$$

(2.95)

Water vapor will start to condense into one-nm radius pores at a relative humidity of 0.34. At higher relative humidity, larger pores will start to fill, or when the pores fill to the top edge, further filling can occur as p is increased. All pores are completely filled only when p_∞^s is reached.

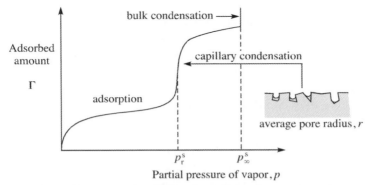

Fig. 2-59: Capillary condensation into a medium of approximately monodisperse pores.

Considerable hysteresis is nearly always found between adsorption and desorption, as shown in Fig. 2-60. One explanation may be contact angle hysteresis, discussed in Chap. 4.B, but a more generally satisfactory explanation is given in terms of pore geometry, specifically, the existence of so-called "ink bottle pores," *i.e.*, pores constricted at the top. They require

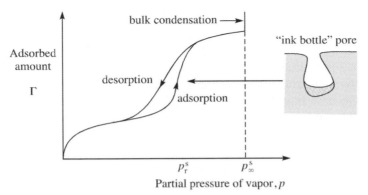

Fig. 2-60: Hysteresis associated with capillary condensation.

higher pressures to fill from the bottom up (for adsorption). For *desorption*, they start full so that the pressure must be reduced further to remove the condensate from the smaller necks. This has received numerous qualitative confirmations by checking that the same back-calculated pore size and shape distribution is predicted for different liquids in the same solid and for the same liquid at different temperatures on the same solid specimen.

Capillary condensation shows promise for the safe storage of natural gas[71] or hydrogen[72] at reasonable pressures. Such technology is of particular currency in view of the need to provide these fuels for the operation of fuel cells. Thus major effort is currently under way to develop high porosity, high surface area materials, and a number of metal-organic framework (MOF) materials have been reported. Currently surface areas of the order of 4,500 m^2/g (!) and pore diameters of roughly 1 nm are being produced.[73]

Capillary condensation also occurs in powders (which are wettable by the condensate liquid) at the points of contact between the powder particles. The condensate then forms liquid bridges between the particles, "gluing" them together. It is thus often found that a powder that is freely flowing on a dry day may clump up and not flow evenly on a humid day. On the other hand, moisture may be introduced into the vapors surrounding the powder to induce their consolidation into clumps (called "spherical agglomeration"). The use of the Kelvin Equation, even though it is based on a macroscopic thermodynamic description of the meniscus, appears to be valid for menisci of nanoscale dimensions. Use of the surface forces apparatus (SFA) (described in Chap. 7.B.4) has revealed that it successfully describes condensate bridges of cyclohexane[74] as small as 4 nm, formed between approaching crossed mica cylinders. Monte Carlo simulations suggest that below this range, capillary condensation is preceded by accumulation of dense vapor between the surfaces, and that the snap-apart event is preceded by a gradual decrease in liquid density.[75] The liquid in the nano-meniscus is found to exist in a layered structure.

4. Nucleation

The phenomenon of phase change by *nucleation* and growth (binodal decomposition) is governed by the Kelvin effect.[76] Consider, for example, the condensation of liquid from a saturated vapor. For phase change to occur by this mechanism, clusters of molecules out of the vapor that subsequently grow into the new liquid phase must be formed. It is first of all clear that *supersaturation* will be required in order for this to happen. Any small cluster of molecules will have a very small radius and hence a very high

[71] Matranga, K. R., Myers, A. L., and Glandt, E. D., *Chem. Eng. Sci.*, **47**, 1569 (1992).

[72] Dillon, A. C., Jones, K. M., Bekkedahl, T. A., Klang, C. H., Bethune, D. S., and Heben, M. J., *Nature*, **386**, 377 (1997).

[73] Hee, K., Chae, H. K., Siberio-Pérez, D. Y., Kim, J., Go, Y., Eddaoudi, M., Matzger, A. J., O'Keefe, M., and Yaghi, O. M., *Nature*, **427**, 523 (2004).

[74] Fisher, L. R., and Israelachvili, J. N., *J. Colloid Interface Sci.*, **80**, 528 (1981).

[75] Stroud, W. J., Curry, J. E., and Cushman, J. H., *Langmuir*, **17**, 688 (2001).

[76] Zettlemoyer, A. C., (Ed.), **Nucleation**, Marcel Dekker, NY, 1969.

vapor pressure, causing it to rapidly re-evaporate. Once droplets of sufficient size do form, condensation occurs readily. Nucleation may be envisioned phenomenologically in the following way, as proposed by Becker and Döring.[77] At any degree of supersaturation, the vapor will contain a population of transitory clusters, as shown in Fig. 2-61, ranging from dimers up to nuclei whose size corresponds to the unstable equilibrium described by the Kelvin Equation. These are called *critical nuclei*, and there must be a

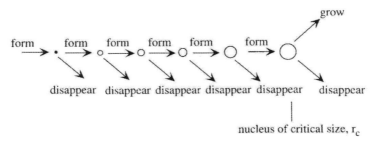

Fig. 2-61: Formation of critical nuclei leading to phase change.

sufficient number of them, *i.e.*, they must be produced at a sufficient rate, for phase change to occur at a finite rate. Only nuclei of this size or larger have a chance to grow under the prevailing conditions. The rate at which critical nuclei form is expressed by a rate equation of the Arrhenius type:

$$J(\text{nuclei/cm}^3\text{s}) = Ae^{-E/kT} , \qquad (2.96)$$

where E is the effective activation energy, and the pre-exponential factor A is dependent on the collision frequency of the vapor molecules. A is thus proportional to p^2, and the value given by simple kinetic theory is approximately: $A(\text{cm}^{-3}\text{s}^{-1}) \approx 10^{23}p^2(\text{mmHg}^2)$.

It is common to use as the activation energy, the change in Helmholtz free energy, ΔF, for the formation of the critical nucleus within a system at constant total volume and temperature. This is given by:

$$\Delta F_{\text{form}} = \Delta G_{\text{form}} - V_{\text{nuc}}\Delta p$$

$$= \Delta G_{\text{phase change}} + \Delta G_{\text{area formation}} - V_{\text{nuc}}\Delta p$$

$$= 0 + \sigma A_{\text{nuc}} - V_{\text{nuc}}\Delta p$$

$$= 4\pi r^2\sigma - \frac{4}{3}\pi r^3\left(\frac{2\sigma}{r}\right) = \frac{4}{3}\pi r^2\sigma , \text{ so that,} \qquad (2.97)$$

$$E \approx (\Delta F)_{\text{form}} = \frac{1}{3}\sigma A_{\text{nuc}} = \frac{4}{3}\pi r^2\sigma . \qquad (2.98)$$

[77] Becker, R., and Döring, W., *Ann. Physik.*, **24**, 719 (1935).

The pressure of the vapor at a given T is p, while the equilibrium vapor pressure (over a flat surface) is p_∞^s. The degree of supersaturation is thus (p/p_∞^s), a quantity designated as x. The critical nucleus for these conditions is the one whose radius satisfies the Kelvin Equation, $i.e.$, the value of r for which $p = p_r^s$. Solving the Kelvin Equation for the radius of the critical nucleus gives

$$r = r_c = \frac{2v^L\sigma}{RT\ln x},$$
(2.99)

and substituting into the expression for E:

$$E = \frac{16\pi}{3}\frac{(v^L)^2\sigma^3}{R^2T^2\ln^2 x}.$$
(2.100)

The presumption is that as the degree of supersaturation increases at a given T, J will increase until it reaches a value large enough to produce critical nuclei at a "sufficient rate," say one nucleus per cm^3 per second, such that $\ln J \approx 0$. (Alternatively, one may hold p constant and decrease T.) A numerical example is illustrative. Consider water vapor at 20°. The needed properties are: $v^L = 18$ cm^3/mole; $\sigma = 72.7$ dynes/cm; $p_\infty^s = 17.5$ mmHg. For this situation:

- $A \approx 10^{23}p^2(\text{mmHg}^2) = 10^{23}(17.5)^2x^2 = 3\times10^{25}x^2$
(2.101)

- $E = \dfrac{16\pi}{3}\dfrac{(18)^2(72.7)^3}{(8.314\times10^7)^2(293.2)^2\ln^2 x} = \dfrac{3.51\times10^{-12}}{\ln^2 x}$ (erg/nucleus)
(2.102)

Then:

$$\ln J = \ln A - \frac{E}{kT} = 58.7 + 2\ln x - \frac{86.7}{\ln^2 x}.$$
(2.103)

A plot of this function is shown in Fig. 2-62, revealing that critical nuclei are formed at only a vanishingly low rate until the degree of supersaturation reaches about 3. This is in reasonable accord with experiment. It is of interest to note that the radius of the critical nucleus corresponding to these conditions is about one nm (entailing approximately 70 water molecules). The use of the continuum property of surface tension in the description of "droplets" so small would appear questionable, but the a $posteriori$ agreement with experiment suggests its validity. One may similarly examine the situation in which the partial pressure of the vapor is held constant while the temperature is reduced, and similar results are found. For the above case of $p = 17.5$ mmHg, the temperature must be reduced to below 20°C before $\ln J \approx 0$, and condensation ensues. The degree of supersaturation required to boil a liquid, freeze a liquid or melt a solid are found by the same type of analysis. Interestingly, it is found that liquid water at atmospheric pressure may be reduced in temperature to below -30°C (!) before it freezes by the above mechanism. A more detailed discussion of nucleation is given by Defay *et al.*[78] Numerous refinements to Becker-Döring theory have been made, but its essential features have been retained.

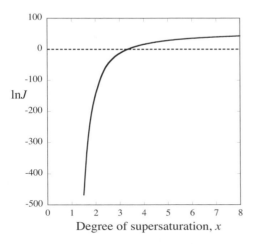

Fig. 2-62: Critical nucleus formation rate as a function of the degree of supersaturation.

What has been described above is *homogeneous nucleation*. Actual phase changes seldom occur in this way and seldom require the indicated degree of supersaturation. More likely, the initial nuclei are formed at imperfections (cracks, crevices, *etc.*) at solid surfaces that bound the system, yielding *heterogeneous nucleation*, as described further in Chap. 4. Not only is the energy required to produce a critical nucleus at such sites considerably less, but such sites may often permanently house nuclei of the new phase. Imperfections in solid surfaces, for example, are sometimes never completely evacuated of gas, so that when the system is heated, the trapped vapor pockets simply grow and are pinched off as bubbles. Boiling chips contain many of these imperfections and therefore provide smooth, even boiling (called *nucleate boiling*), as opposed to the "bumping" associated with higher degrees of superheat.

K. Thin liquid films

1. *Disjoining pressure and its measurement*

When fluids exist in the form of thin films, they are found to have properties differing from those of the same material in bulk, and such films are often unstable. To fix ideas, consider the examples shown in Fig. 2-63: (a) a foam lamella separating gas phases, (b) a liquid film between two flattened liquid droplets being drawn together, and (c) a liquid film supported on a smooth solid surface. (a) and (b) are examples of "free films," and (c) shows a "supported film." Thin films may also exist between

[78] Defay, R., Prigogine, I., Bellemans, A, and Everett, D. H., **Surface Tension and Adsorption**, pp. 310-348, Longmans, London, 1966.

solid surfaces ("confined films"), but discussion of such systems is deferred to Chap. 7, when the interaction between colloidal particles is discussed. Thin fluid films are often unstable in that they spontaneously seek to either thin or thicken themselves. This is because the equilibrium pressure within such a film, p_{film}, differs from the pressure that exists (or would exist) in an adjoining bulk phase of the same fluid under the same thermodynamic

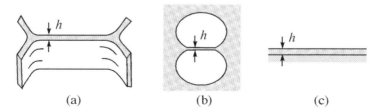

(a)   (b)   (c)

Fig. 2-63: Examples of thin films: (a) foam lamella; (b) liquid film between drops; (c) liquid film on a solid surface.

conditions (without regard to gravity). The difference is termed, originally by Derjaguin, the "disjoining pressure," $\Pi(h)$[79]:

$$\Pi(h) = p_{film} - p_{bulk}, \tag{2.104}$$

and, as indicated, is a function of the film thickness h. The function assumes different forms in different systems under different conditions, but in all cases Π tends to zero as h becomes sufficiently large (usually a few tens to hundreds of nanometers). Disjoining pressure quantifies the driving force for spontaneous thickening (if $\Pi > 0$) or thinning of a film (if $\Pi < 0$) in which, initially, $p_{film} = p_{bulk} = p$. The process of thickening is equivalent to separating (or "disjoining") its bounding surfaces; hence the name. A thermodynamically stable thin film may exist at a particular value (or values) of h in a given case, as described in more detail below, but more typically if a thin liquid film is found to exist over long periods of time (kinetic "stability") it is because free flow of the liquid between the film and an adjoining bulk phase (or potential bulk phase) of the same liquid is somehow impeded. In such cases, the disjoining pressure may be measured.

The immediate question is: what constitutes "thin?" For any particular case, one can distinguish between "bulk" films, for which h is sufficiently large that the film behaves the same as a bulk phase, and $\Pi = 0$, and "thin" films, for which $\Pi = \Pi(h) \neq 0$. Thin films may further be designated as either *thin* thin films or *thick* thin films. (These are designated in the Russian literature as α-films or β-films, respectively.) Thin thin films are adsorbed multilayers, monolayers or sub-monolayers, as examined in more detail in Chap. 3, whose thicknesses are of the order of a single nanometer or less, so that the zones of inhomogeneity of the bounding surfaces overlap. The

---

[79] Derjaguin, B.V., *Kolloid Zh.*, **17**, 205 (1955).

interfacial tensions of these bounding surfaces (generally not measurable properties) must be regarded as functions of film thickness for such films. Thin thin films are more usefully discussed as adsorbed layers, and their "effective thickness" is usually computed as the adsorbed amount Γ(moles/area) divided by the bulk molar density (moles/volume). For thick thin films, our major concern here, the interfacial zones do not overlap, and the interfacial tensions are the same as they would be for a bulk phase, but h is still within the reach of the integrated intermolecular forces. This generally puts thick thin films at a few hundred nm or less.

Extensive direct measurements of disjoining pressure have been reported, as in the classical work of Sheludko and coworkers,[80] who also provided an early perspective on the issue of thin liquid films. Figure 2-64 shows their technique applied to the study of free films (in which the disjoining pressure is generally negative, causing spontaneous thinning), consisting of a biconcave meniscus formed in a circular tube. A flat circular film is formed in the center, while the pressure in the bulk meniscus is

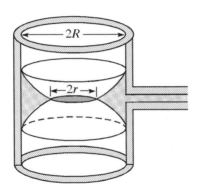

Fig. 2-64: Biconcave meniscus apparatus of Sheludko, *et al.* for measuring disjoining pressure in free thin films.

maintained at a value higher or lower (as needed to impede flow and maintain the film thickness at a desired value h) than the pressure in the film (which is the same as the ambient pressure) and accessed by means of a side port. The thickness of the film is usually measured interferometrically, and the pressure in the bulk liquid at the side required to maintain a given film thickness is recorded. Some results for $\Pi(h)$ obtained by Sheludko are shown in Fig. 2-65 for free films of aniline, and are seen to show that $\Pi(h)$ varies as $-1/h^3$. One of the ways of determining disjoining pressure in supported liquid films is by pushing a bubble of gas against the solid surface, as shown in Fig. 2-66, and monitoring the thickness of the film beneath the bubble as a function of the pressure applied to it by the bubble. This method is generally appropriate only if the disjoining pressure is positive. Reviews of relevant experimental techniques and their interpretation are given by Clunie et al.[81], and by Cazabat.[82]

[80] Sheludko, A., **Colloid Chemistry**, Elsevier, Amsterdam (1966).]
[81] Clunie, J. S., Goodman, J. F., and Ingram, B. T., "Thin Liquid Films," **Surface and Colloid**

2. The molecular origin of disjoining pressure

The origin of disjoining pressure can be traced to intermolecular and surface forces, integrated over finite distances, areas and volumes. As we have seen, it is a function of the film thickness h, and $\Pi(h)$, called the disjoining pressure isotherm (since it is usually obtained at constant temperature), depends on the makeup of the film, the adjoining bulk phases and the interfaces between them. If the film finds itself between two like

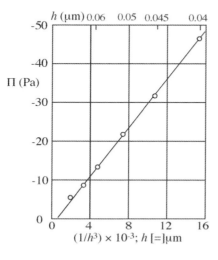

Fig. 2-65: Disjoining pressure isotherm $\Pi(h)$ in a free liquid film of aniline. After [Sheludko, A., **Colloid Chemistry**, pp. 173-207, Elsevier, Amsterdam (1966).]

phases, as in Fig. 2-64, it is usually (but not always) negative, whereas if it is between a condensed phase on one side and a gas on the other, as in Fig. 2-66, it is sometimes (but perhaps more often not) positive, and in other cases may be positive over some ranges of h and negative over others. Further discussion of the formation and properties of thin liquid films are given by Derjaguin et al.[83] and de Gennes.[84, 85]

As an example, consider the case of a thin supported film of non-volatile liquid on a smooth, horizontal solid surface, and a gas above it, as shown in Fig. 2-63(c). One may compute the disjoining pressure isotherm $\Pi(h)$ for the case when the origin of the disjoining pressure is the attractive van der Waals interaction of the molecules in the liquid film with the solid substrate, which can be regarded as a semi-infinite block. As seen earlier, the van der Waals interaction between a single molecule in the liquid L (a

Science, Vol. 3, E. Matijevic (Ed.), pp. 167-239, Wiley-Interscience, New York, 1971.

[82] Cazabat, A. M., in **Liquids at Interfaces**, J. Charvolin, J.F. Joanny and J. Zinn-Justin (Eds.), pp. 372-414, Elsevier, Amsterdam, 1990.

[83] Derjaguin, B. V., Churaev, N. V., and Muller, V. M., **Surface Forces**, V. I. Kisin, Trans., J. A. Kitchner (Ed.), Consultants Bureau, New York, 1989.

[84] de Gennes, P., *Rev. Mod. Phys.*, **57**, 827 (1985).

[85] de Gennes, P., in **Liquids at Interfaces**, J. Charvolin, J.F. Joanny and J. Zinn-Justin (Eds.), pp. 273-291, Elsevier, Amsterdam, 1990.

distance z from the solid surface) and the solid S half-space per unit area is, in accord with Eq. (2.11):

$$\Phi^\sigma_{\text{molec-solid}} = -\frac{\pi B_{\text{SL}}\rho_{\text{S}}}{6z^3},$$

(2.105)

where the B_{SL} is the cross van der Waals interaction constant between the molecule of the liquid and the molecules of the solid, and ρ_{S} is the molecular density of the solid. To obtain an expression for the total interaction energy between all the molecules in the liquid film (per unit area) and the solid, $\Phi^\sigma_{\text{SL}(h)}$, one must integrate over all the molecules in the liquid (per unit area):

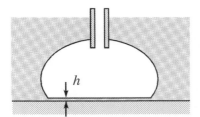

Fig. 2-66: Measurement of positive disjoining pressure in a supported liquid film.

$$\Phi^\sigma_{\text{SL}(h)} = -\frac{\pi B_{\text{SL}}\rho_{\text{S}}\rho_{\text{L}}}{6}\int_{D_0}^h \frac{dz}{z^3} = \frac{\pi B_{\text{SL}}\rho_{\text{S}}\rho_{\text{L}}}{12}\left(\frac{1}{D_0^2} - \frac{1}{h^2}\right)$$

(2.106)

where "D_0", as before, is the distance of closest approach between the molecules of the liquid and the solid. It is evident that if h is large, $1/h^2 \to 0$, and $\Phi^\sigma_{\text{SL}(h)}$ is a constant with h. But for thin films, the second term is not negligible, i.e., even the outermost molecules of the film "feel" the influence of the force field of the solid. The prefatory constant in Eq. (2.106) may be written in terms of the cross Hamaker constant, A_{SL} (see Eq. (2.23)) to give

$$\Phi^\sigma_{\text{SL}(h)} = \frac{A_{\text{SL}}}{12\pi}\left(\frac{1}{D_0^2} - \frac{1}{h^2}\right).$$

(2.107)

To obtain the *excess* energy of molecular interactions in the film resulting from its contact with the solid, one must subtract the energy of interactions within the film itself, $\Phi^\sigma_{\text{LL}(h)}$ to obtain

$$\Phi^\sigma_{\text{E}(h)} = \frac{(A_{\text{SL}} - A_{\text{LL}})}{12\pi}\left(\frac{1}{D_0^2} - \frac{1}{h^2}\right) = -\frac{A_{\text{eff}}}{12\pi}\left(\frac{1}{D_0^2} - \frac{1}{h^2}\right),$$

(2.108)

where $A_{\text{eff}} = (A_{\text{LL}} - A_{\text{SL}})$, the effective Hamaker constant for the system. Treating $\Phi^\sigma_{\text{E}(h)}$ as a free energy, the excess pressure in the film relative to that in a bulk layer, i.e., Π, may be obtained for van der Waals materials as

$$\Pi(h) = -\frac{\partial(\Phi^\sigma_{\text{E}(h)}\mathcal{A})}{\partial V} \equiv -\frac{\mathcal{A}}{\mathcal{A}}\left(\frac{\partial\Phi^\sigma_{\text{E}(h)}}{\partial h}\right) = -\frac{A_{\text{eff}}}{6\pi h^3},$$

(2.109)

where $\mathcal{A}$ is the surface area. The form of Eq. (2.109) is in agreement with the data shown in Fig. 2-65 for free films of aniline, but in this case, since no solid substrate is present, $A_{eff} = A_{LL}$. The Hamaker constant A_{LL} for a van der Waals liquid is always positive, so the disjoining pressure is negative for all h-values. Therefore a free film of van der Waals liquid will spontaneously thin.

As indicated in Eq. (2.108), for a liquid film supported on a solid substrate, the effective Hamaker constant of the film is given by

$$A_{eff} = A_{LL} - A_{SL}. \tag{2.110}$$

For van der Waals materials, A_{SL} is given by the geometric mean mixing rule:

$$A_{SL} = \sqrt{A_{SS}A_{LL}}, \text{ so that} \tag{2.111}$$

$$A_{eff} = \sqrt{A_{LL}}\left(\sqrt{A_{LL}} - \sqrt{A_{SS}}\right). \tag{2.112}$$

Thus A_{eff} is either negative (if $A_{SS} > A_{LL}$), yielding a fully wetting film of finite thickness, or positive (if $A_{SS} < A_{LL}$), causing the film to spontaneously thin itself.

It has been argued that as the film thickness h approaches zero, $\Pi(h)$ should approach the spreading coefficient, $S_{L/S}$, as in Eq. (2.49), [86] i.e.,

$$\Pi(h)\big|_{h \to 0} = S_{L/S}. \tag{2.113}$$

It is useful here to digress briefly to consider the general situation in which a fluid film (1) is separated by phases of different materials (2) and (3). The latter may be gases, liquids immiscible with (1) or solids, or any combination thereof. The effective Hamaker constant of the film for computation of the disjoining pressure isotherm is given by[87]

$$A_{eff} = A_{23} + A_{11} - A_{21} - A_{31}. \tag{2.114}$$

The last two terms account for the interaction of the film with its adjoining phases, while the first two account for the interactions of the adjoining phases with each other and the film molecules with themselves. Applying the geometric mean mixing rule, Eq. (2.111), yields:

$$A_{eff} = \left(\sqrt{A_{33}} - \sqrt{A_{11}}\right)\left(\sqrt{A_{22}} - \sqrt{A_{11}}\right), \tag{2.115}$$

from which it is easy to see the various combinations that lead to either positive or negative values for A_{eff}. In applying Eq. (2.115), the Hamaker

[86] Brochard-Wyart, F., di Meglio, J.-M., Quéré, D., and de Gennes, P.-G., *Langmuir*, **7**, 335 (1991).
[87] Israelachvili, J. N., **Intermolecular & Surface Forces**, 2nd Ed., p. 200, Academic Press, London, 1992.

constant is generally taken to be zero for gas phases, due to the low molecular density in such media. Thus if the film is bound on both sides by gases, $A_{22} = A_{33} = 0$, and $A_{\text{eff}} = A_{11}$. If it is bound on one side by a condensed phase (2) and on the other by a gas (3), $A_{\text{eff}} = \sqrt{A_{11}}\left(\sqrt{A_{11}} - \sqrt{A_{22}}\right)$, as in Eq. (2.112).

In the above it has been tacitly assumed that only van der Waals interactions are relevant. Other intermolecular forces however (hydrogen bonds or other donor-acceptor interactions, solvent structuring effects, anomalous density profiles near the wall, *etc.*), may also contribute to the disjoining pressure, adding additional terms to the $\Pi(h)$ function. These may be especially important as $h \to 0$. Hydrogen bonding in water yields a term of the form

$$\Pi_{\text{H}}(h) = \frac{C_{\text{H}}}{h}, \tag{2.116}$$

where C_{H} is a constant. Solvent structuring effects are manifest near solid boundaries, in which there may be a strong, essentially chemical, affinity of the liquid for the solid surface. This is manifest at the boundary between water and strongly hydrophilic surfaces, such as quartz. The squeezing out of this final layer (often a monolayer) is strongly resisted, and has been approximated as[88]

$$\Pi_{\text{s}}(h) = K_{\text{s}} \exp(-h/\lambda_{\text{s}}), \tag{2.117}$$

where K_{s} is very large ($\approx 10^7$ N/m^2), and λ_{s} is of the order of a few Å (very short-ranged). In the absence of hydration, a strong "solvent structuring effect," for which $\lambda_{\text{s}} \to 0$, just a manifestation of Born repulsion, will be evident. Thus every disjoining pressure isotherm for a film bounded at least on one side by a solid will exhibit a steep positive branch as $h \to 0$. For a confined thin liquid film between two solid surfaces, successive layers of liquid must be squeezed out as the film thins. This may in principle lead to an oscillatory disjoining pressure, as shown in Fig. 2-67, in which the wavelength of the oscillation is the effective molecular diameter.

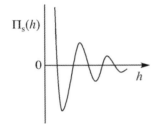

Fig. 2-67: Solvent structuring term of the disjoining pressure for a confined liquid film between smooth solid surfaces. The wavelength of the oscillations corresponds to the diameter of the solvent molecules.

---

[88] Derjaguin, B. V., and Churaev, N. V., *Langmuir*, **3**, 607 (1987).

Another possible contributor to the disjoining pressure is electrostatic forces. These may exist, for example, in an aqueous thin film containing an ionic surfactant adsorbed to both the bounding surfaces, as pictured in Fig. 2-68. Even a clean water surface against air possesses a negative surface charge. If the opposing surfaces possess diffuse electrical double layers (see Chap. 6), *i.e.*, adjacent clouds of ions opposite in charge to that of the surfaces, their overlap contributes to the total disjoining pressure with a term of the form:

$$\Pi_{\text{el}}(h) = K_{\text{el}}\exp(-\kappa h),\tag{2.118}$$

where K_{el} and κ are constants dependent primarily on the charge density at the surfaces and the ionic content and dielectric constant of the film. The bounding surfaces may also possess dissolved polymer adlayers, leading to

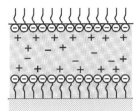

Fig. 2-68: Electrostatic forces due to presence of charges at the surfaces of the film. Here they are due to adsorbed ionic surfactant.

"steric" repulsive interactions as the adlayers overlap. Still other factors may contribute. For example, the Helfrich Equation for the free energy associated with the bending of an interface, Eq. (2.59), has been used to compute forces arising from the mutual undulation of opposing structured surfaces of thin films. Such undulation forces are found to be proportional to the Helfrich first modulus k_1, and inversely to the square of the film thickness. This bending force contribution, when relevant, must be added to the disjoining pressure expression.

3. *The disjoining pressure isotherm*

The disjoining pressure may be regarded as the sum of the various contributions to it, as described above.

$$\Pi_{\text{tot}}(h) = \Pi_{\text{vdW}}(h) + \Pi_{\text{s}}(h) + \Pi_{\text{H}}(h) + \Pi_{\text{el}}(h) + \Pi_{\text{steric}}(h) + \dots\tag{2.119}$$

Depending on the various contributions to $\Pi(h)$, the function may take on a variety of different forms, but such isotherms are usually one of the four types shown in Fig. 2-69, with inserts shown for liquid films supported on a solid substrate. If the film is described entirely in terms of van der Waals interactions, the disjoining pressure curve takes the form shown in Fig. 2-69(a), Type I, if the effective Hamaker constant is negative, and (b), Type II, if it is positive. In the latter case, the steep repulsive force associated with squeezing out the last monolayer or so of the film is included. Even for a free film, there may be such repulsion as the final molecules in the film jockey for position. If not, the curve follows the dashed line. Type I isotherms are exemplified by supported films on solid substrates, if the

Hamaker constant for the solid is greater than that for the liquid, so that A_{eff} is negative, in accord with Eq. (2.112). Examples would be alkanes, for example octane, $A_{LL} = 4.50 \times 10^{-20}$ J, on fused silica, $A_{SS} = 6.55 \times 10^{-20}$ J, so $A_{eff} = \sqrt{4.50}\left(\sqrt{4.50} - \sqrt{6.55}\right) \times 10^{-20} = -0.93 \times 10^{-20}$ J. Type II films are exemplified by free liquid films in air, such as the aniline film shown in Fig. 2-65, or supported films such as the case of octane on a Teflon substrate (with $A_{ss} = 3.80 \times 10^{-20}$ J).

For the case with both van der Waals (with a positive effective Hamaker constant) and electrostatic effects:

$$\Pi(h) = \Pi_s(h) + \Pi_{vdW}(h) + \Pi_{el}(h) = K_s \exp(-h/\lambda_s) - \frac{A_{eff}}{6\pi h^3} + K_{el}\exp(-\kappa h), \quad (2.120)$$

in which the ever-present solvent structuring term is included. This can lead to disjoining pressure curves of the type shown in Fig. 2-69 (c), Type III, or (d), Type IV, as exemplified by films of water on quartz.[89]

The fact that disjoining pressure isotherms of the type shown in Fig. 2-69 can be computed does not mean that the complete curves can all be observed in the laboratory. If a liquid film is in unimpeded contact with a reservoir of bulk liquid, it will spontaneously thicken or thin until it reaches a state of stable equilibrium. Such states correspond to local minima in the free energy of the system. The (Helmholtz) free energy of the thin film system, per unit area, may be written as

$$F^{\sigma}(h) = F_0^{\sigma} + \Phi_{E(h)}^{\sigma}(h), \quad (2.121)$$

where F_0^{σ} is a constant, and it is recalled from Eq. (2.108) that $\Phi_{E(h)}^{\sigma}(h)$ is the excess energy of the system due to its thin film status. In accord with Eq. (2.109), the disjoining pressure is obtained as the negative derivative, in this case, of $F^{\sigma}(h)$:

$$\Pi(h) = -\left[\frac{\partial F^{\sigma}(h)}{\partial h}\right]_{T,p}. \quad (2.122)$$

Equilibrium states are those for which the free energy derivative is zero, and hence disjoining pressure is zero. *Stability* then requires that the second derivative of the free energy be positive, *i.e.*,

$$\left[\frac{\partial^2 F^{\sigma}(h)}{\partial h^2}\right]_{T,p} = -\frac{d\Pi(h)}{dh} > 0, \text{ or } \frac{d\Pi(h)}{dh} < 0. \quad (2.123)$$

Thus thin film stable equilibrium states are limited to those where the disjoining pressure isotherm crosses or touches the $\Pi = 0$ line with a negative slope. These states are identified in Fig. 2-69.

[89] Derjaguin, B. V., and Churaev, N. V., "Properties of Water Layers Adjacent to Interfaces, " in **Fluid Interfacial Phenomena**, C. A. Croxton (Ed.), Chap. 15, Wiley, New York, 1986.

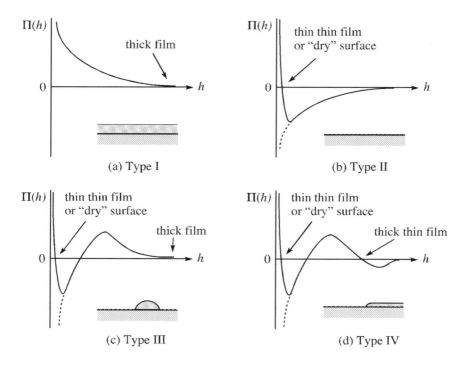

Fig. 2-69: Examples of disjoining pressure isotherms. (a) Type I: Spontaneously thickening film of van der Waals liquid on a solid or liquid substrate; (b) Type II: Spontaneously thinning film of van der Waals liquid on a solid or liquid substrate, or free liquid film; (c) Liquid film with both positive and negative contributions to the disjoining pressure, leading to possible coexistence of a thin thin film with bulk liquid; (d) Liquid film with both positive and negative contributions to the disjoining pressure, leading to possible coexistence of a thin thin and a thick thin film.

If the solid surface is finite in lateral extent, and the disjoining pressure is positive, as in Fig. 2-69 (a), the film will spontaneously thicken until $\Pi(h) \to 0$, and then level itself (even in the absence of gravity) due to surface tension, except at the edges of the solid. If $\Pi(h) < 0$, as in Fig. 2-69 (b), the film will *thin* itself until $\Pi(h) \to 0$ (essentially to $h \approx 0$), with any excess liquid left in the form of drops making a distinct contact angle with the "dry" solid. For the more complex disjoining pressure functions shown in Fig. 2-69 (c) and (d), it is possible to have an equilibrium, flat, horizontal thin film coexisting with bulk liquid, or even flat films of different thicknesses coexisting with each other.

In experiments such as those suggested in Figs. 2-64 and 2-66, as well as others,[90] particularly those in which a drop or bubble is pushed against a wetted solid substrate using the apparatus of atomic force microscopy,[91]

---

[90] Bergeron, V. B., Fagan, M. E., and Radke, C. J., *Langmuir*, **9**, 1704 (1993); Bergeron, V. B., and Radke, C. J., *Colloid Polym. Sci.*, **273**, 165 (1995).
[91] Basu, S., and Sharma, M. M., *J. Colloid Interface Sci.*, **181**, 443 (1996).

most of the isotherms of the type shown in Fig. 2-69 may in principle be observed and measured, either in a dynamic experiment or one in which the flow of liquid between the film and its adjacent bulk liquid is restricted or controlled. An interesting exception occurs for Types III or IV if one attempts to either reduce the thickness of the film or increase its internal capillary pressure beyond the point where the local maximum in disjoining pressure equilibrium exists, the film will undergo a spinodal decomposition to a thin thin film in coexistence with either bulk liquid droplets or thick thin films.

4. *The augmented Young-Laplace Equation*

If one considers gravity in the case of a wetting film on a solid surface that is not horizontal, there may exist a final equilibrium situation in which a thin film coexists with bulk liquid, even for a Type I disjoining pressure isotherm. For example, consider the meniscus of a wetting liquid ($\theta = 0°$) against a vertical flat wall, as shown in Fig. 2-70. Considering disjoining pressure, this is somewhat more complicated than has been described earlier (in Fig. 2-31). In describing the meniscus shape, one must now write:

$$p_{liq} - p_{vap} \equiv \Delta p = \pm \sigma \kappa - \Pi(h), \tag{2.124}$$

which takes account of both the curvature and the disjoining pressure. First introduced by Derjaguin,[92] it is known as the *augmented* Young-Laplace Equation. For the case shown (for a wetting van der Waals liquid on a vertical substrate), it takes the form:

$$\rho g y = \pm \sigma \frac{y''}{\left[1 + (y')^2\right]^{3/2}} - \frac{A_{eff}}{6\pi x^3}, \tag{2.125}$$

The complete solution of this equation yields the detailed shape of the meniscus in the region of the nominal interline, but it is found to differ only microscopically from the solution obtained ignoring disjoining pressure, shown in Fig. 2-31 for a vertical surface. A thin film, however, extends far higher. In this region the meniscus becomes essentially flat so that the curvature term in the augmented Young-Laplace Equation may be neglected, leading to:

$$x = h \approx \left(\frac{A_{eff}}{6\pi \rho g y}\right)^{1/3}. \tag{2.126}$$

As an example, for octane against quartz, $A \approx 6 \times 10^{-20}$ J and $\rho = 0.7$ g/cm^3, so that at a distance of $y = 1$ cm above the flat liquid level, a film of thickness $h \approx 20$ nm exists. The boundary region between films of different

---

[92] Derjaguin, B. V., Churaev, N. V., and Muller, V. M., **Surface Forces**, Plenum Press, New York, 1987.

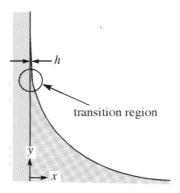

Fig. 2-70: Wetting film against
a vertical wall.

thickness is in general described by the solution to the augmented Young-Laplace Equation. Further details are given in numerous references.[93,94,95]

Supported thin films are discussed further in the context of spreading phenomena in Chap. 4, and free thin films get more attention in Chap. 8 in the description of emulsions and foams. Confined films, *i.e.*, fluids between solid surfaces, are discussed further in the context of interactions between colloid particles in Chap. 7.

[93] Davis, H. T., **Statistical Mechanics of Phases, Interfaces, and Thin Films**, pp. 370-377, VCH, New York, 1996.

[94] Starov, V. M., Velarde, M. G., and Radke, C. J., **Wetting and Spreading Dynamics**, Chap. 2, CRC Press, Boca Raton, 2007.

[95] Hirasaki, G. J., *J. Adhesion Sci. Tech.*, **7**, 285 (1993).

Some fun things to do:
Experiments and demonstrations for Chapter 2

At the end of this and the remaining Chapters, are suggestions for simple experiments to help illustrate the concepts of the chapter. Little or no instrumentation, or expensive or hard-to-get materials are required. Most experiments are suitable for classroom demonstration, either with direct observation or by projection onto a screen using an overhead projector or a camcorder connected to a video projector.

1. *The thinness of clean interfaces*

According to Fresnel's Law, light reflected at or near Brewster's angle (53° for water) from an interface *that is thin relative to its wavelength* (≈ 600 nm) will be plane polarized. This is why polarized sunglasses are able to block reflected glare from horizontal surfaces. The thinness of a clean water surface is demonstrated by reflecting the beam from a laser pointer from the surface of water in a shallow Petri dish. The laser beam from the pointer is plane polarized, so as it is rotated, an orientation will be obtained in which little or no reflection will occur.

Materials:
- laser pointer (A Class IIIa red laser pointer with output power 5 mW and beam diameter 4.5 mm at aperture is recommended. Care should be exercised not to shine this near anyone's eyes.)
- small Petri dish with a flat black piece of paper at the bottom (to avoid reflections from the bottom surface) and filled with clean water.
- small ring stand and clamp
- white cardboard screen

Procedure:
Mount the pointer in the ring stand clamp as shown in Fig. E2-1, and direct the beam at the surface at ≈ 50° and note the intensity of

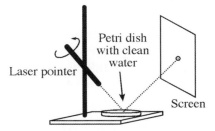

Laser pointer / Petri dish with clean water / Screen

Fig. E2-1: Demonstration of Fresnel's Law for a clean water interface.

the reflected spot on the screen as the pointer is rotated. It may be necessary to use tape to keep the pointer turned on as this is done. Note the beam spot nearly disappears at a particular angle, showing the polarizing ability of the surface and hence its thinness.

2. Soap bubbles and films[96]

Soap films are layers of water stabilized by surfactant on each side. Their thickness is of the order of a few micrometers to two mm, small enough that structures composed of them are largely free of gravitational forces. They can therefore be used to produce surfaces of constant curvature.

Materials:

- 250 mL of soap solution: a 50/50 v/v solution of dishwashing detergent (*e.g.*, Joy®, Dawn®, Palmolive®) in a 250 mL beaker
- 3×3×1/4 in. glass plate (with edges smoothed)
- 4 in. piece of ½ in. Tygon® tubing
- 2 in. or larger diameter plastic ring (from a bubble toy kit)
- 10×5×1/8 in. plastic (Plexiglas®) sheet bent 180°, as shown in Fig. E2-2, to yield a spacing of ≈ 2 cm.
- cubical wire frame, about 2×2×2 in. This can be soldered together from bent pieces of wire. 16 AWG (American Wire Gauge) (≈ 1.3 mm diameter) galvanized bailing wire is about right.
- capful of rubbing alcohol (70% isopropyl alcohol)

Procedure:

1) Dip the plastic ring in the beaker of soap solution and gently blow, as shown in Fig. 2-14, to demonstrate requirement of a pressure jump Δp required to sustain a non-zero curvature.

2) Pre-moisten the surface of the glass plate with the soap solution; then use the Tygon® tube to blow a bubble onto the glass surface. It will create a perfect hemisphere, as shown in Fig. E2-2.

3) Pre-moisten the inside surfaces of the bent plastic sheet, and use the Tygon® tube to blow a bubble that spans the gap between the plate surface to create a perfect cylindrical bubble, as shown in Fig. E2-2. Multiple bubbles can be blown into this space to create a variety of right cylindrical surfaces.

4) Dip the cubical wire frame in the beaker of soap solution. If carefully withdrawn, it produces a pattern of soap films as shown in Fig. 2-37. These can be selectively broken carefully using your fingers or using a sharpened pencil tip dipped in the capful of rubbing alcohol

---

[96] A wealth of additional experiments on soap films and bubbles can be found in:
1) Boys, C. V., **Soap Bubbles, Their Colours and the Forces Which Mould Them**, Dover Pub., New York, 1959.
2) Mysels, K. J., Shinoda, K., and Frankel, S., **Soap Films: Studies of their Thinning**, Pergamon Press, New York, 1959.
3) Isenberg, C., **The Science of Soap Films and Soap Bubbles**, Dover Pub., New York, 1992.

to produce a variety of compound and single saddle shaped surfaces of zero mean curvature, illustrating solutions to Plateau's problem.

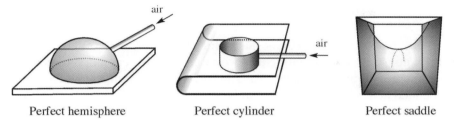

| Perfect hemisphere | Perfect cylinder | Perfect saddle |

Fig. E2-2: Some soap film structures.

3. *Liquid bridges*

A small liquid drop may exert a considerable adhesive force between solid objects that it wets, as shown schematically in Fig. 2-51 for a pair of flat plates. Equation (2.76) shows that the principal force of attraction depends directly on the cosine of the contact angle, θ. For water on glass, $\theta \rightarrow 0°$, the force is maximum. A single drop of water will produce an adhesive force that makes the plates difficult to pull apart. (It must be noted that part of the attractive force is the viscous resistance to thickening of the water film.) For water on Teflon®, $\theta \approx 110°$, $\cos\theta < 0$ so the adhesive force has a negative pressure component and only a weak positive interline force, and a net value near zero. For a fair test, make sure the Teflon® surfaces are as smooth as possible. The water bridge between glass and Teflon has a zero pressure component and only a very weak attractive interline force.

Materials:

- two clean glass plates, 3×3×1/4 in. (with edges smoothed)
- two smooth Teflon® plates, 3×3×1/4 in.
- small container of water, with drop-dispensing tip

Procedure:

Place a single drop of water at the center of one of the glass plates, and place the second plate on top. Try to pull them apart, as in Fig. E2-3. Do the same thing with the Teflon® plates, as well as one glass and one Teflon plate. The differences will be very noticeable. More interesting results can

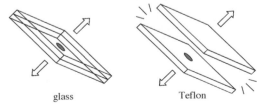

glass   Teflon

Fig. E2-3: Liquid bridge adhesion of water between glass *vs.* Teflon® plates.

be obtained by substituting rubbing alcohol (70% isopropyl alcohol in water) for water.

4. *Shared menisci*

Equation (2.88) suggests that solid objects of like wettability, either wet or non-wet, will be attracted to one another by capillary forces when they share a liquid meniscus. Thus particles of the same type floating or suspended at a water surface should be drawn together and stick. Particles of opposite wettability characteristics will be repelled from one another. Particles will also be either attracted or repelled from the meniscus at the container wall depending on whether the meniscus at the wall shows the same shape, *i.e.* concave upward or downward, as it does against the particles.

Materials:

- small Petri dish half filled with clean water.
- cork particles, approximately 2-4 mm diameter
- Teflon® shavings, approximately 2-4 mm diameter
- forceps
- plastic water bottle with delivery tube.
- overhead projector

Procedure:

Gently place a few cork particles on the surface of water in the Petri dish at least several mm apart from one another. This can be done with the Petri dish placed on an overhead projector focused on the particles and projected on the screen. Then gently blow on the surface to bring the particles close to one another, and watch them snap together. Note also how they are drawn to the wetted glass edge of the dish. Then gently place a few Teflon particles on the surface. (This must be done carefully as Teflon® is heavier than water and will sink if submerged.) These particles also clump together, but are seen to repel the cork particles and to stay away from the glass meniscus. Next, gently increase the water level in the dish until it bulges over the rim, creating a concave downward meniscus. Observe the cork particles leaving the meniscus and the Teflon particles moving into it, as shown in Fig. 2-56 and Fig. E2-4.

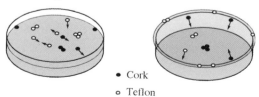

• Cork

o Teflon

Fig. E2-4: Wettable (cork) and unwettable (Teflon) particles on a water surface.

Chapter 3

THERMODYNAMICS OF INTERFACIAL SYSTEMS

A. The thermodynamics of simple bulk systems

1. *Thermodynamic concepts*

Thermodynamics is in general useful for providing over-arching "rules" that govern the descriptions of macroscopic systems in terms of their properties and their interactions with other systems. The systems usually encountered in textbooks for scientists and engineers are pieces of infinite systems, as pictured schematically in Fig. 3-1(a). Their "boundaries" are completely conceptual and serve only to delineate the extent of the system. Furthermore, it is assumed in formulating thermodynamic descriptions that

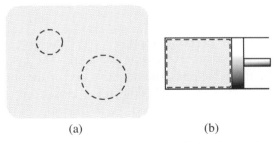

(a)   (b)

Fig. 3-1: (a) Thermodynamic systems as finite pieces of an infinite homogeneous system, (b) Simple-compressible system as a fluid in a piston-cylinder arrangement.

the systems are in internal equilibrium so that, in the absence of external fields, they are homogeneous with respect to their relevant intensive properties. Specifically, since they are assumed to be in internal thermal, mechanical and diffusional equilibrium, they are uniform with respect their temperature, pressure and composition. This requirement poses problems for solids, in which full internal equilibrium often does not exist. In fluids, molecular mobility often (but not always) guarantees internal equilibrium, whereas in solids, non-equilibrium structures, and hence non-uniform stress fields and compositions, can be frozen in place over time scales of practical interest. (This is clearly the case for many of the purpose-built micro or nano constructions of interest in nanoscience and nanotechnology.) "Textbook chemical thermodynamics" thus focuses on fluids, and the systems commonly encountered are fluid masses delineated by boundaries that are close to, but not coincident with their actual physical boundaries, as exemplified by the gas contained within the piston-cylinder arrangement of Fig. 3-1(b). Since the physical interfacial region between the fluid and the cylinder wall is only a few Å thick, essentially all of the mass of the fluid is captured by drawing the model system boundary just far enough away from the wall to exclude all the inhomogeneity of the interfacial layer.

2. The simple compressible system

In the absence of external fields, the systems of chemical thermodynamics are subject only to the single work mode of compression ("p-V work"), and energy added to the system as work is given by:

$$\delta W = -pdV. \tag{3.1}$$

This defines a *simple-compressible system*, and the only purely physical processes to which it is subject are those of compression-expansion and heating-cooling. (Mixing or de-mixing can always be deconstructed into compression-expansion processes.) The list of allowable processes can be extended, however, to include ones with respect to which the system is not necessarily in equilibrium, *viz.*, chemical reactions or phase changes. The conditions of internal *physical* equilibrium can be envisioned to continue to obtain for individual phases while the system is undergoing one or more of the above physicochemical processes.[1]

The first step in describing purely physical processes consists of writing down the appropriate expression for work. The next step makes use of the State Postulate, which asserts the existence of the internal energy, U, as a property of the system, expressible for a closed system (a system of given mass or set of mole numbers) as a function of two independent relevant variables (one more than the number of independent work modes), *e.g.*, using the independent variables T and V:

$$U = U(T,V). \tag{3.2}$$

The First Law of thermodynamics asserts that any change in system energy is given by:

$$dU = dQ + dW, \tag{3.3}$$

where Q is *heat*, defined as energy transfer to the system unaccounted for in macroscopic evaluations of work. The First Law statement is a useful, as opposed to a trivial, statement of the conservation energy because it can be shown that the heat effect Q can be measured independently by *calorimetry*. For systems not undergoing physicochemical processes (phase changes or chemical reactions), the heat effect is associated with a temperature change of the system, *i.e.* $dQ = CdT$, where C is the heat capacity of the system

---

[1] Prigogine, I., and Defay, R., **Chemical Thermodynamics**, Longmans Green and Co., London, 1954.

dependent on the thermodynamic state of the system and the nature of the process (constant V, constant p, *etc.*) during which heat is added. Appropriate scales for both temperature and heat capacity (for reference substances) were established by making use of the mechanical equivalence of heat. Putting the above relationships together allows the development of an explicit expression for the heat effect, Q, accompanying any such process, *e.g.*

$$\delta Q = \delta U - \delta W = \left(\frac{\partial U}{\partial T}\right)_V dT + \left(\frac{\partial U}{\partial V}\right)_T dV + pdV$$

$$= C_V dT + T\left(\frac{\partial p}{\partial T}\right)_V dV, \tag{3.4}$$

where the heat capacity at constant volume, $C_v = \left(\frac{\partial U}{\partial T}\right)_V$, and the form of the coefficient in the second term of Eq. (3.4) is obtained by ordinary thermodynamic reductions; in this case:

$$\left(\frac{\partial U}{\partial V}\right)_T = T\left(\frac{\partial p}{\partial T}\right)_V - p. \tag{3.5}$$

One may proceed from this point to the expression for entropy change, dS, for a quasi-static process, *viz.* dQ/T, substituting from Eq. (3.4):

$$dS = \frac{\delta Q_{rev}}{T} = \frac{C_v}{T} + \left(\frac{\partial p}{\partial T}\right)_V dV, \tag{3.6}$$

and thence to the Helmholtz free energy function: $F = U - TS$:

$$dF = d(U - TS) = -SdT + \delta w = -SdT - pdV. \tag{3.7}$$

The Second Law of Thermodynamics states that for the spontaneity of a proposed process in an isolated system, $dS \geq 0$, or $dF \leq 0$ for systems constrained to constant T and V. Independent variables T and p may be chosen instead of T and V, in which case it is convenient to introduce the system enthalpy, $H = U + pV$ and the Gibbs free energy (or free enthalpy), $G = H - TS$. The enthalpy is a useful function for describing heat and work effects accompanying constant pressure processes, and the Gibbs free energy is a useful function for describing system equilibria and stability with respect to processes at constant temperature and pressure. In this set of variables, the descriptive equations become:

$$\delta W = -p\left(\frac{\partial V}{\partial T}\right)_p dT - p\left(\frac{\partial V}{\partial p}\right)_T dp \tag{3.8}$$

$$\delta Q = C_p dT - T\left(\frac{\partial V}{\partial T}\right)_p dp \tag{3.9}$$

$$dH = d(U + pV) = C_p dT + Vdp \tag{3.10}$$

$$dS = \frac{\delta q_{rev}}{T} = \frac{C_v}{T} + \left(\frac{\partial p}{\partial T}\right)_V dV = \frac{C_p}{T} - \left(\frac{\partial V}{\partial T}\right)_V dp \tag{3.11}$$

$$dG = d(H - TS) = -SdT + Vdp \tag{3.12}$$

In all cases, mathematical reductions have been made which put the expressions in a form such that the coefficients can be evaluated from quantities obtainable in the laboratory.[2] For systems modeled as "simple-compressible," the list of such quantities includes only:

- volumetric data (p-V-T equations of state)
- calorimetric data (heat capacities, latent heats)
- composition (whose changes allow one to follow a physicochemical process)

From the above type of development, much can be done toward describing the behavior of real systems. In particular, expressions for heat and work effects are made available, and expressions for the "driving forces," whose sign and magnitude determine the spontaneity of various processes, are derived.

B. The simple capillary system

1. The work of extension

In constructing a model for systems that includes their interface(s) with adjacent systems, it is first recognized that if it is to be a piece of an infinite system, the piece must include one or more interfaces within its conceptual boundaries. Since the latter are finite, the interfacial area in the system is also finite,[3] and the extensive properties of the system can no longer be obtained by simply multiplying the corresponding intensive properties by the system mass. Such a system will be termed in general an *interfacial system*, and if the bulk phase states involved are fluids, it is a *capillary system*. As stated earlier, because of their greater simplicity, it is useful to first set forth the thermodynamic description applicable to capillary systems and to point out and discuss later the ramifications of extending the description to fluid-solid interfacial systems.

---

[2] Recall that absolute entropy, S, is obtainable (using the Third Law of Thermodynamics) calorimetrically or spectroscopically.

[3] Hill, T. L., **Thermodynamics of Small Systems**, Part I, W. A. Benjamin Publ., New York (1963).

The simplest model one may use is that of the *simple capillary system*, pictured in Fig. 3-2. It consists of three parts in internal equilibrium: two portions of bulk phase, of volumes V' and V'', which are "simple compressible," and the interface itself, of area A, regarded mechanically as a membrane of zero thickness in uniform, isotropic tension. It is not possible to choose any simpler model, such as the interfacial layer by itself, because the latter is not "autonomous," *i.e.* it is inextricably connected to at least small adjacent portions of the bulk phase on either side.[4]

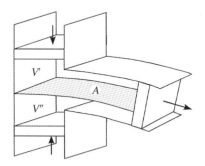

Fig. 3-2: The simple capillary system.

Work may be done on a simple capillary system in accord with:

$$dW = -p'dV' - p''dV'' + \sigma dA, \qquad (3.13)$$

where the last term on the right is the "work of area extension."[5] In writing Eq. (3.13), it is clear that the membrane model of Young has been incorporated into the simple capillary system model. Recall that the difference between p' and p'' is related to the interfacial tension and the local curvature of the interface, κ, in accord with the Young-Laplace Equation, Eq. (2.29):

$$p'' - p' = \sigma\kappa. \qquad (3.14)$$

The pressure difference is usually not significant for purposes of computing work unless the surface is of high curvature ($R_m = 1/\kappa \le 1$ μm). One thus generally uses the approximation of a flat surface system, so that:

$$dW \approx -pdV + \sigma dA. \qquad (3.15)$$

2. Heat effects; abstract properties; definition of boundary tension

Starting with the above work expression, and using the State Postulate in the form, for a closed simple capillary system:

$$U = U(T, V, A), \qquad (3.16)$$

---

[4] Defay, R., Prigogine, I., Bellemans, A., and Everett, D. H., **Surface Tension and Adsorption**, pp. 2-3, Longmans, London, 1966.

[5] If the interface resists bending deformations, as mentioned in Chap. 2, G.4, an additional work term may be required.

together with the First and Second Laws of thermodynamics, one can derive the equations for the heat effect and the various thermodynamic property changes for simple capillary systems, which are as follows:

$$\delta Q = C_{V,A}dT + T\left(\frac{\partial p}{\partial T}\right)_{V,A} dV - T\left(\frac{\partial \sigma}{\partial T}\right)_{V,A} dA \tag{3.17}$$

$$dU = C_{V,A}dT + \left[T\left(\frac{\partial p}{\partial T}\right)_{V,A} - p\right]dV + \left[\sigma - T\left(\frac{\partial \sigma}{\partial T}\right)_{V,A}\right]dA \tag{3.18}$$

$$dS = \frac{C_v}{T}dT + \left(\frac{\partial p}{\partial T}\right)_{V} dV - \left(\frac{\partial \sigma}{\partial T}\right)_{V,A} dA \tag{3.19}$$

$$dF = -SdT - pdV + \sigma dA \tag{3.20}$$

Integration of the expression for dF at constant temperature and volume gives:

$$F = F_0 + \sigma A, \tag{3.21}$$

where F_0 is a constant with respect to system area changes at constant T and V. Such equations show how one can evaluate heat and work effects and other property changes for simple capillary systems from laboratory data obtainable for such systems, the list of which now consists of

• volumetric data,
• calorimetric data,
• composition, AND
• interfacial tension data (as a function of T and composition)

Comment needs to be made regarding the derivatives $\left(\frac{\partial p}{\partial T}\right)_{V,A}$ and $\left(\frac{\partial \sigma}{\partial T}\right)_{V,A}$ in Eqs. (3.17)–(3.19) above. A *pure-component* simple capillary system is just a liquid against its equilibrium vapor, so that the pressure is the vapor pressure at the given temperature, and the surface tension is also fixed when the temperature is fixed. Thus in this case, both expressions should be written as total derivatives. For closed multicomponent capillary systems, both the system pressure and surface tension have to be regarded as functions of volume and surface area as well as temperature because changes in those variables at constant temperature will result in a redistribution of the components amongst the system parts, *i.e.*, between the liquid phase, the vapor phase and the interface.

It may be desirable to change the independent variable set to (T, p, A) or (T, p, σ), but for single-component simple capillary systems this is not possible because once temperature is fixed, p and σ are fixed, as mentioned above. These variable sets may be used for multicomponent capillary systems, in which variations in p and σ would be accompanied by redistribution of the components between the various parts of the capillary system. Multicomponent as well as open capillary systems are discussed more explicitly below.

The above equations have been developed under the assumption of internal equilibrium. Internal diffusional equilibrium in a simple capillary system means that during any process, an equilibrium distribution of the components between the various parts of the system (the two bulk phase portions and the interface) exists. This amounts to maintaining phase equilibrium and adsorption equilibrium, as discussed in more detail below.

The important new terms in the above expressions express the consequences of making interfacial area changes, viz.

- work of extension: $\quad\quad\quad\quad\quad \sigma\, dA$

- heat of extension: $\quad\quad\quad\quad -T\left(\dfrac{d\sigma}{dT}\right)dA$, and

- entropy of extension: $\quad\quad -\left(\dfrac{d\sigma}{dT}\right)dA$, etc.

In particular, the free energy of extension at constant (T,V), i.e., $-\left(\dfrac{\partial F}{dA}\right)_{T,V}$, provides a thermodynamic definition of the boundary tension, σ.

What one learns upon inserting realistic numbers from surface or interfacial tension measurements into the above equations for δQ and δW is that the heat and work effects associated with interfacial area changes, such as might accompany the subdivision of ordinary bulk matter into particles in the colloid size range, are quite small. For example, if at 300°K a droplet of diameter 1 cm (with $A = 3.14$ cm^2) is broken up into droplets of diameter one μm (there will be 1.33×10^{12} such droplets), the surface area will have increased to $A = 4.19$ m^2. If the surface tension is $\sigma = 35$ erg/cm^2, and its temperature derivative is taken as -0.1 erg/cm^{2}K, we have:

$$W = \int_{3.14}^{4.19 \cdot 10^4} \sigma\, dA = (35)(4.19 \cdot 10^4) = 1.47 \cdot 10^6 \text{ erg} = 0.035 \text{ cal}, \tag{3.22}$$

and

$$Q = -\int_{3.14}^{4.19 \cdot 10^4} T\left(\frac{d\sigma}{dT}\right)dA = -(300)(-0.1)(4.19 \cdot 10^4) \text{ erg} = 0.030 \text{ cal} \tag{3.23}$$

These values are almost negligibly small. One might have anticipated the smallness of these effects as they were not taken into account in the early experiments of Clausius, Rumford, Joule, etc., leading to the establishment of the First Law principle, i.e., a statement of the conservation of energy.

The failure to account for them never led to any apparent violation of the principle.

The above type of development also leads to the expression of important driving forces affecting the spontaneous behavior of capillary systems. Most importantly, it shows from the thermodynamic point of view, that systems with positive values of σ will always tend to spontaneously contract their interface. A spontaneous process will occur (at constant T,V) if F can thereby be made to decrease, *i.e.*, dF is negative. This will obviously occur if the change in area, A, is negative.[6]

It is to be noticed that the key properties involved in describing the interfacial effects in pure-component systems or systems of constant composition are the boundary tension and its temperature derivative, the experimental determination of which we have already discussed.

C. Extension to fluid-solid interfacial systems

1. *The work of area extension in fluid-solid systems*

Next consider how fluid-*solid* interfacial systems differ from the fluid-fluid (capillary) systems described above. The first observation is that solids, and in particular interfacial zones between solids and fluids, are often not in states of full internal equilibrium. Stress relaxation times may be large relative to times scales of practical interest so that, dependent on their history, solid specimens may be supporting significant residual stresses. Nonetheless, provided appropriate precautions are taken, such systems can often be qualitatively and even quantitatively described in terms of the thermodynamics of capillary systems.[7] The "tension" σ at a solid-fluid interface represents, analogous to that in a capillary system, the integral effect of a reduced lateral stress component across the zone of inhomogeneity between the fluid and solid phases. The difference is that the stress field in the solid is in general not known and may not be in a relaxed, *i.e.*, internal equilibrium, state. Even in the absence of residual stresses, the boundary tension at a fluid-solid interface can be defined only when the total free energy density tensor in the solid phase reduces to a uniform value. Fortunately, this restriction may not be critical, because the elastic (stress)

---

[6] There are cases where this statement must be amended. The process of breaking up a mass of liquid into droplets, as described above, produces a *configurational* entropy change in addition to the term: $-\int \left(\dfrac{d\sigma}{dT}\right) dA$, as discussed in Chap. 9 in the context of emulsification. This also affects the free energy changes, which have entropy content.

[7] Hering, C., *in* **Structure and Properties of Solid Surfaces**, R. Gomer and C. S. Smith, Eds., pp. 5-82, Univ. Chicago Press, Chicago (1953).

energy density is usually only a small (often negligible) component of the total energy density.[8]

The evaluation of the work associated with area extension for fluid-solid interfacial systems depends on how the area is created. One way this may be effected is through direct interface creation, as occurs, for example when new solid material precipitates out of a solution. In this case, the (reversible) work effect is given by

$$\delta W = dF = \left(\frac{\partial F}{\partial A}\right) dA = \sigma dA,$$ (3.24)

the same result as for capillary (*i.e.*, fluid interface) systems. On the other hand, new solid-fluid interface may be produced by the mechanical *stretching* of a pre-existing solid-fluid interface, a process that changes the structure and properties of the interfacial layer. In capillary systems such changes are instantly relaxed out, and the work expression is the same as that for interface creation, but in solid-fluid systems they may be frozen in indefinitely, or at least for significant periods of time. The isothermal quasi-static mechanical work required to stretch the interfacial area is given in general by[9]

$$\left(\frac{\delta W}{dA}\right)_{\text{int. dilation}} = \sigma_{\text{m}} = \frac{dF}{dA} \equiv \frac{d(f^{\sigma} A)}{dA} = f^{\sigma} + A\frac{df^{\sigma}}{dA} = \sigma + A\frac{d\sigma}{dA},$$ (3.25)

where σ_{m} is the mechanical boundary tension (or total stretching tension), and f^{σ} is the surface free energy per unit area, equal to the thermostatic "tension," σ, defined above. The last term must be included because the structure of the surface is being changed. Finally, work may be associated with both bulk and surface shear strain in the solid. The work terms associated with these processes would be[10]:

$$\text{Work of bulk shear strain} = V(\mathbf{s}^{\beta} : \delta\varepsilon^{\beta}), \text{ and}$$ (3.26)

$$\text{Work of surface shear strain} = A(\mathbf{s}^{\sigma} : \delta\varepsilon^{\sigma}),$$ (3.27)

where $\mathbf{s}^{\beta}$ and $\mathbf{s}^{\sigma}$ are the bulk and surface stress tensors, respectively, and $\delta\varepsilon^{\beta}$ and $\delta\varepsilon^{\sigma}$ are the corresponding bulk and surface displacement tensors. In processes involving surface area extension in fluid-solid interfacial systems, the work of surface shear strain is usually negligible, but the work associated with the accompanying bulk shear strain is often the dominant component of the actual total work effect. It is generally not possible to calculate this effect

---

[8] Defay, R., Prigogine, I., Bellemans, A., and Everett, D. H., **Surface Tension and Adsorption**, pp. 286ff, Longmans, London, 1966.

[9] Tabor, D., **Gases, Liquids and Solids and other States of Matter**, 3rd Ed., pp. 166-168, Cambridge Univ. Press, Cambridge, UK (1996).

[10] Benson, G. C., and Yun, K. S., *in* **The Solid-Gas Interface**, Vol. 2, E. A. Flood, Ed., pp. 203-269, Marcel Dekker, New York (1967).

quantitatively, but it is one of the reasons that the energies associated with size reduction (crushing and grinding) of solids are usually much larger than those suggested by the above equation for ($\delta W/dA$).

2. Compound interfacial systems; Young's equation

Another method for changing the area of a given fluid-solid interface is by advancing or retracting the fluid phase across the solid, as when a liquid drop is advanced or retracted across a solid surface, pictured in Fig. 3-3. In such a case, a given fluid-solid interface is created or destroyed at the expense or benefit of another fluid-solid and fluid- fluid interface. What is required to describe such processes are *compound* capillary or interfacial systems, *i.e.*, systems with more than a single type of interface. Consider the (reversible) work associated with the process shown in Fig. 3-3, whereby the liquid drop expands its base at constant temperature and volume (in the absence of gravity) to cover more of the solid surface. This is given by

$$\delta W = dF = \left(\frac{\partial F}{\partial A_{SG}}\right)_{T,V} dA_{SG} + \left(\frac{\partial F}{\partial A_{SL}}\right)_{T,V} dA_{SL} + \left(\frac{\partial F}{\partial A_{LG}}\right)_{T,V} dA_{LG}$$

$$= (\sigma_{SG} - \sigma_{SL})dA_{SG} + \sigma_{LG}dA_{LG} \tag{3.28}$$

Expressions of this type are the thermodynamic basis for the computation of the work associated with such processes as adhesion and wetting discussed in the next chapter.

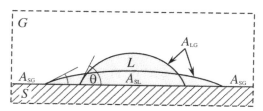

Fig. 3-3: A compound capillary system consisting of a drop of liquid (L) resting on a solid (S) surface in the presence of a gas (G). If the liquid drop is flattened, its interfacial area against the solid, A_{SL}, increases while that of the solid surface against the gas, A_{SG}, decreases by a corresponding amount, and the liquid-gas interface, A_{LG}, increases.

The thermodynamically preferred configuration for a compound capillary system of the type shown in Fig. 3-3 is obtained from the minimization of the free energy, F, with respect to variation of the interfacial areas in an isothermal, constant volume process of the type suggested in the figure[11]:

$$F = F_0 + \sigma_{LG}A_{LG} + \sigma_{SL}A_{SL} + \sigma_{SG}A_{SG}, \tag{3.29}$$

[11] Sheludko, A. **Colloid Chemistry**, pp. 90-92, Elsevier, Amsterdam, 1966.

where F_0 is a constant. If the drop is sufficiently small that gravity can be neglected ($Bo \rightarrow 0$), it may be regarded as spherical cap,[12] allowing its liquid-gas surface, A_{LG}, its volume V and the contact angle θ to be expressed in terms of its height h and its base radius R. Consider the construction shown in Fig. 3-4, with the slice dV of the spherical segment of radius R shown. The element of segment area dA is the width of the slice, $Rd\psi$, times its circumference, $2\pi r' = 2\pi R\sin\psi$. Thus, integrating over the segment, i.e., from $\psi = 0$ to $\psi = \theta$, one obtains:

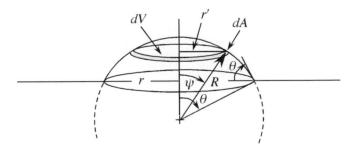

Fig. 3-4: Diagram showing computation of the area and the volume of a spherical segment.

$$A_{LG} = \int dA = \int_0^\theta 2\pi R^2 \sin\psi d\psi = 2\pi R^2 (1 - \cos\theta). \tag{3.30}$$

The volume of the slice dV is its area: $\pi r'^2 = \pi R^2 \sin^2\psi$, times its thickness, $R\sin\psi d\psi$, which upon integration gives:

$$V = \int dV = \int_0^\theta \pi R^3 \sin^3 \psi d\psi = \frac{1}{3}\pi R^3 \left[2 - \cos\theta(\sin^2\theta + 2)\right]. \tag{3.31}$$

Then both $\cos\theta$ and R may eliminated in terms of r and h using the relationships:

$$r = R\sin\theta \quad \text{and} \quad h = R(1 - \cos\theta), \tag{3.32}$$

as suggested by Fig. 3-5. After some algebra, one obtains:

$$A_{LG} = \pi\left(r^2 + h^2\right), \tag{3.33}$$

$$V = \frac{1}{6}\pi h\left(3r^2 + h^2\right), \quad \text{and} \tag{3.34}$$

$$\cos\theta = \frac{(r^2 - h^2)}{(r^2 + h^2)}. \tag{3.35}$$

---

[12] If gravity is to be taken into account, the situation is more complex in that the shape is less easily expressible, and gravitational potential energy must be added in. The derivation follows the same lines, however, and the results are identical.

The Helmholtz free energy of the system may then be expressed as

$$F = F_0 + \pi\left(r^2 + h^2\right)\sigma_{LG} + \pi r^2\sigma_{SL} + \left(A_{S\text{-total}} - \pi r^2\right)\sigma_{SG} \qquad (3.36)$$

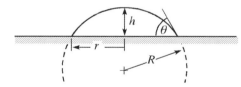

Fig. 3-5: Diagram for derivation of Young's equation by minimization of system free energy.

If F is to be minimum, any variation of it upon variation of r and h (subject to the constraint of constant V) must vanish, $i.e.$,

$$\delta F = \pi(2r\delta r + 2h\delta h)\sigma_{LG} + 2\pi r\delta r\sigma_{SL} - 2\pi r\delta r\sigma_{SG} = 0, \text{ and} \qquad (3.37)$$

$$\delta V = \frac{\pi}{6}(6rh\delta r + 3h^2\delta h + 3r^2\delta h) = 0, \text{ or, simplifying:} \qquad (3.38)$$

$$h\sigma_{LG}\delta h + r(\sigma_{LG} + \sigma_{SL} - \sigma_{SG})\delta r = 0, \text{ and} \qquad (3.39)$$

$$2rh\delta r + (r^2 + h^2)\delta h = 0. \qquad (3.40)$$

Using the Lagrange method for finding the extremum of a function subject to auxiliary constraints, we multiply Eq. (3.40) by the arbitrary constant $-\lambda$, add it to Eq. (3.39), and set the resulting coefficients of both dr and dh equal to zero, giving:

$$\sigma_{LG} + \sigma_{SL} - \sigma_{SG} - 2\lambda h = 0, \text{ and} \qquad (3.41)$$

$$\sigma_{LG}h - \lambda(r^2 + h^2) = 0. \qquad (3.42)$$

Elimination of λ between the above equations, and substituting for $\cos\theta$, gives:

$$\cos\theta = \frac{(\sigma_{SG} - \sigma_{SL})}{\sigma_{LG}}. \qquad (3.43)$$

This important result is known as *Young's Equation*,[13] and it appears to confer upon the contact angle the status of a *thermodynamic property*, $i.e.$, it should have a fixed, unique value for a given physical system at a given temperature and pressure at equilibrium. As seen in Chap. 4, however, experimental values for it depend upon many *non*-thermodynamic quantities, and its use as a thermodynamic property (*e.g.*, to infer "surface energies") must be made only with caution.

A compound capillary system is also exemplified by an oil drop floating at a water-air interface. The total system is made up of three simple capillary systems, involving the water-air, oil-air and oil-water interfaces,

---

[13] Young, T., *Phil. Trans.*, **95**, 65, 82 (1805).

respectively, in mutual equilibrium, as shown in Fig. 2-28. Minimization of the system's free energy in this case leads to the force balance of Eq. (2.48):

$$\underline{\sigma}_w + \underline{\sigma}_o + \underline{\sigma}_{o/w} = 0, \tag{3.44}$$

i.e., Neumann's triangle, where $\underline{\sigma}_w, \underline{\sigma}_o$, and $\underline{\sigma}_{o/w}$, the boundary tensions of the water-air, oil-air and oil-water interfaces, respectively, are treated as vectors.

For a sessile drop on a flat *solid* surface, Neumann's triangle reduces to the set of scalar equations:

$$\sigma_{SG} - \sigma_{SL} = \sigma_{LG} \cos\theta, \text{ and} \tag{3.45}$$

$$\sigma_{LG} \sin\theta = E_S, \tag{3.46}$$

where θ is the contact angle. The first equation above is the horizontal component of the force balance, *viz.* Young's Equation, while the second shows that the vertical component of the interline force must be balanced by an elastic force E_S in the solid phase.

As pointed out in Chap. 2, three-phase interlines that exist in a compound interfacial system may be imbued with a line tension, τ_ℓ. For the case of a sessile drop on a flat surface, the horizontal force must then be augmented by τ_ℓ/R (where R is the base radius of the drop), so that Young's Equation becomes

$$\cos\theta = \frac{\sigma_{SG} - \sigma_{SL}}{\sigma_L} - \frac{\tau_\ell}{R\sigma_L}. \tag{3.47}$$

With a maximum plausible value of $\approx 10^{-9}$ N for τ_ℓ, the final term is seen to be generally negligible, except possibly for micro or nano droplets that may be encountered in hetero-nucleation processes or capillary condensation into micro or nano pores.

D. Multicomponent interfacial systems

1. *The Gibbs dividing surface and adsorption*

To study the effects of composition on the behavior of multi-component capillary systems (*e.g.*, the dependence of surface tension on composition, *etc.*), one must first recognize that the various components in such a system do not distribute themselves uniformly at equilibrium amongst the various parts of the system. An example is shown schematically in Fig. 3-6, where the circles represent a solute component. It is evident that the concentration of the component in the lower bulk phase is higher than it is in the upper bulk phase, and that in this case it is especially high in the interfacial region. The figure also shows a representation of the solute concentration profile through the interfacial region. If we wish to examine the distribution of this component or others within the system, the simple capillary system model must be developed further. It thus far considers the system *only as a whole* (with the assumption that there is maintenance of component distribution equilibrium) as it undergoes various processes.

As mentioned earlier, an attempt to actually "split up" the system into its parts to identify properties with these parts alone, leads to quantities difficult (or impossible) to address in the laboratory, or even to define unambiguously. What can be done, however, is to develop an extension of the model of a simple capillary (or interfacial) system that reflects all of its measurable properties. A number of devices have been proposed for this, but the one most commonly used is that of the *Gibbs Dividing Surface*. Gibbs replaced the interfacial layer by a (zero-thickness) "dividing surface,"

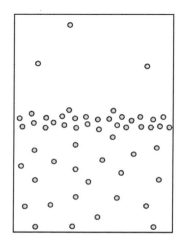 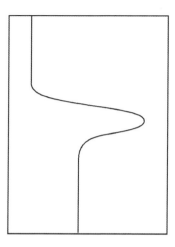

Fig. 3-6: Schematic representation of the variation of a solute concentration in moving from one bulk phase, through the interfacial zone into the adjacent bulk phase.

oriented normal to the density gradient in the interfacial zone, up to which *all* of the intensive variables describing the bulk phase portions of the system are taken to be uniform at their bulk-phase values. This is pictured schematically in Fig. 3-7 for the concentrations of the solvent (1) and solute (2) in a binary simple capillary system. (Once again, the solute is shown to be concentrated in the interfacial zone, but this of course need not be the case.) The bulk phase concentrations of the species are C_1', C_1'', C_2', and C_2''. Depending on the dividing surface location, the model requires an addition (or subtraction) of a particular number of moles of each species to the surface, n_1^σ and n_2^σ, respectively, to exhibit mass equivalence with the real system, *i.e.*,

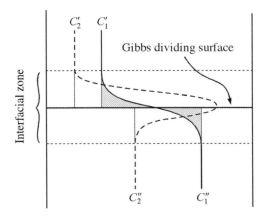

Fig. 3-7: Gibbs Dividing Surface drawn in a binary liquid-gas capillary system.

$$n_1^\sigma = (n_1)_{\text{real}} - (n_1)_{\text{model}}, \text{ and} \qquad\qquad (3.48)$$

$$n_2^\sigma = (n_2)_{\text{real}} - (n_2)_{\text{model}}. \qquad\qquad (3.49)$$

The terms n_1^σ and n_2^σ are called "surface excesses" and are examples of "properties of the dividing surface," as distinct from properties of the capillary system. They are the numbers of moles of the components *ascribed* to the dividing surface. One may define surface excesses for other extensive thermodynamic properties of the system as well, such as "surface excess enthalpy," "surface excess entropy," *etc.*

The location of the dividing surface (for a quasi-flat surface) is arbitrary, but it is clear that both the magnitude and the sign of the n_i^σs are extremely sensitive to where it is placed. (When the surface is curved, mechanical equilibrium considerations require in principle that the dividing surface be located at a specific position called the *surface of tension*.[14] This cannot, in any case, be located experimentally.) A shift of the dividing surface location by just a few Å may change the sign of any surface excess, and may alter its value by many orders of magnitude. The Gibbs dividing surface is clearly of no value for describing mass distribution in the case of a pure-component system, but for a binary system, one may first notice that while the surface excess for either component can be made any value desired by judicious location of the dividing surface, the *relative positions* of the actual concentration profiles of different components are not at all arbitrary. The relationship is fixed by Nature, even if the details of the profiles are not known. One seeks to define a quantity that expresses the *relative* surface excess of one component (usually the solute, 2) *with respect to another* (usually the solvent, 1). *This* quantity is independent of dividing surface

[14] Defay, R., Prigogine, I., Bellemans, A., and Everett, D. H., **Surface Tension and Adsorption**, p. 3, Longmans, London.

location and should, at least in principle, be subject to unambiguous determination in the laboratory. Such a quantity can be derived for a binary system by writing material balances for the solvent and solute components with reference to Fig. 3-8:

1. $n_1^\sigma = n_1 - V'C_1' - V''C_1''$ (substitute $V' = V - V''$)

$$= n_1 - VC_1' + V''(C_1' - C_1''),\ \text{and} \qquad (3.50)$$

2. $n_2^\sigma = n_2 - VC_2' + V''(C_2' - C_2'') \qquad\qquad (3.51)$

The quantities n_1^σ, n_2^σ and V'' are dependent upon dividing surface location, but the other quantities are independent of it and measurable. Eliminating V'' between the above two equations, and putting both the surface excesses on the left side of the equation gives:

$$n_2^\sigma - n_1^\sigma \left[\frac{C_2' - C_2''}{C_1' - C_1''} \right] = \left\{ (n_2 - VC_2') - (n_1 - VC_1') \left[\frac{C_2' - C_2''}{C_1' - C_1''} \right] \right\}. \qquad (3.52)$$

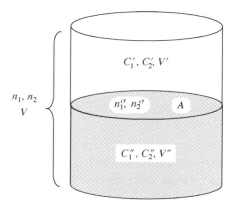

$C_1',\ C_2',\ V'$

$n_1^\sigma,\ n_2^\sigma \qquad A$

$n_1,\ n_2$
V

$C_1'',\ C_2'',\ V''$

Fig. 3-8: Binary simple capillary system.

Finally, dividing through by the interfacial area A gives:

$$\Gamma_{2,1} = \Gamma_2 - \Gamma_1 \left[\frac{C_2' - C_2''}{C_1' - C_1''} \right] = \frac{1}{A} \left\{ (n_2 - VC_2') - (n_1 - VC_1') \left[\frac{C_2' - C_2''}{C_1' - C_1''} \right] \right\}. \qquad (3.53)$$

where the surface excesses on a per-unit-area basis, Γ_1 and Γ_2, are known as the *adsorptions* of those components, and $\Gamma_{2,1}$ is known as the *relative adsorption of 2 with respect to 1*. This quantity meets the requirements called for above, since all of the quantities on the right hand side of the equation are independent of dividing surface location and, in principle, measurable in the laboratory. (The extension to solutions of more than one solute is evident. One may express, for example, the relative adsorption of any of the solutes with respect to the solvent.) Its physical significance may be better appreciated by noting that it is *the adsorption of component 2 (regarded as a solute) when the dividing surface is located in such a way as to make the adsorption of component 1 (the solvent) equal to zero.* One may also appreciate it better by considering the situation when one of the phases (say phase') is a gas. In this event, the concentrations of both components in the gas are generally very small compared with those in the liquid (or solid), and may be neglected. Thus in such a case

$$C_1' << C_1'', \text{ and } C_2' << C_2'', \text{ so that} \tag{3.54}$$

$$\Gamma_{2,1} \approx \Gamma_2 - \Gamma_1\left(\frac{C_2''}{C_1''}\right). \tag{3.55}$$

Dividing through by Γ_1 gives:

$$\frac{\Gamma_{2,1}}{\Gamma_1} = \frac{\Gamma_2}{\Gamma_1} - \left(\frac{C_2''}{C_1''}\right), \tag{3.56}$$

from which it can be seen that the relative adsorption of 2 to 1 is zero when the components are in the same proportion in the interface as they are in the bulk. When component 2 is relatively *enriched* in the interface,

$$\Gamma_{2,1} > 0 \quad (\text{termed } positive \text{ adsorption}), \tag{3.57}$$

and when it is relatively *less* concentrated in the interface than in the bulk,

$$\Gamma_{2,1} < 0 \quad (\text{termed } negative \text{ adsorption}). \tag{3.58}$$

The relative adsorptions of the various solutes in a solution thus describe the distribution of components between the interface and bulk phases.

2. Immiscible interfacial systems

One of the most important simplifications of the current model is that of the *immiscible* simple interfacial system, which occurs when the phases forming the system are completely immiscible, as pictured in Fig. 3-9. This simplification applies in particular to most of the fluid-solid interfacial systems encountered in practice, and amounts to requiring that the solid be non-volatile or insoluble in the adjoining fluid, and that none of the components of the fluid phase dissolve into the solid. Capillary system examples would also include aqueous solutions in contact with an oil or with mercury, in which all components of the solution are assumed to be completely insoluble, or a solution of completely non-volatile components in contact with a gas which does not dissolve in the solution. The most important immiscible interfacial system is that of a solution in contact with an insoluble solid that itself cannot dissolve any of the components of the solution. For immiscible interfacial systems, the Gibbs dividing surface may be located without ambiguity so as to separate all of the atoms or molecules

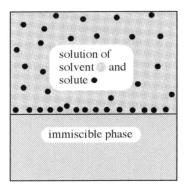

Fig. 3-9: An immiscible interfacial system. The bulk phase portions are completely immiscible, and Gibbs Dividing Surface is made coincident with the actual boundary between the phases. In the system shown, the upper phase is a binary solution, whose components compete for space at the phase boundary.

of the immiscible phases from one another, and the relative adsorption of the solute is given by

$$\Gamma_{2,1} = \Gamma_2 - \Gamma_1\left(\frac{C_2}{C_1}\right),\tag{3.59}$$

where the ($'$) and ($''$) may be dropped from the concentration variables since the components appear in only one phase. It is evident from the above equation as well as Fig. 3-9 that the solute and solvent *compete* for space on the adsorbent surface. Under conditions in which the solution is dilute (*i.e.*, $C_2 \to 0$) or for adsorption of a component from a non-adsorbing gas (so that $\Gamma_1 \to 0$), one may identify the relative adsorption of the solute with its actual adsorption, *i.e.*, $\Gamma_{2,1} \approx \Gamma_2$.

A special case of an immiscible interfacial system occurs when the "solute" is not soluble in *either* bulk phase portion of the system. This is exemplified by the insoluble or Langmuir monolayers discussed briefly in Chap. 2. The monolayer component is spread at the surface. This may also pertain to a spread non-volatile (or insoluble) monolayer at a solid-fluid interface. In such cases, $C_2 = 0$, and

$$\Gamma_{2,1} \approx \Gamma_2.\tag{3.60}$$

3. *The measurement of adsorption*

To measure adsorption, first consider the right hand side of the general expression for it given earlier in Eq. (3.53):

$$\Gamma_{2,1} = \frac{1}{A}\left\{(n_2 - VC_2') - (n_1 - VC_1')\left[\frac{C_2' - C_2''}{C_1' - C_1''}\right]\right\}.\tag{3.61}$$

It at first seems a straightforward thing to do, since all the quantities on the right hand side are measurable. In reality, however, this calls for computing the difference in two terms that are so close to the same value that it cannot in general be determined with any certainty. This derives from the fact that proportionally so little of the solute inventory resides at the interface. The most successful direct attempt at measuring the relative adsorption in a capillary system has been the *microtome* method of McBain and coworkers[15], shown schematically in Fig. 3-10. A very thin slice (0.05 - 0.1 mm thick, 1 - 2 m^2 in area) of solution was scraped from the surface by a knife traveling along a set of rails. Some results from their experiments are shown in Table 3-1. Interferometry was used to measure the ratio of (n_2/n_1) in the scraped-off layer and compared with the ratio of concentrations for a sample taken from the bulk. The idea was that the scraped-off layer was sufficiently thin that its composition would reflect the different ratio of components at the interface, and the relative adsorption could be determined from the general expression (with C_1' and $C_1'' \approx 0$):

$$\Gamma_{2,1} \approx \frac{1}{A}\left[n_2 - n_1\left(\frac{C_2''}{C_1''}\right)\right].$$

(3.62)

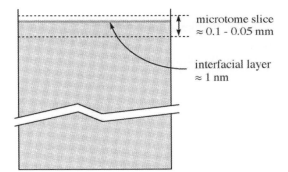

microtome slice
$\approx 0.1 - 0.05$ mm

interfacial layer
≈ 1 nm

Fig.3-10: Schematic of McBain's microtome method for measuring relative adsorption at a gas-liquid interface.

The measured values give the order of magnitude of typical adsorption. Radioactive tracers[16] and bubble fractionation techniques[17] have also been used to measure $\Gamma_{2,1}$, although they are both fraught with difficulties in interpretation. At any rate, one may consider $\Gamma_{2,1}$, to be a measurable (albeit sometimes with difficulty), unambiguous property of fluid interface systems.

Table 3-1: Relative solute adsorption from aqueous solutions obtained by the microtome method. 1 μmol/m^2 = 10^{-10} mol/cm^2.

Solute	C_2 (M)	$\Gamma_{2,1}$ (μmol/m^2)
Phenol	0.218	4.4
n-Hexanoic acid	0.0223	5.8
NaCl	2.0	- 0.74

[15] McBain, J. W. and Humphreys, C. W., *J. Phys. Chem.*, **36**, 300 (1932); McBain, J. W., and Swain, R. C., *Proc. Roy. Soc. (London)*, **A154**, 608 (1936).

[16] Adamson, A. W., **Physical Chemistry of Surfaces**, 4th Ed., pp. 80 ff, Wiley-Interscience, New York, 1982.

[17] Adam, N. K., **The Physics and Chemistry of Surfaces**, pp. 113 ff, Dover Publ., New York, 1968.

For some immiscible interfacial systems, however, determination of adsorption can be made fairly easily, as in the case of adsorption onto the surface of finely divided, high-surface-area solids from either a gas mixture or a solution. These systems are usually immiscible so that C_1'' and C_2'' (concentrations in the solid phase) are both 0, and one may simplify the expression for $\Gamma_{2,1}$ to:

$$\Gamma_{2,1} = \frac{1}{A}\left\{(n_2 - VC_2) - (n_1 - VC_1)\left[\frac{C_2}{C_1}\right]\right\}. \tag{3.63}$$

The second term in the brackets is negligible, following from the assumption that the number of moles of sol*vent* 1 adsorbed, n_1^σ, is negligible relative to the total number of moles of solvent, n_1, in the system. Then one has:

$$C_1 = \frac{n_1 - n_1^\sigma}{V - \bar{v}_1 n_1^\sigma} \approx \frac{n_1}{V}, \tag{3.64}$$

from which the second term in the brackets above is zero. For the case of adsorption out of a liquid solution, a common method is to measure the concentration of solute in the bulk phase before, $C_2^{\,0}$, and after, C_2, adsorption equilibrium has been established, as shown in Fig. 3-11(a). Then:

$$\Gamma_{2,1} \approx \frac{1}{A}\left[n_2 - VC_2\right] = \frac{V}{A}\left[C_2^0 - C_2\right]. \tag{3.65}$$

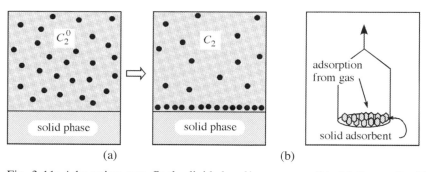

Fig. 3-11: Adsorption onto finely divided and/or porous solids (a) from a liquid phase, or (b) from a gas phase.

This will be a measurable difference if the total amount of adsorption is sufficiently large, a condition which often exists because the area A of the solid-fluid interface is large - often several hundred m²/gram of solid adsorbent.

For adsorption of a component 2 from a gas mixture with a non-adsorbing component 1, measurement can be made directly from the increase in weight (Δw) of the adsorbent (of molecular weight *MW*) upon adsorption, as shown in Fig 3-11(b). The computation for the relative adsorption is:

$$\Gamma_{2,1} = \frac{1}{A} \frac{(\Delta w)}{(MW)g},$$ (3.66)

where g is the gravitational constant.

Chromatography also provides a powerful means for the measurement of adsorption at solid-fluid interfaces, often even when the specific area of the adsorbent is not especially large. High performance liquid chromatography (HPLC)[18] and gas or inverse gas chromatography (IGC)[19] are used for adsorption from solution and from gas mixtures, respectively.

Another case in which "adsorption" is readily determined is that of an insoluble (Langmuir) monolayer spread on a liquid surface. In such a system, we have:

$$\Gamma_{2,1} = \Gamma_2 = \frac{n_2}{A}.$$ (3.67)

A given number of moles of the surfactant, n_2, is deposited on the surface using a micrometer syringe containing a dilute solution of the compound in a volatile "spreading solvent," which evaporates from the surface quickly after it is spread.

Both adsorption at the solid-liquid interface and Langmuir monolayers are discussed in more detail later in this chapter.

4. The phase rule; descriptive equations for binary interfacial systems

Capillary systems, divided up into parts as we have done, may appear to have more variables describing them than do "ordinary systems." For a two-phase, C-component system we have in addition to p, T and C - 1 concentration variables, the surface tension and C - 1 relative adsorptions, $\Gamma_{1,i}$ ($i = 2 \to C$), i.e., a total of C additional variables. This does not mean, however, that the variance of the system has increased, because there is a new equation for each new dependent variable, viz., the adsorption equilibrium equation, $\mu_i^{ads} = \mu_i^{bulk}$, for each component, the μ_i's being the chemical potentials of the components.

Consider a binary two-phase system consisting of solvent (1) and solute (2). The phase rule is:

$$F = C - p + 2 = 2 - 2 + 2 = 2,^{20}$$ (3.68)

---

[18] Sharma, S. C., and Fort, T., *J. Colloid Interface Sci.*, **43**, 36 (1973);
Wang, H. L., Duda, J. L., and Radke, C. J., *J. Colloid Interface Sci.*, **66**, 153 (1978).

[19] Lloyd, D. R., Ward, T. C., and Schreiber, H. P., Eds., **Inverse Gas Chromatography**, ACS Symposium Ser. 391, Washington, DC, 1989.

[20] The variance goes up by one if the curvature of the interface is high ($R_m \leq 1$ μm), with the additional variable being a measure of the curvature.

where in general c is the number of components, p is the number of phases, and F is the "variance," or number of independent intensive variables required to fix the intensive state of the system. If one fixes, for example, T and the mole fraction of the solute in the liquid, x_2, the state of the system is fixed, specifically the values of σ and $\Gamma_{2,1}$ are fixed. More generally, for a binary *isothermal* system, one can express any variable as a function of any other independent variable. At constant T, e.g., $p = p(x_2)$; $y_2 = y_2(x_2)$, etc., or bringing in the capillary properties, we have:

$$\sigma = \sigma(x_2) \quad \rightarrow \quad \text{the } \textit{surface tension equation} \tag{3.69}$$

$$\Gamma_{2,1} = \Gamma_{2,1}(x_2) \quad \rightarrow \quad \text{the } \textit{adsorption isotherm} \tag{3.70}$$

$$\sigma = \sigma(\Gamma_{2,1}) \quad \rightarrow \quad \text{the } \textit{surface equation of state} \tag{3.71}$$

These three types of relationships (named as indicated) each describe the binary capillary system. What is observed, however, is that for a given type of system, usually only one of them is readily accessible in the laboratory. The surface tension equation is easily obtained for fluid interface systems, since both σ and x_2 (or C_2) are readily measured. $\Gamma_{2,1}$, however, is not easily measured for these kinds of systems (recall the microtome method), and therefore neither the adsorption isotherm nor the surface equation of state is generally obtainable directly. The adsorption isotherm is readily determined for solid-liquid or solid-gas systems (if the specific area is large), since $\Gamma_{2,1}$ and C_2 are both accessible. The surface tension, however, cannot be measured for these systems, and so the surface tension equation and surface equation of state are unobtainable. The surface equation of state is readily determined for insoluble monolayers, since for such systems both σ and $\Gamma_{2,1}$ are measurable, but the immeasurability of C_2 precludes getting the other two types of relationships.

E. The Gibbs Adsorption Equation

We are often interested in one of the "inaccessible" relationships for a given system. For example, we may wish to obtain the adsorption isotherm for a liquid-gas system, or the surface pressure (equilibrium spreading pressure, π) of an adsorbing solute at a solid-liquid interface. Thermodynamics solves this problem by providing a rigorous relationship amongst the above equations. (This is one of the most important general things that thermodynamics does for us, *i.e.*, yielding quantities we cannot measure in terms of quantities we *can* measure.) The rigorous relationship referred to is the capillary system analogue of the Gibbs-Duhem Equation. It is called the *Gibbs Adsorption Equation*. At constant T it takes the form:

$$d\sigma = -\sum_{i=2}^{m} \Gamma_{i,1} d\mu_i = -\Gamma_{2,1} d\mu_2 - \Gamma_{3,1} d\mu_3 - \dots \tag{3.72}$$

for a solution consisting of a solvent (1) and any number of solutes (2), (3), *etc.*

Its general derivation follows. The objective is to develop a relationship amongst the variables describing a capillary system in full adsorption equilibrium. To do this, one starts by generalizing the thermodynamic description to *open* multicomponent simple capillary systems. Specifically:

$$dF = -SdT - p'dV' - p''dV'' + \sigma dA + \sum \mu_i dn_i, \tag{3.73}$$

where

$$\mu_i = \left(\frac{\partial F}{\partial n_i}\right)_{T,V',V'',A,n_{j\neq i}}.$$

As is done in bulk phase thermodynamics, we define a potential function whose differential is expressed in terms of differentials of *intensive* properties. In this case, such a function is

$$G = F + p'V' + p''V'' - \sigma A \tag{3.74}$$

Then:

$$dG = -SdT + V'dp' + V''dp'' - Ad\sigma + \sum \mu_i dn_i. \tag{3.75}$$

This can be integrated under conditions such that the intensive state of the system remains constant, *i.e.*, constant T, p', p'', σ and μ_i, leading to:

$$G = \sum \mu_i n_i, \tag{3.76}$$

for which a general differentiation yields:

$$dG = \sum \mu_i dn_i + \sum n_i d\mu_i. \tag{3.77}$$

Comparison of the two expressions for dG gives:

$$SdT - V'dp' - V''dp'' + Ad\sigma + \sum n_i d\mu_i = 0. \tag{3.78}$$

This is the form of the Gibbs-Duhem Equation for simple capillary systems. We may subtract from it the Gibbs-Duhem Equation for each of the bulk phase portions (in the Gibbs model), *viz.*,

i) $\quad S'dT - V'dp' + \sum n_i'd\mu_i = 0$, and $\tag{3.79}$

ii) $\quad S''dT - V''dp'' + \sum n_i''d\mu_i = 0$, leaving: $\tag{3.80}$

$$\underbrace{(S - S' - S'')}_{S^\sigma}dT - \underbrace{(V' - V')}_{0}dp' - \underbrace{(V'' - V'')}_{0}dp'' + Ad\sigma + \sum \underbrace{(n_i - n_i' - n_i'')}_{n_i^\sigma}d\mu_i = 0 \tag{3.81}$$

and dividing through by A:

$$s^o \, dT + d\sigma + \sum \Gamma_i d\mu_i = 0, \text{ or} \tag{3.82}$$

$$d\sigma = -s^o \, dT - \sum \Gamma_i d\mu_i, \tag{3.83}$$

where $s^o = S^o / A$. By virtue of the Gibbs-Duhem Equations for the bulk phase portions (which we have so far just used once, *i.e.*, their sum), not all of the μ_i's are independent. Dividing Eqs. (3.79) and (3.80) through by V' and V'', respectively, and subtracting yields:

$$C's' dT - dp' - \sum C'_i d\mu_i = 0$$

$$C''s'' dT - dp' - \sum C''_i d\mu_i = 0$$

$$\overline{}$$

$$(C's' - C''s'') dT - 0 - \sum (C'_i - C''_i) d\mu_i = 0 \; . \tag{3.84}$$

C' and C'' refer to the total molar concentrations of phase ' and ", and s' and s'' refer to molar entropies. In carrying out the above subtraction, it has been assumed that the interface is quasi-flat, so that $dp' \approx dp''$. We may now solve for $d\mu_1$ in terms of the remaining variables:

$$d\mu_1 = \left[\frac{C's' - C''s''}{C'_1 - C''_1} \right] dT - \sum_{i=2}^{m} \left[\frac{C'_i - C''_i}{C'_1 - C''_1} \right] d\mu_i, \tag{3.85}$$

and substituting into Eq. (3.83) gives:

$$d\sigma = -s^o \, dT - \sum_{i=1}^{m} \Gamma_i d\mu_i$$

$$= -s^o \, dT - \Gamma_1 \left\{ \left[\frac{C's' - C''s''}{C'_1 - C''_1} \right] dT - \sum_{i=2}^{m} \left[\frac{C'_i - C''_i}{C'_1 - C''_1} \right] d\mu_i \right\} - \sum_{i=1}^{m} \Gamma_i d\mu_i$$

$$= -\left[s^o - \Gamma_1 \left(\frac{C's' - C''s''}{C'_1 - C''_1} \right) \right] dT - \sum_{i=2}^{m} \left[\Gamma_i - \Gamma_1 \left(\frac{C'_i - C''_i}{C'_1 - C''_1} \right) \right] d\mu_i$$

$$= -(s^o)_1 \, dT - \sum_{i=2}^{m} \Gamma_{i,1} \, d\mu_i. \tag{3.86}$$

In the above, $(s^o)_1$ is the *relative* surface entropy (with respect to the adsorption of component 1), and $\Gamma_{i,1}$ is the *relative* adsorption of component i (with respect to that of component 1). At constant temperature, we recover the Gibbs Adsorption Equation, as stated in Eq. (3.72). For a binary system, it takes the form:

$$d\sigma = -\Gamma_{2,1} d\mu_2. \tag{3.87}$$

Since $d\mu_2 = RT d\ln\gamma_2 x_2$, where x_2 is the mole fraction and γ_2 is the activity coefficient, one may write out the Gibbs Adsorption Equation for a binary, Eq. (3.87), as:

$$\Gamma_{2,1} = -\frac{1}{RT}\frac{d\sigma}{d\ln\gamma_2 x_2} = -\frac{1}{RT}\frac{d\sigma}{d\ln\gamma_2 + d\ln x_2}. \tag{3.88}$$

Multiplying and dividing the denominator by $d\ln x_2$ (and recalling the condition of constant T), yields:

$$\Gamma_{2,1} = \frac{-\dfrac{x_2}{RT}\left(\dfrac{\partial\sigma}{\partial x_2}\right)_T}{\left[\left(\dfrac{\partial\ln\gamma_2}{\partial\ln x_2}\right)_T + 1\right]}. \tag{3.89}$$

For ideal or ideal-dilute binary solutions ($\gamma_2 \to 1$, or $\gamma_2^H \to 1$, resp.) Eq. (3.89) becomes:

$$\Gamma_{2,1} = -\frac{x_2}{RT}\left(\frac{\partial\sigma}{\partial x_2}\right)_T, \tag{3.90}$$

or in terms of molar concentration:

$$\Gamma_{2,1} = -\frac{C_2}{RT}\left(\frac{\partial\sigma}{\partial C_2}\right)_T. \tag{3.91}$$

Thus it may be seen how the adsorption isotherm can be obtained from surface tension data (together with bulk phase activity data, if the solution is non-ideal). One can similarly interchange amongst the various equations for capillary systems, such as obtaining a surface tension equation from an adsorption isotherm, or a surface equation of state, *etc.* It is seen from any of Eqs. (3.89)-(3.91) that in general, a surface tension reduction implies positive adsorption, and *vice versa*. Recalling Fig. 2-4, one may note that most organic solutes are positively adsorbed at the water surface, while ionizing salts are negatively adsorbed. It is to be noted also that there are interesting cases of extrema in surface tension, corresponding to zero adsorption, or "surface azeotropy."

Figures 3-12 and 3-13 show the derivation of an adsorption isotherm from surface tension data for the acetone-water system, a case when bulk phase non-ideality must be taken into account. The isotherm computed on the assumption of an ideal solution is also shown for comparison, and it is noted that the strong positive deviations from ideality in the acetone-water solution strongly enhance the relative adsorption of acetone at intermediate concentrations. At very low concentrations, the effects of non-ideality are less severe. The reduction in *relative* adsorption at higher solute concentrations is commonly observed. Our greatest interest in what follows will be dilute solutions, particularly those of surface active agents in water.

The cases studied by McBain and coworkers generally corresponded to sufficiently dilute solutions that the ideal form of the Gibbs Adsorption Equation was applicable. (For ionic solutes, some special considerations must be taken, as discussed later.) We may now compare the derived adsorptions with those measured by the microtome method, as shown in Table 3-2. Agreement is seen to be reasonable.

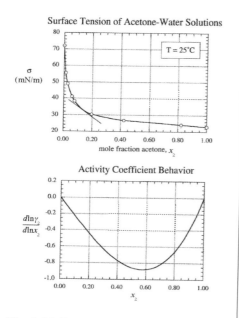

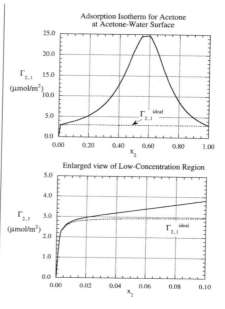

Fig. 3-12: Data needed to derive adsorption isotherm for acetone-water system. Surface tension data from [Howard, K. S., and McAllister, R. A., *AIChE J.*, **3**, 325 (1957).] Activity data from [Gmehling J., Onken, U, and Arlt, W, **Vapor-Liquid Equilibrium Data Collection Vol. 1, Part 1a**, p. 193, Dechema. Frankfurt, Germany, 1981.]

Fig. 3-13: Derived adsorption isotherm for acetone at the acetone-water surface.

Table 3-2: Comparison of relative adsorption from aqueous solutions obtained by the microtome method (obs) and computed using the Gibbs Adsorption Equation (calc).

Solute	C_2 (M)	$\Gamma_{2,1}$ (μmol/m^2)^{obs}	$\Gamma_{2,1}$ (μmol/m^2)^{calc}
Phenol	0.218	4.4	5.2
n-Hexanoic acid	0.0223	5.8	5.4
NaCl	2.0	-0.74	-0.64

It is now evident how, through the use of the Gibbs Adsorption Equation, one may obtain the surface pressure π of an adsorbate at a solid-fluid interface by integration of the appropriate adsorption isotherm. For example, the surface pressure of an adsorbing vapor at a solid-gas interface, as might be required in the interpretation of contact angle data, is computed from the adsorption isotherm, $\Gamma_{2,1} = \Gamma_{2,1}(p_2)$, where p_2 is the partial pressure of component 2 in the gas (assuming ideality in the gas phase), as follows:

$$d\pi = -d\sigma = \Gamma_{2,1}RT\,d\ln\,x_2 = \Gamma_{2,1}RT\,d\ln\,p_2, \text{ so that} \tag{3.92}$$

$$\pi = RT \int_0^{p_2} \Gamma_{2,1} \frac{dp_2'}{p_2'} . \tag{3.93}$$

The gas-solid isotherm, $\Gamma_{2,1}(p_2)$, is conveniently obtained by the direct weight-gain measurements described above or by using inverse gas chromatography, as described in Chap. 4. It should be noted that the surface pressure is not the surface tension (or energy) σ, but merely its reduction upon adsorption, *i.e.*

$$\sigma = \sigma_0^1 - \pi, \tag{3.94}$$

where σ_0^1 in immiscible interfacial systems is the "tension" (or energy) of the solute-free interface between the substrate phase and the pure solvent component phase 1.

Similar to the above, the surface pressure (interfacial free energy reduction) for detergents adsorbed at solid-liquid interfaces, believed to be responsible for the roll-up mechanism in detergency, may be obtained from the measured adsorption isotherm, $\Gamma_{2,1}(C_2)$, which is integrated using the Gibbs Adsorption Equation, *viz.*:

$$\pi = RT \int_0^{C_2} \Gamma_{2,1} \frac{dC_2'}{C_2'} . \tag{3.95}$$

Certain caveats must be observed in applying the Gibbs Adsorption Equation to solid-fluid interfaces.[21] First, the adsorption process cannot be accompanied by any irreversible interfacial stretching, as described earlier. More importantly, it cannot, in the form derived, describe *chemisorption*, *i.e.*, situations in which an adsorbate-adsorbent covalent bond is formed, or even most cases in which an acid-base interaction or electrostatic attraction provides the dominant driving force for adsorption. This has been identified later in this chapter as *amphiphilic adsorption,* and the result of such a process may be an *increase* in the solid-liquid interfacial energy.

---

[21] Molliet, J. L., Collie, B., and Black, W., **Surface Activity**, pp. 94-98, van Nostrand, Princeton, NJ, 1961.

F. Surface tension of solutions

1. *Ideal-dilute capillary systems*

Most solutions whose interfacial properties are of interest are dilute. Recall that for bulk solutions, it is convenient to define *ideal-dilute solutions* and then refer real solution behavior to them. Ideal dilute solutions are defined as those obeying Henry's Law

$$f_2 = x_2 \mathcal{H}_{2,1}, \tag{3.96}$$

where f_2 is the fugacity of component 2 in the solution, and $\mathcal{H}_{2,1}$ is Henry's constant. For the case in which the solution is in equilibrium with an ideal gas mixture:

$$p_2 = x_2 \mathcal{H}_{2,1}, \tag{3.97}$$

where p_2 is the partial pressure of component 2 in the gas. Real solutions could be described by inserting an activity coefficient, γ_2^H, so for example:

$$p_2 = x_2 \gamma_2^H \mathcal{H}_{2,1} . \tag{3.98}$$

All solutions obey Henry's Law when they are sufficiently dilute, *i.e.*, $\gamma_2^H \rightarrow$ 1 as $x_2 \rightarrow 0$.

In a similar fashion, one may define an *ideal-dilute capillary system* as one obeying Henry's Law as above, both in the bulk solution *and* in the interfacial layer. Such a system exhibits linear surface tension behavior, *viz.*

$$\sigma = \sigma_o - \alpha C_2; \quad \text{where} \quad \alpha = -\left(\frac{\partial \sigma}{\partial C_2}\right)_{C_2 \to 0}. \tag{3.99}$$

$\alpha > 0$ refers to positive adsorption. For cases of positive adsorption, a greater degree of dilution is required for a system to be ideal-dilute with respect to capillary behavior than to obey Henry's Law in the bulk solution because of the greater concentration in the surface layer, *i.e.*, greater mutual proximity of solute molecules.

One may deduce the corresponding isotherm for an ideal-dilute capillary system, using the Gibbs Equation.

$$\Gamma_{2,1} = -\frac{C_2}{RT}\left(\frac{\partial \sigma}{\partial C_2}\right)_T = -\frac{C_2}{RT}(-\alpha) \equiv \left(\frac{\alpha}{RT}\right)C_2, \tag{3.100}$$

termed Henry's Isotherm. The surface equation of state, $\sigma = \sigma(\Gamma_{2,1})$ is obtained by eliminating C_2 from the isotherm using the surface tension equation to get

$$\sigma_0 - \sigma \equiv \pi = RT\Gamma_{2,1}. \tag{3.101}$$

Note that α disappeared, and it is therefore evident that all ideal dilute capillary systems have the same surface equation of state. It can be put into a familiar form by defining the "specific area" of the solute as: $a_{2,1} \equiv 1/\Gamma_{2,1}$, so that

$$\pi a_{2,1} = RT, \tag{3.102}$$

a two-dimensional analogue of the ideal gas law. Equation (3.102) is also analogous to the osmotic pressure equation in ideal dilute solutions, *i.e.*, the van't Hoff equation.

It is evident that if one started with the surface equation of state, it would not be possible to regenerate either the surface tension equation or the adsorption isotherm, except to within an additive constant of integration.

2. Moderately dilute capillary systems

As solution concentrations increase beyond the range of the "ideal-dilute," for capillary systems showing positive adsorption, the initial linear surface tension decrease moderates to one that decreases less steeply. This is illustrated by some data for carboxylic acids shown in Fig. 3-14. These data were found by Szyszkowski[22] to be well fit by the two-parameter equation:

$$\pi = RTB\ln(1 + C_2/a), \tag{3.103}$$

now known as the *Szyszkowski Equation*. It reduces, as $C_2 \to 0$, to the linear surface tension equation, with $\alpha = RTB/a$. For homologous series' of

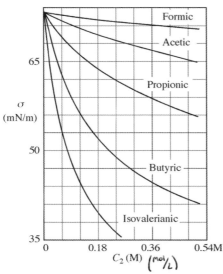

Fig. 3-14: Szyszkowski surface tension data for a homologous series of carboxylic acid solutions. Data from [Szyszkowski, B., Z. Physikal. Chem., **64**, 385 (1908).]

compounds, the constant B was found to be the same for the series, while the parameter "a" was characteristic of the particular member. Adherence to the Szyszkowski equation format at modest solute concentrations is the basis for

---

[22] Szyszkowski, B., Z. Physikal. Chem., **64**, 385 (1908).

the definition of what might be termed "moderately dilute" capillary systems. These are solutions that are sufficiently dilute to obey Henry's Law with respect to the bulk solution properties, but not dilute enough to be ideal dilute in the surface layer. The surface tension behavior of dilute aqueous solutions of many surface active agents are described by this model. A summary of Szyszkowski parameters for a large number of compounds is given by Chang and Franses.[23]

The Szyszkowski equation can be converted, using the ideal solution form of the Gibbs Adsorption Equation, into the corresponding adsorption isotherm and the equation of state for what might be termed *moderately dilute capillary systems*, as shown in Fig. 3-15. The resulting isotherm is in the form of the well-known *Langmuir Isotherm*,

$$\Gamma_{2,1} = \frac{BC_2}{a + C_2}. \tag{3.104}$$

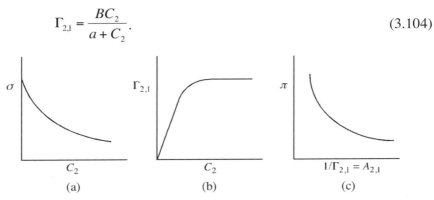

Fig. 3-15: Functional form of thermodynamic equations for a binary moderately dilute capillary system: (a) Szyszkowski surface tension equation, (b) Langmuir adsorption isotherm, (c) Frumkin surface equation of state.

At low C_2, Henry's adsorption isotherm is recovered, with: $\Gamma_{2,1} \rightarrow (B/a)\, C_2$, and at sufficiently high concentration, the surface becomes saturated, *i.e.*, $\Gamma_{2,1} \rightarrow B = \Gamma_\infty$, yielding the maximum molar packing density Γ_∞ of a monolayer of solute at the surface. The equation of state that is derived from the Szyszkowski Equation is termed the Frumkin surface equation of state:

$$\pi = -RT\,B\ln\!\left(1 - \frac{\Gamma_{2,1}}{B}\right) \equiv -RT\,B\ln\!\left(1 - \frac{\Gamma_{2,1}}{\Gamma_\infty}\right), \tag{3.105}$$

from which it is seen, the constant "a" has disappeared. This suggests that any member of a given homologous series obeying the Szyszkowski Equation has the same surface equation of state.

The type of "surface phase" non-ideality exhibited by systems described by the Szyszkowski Equation *et seq.* is that which is attributable

[23] Chang, C.-H., and Franses, E. I., *Colloids Surfaces A*, **100**, 1 (1995).

solely to the space occupied by the adsorbing molecules. It does not account for lateral attractive or repulsive interactions between them (*i.e.*, cooperative effects). While this simple description appears to be satisfactory for many alkane-chain (up to C_{18}) ionic surfactants at both the water/air and water/oil interfaces,[24] it appears to fail for many nonionics. Frumkin[25] had earlier proposed a modification to the Langmuir Isotherm to account for such interactions, *viz.*

$$\Gamma_{2,1} = \frac{\Gamma_\infty C_2 \exp\left(A\frac{\Gamma_{2,1}}{\Gamma_\infty}\right)}{a_F + C_2 \exp\left(A\frac{\Gamma_{2,1}}{\Gamma_\infty}\right)},$$ (3.106)

The bulk concentration has been augmented by the factor $\exp\left(A\frac{\Gamma_{2,1}}{\Gamma_\infty}\right)$, in which "$A$" is an empirical constant accounting for lateral, or solute-solute, interactions in the surface. For $A > 0$, the adsorption is influenced by favorable interactions, such as chain-chain cohesion, whereas for $A < 0$, adsorption is disfavored by solute-solute repulsion in the surface. The implicit nature of the Frumkin Isotherm (as it is termed) precludes the derivation of an analytical expression for the corresponding surface tension equation, but it can be obtained numerically. The surface equation of state can be obtained in closed form. This three-parameter representation is adequate for the representation of data for most dilute surfactant solutions, still in the context of moderately dilute capillary systems.

G. Surface active agents (surfactants) and their solutions

1. *The structure of different types of surface active agents*

Surface active agents, as defined in Chap. 2, are compounds which, when present in very small amounts (≤ 0.01 M), reduce the surface tension of water by a significant amount (≥ 30 mN/m). Surface activity also exists in non-aqueous media, but the extent of surface tension reduction is generally much less. The discussion that follows concerns aqueous systems. A brief aside on nomenclature is useful at this point. The term "surface active agent" was shortened by Langmuir to "surfactant," and is synonymous with it. Surfactants are also referred to inter-changeably as "amphiphiles" (which reflects the nature of their structure). The term "lipid" refers to long chain aliphatic hydrocarbons, or fats, and derivatives originating in living cells. Only some, such as fatty acids, are also surfactants, although the term is often used as synonymous with "surfactant." The word "soap" refers to salts of fatty acids, although it too is often used more broadly. A "detergent"

[24] Lucassen-Reynders, E. H., *J. Phys. Chem.*,**70**, 1777 (1966).

[25] Frumkin, A., *Z. Phys. Chem.*, **116**, 466 (1925).

generally refers to a synthetic surfactant such as a fatty sulfate, sulfonate, or long chain quaternary ammonium salts, or it may reference a commercial cleaning mixture containing detergents as well as other compounds, such as "builders," bleaches, enzymes, fragrances, *etc*.

An enormous array of different substances may be surface active in aqueous media, but all such materials have certain features in common. Their molecules are composed of at least two portions *segregated* from one another, one being hydrophilic (such as a highly polar or ionized functional group) and the other hydrophobic, such as a medium-to-long aliphatic chain (usually in the range of $C_6 - C_{20}$). Surfactants thus generally present themselves as homologous series' of compounds. The hydrophilic portion is referred to as the *head group*, while the hydrophobic portion is called the *tail*. In addition to straight aliphatic chains, the hydrophobes may consist of branched chains, chains with double or triple bonds, or with aromatic groups, dual chains (such as in the dipalmitoyl lecithin lung surfactant shown in Fig. 2-6), chains with halogen (particularly fluorine) substitution, siloxane chains,[26] *etc*.

An enormous variety of hydrophilic head groups is available, and a sample is listed in Table 3-3. Each type of head group yields a different family of surfactants. Surfactants seek the surface because there alone can they orient themselves to satisfy the solubility characteristics of both portions of their structure. The adsorption of surfactant from aqueous solution is pictured (very) schematically in Fig. 3-16. In solution, the hydrophobic tail is believed to be surrounded by an "iceberg" of structured

Table 3-3: Some typical surfactant head groups.

-OH	Hydroxyl
-COOH (low pH)	Carboxyl
-COO⁻ (high pH)	Carboxylate
$-SO_4^-$	Sulfate
$-SO_3^-$	Sulfonate
$-H_2PO_4^-$ (mod. pH)	Phosphate
$-NH_2$ (high pH)	Amino
$-NH_3^+$ (low pH)	Ammonium
$-N(CH_3)_3^+$	Trimethylammonium, or Quaternary ammonium
$-(OCH_2CH_2)OH$	Polyoxyethylene, or polyethylene oxide

water,[27] which, upon adsorption to either the water-air water-solid interface, is released into the solution, accompanied by a large increase in entropy. The free energy of adsorption:

$$\Delta G_{ad} = \Delta H_{ad} - T\Delta S_{ad} \tag{3.107}$$

---

[26] Hill, R. M. (Ed.), **Silicone Surfactants**, Surf. Sci. Ser. 86, Marcel Dekker, New York, 1999.

[27] Frank, H. S., and Evans, M. W., *J. Chem. Phys.*, **13**, 507 (1945).

is large and negative (thus favoring the process) due largely to the entropic term. The enthalpy of adsorption is often small, and may be either positive (due, for example, to electrostatic repulsion between charged head groups) or negative (due, for example, to tail-tail attractive van der Waals interactions). The apparent attraction of hydrophobic moieties in water for surfaces where they may be either expelled or sequestered is referred to as *hydrophobic bonding* or the *hydrophobic effect*.[28] Its primary origin is the entropic effect described above.

The key feature of surfactant molecular structure is the *segregation* of its hydrophobic and hydrophilic moieties. Thus, *e.g.*, glucose is not surface active because its hydrophilic hydroxyl groups are not segregated from the

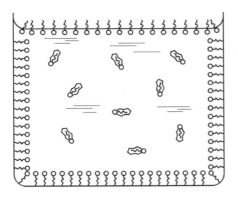

Fig. 3-16: Schematic of surfactant adsorption from aqueous solution, both to the air-water surface and the water-solid interface of the container wall. The "iceberg" of structure water surrounding the hydrophobic tails in solution is depicted.

hydrocarbon structure, as seen in Fig. 3-17, and in fact, glucose appears to produce a small *increase* in the surface tension of water, suggesting *negative*

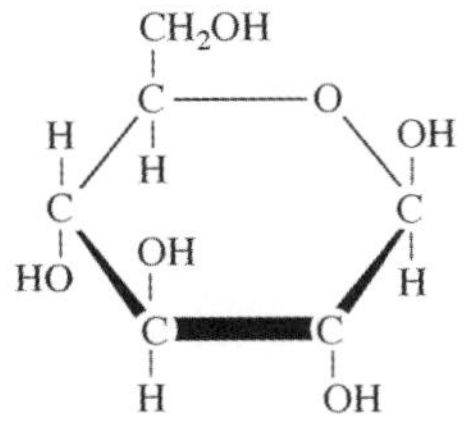

Fig. 3-17: Molecular structure of glucose.

adsorption. Surfactants are used in many ways in our everyday lives. The use of soaps and detergents for cleaning things is just one example. Surfactant compounds, like polymers, can often be tailor-made to suit the needs of a variety of specific applications, as listed in Table 3-4. Much more extensive accounts of the applications of the various types of surfactants can be found in Rosen,[29] Myers[30], Molliet, Collie and Black,[31] or in the detailed

[28] Tanford, C., **The Hydrophobic Effect: Formation of Micelles and Biological Membranes**, 2nd Ed. , Krieger, Malabar, FL, 1991.

[29] Rosen, M. J., **Surfactants and Interfacial Phenomena**, 2nd Ed., pp. 1-32, Wiley, New York, 1989.

compilations of commercially available surfactants, such as in McCuthcheon's Handbooks.[32] The surfactant properties needed to meet various applications are elucidated as the particular applications are discussed. Some of these have already been mentioned, while others can be appreciated only after getting further into the subject of this text.

Table 3-4: Some uses of surface active agents.

---

1. Modification of wetting behavior

   • Promotion of wetting (coating, cleaning,,…)

   • Reduction of wetting (waterproofing, soil release, flotation,…)

2. Stabilization of thin films

3. Formation and stabilization of emulsions

4. Solubilization of oil in water

   • Soil removal, detergency

   • Reactions; emulsion polymerization

5. Formation and stabilization of colloidal dispersions

6. Grinding aids

7. Lubrication

Aqueous surfactants are broadly classified into four categories, based on the charge structure of their hydrophilic "head groups," as follows:

1) **Anionic surfactants**:

"Anionics" are ionized salts in which the anion (- charge) possesses the long hydrophobic chain. Examples are:

Sodium stearate $CH_3(CH_2)_{16}COO^- Na^+$

Sodium dodecyl ("lauryl") sulfate $CH_3(CH_2)_{11}SO_4^- Na^+$

Sodium dodecyl benzene sulfonate $CH_3(CH_2)_{11}SO_3^- Na^+$

The first of these, sodium stearate ("Grandma's lye soap") is an example of a "soap," the salt of a fatty acid. Soaps are usually produced by the reaction (saponification) of fatty acids or esters with alkali, usually sodium or potassium hydroxide. The latter two are examples of "detergents," *i.e.*, synthetic fatty sulfates or sulfonates. They are synthesized by the reaction of

---

[30] Myers, D., **Surfactant Science and Technology**, VCH Publ., Weinheim, West Germany, 1988.

[31] Molliet, J. L., Collie, B., and Black, W., **Surface Activity**, van Nostrand, Princeton, NJ, 1961.

[32] **McCutcheon's: Vol 1. Emulsifiers and Detergents; Vol. 2. Functional Materials**, published annually by McCutcheon's Division. MC Publishing Co., Glen Rock, NJ.

sulfuric acid with unsaturated fats or fatty alcohols (sulfation or sulfonation). Commercial detergents usually contain traces of the unreacted fats, which may have an important influence on their properties. Sodium dodecyl sulfate (SDS) is a major component of bar soap, while sodium dodecyl benzene sulfonate (SDBS) is a major constituent of laundry detergent. They are fully ionized over the whole practical *pH* range. An important di-tail anionic surfactant is Aerosol OT® (Cytec Industries): the di(ethylhexyl) ester of sulfosuccinic acid. It is an especially effective wetting agent and dispersant. Anionics, primarily the detergents, comprise about 70% of the surfactant market. They are inexpensive and make good wetting agents and cleaning compounds. This is accomplished primarily through a reduction in the surface tension of water and a reduction in the solid-liquid interfacial energy. Recalling Young's Equation, Eq. (3.43), we see that the contact angle, θ, is reduced ($\cos\theta$ is increased) when the liquid surface tension, σ_L, and/or the solid-liquid interfacial energy, σ_{SL}, is reduced. Under water, adsorption of the surfactant leads to reductions in both the substrate-water and the dirt-water interfacial energies and eventually to "roll up" of the dirt from the substrate, as described later in Chap. 4, H.1 under "detergency."

Anionics tend to shun dense adsorption onto solid surfaces from water because these surfaces are often negatively charged (as described in more detail in Chap. 6). Adsorption onto the solid under these conditions generally results in a "tail-down" configuration such that the hydrophilic head groups are exposed to the water, increasing the hydrophilicity of the solid surface. The driving force for such adsorption is *hydrophobic bonding*, *i.e.*, the non-specific desire of the hydrophobic portions of the molecules to "get out" of the water.

2) **Cationic surfactants**:

"Cationics" are ionized salts in which the cation (+ charge) is the surfactant species. The quaternary ammonium compounds are fully ionized over whole practical *pH* range. They have good bactericidal properties and a large number of other special applications, for example as waterproofers and anti-stats, owing to their head-down, tail-out adsorption at most solid-water interfaces, which generally carry a negative charge. Most are nitrogen-containing compounds. Some specific examples include:

Hexadecyl (cetyl) trimethylammonium bromide,

(CTAB), $CH_3(CH_2)_{15} N^+(CH_3)_3 Br^-$

Dodecyl pyridinium chloride, $CH_3(CH_2)_{11} \bigcirc N^+ Cl^-$

Dodecyl ammonium hydrochloride, $CH_3(CH_2)_{11} NH_3^+ Cl^-$

Quaternary ammonium compounds are usually synthesized through the reaction of ammonia with fatty alcohols.

3) **Nonionic surfactants**:

"Nonionics" usually refer to the various alkyl poly(ethylene oxide)'s derived from the condensation of ethylene oxide with fatty acids or alcohols. The simplest example is the Brij® series (ICI Americas):

$$C_nH_{2n+1}O\text{-}(CH_2CH_2O)_mH,$$

where m and n can have different values to produce different properties. The structure of such a surfactant is often abbreviated as: C_nE_m. As the m/n ratio increases, the surfactants become more hydrophilic. Nonionics can be tailor-made for many applications in this way. They are compatible with other types of surfactants, resistant to hard water, generally low foamers, and may be soluble in organic solvents. Because of the method of their synthesis, any given member of the series is polydisperse to a certain, often significant, extent. Some important examples of nonionic surfactant series are:

The Triton X® series (Union Carbide): polyethoxylated alkyl phenols, of which Triton X-100,

$$C_8H_{17} \text{—}\bigcirc\text{—} (OC_2H_4)_{9.5} OH$$

is a common wetting agent and dispersant. Other families of nonionics include:

The Tergitol® series (Union Carbide): di-alkyl, di-poly (ethylene oxide)s, useful, for example in formulating injet printing inks.

The Pluronic® series (BASF Corp.): triblock copolymers of poly (ethylene oxide) and poly (propylene oxide), and

The Spans® and Tweens® (ICI Surfactants): various polyoxyethylene-containing sorbitan derivatives. These are important classes of emulsifiers.

4) **Amphoteric surfactants**:

"Amphoterics" can be either anionic or cationic, depending on the *pH* of the solution. Examples include the N-alkylaminoacids, *e.g.*:

$$
\begin{array}{ll}
\text{CH}_2\text{-CH}_2\text{-COOH} & \text{CH}_2\text{-CH}_2\text{-COO}^- \\
\quad/ & \quad/ \\
\text{C}_{12}\text{H}_{25}\text{ - NH}^+ \text{ (at low } pH\text{), or} & \text{C}_{12}\text{H}_{25}\text{ - N} \qquad \text{(at high } pH\text{)} \\
\quad\backslash & \quad\backslash \\
\text{CH}_2\text{-CH}_2\text{-COOH} & \text{CH}_2\text{-CH}_2\text{-COO}^-
\end{array}
$$

Many proteins and other natural surfactants are of this nature.

5) **Zwitterionic surfactants**:

The ionic head group in "zwitterionics" contains both positive and negative charges in close proximity. The phosphoryl choline group of dipalmitoyl lecithin (shown in Fig. 2-6) is an example. An interesting type of di-chain, zwitterionic surfactant is formed from the mixture of anionic and cationic surfactants, which combine in aqueous solution[33] to produce *catanionic* surfactants.

Other, more exotic, types of structures have also been identified,[34] such as *gemini* surfactants, consisting of two single-tail surfactants whose heads are connected by a spacer chain which may be either hydrophilic or hydrophobic.[35] Another example is the *telechelic surfactant*, in which two hydrophobic groups are connected by a hydrophilic chain.[36] *Boloform* surfactants, on the other hand, are hydrophobic chains with hydrophilic groups on each end.

Amphoteric surfactants and nonionics with PEO hydrophilic head groups are example of "function-changing surfactants." Their character as surfactants can be made to change with changes in controllable external variables such as *pH* or temperature. Such changes can lead to wetting transitions, as discussed in Chap. 4, or to the inversion or breaking of emulsions, as discussed in Chap. 9. Another example of function-changing is the conversion of an anionic surfactant with the addition of divalent cations, such as Ca^{+2} or Mg^{+2} to the solution. These will each associate with a pair of surfactant anions to produce an undissociated, hydrophobic di-tail surfactant. Another important class of compounds is that of *cleavable surfactants*.[37] These are intentionally designed with a weak linkage, usually between the hydrophilic and hydrophobic portions of the molecule, that is susceptible to cleavage under high *pH* conditions (*e.g.*, normal esters or carbonates) or low *pH* conditions (*e.g.*, ortho esters or ketals). Ester, amide and carbonate linkages have been investigated with respect to enzymatic hydrolysis, and at least one type of surfactant based on the incorporation of Diels-Alder adducts is found to cleave upon increasing the temperature to approximately 60°C. Cleavability is motivated in part by the desire to achieve biodegradability, but also as an aid to the removal of surfactants from a system once they have performed their function.

Many of the surfactants described above can be produced in polymeric form, with the monomer surfactants as repeat units in the polymers. Some are nonionic, like polyvinyl alcohol or polyethylene oxides, while others are polyelectrolytes (polyanions or polycations). In addition, there are a number of naturally occurring macromolecules that often act as polymeric

---

[33] Kaler, E. W., Murthy, A. K., Rodriguez, B. E., and Zasadzinski, J. A. N., *Science*, **245**, 1371 (1989).

[34] Holmberg, K. (Ed.), **Novel Surfactants**, Surf. Sci. Ser. 74, Marcel Dekker, New York, 1998.

[35] Zana, R., Talmon, Y., *Nature*, **362**, 228 (1993).

[36] Semenov, A. N., Joanny, J. F., Khokhlov, A. R., *Macromolecules*, **28**, 1066 (1995).

[37] Stjerndahl, M., Lundberg, D., and Holmberg, K., "Cleavable surfactants," pp. 317-45, in : **Novel Surfactants**, 2nd Ed. (K. Holmberg, Ed.) Marcel Dekker, New York, 2003; Tehran—Bagha, A., and Holmberg, K., *Curr. Opinion Colloid Interface Sci.*, **12**, 81 (2007).

surfactants. These would include polypeptides (proteins), which chains of amino acid groups folded into tertiary structures, as well as nucleic acids and polysaccharides.

The discussion above pertains to aqueous interfaces. Surface activity in non-aqueous systems refers to the ability of a solute or an insoluble monolayer to reduce the surface tension of a non-aqueous liquid, but as mentioned earlier, such reductions are much less than in aqueous media. They are brought about principally by various fluorocarbons or silicones.

A large and growing literature on surfactants and their applications, in addition to the references cited earlier, is provided by Marcel Dekker's *Surfactant Science Series*,[38] beginning in 1967 with Vol. 1: **Nonionic Surfactants**, and up to Vol. 142 in 2008.

2. *Solutions of non-electrolyte surfactants*

Dilute aqueous solutions of surfactants, whether they are electrolytes or non-electrolytes, show remarkably similar surface tension behavior. They all, of course, show linear behavior at great enough dilution, and many of show agreement with the Szyszkowski Equation over the whole range for which surface tension changes significantly with concentration. Their surface tension behavior is thus often expressible as:

$$\sigma_0 - \sigma \equiv \pi = RTB\ln\left(1 + \frac{C_2}{a}\right). \tag{3.108}$$

Over much of the concentration range of interest, $C_2/a \gg 1$, so that (in this range):

$$\sigma_0 - \sigma \equiv \pi = RTB\ln\left(\frac{C_2}{a}\right). \tag{3.109}$$

It is useful to plot surface pressure (or surface tension) against the logarithm of C_2, or to use semi-log coordinates, as shown in Fig. 3-18.

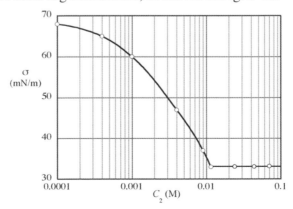

Fig. 3-18: Schematic of surface tension behavior of aqueous surfactant solutions.

---

[38] A listing of volume titles through 2006 is given at: http://liv.ed.ynu.ac.jp/senmon/book02.html

"Linear" Szyszkowski behavior, as given by Eq. (3.109) results in a straight line, often observed over one-to-two decades of C_2. The surface tension decrease with concentration comes to an abrupt end, either when the solubility limit of the surfactant is reached, or when further increases in bulk concentration lead to no further changes. The latter behavior is discussed further below.

Szyszkowski behavior in a homologous series of non-electrolyte compounds, *e.g.*, the normal alcohols, is shown in Fig. 3-19. Note that the

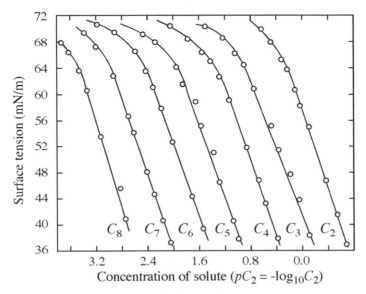

Fig. 3-19: Surface tension dependence on concentration for aqueous solutions of the normal alcohols ($C_2 - C_8$) at 25°C. Data from [Posner, A.M., Anderson, J.R. and Alexander, A.E., *J. Colloid Sci.*, **7**, 623 (1952).]

convenient composition variable, pC_2, defined analogously to the pH, has been introduced. There are two important features to be noticed in Fig. 3-19. The first is that all curves have about the same slope, *i.e.*, $d\pi/d(pC_2) \approx$ constant ($\approx$ the same for each compound). This can be interpreted with the Gibbs Adsorption Equation:

$$\Gamma_{2,1} = \frac{C_2}{RT}\frac{d\pi}{dC_2} \; ; \; \text{or} \quad \frac{d\pi}{d(pC_2)} = -2.303..\,RT\Gamma_{2,1} = \text{const.} \quad (3.110)$$

Equation (3.110) implies that in this region, $\Gamma_{2,1}$ is a constant, or that the surface has been saturated. Such a surface is referred to as a "Gibbs monolayer." We note that the slope, hence $\Gamma_{2,1}$, is the same regardless of chain length (just occurring in different concentration ranges), and furthermore, from the magnitude of that slope, the molecular area turns out to be 30 - 35 Å^2/molecule. Putting this information together implies that the molecules are forming a more-or-less compact monolayer of approximately vertically oriented molecules, the value of $a_{2,1} = 1/\Gamma_{2,1}$ being only about 25-

30 percent higher than the known cross-sectional area of a hydrocarbon chain. The slope gives one the constant "B," which is characteristic of the homologous series, while "a" is characteristic of the particular member of the series. This has been shown for a number of series' of aliphatic surfactants.

A second observation is that the curves are approximately equally spaced from one another, about 0.5 pC_2 units in this diagram. Thus the concentration required to produce a given surface tension reduction π is less, as the chain length increases, by a constant factor, viz., for a given π:

$$\log(C_2)_n \approx \log(C_2)_{n+1} + 0.5, \tag{3.111}$$

where n = number of carbon atoms. Thus:

$$(C_2)_n /(C_2)_{n+1} \approx 10^{0.5} \approx 3. \tag{3.112}$$

This result stated in words is: For dilute solutions of a homologous series of straight chain aliphatic (nonionic) surfactants, the concentration at which a given surface pressure is obtained diminishes by a factor of three for each additional -CH_2- group in the chain. This is known as *Traube's Rule*, first proposed in 1891[39] and still a useful rule of thumb for extrapolating data within homologous series' of surfactants. The thermodynamic basis for Traube's Rule is readily understood in terms of the hydrophobic effect described earlier. Each -CH_2- addition to the hydrophobic chain of the surfactant would be expected to add a given number of water molecules to the "iceberg" surrounding the hydrophobic chain in the water, producing a given increase in entropy upon their release.

A useful measure of the *efficiency* of a surfactant in reducing the surface tension is the *pC*-value at which it gives $\pi = 20$ mN/m, denoted pC_{20}. From Fig. 3-19, one can deduce, for example, that pC_{20} for octanol ≈ 3.1. For most soluble surfactants (ionic or nonionic), pC_{20} is between 2 and 6, but there are some higher. Traube's Rule suggests that pC_{20} should increase by $\approx$ 0.5 for each additional -CH_2- group, and for nonionic surfactants, this is generally the case. Thus, the longer the hydrocarbon chain length, the more *efficient* the surfactant. However, pC_{20} may not be attainable because of limited solubility or if a "break" in the curve of σ vs. C_2 occurs (as shown in Fig. 3-18) before $\pi = 20$ is reached. A few *surfactant efficiencies*, so defined, are listed in Table 3-5. Similar surface tension behavior (Traube's Rule, *etc.*) holds for ionic surfactants as well as nonionics. For these cases, at constant ionic strength, the spacing is approximately 0.3 for each additional -CH_2- group, rather than the 0.5-value observed for nonionics, suggesting that the concentration at which a given surface pressure is obtained is diminished by a factor of approximately two for each additional -CH_2- group in the chain.

---

[39] Traube, I., *Annalen*, **265**, 27 (1891).

Table 3-5: Some values of surfactant efficiencies. From [Rosen, M.J., **Surfactants and Interfacial Phenomena**, 2nd Ed., pp. 70 ff, Wiley, New York, 1989.]

Compound (T, in °C)	pC_{20}
$C_{10}H_{21}SO_4^-Na^+$ (27)	1.89
$C_{12}H_{25}SO_4^-Na^+$ (25)	2.57
$C_{12}H_{25}SO_4^-Na^+$ (60)	2.24
$C_{16}H_{33}SO_4^-Na^+$ (25)	3.70
$C_{16}H_{33}SO_4^-Na^+$ (25) (0.01 N NaCl)	5.24
p- $C_{12}H_{25}C_6H_4SO_3^-Na^+$ (70)	3.10
$C_{12}H_{25}N(CH_3)_3^+Cl^-$ (25) (0.10 N NaCl)	3.68
$C_{16}H_{35}N(CH_3)_3^+Cl^-$ (25) (0.10 N NaCl)	5.00
$C_{16}H_{33}(C_2H_4O)_6$ (25)	6.80

3. *Solutions of electrolyte surfactants*

The use of the Gibbs Adsorption Equation in the form of Eq. (3.91) applies only to non-electrolyte systems. The treatment of ionic surfactant solutions involves additional complications. Consider a situation in which the surfactant solute is fully ionized, *e.g.*, a solution of sodium dodecyl sulfate (SDS): $CH_3(CH_2)_{11}SO_4^-Na^+ = DS^- + Na^+$, in which the adsorbed surfactant species is the DS^- ion. Electrical neutrality in the surface layer requires that an equal adsorption of Na^+ *counterions* be present in the surface layer, so that

$$\Gamma_{Na^+} = \Gamma_{DS^-}.$$
(3.113)

Both species must be considered in using of the Gibbs Adsorption Equation:

$$d\sigma = -RT\Gamma_{Na^+} d\ln[Na^+] - RT\Gamma_{DS^-} d\ln[DS^-],$$
(3.114)

where $[Na^+]$ and $[DS^-]$ refer to the concentrations of the respective ions, both equal to the *nominal* concentration of the SDS, [NaDS]. Similarly, Γ_{Na^+} and Γ_{DS^-} are *each* equal to the nominal adsorption of SDS, Γ_{NaDS}. Thus:

$$d\sigma = -2RT\Gamma_{NaDS} d\ln[NaDS], \text{ and}$$
(3.115)

$$\Gamma_{NaDS} = -\frac{[NaDS]}{2RT} \frac{d\sigma}{d[NaDS]}.$$
(3.116)

A factor of 2 thus appears in the denominator of the Gibbs Adsorption Equation.

If a salt containing a common ion with the surfactant, say NaCl, is present in excess, then $d\ln[Na^+]$ will be essentially zero with (the relatively small) changes in [NaDS], and one returns to the original form of the Gibbs Adsorption Equation (*i.e.*, without the factor of 2):

$$\Gamma_{NaDS} = -\frac{[NaDS]}{RT}\frac{d\sigma}{d[NaDS]} \quad \text{(excess Na}^+ \text{ ion).} \quad (3.117)$$

When the additional (common ion) salt is present, but *not* in excess, the factor in the denominator becomes[40]

$$1 + \frac{[NaDS]}{[NaDS]+[NaCl]}. \quad (3.118)$$

Other problems with the Gibbs Adsorption Equation arise if the ionic dissociation is not complete. In these cases, the undissociated surfactants must be considered as a separate species from the dissociated species in the Gibbs equation. When the undissociated and the dissociated electrolyte surfactants are present in the same solution, the undissociated form is almost always the more surface active, and will be preferentially adsorbed. Thus, even if only a small fraction of the surfactant is undissociated in the bulk solution, a large proportion of the adsorbed layer may be of the undissociated form. As such, the surface properties of partially dissociated surfactants may be strongly sensitive to small *pH* changes or the presence of common electrolyte ions, since the extent of association (and hydrolysis) is strongly sensitive to these changes. There are other difficulties as well associated with the use of the Gibbs equations for ionic, as well as many nonionic, surfactants due to the formation of aggregates in solution, as described below.

Another difference in dealing with electrolyte solutions is that the degree of dilution required to yield Henry's Law behavior in the bulk solution is often orders of magnitude greater than that required for non-electrolyte solutes. (This is a consequence of the long-range Coulombic interactions between ions.) One should in principle, therefore, work with *activities* instead of concentrations.[41] Fortunately for the most important cases with no added salt or excess added salt, the concentration terms in the Gibbs Adsorption Equation for the relative adsorption appear in both the numerator and denominator so that the effects of solution non-ideality effectively cancel out.

H. Self-assembly of surfactant monomers in solution

1. *Formation of micelles: critical micelle concentration (CMC)*

As the concentration of most of the strongly surface active solutes is increased, a point is reached beyond which surface tension shows essentially no further decrease, producing the sharp change in slope seen in Fig. 3-18.

---

[40] Matijevic, E., and Pethica, B. A., *Trans. Faraday Soc.*, **54**, 1382 (1958).

[41] Prausnitz, J. M., Lichtenthaler, R. N., and Gomez de Azevado, E., **Molecular Thermodynamics of Fluid-Phase Equilibria**, 3rd Ed., Chap. 9, Prentice Hall, Upper Saddle River, NJ, 1999.

At about the same concentration, many other properties of the solution also show sharp breaks in slope, as shown schematically in Fig. 3-20. These

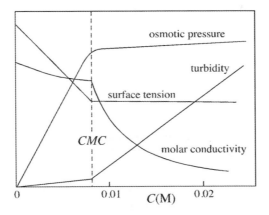

Fig. 3-20: Schematic diagram of physical property changes of aqueous solutions of sodium dodecyl sulfate at 25°C as a function of concentration.

sudden changes are attributable to the formation of aggregates of the surfactant in solution, termed *micelles*, pictured in Fig. 3-21. The spontaneous formation of micelles is an illustration of a general phenomenon associated with amphiphiles, termed *self-assembly*. The ordered arrays of molecules produced in adsorbed surfactant monolayers are also examples of such a process. The precise size and shape of the micellar aggregates differ from case to case, but quite often they are spherical and contain a few tens to hundreds of monomer units each. They form "back-to-back" so as to shield the hydrophobic moiety from the aqueous medium. The tendency to form micelles is a manifestation of the same hydrophobic effect that leads to their adsorption at interfaces. Figure 3-21 also suggests that equilibria exist between the monomers, micelles and adlayers at both the water-air and water-solid interfaces. As the concentration of the surfactant is increased, the formation of micelles begins to occur rather suddenly, or at least in a very narrow concentration range. The concentration at which they begin to form is termed the "critical micelle concentration" (*CMC*) and is characteristic of the particular surfactant and the thermodynamic state, *i.e.*, (T, p, $C_{\text{additional solutes}}$). The micelle size (aggregation number) is also characteristic, and the distribution of micelle sizes for a given case may be quite narrow. The subject of micelle formation is treated in more detail in Tanford,[42] Israelachvili,[43] Rosen,[44] Moroi,[45] and elsewhere.

[42] Tanford, C., **The Hydrophobic Effect: Formation of Micelles and Biological Membranes**, 2nd Ed., Krieger, Malabar, FL, 1991.

[43] Israelachvili, J. N., **Intermolecular and Surface Forces**, 2nd Ed., pp. 341ff, Academic Press, New York, 1992.

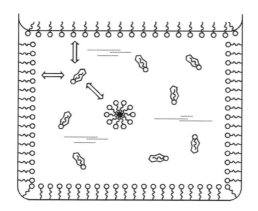

Fig. 3-21: Micelle formation in a surfactant solution.

The formation of micelles can be described using a mass-action model, and the examination of the consequences of such a model illuminates the nature of the process of micellization. Micelles of an electrically neutral species A may be formed in accord with:

$$\text{N } A_1 = A_N,$$ (3.119)

where N is the aggregation number; A_1 refers to the isolated monomer, and A_N is the aggregate. Actually, a family of such "reactions" occurs for N = 2, 3, ... If equilibrium exists for the formation of N-mers from monomers, and Henry's Law holds for all species in solution, we may write for each N:

$$K_N = \frac{\left(\dfrac{x_N}{N}\right)}{x_1^N},$$ (3.120)

where K_N is the equilibrium constant for the formation of N-mers; x_1 is the mole fraction of the surfactant as free monomer, while x_N is the ~~mole~~ fraction of the total # of monomers which find themselves in N-mers. (x_N/N) is thus the mole fraction of N-mers themselves. (We may just as well have used molar concentrations to express the system composition; at the degrees of dilution of interest, the quantities are proportional.) With knowledge of K_N as a function of N (for all N), one may complete the description of the system by noting that the sum of the mole fractions of the surfactant in *all* of the aggregates must equal the nominal mole fraction of monomer in the solution, x_0:

$$x_0 = x_1 + x_2 + x_3 + \dots$$ (3.121)

x_0 is a known quantity, and if K_N (N) is known, one may compute the distribution of aggregates to be expected for any given value of x_0.

[44] Rosen, M. J., **Surfactants and Interfacial Phenomena**, 2nd Ed. Chaps. 3-4, Wiley, New York, 1989.

[45] Moroi, Y., **Micelles: Theoretical and Applied Aspects**, Plenum, New York, 1992.

It may be assumed, on an *ad hoc* basis for now, that aggregates of only a *single* size n can form in significant amounts. This of course does not assume that the micelles are formed via an n-body interaction, but rather that it occurs through the formation of a sequence of multimers, all of which at equilibrium are present in vanishingly small amounts, except for the preferred n-mer. For this case, one has: $x_{i \neq 1, n} \approx 0$, and therefore $x_0 = x_1 + x_n = (1 - \beta)x_0 + \beta x_0$, where β is the fraction of the solute monomers that are present in the aggregates. β may be computed as a function of the nominal surfactant mole fraction, x_0, using the equilibrium expression:

$$K = \frac{\left(\dfrac{x_n}{n}\right)}{x_1^n} = \frac{\left(\dfrac{\beta x_0}{n}\right)}{\left[(1 - \beta)x_0\right]^n}. \tag{3.122}$$

The abruptness of micelle formation as x_0 is increased is demonstrated with the above expression. A typical aggregation number n would be 100. Thus, due to the smallness of x_0, the expected value for K would be enormous. In fact, a reasonable value may be 10^{200}! Using these values produces the results shown in Fig. 3-22, in which the sharpness of the transition to micelles is evident. The *CMC* in this case is in the vicinity of $x_0 = 0.0085$. Presumably, when the concentration is made to exceed the *CMC*, more micelles of the same size and shape are formed, and the monomer concentration remains essentially constant at the *CMC*. Thus, as the *total* concentration increases, only the concentration of the micelles increases.

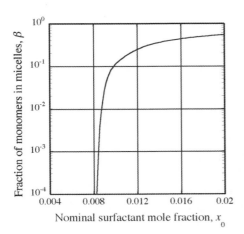

Fig. 3-22: Results of a mass-action model for micellization showing the fraction of the total surfactant monomers in micelles as a function of the nominal surfactant mole fraction.

If the micelles are formed from ionic surfactant, the effective "reaction" by which they are formed might be written as (assuming they are anions):

$$nA_1^- + ynC^+ = A_n C_{yn}^{-(1-y)n}, \tag{3.123}$$

where C^+ refers to a (monovalent) counterion, and y is the fraction of anionic surfactant monomers in the micelles to which such a counterion is *bound*.

The binding of counterions to some of the monomers in the micelles, effecting their partial neutralization, always occurs to some extent. The value of y may typically be between 0.3 and 0.8. The extent of counterion binding is only weakly dependent on the electrolyte concentration, but increases sharply with counterion valence. Often polyvalent counterions will bind to the free monomers, causing their precipitation. The micelle which is formed when there is monovalent counterion binding thus carries a "valence" of $-(1-y)n$. When the surfactant is fully ionized and is the only source of cations (counterions), and the solution continues to obey Henry's Law (a more suspect assumption than in the case of solutions of neutral surfactants), one may set the activity of the counterions equal to the nominal mole fraction of the surfactant: $x_c = x_0$, and the equilibrium equation becomes:

$$ K = \frac{\left(\dfrac{x_n}{n}\right)}{x_1^n x_c^{yn}} = \frac{\left(\dfrac{x_n}{n}\right)}{x_1^n x_0^{yn}} = \frac{\left(\dfrac{\beta x_0}{n}\right)}{(1-\beta)^n x_0^{n(1+y)}}. \tag{3.124} $$

The sharpness of the transition to micelles in this case is even greater than that shown for neutral surfactants.

The narrowness of the micelle size distribution around the value n may be anticipated by a thermodynamic analysis of the micellization process, $i.e.$, the process of the association of one mole of monomers into aggregates of size n. For this case, the standard (Gibbs) free energy of micellization is given by:

$$ \Delta G_{mic}^{\ominus} = -\frac{RT}{n} \ln K. \tag{3.125} $$

The value of $N = n$ to be expected is the one for which $K(N)$ takes on the largest value; hence $\Delta G_{mic}^{\ominus}$ takes on the largest negative value. The aggregation of monomers is a *cooperative* process (*i.e.*, its tendency to occur depends in part on how much it has already occurred). When the first N-mers start to form, the larger they are, the greater the resulting drop in the system free energy when an *additional* monomer is added, since the hydrophobes are more effectively able to shield themselves from the water. This effect does not continue indefinitely, however, or even level off (as would be the case for bulk phase change) because as more monomers are added, the head groups become more crowded at the surface of the aggregate. This crowding is especially acute if the head groups are ionized, and the like-charged entities are obliged to reside next to one another. These opposing effects come into balance at a particular value of N that yields the largest negative $\Delta G_{mic}^{\ominus}$. It has been shown (for the case in which the micelles are spherical in shape)[46] that even when $-\Delta G_{mic}^{\Theta}$ goes through a rather shallow maximum with N, the resulting micellar size distribution is narrow.

The effect that reaching the *CMC* has on many of the solution properties were known for some time before McBain first explained them (in 1911)[47] in terms of micelle formation. Consider the properties of surface tension, specific electrical conductivity, turbidity and osmotic pressure (and other colligative properties). All show dramatic slope changes in the vicinity of the *CMC* that can be explained by the formation of micelles, and all of these changes serve as methods for detecting the *CMC* in the laboratory. Surface tension halts its decrease with concentration as additional surfactant goes into the formation of additional micelles rather than increasing the bulk concentration of monomers or increasing the packing of surfactant into the surface (which is full). Surface tension measurement is the most common method for locating the *CMC*. Specific (molar) electrical conductivity (for ionic surfactants) decreases sharply with micellization primarily due to the binding of counterions to the micelle "interface," thus neutralizing a portion of the electrolyte present. Conductivity measurement thus provides a useful method for the determination of the *CMC* of ionic surfactants. Turbidity increases sharply beyond the *CMC* because micelles (diameter $\geq$ 20 Å) scatter light much more strongly than the monomers (see Chap. 5). The *CMC* may thus be detected using a turbidimeter or spectrophotometer. Osmotic pressure, Π_{OS} (and other colligative properties) depend directly on the number of particles in a unit volume of solution, which increases much more slowly with total surfactant concentration when additional surfactant molecules aggregate into micelles (which count only as one particle each). At least roughly, the slope of the Π_{OS} *vs.* x_0 curve should decrease by a factor of n (the micellar aggregation number) when the *CMC* is exceeded, so that osmotic pressure measurements may be used not only to locate the *CMC*, but also to determine the aggregation number. Although noting the extent of the change in colligative properties of the solution can give information about the size of the micelles, light scattering is probably the commonest method used for this. Other properties are similarly affected, and many of them also serve as methods for locating the *CMC* experimentally.

A large tabulation of *CMC*-values, together with a discussion of the various experimental methods used for obtaining them, is given in Mukerjee and Mysels.[48] Some typical values (together with aggregation numbers) are shown in Table 3-6. These are taken from the text by Rosen, which provides a rich source of property data for surfactants and their solutions.

[46] Mukerjee, P., in **Physical Chemistry, Enriching Topics from Colloid and Surface Science**, H. van Olphen and K. Mysels, Eds., Chap. 9, Theorex, La Jolla, CA, 1975.

[47] McBain, J. W., **Frontiers in Colloid Chemistry**, Interscience, New York, 1949.

[48] Mukerjee, P. and Mysels, K. J., **Critical Micelle Concentrations of Aqueous Surfactant Systems**, Nat. Bur. Stds., NSRDS-NBS 36, U.S. Govt. Printing Office, Washington, DC, 1971.

Table 3-6: Some values for *CMC* and aggregation number. From [From Rosen, M. J., **Surfactants and Interfacial Phenomena**, 2nd Ed., pp. 108ff, Wiley, New York, 1989.]

Surfactant	Medium	$T(°C)$	*CMC* (mM)	Agg. No., n
$C_{10}H_{21}SO_4^-Na^+$	H_2O	40	33	$40_{30°C}$
$C_{12}H_{25}SO_4^-Na^+$	H_2O	40	8.6	54
$C_{14}H_{29}SO_4^-Na^+$	H_2O	40	2.2	
$C_{12}H_{25}SO_4^-Na^+$	H_2O	25	8.2	
$C_{12}H_{25}SO_4^-Na^+$	0.01 M NaCl	21	5.6	
$C_{12}H_{25}SO_4^-Na^+$	0.03 M NaCl	21	3.2	
$C_{12}H_{25}SO_4^-Na^+$	0.10 M NaCl	21	1.5	$90_{20°C}$
$C_{14}H_{29}SO_4^-Na^+$	0.01 M NaCl	23		138
$C_{12}H_{25}SO_4^-Na^+$	3M urea	25	9.0	
$C_{12}H_{25}N(CH_3)_3^+Br^-$	H_2O	25	1.6	$50_{23°C}$
$n\text{-}C_{12}H_{25}(C_2H_4O)_7OH$	H_2O	25	0.05	
$n\text{-}C_{12}H_{25}(C_2H_4O)_6OH$	H_2O	25	0.05	
$n\text{-}C_{12}H_{25}(C_2H_4O)_{14}OH$	H_2O	25	0.055	
$C_{16}H_{33}(C_2H_4O)_6OH$	H_2O	25		2,430
$C_{16}H_{33}(C_2H_4O)_6OH$	H_2O	34		16,600

The *CMC* is of course one of the most important properties of surfactant solutions, and the large database that exists for it lets us examine how its magnitude depends on system properties. It is possible to make some useful generalizations about the magnitude of the *CMC* and its trends with various factors, and some results are summarized below[49]. Many of the trends for the *CMC* can be rationalized using a simple thermodynamic analysis, as follows. Substitution of the expressions for K into the expression given above for $\Delta G_{mic}^{\ominus}$, *i.e.* Eq. (3.125), gives

for nonionic micelle formers:

$$\Delta G_{mic}^{\ominus} = -\frac{RT}{n} \ln\left(\frac{\beta x_0}{n}\right) + RT\ln\left[x_0(1-\beta)\right], \qquad (3.126)$$

and for ionic micelle formers:

$$\Delta G_{mic}^{\ominus} = -\frac{RT}{n} \ln\left(\frac{\beta x_0}{n}\right) + RT\ln\left[x_0(1-\beta)\right] + yRT\ln(x_0). \qquad (3.127)$$

Right at the *CMC*, β (the fraction of the monomers in micelles) ≈ 0, and $x_0 \approx$ *CMC* (expressed as mole fraction). Thus

$$\Delta G_{mic}^{\ominus}\big|_{nonionic} = RT\ln(CMC), \text{ or} \qquad (3.128)$$

---

[49] Osipow, L. I., **Surface Chemistry**, pp. 168 ff, ACS Monograph Series, No. 153, Reinhold, New York, 1962.

$$\Delta G_{mic}^{\ominus}\big|_{ionic} = (1+y)RT\ln(CMC),\qquad\qquad(3.129)$$

so that for either nonionic or ionic micelles, $\ln(CMC) \propto \Delta G_{mic}^{\ominus}/RT$. $\Delta G_{mic}^{\ominus}$ must of course be negative, and the more negative it is, the more favorable the conditions for aggregation and the *lower* the *CMC*. $\Delta G_{mic}^{\ominus}$ may be broken down into contributions from the hydrophobic chain and hydrophilic head group:

$$\Delta G_{mic}^{\ominus} = \Delta G_{mic}^{\ominus}(hc) + \Delta G_{mic}^{\ominus}(head)\qquad\qquad(3.130)$$

The contribution of the hydrophobic group is negative (due to the entropic hydrophobic effect described earlier), and for straight-chain hydrocarbons may be further broken down:

$$\Delta G_{mic}^{\ominus}(hc) = \Delta G_{mic}^{\ominus}(CH_3) + (n_c - 1)\Delta G_{mic}^{\ominus}(-CH_2-).\qquad(3.131)$$

One would expect $\Delta G_{mic}^{\ominus}(-CH_2-)$ to be constant for any surfactant (*i.e.*, head group) type, leading to a systematic decrease of *CMC* with chain length, *i.e.*, Traube's Rule. The contribution of the head group $\Delta G_{mic}^{\ominus}(head)$, on the other hand, is positive, and results from the steric and electrostatic repulsion encountered in bringing the head group into the micelle "interface." It depends inversely on the effective head group area, a_0, in the micelle. For ionic surfactants, the electrostatic contribution to the effective head group area should be proportional to the radius of the "ionic atmosphere" around an ion as given by the Debye-Hückel theory.[50] This in turn is inversely proportional to the total molar concentration of the counterions in the solution. It should be independent of the hydrophobe and only weakly dependent on temperature. Some of the important trends observed experimentally are as follows:

1) Increasing the hydrophobic chain length, n_c (of aliphatic surfactants) decreases the *CMC* in a regular manner. For a given homologous series of ionic surfactants (usually up to $n_C \approx 18$), adding a single -CH_2- group approximately *halves* the *CMC*. For nonionics (of the polyethylene oxide type), the *CMC* is dropped by closer to a factor of 3 for each -CH_2- addition to the chain. It is found that for many types of micelle-formers, this dependence can be accurately expressed by the equation suggested by Klevens[51] (for straight chain surfactants):

$$\log(CMC) = A - Bn_C.\qquad\qquad(3.132)$$

(*CMC*) [=] mole/liter; n_C = number of carbon atoms in the chain. Table 3-7 shows "Klevens constants" for various ionic surfactants. Such relationships,

[50] Lewis, G. N., Randall, M., Pitzer, K. S., and Brewer, L., **Thermodynamics**, 2nd Ed., pp. 332-243, McGraw-Hill, New York, 1961.

[51] Klevens, H. B., *J. Phys. Colloid Chem.*, **54**, 283 (1950).

Table 3-7: Klevens constants for various surfactant types. From [Rosen, M., **Surfactants and Interfacial Phenomena**, 2nd Ed., p. 136, Wiley, New York, 1989.]

Surfactant series	T (°C)	A	B
Na carboxylates (soaps)	20	1.8_5	0.30
K carboxylates (soaps)	25	1.9_2	0.29
Na (K) n-alkyl 1-sulfates or -sulfonates	25	1.5_1	0.30
Na n-alkane-1-sulfonates	40	1.5_9	0.29
Na n-alkane-1-sulfonates	55	1.1_5	0.26
Na n-alkane-1-sulfonates	60	1.4_2	0.28
Na n-alkane-1-sulfates	45	1.4_2	0.30
Na n-alkane-1-sulfates	60	1.3_5	0.28
Na n-alkane-2-sulfates	55	1.2_8	0.27
Na p-n-alkylbenzenesulfonates	55	1.6_8	0.29
Na p-n-alkylbenzenesulfonates	70	1.3_3	0.27
n-Alkylammonium chlorides	25	1.2_5	0.27
n-Alkylammonium chlorides	45	1.7_9	0.30
n-Alkyltrimethylammonium bromides	25	1.7_2	0.30
n-Alkyltrimethylammonium chlorides (in 0.1 M NaCl)	25	1.2_3	0.33
n-Alkyltrimethylammonium bromides	60	1.7_7	0.29
n-Alklpyridinium bromides	30	1.7_2	0.31
n-$C_nH_{2n+1}(OC_2H_4)_6OH$	25	1.8_2	0.49

which are another manifestation of Traube's Rule, are of great practical value in extrapolating and interpolating experimental data on CMC's. The B-value of ≈ 0.3 suggests a molar free energy contribution of $\approx 0.7RT$ for each -CH_2- group. For nonionics (at least of the PEO type), the B-value of ≈ 0.5 (see bottom entry of Table 3-7) yields a free energy contribution of $\approx 1.2RT$ for each -CH_2- group, suggesting in this case a more nearly complete transfer of the hydrophobe from the water into the micelle. This is consistent with the idea that the greater spacing required between the head groups of ionic surfactants (due to electrostatic repulsion) permits greater contact of the hydrophobes in the micelle with water in that case. The nature of $\Delta G_{mic}^{\ominus}$(hc) is further understood by considering the weak temperature dependence of the CMC for ionic surfactants. Applying Gibbs-Helmholtz analysis:

$$\frac{d\ln(CMC)}{dT} = \frac{d\left(\dfrac{\Delta G_{mic}^{\ominus}}{RT}\right)}{dT} = -\frac{\Delta H_{mic}^{\ominus}}{RT^2} = \text{small,} \tag{3.133}$$

it is seen that the enthalpy of micellization is small, a fact confirmed by independent calorimetric studies of the process. Thus, in view of the general relationship

$$\Delta G^{\ominus}_{mic} = \Delta H^{\ominus}_{mic} - T\Delta S^{\ominus}_{mic}, \tag{3.134}$$

it is seen again that the principal driving force for micellization is *entropic*. This may at first be counter-intuitive, because the monomers would be expected to suffer an entropy *decrease* upon confinement to micelles, but this is far outweighed by the entropy increase of the water.

2) Nonionic surfactants (of the PEO type) have lower *CMC's* (for the same hydrophobic chain length and temperature) than ionic ones. This is due primarily to the absence of electrostatic repulsion between adjacent head groups at the micelle "interface." In fact, nonionics have *CMC's* usually an order of magnitude less than ionics with the same hydrophobe. For example (in contrast to the C_{12} ionic surfactants above, which under low salt conditions give $CMC \approx 8$ mM) a C_{12} nonionic surfactant with a head group consisting of six ethylene oxide units, *i.e.*, $C_{12}E_6$, has a *CMC* of 0.087 mM. As the polyethylene oxide chain length is increased for a given hydrophobe, the *CMC* is increased in a regular fashion, in accord with

$$\log(CMC) = A' + B'm_E, \tag{3.135}$$

where m_E is the number of ethylene oxide units. Some values for the constants are listed in Table 3-8.

Table 3-8. Klevens constants for nonionic (PEO) surfactants. From [Rosen, M. J., **Surfactants and Interfacial Phenomena**, 1st ed., p. 103, Wiley-Interscience, New York, 1978.]

Surfactant series	T (°C)	A'	B'
n-$C_{12}H_{25}(OC_2H_4)_x$OH	23	-4.4	+0.046
n-$C_{12}H_{25}(OC_2H_4)_x$OH	55	-4.8	+0.013
p-t-$C_8H_{17}C_6H_4(OC_2H_4)_x$OH	25	-3.8	+0.029
$C_9H_{19}C_6H_4(OC_2H_4)_x$OH	25	-4.3	+0.020
n-$C_{16}H_{33}(OC_2H_4)_x$OH	25	-5.9	+0.024

3) For ionic micelle-formers, anything tending to reduce the electrostatic repulsion between head groups decreases the *CMC*. This may be accomplished by the addition of non-surfactant electrolyte, which has a *screening effect* due to an increased concentration of counterions, as discussed above. For a given compound, the *CMC* depends directly on the total concentration of counterions in accord with

$$\log(CMC) = -a\log(C_C) - b, \tag{3.136}$$

where C_C is the total molar concentration of counterions (both from the surfactant itself and any additional electrolyte present), and "a" and "b" are characteristic constants. A couple of examples given by Osipow [52] are:

K dodecanoate: $\log(CMC) = -0.570 \log C_C^+ - 2.617$, and   (3.137)

Na dodecyl sulfate: $\log(CMC) = -0.458 \log C_C^+ - 3.248$,   (3.138)

where the $^+$ is a reminder that the counterions are the cations in these cases. Under swamping electrolyte conditions, the CMC-values approach those associated with nonionic surfactants. Electrolytes can also affect the CMC for nonionics and zwitterionics, but not in the same way and to the same degree as stated above. In this case, the influence of added electrolyte may be to alter the monomer solubility, $i.e.$, to "salt in" or "salt out" the hydrophobic portion of the monomer. Salting out is more common, and refers to a decreasing of the monomer solubility and hence the CMC. It usually requires a higher level of salt concentration than that needed to affect the CMC of ionic surfactants.

4) Temperature effects on the CMC can be quite complex, as suggested by Fig. 3-23. For ionic micelles, the dependence may be fairly weak over significant ranges, as stated earlier, but for nonionics ($i.e.$, polyethoxylates), increasing temperature always decreases the CMC sharply due to the progressive dehydration of the PEO groups.

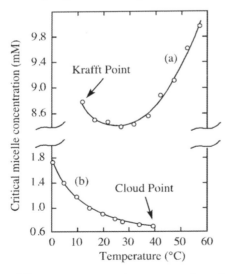

Fig. 3-23: Variation of CMC with temperature for:
 (a) sodium dodecyl sulfate;
 (b) $CH_3(CH_2)_9(C_2H_4O_5)OH$
After [Elworthy, P. H., Florence, A. T., and Macfarlane, C. B., **Solubilization by Surface Active Agents**, Chapman and Hall, London, 1968.]

There are two important benchmark temperatures for surfactant solutions, one for ionic surfactants and one for PEO type nonionics: the *Krafft point* and the *cloud point*, respectively.

[52] Osipow, L. I., **Surface Chemistry**, ACS Monograph Series, No. 153, p. 168, Reinhold, New York, 1962.

Most ionic surfactants can form micelles if the hydrophobic portion is large enough that there is a significant decrease in free energy when they form. Below that, some loose association, "incipient micelle formation," occurs, *i.e.*, dimerization, *etc.* The hydrophobic portion must not be *too* large, however, so that the monomer itself has insufficient solubility. The total solubility of a typical ionic surfactant as a function of T gives the results shown in Fig. 3-24, which shows the definition of the *Krafft Point* as the temperature where the solubility equals the *CMC*, and solubility undergoes a sudden increase. Nonionic surfactants (in particular, the PEO type) have more complex solubility behavior. A monomer solution or solution of micelles of ordinary size may exist at a given temperature, but as temperature is raised, the PEO chains are progressively dehydrated until a point is reached where very large aggregates are formed, producing visible turbidity. The temperature at which this occurs is quite sharp for a given surfactant and is termed its *cloud point*. The transition occurring as the cloud point is traversed is usually regarded as a macroscopic phase change. Some cloud point data for dodecyl ether surfactants are shown as a function of the PEO chain length × in Fig. 3-25.

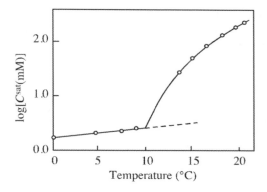

Fig. 3-24: Logarithm of the solubility of sodium dodecyl sulfate as a function of temperature. The dashed line represents the expected behavior in the absence of micelle formation. After [McBain, M.E.L. and Hutchinson, E., **Solubilization and Related Phenomena**, Academic Press, New York, 1955.]

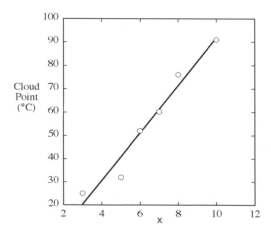

Fig. 3-25: Cloud points for dodecyl ether nonionic surfactants:

$$n\text{-}C_{12}H_{25}(OC_2H_4)_x$$

Data from [Karabinos, J.V., Hazdra, J.J., and Kapella, G.E., *Soap and Chemical Specialties*, **7**, April (1955).]

5) Organic additives may have a strong effect on the *CMC*. Large amounts of such materials may associate themselves with the micelles in the phenomenon termed *solubilization*, discussed below. This may greatly change the size and structure of the micelles, and reduces the *CMC*. Another class of additives is that of solutes whose presence in the solution affects the solubility of the monomer. They are thought of as "structure breakers" or "structure makers." Urea and formamide are examples of structure breakers, which disrupt the water structure and hence the interaction of water with the hydrophobes. Their presence increases the *CMC*. Xylose and fructose are examples of structure makers, which decrease the *CMC*.

6) Most factors that tend to *decrease* the *CMC* also tend to *increase* micelle size, or aggregation number. These are often in the range of 30-300 for ionic micelle-formers and 200-20,000 for nonionic (PEO) surfactant, the latter strongly dependent on temperature.

2. Solubilization

One of the most important consequences of the micellization phenomenon is the fact that micelles provide a hydrocarbon, or more generally an *apolar*, environment within an aqueous medium. Not only does the formation of micelles greatly enhance the solubility of the surfactant itself beyond what it would be if micelles did not form, but the solubility of *other* sparingly soluble organic materials is greatly enhanced as well.

Apolar materials are incorporated into the interior of the micelle as shown in Fig. 3-26, while long-chain, and/or polar molecules may penetrate the outer shell of the micelle (called the palisade layer) in an oriented

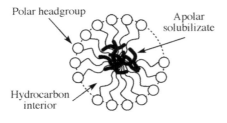

Fig. 3-26: Solubilization of apolar material into the interior of a micelle.

fashion, possibly to form *co-micelles*. Finally, other ionic or highly polar materials may adsorb on the outside of the micelles. Regardless of which mechanism is responsible for the enhanced solubility (by factors of 10^3-10^4 or higher are typical) it is termed *solubilization*.

The cleansing action of soaps and detergents is due in part to solubilization of oily dirt so that it may be swept away. Detergency is a combination of enhanced wetting, detachment of dirt from the substrate surface (termed "roll up") and solubilization. Dry cleaning depends in part on the solubilization of hydrophilic soils into inverted micelles, discussed below.

Solubilization leads to an easy way to detect the *CMC*, by solubilizing a dye.[53] In dilute aqueous solution, the dye would be nearly invisible, but when micelles are formed, most of it would be solubilized, thus concentrated into micelles, and would show visible color. Thus, *CMC*'s could be determined very quickly by titration. It should be noted, however, that the solubilization of the dye itself will at least slightly decrease the *CMC*.

A similar phenomenon explains an anomaly that puzzled surface chemists for many years. When studying the surface tension behavior of many anionic surfactants, there was a *minimum* in the surface tension, as shown in Fig. 3-27, for solutions of sodium dodecyl sulfonic acid and other

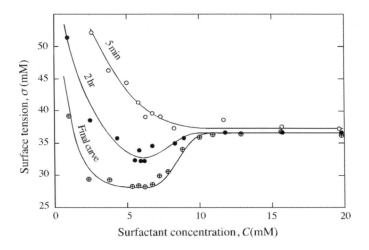

Fig. 3-27: Surface tension behavior of dodecyl sulfonic acid solutions as a function of concentration, measured at different rates. After [McBain, J. W., Vinograd, J. R., and Wilson, D. A., *J. Amer. Chem. Soc.*, **62**, 244 (1940).]

detergents. The Gibbs Adsorption Equation would seem to dictate that the relative adsorption of the surfactant be zero at any minimum in such a curve, and *negative* just beyond it. Such behavior seemed unlikely, but McBain and coworkers documented a number of examples of the apparent paradox. His microtome measurements on a number of these systems,[54] was motivated to check the validity of the Gibbs Adsorption Equation, but revealed $\Gamma_{2,1}$ to have significant positive values throughout the region. The explanation was given finally by Crisp,[55] as sketched in Fig. 3-28(a)-(c). The dodecyl sulfonic acid used (as an example) contained a very small impurity of dodecyl (lauryl) alcohol (the material from which the surfactant was synthesized) (component 3), a non-ionized compound with significantly greater surface activity than the main component. It was this impurity that

[53] Corrin, M. C., and Harkins, W. D., *J. Amer. Chem. Soc.*, **69**, 679 (1947).

[54] McBain, J. W., and Wood, L. A., *Proc. Roy. Soc. A*, **174**, 286 (1940).

[55] Crisp, D. J., *Trans. Faraday Soc.*, **43**, 815 (1947).

predominated at the surface at low overall concentrations, given enough time
to get there, and was primarily responsible for the observed surface tension
reduction. Thus, as C_2 was increased, more of component 3 (the alcohol)
adsorbed, until finally the *CMC* of the dodecyl sulfonic acid was reached.
Then the lauryl alcohol was drawn into the micelles, even away from the
surface, leaving behind SDS. This produced an increase in the surface
tension. In terms of the Gibbs Adsorption Equation for a *ternary* system:

$$\frac{d\sigma}{dC_2} = -\Gamma_{2,1}\left(\frac{\partial\mu_2}{\partial C_2}\right) - \Gamma_{3,1}\left(\frac{\partial\mu_3}{\partial C_2}\right). \tag{3.139}$$

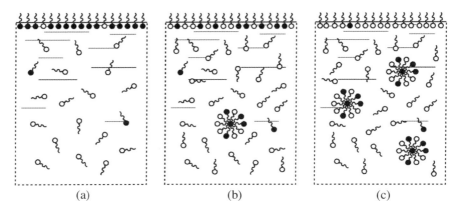

(a)   (b)   (c)

Fig. 3-28: Explanation of surface tension with concentration. Dark head groups represent
fatty alcohols, and light head groups represent sulfate anions.

Above the *CMC* of component 2, $(\partial\mu_2/\partial C_2) \approx 0$ (but *not* $\Gamma_{2,1}$) and $(\partial\mu_3/\partial C_2)$
is *negative*, since as C_2 increases, there are more micelles which *attract* the
alcohol, decreasing its monomer concentration and hence its chemical
potential. Thus:

$$\left(\frac{\partial\mu_2}{\partial C_2}\right) = \frac{RTd\ln(C_2)_{\text{mono}}}{d(C_2)_{\text{total}}} \approx 0, \text{ and} \tag{3.140}$$

$$\left(\frac{\partial\mu_3}{\partial C_2}\right) = \frac{RTd\ln(C_3)_{\text{mono}}}{d(C_2)_{\text{total}}} < 0. \tag{3.141}$$

When carefully purified surfactant was used, there was no minimum in the
surface tension, and conversely, when small (0.1%) amounts of
contaminants like lauryl alcohol were added, the minimum reappeared. A
practical implication of the above is that when *CMC* determinations are to be
made by surface tension measurements for ionic surfactants containing
traces of nonionic precursor surfactants (such as fatty alcohols), the
measurements should be made quickly enough to avoid the effects of these
contaminants.

Solubilization has a great potential for application in chemical reactions, such as in emulsion polymerization, discussed briefly in Chap. 5.C.3. Reactions which occur between (or involve) both water-soluble and water-*insoluble* components (as acid- or base-catalyzed hydrolysis) could have their rates drastically increased by incorporating the water-insoluble ester into micelles in acidic or basic aqueous solutions. Enhancing such a reaction by solubilization of one of the reactants into micelles is termed "secondary valence catalysis" or "micellar catalysis."[56] An example is provided by the Cannizzaro reaction in which benzaldehyde decomposes in the presence of a strong base to benzyl alcohol and the salt of the corresponding carboxylic acid, *viz.*:

$$2\,C_6H_5CHO + KOH \rightarrow C_6H_5CH_2OH + C_6H_5CHOO^-K^+,$$

The alkali is insoluble in the water-immiscible benzaldehyde, dooming the reaction to the water-oil interface. Incorporation of the benzaldehyde into cationic micelles, however, greatly enhances the rate of the reaction as the hydroxyl ions are attracted to the vicinity of the micellar "interface," whereas the use of anionic micelles effectively shuts the reaction off, as pictured in Fig. 3-29. In the latter case the hydroxyl ions are repelled from the micelle "interface." A summary tabulation of reactions mediated by secondary valence catalysis has been given by Fendler.[57]

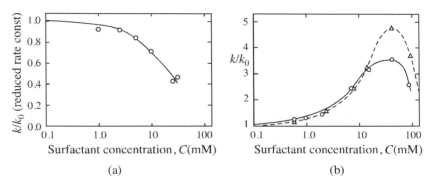

Fig. 3-29: Secondary valence catalysis: (a) the effect of an anionic surfactant (potassium palmitate) on the rate of the Cannizzaro reaction; (b) the effect of cationic surfactants (Δ eicosanyltroimethylammonium bromide, O octadecyl trimethylammonium bromide) on the rate of the same reaction. After [Cramer, L.R., and Berg, J.C., *J. Phys. Chem.*, **72**, 3686 (1968).]

The ultimate amount of solubilizate that can be incorporated in a given case, known as the maximum additive concentration (*MAC*) is limited by the size of the original micelle, and the incorporation of the solubilizate may change the nature and the shape of the micelles. Under certain

[56] Fendler, J., and Fendler, E., **Catalysis in Micellar and Macromolecular Systems**, Academic Press, New York, 1975.

[57] Fendler, J. H., **Membrane Mimetic Chemistry**, pp. 341-409, John Wiley & Sons, New York, 1882.

circumstances, micelles can be swollen by solubilization to hundreds of times their original size (diameters up to 100 nm or more), and such structures are termed *microemulsions*.[58] This high degree of swellability is often facilitated by the addition of medium-chain-length alcohols or other amphiphiles (termed co-surfactants), as shown in Fig. 3-30. Microemulsion droplets are considered to be dissolved species and as such, form thermodynamically stable systems. Microemulsions, and the situations

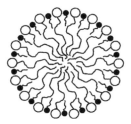

Fig. 3-30: Schematic of a highly swollen micelle formed by a mixture of surfactant and co-surfactant capable of imbibing a large amount of solubilizate to form a microemulsion "droplet."

leading to their spontaneous formation, are discussed in more detail in Chap. 9.D in the context of the treatment of emulsions in general. They are finding applications that parallel the exploitation of solubilization with ordinary micelles, but their requirement for large amounts of (expensive) surfactant to some extent limits their practical use.

I. Micelle morphology, other self-assembled structures, and concentrated surfactant solutions

1. *Micellar shape and the Critical Packing Parameter (CPP)*

Up to this point it has been tacitly assumed that micelles are spherical in shape. This is often not the case. Figure 3-31 suggests that in order for surfactant monomers to be able to form a space-filling sphere, there are

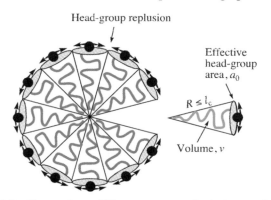

Head-group replusion

Effective head-group area, a_0

$R \le l_c$

Volume, v

Fig. 3-31: Micelles as shape-filling structures, depend upon the size and shape characteristics of the monomer.

---

[58] Bourrel, M., and Schechter, R. S., Eds., **Microemulsions and Related Systems**, Dekker, New York, 483 pp., 1988.

geometric constraints on their hydrophilic "heads" and hydrophobic "tails." It may also be evident that only a narrow range of aggregation numbers will be able to satisfy these requirements for a given monomer. The head groups occupy an effective surface area "a_0" (per molecule) at the aggregate-solution "interface," while the hydrophobic tails have a volume v and a maximum length l_C (the contour length). l_C is the largest chain extension possible while maintaining liquid-like properties. For a spherical micelle of radius R, the aggregation number n may be expressed either in terms of the micelle area occupied by the monomer head groups, or the micelle volume, assumed to consist of the hydrophobic tails:

$$n = \frac{4\pi R^2}{a_0} = \frac{4/3\pi R^3}{v}, \text{ so that}: \quad R = \frac{3v}{a_0} \tag{3.142}$$

Since R must be $\leq l_C$, spherical micelles can form only when $3v/a_0 < l_C$, or when

$$\left[\frac{v}{a_0 l_C}\right] \leq \frac{1}{3}. \tag{3.143}$$

The quantity in brackets above is called the "critical packing parameter" (CPP), or the "surfactant number,"[59] and its value determines the expected shapes of the aggregates that form[60], as discussed below. For spherical micelles to form, the monomer is envisioned to have a conical morphology. While the packing parameter may be less than 1/3 for spherical micelles, if we assume the material of the aggregate to be incompressible, it is clear that it cannot be *much* less than 1/3 or there would be an unfilled region in the core. Estimates of v and l_C may be obtained for aliphatic chains with n_C carbon atoms as[61]

$$v \approx (27.4 + 26.9n_c) \times 10^{-3} \; [=] \; nm^3, \text{ and}$$
$$l_c \approx l_{max} \approx 0.15 + 0.1265n_c \; [=] \; nm \tag{3.144}$$

The effective head group area a_0 is not generally calculable *a priori* and depends not only on the physical size of the head group, but also on its state of hydration, ionization, ionic environment, *etc.* The common single-tail detergents, such as SDS or CTAB, form spherical micelles under low salt conditions.

When the head group area is too small relative to the bulk of the hydrophobe to allow formation of spheres, the monomer morphology more closely resembles that of a truncated cone, and the self-assembled structures

[59] Evans, D.F., Wennerström, H., **The Colloidal Domain**, pp. 12ff, VCH, New York (1994).

[60] Israelachvili, J. N., Mitchell, D. J., and Ninham, B, W., *J. Chem. Soc. Faraday Trans.*, **72**, 1525 (1976).

[61] Tanford, C., **The Hydrophobi Effect: Formation of Micelles and Biological Membranes**, 2nd Ed., p. 52, Krieger, Malabar, FL, 1991.

observed are *cylindrical* micelles, pictured in Fig. 3-32(b), shown in comparison with spherical micelles. Cylindrical micelles are expected in the range $1/3 < CPP < 1/2$. The head group area for ionic surfactants may be reduced by simply increasing the salt concentration of the medium, reducing the degree of head-to-head electrostatic repulsion, so that the common detergents like SDS or CTAB produce cylindrical micelles under high salt conditions. Single hydrophobic chain nonionics with small head groups also

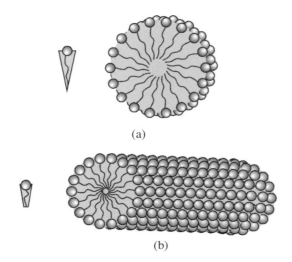

(a)

(b)

Fig. 3-32: (a) Spherical micelles: $CPP \le 1/3$; (b) Cylindrical micelles: $1/3 < CPP < 1/2$.

produce cylindrical micelles. Near the lower end of the above *CPP* range, presumably intermediate structures such as short cylinders with hemispherical caps, form, while higher *CPP* values produce longer cylinders that are polydisperse with respect to their length, with their average length dependent upon the total surfactant concentration. At higher concentrations, they may become very long (> 1 μm), *i.e.*, "wormlike," resembling linear polymers in solution. They are dynamical structures that are able to break and re-form reversibly upon shearing, and have become common model systems for rheological studies.[62]

2. Beyond micelles: other self-assembled structures

When the critical packing parameter becomes still larger, an array of other structures may be formed (self-assembled) from surfactant monomers in solution.[63] When the monomer has a double rather than a single hydrophobic tail, the ratio of hydrophobe volume to head group area becomes quite large so that the monomer shape is still that of a truncated cone, but with a smaller aspect ratio. Such molecules are exemplified by a

---

[62] Keller, S. L., Boltenhagen, P., Pine, D. J., and Zasadzinski1, J. A., *Phys. Rev. Lett.*, **80**, 2725 (1998).

[63] Israelachvili, J. N., **Intermolecular and Surface Forces**, 2nd Ed., pp. 381, Academic Press, New York, 1992.

variety of naturally occurring phospholipids, including the dipalmitoyl lecithin (lung surfactant) pictured in Fig. 2-6, as well as synthetic di-tail surfactants such as Aerosol OT®. With *CPP*-values between 1/2 and 1, such monomers tend to form curved bilayer structures, as pictured in Fig. 3-33(a). These may produce bilayer spheres, termed *vesicles* (or *liposomes*), and under some conditions, bilayer tubules, as shown in Fig. 3-33(b), or even toroids (doughnuts).[64] The structures may be unilamellar or multilamellar.

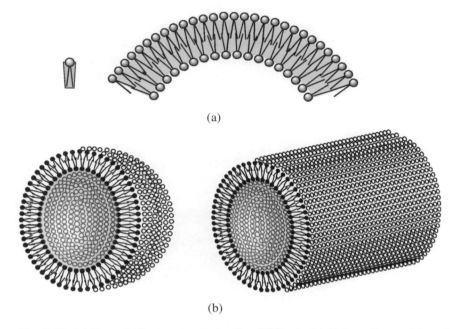

(a)

(b)

Fig. 3-33: (a) Curved bilayers structures: ½ < *CPP* < 1; (b) Unilamellar vesicle and tubule.

Unilamellar vesicles of dipalmitoyl lecithin (lung surfactant) as small as approximately 30 nm diameter may be formed, while multi-lamellar vesicle structures as large as one micrometer or greater in diameter are commonly produced from egg lecithin. The formation of vesicles appears in some cases to require some energy input (stirring, sonication), but in most cases is effectively spontaneous. For example, a polydisperse vesicular dispersion may be formed by gently stirring a mixture of egg lecithin and water at room temperature. The mixture may be fractionated into various sizes by sieving or dialysis through membranes. Vesicles currently show promise as controlled-release drug delivery devices[65,66] or in the synthesis of highly uniform-sized mineral particles with applications in catalysis, or in

---

[64] Mutz, M., and Bensimon, D, *Phys. Rev. A*, 43(8), 4525 (1991);
Fourcade, B., Mutz, M., and Bensimon, D. *Phys. Rev. Lett.*, 68(16), 2551 (1992).

[65] Uster, P. S., "Liposome-Based Vehicles for Topical Delivery," in **Topical Drug Delivery Formulations**, Drugs and the Pharmaceutical Sciences, Vol. 42, Marcel Dekker, New York (1990).

[66] Virden, J. W., and Berg, J. C., *J. Colloid Interface Sci.*, **153**, 411 (1992).

the production of magnetic or electronic devices[67]. The unilamellar form of these and other bilayer structures have much in common with natural membranes, and their study has been termed "membrane mimetic chemistry."[68] When the *CPP* rises to ≈ 1, the monomer morphology is near that of a cylinder, and lamellar micelles, bilayer fragments or even bicontinuous structures may be formed, as shown in Fig. 3-34. The bilayers sheets in any case are of effectively zero mean curvature. The may occur as single wall or multi-wall structures. The monomers are most often di-tail surfactants with small head groups, but may also be single-tail compounds if they contain multiple double bonds or double bonds near the center of the chain (such as oleates). Video enhanced microscopy suggests many other structures can form as well, such as bilayer chains, filaments, and multilayer sandwiches.

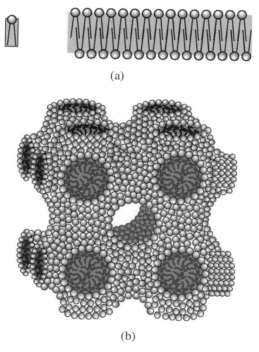

(a)

(b)

Fig. 3-34: (a) Flat bilayer fragment: $CPP \approx 1$; (b) Bicontinuous structure of zero mean curvature.

Values of the critical packing parameter in excess of unity imply the formation of *inverse*, or *reverse* micelles, *i.e.*, aggregates of surfactant molecules in *non*-aqueous media, with the head groups together in the core of the micelle and the hydrophobes extended.[69] These in fact do form under

[67] Mann, S., and Hannington, J. P., *J. Colloid Interface Sci.*, **122**, 326 (1988).

[68] Fendler, J. H., **Membrane Mimetic Chemistry**, John Wiley & Sons, New York (1982).

[69] Luisi, P. L., and Straub, B. E., Eds., **Reverse Micelles**, Plenum, New York, 1984;

the right conditions. The hydrophobic effect contributes little or nothing to the driving force for their formation, but instead it is largely attractive interactions (dipolar, hydrogen bonding, or more general acid-base effects) between the head groups that are responsible for micelle formation. Their formation is believed to be facilitated by or even to require the presence of small (almost undetectable) amounts of water in the organic medium.[70] Water, which acts as both an acid and a base, clearly facilitates donor-acceptor interactions. Reverse micelles are often but not always spherical and more often than not smaller than ordinary micelles, sometimes with aggregation numbers less than ten. One of the most commonly used reverse micelle formers is Aerosol OT®, which has been shown to form trimers in hydrocarbon solutions above a critical concentration (*CMC*).[71] Reverse micelles are swollen through the solubilization of water, and may evolve from spherical shapes to cylinders, and in some circumstances reverse microemulsions are formed. Many of the compounds found to form reverse micelles in low dielectric media are ionic surfactants in aqueous media, but with sufficient oil solubility. They have found application as solubilizing vehicles important in dry cleaning and as agents for secondary valence catalysis in apolar media.[72] Their widest use arises from their ability to stabilize electric charges in apolar media. They may thus be used to confer sufficient electrical conductivity to liquid fuels to prevent the dangerous buildup of static charges. They also provide a mechanism for charging surfaces, such as those of colloidal particles, in contact with low dielectric media, as discussed further in Chap. 6. This in turn provides a contribution to their stabilization against aggregation, as occurs in the buildup of carbon sludges in motor oil. Charged colloids in apolar media are also the basis for the rapidly developing technology of electrophoretic inks for electronic displays.[73] In addition to micelles, reverse vesicles[74] and toroidal reverse vesicles[75] have also been reported.

In summary, it must be noted that while the use of the critical packing parameter of a surfactant monomer as a predictor of self-assembled structures in solution appears to be very powerful, its use is hampered by the fact that one seldom knows the appropriate value for the head group area, a_0. It is often the *post hoc* observation of structure that gives clues to the appropriate value for this parameter.

Morrison, I. D., *Colloids Surfaces A*, **71**, 1 (1993).

[70] Nelson, S. M., and Pink, R. C., *J. Chem. Soc.*, **1952**, 1744 (1952).

[71] Denal, A., Gosse, B., and Gosse, J. P., *Rev. Phys. Appl.*, **16**, 673 (1981).

[72] Kitihara, A., *Adv. Colloid Interface Sci.*, **12**, 109 (1980).

[73] Cominsky, B., Albert, J. D., Yoshizawa, H., and Jacobson, J., *Nature*, **394**, 253 (1998).

[74] Kunieda, H., Shigeta, K., Nakamura, K., and Imae, T., *Progr. Colloid Polym. Sci.,* **100**, 1 (1996).

[75] Murdan, S., Gregoriadis, G., and Florence, A. T., *Intl. J. Pharmaceutics,* **183** [1], 47 (1999).

3. *Concentrated surfactant solutions; liquid crystalline mesophases*

When the concentration of surfactant in aqueous solution is made to exceed about 10% by weight, micelle-micelle interactions become significant, and the spherical or finite cylindrical structures generally undergo conversion to long cylinders or multi-bilayers, as pictured in Fig. 3-35. The cylinders usually form first (as concentration is increased) and pack themselves into hexagonal arrays yielding a liquid crystalline mesophase called the "normal hexagonal phase" or the "middle phase." This is a nematic liquid crystalline phase, and it exhibits optical birefringence (opalescence). Further concentration often produces the multilayer structure known as the "lamellar" or "neat" phase, an example of a smectic liquid crystal. Other liquid crystalline phases may also occur, and in general, the phase behavior of concentrated surfactant solutions is quite complex. It is well described in the monograph by Laughlin.[76]

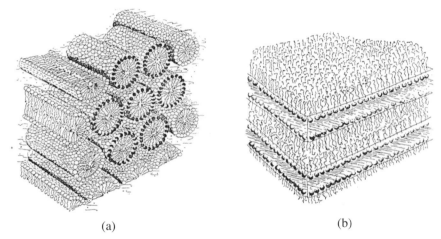

(a)   (b)

Fig. 3-35: Structures forming in concentrated surfactant solutions: (a) normal hexagonal (or "middle") phase; (b) lamellar (or "neat") phase. Other structures also can form in these solutions under different conditions. From [Bourrel, M., and Schechter, R. S., **Microemulsions and Related Systems**, Surf. Sci. Ser. 30, pp. 100 and 102, Marcel Dekker, New York, 1988.]

The reason for the occurrence of the above structural changes with concentration is the long-range repulsive forces that exist between the aggregates. They consist chiefly of electrostatic and hydration forces (solvent structuring around the hydrophilic groups). These can be ignored in the dilute systems (as we have done), but not in more concentrated solutions. The micelle surfaces try to get as far apart as possible, but their ability to do so is limited. Converting to a structure of quasi-infinite cylinders increases the average inter-surface distance, and subsequent conversion to lamellae increases the distance still further.

---

[76] Laughlin, R. G., **The Aqueous Phase Behavior of Surfactants**, Academic Press, London (1994).

One of the important recent applications of the various fluid micro- and nano-structures that can be created from the spontaneous assembly of surfactants in solution is their use as templates for mineralization via inorganic precursors to produce micro- or nano-structured solids (monoliths). The latter are examples of "nanocomposites."[77] A virtually inexhaustible array of new materials including high-performance coatings and catalysts, as well as electronic, photonic, magnetic and bio-active materials, have either been produced in this way, or are contemplated.

4. *Kinetics of micellization and other self-assembly processes*

It has been tacitly assumed to this point that surfactant solutions are in instantaneous equilibrium with respect to micelle or other structure formation when changes in conditions (degree of dilution, temperature, salt addition, *etc.*) are made. Indeed it has been shown that relaxation times for the exchange of single monomers in solution with simple micelles is generally of the order of microseconds, while the time required for complete breakdown and reformation of a micelle if of the order of one millisecond.[78] The latter should correspond to the time required for solubilizate to distribute itself amongst the micelles in a solution, provided it is well mixed. The times required for the formation or disruption of the equilibrium phase structure in concentrated surfactant solutions are often much greater.[79] For example, the dissolution of sufficient sodium dodecyl sulfate (SDS) in water to form the birefringent hexagonal phase described earlier may require several days. The kinetics of annealing in such systems of long-range order could benefit from further study.

J. Dynamic surface tension of surfactant solutions

1. *Diffusion-controlled adsorption*

We must return to consideration of the surface tension of surfactant solutions and recognize that in many, perhaps even most, of the situations of practical interest, the system is not in a state of adsorption equilibrium. While pure liquids do not show a surface tension time-dependence in any measurable time range, dilute surfactant solutions show equilibration times ranging from less than 1 ms (the shortest that can currently be measured) to several minutes or even hours.[80] In spray coating operations of various kinds, surfactants are added to the liquid in part to assist in spreading as the drops impact the target surface. This is accomplished only if sufficient

---

[77] Dagami, R., *Chem. & Eng. News*, **77** [23], 25 (1999).

[78] Aniansson, E. A. G., Wall, S. N., Almgren, M., Hoffmann, H., Kielmann, I., Ulbricht, W., Zana, R., Lang, J., and Tondre, C., *J. Phys. Chem.*, **80**, 905 (1976).

[79] Zana, R., **Dynamics of Surfactant Self-Assemblies: Micelles, Microemulsions, Vesicles and Lyotropic Phases**, Surf. Sci. Ser. 125, Marcel Dekker, New York, 2005.

[80] Dukhin, S. S., Kretzschmer, G., Miller, R., **Dynamics of Adsorption at Liquid Interfaces**, Elsevier, Amsterdam (1995).

adsorption has occurred over the lifetime of the spray droplets, which in the case of inkjet printing is less than one millisecond. It is thus important to be able to calculate the time course of adsorption and surface tension, as well as to be able to measure dynamic surface tension, as discussed in Chap. 2. The maximum bubble pressure technique (Fig. 2-46) can now be used[81] for surface ages in the millisecond and sub-millisecond range, supplanting the more difficult oscillating jet method (Fig. 2-48).[82]

In the simplest case, the time effect is controlled by the rate of surfactant diffusion to the surface, as pictured in Fig. 3-36. For some cases, there may be a kinetic barrier to adsorption, or surface re-orientation of the adsorbate molecules, but if the solution is sufficiently dilute, one would expect the time effect to be due to diffusion alone. In that case, transport of the monomers through the solution is described by the diffusion equation:

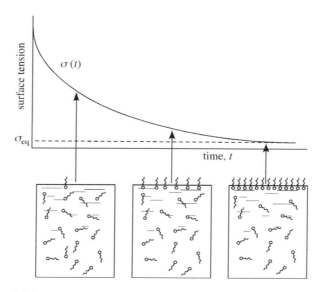

Fig. 3-36: Dynamic surface tension. For dilute solutions (particularly of surface active agents) the surface tension decreases with surface age as the solute diffuses to the surface and adsorbs.

$$\frac{\partial C}{\partial t} = D\frac{\partial^2 C}{\partial z^2},\tag{3.145}$$

where C is the concentration of surfactant, D is its diffusivity, and z is the distance measured *away* from the surface (hence the positive sign on the right hand side). Before the surface is formed, the surfactant is at the uniform concentration C_∞. The adsorption Γ is a function of time as the

[81] Miller, R., Fainerman, V. B., Schano, K.-H., Hoffmann. A., Heyer, W., *Tenside Surf. Det.*, **34** [5], 357 (1997).

[82] Defay, R., and Pétré, G., "Dynamic Surface Tension," in **Surface and Colloid Science**, Vol. 3, pp. 27-81, E. Matijevic, Ed., Wiley-Interscience, New York (1970).

surface "fills," and during the course of filling is assumed to be in local equilibrium with the sublayer immediately beneath it, where $C = C_s$. The initial conditions are:

$$t = 0(-); \quad C = C_\infty(\text{all } z)$$
$$t = 0(+); \quad C_s = 0 \qquad , \qquad \qquad (3.146)$$
$$t = 0; \quad \Gamma = 0$$

and the boundary conditions are:

$$z \to \infty; \quad C \to C_\infty \text{ and} \qquad (3.147)$$

$$z = 0; \quad C = C_s(\Gamma), \text{ the adsorption isotherm.} \qquad (3.148)$$

Then the time course of the adsorption, $\Gamma(t)$ is given by

$$\frac{d\Gamma}{dt} = D\left(\frac{\partial C}{\partial z}\right)_{z=0}. \qquad (3.149)$$

The solution for $\Gamma(t)$, obtained first by Ward and Tordai[83], takes the form of the integral equation:

$$\Gamma(t) = 2\sqrt{\frac{D}{\pi}}\left[C_\infty\sqrt{t} - \int_0^{\sqrt{t}} C_s(t - \tau)d(\sqrt{\tau})\right], \qquad (3.150)$$

where τ is a delay time. The first term on the right reflects the rate of diffusion of solute toward the surface, while the second term expresses the rate of back-diffusion from the surface as it fills. As is evident, the solution to this equation requires knowledge of the adsorption isotherm, which relates C_s to Γ at equilibrium, and the diffusivity D. The surface tension $\sigma(t)$ is then given by the surface equation of state. For many cases, the Langmuir isotherm and the corresponding Frumkin equation of state are adequate. These involve only two parameters, which together with the diffusion coefficient, suffice to predict the dynamic surface tension. The predicted behavior of solutions of the nonionic surfactant Triton X-100 obtained on this basis is shown in Fig. 3-37, and it is found to agree reasonably well with experimental data obtained by the pendant bubble technique for this case. Solutions to the Ward and Tordai equation must in general be obtained numerically, although analytical solutions exist for very short times (when back diffusion can be neglected, and Henry's Law adsorption may be assumed)[84] or very long times, as equilibrium is approached.[85]

The Ward and Tordai analysis using Langmuir-Frumkin equilibrium can be nondimensionalized using a characteristic length h, the "adsorption

[83] Ward, A. F. H., and Tordai, L., *J. Chem. Phys.*, **14**, 453 (1946).

[84] Hansen, R. S., *J. Colloid Sci.*, **16**, 549 (1961).

[85] Daniel, R., and Berg, J. C., *J. Colloid Interface Sci.*, **237**, 294 (2001).

depth," and a characteristic time τ_D. The adsorption depth is defined as the depth of solution of concentration C_∞ needed to supply a unit area of surface

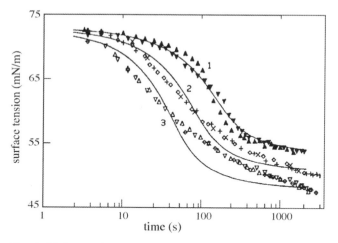

Fig. 3-37: Dynamic surface tension of aqueous solutions of Triton X-100 calculated using Ward and Tordai analysis with $D = 2.6 \times 10^{-6}$ cm^2/s. Data are shown for C_∞ = (1) 9.89, (2) 15.5, and (3) 23.2 mM. From [Lin, S.-Y., McKeigue, K., and Malderelli, C., *AIChE J.*, **36**, 1785 (1990).]

with its equilibrium inventory of surfactant, $\Gamma(C_\infty)$. It is thus expressible in terms of the adsorption equilibrium parameters, *viz.*

$$h = \frac{\Gamma(C_\infty)}{C_\infty} = \frac{B}{a + C_\infty},$$ (3.151)

where B and a are the Szyszkowski-Langmuir parameters defined earlier. The characteristic diffusion time is then defined as[86]

$$\tau_D = \frac{h^2}{D}.$$ (3.152)

When the surface tension is nondimensionalized as

$$\Theta = \frac{\sigma(t) - \sigma_{eq}}{\sigma_0 - \sigma_{eq}},$$ (3.153)

where σ_{eq} is the final equilibrium surface tension (at C_∞), and σ_0 is the surface tension of pure water. When plotted against (t/τ_D), a family of curves is obtained for varying values of the parameter $k = C_\infty/a$. $\Theta(t/\tau_D)$ drops from 1 to 0 as the system equilibrates, but for all values of k, falls to ½ at $(t/\tau_D) = 1$. τ_D is thus the characteristic diffusion time for the system. The time

---

[86] Ferri, J. K., Stebe, K. J., *Adv. Colloid Interface Sci.*, **85**, 61 (2000).

required, in terms of t/τ_D, for, say 90% equilibration, depends on k, being longer for low-k (low concentration) systems. Even for k as low as 0.1, however, 90% equilibration is achieved at a time of approximately $10\tau_D$. The diffusivities of ordinary monomeric surfactants in water are mostly in the range of $1.0 - 10.0 \times 10^{-6}$ cm^2/s, so if the equilibrium adsorption parameters are known for a given surfactant, the diffusion time can readily be estimated for a solution of given concentration. For example, for Triton X-100, the Szyszkowski-Langmuir parameters are: $B = 2.91 \times 10^{-10}$ mol/cm^2; $a = 0.662 \times 10^{-9}$ mol/cm^3, and the diffusivity is 2.6×10^{-6} cm^2/s. For a solution of $C_\infty = 9.89$ μM $(= 9.89 \times 10^{-9}$ mol/cm^3), we may compute: $h = 0.0276$ cm, and $\tau_D = 293$ s, which may be compared with the result shown in Fig. 3-37. At the CMC for Triton X-100, 2.3×10^{-7} mol/cm^3, the characteristic diffusion time is computed to be $\tau_D = 0.70$ s. As another example, for sodium dodecyl sulfate (SDS) at 10^{-4}M, one may compute $\tau_D \approx 0.5$ ms.

2. Finite adsorption-desorption kinetics

With the exception of some pure nonionic surfactants at very low concentrations, and certain alcohols, many surfactants show slower adsorption (in any case, slower surface tension decrease) than that predicted by the diffusion-control model, suggesting the importance of adsorption barriers for many cases,[87] so that calculated diffusion times represent upper limits for the rate of surface tension relaxation.

The very long-time dependence of surface tension that is sometimes observed is usually associated with the presence of small traces of impurity more surface active than the main component, as encountered when a minimum in the surface tension curves is observed, as in Fig. 3-27. Such behavior is often used as an indicator of the presence of such impurities. For solutions well above the CMC, time effects are shorter, since the micelles provide a large reservoir of monomers near the surface, and a simplified analysis of adsorption kinetics for this situation has been suggested.[88] The kinetic model consistent with Langmuir adsorption, cf. Eq. (3.104) is

$$\frac{d\Gamma}{dt} = k_1 C_\infty (B - \Gamma) - k_{-1}\Gamma, \tag{3.154}$$

where k_1 and k_{-1} are the adsorption and desorption rate constants. At $t \rightarrow \infty$, $d\Gamma/dt = 0$, from which it may be seen that $k_{-1}/k_1 =$ the adsorption parameter a. Solution to the rate equation gives

$$\Gamma = \Gamma_{eq}\left[1 - e^{-k_1 a(1+k)t}\right], \tag{3.155}$$

[87] Eastoe, J., Dalton, J. S., Adv. Colloid Interface Sci., 85, 103 (2000).

[88] Daniel, R. C., and Berg, J. C., J. Colloid Interface Sci., 260, 244 (2003).

where it is to be recalled that $k = C_\infty/a$. $\sigma(t)$ is then given by the Frumkin equation. While adjustment of the rate parameter k_1 can sometimes produce better fits to the shape of the $\sigma(t)$ curve than adjustment of the diffusivity in the Ward and Tordai model, *a priori* or independent determination of k_1 is in general not possible.

K. Insoluble (Langmuir) monolayers

1. *Formation of monolayers by spontaneous spreading*

Insoluble monomolecular films are formed at the air-water interface by water-insoluble compounds that spontaneously spread, *i.e.*, have a positive spreading coefficient,[89] as defined in Chap. 2:

$$S_{o/w} = \sigma_w - (\sigma_o + \sigma_{o/w}) > 0. \tag{3.156}$$

The spreading coefficient for any liquid at any fluid-fluid or fluid-solid interface may be computed in the same way. It is thus possible to predict "immiscible displacement" at a solid-liquid interface, if the solid-vapor interfacial energy is known. If $S_{o/w}$ is positive, spreading should occur, and if negative, it should not. When spreading *does* occur, it proceeds until the entire available surface is covered with monolayer, with the excess material existing as unspread bulk lenses or solid particles. Under such conditions, the surface tension reduction achieved is the equilibrium spreading pressure, π, of the material at the given interface.

When mineral oil (*e.g.*, Marcol-70) is deposited at the water/air interface, one may compute the spreading coefficient from the following data: $\sigma_w = 72.8$, $\sigma_o = 31.0$, and $\sigma_{o/w} = 50.0$ mN/m, respectively, yielding: $S_{o/w} = 72.8 - (31.0 + 50.0) = -8.2$ mN/m. Thus one would not expect spreading, and such behavior is typical of all the higher hydrocarbons at the water/air interface. A lower hydrocarbon, such as benzene, is an interesting case to examine. For this case, $S_{b/w} = 72.8 - (28.9 + 35.0) = +8.9$ mN/m, a positive value, suggesting that spreading should occur. A demonstration, using talc particles at the surface to render the spreading visible, shows that spreading does indeed occur, but once the film is formed, it quickly retreats backward into lenses. The explanation for such behavior, typical of the light hydrocarbons at the water/air interface, is that when the oil and water equilibrate, there is a slight amount of oil dissolved in the upper layer of water in contact with the oil, and a small amount of water dissolved in the oil. The former effect in particular, changes the equilibrium tensions involved. At equilibrium between benzene and water, $\sigma_{w(b)} = 62.2$ mN/m and $\sigma_{b(w)} = 28.8$ mN/m. The interfacial tension in this case remains unchanged. The final *equilibrium* spreading coefficient is thus $62.2 - (28.8 + 35.0) = -1.6$ mN/m. We thus distinguish between *initial* and *final* spreading coefficients,

[89] Gaines, G. L., Jr., **Insoluble Monolayers at Liquid-Gas Interfaces**, pp. 136ff, Interscience, New York, 1966.

and while a number of materials may have positive initial spreading coefficients at the water/air interface, only those we have recognized as surface active agents have positive *final* spreading coefficients. For example, the data for oleic acid yield $S_{o/w} = 72.8 - (32.5 + 15.7) = 24.6$ mN/m.

Solids as well as liquids may be observed to spread at interfaces. Camphor, for example, is surface active due the carbonyl group in its structure, as shown in Fig. 3-38. It is a crystalline material with slightly

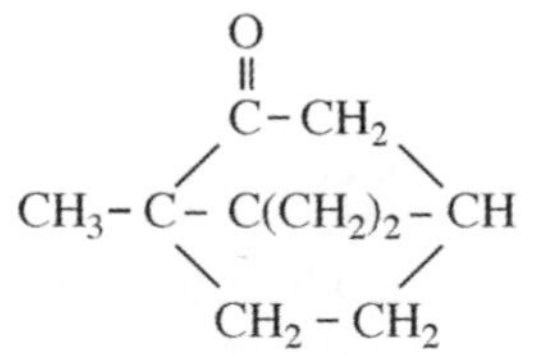

$$
\begin{array}{c}
O \\
\parallel \\
C-CH_2 \\
\diagup \qquad \diagdown \\
CH_3-C- C(CH_2)_2-CH \\
\diagdown \qquad \diagup \\
CH_2 - CH_2
\end{array}
$$

Fig. 3-38: Molecular structure of camphor.

different values of the spreading coefficient from the different crystal faces. The particles are thus observed to spin while spreading, leading to the well-known "camphor dance."[90] The average spreading coefficient for camphor is 16 mN/m. Thus when oleic acid is added to a surface upon which camphor is spreading, its larger spreading coefficient suppresses the spreading, and brings the "dance" to a halt.

The kinetics of spreading of *liquid* surfactants is governed primarily by the viscous drag of the water,[91] and the hydrodynamics of the event is quite complex.[92] Spreading from solids may be governed, however, by a much slower rate process at the solid-fluid interline. Stearic acid at room temperature, for example, has a large positive spreading pressure at the water/air interface, but shows little spreading over a period of several days.

2. *Hydrodynamic consequences of monolayers: Gibbs elasticity*

The presence of an insoluble surfactant monolayer at a fluid interface has a profound influence on the properties of that interface. Many of the changes are hydrodynamic in nature and are dealt with in more detail in Chap. 10. Simple laboratory demonstrations reveal that a clean water/air interface is quite fragile with respect to mechanical, thermal or chemical disruption. Gently blowing on a talc-covered surface shows that the particles may be moved around quite easily in this way. When a monolayer is present, however, the particles are rigidly held in place. Similarly, if a clean surface is heated locally, the surface is seen to dilate itself at that point, and if an adsorbing vapor such as acetone (decreasing surface tension) is brought near the surface, it is also seen to dilate. The presence of a monolayer suppresses such dilation in both cases. The interface with an insoluble monolayer acts

[90] Lord Rayleigh (J. W. Strutt), *Proc. Roy. Soc.*, **47**, 364 (1890).

[91] Di Pietro, N. D., Huh, C., and Cox, R. G., *J. Fluid Mech.*, **84**, 529 (1978);
     Foda, M. A., and Cox, R. G., *J. Fluid Mech.*, **101**, 33 (1980).

[92] Camp, D. W., and Berg, J. C., *J. Fluid Mech.*, **184**, 445 (1987).

like an elastic membrane resisting distortion, and the mechanism for such action is shown in Fig. 3-39. A local dilation sweeps away surfactant

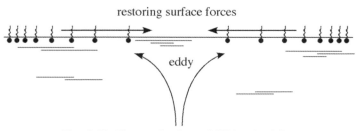

Fig. 3-39: The mechanism of Gibbs elasticity.

leading to a strong gradient in surface tension tending to resist the dilation. This is the phenomenon of *Gibbs elasticity* to be described in more detail in Chap. 10. It is responsible for the legendary ability of oil "to calm troubled waters," as mentioned in ancient times in the writings of Roman historians Pliny the Elder and Plutarch,[93] and exemplified in the photograph of the mirror-like surface of Loch Laggan in Fig. 1-4. It was also the subject of a report by Benjamin Franklin to the Royal Society in 1774 documenting experiments in which he deposited a teaspoon of olive on the pond at Clapham Common in London, and observed it "spread amazingly…making perhaps half an acre as smooth as a looking glass."[94] A schematic of the mechanism for the damping of capillary waves is shown in Fig. 3-40.[95]

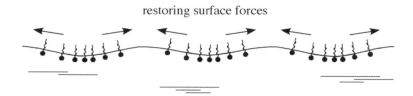

Fig.3-40: Damping of capillary waves by surface active agents.

3. π-A isotherms of insoluble monolayers

Insoluble monolayers are studied using Langmuir troughs, as shown in Fig. 2-42 or Langmuir-Wilhelmy troughs, in Fig. 2-43. The films are generally deposited by means of a micrometer syringe in carefully measured amounts using dilute solutions of the surfactant in a volatile *spreading solvent*. The latter quickly evaporates leaving the monolayer behind. The

[93] Gaines, G. L., Jr., **Insoluble Monolayers at Liquid-Gas Interfaces**, p. 1, Interscience, New York, 1966.

[94] **The Complete Works of Benjamin Franklin**, Vol. V, p. 253, J. Bigelow (Ed.), G. P. Putnam's Sons, New York, 1887.

[95] Levich, V. G., **Physicochemical Hydrodynamics**, pp. 609 ff, Prentice-hall, Englewood Cliffs, NJ, 1962.

film is then manipulated (compressed or expanded) by means of movable barriers that form a leak-proof seal between the film-covered and clean portions of the surface. Results of compression-expansion experiments are generally reported as plots of surface pressure, π (surface tension reduction) vs. specific area of the surfactant, A, i.e., area/molecule. An example of π-A behavior (on a highly nonlinear set of axes) is shown schematically in Fig. 3-41. This is a plot of the surface equation of state for the monolayer. The features of the π-A curve (or a family of such curves for different temperatures) for a given surfactant reveal a wealth of information concerning the size, shape and molecular interactions between the molecules that constitute the monolayer. These were broadly exploited by Langmuir and coworkers,[96] and played an important role in Langmuir's attainment of the 1932 Nobel Prize for Chemistry.

Imagine an experiment in which an initially sparsely populated monolayer is gradually compressed. At the highest degrees of expansion, a two-dimensional ideal gas equation is observed:

$$\pi A = kT, \tag{3.157}$$

where A is the specific area (usually expressed in Å^2/molecule). This has some practical utility in providing a method for determining the molecular weight of proteins, which spread as monolayers at the water/air interface and exhibit an accessible range of 2-d ideal gas behavior. Compression to about 1000 Å^2/molecule leads to a first-order phase change, producing islands of a two-dimensional "liquid" within the gaseous film. This occurs at a surface pressure of the order of one mN/m, a two-dimensional "vapor pressure," π^s.

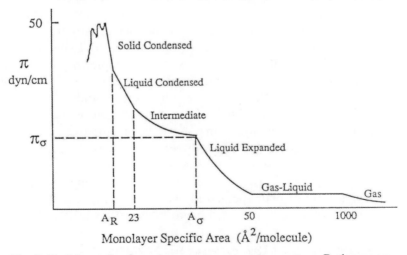

Fig. 3-41: Schematic of a π-A curve at constant temperature. Both axes are greatly expanded near the origin to reveal the phase behavior occurring there.

[96] **The Collected Works of Irving Langmuir**, C. G. Suits and H. E. Way, Eds., Vol. 9: Surface Phenomena, Part 2: Monomolecular Films, Pergamon, New York, 1961.

The temperature dependence of π^s, which requires very accurate data, can be subjected to analysis using the Clausius-Clapeyron equation, to give the enthalpy of surface vaporization:

$$\Delta h^{\text{s-vap}} = RT^2 \left(\frac{d\pi^{\text{s-vap}}}{dT} \right),$$ (3.158)

and the latter yields information on the energetics of tail-tail interactions in the surface liquid phase. The surface pressure remains at π^s until a specific area of approximately 50 Å^2/molecule is reached, when it begins to rise sharply (called "liftoff"). At this point, the monolayer is single-phase and liquid-like, but of greater compressibility than a corresponding three-dimensional liquid and is referred to as a "liquid expanded" phase. Further compression often (but not always) leads to a second-order phase transition corresponding to the onset of the formation of surface aggregates ("surface micelles") of greater density, and noticeable in the $\pi-A$ trace as a characteristic "knee." The surface micelles become more numerous through what is called the "intermediate phase," until the entire film is comprised of such aggregates, leading to the "liquid condensed" phase. Further compression leads to a two-dimensional solid phase as the last lateral water of hydration is squeezed out of the film, producing a highly ordered structure. Eventually, the film collapses (fractures) in a macroscopic failure, sometimes visible in what Langmuir called "crumple patterns." In reality, a more subtle form of collapse generally begins at a slow rate at a much lower degree of compression. This collapse is in the form of the nucleation and growth of bulk phase nuclei of the surfactant.[97] Frozen surface biopsies of Langmuir monolayers have been examined by "cryo-electron microscopy."[98] The foregoing sequence of monolayer structures and phase states as surface pressure is increased was postulated by Langmuir but later buttressed by independent measurements. For example, the use of an air-ionizing electrode in combination with a reference electrode in the subphase measured surface electrical potential, which in turn could be related to the orientation of the molecular dipoles, and thus the closeness of the head group packing. Some techniques that have been employed to detect the orientation of hydrocarbon tails, two-dimensional phase structures and the coexistence of different surface phases include ellipsometry,[99] Brewster angle microscopy (BAM),[100] and fluorescence microscopy.[101] In ellipsometry, a beam of plane-polarized, monochromatic laser light is shone on the surface, producing a reflected beam that is elliptically polarized, *i.e.*,

[97] Smith, R. D., and Berg, J. C., *J. Colloid Interface Sci.*, **74**, 273 (1980).

[98] Berg, J. C., *Proc. Work. Interfacial Phen.*, Seattle, pp. 89-107, NSF/RA-790442, PB80-201551 (1979).

[99] Thompkins, H. G., **A User's Guide to Ellipsometry**, Academic Press, Boston, 1993.

[100] Vollhardt, D., *Adv. Colloid Interface Sci.*, **64**, 143 (1996).

[101] Knobler, C. M., *Science*, **249**, 870 (1990).

has both horizontal and vertical components of polarization. For a given angle of incidence, the state of polarization of the reflected beam depends on the both the refractive index and the effective thickness of the film. It is thus possible to distinguish clean surface area (which produces no ellipticity) from both sparsely covered surfaces and densely packed monolayers. In Brewster angle microscopy the beam is aimed at precisely Brewster's angle ($\approx$ 53°), producing very high contrast in the polarization between regions of condensed and expanded monolayer and therefore yields detailed images of the domain structure. In fluorescence microscopy a small amount of a fluorescent tag, miscible with the monolayer, is incorporated into the film, providing contrast between condensed and expanded regions.

A family of π-A curves for different temperatures shows remarkable similarity to three-dimensional phase behavior, as shown schematically in Fig. 3-42. For example, two-dimensional critical points may be identified for many insoluble monolayers. This figure makes it evident that not all of the phase states pictured in Fig. 3-41 are observed for all Langmuir monolayers. In particular for the fatty acids, the intermediate phase state and the liquid expanded phase eventually disappear as the chain length is increased or the temperature is decreased. In these cases, the surface pressure "lifts off" directly to the liquid condensed state upon compression. On the other hand, when the temperature is raised to a sufficient level, the film is in the gaseous (supercritical) state all the way to collapse.

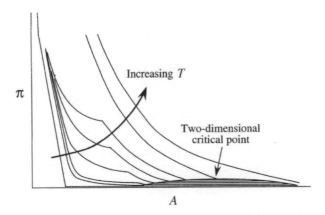

Fig. 3-42: Schematic of π-A isotherms for fatty acid monolayers over a range of temperatures.

One application of insoluble surfactant monolayers is illustrated in Fig. 3-43. When compressed to the point where the molecules are tightly packed together, the measured specific area may be used to determine the size of the molecule by extrapolating the solid-condensed branch of the curve to zero surface pressure. Langmuir used such measurements on straight-chain aliphatic surfactants to obtain the cross-sectional area of a

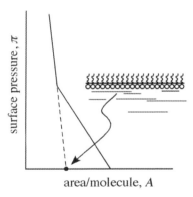

Fig. 3-43: Determination of molecular cross-sectional area from π-A measurements.

hydrocarbon chain.[102] His results were within 15% of the results obtained years later using X-ray diffraction. In the 1920's and 30's, such measurements were used to settle disputes concerning the structure of a variety of molecules, as illustrated in Fig. 3-44. Two suggestions for the structure of cholestanol had been proposed, each with a different implication for the molecular area. Langmuir trough measurements led unambiguously to a decision in favor of the second structure shown.

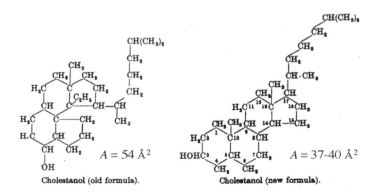

Fig. 3-44: Two competing formulas for cholestanol, each implying a different specific area. Molecular diagrams from [Adam, N. K., **The Physics and Chemistry of Surfaces**, pp. 79-80, Dover, New York (1968).]

4. *Langmuir-Blodgett films*

Another important method for studying insoluble monolayers was pioneered by Langmuir, in collaboration with his coworker, Katherine Blodgett.[103] It is shown in Fig. 3-45 and consists of transferring the monolayer in a highly compressed two-dimensional phase state from the liquid surface to the surface of a solid. This is referred to as the Langmuir-

[102] Langmuir, I., *J. Amer. Chem. Soc.*, **39**, 1848 (1917).

[103] Langmuir, I., and Blodgett, K. B., *Phys. Rev.*, **51**, 964 (1937).

Blodgett (or LB) technique, and by repeated dippings may be used to produce films of up to a thousand monolayers or more. The successive

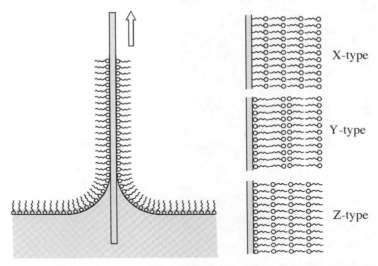

Fig. 3-45: The Langmuir-Blodgett technique. Different types of multiple layers may be produced by successive immersion and emersion of the plate.

layers may be built up in different ways, producing multilayers of different properties. In what is called an X-type multilayer, the original deposition and all subsequent depositions are in the tail-down configuration. In a Y-type multilayer, all monolayers are alternately head-down and tail down, and in the Z-type, the monolayers all in the head-down configuration. While a primary motivation of its inventors was the production of anti-reflective coatings (since the thickness could be so carefully controlled), it later proved invaluable for studying monolayer structure. Monolayers on a *solid* substrate may be treated and examined in the electron microscope. An example, showing a two-phase region of the stearic acid monolayer, is shown in Fig. 3-46. (Note that the condensed-phase islands are perfect circles, suggesting the existence of an "edge tension.") LB films are also amenable to examination by an array of different scanning probe microscopy (SPM) techniques, described in more detail in Chap. 4. The use of the LB technique has aroused interest as a means for producing new materials with unique optical properties (*e.g.*, secondary harmonic generators) that may lead to the development of new types of optical sensors or switches.[104] The most common monolayers used for LB deposition are those of fatty acids and their salts, and a review of these systems, including methods for their preparation and characterization has been given by Peng, *et al.*[105]

---

[104] Möbius, D., Ed., **Langmuir-Blodgett Films 3**, Vols. 1 and 2, Elsevier, London, 1988.

[105] Peng, J. B., Barnes, G. T., and Gentle, I. R., *Adv. Colloid Interface Sci.*, **91**, 163 (2001).

Fig. 3-46: Circular islands of condensed phase, one monolayer thick, in a sea of the uncondensed monolayer phase. Electron micrograph of a Langmuir-Blodgett monolayer of stearic acid at 10 mN/m, originally spread on water. From [Ries, H. E., Jr., and Kimball, W. A., *Nature*, **181**, 901 (1958).]

5. *Transport properties of monolayers*

One of the early applications of Langmuir monolayers was to the suppression of evaporation.[106] Close-packed monolayers of hexadecanol or mixtures of hexadecanol and octadecanol have produced as much as a 90% reduction in the rate of water evaporation in the laboratory and as high as 50% in the field. While a number of practical problems must be overcome, such as the need for continuing repair of damage caused by wind and wave action, the use of these materials is seen as a viable strategy for water conservation. The reduction in evaporation rate is attributed to the sieving effect of the film, *i.e.*, the resistance of the close-packed hydrocarbon layer of the surfactant to penetration by water molecules. Thus there was found a direct relationship between the hydrophobic chain length and the evaporation resistance. Anything leading to a more open structure, such as branching, double bonds or halogen substitution in the chains, destroys the evaporation resistance. A breakthrough in laboratory studies of the phenomenon occurred when it was realized that certain compounds (in particular benzene) used as spreading solvents were retained in the films to a sufficient extent to produce "holes" that allowed for escape of evaporating water.

The intrinsic surface rheology of monolayers may be envisioned in terms of surface constitutive equations relating bending, dilational and shear

---

[106] La Mer, V. K. (Ed.), **Retardation of Evaporation by Monolayers**, Academic Press, New York, 1962.

deformations of the surface to the stresses that induce them. Dilational properties, such as surface dilational viscosity and elasticity, are difficult to separate from the effects of surface compositional changes that inevitably accompany such dilations (or contractions) leading to Gibbs elasticity. Analysis of both transverse and compressional waves produced in Langmuir monolayers, however, has yielded values for their complex surface moduli.[107] The simplest surface rheological parameter to envision is the surface viscosity, μ_s, which may be defined for a one-dimensional surface shear using a two-dimensional form of Newton's Law of viscosity:

$$\tau_{yx}^{\sigma} = -\mu_s \frac{dv_x^{\sigma}}{dy}, \tag{3.159}$$

where x and y are coordinates drawn in the surface, v_x^{σ} is the velocity in the surface in the x-direction (a function of y), and τ_{yx}^{σ} is the tangential force per unit length on a line in the surface perpendicular to the y-axis. It is evident that the units of μ_s will be mass/time (compared with bulk viscosity units of mass/length·time). Attempts to measure the surface viscosity are generally confounded with the need to separate out the effect of the underlying subphase, to which the monolayer is attached by virtue of the hydrodynamic no-slip condition. A successful device, however, is the deep channel viscometer of Burton and Mannheimer,[108] in which an annular surface is produced by dipping a pair of concentric circular cylindrical surfaces into a circular dish that is rotated while the annular walls are held stationary. A parabolic flow is established in the channel, as shown in Fig. 3-47. The

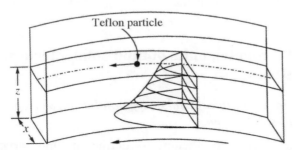

Fig. 3-47: Flow produced in the Burton-Mannheimer deep channel surface viscometer.

measurement consists of tracking the circuit time for a small Teflon particle deposited at the center of the channel. A slight concave-upward shape to the meniscus guarantees that the particle stays in the center. The circuit time measured when a monolayer is present is compared to that obtained when the surface is clean, and the ratio permits deduction of the surface viscosity. Values obtained for close-packed monolayers are in the range of 10^{-3}–1

---

[107] Lucassen, J., and van den Tempel, M., *Adv. Colloid Interface Sci.*, **41**, 491 (1972).

[108] Burton, R. A., and Mannheimer, R. J., **Adv. Chem. Ser. 3**, p. 315, 1967.

surface Poise (g/s). To obtain the same result with a bulk layer of thickness ≈ 1 nm (typical for a surfactant monolayer) would require a viscosity of 10^4– 10^7 Poise, approximately the viscosity of butter at room temperature.

Surface diffusion in monolayers has been measured using radioactive tracers.[109] The surface of a Langmuir trough was divided by a flexible Teflon thread into two parts, one supporting a monolayer of myristic acid, and the other a monolayer of the same compound, but containing a proportion of ^{14}C-tagged material. The surface pressure on the pure monolayer side was adjusted until the disposition of the flexible dividing thread became slightly S-shaped, indicating that it was the same on both sides. The thread was then removed, allowing surface inter-diffusion. The time evolution of the surface concentration profile of the tagged surfactant was monitored with a Geiger-Mueller tube, and the results analyzed using a two-dimensional form of Fick's Law modified to account for dissolution and diffusion in the underlying bulk phase, *viz.*,

$$\frac{\partial \Gamma}{\partial t} = D_s \frac{\partial^2 \Gamma}{\partial x^2} - k_{-1}\Gamma + k_1(hC), \qquad (3.160)$$

where D_s is the surface diffusivity, k_1 and k_{-1} are the rate constants for adsorption and desorption of the surfactant, h is the depth of the liquid and C is the bulk concentration of surfactant. Independent measurements provided values for k_1 and k_{-1}, and the above equation was solved simultaneously with the corresponding equation for bulk transport. Values for the surface diffusivity in the intermediate surface phase state were found comparable to those for bulk liquids, but in the liquid-expanded state, about an order of magnitude larger.

L. The thermodynamics of fluid-solid interfacial systems revisited

1. *The concept of interfacial energy and its measurement in fluid-solid systems*

It is useful first to attempt to define an appropriate surface or interfacial energy to be associated with a given interfacial system for purposes of comparing different systems to one another and for formulating expressions for the driving forces for various processes involving interfacial systems. Thus the definition and experimental determination of the interfacial energy of various interfacial systems is one of the major goals of the thermodynamic analysis of such systems. It is evident that more than one "interfacial energy" may be defined. First, it is generally assumed that the

[109] Sakata, E. K., and Berg, J. C., *Ind. & Eng. Chem. Fund.*, **8**, 570 (1969); Chung, S. T., and Berg, J. C., *J. Kor. Chem. Eng.*, **10**, 189, 231 (1972).

terminology refers to interfacial *free* energy, as opposed to interfacial internal energy or enthalpy. One measure of interfacial energy, particularly relevant to capillary systems, is the boundary tension, σ, defining the Helmholtz free energy change associated with unit interfacial area extension under conditions of constant temperature and total volume, and internal equilibrium. For capillary systems, the boundary tension is readily measured, but the "boundary tensions" of solid-fluid interfaces are not so easily accessible by direct measurement. It is important first to recognize that σ is a surface free energy *derivative* rather than the surface energy itself. A surface free energy (per unit area) may be defined with reference to the Gibbs dividing surface as a surface excess quantity, *viz.*

$$\left(\frac{F^{\sigma}}{A}\right) \equiv f^{\sigma} = \frac{1}{A}\left[F - F' - F''\right].\tag{3.161}$$

The above expression can be given for an arbitrary location of the dividing surface in terms of the component adsorptions and chemical potentials, *viz.*[110]:

$$f^{\sigma} = \sigma + \sum_i \Gamma_i \mu_i.\tag{3.162}$$

f^{σ} is clearly a property of the dividing surface, but if it is computed for the dividing surface location such that the adsorption of the solvent 1 is zero, one obtains the *relative* surface free energy:

$$f_1^{\sigma} = \sigma_0^1 + \sum_{i=2} \Gamma_{i,1} \mu_i,\tag{3.163}$$

which is independent of dividing surface location, but of course dependent on the standard state values chosen for evaluating the component chemical potentials. It is clear that the relative surface free energy is equal to the boundary tension for a pure-solvent (component 1) system or for a system in which solute adsorption is zero, designated as σ_0^1, but not otherwise. If the component standard states are chosen appropriately, one may identify f_1^{σ} with σ, so that for capillary systems, the interfacial free energy is identifiable with the measurable boundary tension. The same identification can be made for fluid-solid systems, but determination of the "interfacial tension" for fluid-solid interfacial systems is more challenging, as discussed below.

In examining the various approaches to determining fluid-solid interfacial energies, it is first assumed that the system is immiscible, and it is useful to distinguish three types of fluid-solid interfacial systems:

[110] Defay, R., Prigogine, I., Bellemans, A., and Everett, D. H., **Surface Tension and Adsorption**, p. 288, Longmans, London, 1966.

1) *Pristine* surface systems refer to surfaces formed and kept *in vacuo* (or if the solid has a finite volatility, in contact with the equilibrium vapor). The surface energy for pristine systems is designated as:

$$\left(f^{\sigma}\right)_{\text{prist}} = \sigma_0. \tag{3.164}$$

2) *Clean* surface systems are those in which the surface is formed and kept in a pure "solvent medium", which may be a pure gas or a pure liquid, designated as component 1. One may distinguish two types of "clean" surfaces, *viz.* "pure" and "modified" surfaces. In the case of pure clean systems, component 1 is capable of only physical interaction (*i.e.*, physical adsorption) with the solid. The surface energy for a pure clean system in the presence of fluid 1 is designated as $\sigma_0{}^1$, and is related to the pristine surface energy by

$$\left(f^{\sigma}\right)_{\text{clean}} = \sigma_0^1 = \sigma_0 - \pi_{\text{eq}}^1, \tag{3.165}$$

where π_{eq}^1 is the "equilibrium spreading pressure" of pure liquid 1 physisorbed on the solid surface. It may be obtained by integrating the gas-phase adsorption isotherm for the physical adsorption of component 1 from zero to its saturation concentration (or partial pressure), in accord with the Gibbs Adsorption Equation, *i.e.*,

$$\pi_{\text{eq}}^1 = \int_0^{C_1^{\text{sat}}} \Gamma_1(C_1')d\ln C_1', \ \ (\text{or} \int_0^{p_1^{\text{sat}}} \Gamma_1(p_1')d\ln p_1'). \tag{3.166}$$

In the case of modified clean surface systems, some of component 1 is capable of *chemisorbing* to the solid surface, changing its surface chemistry and hence energy. Most high-energy surfaces sustain such chemisorption. For example, most metal surfaces chemisorb oxygen to form an oxide layer, and most mineral oxide surfaces chemisorb water to form a layer of hydroxyl groups. If the pristine surface is modified by chemisorption either from the adjoining fluid phase 1 (or from some previous contact with a chemisorbing component), one may write:

$$\left(f^{\sigma}\right)_{\text{mod. clean}} = \sigma_0^1 = \sigma_0^{\text{mod}} - \pi_{\text{eq}}^1 = (\sigma_0 - \pi_{\text{chem}}) - \pi_{\text{eq}}^1, \tag{3.167}$$

where σ_0^{mod} is the energy of the surface modified by chemisorption, and π_{chem} is the surface energy reduction caused by the chemisorption. This reduction is often a significant fraction of the original pristine surface energy. The modified clean surface is then capable of sustaining subsequent physical adsorption of component 1, as indicated above.

3) Finally, *practical* surface systems refer to the case where the solid is or has been in contact with "practical," *i.e.*, dirty, multicomponent fluid environments such as gas mixtures or solutions containing physically adsorbable components. The surface free energy for practical surface

systems in which the fluid-phase portion consists of a diluent gas or solvent component 1 and one or more adsorbable components is given by:

$$\left(f^\sigma\right)_{pract} \equiv \sigma_{pract} = \sigma_0^1 - \pi = \sigma_0^1 - \sum_{i=2} \int_0^{C_i} \Gamma_{i,l} d\ln C_i', \qquad (3.168)$$

where π represents the surface pressure resulting from the combined adsorption of all of the solute species.

Pristine surface energies can be measured if fresh solid surface can be created *in vacuo* without surface stretching. This has been accomplished in two ways, as pictured schematically Fig. 3-48. The first, shown in Fig. 3-48(a), is that of cleaving a brittle solid and measuring the total reversible work required to open a unit area of the crack. Subtracting from this the elastic strain energy yields the energy involved in creating new surface. The pristine surface energy is given by:

$$\sigma_0 = \frac{1}{2}\left[\frac{\text{total work}}{\text{crack area}} - \frac{\text{strain energy}}{\text{crack area}}\right]. \qquad (3.169)$$

This method has been applied to ionic and covalent crystalline materials, particularly at cryogenic temperatures at which any ductility in the solid specimen is frozen out. It has also been applied to so-called van der Waals solids under these conditions. These are solids formed of simple molecules held together only by van der Waals forces. For ductile materials, like metals and polymers, a second technique, pictured in Fig. 3-48(b), and known as the "zero-creep" method has been developed. A series of wires of fixed radius r are hung with a range of different weights and brought to a temperature just below the melting point. Those with weights too large will become distended, while those with the smallest weights will contract

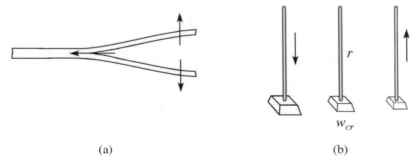

(a)   (b)

Fig. 3-48: Methods for measuring the surface free energy of solids. (a) The fracture method for brittle solids. (b) The zero-creep method for ductile solids.

upward due to the effect of surface tension forces. For a particular critical weight, w_{cr}, there will be zero creep, and this is used to determine the surface tension in accord with[111]:

$$\sigma_0 = \frac{w_{cr}}{\pi r}.$$

(3.170)

Since at "zero creep" one is not stretching the surface, it is the surface free energy that is being measured rather than the "stretching tension" discussed earlier. Clean surface energies may also be obtained by the cleavage and zero-creep methods when the processes are carried out in either a pure gas or liquid. Some qualitative ranges of results of measurements of this type are summarized in Table 3-9. It is reassuring that the numbers for the pristine

Table 3-9: Comparison of ranges of surface energy values for "pristine" surfaces (formed and kept *in vacuo*) and "clean" surfaces (formed and kept in air) for various types of materials.

Solid type	Surface energy range (mJ/m^2)	
	In vacuo (σ_0)	In air (σ_0^1)
Van der Waals	20-60	20-60
Polymers	20-60	20-60
Ionic crystals	100-1000	60-300
Metals	500-2000	60-300
Covalent crystals	3000-9000	300-600

surfaces are in reasonably good accord with theoretical calculations for these surface energies.[112] Such calculations are easier to carry out than the experiments. The large differences in surface energy between those formed *in vacuo* and those formed in air are due to surface modification by chemisorption, particularly of oxygen, onto the surface and only negligibly by subsequent physical adsorption. Another method for the determination of a clean surface energy σ_{SL} of a sparingly soluble solid S in a liquid L is afforded by the Kelvin effect. If the solubility in the form of particles of known radius a is compared with its solubility in macroscopic form, the Kelvin equation (analogous to that for vapor pressure in Chap. 2, and discussed further with regard to the solubility of solids in Chap. 5) yields:

$$\sigma_{SL} = \frac{aRT}{2v_S} \ln \frac{(C_S)_a^{\text{sat}}}{(C_S)_\infty^{\text{sat}}},$$

(3.171)

where a is the radius of the solid particle and v_s is its molar volume.

---

[111] Tabor, D., **Gases, Liquids and Solids and other States of Matter**, 3rd Ed., p. 418, Cambridge Univ. Press, Cambridge, UK, 1996.

[112] Tabor, D., *ibid.*, p. 164.

Calorimetric measurement of the net heat absorbed when particles initially in gas G are dissolved into liquid L yield the difference between the actual heat of solution and the heat evolved due to the destruction of the SG interface. If a mass m of solid of specific area Σ is dissolved, the measured amount of heat absorbed is given by

$$Q_{net} = m\hat{\lambda}_{soln} - m\Sigma u_{SG}, \qquad (3.172)$$

where $\hat{\lambda}_{soln}$ is the heat of solution per unit mass, and $\dot{u}_{SG}$ is the internal energy (per unit area) of the solid-gas interface. It may be related to the free energy by recalling that for an isothermal, constant-volume process

$$u_{SL} \equiv \left(\frac{\partial U_{SL}}{\partial A_{SL}}\right)_{T,V} = \left[\sigma_{SL} - T\left(\frac{d\sigma_{SL}}{dT}\right)\right], \text{ so that} \qquad (3.173)$$

$$Q_{net} = m\hat{\lambda}_{soln} - m\Sigma u_{SG} = m\hat{\lambda}_{soln} - m\Sigma\left[\sigma_{SL} - T\left(\frac{d\sigma_{SL}}{dT}\right)\right]. \qquad (3.174)$$

Calorimetric measurement of the heat evolved when an immiscible solid is immersed in a liquid yields information on the difference between the energy of the SG interface destroyed and the SL interface created:

$$\frac{Q_{imm}}{m\Sigma} = u_{imm} = u_{SL} - u_{SG} = (\sigma_{SL} - \sigma_{SG}) - T\frac{d(\sigma_{SL} - \sigma_{SG})}{dT}. \qquad (3.175)$$

The above are statements of the Gibbs-Helmholtz Law relating energy quantities to the corresponding free energies and their temperature derivatives.

The other common methods for determining surface or interfacial energies are based on wetting, adhesion or wicking measurements, or vapor adsorption measurements, as discussed in Chap. 4. They generally yield information concerning "clean" interfaces (if pure fluids are used and care is taken) or "practical" interfaces.

2. Adsorption of non-polymeric molecules at the solid-liquid interface

Adsorption at the solid-liquid interface differs in fundamental ways from adsorption at a fluid-fluid or a gas-solid interface, and its description is frequently effected in different ways.[113-114] As discussed earlier in the context of describing the direct measurability of such adsorption, it is usually reasonable to assume an *immiscible interfacial system, i.e.,* that there is no mutual solubility between the solid substrate and the components of the

---

[113] Parfitt, G.D., and Rochester, C.H. (Eds.), **Adsorption from Solution at the Solid/Liquid Interface**, Academic Press, London (1983).

[114] Kipling, J. J., **Adsorption from Solutions of Non-Electrolytes**, Academic Press, London, 1965.

liquid solution. For such systems, the dividing surface location is not ambiguous and may be taken as *precisely* at the outer boundary of the solid, *i.e.*, to separate the atoms of the solid lattice from the liquid. The dividing surface located in this way leads to finite values for both Γ_1 and Γ_2 for a binary solution of solute 2 (*adsorbate*) in solvent 1, and the total occupancy of the surface is constant. For example, if the area occupied by one mole of solute is n times that occupied by a mole of solvent molecules, one would have:

$$\Gamma_2 + n\Gamma_1 \; = \; \Gamma_\infty, \tag{3.176}$$

where Γ_∞ is the total number of (moles of) "sites"/area for adsorption on the adsorbent surface.

The relative adsorption is computed as

$$\Gamma_{2,1} = \Gamma_2 - \Gamma_1\left(\frac{C_2}{C_1}\right), \tag{3.177}$$

and if interest is limited to sufficiently dilute solutions (as is often the case), the relative adsorption reduces to Γ_2. It is clear that in general adsorption at the solid-liquid interface is *competitive* between the solute and solvent for the available adsorbent surface area. (This situation contrasts with the case of adsorption at the gas-solid interface, where adsorbate molecules are assumed to occupy otherwise empty space at the surface.) The competition involved in such adsorption is more complex than that of the solute and solvent for the adsorbent area, as suggested by Fig. 3-49. The adsorbent and

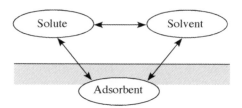

Fig. 3-49: Competition in adsorption at the solid-liquid interface.

solvent may also be thought of as competing for the solute, and the solute and the adsorbent are competing for solvent. The nature of the adsorption that occurs depends on the result of this three-way competition. Considering adsorption is the process whereby a molecule of adsorbate (A) replaces a molecule (or n molecules) of solvent (B) from the adsorbent surface (S), it may be represented in terms of the pseudo chemical reaction:

$$\text{(A)(B)} + \text{(S)(B)} \Longleftrightarrow \text{(A)(S)} + \text{(B)(B)} \tag{3.178}$$

The energy change associated with the process is

$$\Delta\Phi_{\text{ads}} = \Phi_{\text{AS}} + \Phi_{\text{BB}} - \Phi_{\text{AB}} - \Phi_{\text{SB}} \propto -B_{\text{AS}} - B_{\text{BB}} + B_{\text{AB}} + B_{\text{SB}}, \tag{3.179}$$

where B_{AS}, *etc.* refer to the intermolecular interaction coefficients, *cf.* Eq. (2.7). If the interactions are of the dispersion (apolar) type only, these are proportional to the products of the molecular polarizabilities, so that

$$\Delta\Phi_{ads} \propto -\alpha_A\alpha_S - \alpha_B\alpha_B + \alpha_A\alpha_B + \alpha_S\alpha_B = (\alpha_A - \alpha_B)(\alpha_B - \alpha_S). \qquad (3.180)$$

Thus if the relative values of the molecular polarizabilities are such that either

$$\alpha_B < \alpha_A < \alpha_S \text{ or } \alpha_B > \alpha_A > \alpha_S, \qquad (3.181)$$

i.e., the polarizability of the solute A is intermediate to that of the solvent B and the substrate S, $\Delta\Phi_{ads}$ is guaranteed to be negative, and adsorption is thermodynamically favored. Since molecular polarizability is directly proportional to dielectric constant ε (and to the square of the refractive index) the above leads to the useful rule of thumb that adsorption of a solute should occur when the dielectric constant (or the refractive index) of the solute is intermediate to that of the solvent and the adsorbent, an axiom known in the older literature as *Rehbinder's Rule.* The rule applies only when van der Waals forces are dominant, and definitely not when electrostatic effects or acid-base interactions (*e.g.*, H-bonding) are at play. Thus, for example, an acidic solute (like chloroform) in a neutral solvent (like heptane) will adsorb strongly onto a basic adsorbent surface (like calcium carbonate), but not so strongly from a basic solvent like benzene.

If one examines adsorption from a binary solution of miscible components over the whole composition range, as shown in Fig. 3-50, three types of isotherms are observed.[115] In Type I, the solute is positively adsorbed over the whole composition range, while in Type II, the solvent is favored (the solute is negatively adsorbed). In Type III, the adsorption changes sign over the composition range. Another type of isotherm, as

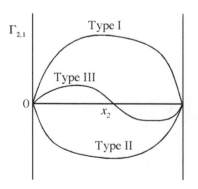

Fig. 3-50: Types of adsorption at the solid-liquid interface.

shown in Fig. 3-51, may be observed for the case in which the adsorbate is a liquid with finite solubility in the solvent. The adsorption often increases sharply as the solubility limit is approached. Another effect may give rise to such behavior when the adsorbent surface is porous. In this case, the steep

[115] Voyutsky, S., **Colloid Chemistry**, p. 156, Mir Publ., Moscow, 1978.

increase in apparent adsorption as the solubility limit is approached may be capillary condensation of the adsorbate liquid into the pores of the adsorbent, as described in Chap. 2.

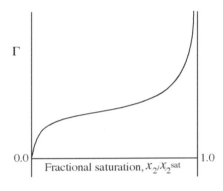

Fig. 3-51: S-shaped adsorption isotherm characteristic of adsorption of a partially miscible adsorbate onto a finely porous adsorbent.

The situations of greatest practical interest are those of positive adsorption from dilute solutions, and are be the focus of the rest of this section. For such cases, particularly with reference to aqueous surfactant solutions, it is useful to classify adsorption at the solid-liquid interface in another way, *viz.*, in terms of the two general mechanisms responsible for it. When adsorption occurs *primarily* because the adsorbing solute (or part of its functionality) would like to escape the solvent (such as a hydrocarbon surfactant seeking to escape water), the driving force is the "hydrophobic effect," and adsorption may be referred to as *amphipathic*. The mechanism of amphipathic adsorption is pictured in Fig. 3-52(a). The hydrophobic (usually hydrocarbon) tails of the surfactant in solution are encased in ice-like water structures. Upon adsorption, these structured water sheaths are released (with the attendant entropy increase), and the hydrophobic moieties of the solute are oriented toward the adsorbent surface. The hydrophobic effect is the same driving force that leads to adsorption of surfactants at the

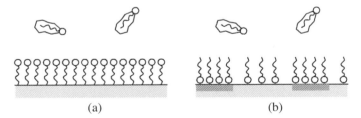

(a)   (b)

Fig. 3-52: Mechanisms of surfactant adsorption from aqueous solutions. (a) Amphipathic adsorption, (b) amphiphilic adsorption.

water/air interface and to the formation of micelles and other fluid microstructures in aqueous media, as discussed earlier.[116] On the other hand,

[116] Tanford, C., **The Hydrophobic Effect: Formation of Micelles and Biological Membranes**, 2nd Ed., Krieger Publ., Malabar, FL, 1991.

if adsorption occurs primarily because of an attraction between the solute and the adsorbent surface, it may be referred to as *amphiphilic*. Its mechanism is suggested in Fig. 3-52(b). These specific attractions may be, for example, electrostatic or may be the result of acid-base interactions (including hydrogen bonding) between the head groups of the solute and functional groups on the solid adsorbent surface. These may be distributed unevenly (often characterized in terms of adsorbent surface energetic heterogeneity) and to monolayer adsorption of uneven density. The orientation of the adsorbate is "head down," exposing the hydrophobic groups to the solution. Adsorption of this type should be regarded as a type of *chemisorption*. Both mechanisms are described in greater detail below, and some examples are sketched in Fig. 3-53.

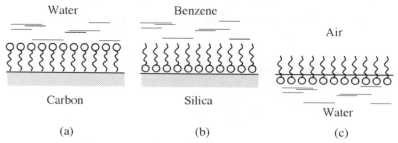

Fig. 3-53: Examples of amphipathic adsorption (a) and (c), and amphiphilic adsorption (b).

Amphipathic adsorption is often well described by the Langmuir adsorption isotherm,

$$\Gamma = \frac{\Gamma_\infty C}{a + C},$$
(3.182)

as shown in Fig. 3-54, which also shows the coordinates used to render such results in the form of a straight line (a "Lineweaver-Burke plot"):

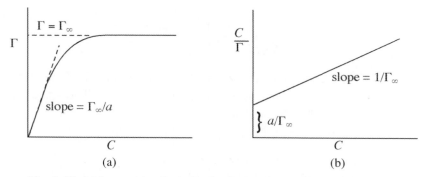

Fig. 3-54: (a) Langmuir adsorption isotherm, characteristic of amphipathic adsorption; (b) Lineweaver-Burke plot.

$$\frac{C}{\Gamma} = \left(\frac{a}{\Gamma_\infty}\right) + \left(\frac{1}{\Gamma_\infty}\right)C, \tag{3.183}$$

for purposes of obtaining the constants from the least squares slope and intercept. Langmuir behavior shows linear adsorption at very low solute concentrations and saturation (assumed to correspond to a close-packed monolayer) as concentration is increased. Once a saturated monolayer is formed, there is little tendency to form a second layer even when solute concentration is increased further. The initial linear slope of the Langmuir isotherm may be interpreted in terms of the standard free energy change of adsorption as follows. For simplicity, assume solute (A) and solvent (B) molecules occupy approximately the same area on the surface. The adsorption process may then be envisioned as a "chemical reaction," in which an A molecule displaces a B molecule from the surface:

$$A^{bulk} + B^\sigma \xrightarrow{\quad K_{ads} \quad} A^\sigma + B^{bulk}. \tag{3.184}$$

Then at equilibrium:

$$K_{ads} = \frac{a_{A^\sigma} a_{B^{bulk}}}{a_{A^{bulk}} a_{B^\sigma}} = \frac{\gamma_A^\sigma \Gamma_A \gamma_B C_B}{\gamma_A C_A \gamma_B^\sigma \Gamma_B}, \tag{3.185}$$

where a_i's are species activities and γ_i's activity coefficients. Noting that $\Gamma_A + \Gamma_B = \Gamma_\infty$, and assuming that the system is sufficiently dilute that all the γ's ≈ 1, and $C_B \approx C_B^{\,0}$, the pure solvent concentration, [=] mole/m^3:

$$K_{ads} = \frac{\Gamma_A C_B^0}{C_A(\Gamma_\infty - \Gamma_A)}, \text{ or} \tag{3.186}$$

$$\Gamma_A = \frac{K_{ads}\Gamma_\infty C_A}{C_B^0 + K_{ads}C_A}, \text{ the Langmuir isotherm.} \tag{3.187}$$

The initial slope ($C_A \to 0$) of the isotherm is:

$$\left(\frac{\Gamma_A}{C_A}\right)_{C_A \to 0} = \left(\frac{\Gamma_\infty}{C_B^0}\right)K_{ads} = \left(\frac{\Gamma_\infty}{C_B^0}\right)\exp\left(-\frac{\Delta G_{ads}^\ominus}{RT}\right), \tag{3.188}$$

where $\Delta G_{ads}^\ominus$ is the standard free energy of adsorption. Thus the steeper the initial slope of the isotherm, the stronger the adsorption. The standard states are conveniently chosen as

for B in solution: pure B, at C_B^0 [=] mole/m^3

for adsorbed B: $\Gamma_B = \Gamma_\infty$ [=] mole/m^2

for A in solution: (Henry's Law solution at $C_A^\ominus = 1$ mole/m^3)

for adsorbed A: (Henry's Law surface solution at $\Gamma_A^\ominus = 1$ mole/m^2)

Amphipathic adsorption is stronger (*i.e.*, the initial slope of the isotherm is steeper) in direct proportion to the hydrocarbon chain length in aliphatic surfactants. Such adsorption thus obeys Traube's Rule, as exemplified in the data of Fig. 3-55 for a homologous series of carboxylic

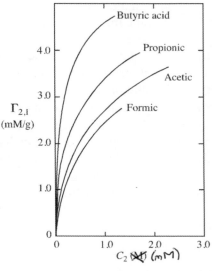

Fig. 3-55: Adsorption of fatty acids onto carbon from aqueous solution, illustrating the applicability of Traube's Rule. For a constant value of Γ, $C(n)/C(n+1) = 2.....3$. Data from [Traube, I., *Annals*, **265**, 27 (1891).]

acids onto carbon. For a constant extent of adsorption, Γ, the required bulk concentration decreases by a constant factor (usually between 2 and 3) for each $-CH_2-$ group added to the hydrophobic chain. When the solute may be regarded as a surfactant, amphipathic adsorption is generally expected for nonionics and for ionic surfactants of charge the same as that of the substrate surface. Anionics are usually amphipathically adsorbed, since most solid surfaces in contact with water bear a negative charge. If micelles form, the *CMC* is usually a concentration slightly less than that corresponding to the knee of the isotherm. For surfactants, the isotherm observed may sometimes exhibit a second (or even a third) plateau, as pictured in Fig. 3-56. This

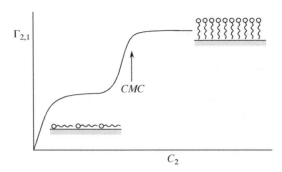

Fig. 3-56: Stepwise isotherm due to adsorbate re-orientation.

behavior is attributed to orientational effects. At low concentrations, the adsorbate molecules are believed to be lying flat on the surface, but as concentration increases, and adsorbate-adsorbate interactions arise, the molecules become more nearly vertically oriented.

Amphiphilic adsorption is often described by a Freundlich isotherm,

$$\Gamma_2 = kC_2^{1/n}, \tag{3.189}$$

where k and n are empirical constants, as shown in Fig. 3-57. It may be plotted on logarithmic coordinates to yield a straight line, as shown. This isotherm is associated with an energetically heterogeneous adsorbent surface. It is steepest at lowest concentrations, since the most energetic sites are covered first, *etc*. The Freundlich isotherm is reproduced when the surface is modeled as consisting of energetically homogeneous patches of exponentially varying adsorption energy, *i.e.*, $\Delta G_{ads}^{\ominus}$, and the adsorption to each "patch" is described by a Langmuir isotherm. Frequently a Freundlich-Langmuir behavior is observed as full monolayer coverage is eventually attained. On the other hand, a second adsorbate layer is often formed (by amphipathic adsorption) on top of the first layer, producing a stepped isotherm similar in shape to that shown in Fig. 3-56, but with a different interpretation. Other and more complex adsorption equilibria may also be observed for amphiphilic adsorption as described below.

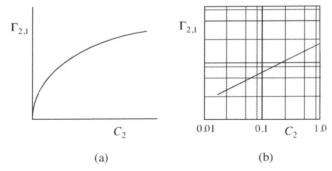

Fig. 3-57: (a) Freundlich adsorption isotherm, characteristic of amphiphilic adsorption; (b) log-log plot of Freundlich isotherm.

Amphiphilic adsorption does not obey Traube's Rule, and in some cases, as shown in Fig. 3-58, may seem to reverse it. This shows data for the adsorption of fatty acids onto silica gel from a toluene solution. In this case the higher members of the series are least adsorbed because the solvent competes more successfully for them. It should be mentioned that this type of adsorption series may also sometimes be observed even for amphipathic adsorption, when the adsorbent is very finely porous. In this case it is the fact that the smaller molecules are able to access the finer pores, and hence more surface area, than the larger molecules.

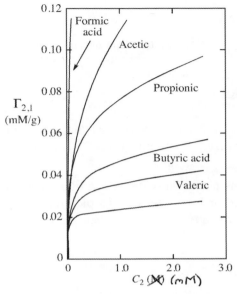

Fig. 3-58: Adsorption of fatty acids onto silica gel from toluene, illustrating what appears to be a "reverse Traube's Rule." Data from [Traube, I., *Annals*, **265**, 27 (1891).]

Amphiphilic adsorption may occur through a variety of mechanisms. Ion pairing and ion exchange are shown in Fig. 3-59. This is the mode of adsorption of cationic surfactants from water onto most solid surfaces (since they are usually negatively charged in contact with water) at low solute concentrations. The adsorption is "head-down" in its orientation and may result in the neutralization (or reversal) of the surface charge. These attributes lead to a number of important applications for cationic surfactants, such as antistats in plastics and textiles, anti-caking agents for powdered or

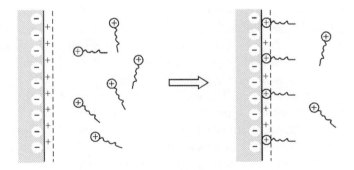

Fig. 3-59: Electrostatic (ion exchange) mechanism of ionic surfactant adsorption.

granular materials, flotation agents, boundary lubricants and corrosion inhibitors. All of these applications rely on the adsorption rendering the surface hydrophobic. At increasing bulk concentration, however, as mentioned above, a second layer of adsorption often occurs as shown in Fig. 3-60. This leads eventually back to a hydrophilic surface. The second layer

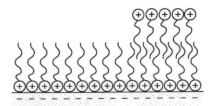

Fig. 3-60: Hemi-micelle formation as
second monolayer adsorbs in amphiphilic
adsorption.

appears to form in patches, which are sometimes called "hemi-micelles" and
may be sites for immobilized solubilization[117] referred to as
adsolubilization.

Other specific mechanisms for amphiphilic adsorption are of the acid-
base type (in either the Brønsted or Lewis senses), as pictured in Fig. 3-61.
This is the mode of adsorption of fatty acids onto silica gel from a toluene
solution, as shown in Fig. 3-58.

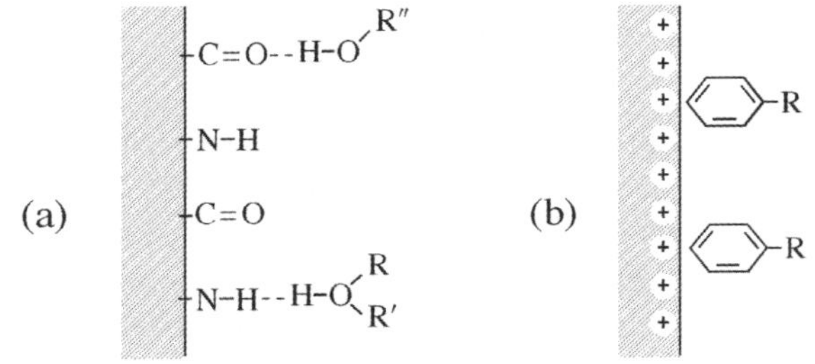

(a)

$$-C=O--H-O^{R''}$$

$-N-H$

$-C=O$

$$-N-H--H-O^{R}_{R'}$$

(b)

Fig. 3-61: Amphiphilic adsorption by acid-base interactions: (a) hydrogen bonding;
(b) polarization of π-electrons.

Another important type of amphiphilic adsorption of surfactants at the
solid-liquid interface is that which leads to the formation of "self-assembled
monolayers," or SAM's. In fact, all adsorption is a form of self-assembly,
but the term SAM refers to highly ordered monolayers that are chemisorbed
to the substrate. Their structure resembles that which is attainable through
Langmuir-Blodgett (LB) dipping, but they are often more robust than such
films. Tri-hydroxy or tri-chloro silane coupling agents (as described in Chap.
4) with long alkane organofunctional groups, chemisorbed to smooth SiO_2
(or other oxide) surfaces provide one example of SAM's. The name has been
more commonly associated, however, with adsorbed layers of alkane thiols,
$HS(CH_2)_nX$ (with $n \geq 7$) adsorbed to smooth, clean surfaces of various
metals, usually gold,[118] but also silver, copper and others ("coinage metals").
The structure is pictured in Fig. 3-62. The sulfur of the thiol coordinates
covalently to the metal surface, in accord with

[117] Sharma, R., Ed., **Surfactant Adsorption and Surface Solubilization**, ACS Symposium
Series 615, ACS, Washington DC, 1995.

[118] Bain, C. D., and Whitesides, G. M., *Science*, **240**, 62 (1988).

$$Au + RSH = AuSR + 1/2 H_2. \tag{3.190}$$

The -CH_2- chains pack in an all-trans conformation, tilted approximately 30° from the perpendicular to the surface. They are most often prepared by dipping the clean metal surface into an ethanol solution of the desired alkanethiol(s) and allowing a few minutes to a few hours for the chains to anneal into the ordered configuration. The end-group X may be of many different types, or mixtures of types, of functional groups. SAM's thus provide a powerful tool for the study of carefully tailored surface chemistry.

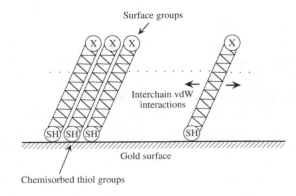

Fig. 3-62: Self-assembled monolayers (SAM's) of long-chain thiols onto a gold surface.

They are well suited to such studies owing not only to their well-ordered and characterizable structure and their ease of preparation, but also to their extraordinary stability. In fact, a terminal =CH_2 group may be functionalized *in situ* under a variety of conditions without disturbing the monolayer. The preparation, properties and experimental characterization of SAM's has been treated in detail by Ulman.[119]

3. *Experimental measurement of small molecule solid-liquid adsorption*

The experimental investigation of adsorption at the solid-liquid interface usually focuses on obtaining the equilibrium adsorption isotherm, and when sufficient solid-liquid interfacial area is available (as is the case when the solid is micro-porous and/or in a finely-divided form), this is conveniently done stoichiometrically by measuring the amount of adsorbate removed (and presumably adsorbed) from a supernatant solution. A variety of analytical techniques may then be used to measure this concentration change. One method for investigating surfactant adsorption from aqueous solutions is that of *soap titration*[120]. A dispersion containing a known amount of dispersoid (with known specific area) is titrated with a "soap

[119] Ulman, A., **An Introduction to Ultrathin Organic Films: from Langmuir-Blodgett to Self-Assembly**, Academic Press, Boston, 1991;
Ulman, A. (Ed.), **Characterization of Organic Thin Films**, Butterworth-Heinemann, Boston, 1995.
[120] Maron, S. H., *et al.*, *J. Colloid Sci.*, **9**, 89, 104, 263, 382 (1954).

solution," where the "soap" is the desired surfactant adsorbate. The surface tension (or sometimes the conductivity) of the dispersion is monitored during the titration until the *CMC* is reached. A comparison of the amount of soap added to that which would have to be added to an equivalent amount of particle-free solvent yields the amount adsorbed at the *CMC*. Knowledge of the surface tension *vs.* concentration behavior of the solution below the *CMC* and its measurement during the titration permits the amount adsorbed at any bulk concentration to be determined using a material balance. If the adsorbed area-per-molecule is known, and the assumption of a fully close-packed monolayer at the *CMC* is made, the technique may be used as a method for determining the specific area of the particles.

Another convenient technique, especially when the specific area of the solid is not large, is frontal analysis solid-liquid chromatography[121]. Adsorption is measured by noting the "break-through" times for solutions of different concentrations in passing through a chromatographic column packed with the adsorbent. First, the dead volume of the packed column is determined by passing pure solvent (or solvent containing a non-adsorbing tracer) through the column. Then, with solvent being pumped through the system at a steady rate, the inlet is switched from solvent to a solution of known concentration at the same flow rate. With the eventual emergence of the adsorbing solute from the column, a detector monitors the outlet concentration until a new steady-state value (corresponding to adsorption equilibrium) is established. The amount of solute adsorbed may then be computed from a material balance. By repeating the procedure for different inlet concentrations, the adsorption isotherm is built up.

A technique closely related to the above is that of *serum replacement*[122]. The particulate dispersion (adsorbent) is confined to a stirred cylindrical cell, one end of which is bounded, for example, by a Nuclepore® or other appropriate membrane with pore size sufficiently small to retain all the particles. After adsorption equilibrium is established between the solid particles and a known amount of adsorbate-containing solution ("serum"), pure solvent is used to flush the adsorbate from the cell. The concentration of adsorbate is monitored in the effluent stream, and a material balance equating the total adsorbate to that remaining in the cell plus that which has been flushed out at any instant gives the adsorption isotherm.

4. Adsorption of polymers at the solid-liquid interface

The adsorption of polymers differs in fundamental ways from the adsorption of lower molecular weight solutes.[123,124] Thermodynamically,

[121] Sharma, S. C., and Fort, T., Jr., *J. Colloid Interface Sci.*, **43**, 36 (1974).

[122] Ahmed, S. M., El-Aasser, M. S., Pauli, G. H., Poehlein, G. W., and Vanderhoff, J. W., *J. Colloid Interface Sci.*, **73**, 388 (1980).

[123] Fleer, G. F., and Lyklema, J., "Adsorption of Polymers," in **Adsorption from Solution at the Solid-Liquid Interface**, Parfitt, G. D., and Rochester, C. H., Eds., Academic Press, New

polymer adsorption occurs when the entropic penalty associated with tethering the polymer to the surface and restricting the number of confirmations it can assume is balanced by a favorable (exothermic) enthalpy of sufficient magnitude associated with establishing segment-surface contact. This is often quite low, *i.e.*, of the order of kT. Thus most polymers adsorb to most surfaces to some extent, and assume a variety of configurations. Linear polymers adsorb in a series of loops, tails and trains, as shown in Fig. 3-63. "Train segments" correspond to those groups in the polymer that are in direct contact with the surface, while "loops" and "tails" dangle out into the solution. When even a single segment of the polymer is in direct contact with the solid, the entire molecule is considered to be

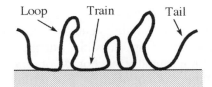

Fig. 3-63: Adsorption of a linear homopolymer to a solid substrate.

adsorbed. The trains generally provide multiple contacts so that polymeric adsorption is often effectively irreversible upon simple dilution. Polymers *can* be desorbed, however, by use of a competitive lower molecular weight adsorbate that "zippers off" the polymer segments. When polymer adsorption is established by increasing the bulk concentration, the observed isotherm is often of the "high affinity" type shown in Fig. 3-64, wherein up to a certain adsorbed amount, all dissolved polymer will be scavenged from the solution, after which the isotherm forms a pseudo-plateau. The plateau level is generally of the order of a few mg/m^2, corresponding to 2-5 equivalent segment monolayers. The bulk composition is usually expressed in terms of polymer mass concentration or volume fraction. Polymer adsorption increases with decreasing solvent quality and with polymer molecular weight in a poor solvent. In a good solvent, adsorption is low and indifferent to polymer molecular weight. Polymer adsorption is slower than that of non-polymeric adsorbates, and if the polymer is broadly polydisperse, there is a slow re-conformation process that occurs as the segments from the higher molecular weight portions gradually displace those from the lower molecular weight molecules that got there first. Desorption, if it occurs at all, proceeds in the reverse order.

For many of the applications in which polymeric adsorption is important (such as steric stabilization of colloids, to be discussed later), one needs to know more than the adsorption isotherm. Structural aspects of the adlayer are also of interest, *e.g.*, the proportion of train segments and the

York, 1983.
[124] Fleer, G. J., Cohen-Stuart, M., Scheutjens, J. M. H. M., Cosgrove, T., and Vincent, B., **Polymers at Interfaces**, Chapman and Hall, London, 1993.

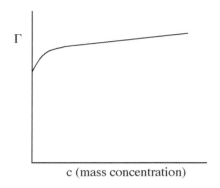

Fig. 3-64: A "high affinity" isotherm, characteristic of polymeric adsorption.

thickness and density of the tails and loops. Of interest in many cases is knowledge of the complete segment density distribution, *i.e.*, the density of segments as a function of distance from the adsorbent surface. Such information may be obtained using small angle neutron scattering (SANS).[125] Many scenarios for adlayer structure are possible, but the most important ones to consider are those of terminally anchored, or grafted, chains and multiply anchored homopolymers or copolymers. For terminally anchored chains at low grafting density, the polymers resemble the Gaussian coils that exist in free solution, with a high segment density near the center dropping to zero density at the edge of the coil. The segment density of the adlayer would rise from a low value at the surface to a maximum at a distance from the surface approximately equal to the radiation of gyration, R_g, of the free coil, and fall off thereafter. As the grafting density is increased, crowding causes the coils to become distended, leading to a more nearly uniform segment density profile. In the extreme case of very high grafting density one obtains a polymer *brush*, consisting entirely of more or less linear tails. Brushes may also be produced by closely spaced side chains on a polymer backbone that adsorbs to the substrate in train segments. Multiply anchored polymers may also produce a fairly uniform segment density, but the segments are a mixture of those occurring in loops and tails. Copolymers consisting blocks that are solvent-incompatible (*anchor blocks*) and solvent-compatible (*buoy blocks*) are very important as steric stabilizers against particle aggregation, to be discussed in Chap. 7. The anchor blocks adsorb as trains separated or terminated by loops or tails, respectively, of buoy blocks. An important simple example is provided by the Pluronic® series of compounds from BASF Wyandotte, which consist of two polyoxyethylene (PEO) chains separated by a block of polyoxypropylene (PPO).[126] The PPO block acts as an anchor in aqueous media, while PEO chains are buoy blocks. An important question for a given adsorbate of this type is what relative proportion (in terms of molar mass) of anchor groups to buoy groups optimizes the adsorbed amount and the adlayer thickness. Both

[125] Cosgrove, T., Crowly, T. L., and Vincent, B., *ACS Symp. Ser.*, **240**, 147 (1984).

[126] Reverse Pluronics are also available, consisting of two PPO blocks separated by a central PEO block.

simulation calculations and data suggest that this occurs at an anchor fraction between 10 and 20%.[127]

A number of theories for modeling polymer adsorption have been developed. Scaling theories[128] have led to expressions proportional to the segment volume fraction dependence on distance from the surface for terminally attached chains, polymer brushes and homopolymers, and Monte Carlo and molecular dynamic simulations have been applied to a wide variety of situations.[129] A number of lattice models have been proposed, first for isolated polymer molecules and later for the general case. These are described and reviewed by Fleer and Lyklema[130] and Fleer, *et al*.[131] The most versatile (powerful) of them is the approach of Scheutjens and Fleer (SF).[132] Polymer segments (from multiple chains) and solvent molecules are assigned to positions in the lattice layer by layer based on their attachments and on extended Flory-Huggins theory.[133] The latter requires the Flory-Huggins χ-parameter for the solution and a corresponding surface parameter χ_S expressing the strength of interaction between a segment of the polymer and the surface. χ_S must be larger than a certain critical values for adsorption to occur. The partition function is computed for each allowable assignment of lattice occupancies, and the arrangement that maximizes it is determined. The theory yields not only the segment density distribution, but gives also the density of segments in any given layer that are in loops or tails, information not at present accessible to measurement. The overall segment density profiles predicted from SF theory are generally found to be in good agreement with SANS data.

Several experimental methods are available for accessing some of the useful descriptors of polymer adlayers short of a full segment density profile. The amount or proportion of train segments (bound fraction), for example, may be determined by monitoring the amount of small-ion adsorbate (counterions) initially present that are ejected into the solution,[134] by spectroscopic methods such as FTIR, by microcalorimetry or solvent

[127] As reported by T. Cosgrove in: Cosgrove, T. (Ed.), **Colloid Science**, pp. 136-139, Blackwell, Oxford, 2005.

[128] de Gennes, P.-G., **Scaling Concepts in Polymer Physics**, Cornell Univ. Press, Ithaca, NY, 1979.

[129] Binder, K., **Monte Carlo and Molecular Dynamics Simulations in Polymer Science**, Oxford Univ. Press, Oxford, 1995.

[130] Fleer, G. J., and Lyklema, J., "Adsorption of Polymers," in: **Adsorption from Solution at the Solid/Liquid Interface**, G. D. Parfitt and C. H. Rochester, Eds., pp. 153-220, Academic Press, London, 1983.

[131] Fleer, G. J., Cohen Stuart, M. A., Scheutjens, J. M. H. M., Cosgrove, T., and Vincent, B., **Polymers at Interfaces**, Chapman & Hall, London, 1993.

[132] Scheutjens, J. M. H. M., and Fleer, G. J., *J. Phys. Chem.*, **83**, 1619 (1979).

[133] Flory, P. J., **Principles of Polymer Chemistry**, pp. 497ff., Cornell Univ. Press, Ithaca, NY, 1953.

[134] Wågberg, L., Ödberg, L., Lindström, T., and Aksberg, R., *J. Colloid Interface Sci.*, **123**, 287 (1988).

NMR relaxation. The effective polymer adlayer thickness may be obtained from viscometric measurements (*cf.* Chap. 8), from measurements of effective particle diffusivity (as with photon correlation spectroscopy,[135] *cf.* Chap. 5), from sedimentation rates, *etc.*, all dependent on the assumption that the effective particle size includes the adlayer. Another popular technique is ellipsometry[136], mentioned in Chap. 2 and earlier in this chapter, but suitable generally only for adsorption onto flat surfaces. Another technique, suitable for studying adsorption onto the surface of a transparent solid, is that of internal reflection spectroscopy.[137] When a light beam enters a transparent medium at a sufficiently oblique angle, it is reflected internally between the back and front surfaces of the solid many times before it finally exits the material. While the light beam does not cross the surface as it is totally internally reflected, an "evanescent" non-propagating electric field is generated above the outside surface of the solid. Decaying exponentially in amplitude with distance from the surface, it penetrates a distance of the order of 100 nm. The evanescent beam may experience absorption characteristic of the thickness and nature of the layer directly adjacent to the external surface of the solid, absorption that is evident in the characteristics of the light beam exiting the solid. Such absorption may thus reveal information on the chemistry, refractive index, thickness, *etc.* of the adsorption layer present.

[135] Baker, J. A., and Berg, J. C., *Langmuir*, **4**, 1055 (1988).

[136] Takahashi, A., Kawaguchi, M., and Kato, T., **Adhesion and Adsorption of Polymers**, p. 729, Plenum, New York, 1980.

[137] Harrick, N. J., **Internal Reflection Spectroscopy**, Interscience Publ., New York, 1967.

Some fun things to do:
Experiments and demonstrations for Chapter 3

1. *The work of surface area extension*

The thermodynamic work of area extension at constant temperature and composition is given by $\sigma \Delta A$. While this is usually very small, it can be demonstrated by noting the need to apply a force to extend the area of a soap film. The soap film has two sides, so the work of extending the film is $2\sigma \Delta A$.

Materials:

- Open wire frame measuring about 2×3×2 inches, with a handle fashioned from a kink in the wire as shown in Fig. E3-1. 16 AWG ($\approx$ 1.3 mm diam.) bailing wire is about right. Tie a flexible string (dental floss is good) about 4 in. in length to each end of the wire. Tie a second string, about 2 in. in length to the center.
- Petri dish large enough to accommodate the wire frame
- Soap solution (50/50 v/v mixture of water and dishwashing liquid) to fill Petri dish

Procedure:

1) Dip the wire frame with string attached into the soap filled dish and withdraw, to produce the film of minimum area shown in Fig. E3-1(b). Then pull down on the string to increase the film area (c), and release to see the minimum area return.

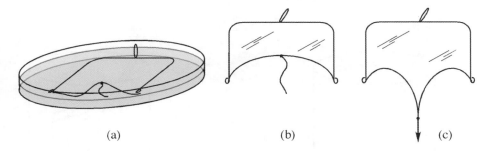

        (a)   (b)   (c)

Fig. E3-1: Demonstrating the work of area extension with a soap film on a wire frame. (a) Dip the frame in a soap solution, (b) withdraw the frame and note that the thread deforms to minimize the area, and (c) when force is applied, the area of the film may be expanded, but as soon as it is removed, the configuration of minimum area returns.

2. Surface tension reduction with surfactant

Surfactants, by definition, drop the surface tension of water from a value of ≈ 70 to ≈ 25 mN/m at low concentrations (usually < 10mM). One way to demonstrate the effect is to "float" a paper clip on water, and then cause it to sink by injecting surfactant into the water beneath it. The paper clip is about eight times as dense as water, but is held up by buoyancy and surface tension forces acting upward around its perimeter. While the exact solution to the problem is rather complex, one can make a *rough* estimate of the surface tension required to maintain the paperclip at the surface as follows. The paperclip may be modeled as a rod of circular cross-section of diameter d against which the water has a contact angle of $\theta \approx 90°$. The metastable equilibrium that exists just before it sinks is shown in Fig. E3.2. The rod is essentially completely submerged, so the buoyancy force/length is: $1/4\pi d^2 \rho_w g$. The upward component of the surface tension force/length is $2\sigma\cos 45° \approx 1.41\sigma$. These forces must balance the weight/length of the rod, $1/4\pi d^2 \rho_{rod} g$, so that for the rod to "float,"

$$\sigma \geq \frac{\pi d^2}{8\cos 45°}(\rho_{rod} - \rho_w)g.$$

For the paperclip floating on water described above, this works out to $\sigma \geq 38.5$ mN/m.

Materials:

- 100-mL beaker, filled to the brim with clean water
- Paperclip and forceps
- Disposable 1 mL polyethylene pipette, filled with soap solution (*e.g.*, Joy® dishwashing detergent)

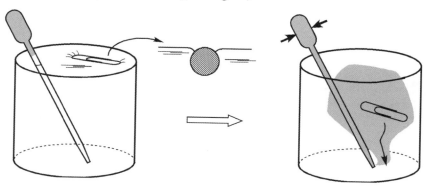

Fig. E3-2: "Floating" a paper clip on water, and sinking it by injection of surfactant from beneath.

Procedure:

1) Use forceps to gently place paperclip on top of the water in the beaker, in which the soap-filled pipette is placed.

2) Squeeze the pipette, injecting the soap into thereby lowering the surface tension and causing the paperclip to sink.

3. CMC determination by dye titration

A rough, but quick and easy, way to determine the critical micelle concentration of a surfactant is the method of dye titration, in which a concentrated surfactant solution (assumed to be above its *CMC*) containing a solubilized dye is progressively diluted until it exhibits a fairly abrupt drop in color intensity (optical absorbance), signaling the loss of micelles. Since the dye itself may participate in the micelle formation, the *CMC* value obtained is likely to be somewhat lower than that for the surfactant by itself. In the experiment described below, the *CMC* of sodium dodecyl sulfate (SDS) is determined using the dye pinacyanol chloride.

Materials:
- 200 mL Erlenmeyer flask
- Magnetic stirrer and stir bar
- Sodium dodecyl sulfate (MW 288.38)
- Pinacyanol chloride (or quinaldine blue) (MW 388.94)
- Spectrophotometer, set for measurement at 615 nm
- 1. 5- and 20-mL pipettes

Procedure:

1) Prepare 100 mL of stock solution containing SDS at 0.0175M (0.505g) and 1L stock solution of pinacyanol chloride at 7.3×10^{-5}M (28.4 mg). Prepare a second dye stock solution by adding 100 mL of the original solution to a 1L volumetric flask and fill to the line with de-ionized water.

2) Pipette 40 mL of the SDS solution, 5 mL of the dye solution and 5 mL of pure water into the 200mL flask placed on the magnetic stirrer.

3) Withdraw a sufficient volume to fill the spectrophotomer cuvette and measure the absorbance, and return the liquid to the mixing flask.

4) Add a succession of aliquots of the diluted dye stock solution according to the following schedule:

Data Pt.	Dye Soln Added	Total Vol.	Conc. of Soln
1	0 mL	50 mL	14.0mM
2	5	55	12.7
3	5	60	11.7
4	5	65	10.8
5	5	70	10.0
6	10	80	8.75
7	10	90	7.77
8	10	100	7.00
9	20	120	5.83
10	20	140	5.00
11	20	160	4.38
12	20	180	3.89

5) Plot the absorbance for each data point, resulting in a graph as shown in Fig. E3-3.

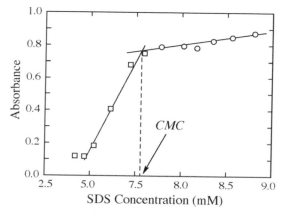

Fig. E3-3: Absorbance data for determination of the *CMC* of SDS by dye titration. (Data from a student lab report)

6) Fit the absorbance data to least squares straight lines, and take the intersection as the location of the *CMC*. For the data shown, the *CMC* is approximately 7.6 mM, slightly lower than the handbook value of 8.0 mM, as expected.

    * If desired, the procedure above may be repeated for various salt concentrations to determine the effect of electrolyte on the *CMC* of this ionic surfactant.

3. *Insoluble monolayers*

Insoluble (or Langmuir) monolayers are formed from the direct spontaneous spreading of an insoluble "oil" from a lens or crystal at the surface. This requires that the spreading coefficient be positive, as expressed by Eq. (3.156):

$$S_{o/w} = \sigma_w - (\sigma_o + \sigma_{o/w}) > 0.$$

where the subscripts "w" and "o" refer to water and "oil" (the candidate spreader), respectively. Mineral oil, with a surface tension of 31.0 mN/m and an interfacial tension against water of 50 mN/m, gives (with the surface tension of water equal to 72.8 mN/m) a spreading coefficient of $S_{o/w} = -8.2$ mN/m at 20°C, and is therefore not expected to form a monolayer. This behavior is typical of most insoluble liquids and solids on water. Those that do spread are materials that have been identified as surface active agents. Oleic acid, for example, has a spreading coefficient on water of 24.6 mN/m. Solid surfactants, such as stearic acid will also spread, but often so slowly that if one wishes to form a monolayer, a spreading solvent is required. Other surfactant solids may spread quickly. Camphor is an interesting example because it is crystalline and has slightly different spreading coefficients from its different crystal faces. This gives the spreading particles a torque that causes them to spin during spreading in what is termed the "camphor dance," shown in Fig. E3-4. Its average spreading coefficient is about 16 mN/m. When a material of higher spreading coefficient is put on a water surface that already has a monolayer of lower spreading coefficient it will displace the original monolayer. All of these spreading phenomena are easily demonstrated in a *clean* Petri dish of *clean* water on which talc particles have been sprinkled, placed on an overhead projector.

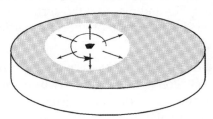

Fig. E3-4: The camphor dance.

When an insoluble monolayer is present, it confers some remarkable properties on the surface. Whereas clean water surfaces are quite fragile to mechanical disturbances, such as air drafts, or temperature variations, or the presence of vapors of a solvent lower surface tension than water, monolayer-covered surfaces are nearly impervious to all these insults, as a direct result of Gibbs elasticity.

Materials:

- 250 mL of *clean* (preferably distilled) water in a plastic squeeze bottle.
- Six clean disposable plastic Petri dishes (10 cm diameter)
- Talc stick (also called "Tailor's chalk")
- Small penknife
- Tweezers
- About 5 g of white mineral oil

- About 5 g of natural camphor (the synthetic variety will not work)
- About 5 g of oleic acid
- About 5 g of acetone (in a tightly capped vial)
- Matches
- Two or three cotton swabs
- Five or six 1-mL disposable PE pipettes
- Large beaker or container for disposal
- Paper towels

Procedure:

1) Show that mineral oil does not spread. Put one of the Petri dishes on the overhead projector and fill it half full with water from the squeeze bottle. Use the penknife to scrape particles from the talc bar and sprinkle them lightly on the surface. Then deposit a small drop of mineral oil on the surface with one of the disposable pipettes and notice that no spreading occurs. Dispose of the Petri dish and contents in the waste container.

2) Demonstrate the camphor dance. Place a clean Petri dish on the overhead projector, half fill it with water, and sprinkle talc particles on the surface. With the tweezers, extract three or four small ($\approx 1 - 3$ mm) pieces of camphor and place on the surface. Observe the particles spinning and the displacement of the talc particles as the camphor spreads.

3) Demonstrate the displacement of one monolayer by another. After the camphor dance has gone on for one or two minutes, place a small drop of oleic acid at the center of water surface. Notice that the oleic acid instantly spreads to the edge of the dish, displacing the camphor film and bringing the "dance" to a halt. Set this Petri dish aside being careful not to let the oleic acid contaminate anything.

4) Demonstrate the fragility of a clean surface *vs.* the rigidity of a monolayer-covered surface by putting another Petri dish on the overhead projector (next to dish with the oleic acid-covered surface) and fill as close to the top as possible without spilling any water. Sprinkle talc particles on the surfaces in both dishes. Gently blow on the clean surface, showing that the talc particles are easily moved around, while on the oleic acid surface it is very difficult to move them. Re-apply particles to the clean surface, and then light a match and hold it close to (but don't touch) the center of the dish, noticing how the particles move away. Do the same thing to the surfactant-covered surface and notice that the particles do not move. Again, re-apply particles to the clean surface if needed. Then dip a cotton sway into the acetone, recap the vial, and bring the swab near (but do not touch) the center of clean surface. Notice how the particles rush away

from where the acetone vapors fall upon the surface. In contrast, the surfactant-covered surface shows no response to the presence of the acetone vapors.

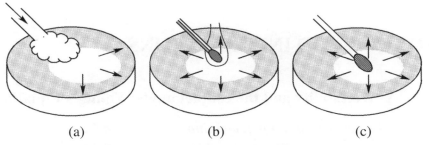

(a)   (b)   (c)

Fig. E3-5: Fragility of clean water surfaces to (a) mechanical disturbances, as being blown upon, (b) heat, as when a lit match is brought near, or (c) chemical disturbances, as when a swab with acetone releases acetone vapor on the surface.

5. *Adsorption from solution by activated carbon*

Activated carbon or charcoal is one of the most commonly used materials for the removal of contaminants from water or other liquid streams, owing to its enormous adsorptive capacity. It typically has a specific area from several hundred to more than one thousand m^2/g. This is easily demonstrated by observing its removal of a dye from a water sample.

Materials:
- 5–10 g activated carbon, coarse-grained (4-8 mesh) or pelletized
- Two or three tea bags
- 250-mL beaker
- Food coloring

Procedure:

1) Sacrifice a tea bag by carefully removing the staple and emptying out the tea, replacing it with about 5 g of activated carbon and re-stapling it together.

2) Add one drop of food coloring to water in the 250-mL beaker, so that it is brightly colored.

3) Hang the tea bag of carbon particles in the solution, and in about 30 min the color will disappear.

Chapter 4

SOLID-FLUID INTERACTIONS

A. Wettability and the contact angle: Young's Equation

1. *Importance of wetting; definition of contact angle*

The physical interaction between a liquid and a solid can vary enormously from case to case. If the solid is a flat, slightly inclined plane, and a drop of liquid is deposited upon it, as shown in Fig. 4-1, it may just ball up and roll off leaving the surface just as it was before contacting the drop. On the other hand, it may partially stick, flowing down the surface as a

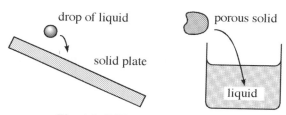

Fig. 4-1: Differences in wettability.

rivulet. Still differently, it may spread out in all directions as a coherent film, perhaps to the extent of a monolayer. If the solid is in the form of a porous matrix, and it is put into contact with a liquid, the liquid may be quickly imbibed into the solid (like water into a sponge) or it may not be. In fact, it may be possible to immerse the porous solid with no liquid entering the pores at all. The nature of the physical interaction between liquids and solids is important in many situations of practical interest, as listed in Table 4-1.

Table 4-1: Importance of wetting behavior in various applications.

Situations where good wetting is desired	Situations where non-wetting is desired	Situations requiring patterned wetting or wetting differences
• Coating • Washing • Adhesion • Absorbency (wicking)	• Water-proofing • Anti-stick surfaces • Anti-soil surfaces • Release coatings	• Printing plates • Ore flotation

One quantitative measure of the liquid-solid interaction is the *contact angle*, θ, made by a liquid when placed against a solid as pictured in Fig. 4-2. The assumption is that the third phase involved is the equilibrium vapor of the liquid, or some inert gas such as air. (For liquid-liquid-solid systems, the liquid in which θ is drawn must be specified.) On rough or inhomogeneous surfaces the three-phase interline may be quite complex. To avoid ambiguity, θ is defined rigorously as the angle made between the *normals* to

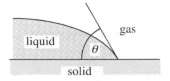

Fig. 4-2: Contact angle θ of a liquid against a solid surface.

the solid and liquid surface (measured in the gas) at the point of interest along the interline, *i.e.*,

$$\theta = \cos^{-1} \underline{n}_1 \cdot \underline{n}_2. \tag{4.1}$$

For sufficiently smooth interlines, this is the same as the angle made between the solid surface and the *tangent* to the liquid-gas surface (drawn in the liquid) in the plane perpendicular to the interline (the interline plane) at the point of interest.

The physical interline between the gas, liquid and solid phases is ultimately made up of individual molecules, and at that level, the contact angle may differ from that which is observed macroscopically. The macroscopic angle θ, as pictured in Fig. 4-2, is the integral result of long-range intermolecular forces between the molecules of the solid, liquid and gas, whereas the microscopic angle θ_m, shown in Fig. 4-3, is determined by short-range interaction forces at the leading edge. θ_m may be defined as the angle between the line connecting the centers of two atoms located in the interline plane in the first two successive molecular layers of the liquid above the solid at the interline.[1] It may be either larger or smaller than θ, as pictured in Fig. 4-3 (a) and (b), respectively, depending on whether short range repulsive or attractive intermolecular forces exist. Even more exotic interface shapes in the immediate vicinity of the interline may occur, depending on the details of the disjoining pressure isotherm.[2] The microscopic angle influences the shape of the liquid interface only out to about 3-4 nm from the interline,[3] and does not affect macroscopic wetting behavior except in the case of $\theta = 0°$, as indicated later in the discussion of spreading. It is thus only the macroscopic contact angle to which we refer in the present context.

[1] Ruckenstein, E., *Colloids Surfaces A*, **206**, 3 (2002).

[2] Brochard-Wyart, F., di Meglio, J.-M., Quéré, D., and de Gennes, P.-G., *Langmuir,* **7**, 335 (1991).

[3] Ruckenstein, E., and Lee, P. S., *Surface Sci.*, **52**, 298 (1975).

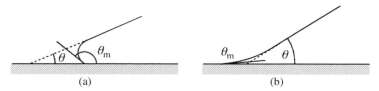

Fig. 4-3: Close-up view of interline region, showing difference between microscopic contact angle, θ_m, and macroscopic contact angle, θ. Case where (a) short-range repulsive forces and (b) attractive forces prevail.

2. Young's Equation revisited; classification of wetting and contact angle values

The contact angle is related to the surface energies, *i.e.*, the σ's, of the three interfaces involved, as shown in Chap. 3, in which system free energy minimization led to Young's Equation:

$$\cos\theta = \frac{(\sigma_{SG} - \sigma_{SL})}{\sigma_{LG}}.$$ (4.2)

This result may also be derived, at least heuristically, by balancing the horizontal forces on an element of the interline, $\delta\ell$, in a manner similar to the construction of Neumann's triangle of forces discussed in Chap. 2. σ_{SL} and σ_{SG} are thus taken as the "effective" boundary tensions of the solid-liquid and solid-gas interfaces, as shown in Fig. 4-4. Consideration of the

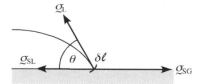

Fig. 4-4: Balance of horizontal forces at the solid-liquid-gas interline.

normal forces on the interline segment requires that the upward capillary force of $\sigma_{LG}\sin\theta$ be balanced by a stress in the underlying solid. The existence of such a force has been demonstrated by experiments with mercury drops on thin (1 μm) mica sheets, which are seen to exhibit slight upward deflections along the interline.[4]

While Young's Equation confers upon the contact angle the status of a thermodynamic property, as noted, measured values of it are notoriously irreproducible and seem to depend upon many "*non*-thermodynamic" quantities. These apparent anomalies are examined shortly, but despite them, the contact angle provides a useful starting point for the classification of liquid-solid interactions, as shown in Fig. 4-5. For the case of $\theta = 0°$, the solid is said to be "wet out" by the liquid. In principle, the liquid can then spread out to form a monomolecular film (with the excess collected in lenses). Wetting out is favored by high σ_{SG} (*i.e.*, high solid-gas surface

---

[4] Bailey, *Proc. 2nd Internat. Congr. Surface Activity*, **3**, 189 Butterworths, London, 1957.

energy), low σ_{SL}, and low liquid surface tension. Examples include organic liquids on most *clean* metals, oxides and many silicate minerals (in general, hard materials). The situation can be deceptive because with polar liquids, a monolayer of the liquid may be present on the apparently dry surface, preventing further bulk spreading, as shown in Fig. 4-6. The excess liquid then collects as drops on the surface, giving an apparent finite contact angle. Such a liquid will not wet out an oriented monolayer of itself, and is referred to as "autophobic." The presence of the monolayer (an adsorbed film) reduces the surface free energy of the solid-gas surface by an amount π_e,

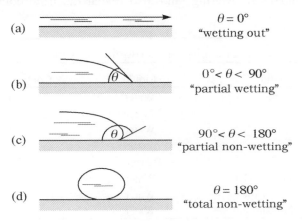

(a) $\theta = 0°$ "wetting out"

(b) $0° < \theta < 90°$ "partial wetting"

(c) $90° < \theta < 180°$ "partial non-wetting"

(d) $\theta = 180°$ "total non-wetting"

Fig. 4-5: Classification of wetting behavior in terms of contact angle.

the equilibrium "surface pressure," or "equilibrium spreading pressure." If π_e is taken into account, Young's Equation must be modified to:

$$\cos\theta = \frac{(\sigma_{SG} - \sigma_{SL} - \pi_e)}{\sigma_L}. \tag{4.3}$$

In general, with a multicomponent liquid, there may be adsorption at all three interfaces, each producing its own surface pressure.

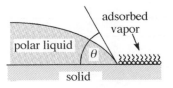

adsorbed vapor

polar liquid θ solid

Fig. 4-6: Autophobic wetting.

In the second case, $(0° < \theta < 90°)$, the liquid is said to "partially wet" the solid. Examples include water on most plastics, waxes and mineral sulfides (*i.e.*, soft-surface materials). In the third case, $(90° < \theta < 180°)$, the liquid is said "not to wet" the solid. Such behavior is observed with high-surface-tension liquids on low-surface-energy solids. Water on very low-surface-energy materials, such as polyethylene or Teflon, exhibits such behavior. A good example would be mercury against most solids, for which in general $\theta \approx 135°$. Lastly, the limiting case of $\theta \to 180°$ is approached by

mercury on Teflon, as well as by water against a number of different micro
or nano-textured surfaces (known as "ultra-hydrophobic surfaces")[5,6] of
hydrocarbons, siloxanes or fluorocarbons. It is also quite commonly
exhibited by water-oil-solid systems, particularly in the presence of
detergents, as discussed later in this chapter in the context of detergency.

Tables of contact angles are generally not to be found in the literature
because of their apparent irreproducibility. Nonetheless, some values can be
found, and a sampling is shown in Table 4-2. These data show examples of
the various classes of wetting, and also reveal the irreproducibility of the
property in several cases.

Table 4-2: Contact angle values (liquids on solids against air)

Solid	Liquid	Contact angle (°)
Glass	Water	0
"	Benzene	0
Silica	Water	0
"	Acetone	0
"	Benzene	0
Anatase (TiO$_2$)	Water	0
Tin oxide (SnO$_2$)	Water	0
Barium sulfate	Water	0
Graphite	Water	86
"	Benzene	60
Stibnite (Sb$_2$S$_5$)	Water	84
"	"	38
Talc	Water	88
"	"	52
"	CH$_2$I$_2$	53
Hexadecyl alcohol	Water	50-72
Paraffin	Water	105
Teflon	Water	110
Glass	Mercury	135
Steel	Water	70-90

B. Contact angle hysteresis

1. *Origins of hysteresis: roughness and heterogeneity*

There are two categories of reasons for the "irreproducibility" of
contact angle data, as indicated in Fig. 4-7. The first refers to the fact that it
is only the uppermost surface layers of the solid that are relevant to
determining the contact angle. Solid surfaces often have coatings or traces of

[5] Öner, D., and McCarthy, T. J., *Langmuir*, **16**, 7777 (2000).
[6] Erbil, H. Y., Demeril, A. L., Avci, Y., and Mert, O., *Science*, **299**, 1377 (2003).

contamination that bear little resemblance to the bulk material beneath and may thus cause differences in the contact angle. Traces of adsorbing solutes in the fluid phases can play a similar role. The second category refers to the method of measurement, *i.e.* to the difference between the contact angle measured when the liquid is being *advanced* over the solid surface as opposed to when it is being *receded* from the surface. In all cases, the advancing angle, θ_A is greater than or equal to the receding angle, θ_R. For example, when a drop of liquid is deposited on a solid surface, θ may

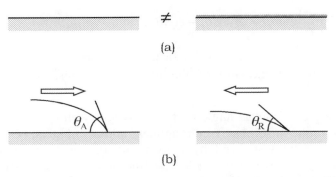

(a)

(b)

Fig. 4-7: "Irreproducible" contact angles: (a) Surface composition is different between nominally identical bulk solids, (b) contact angle depends on whether liquid is advancing or receding across solid surface: hysteresis.

assume some finite value. Then as more liquid is added, the interline remains fixed (or "pinned") so that the apparent contact angle increases. Finally an angle is reached for which the interline "jumps" to a new position. The angle just before the jump occurs is the "static advanced contact angle" and is generally the value reported in the literature. The reverse situation occurs as liquid is withdrawn from the drop, leading to the "static receded contact angle." The magnitude of the difference between θ_A and θ_R is usually quite large and is referred to as contact angle *hysteresis*.

Another manifestation of contact angle hysteresis occurs when a drop of liquid on a flat solid surface remains affixed when the surface is inclined to the horizontal (by an angle β), as shown in Fig. 4-8, in which the drop is

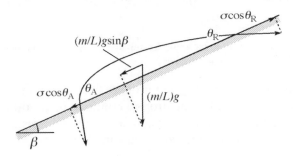

Fig. 4-8: Profile of cylindrical drop (shown in profile) on an inclined plane, showing forces due to contact angle hysteresis preventing it from sliding.

assumed to have a cylindrical shape. If there were no hysteresis, the existence of even the slightest inclination of the plate would cause the drop to slide off, because no net interline forces could be marshaled against gravity. But what is more often observed is that the drop appears to have a larger contact angle along the lower edge and a smaller one at the upper edge. The gravitational force on the drop pulling it downward along the plate is balanced by the net interline force, due to hysteresis, holding the drop in place. The balance of forces yields:

$$\left(\frac{m}{L}\right)g\sin\beta = \sigma\left(\cos\theta_R - \cos\theta_A\right). \tag{4.4}$$

The maximum angle β that can be sustained before the drop runs off the plate is a common measure of the extent of hysteresis, quantified as $\left(\cos\theta_R - \cos\theta_A\right)$.

The origin of contact angle hysteresis is now well understood and can generally be found in solid surface roughness and/or chemical heterogeneity. These can both be visualized by considering the drop of liquid on the inclined plate, as shown in Fig. 4-9. The true or intrinsic contact angle, θ_0, i.e., the one given by Young's Equation, is maintained against the actual surface, but as the interline advances (on a tilted surface), a larger apparent angle is made with the "flat" surface, as shown in Fig. 4-9(a). This produces a larger vertical component of the surface tension force to resist the gravitational pull. Similar reasoning leads to reduction in the apparent angle at the receding edge. As the plate is tilted more steeply, the forward interline advances slightly in an attempt to compensate for the increased load, producing a still higher θ_{app}, until finally the configuration becomes unstable. Beyond that point, θ_{app} starts to *fall* as the interline advances, and it must

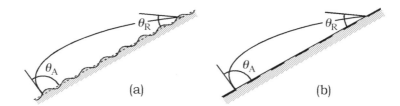

Fig. 4-9: Origins of contact angle hysteresis: (a) Surface roughness, (b) surface chemical (energetic) heterogeneity.

jump to a new position on the next groove. Once the jump occurs, there may be sufficient momentum for the drop to continue its motion and slide off the plate. This type of explanation for contact angle hysteresis has been verified by experiments with uniformly grooved surfaces.[7]

---

[7] Johnson, R. E., Jr. and Dettre, R. H., *J. Phys. Chem.*, **68**, 1744 (1964).

Surface energetic heterogeneity provides the other principal explanation for contact angle hysteresis. Imagine the inclined surface of Fig. 4-9(b) to consist of stripes of wettable and non-wettable surface. The interline will locate itself at the (presumably sharp) edges between the patches of different kinds of surface, where the contact angle is not the sole determiner of the observed angle (closer to the fixed interline case). Again, the magnitude of the maximum and minimum observed angles depends on meniscus stability. Roughly, the forward interline will be unstable at angles exceeding θ_A for the low surface energy material, and the trailing interline will be unstable for angles less than θ_R for the higher energy surface.

2. Complexity of real surfaces: texture and scale

Actual solid surfaces are generally both rough *and* energetically heterogeneous, and the heterogeneity may be both non-uniform and anisotropic. In the latter case, the contact angle may be different depending on the direction of observation. Typically, the combination of irregularities will lead to an irregular interline at the micro level, as pictured in Fig. 4-10.

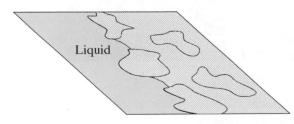

Fig. 4-10: Liquid interline on a heterogeneous surface.

A scanning electron micrograph of ink drops (diameter ≈ 50 μm) against a bond paper is shown in Fig. 4-11. Figure 4-12 shows the effect of surface texture (grooves in a machine-lathed surface of aluminum) and its influence on the anisotropy of wetting of molten poly(phenylether). Thus while the above explanations for contact angle hysteresis are plausible, they do not provide, in general, for the *a priori* prediction of advancing or receding

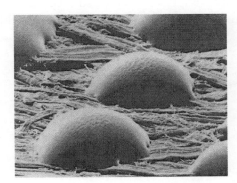

Fig. 4-11: SEM of solidified phase-change ink drops against a bond paper. [Le, H. P., *J. Imaging Sci. Tech.*, **42**, 1 (1998).]

contact angles. As a result of the various heterogeneities, when a fluid interline is either advanced or receded across a solid surface at a finite rate, it generally does so in a series of jumps referred to as "stick-slip motion."

The size scale of the surface heterogeneities is important. When these are large (≥ 10 μm) and regular, the hysteresis may take on a predictable character. The slow motion of a liquid interline across a straight sharp edge was studied by Gibbs and is pictured in Fig. 4-13. In this case, the advancing angle made by the liquid against the upstream surface must be augmented by $(180 - \phi)°$, where ϕ is the angle of the asperity, in order to cross over the

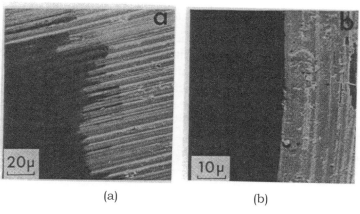

(a)   (b)

Fig. 4-12: Polyphenyl ether drop spreading on machine-lathed aluminum: (a) Along grooves, and (b) against grooves. From [Oliver, J. F., and Mason, S. G., *J. Colloid Interface Sci.*, **60**, 480 (1977).]

edge. This analysis shows that even fully wetting liquids may be confined by sharp edges, provided the latter are free of defects.

When the size scale of the heterogeneities is sufficiently small (generally << 1 μm) and uniform, another type of analysis may be used. Consider a drop on a uniformly roughened solid surface, as shown schematically in Fig. 4-14. The figure also shows computed values of the

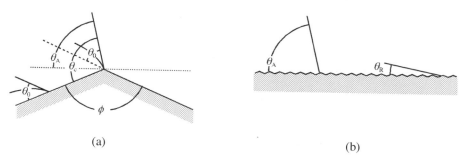

(a)   (b)

Fig. 4-13: Macro roughness: motion of liquid over a macroscopic sharp edge, and observed hysteresis: (a) Close-up view, and (b) long-range view.

system free energy as a function of the nominal (observed) contact angle. Thermodynamics thus suggests that there is a preferred observed contact angle (which minimizes free energy) different from the intrinsic contact angle, θ_0, as given by Young's Equation. The free energy "barriers" between the various interline positions may be sufficiently small that small disturbances (*e.g.*, vibrations) will eventually lead the system to assume the preferred (equilibrium) configuration. Based on such assumptions, treatments have been given to predict the observed equilibrium contact angle in terms of the intrinsic contact angle(s) for both rough and chemically heterogeneous surfaces.

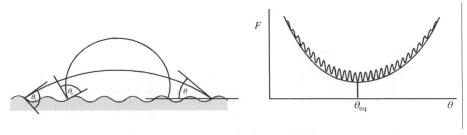

Fig. 4-14: Micro roughness: (a) Two configurations of a drop on a micro-rough surface; (b) free energy as a function of nominal contact angle on a micro-rough surface. After [Johnson, R. E., Jr. and Dettre, R. H., "Wettability and Contact Angles," in **Surface and Colloid Science, Vol. 2**, E. Matijevic, Ed., pp. 85-153, Wiley-Interscience, New York, 1969.]

3. *Wenzel Equation for rough surfaces*

Wenzel[8] considered rough surfaces and supposed that the true solid surface area A could be related to the nominal area, A_{smooth}, by a "rugosity factor," r, where

$$A = r\,A_{smooth} \ ; \ r = A/A_{smooth} > 1. \tag{4.5}$$

Then the minimization of the total system surface free energy, as outlined in Chap. 3, leads to a modified form of Young's Equation in terms of the observed or apparent contact angle, *viz.*,

$$\cos\theta_{app} = r\frac{\left(\sigma_{SG} - \sigma_{SL}\right)}{\sigma_{LG}}, \tag{4.6}$$

known as Wenzel's Equation. Its validity requires not only that the size scale of the roughness be sufficiently small, but also that the liquid mass (drop) must be large relative to the size-scale of the roughness.[9] The roughness of surfaces is discussed in more detail later in this chapter. Although $(\sigma_{SG} - \sigma_{SL})$ is not known independently, Eq. (4.6) is useful in assessing the effect on the apparent (thermodynamically preferred) contact angle of roughening a given

[8] Wenzel, R. N., *Ind. Eng. Chem.*, **28**, 988 (1936).
[9] Wolansky, G., and Marmur, A., *Colloids Surfaces A*, **156**, 381 (1999).

solid surface. What is seen is that one can *increase* the wettability of a surface which is already wet (*i.e.*, $\theta_0 < 90°$) by a given liquid, or *decrease* the wettability of a solid by a liquid that does not wet it. Since $r > 1$ for a roughened surface, if $\cos\theta$, is positive ($\theta_0 < 90°$), the $\cos\theta_{app}$ will be larger, *viz.*

$$\cos\theta_{app} = r\cos\theta_0. \tag{4.7}$$

If the liquid is non-wetting, $\cos\theta_0$ is negative, and $\cos\theta_{app}$ will become more negative if the surface is roughened, as illustrated in Fig. 4-15. The ability to change wetting characteristics by roughening is important in a number of

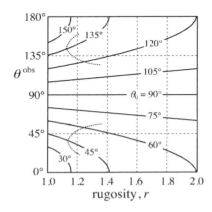

Fig. 4-15: Modification of observed contact angle by micro-roughness Effect of hysteresis shown by dashed lines for intrinsic angles of 45° and 135°.

applications. In particular, if θ_{app} can be made to go to 0°, wetting out will occur. The rugosity factor required for this to occur is $r = \cos^{-1}\theta_0$, and since r-values up to 2.0 can readily be achieved, this is often possible. Examples of the effect of hysteresis, as a result of roughness size scales that are too large for Wenzel analysis to apply, are shown by the dashed lines in Fig. 4-15. In these cases, increased roughness always produces advancing angles higher than θ_0 and receding angles lower than θ_0 as r increases.

4. *Cassie-Baxter analysis of heterogeneous surfaces; composite surfaces and ultra-hydrophobicity*

A treatment similar to Wenzel's has been given by Cassie and Baxter[10] for chemically heterogeneous surfaces. For the case of a surface consisting of two types of patches, for which the intrinsic contact angles are θ_1 and θ_2, respectively, minimization of surface free energy leads to the Cassie-Baxter Equation:

$$\cos\theta_{app} = \phi_1\cos\theta_1 + \phi_2\cos\theta_2, \tag{4.8}$$

where ϕ_1 and ϕ_2 are area fractions for the two types of surface. It is readily extended to the case of many different types of patches. One important case is that in which pores in the surface lead to vapor gaps across which the

---

[10] Cassie, A. B. D., and Baxter, S., *Trans. Faraday Soc.*, **40**, 546 (1944).

liquid does not contact the solid, as pictured in Fig. 4-16. The effective contact angle over such gaps is 180°, and if the area fraction of them is ϕ_2, the Cassie-Baxter Equation, (4.8), becomes:

$$\cos\theta_{app} = \phi_1 \cos\theta_1 - \phi_2. \tag{4.9}$$

It is structures of this type that produce the "ultra-hydrophobic" surfaces ($\theta > 130°$) mentioned earlier in the classification of wetting behavior. Many plant leaves, most notably those of the lotus flower, have ultra-hydrophobic surfaces owing to a very fine surface structure coated with hydrophobic wax crystals only a few nm in diameter. In the lotus plant, the actual contact area

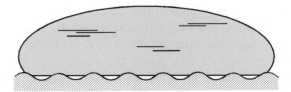

Fig. 4-16: A composite surface with unwetted gas pockets on the rough solid surface.

is only about 2-3%, yielding water contact angles as high as 160°. Beads of water roll off the surface, collecting dirt particles as they go, so that the lotus leaves are self-cleaning in a process now known as the "lotus effect."[11] These properties have been mimicked by researchers at BASF-Mannheim who have recently patented and produced an aerosol product ("Lotus Spray") for the treatment of various surfaces such as wood, paper, masonry, leather, *etc*. Figure 4-17 shows water drops on a wood surface treated by Lotus Spray. The product combines nanoparticles with hydrophobic polymers such as polypropylene, polyethylene and various waxes. It is delivered by a propellant gas, and as the coating dries, it develops a nanostructure through self-assembly.

Fig. 4-17: Water droplets on an "ultra-hydrophobic" surface created by coating wood with "Lotus Spray," an aerosol of hydrophobic nanoparticles. From [http://nanotechweb.org/cws/article/tech/16392/1/0611102.]

---

[11] Marmur, A., *Langmuir*, **20**, 3517 (2004).

Another situation may also arise for a rough surface, as pictured in Fig. 4-16. The small pores on the surface may be pre-filled by capillary condensation of the wetting liquid, and the nominal area fraction of the filled pores may be taken as ϕ_2. Over this area, the contact angle is 0°, so that the Cassie-Baxter Equation takes the form:

$$\cos\theta_{app} = \phi_1 \cos\theta_1 + \phi_2. \tag{4.10}$$

In such situations, one often observes wetting out, *i.e.*, $\theta_{app} \rightarrow 0°$.

The deceptive simplicity of both the Wenzel and Cassie-Baxter Equations depends on the accessibility of the thermodynamically preferred (equilibrium) configuration, which in turn requires that the roughness or the chemical heterogeneities be of sufficiently small size scale that the "barriers" between adjacent configurations are overcome by the general noise level in the system. If this is not the case, one observes widening hysteresis with either increasing roughness, or larger patch sizes for chemically heterogeneous surfaces. The question with regard to the applicability of the Wenzel or Cassie-Baxter analysis is: "How small is small enough?" Recent direct evidence suggests that the critical size scale is in the range of 6-12 nm.[12] It is the ultra-fine scale of the roughness characteristic of the lotus effect described above that all but eliminates hysteresis in those systems and provides for the ease of roll-off. For the more commonly-occurring chemically heterogeneous surfaces, one is likely to obtain results as pictured in Fig. 4-18, where the advancing contact angle curve is in the

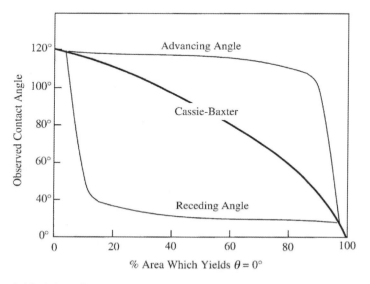

Fig. 4-18: Advancing, receding and equilibrium contact angles on a model heterogeneous surface consisting of areas yielding $\theta = 0°$ and $\theta = 120°$.

[12] Fang, C., and Drelich, J., *Langmuir*, **20**, 6679 (2004).

upper right hand side of the diagram, and the receding angle is shown at the lower left. Both are shown for a surface with high-energy portions ($\theta_2 = 0°$) and low energy portions ($\theta_1 = 120°$) as a function of the fraction of the surface occupied by the high-energy portion. It is to be noted that the advancing angles closely reflect the low surface energy behavior, while the receding angles reflect the high-energy behavior.

5. *The dynamic contact angle; Tanner's Law*

There are two important kinetic effects, one hydrodynamic and one thermodynamic, leading to a distinction between "static" and "dynamic" contact angles. As the liquid interline is advanced or receded across the solid surface at sufficient interline velocity, U, the apparent contact angle becomes a function of U.[13] This can be visualized, as shown in Fig. 4-19, by considering the downward movement of a solid surface into a liquid pool, as might be effected using a plunge tank as illustrated in Fig. 4-29. At sufficiently low velocity, the observed angle will be the intrinsic contact angle, θ_0, or the static advanced angle (actually, there will generally be stick-slip motion). As interline velocity increases, a dynamic advancing contact angle will be established which is a function of the interline velocity. The greater the velocity, the higher will be the contact angle. At a certain rate, known as the rise-canceling velocity, U_c, the dynamic contact angle is 90°. This is a measure of wettability sometimes used in the textile industry. As the interline moves still faster, non-wetting contact angles are obtained, and eventually air entrainment occurs as the dynamic angle approaches 180°, the upper limit for high-speed coating.[14]

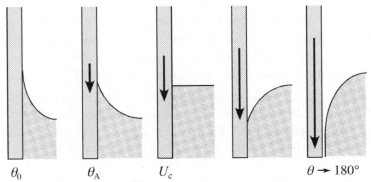

θ_0 θ_A U_c $\theta \rightarrow 180°$

Fig. 4-19: Effect of interline velocity on observed contact angle.

[13] Cox, R. G., *J. Fluid Mech.*, **168**, 169 (1986);
de Gennes, P. G., *Revs. Mod. Phys.*, **57**, 827 (1985);
Hoffman, R. L., *J. Colloid Interface Sci.*, **94**, 470 (1975);
Kistler, S., in **Wettability**, J. C. Berg, Ed., Marcel Dekker, 1992;
Seebergh, J. E., and Berg, J. C., *Chem. Eng. Sci.*, **47**, 4455 and 4468 (1992).
[14] Burley, R., and Jolly, R. P. S., *Chem. Eng. Sci.*, **39**, 1357 (1984).

Hydrodynamic analysis shows that it is primarily viscous effects that are responsible for dynamic wetting behavior, and that these are conveniently described in terms of the Capillary number, Ca,[17, 18] defined as the ratio of viscous forces to surface tension forces:

$$Ca = \frac{U\mu}{\sigma}.$$ (4.11)

The value for Ca below which viscous effects are negligible may depend on the specific geometry involved[15], but is generally between 10^{-5} and 10^{-6}. For the case of the static angle $\theta_0 = 0$, and $Ca \leq 0.1$, dynamic wetting on smooth surfaces is well described by Tanner's Law[16]:

$$\theta_D^3 = (\text{const})Ca,$$ (4.12)

where θ_D is the dynamic contact angle. The database for the case in which the static contact angle is finite is more scattered than that for the case of complete wetting under static conditions, but it appears to be reasonably well correlated in terms of a shifted form of Tanner's Law[17]:

$$\theta_D^3 - \theta_{\text{static}}^3 = (\text{const})Ca.$$ (4.13)

The full range of Capillary numbers is covered by Kistler's correlation,[18] which takes the form (for the case of $\theta_{\text{static}} = 0°$):

$$\theta_D = \cos^{-1}\left\{1 - 2\tanh\left[5.16\left(\frac{Ca}{1+1.31Ca^{0.99}}\right)^{0.706}\right]\right\},$$ (4.14)

shown in Fig. 4-20 together with data from a number of sources for the spreading of various silicone oils on glass. For $Ca < 0.1$, the above equation is well approximated by $\theta_D(°) = 260Ca^{0.353}$ in close agreement with Tanner's Law. The exponent being somewhat larger than the 1/3 prescribed by Tanner's Law may be attributed to a slight roughness of the glass.

The second kinetic effect associated with contact angles concerns thermodynamic of adsorption. The surface energies, *i.e.*, the σ's, can be reduced by the presence of adsorbed materials. Thus the adsorption of vapors at the solid-gas surface or of solutes at the solid-liquid or liquid-gas surfaces influences the contact angle. The kinetic effect results from the time required for the adsorbing solute(s) to diffuse to the interface(s). In some cases, the adsorption process itself, or adsorbed species re-orientation, such as overturning, will be rate determining.

[15] Dussan V, E. B., *Ann. Rev. Fluid Mech.*, **11**, 371 (1979).
[16] Tanner, L. H., *J. Phys. D: Appl. Phys.*, **12**, 1473 (1979).
[17] Seaver, A. E., and Berg, J. C., *J. Appl. Polym. Sci.* **52**, 431(1994).
[18] Kistler, S. F., in **Wettability**, J. C. Berg (Ed.), pp. 311-429, Marcel Dekker, New York, 1993.

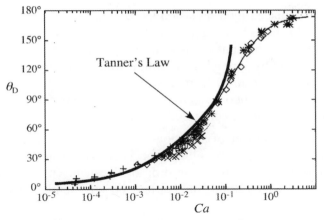

Fig. 4-20: Observed dynamic contact angles for silicone fluids displacing air in capillary tubes and between glass plates, compared with the prediction of Tanner's Law. After [Kistler, S. F., in **Wettability**, J. C. Berg (Ed.), pp. 311-429, Marcel Dekker, New York, 1993]. The static angle in all cases is 0°.

C. Methods for measuring the contact angle

1. *Optical or profile methods: contact angle goniometry*

Because the contact angle is subject to the uncertainties and interpretations discussed above, it is rarely possible to obtain the needed value in a particular circumstance by consulting the literature. Thus it is important to know how to obtain useful values in the laboratory. The direct techniques available are either optical in nature or require the measurement of a force (or related property). Other methods are indirect and subject to simplifying assumptions. Some of the latter do not yield contact angles at all, but do provide some measure of "wettability."

The most common optical method is goniometry. As shown schematically, a sessile drop or captive bubble is formed against the solid in the form of a flat, horizontal surface, as shown in Fig. 4-21. In the simplest

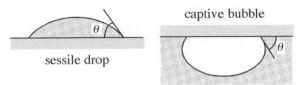

Fig. 4-21: Contact angle goniometry.

devices available, the profile of the drop or bubble is viewed through a telescope equipped with a protractor eyepiece (goniometer) so that the contact angle can be observed directly. The most commonly used device for obtaining drop or bubble profiles, however, is the axisymmetric drop shape analysis (ADSA) system[19], in which the optical component of the manual

---

[19] Li, D., Cheng, P., and Neumann, A.W., *Adv. Colloid Interface Sci.*, **39**, 347 (1992).

goniometer is replaced by a precision CCD camera, attached to a computer, as shown in Fig. 4-22. Tangent lines may be drawn in by eye, or the computer may give the contact angle directly. Both advancing and receding angles are obtained by supplying or withdrawing liquid by a drop-dispensing needle penetrating the drop from above. In either type of instrument, the specimen stage may support an environmental chamber in which a liquid is held and/or temperature is controlled.

There are a number of situations in which the surface shape near the interline is not readily accessible, especially when the meniscus is very small, such as a very small sessile drop or the meniscus in a fine capillary tube. In these cases, the Bond Number may be sufficiently small that the

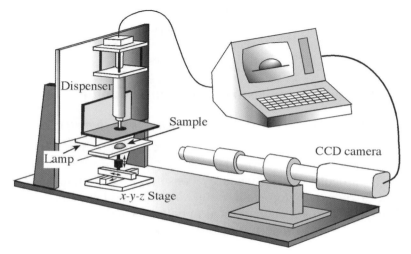

Fig. 4-22: An Axisymmetric Drop Shape Analysis (ADSA) system.

interface shape is a segment of a sphere, and one can calculate the contact angle from other measured features making use of the geometric description of a spherical segment developed earlier. For the sessile drop (or meniscus in a tube), knowledge of two of the three quantities: drop volume (or meniscus volume on the concave side), V, base radius (or tube radius), r, and height, h, suffices to compute θ by the following relations obtained by combining Eqs. (3.32) and (3.34):

$$\tan\frac{\theta}{2} = \frac{h}{r}, \quad \text{and} \tag{4.15}$$

$$\frac{r^3}{V} = \frac{3\sin^3\theta}{\pi(2 - 3\cos\theta + \cos^3\theta)}. \tag{4.16}$$

For the case of a meniscus in a tube, what generally appears in the cathetometer telescope is a black band or "blind zone" of thickness equal to the height of the meniscus. Measurement of this thickness together with knowledge of the tube diameter suffices to calculate θ. For larger drops, full Bashforth-Adams analysis may be applied, as in the axisymmetric drop shape analysis (ADSA) technique.

Attempts to infer θ-values from the shape of symmetrical menisci around fine fibers have been less successful. In this situation, the observed meniscus profile is the net result of two opposing curvatures roughly comparable in magnitude. Interpretation of the observed profile is sometimes ambiguous. Furthermore, the method is limited because when the contact angle exceeds a certain critical value (dependent upon the relative size of the drop to the filament) it flips over to one side of the filament, assuming a "clam shell" configuration.[20]

2. Force methods: contact angle tensiometry

The other principal class of methods for measuring θ does not rely upon optical determinations but instead measures the force on an object, such as the Wilhelmy slide shown in Fig. 4-23, dipping into the liquid. The technique is identical to the Wilhelmy method for measuring surface tension (under the assumption of zero contact angle). In the present case, σ is assumed to be known *a priori*, and the measured force can be used to determine θ. The method is also readily used with fibers. The downward force on the plate, after the weight of the plate in air has been tared out, is given by:

$$(F \downarrow)_{\text{net}} = (\text{perimeter})\sigma\cos\theta - V_{\text{disp}}\rho_{\text{L}}g. \tag{4.17}$$

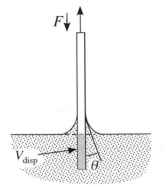

Fig. 4-23: Wilhelmy method for measuring contact angles.

One must know the perimeter of the plate, fiber or filament and sufficient information (density of the liquid, ρ_{L}, and cross-sectional area) to calculate the indicated buoyancy "correction" term, $V_{\text{disp}}\rho_{\text{L}}g$. For very fine fibers ($d <$ 20 μm) this correction is generally negligible. The Wilhelmy method is especially powerful for obtaining contact angle information. The availability of sensitive electrobalances make it possible to resolve contact angle

[20] Carroll, B. J., and Lucassen, J., *Chem. Eng. Sci.*, **28**, 23 (1973).

differences of only one or two degrees against fibers as fine as 10 μm or less. Figure 4-24 is a schematic drawing of a specific system constructed for this purpose.[21] The fiber is suspended from the balance arm of a Cahn Model RG-2000 electrobalance via a hangdown wire and dips into the liquid contained in a dish housed within an environmental chamber. The heater allows control of temperature (to 250°C), and gas composition is controlled by means of the vacuum pump. Optical access is not needed for the measurements, and is available only in the absence of the heater. While the fiber is suspended in null position, the liquid in the Teflon cup is raised or lowered at a carefully controlled rate by means of a Burleigh Instruments Model IW 601-2 Inchworm Translator. Rates between one mm and one μm

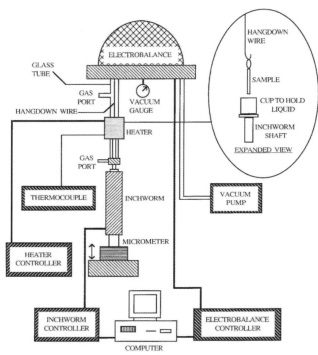

Fig. 4-24: Schematic diagram of dynamic fiber wettability apparatus. [Berg, J. C., in **Composite Systems for Natural and Synthetic Polymers**, L. Salmén, A. de Ruvo, J. C. Seferis, and E, B. Stark (Eds.), p. 23, Elsevier, Amsterdam, 1986.]

per second may conveniently be used. While the liquid is made to move up or down over the fiber at the desired rate, the output force of the electrobalance is fed simultaneously to a computer for display, storage and analysis, so that advancing and receding contact angles can be determined as a function of the interline velocity. A number of fully automated packaged tensiometer devices of the type described above are available commercially.

---

[21]Hodgson, K. T., and Berg, J. C., *Wood Fiber Sci.*, **20**, 3 (1988).

A typical force-readout (corresponding to a Nylon monofilament of diameter 0.71 mm in an aqueous SDS solution) is shown in Fig. 4-25. As the liquid is first brought into contact with the fiber, the force jumps to the advancing interline condition. The subsequent slope of the line results from the buoyancy correction. The stick-slip character of the interline movement is evident in the trace, and the amplitude of this variation is a measure of the fiber surface heterogeneity, both chemical and morphological. A computer may be programmed to average this force trace, compute and subtract out the buoyancy term, and yield an advancing contact angle. It also may compute what might be called the "stick-slip amplitude" (the standard deviation about the mean) and the "stick-slip frequency," a measure of the geometric scale of the surface heterogeneities, all of these data constituting part of the fiber "wetting profile" used to characterize the fiber-liquid interaction. When the liquid level movement is stopped and reversed, and the meniscus is reoriented, the force trace yields the receding contact angle. For the case shown in Fig. 4-25, the advancing force, F_{adv} gives $\theta_{adv} = 82°$, and F_{rec} gives $\theta_{rec} = 15°$. Repeated cycling reveals any changes in surface composition (due to desorption, adsorption, swelling, *etc.*) as well as the time scales associated with those changes. Repeated cycles showing no further changes in wetting may be presumed to have reached equilibrium.

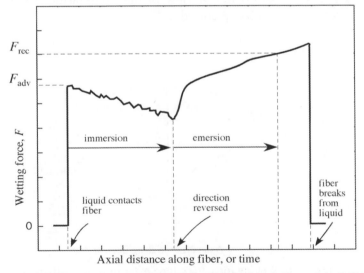

Fig. 4-25: Wetting force trace for Nylon 6 fiber ($d = 0.71$ mm) in $5·10^{-4}$ M SDS solution. Excursion = 2.0 mm.

A comparison of the various forces typically involved in a set of wetting measurements may be used to highlight some of the limitations and problems with this technique, and are shown in Table 4-3. Room vibration and natural convection (if the environmental chamber must be heated) can be serious problems. To minimize the effects of natural convection in a heated environmental chamber, the fiber must be carefully centered, and a small tube should surround the hang-down wire from the balance to the chamber.

Figures 4-26 and 4-27 show wetting traces for two other fiber-liquid systems, revealing some of the differences existing between different fibers. The force scales are approximately the same (scaled to the respective

Table 4-3: Forces encountered in measuring single-fiber wetting ($d = 10$ μm).

Sensitivity of microbalance	± 2.0 N
Maximum surface force ($\theta = 0°$; $\sigma = 30$ mN/m)	10,000
Change of surface force ($0° \rightarrow 5°$)	100
Change of surface force ($40° \rightarrow 45°$)	6,000
Weight of adsorbed film on 1 cm² area	0.1
Buoyancy correction ($h = 2$mm; $\rho_L = 1$ g/cm³)	10
Viscous drag ($\mu = 1$cP; $\dot{\gamma} = 0.1$ s⁻¹; area $= 0.01$cm²)	1
Natural convection in air	≈ 1,000

perimeters). The stick-slip frequency of the Kevlar® fiber is seen to exceed that of the Nylon fiber. The severe heterogeneity of a wood pulp fiber, seen in Fig. 4-27, is not surprising in view of the irregularities of such fibers, a typical example of which is pictured in Fig. 4-28.

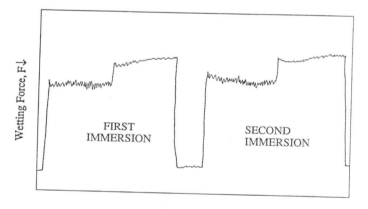

Fig. 4-26: Wetting force trace for Kevlar fiber ($d = 7.0$ μm) in $5 \cdot 10^4$M SDS solution. Excursion: 2.0 mm.

The most important information present in the wetting profiles for the purpose of predicting wicking flow or adhesive performance is the advancing contact angle. For a smooth surface, it is this quantity that is characteristic of the low-energy (and most difficult to wet) portions of the surface. The lowest possible value of the advancing contact angle is sought, with the best choice being the penetrant or adhesive for which it is zero.

Next in importance, if the advancing contact angle is minimized but not zero, is to seek a condition of minimal contact angle hysteresis. In the

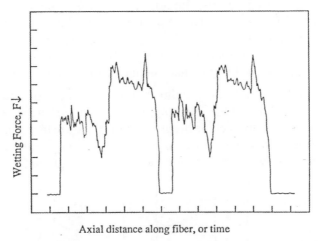

Fig. 4-27: Wetting force trace for self-sized Douglas fir fiber (d = 20.0 μm) in water. Excursion: 1.68 mm.

removal of trapped gas pockets by shearing during the application of the penetrant or adhesive, the resistance of the bubble to being removed is proportional to the hysteresis ($\cos\theta_R$ - $\cos\theta_A$). Finally, the surface heterogeneity, as indicated by the stick-slip amplitude of the wetting force variations, is a good indication of the ability of the surface to trap air pockets.

Fig. 4-28: SEM of a bulky, open pulp sheet of southern pine latewood. Fiber width ≈ 30 μm. From [Parham, R. A., and Kaustinen, H. M., **Papermaking Materials: An Atlas of Electron Micrographs**, p. 23, The Institute of Paper Chemistry, Appleton, WI, 1974.]

3. *Dynamic contact angle measurement*

While the Wilhelmy method provides a convenient means for measurement of dynamic contact angle at relatively low Capillary numbers or for changes occurring as a result of adsorption, other techniques are generally required to access cases of higher interline velocity. Three commonly used methods are pictured in Fig. 4-29.[22] These include the capillary flow technique, in which the interline speed is controlled by the imposed pressure drop on the liquid in the tube. The plunge tank and syringe-needle methods are convenient for use with flat surfaces, which may be moved as rapidly as desired. The syringe-needle method is especially accurate because it allows one to focus sharply on the meridian plane of the curved liquid surface, yielding a sharp image of the contact angle.

Other methods yield only indirect information relative to the contact angle, as described later. One example is the so-called "wipe test," applied generally to polymer surfaces. It involves wiping the surface with a series of liquids covering a range of surface tensions, starting with the liquid of lowest surface tension, and noting for each the time required for the film to

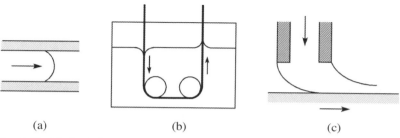

(a)   (b)   (c)

Fig. 4-29: Methods for measurement of dynamic contact angle. (a) Capillary displacement; (b) plunge tank, and (c) syringe-needle extrusion coating.

break up into droplets. Liquids are tried until one is found for which the film lifetime is just two seconds. The surface tension of this liquid is then identified as the surface energy of the plastic. Another example of such a method is the "sink test," used in the textile industry. The wettability of a skein of yarn is determined by measuring its "sinking time" in a vessel of liquid. The wettability of a fine powder might be inferred from the time required for a drop of liquid to sink into a layer of the dry powder. Other types of wicking are used in other situations. Rough values of an effective contact angle can be inferred only after assumptions concerning the pore structure in such materials are made, or by comparison with the behavior of liquids known to wet out the solid, as discussed later in this chapter.

D. Relation of wetting behavior to surface chemical constitution

1. Zisman Plots; the critical surface tension

Contact angle measurements, through the use of Young's Equation, Eq. (4.2), should in principle provide information on the surface energetics of solids, *viz.*,

[22] Kistler, F. S., in **Wettability**, J. C. Berg (Ed.), p. 319, Marcel Dekker, New York, 1993.

$$(\sigma_{SG} - \sigma_{SL}) = \sigma_L \cos\theta. \tag{4.2}$$

The question arises as to what, in terms of solid surface chemistry, leads to specific values for $(\sigma_{SG}- \sigma_{SL})$ or of σ_{SG} or σ_{SL} by themselves. These are important parameters, for example, in predicting adhesive or "abhesive" (*i.e.*, non-stick) performance. High-energy solids, which are wet out by most liquids, allow one to determine only that $(\sigma_{SG}- \sigma_{SL}) > \sigma_L$. Contact angle measurements on low energy solids, such as plastics, polymers, *etc.*, however, can provide more information. These in any case are the types of materials for which quantitative characterization of wetting behavior is needed. Low surface energy solids are materials for which the adsorption of a monolayer of molecules of the liquid at the solid-gas interface, lowering its surface energy, is usually neglected,[23] although there is evidence that this may not be valid.[24,25] Zisman and coworkers at the Naval Research Laboratories[26] provided many contact angle data for low-surface-energy solids using optical goniometry. They noted that if for a given smooth-surface solid one plotted $\cos\theta$ (static advanced angle) against the surface tension of the liquid for a series of liquids of different surface tension, they would fall on a single straight line (except for the higher σ values), as shown schematically in Fig. 4-30. Such lines are known as *Zisman Plots*, and they permit extrapolation to $\cos\theta = 1$ $(\theta = 0°)$.[27] The extrapolated value of the surface tension on a Zisman Plot is termed the "critical surface tension" of the solid, σ_c, and has the practical significance of being the surface tension of a liquid at or below which the solid will be wet out.

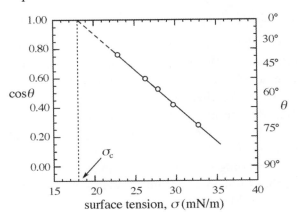

Fig. 4-30: Schematic of a Zisman Plot.

In preparing Zisman Plots, a number of precautions must be observed concerning the probe liquids used: 1) they must not dissolve or swell the solid, 3) they must not interact *specifically* with the solid (no H-bonding or

[23] Fowkes, F. M., McCarthy, D. C., and Mostafa, M. A., *J. Colloid Interface Sci.*, **78**, 200 (1980).

[24] Schröder, J., *Farbe + Lack*, **1/1980**, 19 (1986).

[25] Jacob, P. N., and Berg, J. C., *J. Adhesion*, **54**, 115 (1995).

[26] Zisman, W. A., **Contact Angle, Wettability and Adhesion**, Adv. in Chemistry Ser., No. 43, p. 1, ACS, Washington, DC, 1964.

[27] Fox, H. W., and Zisman, W. A., *J. Colloid Sci.*, **7**, 109 (1952).

other acid-base interactions), and 4) their vapors must not adsorb on the solid. It is also preferred that they be pure (as opposed to solutions). Zisman noted that, provided the above conditions were met, the value of σ_c was independent of the probe liquids used and was therefore taken as a characteristic surface energy of the solid alone.

Under conditions where $\sigma_L \rightarrow \sigma_c$, $\cos\theta \rightarrow 1$, so that:

$$\sigma_C = \sigma_{SG} - \sigma_{SL}. \tag{4.18}$$

For the case in which the molecules of both the solid as well as the liquids used in preparing the plot interact only through dispersion forces (the usual situation), the equation of Girifalco and Good, Eq. (2.26) (used earlier to predict the interfacial tension between immiscible liquids in terms of their surface tensions) may be used to express σ_{SL}, giving:

$$\sigma_C = \sigma_{SG} - \left[\sigma_{SG} + \sigma_L - 2\sqrt{\sigma_{SG}\sigma_L}\right]. \tag{4.19}$$

Substituting $\sigma_L \rightarrow \sigma_c$, and simplifying leads to:

$$\sigma_C = \sigma_{SG}. \tag{4.20}$$

If functional groups on the solid surface interact through more than dispersion forces, but the molecules of the liquids self-interact only via such forces, then $\sigma_L = \sigma_L^d$, and one may apply the Fowkes Equation, Eq. (2.27), to obtain σ_{SL}:

$$\sigma_{SL} = \sigma_{SG} + \sigma_L - 2\sqrt{\sigma_{SG}^d\sigma_L}, \tag{4.21}$$

which leads to: $\sigma_C = \sigma_{SG}^d$, i.e., the Zisman critical surface tension gives the dispersion force (or Lifshitz-van der Waals) component of the solid surface energy. Finally, it should be noted that the σ_C-values obtained from Zisman Plots correspond to σ_{SG} (or to σ_{SG}^d) rather than to what has been termed in Chap. 3.L the "pristine" surface energy of the solid in vacuo, σ_0, which is equivalent to σ_s. σ_{SG} is equivalent to a "clean" surface energy if the gas phase is the equilibrium vapor of the liquid, in which case, σ_{SG} is equal to σ_s reduced by the equilibrium spreading pressure, π_e. σ_{SG} may be a "practical" surface energy, reduced further by the adsorption of additional components from the gas phase. Thus one would expect $\sigma_C \leq \sigma_{SG} \leq \sigma_s$.

2. The wettability series

Critical surface tension values have been found for a wide variety of materials. The range of interfacial composition and structure studied was greatly expanded by the fact that it is only the uppermost surface monolayers that govern wetting behavior. Thus oriented monolayers of various surfactants, most notably the polyfluorinated compounds, were deposited on convenient base materials, and wetting studies done on them. Some Zisman Plots for these coated surfaces are shown in Fig. 4-31. They show the remarkable result that surfaces consisting of close-packed -CF$_3$ groups have a critical surface tension of only 5.6 mN/m!, unwettable by virtually any known liquid at room temperature. Some critical surface tension data for various materials are summarized in Table 4-4. Using extensive data of this type, it was possible to build up generalizations about the surface energy attributable to chemical functionality. In fact it was possible to construct a series of ascending surface energies (a "wetting series") based on the atomic constitution, *viz.*,

$$F < H < Cl < Br < I < O < N. \tag{4.22}$$

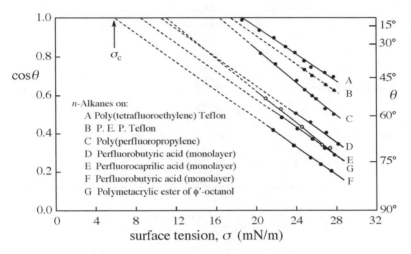

Fig. 4-31: Zisman Plots for various fluorine-rich surfaces. After [Shafrin, E. G., Zisman, W. A., *J. Phys. Chem.*, **64**, 519 (1960).]

The underlying explanation for a series of the above type may be given in part by the how strongly the electrons are held by the atomic nucleus. The more strongly they are held, the lower their polarizability, and the weaker is their interaction with neighboring molecules. In any case, a series such as (4.22) provides valuable information on how one should modify the surface chemistry of a solid in order to achieve desired changes in wetting behavior.

3. *Estimates of surface energies from contact angle data or vice versa*

It is in principle possible to extract surface energies from contact angle data without preparing a full Zisman Plot. The solid-liquid interfacial energy, σ_{SL}, called for in Young's Equation may be eliminated in terms of σ_s and σ_L by using one of the combining equations proposed originally for interfacial tension, *viz.* Eq. (2.26) or (2.27). This gives:

Table 4-4: Critical surface tension of surfaces in terms of their chemistry. From [Osipow, L. I., **Surface Chemistry**, p.251, ACS Monograph Series No. 153, Reinhold, New York, 1962.]

Surface	Chemical structure	σ_c (mN/m)
Perfluorolauric acid, monolayer	CF_3, close-packed	5.6
Perfluorobutyric acid, monolayer	CF_3, less closely packed	9.2
Perfluorokerosene, thin liquid film	CF_2, some CF_3	17.0
Polytetrafluoroethylene, solid	CF_2	18.2
Octadecylamine, monolayer	CH_3, close-packed	22.0
α-Amyl myristic acid, monolayer	CH_3 and CH_2	26.0
2-Ethyl hexyl amine, monolayer	CH_3 and CH_2	29.0
n-Hexadecane, crystal	CH_2, and some CH_3	29.0
Polyethylene, solid	CH_2	31.0
Naphthalene, crystal	⬡⬡, edge only	25.0
Benzoic acid, monolayer	⬡, edges and faces	53.0
2-Naphthoic acid, monolayer	⬡⬡, edges and faces	58.0
Polystyrene, solid	CH_2, some ⬡	32.8- 43.3
Polyethylene terephthalate, solid	⬡, CH_2, ester	43.0
Nylon, solid	CH_2, amide	42.5-46.0

Girifalco-Good: $$\sigma_S = \frac{\sigma_L(1 + \cos\theta)^2}{4\Phi^2}, \text{ or} \qquad (4.23)$$

Fowkes: $$\sigma_S = \frac{\sigma_L^2(1 + \cos\theta)^2}{4\sigma_L^d}. \qquad (4.24)$$

The treatment of Girifalco and Good is of practical value when only dispersion forces are involved (so $\Phi = 1$), and the more general Fowkes treatment shows how it is again the dispersion-force component of the solid surface energy that is obtained. Both results tacitly assume that the spreading pressures, π_e, of the vapors of the liquids used are ≈ 0.

It should be noted that combining rules in addition to those of Eq. (2.26) and (2.27) have been proposed, particularly for use in computing solid-liquid interfacial energy. Wu[28] proposed using a harmonic rather than geometric mean mixing rule to get:

$$\sigma_{SL} = \sigma_S + \sigma_L - \frac{4\sigma_S\sigma_L}{\sigma_S + \sigma_L}. \qquad (4.25)$$

---

[28] Wu, S., *J. Polym. Sci.*, **C34**, 19 (1971).

Neumann and coworkers[29] suggested that an equation should exist, even for polar materials, in the form

$$\sigma_{SL} = f(\sigma_S, \sigma_L),\tag{4.26}$$

and proposed on empirical grounds:

$$\sigma_{SL} = \frac{\left(\sqrt{\sigma_S} - \sqrt{\sigma_L}\right)^2}{1 - 0.015\sqrt{\sigma_S \sigma_L}}.\tag{4.27}$$

Relationships of the type of Eq. (4.23) and (4.24) may be inverted to show explicitly how the contact angle should depend on the surface tension of the liquid and the surface energy of the solid. Specifically, for the case in which only dispersion forces are important, one obtains

$$\cos\theta = -1 + 2\sqrt{\frac{\sigma_S}{\sigma_L}}.\tag{4.28}$$

Equation (4.28) provides an alternate format for the preparation of Zisman Plots, *viz.* plotting $\cos\theta$ vs. $1/\sqrt{\sigma_L}$ and extracting σ_S (presumably the same as σ_c) from the slope. Using this procedure, Fowkes[30] found surface energy values somewhat different from but in the same sequence as those reported in Table 4-4.

When acid-base interactions between the liquid and functional groups on the substrate occur, Eq. (2-28) suggests a relationship of the form

$$\cos\theta = -1 + 2\sqrt{\frac{\sigma_S}{\sigma_L}} + \frac{I^{ab}}{\sigma_L},\tag{4.29}$$

where I^{ab} represents the energy per unit area due to acid-base interactions. While Eqs. (4.28) and (4.29) shouldn't be taken too literally, they do show how wetting behavior is related to the relative values of the liquid surface tension and the surface energy of the substrate, as well as the importance of acid-base interactions to the enhancement of wetting.

For the wetting of solid surfaces with ionizable functional groups by aqueous media, it is evident that the contact angle should be a function of *pH*, suggesting the method of *contact angle titrations*. Bolger and Michaels,[31] who were first to explore this idea, focused on metal oxides. These are populated with –MOH groups that are capable of picking up a proton at low *pH* (to acquire a positive charge) or losing a proton at high *pH* (to acquire a negative charge). The *pH* at which there is a balance between

---

[29] Driedger, O., Neumann, A. W., and Sell, P. J., *Kolloid Z. Z. Polym.*, **201**, 52 (1965).

[30] Fowkes, F. M., **Contact Angle, Wettability and Adhesion,** Adv. in Chemistry Ser., No. 43, p. 99, ACS, Washington, DC, 1964.

[31] Bolger, J. C., and Michaels, A. S., in **Interface Conversion for Polymer Coatings**, Proc., P. Weiss and G. D. Cheevers, Eds., Elsevier, New York, 1968.

the positively and negatively charged groups on the surface was identified as the isoelectric point of the solid (*IEPS*), a quantity also accessible by electrokinetic titrations, as described in Chap. 6. *IEPS* values for a variety of metal oxides are tabulated by Bolger and Michaels and elsewhere. I^{ab} and hence $\cos\theta$ in Eq. (4.29) above should be greatest when the groups are fully ionized, *i.e.*, in proportion to (*IEPS* – *pH*) for *pH* < *IEPS*, or (*pH* – *IEPS*) for *pH* > *IEPS*. One of the primary objectives of Bolger and Michaels was the determination of the conditions under which water would *not* be able to interpose itself between a solid (S) and some organic acidic or basic (usually polymeric) coating (L). This would be expected to be the case if the ionic (acid-base) interactions between L and S exceeded those that would exist between water (at its *pH*) and S. For acidic liquids L, this should occur when pK_A < *pH* < *IEPS*, and for basic liquids when *IEPS* < *pH* < pK_B. A large body of experimental data appears to be explained by these predictions.

Whitesides and coworkers measured the water wettability of surface functionalized poly(ethylene) as a function of *pH*. Most attention was given to carboxyl groups that were implanted at high density by treating the PE with concentrated chromic/sulfuric acid solutions. An example of their contact angle titrations is shown in Fig. 4-32. From curves of this type, they were able to deduce the degree of dissociation α of the surface carboxyl groups:

$$\alpha = \frac{\left[CO_2^-\right]}{\left[CO_2^-\right]+\left[CO_2H\right]}. \tag{4.30}$$

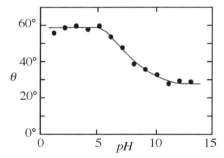

Fig. 4-32: Contact angle titration of carboxyl-functionalized poly-ethylene. After [Whitesides, G. M., Biebuyck, H. A., Folkers, J. P., and Prime, K. L., *J. Adhesion Sci. Tech.*, **5**, 57 (1991).]

Assuming the solid/liquid interfacial energy, σ_{SL}, is linearly related to α, they wrote:

$$\alpha(pH) = \frac{\sigma_{SL}(pH\,1) - \sigma_{SL}(pH)}{\sigma_{SL}(pH\,1) - \sigma_{SL}(pH\,13)}. \tag{4.31}$$

Substituting for σ_{SL} from Young's Equation, and assuming the surface tension to be negligibly affected by changes in *pH*, they obtained:

$$\alpha = \frac{\cos\theta(pH\ 1) - \cos\theta(pH)}{\cos\theta(pH\ 1) - \cos\theta(pH\ 13)} = 2.83\cos\theta - 1.63. \qquad (4.32)$$

Additionally, contact angle titrations of a variety of carbon fibers,[32] and of wood pulp fibers[33] have revealed the presence of several different functional groups on the surfaces. Contact angle titrations appear to provide a simple but powerful technique for the study of wetting phenomena influenced by the ionization of functional groups at solid surfaces in contact with aqueous media. Acid-base effects in wetting are further reviewed elsewhere.[34]

4. Thermodynamics of solid-liquid contact: work of adhesion, work of wetting and work of spreading; the Young-Dupré Equation

It is also useful to consider the free energy of the interaction between a liquid and a solid. Three different expressions for such a quantity have been formulated, depending on how the contacting event is visualized, as pictured in Fig. 4-33. Each of the parameters represents the reversible work associated with a *de*-contacting process, as shown. The first, referring to Fig. 4-33(a), is the "work of adhesion," W_A, defined as the work required to

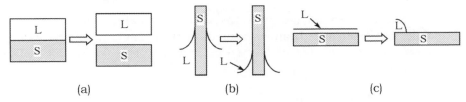

(a)   (b)   (c)

Fig. 4-33: Solid-fluid contacting or disjoining processes: (a) de-bonding or adhesion; (b) de-wetting or wetting; (c) de-spreading or spreading.

disjoin a unit area of the solid-liquid interface, thereby creating a unit area of liquid-vacuum (or liquid-gas) and solid-vacuum (or solid-gas) interface. If the interface is disjoined *in vacuo*, it is

$$W_A = \sigma_S + \sigma_L - \sigma_{SL}, \qquad (4.33)$$

whereas if it is disjoined in a gas, one should write:

$$W_A = \sigma_{SG} + \sigma_{LG} - \sigma_{SL}. \qquad (4.34)$$

The distinction is often neglected, and this can lead to error, especially if the liquid is volatile. In that case, the vapor may adsorb to the solid surface, decreasing its surface energy an amount equal to the equilibrium spreading

[32] Hüttinger, K. J., Höhmann-Wien, S., and Krekel, J., *J. Adhesion Sci. Tech.*, **6**, 317 (1992).

[33] Berg, J. C., *Nordic Pulp Paper Res. J.*, **8** [1], 75 (1993).

[34] Berg, J. C., "*Role of Acid-Base Interactions in Wetting and Related Phenomena*," in **Wettability**, J. C. Berg, Ed., Marcel Dekker, New York, 1993; Sun, C., and Berg, J. C., *Adv. Colloid Interface Sci.* **105**, 151 (2003).

pressure, π_e. Recall the earlier discussion of autophobic wetting, Fig. 4-6 and Eq. (4-3). In this case, we must write

$$\sigma_{SG} = \sigma_S - \pi_e. \tag{4.35}$$

A second useful parameter, referring to Fig. 4-33(b), is the "work of wetting," W_W (or more commonly called the "wetting tension" or the "adhesion tension") is the work expended in eliminating a unit area of the solid-liquid interface, while exposing a unit area of the solid-vacuum (or solid-gas) interface:

$$W_W = \sigma_S - \sigma_{SL}, \text{ (or } \sigma_{SG} - \sigma_{SL}). \tag{4.36}$$

A third parameter, Fig. 4-33©, is the "work of spreading," defined as the work required to expose a unit area of the solid-vacuum (or solid-gas) interface, while destroying a corresponding amount of solid-liquid and liquid-vacuum (or liquid-gas) interface:

$$W_S = \sigma_S - \sigma_L - \sigma_{SL}, \text{ (or } \sigma_{SG} - \sigma_{LG} - \sigma_{SL}). \tag{4.37}$$

It is recognized as identical to the spreading coefficient introduced in Chap. 2, this time applied to a liquid at a fluid-*solid* interface, $S_{L/S}$.

The first two of the wetting parameters defined above may in principle be evaluated from measured contact angles, using Young's Equation for $\sigma_S - \sigma_{SL}$ (or $\sigma_{SG} - \sigma_{SL}$) $= \sigma_L \cos\theta$. This yields the set of relationships:

$$W_A = \sigma_L(1 + \cos\theta), \tag{4.38}$$

known as the Young-Dupré Equation,[35] and

$$W_W = \sigma_L \cos\theta. \tag{4.39}$$

The spreading coefficient would be given by

$$W_S = S_{L/S} = \sigma_L(\cos\theta - 1), \tag{4.40}$$

which shows that W_S is non-negative only when $\theta = 0°$.

There are a number of precautions to be observed in inferring solid surface energies or related quantities from contact angle measurements, particularly in view of hysteresis effects. Another problem with such inferences is the possible importance of equilibrium spreading pressure, π_e, mentioned earlier. If taken into account, the appropriate π_e must be added to the right hand side of each of the above equations. It will in general be different for different probe liquids on a given solid, so the error committed in ignoring it will not be systematic, nor will different probe liquids in general yield the same value for the surface energy of a given solid.

---

[35] Dupré, A., **Théorie méchanique de la chaleur**, Paris, 1869.

It is also useful to express the various energies of solid-liquid interaction in terms of the energies of the solid and liquid surfaces alone, *i.e.*, in terms of σ_S and σ_{SL}. This can be done using the combining rule, Eq. (2.25), to substitute for σ_{SL} for the case in which dispersion forces dominate, leading, for nonvolatile liquids, to:

$$W_A = 2\sqrt{\sigma_S^d \sigma_L^d} = 2\sqrt{\sigma_S \sigma_L}, \tag{4.41}$$

$$W_W = -\sigma_L + 2\sqrt{\sigma_S^d \sigma_L^d} = \sqrt{\sigma_L}\left[2\sqrt{\sigma_S} - \sqrt{\sigma_L}\right], \text{ and} \tag{4.42}$$

$$W_S \equiv S_{L/S} = 2\left[-\sigma_L + \sqrt{\sigma_S^d \sigma_L^d}\right] = 2\sqrt{\sigma_L}\left[\sqrt{\sigma_S} - \sqrt{\sigma_L}\right]. \tag{4.43}$$

5. *The promotion or retardation of wetting: practical strategies*

There are many reasons, related to applications listed in Table 4-1, to want to alter the wetting characteristics of a solid material, and there are a number of procedures available for this purpose, listed in Table 4-5. All can be rationalized with reference to Young's Equation:

$$\cos\theta = \frac{\sigma_{SG} - \sigma_{SL}}{\sigma_L}. \tag{4.2}$$

For good wetting, $\cos\theta$ should be as high as possible (so that $\theta \rightarrow 0°$), and for poor wetting or non-wetting, it should be as large a negative value as possible (so that $\theta \rightarrow 180°$).

Table 4-5: Strategies to promote or to hinder wetting.

TO PROMOTE WETTING
- Roughen surface (if $\theta_0 < 90°$)
- Reduce σ_L with surfactants
- Reduce σ_{SL} with surfactant adsorption to solid/liquid interface
- Chemically modify solid surface (oxygen enrichment)

TO HINDER WETTING
- Roughen surface (if $\theta_0 > 90°$)
- Increase σ_{SL} with surfactant adsorption to solid/liquid interface
- Chemically modify solid surface (fluorine enrichment)

The usual candidates for wettability enhancement are plastics, such as polyolefins, vinyls, fluorocarbons and silicones. These are low surface energy, poorly wettable materials that are difficult to coat or with which to form adhesive bonds. Perhaps the simplest strategy for enhancing their wettability, if the initial contact angle is < 90° (*i.e.*, the right-hand side of Young's Equation is positive) is to roughen the surface. Recalling Wenzel's Equation, Eq. (4.6), roughening multiplies the right-hand side of Young's Equation by the factor $r > 1$, *cf.* Fig. 4-15. Roughening may be accomplished through sanding, grit blasting, chemical etching (such as anodization,[36] in the case of aluminum), or plasma etching.[37] This would not work for improving the water wettability of Teflon, for which the smooth surface exhibits a contact angle of about 110°, so that the right hand side of Eq. (4.6) is negative.

The wettability (particularly by water) may be increased by lowering the surface tension, as might be accomplished by adding surfactants, *i.e.*, "wetting agents." These compounds also adsorb to the solid-liquid interface in a way that promotes wetting, *i.e.*, reduces σ_{SL}. For example, water wetting of most solids is promoted by the use of anionic surfactants. Since most solid surfaces take on a negative charge in contact with water (to be discussed in Chap. 6), anionic surfactants adsorb in a "tail-down" configuration, exposing their hydrated hydrophilic anionic groups to the water, reducing σ_{SL}, as pictured in Fig. 4-34(a). Surfactants that promote water wetting by reduction of σ_L and/or σ_{SL} are termed "wetting agents." Compounds which adsorb at the solid-liquid interface and maintain favorable orientation for promoting wetting when the solid is dried are termed "re-wetting agents.[38]"

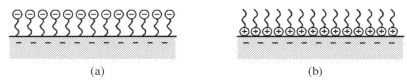

(a)   (b)

Fig. 4-34: Promotion (a) or inhibition (b) of wetting by use of surfactants adsorbed at the water/solid interface.

Wetting may also be promoted by chemically modifying the solid surface in various ways that increase the dry surface energy, σ_S, decrease the interfacial energy σ_{SL} (against the liquid of interest, usually water), or both. In view of the atomic wetting series described earlier, (4.22), increasing σ_S might be achieved by increasing the oxygen (or nitrogen) content of the surface. One approach is to treat the solid surface with a plasma, *i.e.*, a gas under moderate vacuum subject to radio frequency waves which disrupt the molecules into free radicals that react chemically with the solid surface. Oxygen-containing plasmas implant oxygen functionality (usually in the form of carbonyl and hydroxyl groups) in otherwise low-energy surfaces such as those of plastics and increase their surface energy and chemical reactivity. Examples of some results of the treatment of polyolefins with an O_2-plasma are shown in Table 4-6 (left). It is noted that the treatments have

[36] Pocius, A. V., **Adhesion and Adhesives Technology**, pp. 172ff, Hanser/Gardner Publ., Cincinnati, OH, 1997.

[37] Lejeune, M., Lacroix, L. M., Brétagnol, F., Valsesia, A., Colpo, P., and Rossi, F., *Langmuir*, **22**, 3057 (2006).

[38] Berg, J. C., in **Absorbent Technology**, P. K. Chatterjee and B. S. Gupta, Eds., pp. 149-198, Elsevier, Amsterdam, 2002.

limited permanence. Oxygen functionality may also be added to polymeric surfaces by corona discharge treatments, in which the surface to be treated is drawn through an air-ionizing electric arc that also produces oxygen radicals that interact with the surface. Oxidizing flame treatments have a similar effect. The flame induces an ionized air stream by burning an ultra-lean gas mixture whose excess oxygen is rendered reactive by the high temperature. Corona and flame ionization treatments also have somewhat limited permanence.

Table 4-6: Results of plasma treatments of solid surfaces in terms of water wettability.

Example 1: O_2 Plasma treatment of polymers. Data from [Yasuda, H., *in* **Adhesion Aspects of Polymeric Coatings**, K. Mittal, Ed., p. 193, Plenum, NY (1983).]

Polymer	Substrate	θ_{H2O}
Polypropylene	Untreated	100°
	Treated (after 20 min)	57°
	Treated (after 2 days)	68°
	Treated (after 1 month)	85°
Polyethylene	Untreated	76°
	Treated (after 20 min)	33°
	Treated (after 1 month)	35°

Example 2: $C_3F_8/C_2H_4O/O_2$ plasma treatment of glass. From [Ratner, B. D., Haque, Y., Horbett, T. A., Schway, M. B., and Hoffman, A. S., **Biomaterials '84 Trans.**, VII, Washington, DC (1984).]

$C_3F_8(\%)$	$C_2H_4O(\%)$	$O_2(\%)$	σ_C (mJ/m^2)
100	-	-	1 (!)
50	50	-	22
-	100	-	40
-	80	20	45

Another way to get oxygen functionality at the surface of a plastic is to incorporate oxygen-rich compounds as "blooming agents" into the bulk of the apolar material (such as a polyolefin) so that they gradually migrate (or "bloom") to the surface. Finally, a variety of different wet chemical methods may be used to enhance wettability. An example is shown in Fig. 4-35, in

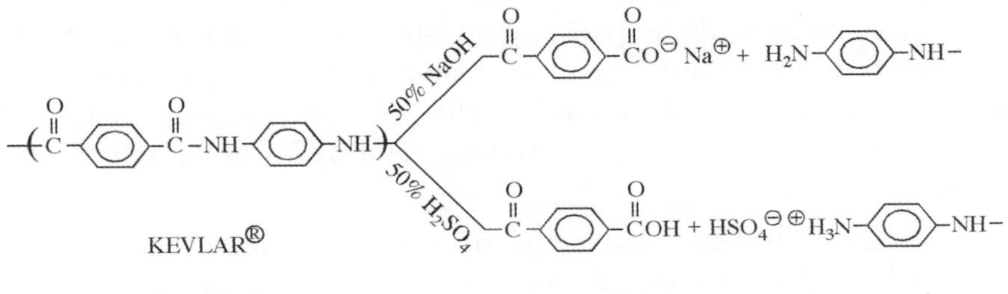

Fig. 4-35: Making Kevlar hydrophilic by acid or base induced hydrolysis of the polyaramid chains. After [Keller, T.S., A.S. Hoffman, B.D. Ratner and B.J. McElroy, in **Physicochemical Aspects of Polymer Surfaces**, Vol. 2, K.L. Mittal, Ed., Plenum Press, New York, NY, p. 861 (1983).]

which Kevlar®, a hydrophobic polyaramid polymer used for making bulletproof vests, is made hydrophilic by treatment with strong bases or acids. The cleavage of the linkage between the carbonyl carbon and the nitrogen leaves ionized groups at the surface in either case. The treatment of polyamides or polyesters with strong acids or bases leads to similar results.

To hinder wetting, many of the above strategies may be applied in reverse. Roughening of surfaces for which the smooth-surface contact angle is > 90° will make the surface even more non-wettable, in accord with the Wenzel effect.

The surfaces for which reduction in wettability is sought are usually those of high dry surface energy, such as metals and ceramics, or those with high water affinity, such as cellulosic and many other natural fibers. Adsorbates, such as cationic surfactants, may be used to render surfaces hydrophobic. They may adsorb at the solid-water interface in a "head-down" configuration, as shown in Fig. 4-34(b), exposing their hydrocarbon moiety to the water, *increasing* σ_{SL}. An increase in σ_{SL} may seem thermodynamically unlikely, but may occur when there is a strong specific interaction between the solid and the adsorbate head group, as discussed in Chap. 3. In the case of cationic surfactants, it is the electrostatic interaction between the head groups and the negative charge that is usually found at a water/substrate interface that governs the adsorption. (Of course, if the water/substrate interface happens to carry a positive charge, anionic surfactants may be used with the same effect.) Similarly, monolayers of straight-chain fatty acids may be chemisorbed to metal surfaces such as steel, where they are often used as "boundary lubricants." Compounds that reduce dry surface energy are sometimes termed "release agents," as they are used in molding and casting to aid in the separation of the mold from the material being molded. In other examples, waterproofing and imparting soil repellency to fabrics are often achieved by the adsorption of such compounds, and other such materials are sprayed on cooking surfaces to prevent food from sticking. Similar materials, called sizing agents are used on paper to reduce water penetration and to enhance printability (*i.e.*, to prevent a blotter effect, or "feathering"). Typical materials used for this purpose are rosin soaps, alkenylsuccinic anhydride (ASA), and alklylketene dimer (AKD). Rosin soaps (bulky anionic surfactants) are bound to negatively charged cellulose surfaces using polyvalent cations (usually Al^{+3}) bridges. ASA and AKD are covalently bonded to the paper, exposing their hydrocarbon moieties to the water.

Figure 4-36 shows an example of an adsorbate used to render surfaces both hydrophobic and oleophobic.[39] It is a comb polymer prepared by copolymerizing acrylate esters of a perfluoroalkyl alcohol with a carboxylic acid functional monomer. In aqueous solution, it self-organizes as the water evaporates, with the perfluoro groups extended out from the aqueous/air

---

[39] Brady, Jr., R.F., *Nature*, **368**, 16 (1994).

interface. The polymer backbone is cross-linked (with a polymer containing pendant oxazoline rings) to give it permanence after it is dried and cured. The final, hard, smooth coating, which may be easily applied to a wide variety of substrates, has a surface energy of only 11-16 mN/m and is unwet by essentially all solvents and adhesives.

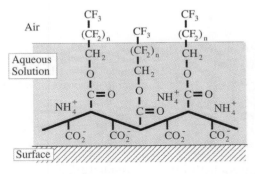

Fig. 4-36: An adsorbing comb polymer that renders a surface both hydro- and oleo-phobic by populating the surface with close-packed CF_3 groups.

Using fluorine-containing plasmas has the effect of decreasing the solid surface energy, and hence the wettability, as illustrated by data in Table 4-6 (right). As the plasma is enriched in perfluoropropane relative to oxygen or ethylene oxide, the Zisman critical surface tension, σ_c, is reduced to lower and lower values. The ultimate value of 1 mJ/m² (!) is of course an extrapolation. Finally, various wet chemical techniques can be employed to plant hydrogen and/or fluorine at the surface (usually in the form of $-CH_3$ or $-CF_3$ groups) to lower the surface energy.

One of the most important generic methods for varying or controlling the chemistry of many surfaces is shown in Figs. 4-37 and 4-38, *viz.*, the use of organo-functional *silanes*.[40,41,42] The most common types of these compounds are the tri-alkoxy silanes, having the formula: Y-Si-$(OR)_3$, or tri-chloro silanes, with the formula: Y-Si-Cl_3. For the tri-alkoxy silanes, R is usually methyl or ethyl. Y is an organo-functional group that can be tailored to whatever application is desired. In general, the silanes are covalently bonded to any surface possessing a population of hydroxyl groups. This includes mineral and other oxides as well as metals (which are superficially oxidized), because their contact with moist air hydroxylates the surface oxygens. Upon contact with water, the alkoxy groups (OR) or the chlorine atoms are replaced with OH groups, as shown in Fig. 4-37. These subsequently condense with other silane molecules and with OH groups on the surface to form a laterally cross-linked layer attached to the surface by ether (oxane) linkages, as shown in the idealized representation of Fig. 4-37. More often many of the multi-alkoxy or multi-chloro silane molecules will

[40] Plueddemann, E. P., **Silane Coupling Agents**, Plenum Press, New York, 1982.

[41] Mittal, K.L., Ed., **Silanes and Other Coupling Agents**, VSP, Utrecht, 1992.

[42] Arkles, B., and Larson, G., (Eds.), **Silicon Compounds: Silanes and Silicones, 2ⁿᵈ Ed.**, Gelest, Inc., Morrisville, PA, 2008.

condense with each other without direct attachment to the surface, producing entangled multilayers with thicknesses of up to several hundred nm. The outermost layer, however, is still populated by the organo-functional groups. Di- or mono-silanes: *e.g.* Y_2-Si-$(OR)_2$ or Y_3-Si-(OR), are also used. The latter are capable of forming only monolayers.

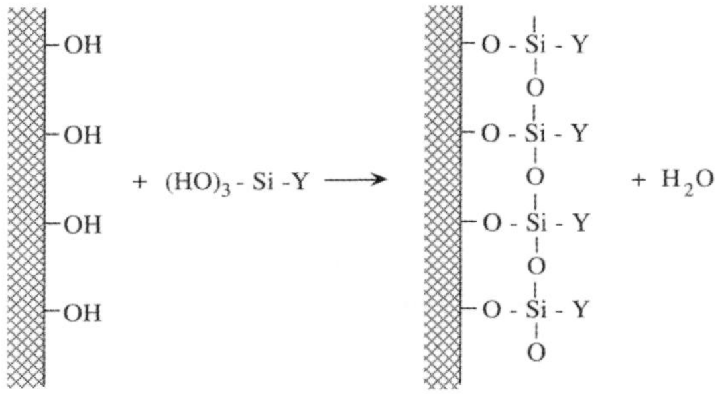

Fig. 4-37: Idealized view of the application of a silane coupling agent to a hydroxylated mineral surface.

If the organo-functional group is of low surface energy, its application to the treated surface renders it hydrophobic. Figure 4-38 shows the hydrophobization of quartz by dichloro dimethylsilane. The quartz (SiO_2) is

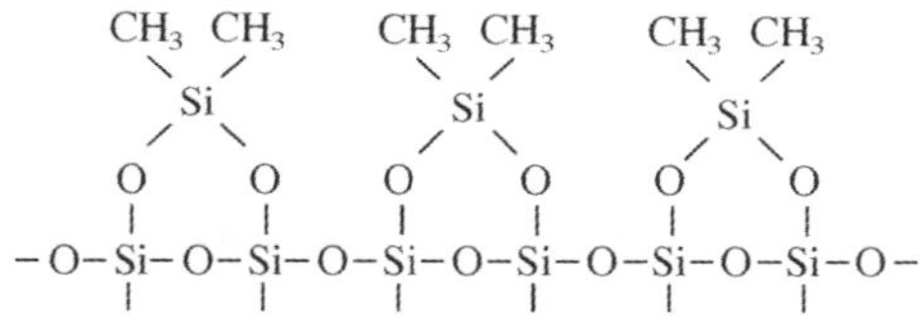

Fig. 4-38: Hydrophobization of quartz by reaction with a dichlorosilane.

superficially hydroxylated by contact with moist air, and the silane reacts with the hydroxyl groups on the surface in accord with:

$$(CH_3)_2SiCl_2 + 2 \text{ M-OH} \rightarrow M_2Si(CH_3)_2 + 2HCl \qquad (4.44)$$

The most common use of silanes is the promotion of adhesion between mineral surfaces and polymeric matrices, in which application they are referred to as "primers" or "coupling agents." Typical organo-functional groups are thus entities such as methacrylates, aminos, epoxies, styrene, *etc.*, chosen to be compatible with the desired polymeric material.

E. Spreading of liquids on solid surfaces

1. *Criteria for spontaneous spreading; spreading morphology*

Spontaneous spreading refers to the process that occurs when the boundary condition expressed by "Neumann's triangle of forces" cannot be satisfied. As noted in Chap. 2.F.4, the spreading of a liquid (B) at a fluid interface (between A and C) is possible only when the spreading coefficient is non-negative, *i.e.*

$$S_{B/AC} = \sigma_{AC} - \sigma_{AB} - \sigma_{BC} \geq 0. \tag{4.45}$$

For the analogous situation of a liquid at a horizontal solid-gas interface, spontaneous spreading may occur when

$$S_{L/S} \equiv W_S = \sigma_S - \sigma_L - \sigma_{SL} \geq 0, \tag{4.46}$$

where the subscript "$_G$" denoting the gas phase has been dropped. The contact angle under such conditions, of course, is 0°. In reality, even when the above condition appears to be met, spontaneous spreading may not be observed, as the advancing interline may become pinned or anchored by roughness elements or other heterogeneities. Even nominally smooth surfaces are generally heterogeneous on the micron or sub-micron scale; *e.g.*, so-called "atomically smooth" surfaces (like freshly-cleaved mica or clean silicon wafers) have defects, dislocations, *etc*. If the liquid mass is large enough, gravitational and/or inertial effects may assist the advancing interline in overcoming such obstacles, but one may argue that that would be "forced," as opposed to spontaneous spreading. In the case of very small droplets, as might be produced in spray coating or ink jet printing, surface forces alone must cause sufficient spreading so that the droplets merge to form a coherent, uniform film. A non-negative spreading coefficient is thus a necessary, but not a sufficient, condition for spontaneous spreading of liquids on solid surfaces.

When spontaneous spreading does occur, one is interested in both the rate of the process and the final equilibrium state. It has long been known[43] that spontaneous spreading occurs through the agency of a "precursor foot," as shown in Fig. 4-39, extending a few millimeters ahead of the observed interline. The foot varies in thickness from case to case from a few Å to

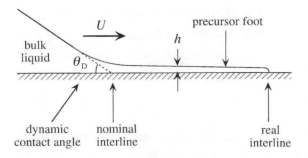

Fig. 4-39: Spreading of a wetting liquid.

[43] Hardy, W., *Phil. Mag.*, **38**, 49 (1919);
    Bascom, W., Cottington, R., and Singleterry, C., **Contact Angle, Wettability and Adhesion**, Adv. Chem. Ser., 43, p. 341, 1964.

perhaps 100-200 nm (a nanofilm), and is therefore generally not visible to the naked eye. One may use suspended particles[44] or interferometry to detect its presence and properties. Generally, however, studies of the rate of spreading of wetting liquids, as described for example by Tanner's Law, pertains to the advance of the *nominal* interline between the bulk liquid and the thin film in front of it. The leading edge of the precursor foot is an abrupt nano-cliff, whose rate of advance is greater than that of the nominal interline.

The final morphology of the liquid on the surface may be addressed by considering the situation in which a small volume V of a liquid is put in contact with a large (infinite), smooth horizontal surface, as pictured schematically in Fig. 4-40. If the spreading coefficient S is positive, it is generally presumed to reach the thinness of an adsorbed layer, *i.e.*, a thin thin film. This is not always the case, however, and the final film may be a *thick* thin film. Since the spreading precursor film is sufficiently thin that disjoining pressure forces come into play, one must return to the

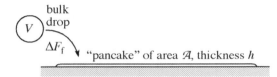

Fig. 4-40: Equilibrium thickness of a spread film.

consideration of thin films begun in Chap. 2 to explore the different possibilities. To do this, consider the (Helmholtz) free energy change, ΔF_f, of forming a film of thickness h, equal to $V/\mathcal{A}$, where $\mathcal{A}$ is the area of the film against the solid surface. ΔF_f is the reversible isothermal work associated with the process pictured in Fig. 4-40, and consists of two terms. The first is the work $-W_S$ required to create new solid-liquid and liquid-gas interfacial area $\mathcal{A}$ while destroying an equivalent solid-gas interfacial area.

This amounts to minus the work of spreading, (*i.e.*, $-S_{L/S}$) times $\mathcal{A}$. The second is the work against disjoining pressure in thinning the film from $z \rightarrow \infty$ to $z = h$. Thus:

$$\Delta F_f = -S_{L/S}\mathcal{A}(h) + \int_{\infty}^{h}\Pi(z)\mathcal{A}(z)dz = -S_{L/S}\frac{V}{h} + V\int_{\infty}^{h}\frac{1}{z}\Pi(z)dz . \qquad (4.47)$$

Equation (4.47) shows that wetting depends on more than just the spreading coefficient, but also involves the nature of the disjoining pressure isotherm, examples of which have been shown in Fig. 2-69. For a van der Waals liquid forming a film with a positive effective Hamaker constant, A_{Heff} (Type I of Fig. 2-69), *i.e.*,

---

[44] Forester, J. E., Sunkel, J. M., and Berg, J. C., *J. Appl. Polym. Sci.*, **81**, 1817 (2001).

$$\Pi(z) = \frac{-A_{Heff}}{6\pi z^3},$$ (4.48)

one obtains

$$\Delta F_f = -\frac{S_{L/S}V}{h} - \frac{A_{Heff}V}{2\pi h^3}.$$ (4.49)

(A subscript "$_H$" has been temporarily added to the notation for the Hamaker constant, to avoid confusion with the interfacial area, $\mathcal{A}$.) At equilibrium one has: $(\partial \Delta F / \partial h) = 0$, which gives an equilibrium thickness, h_e:

$$h_e = \left(\frac{-3A_{Heff}}{2\pi S_{L/S}}\right)^{1/2}.$$ (4.50)

Thus the final thickness of the wetting film depends inversely on the spreading coefficient, $S_{L/S}$. For large values of the spreading coefficient, the film becomes monomolecular, but when the spreading is marginal, $i.e.$, $S_{L/S}$ is very near zero, the final "pancake" may be quite thick. If more solid surface is available than needed to accommodate the equilibrium pancake, the remainder of that surface will be dry.

If the spreading coefficient is positive, but the disjoining pressure isotherm is different from Type I, different results may be obtained. For the case of a van der Waals liquid, but one for which the film's effective Hamaker constant is negative (Type II in Fig. 2-69), it is evident that as the film is thinning, ΔF_f will continue to become more negative until $\Pi(h)$ returns to zero, a point corresponding to a thin thin (adsorbed) film. Any excess of liquid will exist as macroscopic droplets on the surface. If the disjoining pressure isotherm is of Type III or IV, the film will thin until the first stable equilibrium film thickness is achieved. For Type III, that will mean either a pancake (the same as for Type I) or droplets surrounded by thin thin (adsorbed) film. Type IV may produce droplets surrounded by thick thin film, a metastable state, or a thin thin film.

The situation given by Type II, III or IV disjoining pressure isotherms, is shown in Fig. 4-41. Thin liquid films are in a state of tension, the "film tension," σ_f, that is not only the sum of the tensions of the bounding surfaces, but accounts also for the work required to bring the film to its thickness h against the disjoining pressure forces. The film tension for a supported thick thin film is thus given by

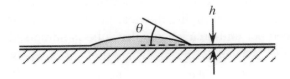

Fig. 4-41: Droplet in equilibrium with a supported thin film.

$$\sigma_f = \sigma_{LG} + \sigma_{SL} - \int_\infty^h \Pi(z)dz. \tag{4.51}$$

The angle θ that the droplet makes against the thin film is obtained from a horizontal force balance in which the film tension (pulling to the right in Fig. 4-41) is set equal to the sum of the horizontal boundary tensions pulling to the left:

$$\sigma_{LG} + \sigma_{SL} - \int_\infty^h \Pi(z)dz = \sigma_{LG}\cos\theta + \sigma_{SL}, \text{ or} \tag{4.52}$$

$$\cos\theta = 1 - \frac{1}{\sigma_{LG}}\int_\infty^h \Pi(z)dz. \tag{4.53}$$

For the case of a thin thin film, the disjoining pressure integral becomes:

$$\int_\infty^h \Pi(z)dz = \int_\infty^{\Gamma/\rho_N} \Pi(z)dz = -(S_{L/S} - \pi_e), \tag{4.54}$$

where π_e is the equilibrium spreading pressure. This will be present only if the liquid is volatile, giving finally:

$$\cos\theta = \frac{\sigma_{SG} - \sigma_{SL} - \pi_e}{\sigma_{LG}}, \tag{4.55}$$

the same as Eq. (4.3), *i.e.*, Young's Equation, modified to account for the presence of an adsorbate at the solid-gas interface.

2. Temperature effects of wetting; heats of immersion and wetting transitions

Differentiation of Young's Equation, with respect to temperature leads to the awkward result:

$$\frac{d\cos\theta}{dT} = \frac{\cos\theta}{\sigma_L}\frac{d\sigma_L}{dT} + \frac{1}{\sigma_L}\frac{d(\sigma_S - \sigma_{SL})}{dT}. \tag{4.56}$$

Recalling that the temperature derivative of the liquid surface tension is of the order of -0.1 mJ/m°K, it may be anticipated that the temperature dependencies of σ_S and σ_{SL} are of similar magnitudes. Thus while Eq. (4.56) may be integrated to give the explicit dependence of the contact angle on temperature, the smallness of the surface energy temperature derivatives suggests that wetting depends only rather weakly on temperature. Nonetheless, measurement of the T-dependence of the contact angle provides a means for determining the heat of immersion, ΔH^σ_{imm}, defined as the heat absorbed per unit area when a dry solid is immersed in a liquid, *i.e.*, $\Delta H^\sigma_{imm} = H^\sigma_{SL} - H^\sigma_S$. Usually a small amount of heat is evolved upon immersion, so that ΔH^σ_{imm} takes on small negative values. Recalling that the surface "tensions": σ_L, σ_S and σ_{SL} are free energies per unit area, one may use the Gibbs-Helmholtz Equation to obtain ΔH^σ_{imm}:

$$d\frac{\dfrac{(\sigma_{SL} - \sigma_S)}{T}}{dT} = -\frac{\Delta H^{\sigma}_{imm}}{T}.$$

(4.57)

Substitution of Eq. (4.56) into (4.57) leads to:

$$\Delta H^{\sigma}_{imm} = -\sigma_L \cos\theta + T\cos\theta\frac{d\sigma_L}{dT} + T\sigma_L\frac{d\cos\theta}{dT}.$$

(4.58)

Comparison of the heat of immersion determined using Eq. (4.58) with the value determined directly by calorimetry provides a confirmation of the validity of Young's Equation.

An exception to the weak dependence of wetting behavior on temperature occurs when a system consisting initially of a liquid that only partially wets a given surface, as pictured in Fig. 4-5(b), transforms itself spontaneously into a wetting film (Fig. 4-5(a)). It is said to undergo a *wetting transition*, and such an event may be the result of a temperature change, but may also result from changes in chemical composition, *etc.* The temperature at which the process occurs is T_w, the critical temperature for the wetting transition. The initial state may be droplets in contact with a bare surface (if the liquid is non-volatile), but more often droplets coexisting with a monolayer, sub-monolayer or thin film of the droplet liquid, as described above. The origin of such changes can be traced to changes in the relevant disjoining pressure isotherm. The transition may of course occur in the reverse direction, *i.e.*, a thick thin film may break up into droplets at a given temperature. The temperatures for the two processes will in general be different owing to the same factors responsible for contact angle hysteresis.[45]

3. The kinetics of spreading on smooth surfaces

The rate of spreading is influenced by gravitational and inertial effects (if the spreading mass is sufficiently large), by viscosity (principally of the underlying liquid, if the substrate is a liquid), and by the magnitude of the spreading coefficient. Taking the rate of spreading (on a solid substrate) to be the rate of movement of the nominal interline between the bulk liquid and the precursor film, one obtains an expression for it starting with Tanner's Law under low Capillary number conditions ($Ca = U\mu/\sigma \leq 0.1$):

$$\theta_D^3 \propto Ca.$$

(4.59)

Consider a small ($Bo \ll 1$) spreading drop, hence a spherical cap, with radius r and height h, as shown in Fig. 4-42. Consider θ_D to be sufficiently small that its volume is given to a rough approximation by: $V \approx \pi r^2 h$, with h given by $h \approx 1/2 r\theta_D$ (θ_D in radians). Substituting into the expression for V and rearranging: $\theta_D \approx 2V/\pi r^3$. Finally, substitution into Tanner's Law gives:

---

[45] Wang, J. Y., Crawley, M., and Law, B. M., *Langmuir*, **17**, 2995 (2001).

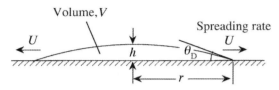

Fig. 4-42: Figure for derivation of spreading rate law.

$$\theta_D^3 \approx \frac{8V^3}{\pi^3 r^9} \propto Ca = \frac{U\mu}{\sigma} = \frac{\mu}{\sigma}\frac{dr}{dt}, \text{ or} \tag{4.60}$$

$$U \equiv \frac{dr}{dt} \propto \frac{V^3\sigma}{\mu} \cdot \frac{1}{r^9}, \tag{4.61}$$

which upon integration yields:

$$r \propto t^{1/10}. \tag{4.62}$$

The latter form of spreading rate law has been confirmed by experiment.[46] The smallness of the exponent on time shows that while spreading may be rapid at first, it quickly slows with time.

What is interesting (and perhaps puzzling!) about the above result is the apparent non-dependence of the spreading rate on the spreading coefficient. This apparent contradiction has been explained by de Gennes[47] as follows. The total hydrodynamic driving force for the spontaneous spreading of a liquid at any instant is the uncompensated "Young's force" (per unit of interline):

$$\mathcal{F}_Y = \sigma_{SG} - \sigma_{SL} - \sigma\cos\theta_D = S + \sigma(1 - \cos\theta_D) \approx S + \frac{1}{2}\sigma\theta_D^2, \tag{4.63}$$

the latter geometric simplification being valid for small θ_D. $\mathcal{F}_Y$ times the rate of interline movement U is the spreading power generated by the disequilibrium of the system. It is balanced by the total rate of viscous dissipation in the film, $\dot{\Delta}$. What de Gennes showed was that this dissipation could be split into two terms, corresponding to dissipation in the precursor film and in the bulk film, and that these could be identified one-to-one with the respective terms in the spreading power:

$$\dot{\Delta} = \dot{\Delta}_{precursor} + \dot{\Delta}_{bulk} = SU + \frac{1}{2}\sigma\theta_D^2 U. \tag{4.64}$$

The product SU is often over 1000 times the magnitude of the second term on the right, but this larger term consists of energy that is entirely consumed in the spreading of the invisible precursor film. The observed spreading then

---

[46] Marmur, A., *Adv. Colloid Interface Sci.*, **19**, 75 (1983).

[47] de Gennes, P. G., in **Liquids at Interfaces**, J. Charvolin, J.F. Joanny and J. Zimm-Justin (Eds.), p. 273, Elsevier, Amsterdam, 1990.

depends only on the rather small driving force given by the second term, which is independent of the spreading coefficient. A full description of the hydrodynamics of spreading of pure liquids has been given by Teletzke *et al.*[48]

4. *Spreading agents; super-spreaders*

One of the most effective methods of promoting wetting or spreading is through the use of wetting or spreading "agents." When the liquid whose spreading is sought is water, these are typically surface active agents. It is evident how they may promote wetting by reducing the surface tension of the water and/or by decreasing the solid-liquid interfacial energy, as discussed earlier. Not all surfactants are effective in doing either of these, however, and in fact their effectiveness depends upon the details of their structures in solution, which are not yet fully understood. Comprehensive studies of Stoebe *et al.*,[49] and the references they contain, give a picture of the state-of-the-art regarding spreading agents. A number of surfactants, usually of low water solubility, at concentrations somewhat above their saturation solubilities, have been found to produce spreading of water over otherwise quite hydrophobic surfaces (θ up to 90°), but a new class of compounds, called "super-spreaders," have been found which will cause water to spread over virtually any surface.[50] These materials are typically polyethoxylated trisiloxanes, with the structure shown in Fig. 4-43. An example is the commercial product Silwet-77® from Crompton Corp. USA, in which the number of ethylene oxide units, $n = 7$-8. Variations on this structure have been tested, particularly for the difficult task of wetting

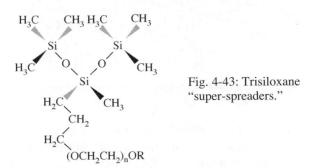

Fig. 4-43: Trisiloxane "super-spreaders."

hydrophobic foliage.[51] They are believed to act by forming very large aggregate structures in solution which readily break up in the vicinity of the advancing interline, setting up large surface tension gradients which pull the

[48] Teletzke, G.F., Davis, H.T., and Scriven, L.E., *Chem. Eng. Comm.*, **55**, 41 (1986).

[49] Stoebe, T., Lin, Z., Hill, R.M., Ward, M.D., and Davis, H.T., *Langmuir*, **13**, 7270; 7276; 7282 (1997).

[50] Ananthapadmanabhan, K.P., Goddard, E.D., Chandar, P., *Colloids Surfaces*, **44**, 281 (1990); Zhu, X., Miller, W.G., Scriven, L.E., and Davis, H.T., *Colloids Surfaces*, **90**, 63 (1994).

[51] Zhang, Y., Zhang, G., and Han, F., *Colloids Surfaces A*, **276**, 100 (2006).

interline forward. This is an example of the "Marangoni effect" to be discussed in Chap 10. Super-spreaders are also effective in causing water to spread over hydrocarbon liquids, but trisiloxanes are not unique in this respect. Such spreading is also accomplished with polyfluorinated surfactants, used to produce the 3M Company's "Light Water™," used to extinguish oil fires.

F. The relationship of wetting and spreading behavior to adhesion

1. *Definition of adhesion; adhesion mechanisms*

Adhesion refers to the sticking together of solid surfaces by *molecular* attraction across their common interface.[52] There is, however, little direct molecular contact between real solid surfaces, due to their roughness and the presence of adsorbed contamination, as indicated in Fig. 4-44, since there are relatively few points of contact where molecular interactions (which are very short-ranged) between the solids can occur. The extent of contact will depend on the softness of the materials and the compressive load applied, but in general sufficient intimacy can be achieved only if one of the materials, the *adhesive*, is in liquid form. The solid surface to which the adhesive is applied is termed the *adherend*. In the ideal case, the liquid adhesive may be envisioned as completely contacting the solid over all its

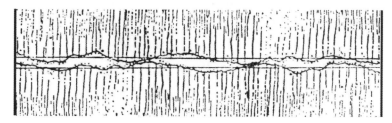

Fig. 4-44: Minimal molecular contact between solid surfaces, due to roughness and contamination.

area, and displacing any physisorbed contamination in the process. One may further idealize by assuming that as the liquid solidifies (cures), this full contact is maintained. Attention is then focused on either one of the adhesive-adherend interfaces or on the "sandwich" as a whole, referred to as a "bondline." When the adhesive joint is pulled apart, the locus of failure may be right at or near one of the actual interface(s), *i.e.*, "adhesive failure," or it may be further away from the interface in either the cured adhesive or bulk adherend phase, yielding "cohesive failure." The concern here is with *ad*hesive failure.

---

[52] Kendall, K., **Molecular Adhesion and Its Applications: The Sticky Universe**, Kluwer/Plenum, New York, 2001.

Wetting and spreading play an enormous role in the phenomenon of adhesion,[53] but there is much more to it than that. Comprehensive monographs are available,[54] so suffice it here to point out what are believed to be the four principal mechanisms of adhesion, pictured in Fig. 4-45. The first is *contact adhesion*, in which molecular interactions (physical, acid-

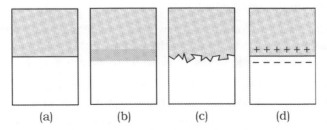

(a)   (b)   (c)   (d)

Fig. 4-45: Four mechanisms of adhesion. (a) Contact adhesion, (b) diffusion interphase adhesion, (c) mechanical interlock, (d) electrostatic adhesion.

base or covalent) occur across a more-or-less smooth, defined interface. It is to this mechanism that most of what follows applies. The second is *diffusion interphase adhesion*, generally applicable to the bonding between polymeric adhesives and adherends. If there is adequate thermodynamic compatibility (mutual solubility) between the polymers, the polymer molecules are mobile (not locked into a tightly cross-linked or crystallized structure), and there is adequate contact time, the polymers will inter-diffuse, or inter-digitate, leading to the formation of an interphase of some thickness. The third mechanism is *mechanical interlock*. The liquid adhesive penetrates the cavities of a rough or porous adherend, and upon solidification forms effective hooks holding the phases together. The final mechanism proposed is that of electrostatic interactions across the interface. The contribution of this mechanism is generally thought to be small. In many situations, two or more of these mechanisms will be important. It appears likely that while wetting behavior may directly relate to contact adhesion, it also must play some role in the remaining mechanisms, because if good physical intimacy between the phases is not achieved, it will be difficult for the remaining three mechanisms to be effective.

2. The "Laws of Molecular Adhesion"

While the practical consequences of adhesion, and the general approaches used to describe it, are macroscopic, its origin is molecular

[53] Chaudhury, M. K., *Mater. Sci. Eng.*, **R16**, 97 (1996).
[54] Kinloch, A. J., **Adhesion and Adhesives**, Chapman and Hall, London, 1987;
Pocius, A. V., **Adhesion and Adhesives Technology**, Hanser/Gardner, Cincinnati, 1997;
Adhesion Science and Engineering, Vols. I and II, (A. V. Pocius, Series Ed.):
Vol. I: **The Mechanics of Adhesion**, D. A. Dillard and A. V. Pocius (Eds.)
Vol. II: **Surfaces, Chemistry & Applications**, M. Chaudhury ands A. V. Pocius (Eds.),
Elsevier, Amsterdam, 2002.

interaction. At the molecular level, adhesion behavior often appears counter-intuitive, as pointed out by Kendall[55], who suggests three "Laws of Molecular Adhesion" as follows:

1. All atoms and molecules adhere with considerable force. If two solid bodies approach to nanometer separations (across a vacuum), they will jump into contact.

2. The effect of contaminant "wetting" molecules is to reduce adhesion, or even to make the bodies repel each other. It would thus appear that "adhesives" *reduce* molecular adhesion.

3. Molecular adhesion forces are of such short range that various factors, such as roughness, Brownian motion, cracking, plastic deformation, *etc.* can have large effects on macroscopic adhesion while molecular adhesion remains the same.

The first two of these "laws" can readily be understood in terms of considerations developed earlier. Considering a thin fluid film separating two solid surfaces a distance h apart, Eq. (2.109) gives the adhesion force per unit area (equivalent to the negative of the disjoining pressure) as

$$\frac{F_{adhesion}}{Area} = -\Pi(h) = \frac{A_{eff}}{6\pi h^3},$$ (4.65)

where A_{eff} is the effective Hamaker constant. Equation (4.65) assumes the existence of the universally present dispersion forces between the molecules. The effective Hamaker constant is given by the appropriate form of Eq. (2.115) for material (2) interacting across a film of liquid (1) with a material (3):

$$A_{eff} = \left(\sqrt{A_{33}} - \sqrt{A_{11}}\right)\left(\sqrt{A_{22}} - \sqrt{A_{11}}\right).$$ (4.66)

Equation (4.66) illuminates two facts that are equivalent to the first two laws of molecular adhesion. First, the effective Hamaker constant for the interaction between two materials of the same type (where $A_{22} = A_{33}$) is always positive, yielding positive adhesion forces. Second, the presence of any intervening medium, with $A_{11} > 0$, will reduce A_{eff} and the strength of the adhesion forces. A_{eff} may even become negative for the adhesion force between two different materials if the Hamaker constant of the intervening medium is intermediate to that of the two materials, *i.e.*, if

$$A_{22} < A_{11} < A_{33} \text{ or } A_{33} < A_{11} < A_{22}.$$ (4.67)

The third law of molecular adhesion addresses some of the features illustrated in Fig. 4-44, *viz.* roughness and contamination of actual macroscopic surfaces that prevent intimacy of contact, vital since the range of action is so short. If the system pictured in Fig. 4-44 is subjected to stress

[55] Kendall, K., **Molecular Adhesion and Its Applications: The Sticky Universe**, Kluwer/Plenum, New York, 2001.

(say a shear stress), micro-cracks will develop at the points of molecular contact. The Brownian motion or diffusion of molecules or nano-sized portions of material at the crack tip will continually probe the crack configuration. The existence of stresses biases the system against re-healing when microscopic motion is in the direction of the advancing the crack.

3. *"Practical adhesion" vs. "thermodynamic adhesion"*

Macroscopic adhesion is quantified in practice by the strength or durability of the bond between two solids as measured by some mechanical test. This "mechanical adhesion" or *practical adhesion*, is what is of interest for actual applications, but is a difficult property to pin down, depending always on the method used for its measurement and the conditions of that measurement, as suggested in Fig. 4-46, which shows from left to right: peel tests (90° and 180°), a lap shear test, and a fiber pull-out test. It is essential to keep in mind that adhesion involves many things besides the chemistry and physics of the interface (or often an "inter*phase*" of finite thickness in which the adhesive and the adherend have inter-diffused[56]). These include

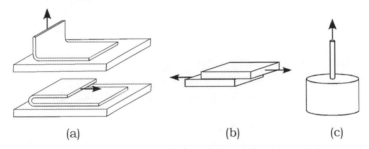

(a)   (b)   (c)

Fig. 4-46: Tests for measurement of practical (mechanical) adhesion: (a) Peel tests, (b) lap shear test, (c) fiber pull-out test.

roughness and mechanical interlocking, electrostatic effects, and the mechanics of the bulk phases near the boundary (*e.g.*, the existence and level of residual stresses), and the rheology of the disjoining event.[57,58,59]

Another concept of adhesion, based purely on thermodynamics and not subject to the ambiguities of practical adhesion, is the work of adhesion, W_A, defined earlier, Fig. 4-33(a). Furthermore, if the contact angle of the adhesive against the adherend is finite, the Young-Dupré Equation, Eq. (4.38), may be used to measure it in the laboratory. One must ask, however, if there is a meaningful relationship between the above two concepts of "adhesion." Any such relationship between W_A and practical adhesion is in general complex and would be limited to contact adhesion, *i.e.*, cases in which an extensive inter*phase* is not involved. It should first be noted that W_A is an energy quantity rather than a force. The quantity that should be

[56] Sharpe, L. H., *J. Adhesion*, **67**, 227 (1998).

[57] Kinloch, A. J., **Adhesion and Adhesives**, Chapman and Hall, London, 1987.

[58] Wu, S., **Polymer Interface and Adhesion**, Marcel Dekker, New York, 1982.

[59] Pocius, A. V., **Adhesion and Adhesives Technology**, Hanser/Gardner, Cincinnati, 1997.

compared with W_A is a mechanical adhesive failure *energy*, w_m, *i.e.*, the work required (per unit area) to disrupt the bond. It is likely, at least, that the maximum adhesive force is proportional to w_m. In accord with the Young-Dupré Equation, Eq. (4.38), for a given adhesive (*i.e.*, given σ_L), thermodynamic adhesion should vary as $(1+\cos\theta)$. As an example, Fig. 4-47 shows such a comparison of the mechanical adhesive strength for ice against various solids[60] against thermodynamic adhesion. The result is typical, and

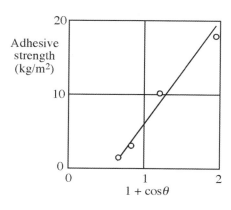

Fig. 4-47: Relationship between measured mechanical adhesion strength and $(1+\cos\theta)$ for ice against various solids. After [Mittal, K. L., "Surface Chemical Criteria for Maximum Adhesion and Their Verification Against the Experimentally Measured Adhesive Strength Values," in **Adhesion Science and Technology**, Vol. 9A, L.-H. Lee, Ed., pp. 129-171, Plenum Press, New York, 1975.]

the literature contains many such comparisons. An expression relating W_A and the adhesive fracture energy w_m, based on continuum fracture mechanics, has been derived for elastic solids and takes the simple form:[61]

$$w_m = C(\dot{a}, T, \varepsilon) \cdot W_A, \tag{4.68}$$

where the factor C depends on crack growth rate $\dot{a}$, temperature T, and the stored strain energy level ε. Thus while w_m may be orders of magnitude larger than W_A, the proportionality should hold for a series of tests on different materials conducted in the same way. A convincing demonstration is given in Gent and Schulz,[62] who performed peel tests disjoining bonds formed by a molten polybutadiene adhesive against a Mylar adherend in air and in various liquids. Changing the fluid medium had the effect of changing only W_A from case to case, without changing C. At a given peel rate it was found that there was a good correlation between mechanical peel energy and W_A. Thus while mechanical adhesion depends upon much more than interfacial thermodynamic properties, good thermodynamic adhesion appears to be a *necessary* (if not sufficient) condition for good practical adhesion.[63] In terms of wetting, the best thermodynamic adhesion occurs when $\theta \to 0°$. In this case, $W_A = 2\sigma_L =$ work of *cohesion*, W_C, so that the

[60] Raraty, L. E., and Tabor, D., *Proc. Roy. Soc.*, **245A**, 184 (1958).

[61] Gent, A. N., and Kinloch, A. J., *J. Polym. Sci.*, **A2**, 659 (1971);
    Andrews, E. H., and Kinloch, A. J., *Proc. Roy. Soc.*, **A332**, 385, 401 (1973).

[62] Gent, A. N., and Schulz, J., *J. Adhesion*, **3**, 281 (1972).

[63] Berg, J. C., in **Composite Systems from Natural and Synthetic Polymers**, L. Salmén, A. de Ruvo, J. C. Seferis, and E, B. Stark (Eds.), pp. 23-44, Elsevier, Amsterdam, 1986.

locus of failure is as likely to be in the bulk phase as at the interface. This suggests that the optimal adhesive should have σ_L just low enough to make θ = 0°. The bulk adhesive in the solid state, however, generally has a much greater cohesive energy than $2\sigma_L$, due usually to internal cross-linking.

4. *The importance of wetting (contact angle) to practical adhesion*

The minimization of the contact angle between adhesive and adherend is important for reasons other than the maximization of W_A. Often the first step in the formation of a bond between two solid surfaces by means of an adhesive is the establishment of a liquid bridge, as discussed in Chap. 2. The strength of this "capillary adhesion" is directly proportional to $\cos\theta$. When the liquid bridge is cured to a solid, the mechanical properties of the bondline are also dependent on the contact angle. Figure 4-48 shows that the magnitude of the stress concentration factor (local stress/applied stress) in a stressed lap joint increases quite sharply with contact angle beyond about

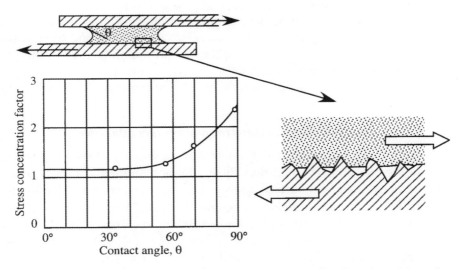

Fig. 4-48: Magnitude of maximum stress concentration as a function of contact angle in a strained lap joint. Location of maximum stress moves from center of bondline toward the interline as contact angle increases. Inset shows vapor void in a rough surface, at edges of which high stress concentrations may develop. After [Zisman, W.A., *in* **Contact Angle, Wettability and Adhesion**, F.M. Fowkes, Ed., pp. 1-51, ACS Adv. In Chem. Ser. 43, ACS, Washington, DC (1964).]

30°.[64] Furthermore, the locus of the stress concentration moves out to the edge of the adhesive layer when the contact angle increases.[65] Such stress concentrations are not unlike those that exist at the sites of vapor inclusions or voids between adhesive and rough porous surfaces, as shown in the inset to Fig. 4-48. The presence of a row of such voids can lead to a zippering type of failure. Voids are also a serious problem in the formation of fiber-

[64] Tabor, D., *Rep. Progr. Appl. Chem.*, **36**, 621 (1951).
[65] Mylonas, C., **Proc. VII Intern. Congr. Appl. Mech.**, London, 1948.

matrix composite materials,[66,67] as shown in Fig. 4-49. The presence of such voids is minimized when the contact angle is minimized, and the best condition is, of course, $\theta = 0°$. It is clear from the above that roughness of the adherend may be either favorable or unfavorable to good practical adhesion, depending on wettability. If wetting is complete, roughness will

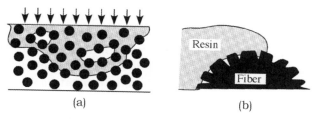

(a)   (b)

Fig. 4-49: Void formation due to poor wetting during the preparation of fiber-reinforced polymeric composite prepregs.

lead to greater molecular contact area, and possibly to mechanical interlock. On the other hand, if wetting is incomplete, roughness may produce air pockets or voids leading to reduced molecular contact area and to sites for high local stress concentrations. Given the central importance of wetting to the establishment of adhesive bonds, many of the practical strategies discussed earlier for optimizing wetting are also appropriate for maximizing adhesion.

5. The optimization of thermodynamic contact adhesion

For a given adherend, it is of interest to know what adhesive (in terms of its surface tension, σ_L) will maximize adhesion strength, or for a given adhesive, what adherend surface energy, σ_S, will maximize bond strength. The simple answer would seem to be that W_A should be maximized, and there indeed appears to be evidence for this. In one such study, silica spheres were loaded into specimens of poly(vinyl butyral) (PVB) that were subsequently subjected to three-point bend tests. The silica spheres were treated to various levels with methyl silane and octyl silane to vary their surface energy, σ_S, which was determined independently in each case using inverse gas chromatography, IGC (to be described later in this chapter). The yield stress of the specimens, assumed to be proportional to the strength of particle-matrix adhesion, was plotted against the work of adhesion, as given by Eq. (4.41), $W_A = 2\sqrt{\sigma_S \sigma_L}$. The surface energy of the polymer, σ_L, was also determined by IGC. A direct correlation between the yield stress and W_A was found, as shown in Fig. 4-50.

[66] Lee, W. J., Seferis, J. C., and Berg, J. C., J. Polym. Composites, **9**, 36 (1987).
[67] Conner, M., Harding, P., Månson, J.-A., and Berg, J. C., J. Adhesion Sci. Tech., **9**, 983 (1995).

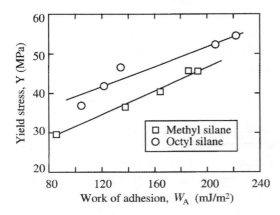

Fig. 4-50: PVB composite yield stress *vs.* work of adhesion between the filler and the matrix. W_A is estimated from IGC measurements on both the matrix polymer and the silane-treated silica filler. After [Harding, P. H., and Berg, J. C., *J. Adhesion Sci. Tech.*, **11**, 471 (1997).]

Sharpe and Schonhorn[68] report convincing data, however, showing that more than W_A must be involved in maximizing adhesion. In an experiment shown in Fig. 4-51, involving epoxy resin and polyethylene, they found the adhesion to be excellent when the epoxy was first cured, and molten polyethylene was applied to it as the adhesive liquid (a), but very poor when (uncured) epoxy resin was deposited as an adhesive to the polyethylene substrate (b). Since the same interface was presumably created in both cases, one would assume that the work of adhesion would be the same, but the results differed sharply. The difference is that the polyethylene solid, with its low surface energy, is not wet by the epoxy liquid ($\theta > 0°$), which must then be coated onto the substrate, while the low surface tension

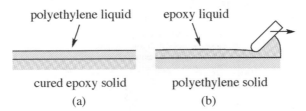

Fig. 4-51: The importance of spreading on adhesion: (a) liquid polyethylene spontaneously spread on an epoxy solid to produce good adhesion, while (b) liquid epoxy coated (but not spontaneously spread) on polyethylene solid produced poor adhesion.

molten polyethylene spontaneously spreads over the cured epoxy adherend ($\theta = 0°$). It was argued that in the absence of spontaneous spreading, so that forced spreading is required to apply the adhesive, only incomplete contact (micro air pockets?) between the adhesive and the adherend would be achieved, leading to poor adhesion. This premise needs further research, but if it is accepted, one may re-state the thermodynamic criterion for the optimization of contact adhesion between an adhesive and a solid surface as follows: Maximum contact adhesion is achieved when the work of adhesion

[68] Sharpe, L. H., and Schonhorn, H., in **Contact Angle, Wettability and Adhesion**, Adv. Chem. Ser., 43, p. 189, 1964.

is maximized subject to the requirement of spontaneous spreading of the adhesive on the substrate. Spreading requires that the spreading coefficient be positive (or at least, non-negative), which in terms of the surface energies of the adhesive and adherend, Eq. (4-43) gives:

$$S_{L/S} = 2\sqrt{\sigma_L}\left[\sqrt{\sigma_S} - \sqrt{\sigma_L}\right] \geq 0,$$ (4.69)

i.e., the surface energy of the solid adherend must exceed the surface tension of the adhesive, $\sigma_S > \sigma_L$. ~~The maximum of W_A under these conditions is~~ *Thus for a given adherend (ie given σ_S), the* ~~achieved~~ *maximum in W_A occurs* when $\sigma_S = \sigma_L$. The literature contains evidence that this is in fact true, an example of which is shown in Fig. 4-52). Dykerhoff and Sell[69] investigated the bonding of various adhesive resins with different surface tensions to polished steel that had been treated with different carboxylic acids to produce adherends of varying surface energy. The adhesive strength was measured using butt-joint tests. The surface tension of the treated steel surfaces was determined by measuring the contact angles of water, formamide and ethylene glycol against them and computing the solid surface energy from the equation resulting from substitution of the combining rule, Eq. (4.27), into Young's Equation, *viz.*

$$\cos\theta = \frac{(0.015\sigma_S - 2)\sqrt{\sigma_S\sigma_L} + \sigma_L}{\sigma_L\left(0.015\sqrt{\sigma_S\sigma_L} - 1\right)}.$$ (4.70)

Results are shown in Fig. 4-52, which clearly shows a maximum in adhesion strength correlating with $\sigma_S = \sigma_L$.

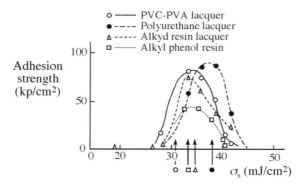

Fig. 4-52: Data showing the adhesion strength of various resins against surface-treated steel as determined using butt-joint tests. The surface tension of each resin, σ_L, is indicated by the arrow crossing the abscissa through the identifying plot symbol. After [Dyckerhoff, G. A., and Sell, P. J., *Angew. Makromol. Chem.*, **21(312)**, 169 (1972).]

[69] Dyckerhoff, G. A., and Sell, P. J., *Angew. Makromol. Chem.*, **21(312)**, 169 (1972).

It may be argued that this criterion is too severe.[70] For the situation in which the adhesive is to form first a liquid bridge between two solid surfaces, as pictured in Fig. 4-48, the free energy per unit bondline area associated with the complete disjoining of the bondline, as shown in Fig. 4-53, would be the appropriate driving force for spontaneous spreading of the

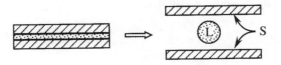

Fig. 4-53: Disjoining of a bondline.

adhesive between the two solid surfaces. This can be denoted as $S_{S/L/S}$, and is given by:

$$S_{S/L/S} = 2(\sigma_S - \sigma_{SL}),\qquad(4.71)$$

recognizable as twice the adhesion tension, defined by Eq. (4.36). We must then have $S_{S/L/S} > 0$ for spontaneous "spreading" of the bondline. (This is actually a wicking event, as described later in this chapter.) Substitution of the Young-Dupré Equation, Eq. (4.38), gives

$$S_{S/L/S} = 2(\sigma_S - \sigma_L) = 2\sigma_L \cos\theta > 0,\qquad(4.72)$$

a criterion satisfied if $\theta < 90°$. Equation (4.72) then suggests that the adhesive surface tension, σ_L, may be as large as $4\sigma_S$ without violating the "spreading" criterion.

6. Acid-base effects in adhesion

To extend the treatment to more general cases, one may use the Fowkes combining equation for σ_{SL}, modified to include the possibility of interactions (I) across the interface in addition to those of London dispersion forces, Eq. (2.28):

$$\sigma_{SL} = \sigma_S + \sigma_L - 2\sqrt{\sigma_S^d \sigma_L^d} - I.\qquad(4.73)$$

Upon substitution, the expression for W_A becomes:

$$W_A = 2\sqrt{\sigma_S^d \sigma_L^d} + I.\qquad(4.74)$$

Owens and Wendt[71] and Kaelble[72] suggested that surface energies consist of contributions due to dispersion forces plus those due to polar effects, i.e.: $\sigma = \sigma^d + \sigma^p$, and that the latter may act across the interface with the same

[70] Paul, C. W., *J. Adhesion Sci. Tech.*, **22**, 31 (2008).

[71] Owens, D. K., and Wendt, R. C., *J. Appl. Polym. Sci.*, **13**, 1741 (1969).

[72] Kaelble, D. H., **Physical Chemistry of Adhesion**, Wiley-Interscience, New York, 1971.

geometric mean mixing rule as is valid for dispersion forces. (Neither assumption is theoretically quite right.) This yields:

$$W_A = 2\sqrt{\sigma_S^d \sigma_L^d} + 2\sqrt{\sigma_S^p \sigma_L^p},$$ (4.75)

an expression that is maximized when the "fractional polarity" of the adhesive and the adherend are equal, *i.e.*:

$$\frac{\sigma_S^p}{\sigma_S^d} = \frac{\sigma_L^p}{\sigma_L^d}.$$ (4.76)

This so-called "polarity matching" condition is consistent with "de Bruynes' Rule," which states that optimum adhesives for nonpolar materials are nonpolar and for polar materials are polar[73]. The adherend surface ("treated" or otherwise) may be "characterized," *i.e.*, given values of σ_S^d and σ_S^p, by using a pair of *probe liquids*, each with known σ_L^d and σ_L^p values. These are widely tabulated, as in Table 2-2,[74] or may in principle be determined from interfacial tension measurements against immiscible apolar liquids. The work of adhesion values of these probe liquids against the solid are measured from their contact angles against the solid and their surface tensions:

$$W_{Ai} = \sigma_i(1 + \cos\theta_i) = 2\sqrt{\sigma_i^d \sigma_S^d} + 2\sqrt{\sigma_i^p \sigma_S^p},$$
$$W_{Aj} = \sigma_j(1 + \cos\theta_j) = 2\sqrt{\sigma_j^d \sigma_S^d} + 2\sqrt{\sigma_j^p \sigma_S^p},$$ (4.77)

providing two equations in the two unknowns σ_S^d and σ_S^p. As an example, the method has been applied to the assessment of various surface treatments of cellulose fibers.[75]

An alternative view of the work of adhesion was proposed by Fowkes[76] beginning in 1978, and developed significantly since that time.[77] What had been identified above as "polar" interactions are more likely to be due to H-bonding, or more generally, to "acid-base interactions." Recognition of the importance of acid-base interactions means that the polarity-matching requirement is an oversimplification. Further investigation has revealed that the interactions due to the presence of permanent dipoles is very small and easily lumped together with the dispersion forces into a term which might more properly be designated as "Lifshitz-van der Waals" interactions. (The Lifshitz approach to the computation of molecular interactions between macroscopic bodies is discussed in Chap. 7.) In addition to Lifshitz-van der Waals effects, which are often termed simply

[73] de Bruyne, N. A., *The Aircraft Engineer* (supplement to flight), 18(12), 51 (1939).
[74] σ^p values would be computed as $(\sigma - \sigma^d)$.
[75] Westerlind, B. S., and Berg, J. C., *J. Appl. Polym. Sci.*, **36**, 523 (1988).
[76] Fowkes, F. M., *J. Adhesion Sci. Tech.*, **1**, 7 (1987).
[77] Jensen, W. B., **The Lewis Acid-Base Concepts: An Overview**, Wiley, New York, 1980; Berg, J. C., in **Wettability**, J. C. Berg, Ed., 75-148, Marcel Dekker, New York, 1993.

"dispersion" interactions, there may be those due to metallic bonding or to specific chemical interactions. Virtually all of the latter may be lumped into "acid-base" interactions, when they are considered in the most general "Lewis" sense. A Lewis acid-base interaction occurs whenever an electron pair from one of the participants is shared and may vary from an ionic interaction in one extreme to a covalent bond in the other. With these ideas, the work of adhesion may be expressed as (suggested by Fowkes):

$$W_A = 2\sqrt{\sigma_S^d \sigma_L^d} + f \cdot n \cdot (-\Delta H^{ab}), \qquad (4.78)$$

where ΔH^{ab} is the enthalpy (per mole) of acid-base adduct formation between the acid or base functional groups on the adherend and in the adhesive, n is the number (moles) of accessible functional groups per unit area of the adherend, and f is an enthalpy-to-free energy correction factor (which Fowkes assumed to be ≈ 1). In view of the above equation, what should really be sought for maximum adhesion is not a "polarity match," but rather complementarity, i.e., acidic adhesives for basic surfaces and vice versa. Many materials have both an acid *and* a base character, and are sometimes referred to as "bipolar."[78] Examples include water, alcohols, amides, carboxylic acids and many more, and when these are liquids, they are self-associated. Only a minority of materials may be regarded as "monopolar," i.e., exclusively acids (such a chloroform) or exclusively bases (such as most ketones). Thus whenever a monopolar or bipolar adhesive interacts with a solid with bipolar functional groups, or vice versa, there will be an acid-base interaction. Materials capable of entering into acid-base interactions are often thought of as "polar," (even though their dipole moments may be nil), and it is easy to see the origin of the fortuitous success of the Owens-Wendt-Kaelble approach. That approach misidentified the acid-base interaction: $f \cdot n \cdot (-\Delta H^{ab})$ as a "polar" interaction: $\sqrt{\sigma_S^p \sigma_L^p}$, and the probe liquids used and/or the adherends were generally bipolar, so that *some* interaction was detected. Consistent with the acid-base theory, the surface free energies of materials should be decomposed into contributions from dispersion forces and from those of acid-base *self*-association. It is the latter term that was misidentified as the "polar contribution" in the earlier work. Conspicuous, i.e. qualitative, failures of the old theory would occur only for the relatively uncommon cases when a monopolar acid interacted with a monopolar base.

In light of acid-base interactions, the criteria for adhesive optimization may be re-formulated as follows. The spreading criterion becomes:

$$S_{L/S} = \sigma_S - \sigma_L - \sigma_{SL} = 2\sqrt{\sigma_S^d \sigma_L^d} - 2\sigma_L + I^{ab} > 0, \qquad (4.79)$$

so that the maximum allowable surface tension of the adhesive, σ_L^{max}, is now seen to depend upon the extent of acid-base interaction across the interface.

---

78 van Oss, C. J., Good, R. J., and Chaudhury, M. K., *J. Colloid Interface Sci.*, **111**, 378 (1986).

It may be considerably larger than σ_s, thus increasing the possible value of the work of adhesion.

The acid-base theory of adhesion has met with qualitative success, two examples of which are pictured in Figs. 4-54 and 4-55, taken from

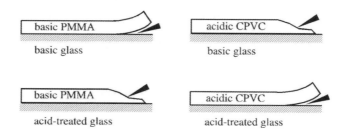

Fig. 4-54: Films of the basic polymer PMMA cast from solution onto ordinary (basic) glass are easily peeled off, but cast films of the acidic polymer *CPVC* cannot be peeled off. However, after the glass is rinsed with HCl, it allows the *CPVC* to be peeled off, but not the PMMA. After [Fowkes, F. M., *J. Adhesion Sci. Tech.*, **1**, 7 (1987).]

Fowkes. When a basic polymer film, poly(methyl methacrylate), PMMA, was cast against a basic soda glass substrate, poor adhesion was obtained, but when the acidic polymer, post-chlorinated poly(vinyl chloride), CPVC, was cast against the basic substrate, adhesion was very strong. W_A based on

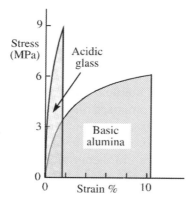

Fig. 4-55: Instron stress-strain results for 250 μm films of an acidic polymer poly(vinyl butyral) filled with 88 wt % of an acidic glass powder or of a basic alumina powder. Area under curve is proportional to toughness. After [Fowkes, F. M., *J. Adhesion Sci. Tech.*, **1**, 7 (1987).]

dispersion interactions was about the same for both cases. The situation was permuted by converting the soda glass to an acidic substrate by soaking it in a concentrated HCl solution, producing the results shown. In another example, Fig. 4-55 compares the very low toughness of an acidic polymer, poly(vinyl butyral), PVB, reinforced with an acidic glass polymer with the high toughness of the same polymer filled with basic alumina particles. There has been hope that relations such as Eq. (4.78) might eventually achieve semi-quantitative status, but that is not the case at present. Among the difficulties are that the factor n is not known for any given solid of interest, and other evidence suggests that the enthalpy-to-free energy factor, f, *cannot* be assumed to be unity.[79] Nonetheless, some guidance in optimizing acid-base interactions may be obtained by examining the strength of the relevant acid-base enthalpies, (ΔH^{ab}). These have been correlated in a number of ways, two of the most common being the Gutmann *Donor Number* and *Acceptor Number* scales and the Drago "*E & C*" formulation,[80] some values of which are shown in Table 4-7. The Donor Number, *DN*, quantifies the Lewis basicity of a variety of solvents and was defined for a given base B as the exothermic heat of its reaction with the reference acid, antimony pentachloride, in a 10^{-3} M solution in a neutral solvent (1,2-dichloroethane). The units are (kJ/mol). Specifically:

$$DN_B = -\Delta H(SbCl_5 : B). \qquad (4.80)$$

The dimensionless Acceptor Number, *AN*, ranked the acidity of a solvent and was defined for an acidic solvent A as the relative ^{31}P NMR downfield shift $(\Delta \delta)$ induced in triethyl phosphine when dissolved in pure A. A value of 0 was assigned to the shift produced by the neutral solvent hexane, and a value of 100 to the shift produced by $SbCl_5$. Gutmann suggested that the enthalpy of acid-base adduct formation be written as:

Table 4-7: Acid-base interaction parameters.

Gutmann Donor Numbers (*DN*) [Gutmann, V., **The Donor-Acceptor Approach to Molecular Interactions**, p. 20, Plenum Press, New York, 1978] and Acceptor Numbers (*AN*) [Riddle, F. L., Jr., and Fowkes, F. M., *J. Amer. Chem. Soc.*, **112**, 3259 (1990)]

Bases	(DN) (kJ/mol)	Acids	(AN)
Acetone	17.0	Acetonitrile	16.3
Dioxane	14.8	Chloroform	18.7
Methyl Acetate	16.5	2-Propanol	31.5

Drago *E* and *C* constants (kcal/mol)$^{1/2}$ From [Drago, R. S., Wong, N., Bilgrien, C., and Vogel, G. C., *Inorg. Chem.*, **26**, 9 (1987)]

Bases	E	C	Acids	E	C
Triethylamine	0.99	11.09	Phenol	4.33	0.44
Pyridine	1.17	6.40	Chloroform	3.31	0.16
Ethyl acetate	0.97	1.74	Water	2.45	0.33
PMMA	0.68	0.96	Silica	4.39	1.14

$$-\Delta H^{ab} = \frac{(AN_A)(DN_B)}{100}, \qquad (4.81)$$

[79] Vrbanac, M. D., and Berg, J. C., *J. Adhesion Sci. Tech.*, **4**, 255 (1990).

[80] Drago, R.S., and Wayland, B. B., *J. Amer. Chem. Soc.*, **87**, 3571 (1965); Drago, R. S., Dadmun, A. P., and Vogel, G. C., *Inorg. Chem.*, **32**, 2473 (1993); Vogel, G. C., and Drago, R. S., *J. Chem. Ed.*, **73**, 701 (1996).

where the factor 100 converts the tabulated *AN* to a decimal fraction of the SbCl$_5$ value. A considerable database for donor and Acceptor Numbers (corrected for solvent effects) is available[92].

Drago, on the other hand, expressed the enthalpy of acid-base complex formation in apolar organic solvents as:

$$-\Delta H^{ab} = E_A E_B + C_A C_B,$$ (4.82)

where E_A and E_B represent the "hardness" or electrostatic contribution, and C_A and C_B the "softness" or covalent contribution to the exothermic heat. The *E* and *C* constants are obtained experimentally from calorimetric or other measurements, with iodine chosen as a reference acid having $E = 1.00$ and $C = 1.00$ (kcal/mol)$^{1/2}$. Materials with *E* much larger than *C* are termed *hard* acids or bases, while those with *C* much larger than *E* are termed *soft* acids or bases. *E* and *C* constants for many liquids are presently tabulated, but those for a given solid of interest must generally be obtained experimentally.

The above considerations are all relevant to guidelines for the selection of an optimum adhesive in a given situation, or for assessing the effect of various adherend surface treatments. Examples of solid surface treatments to promote adhesion include: chemical grafting (coupling agents), plasma treatment (CASING = cross-linking by activated species of inert gas), acid etching, corona discharge, transcrystalline growth (promotion of surface crystallinity by solidification against a particular solid), anodization (metal such as aluminum treated by making it the anode of an electrolytic cell and thereby producing surface porosity), photochemical treatment, and UV laser treatment. While the thermodynamic considerations and criteria apply directly only to contact adhesion, they must also be favorable if the mechanisms of adhesive-adherend interpenetration or successful mechanical interlocking are to occur. They remain as important screening criteria in nearly all cases.

7. Contact mechanics; the JKR method

As evidenced above, it is difficult to obtain values for the work of adhesion, W_A, for solid-liquid or solid-solid interfaces or for the solid-fluid interfacial energies such as σ_{SG} or σ_{SL} from mechanical measurements without uncomfortable assumptions and extrapolations. One technique has been developed, however, which makes this possible.

Consider the situation when two spheres, or a sphere and a plate as shown in Fig. 4-56, are pressed together under a compressive load *P*. Assuming elastic materials, Hertz[81] gave a complete description of this system, including the surface deformations and the internal stress fields. For the case of a sphere of radius *R*, Young's modulus *E*, and Poisson's ratio *v*

---

[81] Hertz, H., *J. Reine Angew. Math.*, **92**, 156 (1882).

pressed against a rigid (undeformable) flat plate with a load P, they obtained for the radius of the circle of contact, a:

$$a^3 = \frac{PR}{K}, \qquad \text{where} \tag{4.83}$$

$$\frac{1}{K} = \frac{3}{4}\left[\frac{1-\nu^2}{E}\right]. \tag{4.84}$$

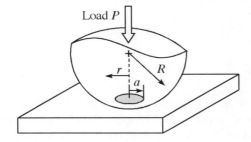

Load P

Fig. 4-56: Schematic of the Hertz experiment, in which an elastic sphere is pressed against a rigid flat plate.

The extent of the sphere flattening (the central displacement), δ, was

$$\delta = \frac{P^2R}{K} = \frac{a^2}{R}, \tag{4.85}$$

and the normal stress distribution within the contact circle was given by

$$S(r) = \frac{3Ka}{2\pi R}\left[1-\left(\frac{r}{a}\right)^2\right]^{1/2}. \tag{4.86}$$

The analysis assumed the contact was frictionless, so that no shear stresses could be sustained at the interface, and that no adhesion occurred between the bodies. Upon unloading, the contact circle, central displacement and normal stress all reversibly disappeared, and detachment required no tensile force.

Johnson, Kendall and Roberts,[82] in their landmark paper, considered the situation where adhesion existed across the contact circle and was sufficiently strong to deform the shape of the sphere in its vicinity, as pictured in Fig. 4-57. A finite circle of contact would therefore exist even under conditions of no applied load, and a finite tensile load would be required to detach the sphere. Their solution added the adhesion force, as dictated by the work of adhesion, to the normal stress distribution, to obtain:

$$S(r) = \frac{3Ka}{2\pi R}\left[1-\left(\frac{r}{a}\right)^2\right]^{1/2} - \left(\frac{3W_AK}{2\pi a}\right)^{1/2}\left[1-\left(\frac{r}{a}\right)^2\right]^{-1/2}. \tag{4.87}$$

[82] Johnson, K. L., Kendall, K., and Roberts, A. D., *Proc. Roy. Soc. London A*, **324**, 301 (1971); Johnson, K. L., **Contact Mechanics**, Cambridge Univ. Press, New York, 1987.

Fig. 4-57: Schematic of JKR experiment.

The result for the radius of the enlarged contact spot was:

$$a^3 = \frac{R}{K}\left[P + 3\pi W_A R + \sqrt{6\pi W_A R P + \left(3\pi W_A R\right)^2}\right].$$

(4.88)

For the case of zero applied load ($P = 0$), Eq. (4.88) reduces to:

$$a_0^3 = \frac{6\pi W_A R^2}{K}.$$

(4.89)

When a tensile load is applied ($P < 0$), a is reduced until the point is reached where the expression under the radical in Eq. (4.88) reaches 0, beyond which there is no solution, and the sphere detaches. The tensile (pull-off) force is:

$$P_S = \frac{3}{2}\pi W_A R.$$

(4.90)

The pull-off force is thus clearly a measurable quantity relating directly back to the work of adhesion. The radius of the contact spot at the instant of detachment is given by substituting $P = -P_S$ from Eq. (4.90) into Eq. (4.88) to get

$$a_S^3 = \frac{3\pi W_A R^2}{2K}.$$

(4.91)

If either the plate or the sphere is transparent, the course of the contact radius may be followed as the applied load is varied. This is the basis of the "JKR method." The ability to fit data for the contact spot size variation with load as well as the pull-off force with a single value for W_A provides an important check on internal consistency. Figure 4-58 shows a set of results redrawn from the original JKR paper for the contact of gelatin spheres of various radius from a few mm to a few cm against a poly(methylmethacrylate) (PMMA) plate. The data showed no hysteresis between loading and unloading, and were fit as shown when the work of adhesion, W_A, was adjusted to a value of 105 ± 10 mJ/m^2.

In other experiments, Johnson *et al.* addressed themselves to a verification of Young's Equation, Eq. (3-43) or (4.2):

$$\cos\theta = \frac{\sigma_{SG} - \sigma_{SL}}{\sigma_L}$$

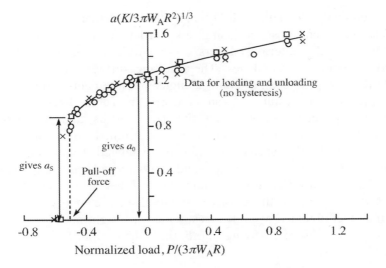

Fig. 4-58: Data for the contact of gelatin spheres against a PMMA plate. Solid line is the data fit using $W_A = 105\pm10$ mJ/m². After [Johnson, K. L., Kendall, K., and Roberts, A. D., *Proc. Roy. Soc. London A*, **324**, 301 (1971).]

The dry contact of two rubber spheres led to a work of adhesion of $(W_A)_{air} = 71\pm4$ mJ/m², and contact under water gave $(W_A)_{water} = 6.8\pm0.4$ mJ/m². Since:

$$(W_A)_{air} = 2\sigma_{SG} - \sigma_{SS}, \text{ and } (W_A)_{water} = 2\sigma_{SL} - \sigma_{SS}, \tag{4.92}$$

it is seen that

$$\frac{1}{2}\left[(W_A)_{air} - (W_A)_{water}\right] = (\sigma_{SG} - \sigma_{SL}) = \sigma_L \cos\theta. \tag{4.93}$$

Using the two measured values of W_A, together with the measured surface tension of water, $\sigma_L = 72$ mJ/m², led to $\theta = 64°$, which compared with the independently measured contact angle of water against the dry rubber of $\theta = 66°$.

Chaudhury and Whitesides[83] repeated the JKR experiments using small ($R \approx 1$–20 mm) hemispherical lenses and flat sheets of poly(dimethyl siloxane) (PDMS). The hemispheres were cast as liquid drops of varying size on non-wetting surfaces before being cured into solid form. Both the hemispheres and the flat sheets were found to be stress free (by examination under cross-polarizers) and smooth (as evidenced by high-resolution electron micrographs and by the near absence of contact angle hysteresis). Measurement of the pull-off force for PDMS in air using spheres of varying radius against flat surfaces led to a work of adhesion equal to 45.2 mJ/m², giving a value of the surface energy of 22.6 mJ/m², consistent with expectations for PDMS and with the value obtained by fitting the $a(P)$ data. These authors went on to plasma oxidize the PDMS surfaces, converting

[83] Chaudhury, M. K., and Whitesides, G. M., *Langmuir*, **7**, 1013 (1991).

them chemically to silica, following which further surface chemical modifications were possible using alkyl trichlorosilanes. The chemical composition of the outermost layers was controlled by functionalizing the tail group region of the silanes. The resulting surfaces were studied with both JKR tests and separate contact angle measurements with a series of probe liquids, and results are summarized in Table 4-8.[84] Reasonable agreement was achieved between the JKR results and the dispersion component of the surface energies determined by contact angle measurements. If the surface energy is evaluated from contact angle measurements using probe liquids and decomposed into dispersion and "polar" (actually a reflection of H-bonding or other acid-base interactions) components in accord with Eq. (4.75), the total surface energy is found to be larger than that obtained by JKR measurements. Insight into this is provided by noting that cases involving more than dispersion force interactions produce hysteresis

Table 4-8: Surface energies of surface functionalized PDMS as determined by JKR experiments and contact angle measurements, as analyzed by Eq. (4.75). From [Mangipudi, V. S., and Falsafi, A., "Direct estimation of the adhesion of solid polymers," in **Adhesion Science and Engineerig, Vol. II, Surfaces, Chemistry & Applications**, M. Chaudhury and A. V. Pocius, Eds., pp. 75-138, Elsevier, Amsterdam, 2002.]

System	σ_{SG}^{JKR}	σ_{SG}^{d}	σ_{SG}^{p}	σ_{SG}^{tot}
-CF$_3$	16.0	15.0	0.8	15.8
-CH$_3$	20.8	20.6	0.09	20.7
-OCH$_3$	26.8	30.8	6.4	37.2
-CO$_2$CH$_3$	33.0	36.0	6.4	42.4
-Br	36.8	37.9	1.7	39.6

between the loading and unloading traces for the contact radius *vs.* time, and that the degree of hysteresis depended on the dwell time under compressive load. Kim *et al.*,[85] who performed JKR measurements between PDMS and both Si-OH and –COOH surfaces reasoned that the number of H-bonds increased by pressure-induced reorganization of the PDMS network near the interface. Further study of JKR hysteresis is providing insight into other processes important to adhesion, such as viscoeleastic or plastic deformation, interdigitation, and reptation or inter-diffusion.

As powerful as it is, there are limitations to the JKR method. Examination of Eq. (4.87) reveals that the interfacial stress exhibits a singularity at the edge of the contact circle, $r = a$. This can be attributed to

---

[84] Chaudhury, M. K., and Whitesides, G. M., *Science*, **255**, 1230 (1992):
    Chaudhury, M. K., *J. Adhesion Sci. Tech.*, **7**, 669 (1993).
[85] Kim, S., Choi, G. Y., Ulman. A., and Fleischer, C., *Langmuir*, **13**, 6850 (1997).

the JKR neglect of any molecular interactions outside the direct contact zone. These were accounted for in the analysis of Derjaguin, Muller and Toporov (DMT),[86] who assumed, however, a Hertzian shape profile for the interacting bodies. Among other differences, the DMT analysis led to a predicted pull-off force of

$$P_S^{DMT} = 2\pi W_A R,$$ (4.94)

in contrast to the JKR result, for which the prefatory constant is 3/2. DMT theory is more appropriate to hard solids of small radius and low adhesion, while JKR analysis is applicable to soft solids, larger radius and higher adhesion. Maugis has constructed a smooth transition between the two cases.[87] A basic assumption of JKR analysis is that the deformations remain small compared with the dimensions of the sample so that the bodies may be regarded as semi-infinite elastic media. This is especially important considering the convenient small-lens ($R \approx 1$ mm) technique developed by Chaudhury and Whitesides, and implemented in a commercial JKR instrument.[88] The surface forces apparatus described briefly in Chap. 1, is also used for JKR studies and effects pull-off force measurements between crossed cylinders (of radii of a few mm), providing geometrically the same as a sphere-plate contact. It is also common to use AFM pull-off forces (to be described later in this chapter) between very small ($R \approx 10$ μm) spheres attached to the cantilever tip and the substrate to quantify adhesion. It has been demonstrated for the study of a hard sphere pushed against an elastomeric plate, the thickness of the latter should exceed 700 μm, and for a small elastomeric lens against a rigid surface, the lens may be as small as $R \approx 1$ μm for $a(P)$ determinations, but not for penetration depth studies.[89] In another study, the validity of pull-off force measurements for spheres as small as 300 μm has been demonstrated.[90] Finally, the possible presence and consequences of capillary condensation of a liquid bridge surrounding the circle of contact has been taken into account.[91,92]

G. Heterogeneous nucleation

Homogeneous nucleation, in which embryos of a new phase are created within the bulk of a parent phase, was treated in Chap. 2. It was noted that high degrees of super-saturation were generally required for the process to occur. Heterogeneous nucleation, in which nuclei of the new phase are created at the surface(s) of the system, such as container walls, dust particles, boiling chips, *etc.*, occur at far lower free energy cost and

[86] Muller, V. M., Derjaguin, B. V., and Toporov, Y. P., *Colloids Surfaces*, **7**, 251 (1983).

[87] Maugis, D., *J. Colloid Interface Sci.*, **150**, 243 (1992).

[88] www.aak.se/MAMA

[89] Deruelle, M., Hervet, H., Jandeau, G., and Léger, L., *J. Adhesion Sci. Tech.*, **12**, 225 (1998).

[90] Briscoe, B. J., Liu, K. K., and Williams, D. R., *J. Colloid Interface Sci.*, **200**, 256 (1998).

[91] Fogden, A., and White, L. E., *J. Colloid Interface Sci.*, **138**, 414 (1990).

[92] Maugis, D., and Gauthier-Manuel, B., *J. Adhesion Sci. Tech.*, **8**, 1311 (1994).

hence at lower degrees of super-saturation. The central idea, as in homogeneous nucleation, is that for the process to occur at a finite rate, critical nuclei must be formed at a finite rate, J, which has been arbitrarily set at one nucleus/cm³·s. Critical nuclei are those of sufficient size to satisfy the Kelvin Equation, Eq. (2.92) at the prevailing degree of super-saturation.

To fix ideas, consider the condensation of a vapor to liquid through the formation of nuclei at a flat solid wall bounding the vapor, as pictured in Fig. 4-59, a re-labeled version of Fig. 3-5. The critical nucleus may be

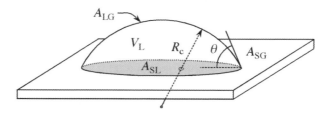

Fig. 4-59: Critical nucleus of condensate liquid at a vapor/solid interface.

considered a spherical cap, and its size , in terms of its radius of curvature, $R = R_C$, is given by the Kelvin Equation as [cf. Eq. (2.99)]:

$$R_C = \frac{2v^L \sigma}{RT \ln x},$$

(4.95)

where $x = p/p_\infty^s$, the degree of super-saturation, $i.e.$, the ratio of the partial pressure to the handbook vapor pressure at the temperature of the system, v^L is the molar volume and σ the surface tension of the condensate liquid. As in homogeneous nucleation, cf. Eq. (2.96), the rate of nucleation is given by an Arrhenius expression:

$$J(\text{nuclei/cm}^3\text{s}) = Ae^{-E/kT},$$

(2.96)

where the pre-exponential factor A depends on the collision frequency of the vapor molecules with the wall, and the activation energy E, is equal to the free energy of formation of the critical nucleus, ΔF_{form}. It is in the expression for ΔF_{form} that heterogeneous nucleation differs from that of the homogeneous process.

ΔF_{form} consists of the free energy change involved in creating new vapor-liquid and solid-liquid interfacial area, destroying vapor-solid area, and creating new volume, V_L, against the pressure difference $\Delta p = (p - p_\infty^s)$:

$$\Delta F_{\text{form}} = \sigma_L A_{\text{LG}} + (\sigma_{\text{SL}} - \sigma_{\text{SG}}) A_{\text{SL}} - V_L \Delta p.$$

(4.96)

Upon substitution of Young's Equation, Eq. (3.43), into the second term on the right, and the Young-Laplace Equation for a spherical surface, Eq. (2.44), into the final term of Eq. (4.96), one gets:

$$\Delta F_{\text{form}} = \sigma_L A_{LG} - \sigma_L A_{SL} \cos\theta - 2\frac{V_L \sigma_L}{R_C}. \tag{4.97}$$

From the geometric description of a spherical cap, Eqs. (3.31) – (3.33):

$$A_{LG} = 2\pi R_C^2 (1 - \cos\theta), \tag{4.98}$$

$$A_{SL} = \pi R_C^2 \sin^2\theta, \text{ and} \tag{4.99}$$

$$V_L = \frac{1}{3}\pi R_C^3 \left[2 - \cos\theta(\sin^2\theta + 2)\right]. \tag{4.100}$$

Substitution of Eqs. (4.98)–(4.100) into Eq. (4.97), and some algebra, leads to

$$\Delta F_{\text{form}} = \frac{1}{3}\pi R_C^2 \sigma_L \left[2(1 - \cos\theta) - \cos\theta\sin^2\theta\right]. \tag{4.101}$$

It is evident that for $\theta \to 180°$, $\Delta F_{\text{form}} \to (4/3)\pi R_C^2 \sigma_L$, the same result as obtained for homogeneous nucleation, Eq. (2.98). Also, if $\theta \to 0°$, $\Delta F_{\text{form}} \to 0$, so that if the condensate liquid wets out the solid, it is impossible to support any super-saturation of the vapor. For intermediate condensate contact angles, ΔF_{form}, and hence the activation energy for heterogeneous nucleation, depends on θ and is always less than that for homogeneous nucleation. More details may be found in Defay *et al.*[93]

H. Processes based on wettability changes or differences

1. *Detergency*

Examples of the many processes controlled by wettability changes and differences are detergency, flotation, selective agglomeration and offset lithographic printing. Detergency refers to the complex set of processes by which soil is removed from surfaces by flushing them with water containing *detergents*, *i.e.*, surface active solutes. An important part of detergent action is increasing the contact angle between a liquid (the soil) and a solid (the substrate to be cleaned), as pictured in Fig. 4-60. Dirt or soil may initially cling to a solid because its contact angle (under water) against the solid is low. A detergent has the major effect of adsorbing at the solid-water interface and decreasing the water-solid interfacial energy, σ_{SW}, often effectively to zero, while adsorption at the dirt-water interface, reduces the tension, σ_{DW}. The result is that $(\sigma_{SW} - \sigma_{SD})/\sigma_{DW} < -1$. This leads to a contact angle of 180°, *i.e.*, "roll-up" and eventual detachment of the dirt particle from the surface, as illustrated in Fig. 4-60. Other aspects of detergency include the "solubilization" (described in Chap. 3) or sequestration of the detached dirt so that it does not become re-deposited on the cleaned surface.

---

[93] Defay, R., Prigogine, I., Bellemans, A., and Everett, D. H., **Surface Tension and Adsorption**, pp. 335ff, Longmans, London, 1966.

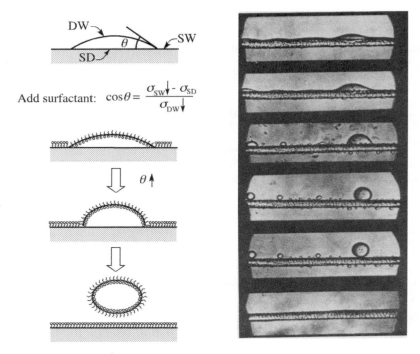

Add surfactant: $\cos\theta = \dfrac{\sigma_{SW}\downarrow - \sigma_{SD}}{\sigma_{DW}\downarrow}$

Fig. 4-60: The role of wettability change in detergency. The reduction of the energy of the solid-water interface by detergent is the most decisive, leading to a sign change of the right hand side of Young's Equation, and $\theta \rightarrow 180°$ ("roll-up"). The right figure shows stages in roll-up of oil on a wool fiber after addition of detergent over a period of approximately 30 s, from [Adam, N.K., and Stevenson, D.G., *Endeavor* **12**, 25 (1953).]

2. Flotation

Another important situation in which de-wetting is desired is in the process of froth flotation[94], as pictured in Fig. 4-61. In this process, valuable ores (usually sulfides) are separated from gangue (usually siliceous material) by sparging gas bubbles through a mixed slurry of the two substances. The bubbles stick to the substance for which the contact angle is high and will not stick to material (such as silica or silicates) for which it is 0°. Often flotation agents, referred to as "collectors" (adsorbates for the ore) are added to enhance the non-wetting characteristics of the material to be floated, while other materials ("frothers") are added to promote frothing. The ores to be extracted are thus separated by "floating" them to the surface (attached to the bubbles) where they are trapped in a froth, which can be removed from the top of the slurry by mechanical means. This is an important separation process, based on differences in wettability. It has found important application not only in the purification of ores but also in other separation processes, such as the de-inking of recycled paper slurries.

---

[94] Leja, J., **Surface Chemistry of Froth Flotation**, Plenum, New York, 1982.

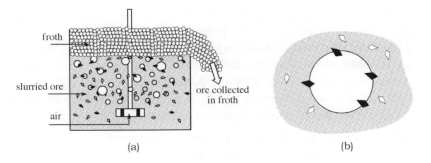

Fig. 4-61: Froth flotation, a separation process based on differences in particle wettability: (a) Schematic of a flotation cell; (b) Close-up of a bubble, showing the attachment of un-wetted particles.

3. Selective or "spherical" agglomeration

An analogous situation to flotation is one in which oil droplets are injected into an aqueous slurry, leading to selective agglomeration. Particles that are not fully water-wet will cling to the oil droplets that grow into large aggregates held together by liquid bridges formed by the oil, as pictured schematically in Fig. 4-62. The process is often referred to as "spherical agglomeration." Such oil-assisted agglomeration[95] is used to remove and consolidate coal fines from aqueous slurries and more recently to remove ink

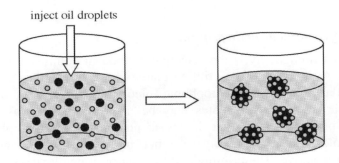

inject oil droplets

Fig. 4-62: Schematic of the spherical (or oil-assisted) agglomeration process.

from re-pulped slurries encountered in the recycling of paper.[96] They have also shown promise for the removal of "stickies" or other oleophilic contaminants from paper slurries. In general, the nature of the aggregates formed will depend on the volume ratio of the oil to particulates being collected. For low oil-to-particle ratios, tenuous liquid bridges will form between the particles, leading to what are termed pendular aggregates. These are fairly easily broken up by shearing action. At somewhat higher oil-to-particle ratios, the number of bridges is increased, leading to stronger, more consolidated structures called funicular aggregates. Finally, when a high

95 Pietsch, W., **Enlargement by Agglomeration**, p. 41, Wiley, Chichester, UK, 1990.
96 Snyder, B. A., and Berg, J. C., *AIChE J.*, **43**, 1480 (1997).

enough ratio of oil is present, all of the inter-particle spaces are filled with oil, and the structure is held together by capillary forces at the aggregate/water interface. This is the capillary regime, and the structures are strongly resistant to breakup by shear forces. If still more oil is added one obtains macroscopic oil drops with the particles dispersed within them. The first three regimes are illustrated in Fig. 4-63 for a dispersion of ink toner particles dispersed in water after injection of mineral oil in varying amounts.

More generally, spherical agglomeration represents a variety of processes for size enlargement by agglomeration.

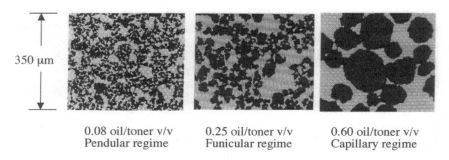

350 μm

| 0.08 oil/toner v/v | 0.25 oil/toner v/v | 0.60 oil/toner v/v |
| Pendular regime | Funicular regime | Capillary regime |

Fig. 4-63: Liquid-bridge agglomerate structures formed by ink toner particles and mineral oil in water. From [Snyder, B. A., and Berg, J. C., *Tappi Journal.* **77** [5], 79 (1994).]

4. *Offset lithographic printing*

Offset lithography is the principal method of commercial printing in use today. At its heart is the printing plate, whose surface has the image to be printed as a pattern of hydrophobic (oleophilic) regions and hydrophilic (oleophobic) regions. The plates may be made of metal, plastic, rubber, paper, or other materials. The image is put on the printing plates using photochemical, laser engraving or other processes. The image may be positive or negative. The printing and the non-printing elements are parts of a single continuous surface (not recessed or raised) and differ only on the basis of their wettability. Oil-based ink is applied to the plate, but adheres only to the hydrophobic "print areas," while the non-image areas are hydrophilic and covered with an aqueous solution called a "fountain solution." The process is carried out as indicated schematically in Fig. 4-64. Fountain solution is first applied to the printing plate on the plate cylinder, followed by the application of the ink. The inked image on the printing plate is next printed onto the rubber blanket (cylinder) and finally transferred (*i.e.,* offset) to paper or other print medium. Excess water is squeezed from the print medium as it passes between the blanket cylinder and impression cylinder.

Most commonly, the printing plates are finely brushed or roughened hydrophilic flexible aluminum sheets coated with a thin photosensitive layer ("emulsion"). A photographic negative of the desired image is placed in contact with the emulsion surface, which is then exposed to light. The light induces a chemical change (cross-linking) in the emulsion layer that allows the exposed areas to remain adherent to the substrate during chemical development, while the unexposed areas of the emulsion are washed away. The hydrophobic emulsion constitutes the image area, while the exposed substrate is the non-print area.

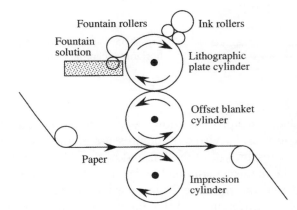

Fig. 4-64: Schematic of the offset lithographic printing process.

The fountain solution is mostly water with *pH* adjusted to the acid side, isopropyl alcohol (as a wetting agent), and gum Arabic or other water-compatible polymer (such as carboxymethylcellulose) to act as an anti-fouling agent for the non-image areas. The inks vary widely in their recipes but generically consist of colorant pigment (such as carbon black), a carrier solvent (such as mineral oil), a binder (various resins), and additives (driers, antioxidants, *etc.*).

The basic requirements for successful printing are that the disjoining pressure in aqueous films between the image area of the plate and the ink, and ink films between the non-image area of the plate and the aqueous fountain solution be negative, so that they are spontaneously squeezed out. Conversely, the disjoining pressure in the ink films between the image area of the plate and the fountain solution, and aqueous films between the non-image area of the plate and the ink must be positive. In terms of contact angle, the fountain solution must wet out ($\theta = 0°$) the non-image area in preference to the ink, and the ink must wet out the image area in preference to the fountain solution. The demarcations between the two areas on the printing plate must be sharp, and resolutions down to 10 μm are routine.

I. Wicking flows (capillary action) and absorbency

1. *Wicking into a single capillary tube*

Next to be examined is the interaction of liquids with porous solids. To start, consider again the movement of a liquid inside a capillary tube, as discussed in Chap. 2. The concern this time, however, is with the *rate* of capillary imbibition. Consider first that the tube is oriented horizontally, as shown in Fig. 4-65, and assume also that Bo << 1, so that the meniscus

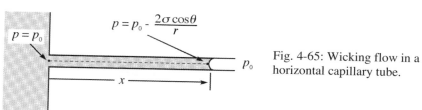

Fig. 4-65: Wicking flow in a horizontal capillary tube.

shape is the segment of a sphere. In a process known as *wicking*, or *capillary action*, the meniscus moves outward by virtue of the pressure gradient which is the difference between the pressure p_0, at the left end of the tube, and the pressure inside the liquid at the meniscus, divided by the length of the liquid column, x. If it is assumed that the pressure in the gas phase outside the meniscus is also p_0, the pressure difference will be equal to just the Young-Laplace capillary pressure, *i.e.*,

$$\Delta p = \frac{2\sigma\cos\theta}{r}, \tag{4.102}$$

where θ is the contact angle of the liquid against the tube wall and r is the radius of the tube. Assuming quasi-steady state conditions, the linear velocity of the liquid in the capillary is given by the Hagen-Poiseuille Law for laminar flow in a round tube[97]:

$$v = \frac{r^2}{8\mu}\left(\frac{\Delta p}{x}\right). \tag{4.103}$$

Substituting Eq. (4.102) into (4.103) gives:

$$v = \frac{dx}{dt} = \frac{r\sigma\cos\theta}{4\mu x}, \tag{4.104}$$

where θ is the advancing contact angle. Integration of Eq. (4.104) from $t = 0$ gives the position of the meniscus as a function of time:

$$x = \left[\frac{r\sigma\cos\theta}{2\mu}\right]^{1/2} t^{1/2} = kt^{1/2}. \tag{4.105}$$

---

[97] Bird, R. B., Stewart, W. E., and Lightfoot, E. N., **Transport Phenomena**, 2nd Ed., pp. 48-53, Wiley, New York, 2007.

The above is known as the Lucas-Washburn Equation, or simply the Washburn Equation, after the investigators who first derived it.[98] It is used as a starting point to describe the motion of liquids through porous media as a result of surface tension forces.

It is evident that the mechanical requirement for spontaneous wicking into a capillary tube is that the advancing contact angle be less that 90°, so that its cosine will be positive. The thermodynamic requirement is that the free energy of replacing solid-gas surface area with solid-liquid interfacial area be negative, or that the "work of wetting," defined in Eq. (4.36) be positive.

Before generalizing to flow in porous media, consider the effect of gravity on wicking. When the capillary tube is oriented such that the direction of flow makes an angle β with the gravitational acceleration, g, the pressure drop across the tube is

$$\Delta p = \frac{2\sigma\cos\theta}{r} + \rho g y \cos\beta, \tag{4.106}$$

where y is now the distance along the tube. If the meniscus is rising vertically upward, $\beta = 180°$, and $\cos\beta = -1$. Under these circumstances, the Hagen-Poiseuille Equation integrates to an implicit relationship for y:

$$t = \frac{8\mu h}{\rho g r^2}\left[-\ln\left(1-\frac{y}{h}\right)-\frac{y}{h}\right], \tag{4.107}$$

with h equal to the equilibrium rise height:

$$h = \frac{2\sigma\cos\theta}{\rho g r}. \tag{4.108}$$

This can be cast into dimensionless form to give the result shown in Fig. 4-66, expressed in "Washburn coordinates," i.e., y vs. $\sqrt{t}$. The time scale factor, t^*, gives the time required for the meniscus to reach approximately 83% of the equilibrium rise height, and is a measure of the minimum time required to obtain a surface tension value by the capillary rise method. For a 0.4-mm diameter capillary, this time amounts to 1.5 and 0.5 s, respectively, for water and benzene at 20°C, but more than 20 min for the viscous liquid glycerol. The method is thus seen to be inappropriate for viscous liquids. Examination of Fig. 4-66 also reveals that gravity will not play a significant role in the rate of capillary rise until the meniscus is above approximately 35% of the equilibrium rise height. Up to this point, it follows the simple Washburn Equation derived for horizontal tubes, Eq. (4-104). When r is sufficiently small (say 10 μm or less) the indicated equilibrium rise heights

[98] Washburn, E. W., *Phys. Rev.*, **17**, 273 (1921);
Lucas, R., *Kolloid Z.*, **15**, 23 (1918).

may be several meters, so that in fine capillaries, gravity may be neglected completely.

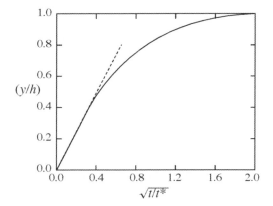

Fig. 4-66: Capillary rise in a vertical tube.

$$t^* = \frac{8\mu h}{\rho g r^2}$$

$$h = \frac{2\sigma}{\rho g r}$$

2. Wicking in porous media

The format of the Washburn Equation (or its generalization to include gravity) for flow in round tubes, often describes very well the rate of capillary penetration, or wicking, of liquids into complicated porous media.[99] Consider a piece of filter paper dipping into water, as pictured in Fig. 4-67(a). The liquid rises over most of the height of the paper as though

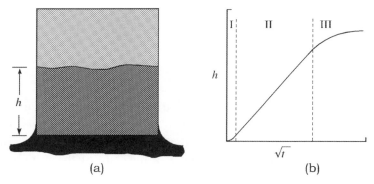

Fig. 4-67: Wicking flow in a porous medium (filter paper dipping into water). Wicking often occurs over three regimes: I - a "wetting delay", usually < 1 s, II – the "Washburn period", and III - fall-off due to approach to the equilibrium rise height (or due to evaporation).

the latter consisted of a bundle of circular capillary tubes, despite the tortuous, non-cylindrical pores involved. Equation (4.103) may be modified by replacing r with r_e, the "wicking equivalent radius" of the pores:

---

[99] Chatterjee, P. K., and Nguyen, H. V., **Absorbency**, P. K. Chatterjee, Ed., pp. 29-84, Elsevier, Amsterdam, 1985.

$$v = \frac{dx}{dt} = \frac{r_e \sigma \cos\theta}{4\mu x}.$$ (4.109)

The value of r_e accounts for the non-uniform, irregular shape of the pores, but there is one other effect that should be shown explicitly. The distance x refers to the distance traveled by the wicking front in the general direction of flow, but due to the tortuosity of the liquid path, the actual distance traveled by the liquid, ℓ, should be written as $\ell = \tau x$, where τ is defined as the tortuosity factor, ℓ/x. In a fibrous medium, it is typically of order 10 or larger. It is useful to re-write Eq. (4.109) as

$$v_{wick} = \frac{dx}{dt} = \frac{r_{pore} \sigma \cos\theta}{4\mu\tau x},$$ (4.110)

where v_{wick} is the rate of motion of the wicking front, and r_{pore} is the effective pore radius, which should bear at least a qualitative relationship to the actual size of the pores in the structure. The wicking-equivalent pore radius is thus $r_e = r_{pore}/\tau$. Integration of Eq. (4.110) gives

$$x = \left[\frac{r_e \sigma \cos\theta}{2\mu}\right]^{1/2} t^{1/2} = \left[\frac{r_{pore} \sigma \cos\theta}{2\mu\tau}\right]^{1/2} t^{1/2} = kt^{1/2}.$$ (4.111)

The advance of the wicking front often appears as shown in Fig. 4-67(b). There is sometimes a "non-Washburn" period right at the beginning of wicking (referred to as a "wetting delay"), which may be due to the time required for the meniscus to establish itself and the flow to reach quasi-steady state, or to other causes. In many cases it is negligibly small, usually a fraction of a second. It is, however, important when the total wicking distance is small, as in wetting in the z-direction in paper. Slowing at large times may be observed, due to the approach of the wicking distance to the equilibrium rise height, or to evaporation of the liquid. Over most of the wicking distance, however, the format of Washburn's Law is obeyed.

The slope of the plot of wicking distance *vs.* $\sqrt{t}$ is termed the "Washburn slope," and according to Eq. (4.111) should vary directly as the quantity $\sqrt{\sigma/\mu}$, and such a dependence has been verified for a number of different systems.[100] Thus a contact angle for a fibrous or a granular material may be inferred from wicking experiments by comparing the Washburn slope of the liquid of interest against the slope obtained for a liquid giving $\theta = 0°$[101]. For a given porous material, such as a powder packed into a tube, an experiment is performed with a liquid which is known to wet out the solid, so that for it $\cos\theta = 1$. A good candidate is hexamethyldisiloxane (HMDS), the 1-cs Dow-Corning 200 fluid, with its low surface tension of 16 mN/m (at 20°C), and volatility about that of water.

[100] Everett, D. H., Haynes, J. M., and Miller, R. J., **Fibre-Water Interactions in Papermaking**, (Fundamental Research Com., Eds.) BPBIF, p. 519, Clowes, London, 1978.
[101] Studebaker, M. L., and Snow, C. W., *J. Phys. Chem.*, **59**, 973 (1955).

Assuming that σ and μ for this liquid are known, the measured slope will yield the wicking equivalent radius, r_e, for the material. The "inferred contact angle" is computed as

$$(\cos\theta)_{\text{inferred}} = \left(\frac{k}{k_{\text{ref}}}\right)^2 \frac{(\sigma/\mu)_{\text{ref}}}{(\sigma/\mu)}, \qquad (4.112)$$

assuming r_e of the medium to be the same for both liquids (e.g., no swelling of the particles or the fibers of the medium). This method has been used[102] to obtain surface energies of carbon fibers for the manufacture of fiber-reinforced composite materials. A bundle of carbon fibers was drawn into a tube, and the weight of liquid drawn up into the skein is recorded over time. Similar studies were performed on various non-woven fibrous media.[103] When single-fiber contact angles were obtained using the Wilhelmy method described earlier, for the same fibers abstracted from the fiber networks, they were found to be in agreement with the contact angles inferred from the Washburn slopes, verifying the predicted $(\cos\theta)^{1/2}$ dependence of the latter.

For most fibrous materials, the Wilhelmy "fiber balance" described earlier is a valuable tool for obtaining single-fiber wetting data. It can be used in conjunction with wicking or absorbency measurements to provide a complete description of the interaction of the material with liquids.[104] Single-fiber wetting measurements are especially useful for explaining the effects of wetting agents, re-wetting agents, surface treatments,[105] etc., in modifying wicking performance because they separate geometric effects from those of altered surface chemistry. For powders and fine granular materials, inferred contact angle techniques may be one of the few practical means that can be used to quantify wetting properties.

Wicking in porous media is important in many processes and in everyday events, some of which are listed in Table 4-9. It is important, for example, in the cleaning, drying, dying, adhesion, and other processes

---

Table 4-9: Examples of wicking flows, or capillary action.

- Cleaning of textiles and other porous materials
- Drying of porous materials
- Adhesion to rough or porous substrates: wood, paper, concrete, etc.
- Moisture movement and retention in soils
- Liquid uptake by absorbent products: paper towels, diapers, etc.
- Liquid fuel transport in kerosene lamps, heaters, space vehicles, etc.
- Flow of working fluid in heat pipes
- Frost heaving in silty arctic soils

---

[102] Cwastiak, S., J. Colloid Interface Sci., 42, 298 (1973).
[103] Hodgson, K. T., and Berg, J. C., Wood and Fiber Sci., 20, 3 (1988).
[104] Hodgson, K. T., and Berg, J. C., J. Colloid Interface Sci., 121, 22 (1988).
[105] Westerlind, B. S., and Berg, J. C., J. Appl. Polym. Sci., 36, 523 (1988).

involving porous materials, such as paper, textiles, wood and concrete, and to the movement of moisture in soils and rock. Absorbent products, on the other hand, are porous materials specifically designed to move or remove liquids by wicking action, and they include towels, sponges, diapers, feminine hygiene pads, *etc.* Similarly, wicks are strips of porous material designed to transport liquids from one location to another, as fuels in lamps and heaters. An interesting application of wicking is in *heat pipes*, used for the rapid transport of heat from one location to another.[106] The basic design is a sealed vessel containing a working fluid, which is a vapor at the temperature of the hot surface but a liquid at the cold surface, as pictured in Fig. 4-68. Heat is transferred to the cold surface as the latent heat of condensation of the working fluid, which is returned by a wick to the hot surface, where it is re-vaporized. Thermal transport rates of heat pipes are enormous. A tubular heat pipe of the type shown in Fig. 4-68, using water as the working fluid and operating between 150°C and 20°C has an effective thermal conductivity several hundred times that of copper.

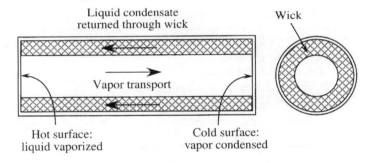

Fig. 4-68: Schematic diagram of a heat pipe.

An important situation involving wicking that occurs in Nature is *frost heaving*,[107] a phenomenon requiring cold temperatures, such as exist in the Far North in winter, and fine or silty soils that might have been produced by millennia of glacial grinding. It is an upward movement of the subgrade soil resulting from the expansion of ice lenses beneath the surface. These are formed originally within the larger soil voids and grow as they are fed by liquid water trapped in the fine, silty soil beneath and around them. By virtue of the Kelvin effect, moisture exists as liquid water well below the nominal freezing point in these soils of micron-size particles. It is transported to the growing lenses by capillary action. As the ice lenses grow, the overlying soil "heaves," often exerting pressures great enough to break up roads, foundations, airstrips, *etc.* Frost heaving occurs only in soils containing fine enough particles. Many agencies classify materials as being frost susceptible

[106] Dunn, P., and Reay, D. A., **Heat Pipes**, Pergamon Press, Oxford, 1976.
[107] Berg, R. L., **Frost Action and Its Control**, Amer. Soc. Civil Engineers, 1984.

if 10 percent or more passes a No. 200 sieve (≤ 74 μm) or 3 percent or more passes a No. 635 sieve (≤ 25 μm).

3. *Practical strategies for promoting absorbency*

Absorbency may be characterized in terms of the *volumetric* rate of wicking, Q, per unit area, A, of the wicking medium at a given distance x, from the liquid reservoir. This may be expressed, using Eq. (4.110) as

$$Q/A = v_{\text{wick}}\phi = \frac{r_{\text{pore}}\phi\sigma\cos\theta}{4\mu\tau x}, \tag{4.113}$$

where ϕ is the porosity (void fraction) of the medium. Maximization of absorbency is seen to be a combination of structural factors: r_{pore}, ϕ, and τ ; fluid properties: σ and μ; and the wettability of the structure by the wicking liquid, as expressed by $\cos\theta$. Some strategies for promoting wicking or absorbency are listed in Table 4-10.

Table 4-10: Strategies for enhancing the rate of wicking.

- Increase the pore size in the medium, r_{pore}
- Increase the porosity, ϕ
- Minimize the pore tortuosity, τ
- Increase the surface tension of the liquid, σ
- Decrease viscosity of liquid, μ
- Improve wetting of the medium: decrease the contact angle (*i.e.*, increase $\cos\theta$) (decrease σ_{SL})

Increasing the pore size r_{pore} should be effective, but in consideration of gravity, can be done only to a certain point. For vertical wicking in large pores, there will be meniscus flattening, as discussed in Chap. 2, and the driving force for the process will be reduced. The appropriate driving force is

$$\Delta p = \frac{2\sigma}{b}, \tag{4.114}$$

where b is the radius of curvature of the meniscus at its base, dependent on both the Bond number and the contact angle. When the contact angle is 0°, b may be computed for a given liquid in a tube of given diameter using the Sugden Tables, Table 2-6. Secondly, as r_{pore} increases, the equilibrium rise height will be reduced. The appropriate equilibrium rise height is to be computed using r_{pore} rather that the wicking equivalent pore radius, r_{e}. Maximizing porosity can also be carried only to a certain point because wicking requires enclosed channels. Minimizing the pore tortuosity τ would occur in fibrous media if the fibers were aligned as much as possible along the direction of flow.

With respect to fluid properties, the surface tension should be as high as possible, without increasing the contact angle, and the viscosity should be as low as possible. Neither of these can generally be controlled, because the liquid to be absorbed is given, and additives are not available to raise the surface tension or lower the viscosity. One method to enhance absorbency might appear to be the use of a surfactant to improve the wetting (increase $\cos\theta$), but it is quickly to be noted that this will also reduce the surface tension, so that the effects may cancel each other out. This is made clear when Young's Equation (which is valid if the contact angle is greater than zero) is substituted into Eq. (4.113), giving:

$$Q/A = \frac{r_{pore}\phi(\sigma_{SG} - \sigma_{SL})}{4\mu\tau x}.$$

(4.115)

Thus as long as the contact angle is finite, any surfactant that has the sole effect of reducing the surface tension of the liquid will have little effect on the wicking rate. If the contact angle is 0°, the addition of surfactant will only reduce surface tension, making the wicking rate slower. The only way a surfactant can assist in wicking is through its adsorption at the solid/liquid interface, and producing a reduction in σ_{SL}, and thus an increase in the wetting tension ($\sigma_{SG} - \sigma_{SL}$). Such observations are borne out by experiment.[108] Further discussion of the effects of surfactants on absorbency is given elsewhere.[109,110] Of particular importance in the use of surfactants for enhancing absorbency are the dynamics of the process. For the surfactant to increase the wetting tension it must be able to reach the advancing meniscus and adsorb. As the liquid meniscus advances through the porous medium it is depleted of surfactant by adsorption to the pore walls of the medium. It can be replenished only by diffusion from the liquid behind the meniscus. The advancement of the wicking front goes as $k\sqrt{t}$, and the diffusion path length varies as $C\sqrt{Dt}$, so depending on the relative values of the Washburn slope, k, and the concentration C and diffusivity D of the surfactant, the meniscus will be out of equilibrium with respect surfactant adsorption by a constant amount during the wicking process. While the Washburn format would be thus be expected for the wicking process, the effective value of the wetting tension would differ by a constant amount from the equilibrium value assumed in Eq. (4.115). This is what has been observed for the wicking of various surfactant solutions into strips of various types of cellulosic fibers.[103]

108 Pyter, R. A., Zografi, G., and Mukerjee, P., *J. Colloid Interface Sci.*, **89**, 144 (1982).

109 Hodgson, K. T., and Berg, J. C., *J. Colloid Interface Sci.*, **121**, 22 (1988).

110 Berg, J. C., "The Role of Surfactants," pp. 149-198, **Absorbent Technology**, P. L. Chatterjee and B. J. Gupta, (Eds.), Elsevier, Amsterdam, 2002.

4. *Immiscible displacement*

The Washburn analysis may be generalized to the displacement of one viscous liquid from a capillary tube (or porous medium) by another and also to the case in which an external pressure drop is imposed either with or against the capillary flow. This is pictured in Fig. 4-69. Pressure flow through a capillary medium in the absence of capillary effects is termed Darcy (originally, d'Arcy) flow, so the combined result is referred to as the Washburn-Darcy Equation

$$v = \frac{r^2\left[(p_A - p_B) + \dfrac{2\sigma\cos\theta}{r}\right]}{8\left[\mu_A x + \mu_B(L - x)\right]}. \tag{4.116}$$

The viscous resistance to flow has been apportioned in accord with the lengths of the tube occupied by each fluid.

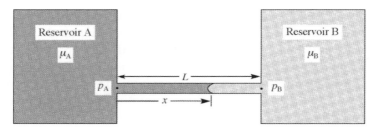

Fig. 4-69: Washburn-Darcy flow (immiscible displacement).

The application of Eq. (4.116) to the immiscible displacement of one liquid by another in a porous medium represents a great simplification, perhaps an over-simplification. In reality the process is a set of complex pore-scale flow phenomena, including viscous fingering, pinch-off of liquid threads, entrapment of ganglia, flow of wetting films, *etc.*[111] Despite the complexity of these phenomena, their description can be given in terms of the equilibrium contact angle, capillary number, *Ca* (which would yield a dynamic contact angle), and the viscosity ratio, as suggested at least qualitatively by the Washburn-Darcy Equation. It gives a reasonable zeroeth approximation to the movement of the front through the medium.

5. *Mercury porosimitry*

An important example of a situation in which a pressure driving force opposed to capillary flow is imposed is the technique of *mercury porosimetry,*[112] used for determining the pore size distribution in a solid

---

[111] Payatakes, A. C., and Dias, M. M., *Rev. Chem. Eng.*, **2**, 85 (1984).
   Homsy, G. M., *Ann. Rev. Fluid Mech*, **19**, 271 (1987).
   Sahimi, M., *Rev. Mod. Phys.*, **65**, 1393 (1993).
[112] Ritter, H. L., and Drake, L. C., *Ind. Eng. Chem. Anal. Ed.*, **17**, 782 (1945).

specimen and shown in Fig. 4-70. Mercury is strongly non-wetting to most solids and generally exhibits a contact angle of 130-140°. It thus requires pressure to force mercury into pores. At equilibrium with respect to a given pore size, one may set the Washburn velocity ≈ 0 and solve for the pore radius r_{pore} in terms of the required pressure, $p_A (= p)$. The result is

$$p = -\frac{2\sigma\cos\theta}{r},$$ (4.117)

which, upon substitution of the properties of mercury ($\sigma = 485$ mN/m, $\theta \approx 135°$) gives the pore radius in μm as approximately $60/p$ (in psi). In a

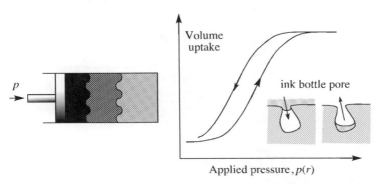

Fig. 4-70: Schematic of mercury porosimetry for measurement of pore size distribution in a porous solid. The ink bottle pore configuration provides one explanation for the hysteresis usually observed.

porosimetry experiment, the volume of mercury forced into the pores (uptake) versus the pressure required can thus be interpreted to yield a volumetric pore size distribution. Pores down to radii of less than 10 nm can be determined this way. Usually hysteresis is observed as the pressure is released, and this is most often interpreted in terms of the ink bottle pore configuration discussed in the earlier consideration of capillary condensation.

6. Motion of liquid threads

Consider next the possible motion of small *isolated* liquid masses (known as liquid threads or "indices") in pores open to the same gas pressure on both sides. Under certain circumstances, these will move spontaneously. All that is required in principle is that a pressure difference develops along the length within the thread. For the case of a capillary tube with changing pore diameter, as shown in Fig. 4-71(a), a pressure difference develops due to the difference in curvature of the opposing menisci and the Young-Laplace requirement. If the Bond Number is assumed small, the pressure in the liquid at Point A is: $p_A = p_0 - 2\sigma/R_A$, where p_0 is the pressure outside the pore and R_A is the effective meniscus radius at point A. Similarly at Point B: $p_B = p_0 - 2\sigma/R_B$, so that

$$p_A - p_B = 2\sigma\left(\frac{1}{R_B} - \frac{1}{R_A}\right) > 0, \qquad (4.118)$$

and the thread will want to spontaneously move from left to right. Liquids that wet the solid will spontaneously move into the smaller pores, while those that do not will move out into the larger pores, and perhaps out of the entire porous mass. Another way to develop pressure differences is shown in Fig. 4-71(b). By an analysis similar to that leading to Eq. (4.118):

$$p_A - p_B = 2\sigma\left(\frac{1}{R_B} - \frac{1}{R_A}\right) = \frac{2\sigma}{R}(\cos\theta_B - \cos\theta_A) > 0, \qquad (4.119)$$

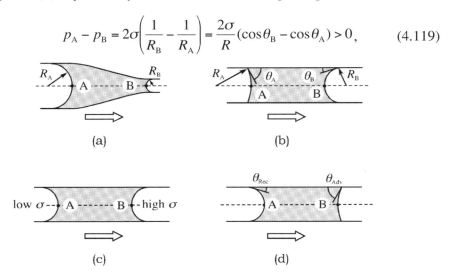

(a)   (b)

(c)   (d)

Fig. 4-71: Liquid threads in porous media.

where R is the pore radius, so again the liquid will want to move from left to right, *i.e.*, liquids will move from less wettable to more wettable regions. In Fig. 4-71(c), the surface tensions at A and B are different, and

$$p_A - p_B = \frac{2\sigma_B}{R} - \frac{2\sigma_A}{R} = \frac{2}{R}(\sigma_B - \sigma_A) > 0, \qquad (4.120)$$

so it is seen that the liquid will move from low-σ to high-σ regions. A household application of this effect is described by Morrison and Ross:[113] A grease spot may be removed from cloth by placing a paper towel over the spot, turning it over and applying a hot iron to the opposite side. The surface tension difference, due to the temperature difference, drives the grease spot into the paper towel.

Often even when a pressure driving force is developed in a liquid thread, motion is *not* observed. This is because in order to move, the liquid must overcome contact angle hysteresis, as shown in Fig. 4-71(d). This can be an especially large effect if several such threads are present in series in

---

[113] Morrison, I. D., and Ross, S., **Colloidal Dispersions**, p. 155, Wiley-Interscience, New York, 2002.

the same pore, referred to as the *Jamin effect*.[114] Thus it is that contact angle hysteresis plays an important role in the ease of removing either trapped gas bubbles or trapped liquid masses from porous media.

7. *Surface wicking; spreading over rough or porous surfaces*

The spreading of liquids on rough, as opposed to smooth, surfaces may be quite different from that on smooth surfaces and is best understood in the context of wicking. Surface roughening may induce spreading of an otherwise incompletely wetting liquid, and in the case of a spreading liquid, may considerably enhance the spreading rate. Roughening in a given direction to provide texture in the form of grooves, as suggested by Fig. 4-12(a), allows one to control the direction of spreading. When the underlying surface is horizontal, a liquid will in principle spread indefinitely by "surface wicking" into the grooves. A schematic representation of liquid in a single groove is shown in Fig. 4-72. It is easy to show that the critical condition

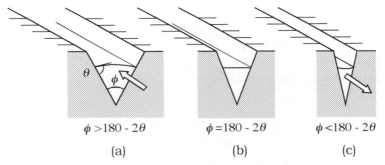

$$\phi > 180 - 2\theta \qquad \phi = 180 - 2\theta \qquad \phi < 180 - 2\theta$$

(a)   (b)   (c)

Fig. 4-72: Surface wicking. θ and ϕ [=] °

for surface wicking in a horizontal groove is that the groove angle ϕ be less than $(180 - 2\theta)°$, where θ is the contact angle. When $\phi = (180 - 2\theta)°$, it is evident from the geometry that the liquid interface is flat. For larger values of ϕ, the meniscus will be concave upward, requiring a reduced pressure in the liquid in the groove relative to that in the air in the dry groove in front of it, causing the liquid to retreat. Surface wicking from a pool of liquid into a groove on a surface slanted at the angle β relative to the horizontal, will come to a halt when the gravitational force on the liquid in the groove balances the capillary force due to the curved meniscus, analogous to capillary rise in a round tube described in Chap. 2. A classic example of the latter is the wicking of a liquid into the vertical groove provided by the edge contact of two glass plates, as shown in Fig. 4-73. A wetting liquid ($\theta = 0°$) will rise into the groove in accord with the gap provided by the groove angle ϕ. If the gap is small enough, the Bond Number ≈ 0, and the meniscus elevation y above the liquid in the dish is given by

---

[114] Schwartz, A.M., Rader, C.A., and Huey, E., **Contact Angle, Wettability and Adhesion**, Adv. Chem. Ser., No. 43, p. 250, ACS, Washington DC, 1964.

$$\Delta p = \rho g y = \frac{\sigma}{R}, \tag{4.121}$$

where the radius of curvature of the meniscus may be taken as half the plate spacing, *i.e.*, $R = \phi x/2$ (ϕ in radians), so that the profile of the meniscus is given by

$$y(x) = \frac{2\sigma}{\rho g \phi x}. \tag{4.122}$$

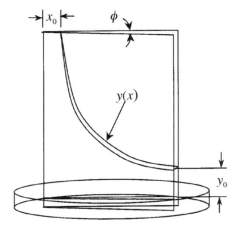

Fig. 4-73: Capillary rise between glass plates in contact at a shallow angle, ϕ.

Spontaneous wetting or de-wetting in other situations, as a result of a variety of different geometries, is described by Schwartz.[115]

The spreading of liquids on highly rough surfaces may be characterized as surface wicking. Spreading via unidirectional grooves on horizontal surfaces should exhibit a rate law analogous to that of the Washburn Equation, Eq. (4.105), *i.e.*, the interline position should advance as $\sqrt{t}$. Spreading of a drop on a highly rough surface could proceed with a rate law of the form

$$r \propto t^{\kappa}, \tag{4.123}$$

where the exponent κ takes on different values dependent on the model assumptions made,[116] but might approach a value as high as 0.5, in contrast to spreading on a smooth surface, as described by Eq. (4.62), with an exponent of 0.1.

[115] Schwartz, A. M., *Ind. & Eng. Chem.*, **61**[1], 10 (1969).
[116] Chaudhury, M. K., and Chaudhury, A., *Soft Matter*, **1**, 431 (2005)

J. Particles at interfaces

1. *Particles at solid-fluid interfaces: effects on wetting and spreading*

Small particles adhere to solid surfaces by virtue of long-range van der Waals or lyophobic forces (described in Chap. 7) and short-range attractions that may be electrostatic, acid-base or ultimately covalent interactions. One of the primary consequences of their presence is to impart roughness to the solid interface. This may promote wetting or spreading (surface wicking) of a liquid over the surface if the particles are wet by the liquid. If the contact angle of the liquid against the particle surfaces is greater than 90°, they will retard wetting or even produce what has been described as ultra-hydrophobicity. Coffee rings, *i.e.*, patterns left by a puddle of particle-laden liquid after it evaporates, are also explained similarly.[117] The evaporating liquid at the edge of the drop deposits particles there, as shown in Fig. 4-74. This ridge of particles pins the interline as the drop

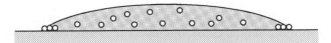

Fig. 4-74: The origin of "coffee rings" from the evaporative deposition of particulates that impede wetting and pin the drop interline.

flattens due to further evaporation. Eventually the apparent contact angle becomes sufficiently low that the interline jumps in to a new position, and the process repeats. If the particles are in the nanometer size range, these effects are described by the Wenzel Equation, Eq. (4.7) or the Cassie-Baxter Equation for composite surfaces, Eq. (4.9) or (4.10).

2. *The disposition of particles at fluid interfaces*

The presence of particles embedded in fluid interfaces may have profound effects on the properties of systems in which they occur. We must first consider the conditions under which particles reside or become trapped at fluid interfaces. In the presence of gravity, particles will always collect at an interface between fluid phase A (on top) and fluid phase B (below) if their density is intermediate to that of the two fluid phases, *i.e.*, $\rho_A < \rho_p < \rho_B$, regardless of their size, shape or relative wettability, or the magnitude of the interfacial tension. The density of the particle relative to that of the fluid phases may be expressed in terms of a dimensionless Density number, Dn:

$$Dn = \frac{\rho_p - \rho_A}{\rho_B - \rho_A},$$
(4.124)

[117] Deegan, R. D., Bakajin, O., Dupont, T. F., Huber, G., Nagel, S. R., and Witten, T. A., *Nature*, **389**, 827 (1997).

and it is seen that the above condition pertains to $0 < Dn < 1$. If the particle is either denser than the lower phase ($Dn > 1$) or lighter than the upper phase ($Dn < 0$), the situation is more complicated.[118] It may be possible to retain, metastably, a particle denser than the lower phase if it is preferentially wet by the upper phase, or vice versa, because in these situations, interline forces, together with buoyancy forces, will be present pulling the particle in the direction opposite to the direction of gravity. Otherwise, the particle (starting from the phase with which its density differs most) has no option but to pass directly through the interface. Solutions for the equilibrium vertical positions of the particles of given shape and the shape of the interface around them, for a given value of Dn and contact angle, can be determined by solution of the Young-Laplace Equation and are found to be a function of the Bond Number. Recall from Chap. 2 that the Bond Number is the dimensional ratio of gravity forces to boundary tension forces acting on the particle, *viz.*

$$Bo = \frac{(\rho_B - \rho_A)gR^2}{\sigma}, \tag{4.125}$$

where R is the particle radius (assuming it to be spherical or cylindrical). Figure 4.75 shows a sequence of images for a sphere in a system of a given

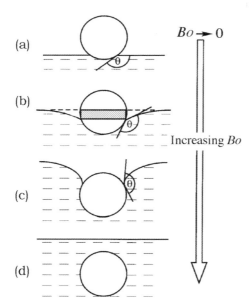

Fig. 4-75: The disposition of a spherical particle at a fluid interface. The particle is denser than the lower fluid phase ($Dn > 1$) and is preferentially wet by the upper phase ($\theta > 90°$). Parts (a) – (d) show the effect of increasing the Bond Number, Bo. When $Bo \approx 0$, there is no interface distortion, but when Bo reaches a critical value, the particle detaches and sinks into the lower phase. The cross-hatched part in (b) shows the volume of liquid whose buoyancy must be taken into account in determining the equilibrium position of the particle.

$Dn > 1$ and given contact angle θ as Bo is increased from 0 to a value beyond which the sphere can be retained at the interface. Analytical solutions for the depth of immersion of long cylindrical particles (neglecting end effects), and numerical solutions for spheres have been reported by

---

[118] Princen, H. M., **Surface and Colloid Science**, Vol. 2, E. Matijevic, (Ed.), pp. 30ff, Wiley-Interscience, New York, 1969.

Princen and others.[119] An important term to be included in the force balance, as recognized by Princen and rediscovered later,[120] is the buoyancy of the volume of liquid equal to the area bounded by the interline times the depth of the interline below the undisturbed surface, and sketched in Fig. 4-75(b). Of practical importance to flotation is the maximum size of particle of given density that can be retained at the water-air interface, $i.e.$ "floated." One may find for example that for spherical particles of specific gravity 7.5 (corresponding to that of lead sulfide, an ore commonly recovered by flotation), which we assume to be completely un-wet by water ($\theta \rightarrow 180°$), the maximum Bo for which a solution exists is approximately 0.25. If the surface tension of the water is taken as 40 mN/m (in view of the presence of frother), this corresponds to a maximum "floatable" particle of radius ≈ 1.0 mm. For the cylindrical waxed needle of density 8 times that of water "floating" on clean water, as mentioned in Chap. 1, the maximum Bo is about 0.3, corresponding to a maximum needle diameter of approximately 1.5 mm.

Of particular interest is the behavior of very small (colloidal) particles, giving small values of the Bond Number. When $Bo < 0.01$, gravity determines only whether or not the particle reaches the interface, when present initially in one of the bulk phases. They will be able to $enter$ the interface if the contact angle of neither fluid against the particle surface in the presence of the other is $0°$, and will then assume a position in the interface without deforming it, as shown in Fig. 4.75(a). The exact elevation will be that which satisfies the contact angle requirement. Under certain circumstances, the presence of a layer (or multilayers) of such particles has been shown to stabilize emulsions ("Pickering emulsions") or foams with respect to coalescence, or to stabilize isolated bubbles ("armored bubbles") against dissolution or size disproportionation. On the other hand, they may bridge emulsion droplets together or de-stabilize them, or may act as foam beakers. They may assist in the spreading of one liquid on the surface of another ("particle-assisted wetting"), or to retard the spreading of liquid drops ("liquid marbles") on solid surfaces. The collection of particles at fluid interfaces provides a powerful pathway to the synthesis of complex particles or to their organization into various structures. Some of these effects are explored below.

3. Particle-assisted wetting

To fix ideas, consider the situation existing when a drop of mineral oil is placed on water. As discussed in Chap. 3, the spreading coefficient in such a situation is negative, so the oil will form a lens. It may be possible, however, that a layer of fine particles adherent to either the oil-water or the oil-air interfaces (or both) may cause the oil to spread out over the water surface. One may inquire how this is possible. The short answer is that the

[119] Vella, D., Lee, D. G., and Kim, H.-Y., $Langmuir$, **22**, 5979 (2006).

[120] Keller, J. B., $Phys. Fluids$, **10**, 3009 (1998).

presence of particles at an interface may reduce the free energy of a fluid interface system, as outlined by Goedel and coworkers.[121] Consider the insertion of small spherical particles, initially in air, into the system consisting initially of an oil lens on water, as shown in Fig. 4-76. A number of possibilities exist, with the preferred scenario being the one(s) of minimum free energy. The free energy of the system may be computed as

$$F = F_0 + \Delta F, \qquad (4.126)$$

where F_0 corresponds to the system of water/oil lens/air with dry particles, and ΔF is taken as the free energy change accompanying one of the scenarios indicated in Fig. 4-76. If a single spherical particle of radius R enters either of the bulk liquids, one obtains

$$\Delta F_W = -4\pi R^2(\sigma_{PA} - \sigma_{PW}) = -4\pi R^2 \sigma_{WA}\cos\theta_{AWP}, \text{ or} \qquad (4.127)$$

$$\Delta F_0 = -4\pi R^2(\sigma_{PA} - \sigma_{PO}) = -4\pi R^2 \sigma_{OA}\cos\theta_{AOP} \qquad (4.128)$$

for the cases in which the particle enters the water or the oil, respectively.

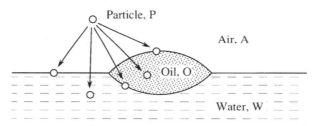

Fig. 4-76: Various scenarios for the insertion of a particle into a system with an oil lens on water.

The second parts of Eqs. (4.127) and (4.128) are obtained by substitution of Young's Equation, Eq. (4.2) in each case. θ_{AWP} is the contact angle at the air-water-particle interline, drawn in the water phase, and θ_{AOP} is the contact angle at the air-oil-particle interline, drawn in the oil phase. Such a substitution is valid only under conditions where Young's Equation applies, *i.e.*, the contact angles are finite. If the particle enters one of the interfaces, we obtain one or more of the situations shown in Fig. 4-77. Consider the case of the particle entering the water/air interface. The free energy change is given by

$$\Delta F_{W/A} = \sigma_{PA}\Delta A_{PA} + \sigma_{PW}\Delta A_{PW} + \sigma_{WA}\Delta A_{WA}, \qquad (4.129)$$

where ΔA_{PA} is the area change of the particle/air interface, *etc*. The following

[121] Goedel, W. A., *Europhys. Lett.*, **62**, 607 (2003);
Xu, H., Yan, F., Tierno, P., Marczewsi, D., and Goedel, W. A., *J. Phys. Cond. Matter*, **17**, S465 (2005).

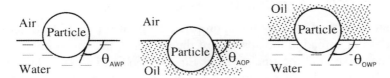

Fig. 4-77: Small spherical particle positioned at an air/water, air/oil or oil/water interface.

area changes accompany the process:

$$\Delta A_{PA} = A_{cap} - 4\pi R^2, \tag{4.130}$$

$$\Delta A_{PW} = 4\pi R^2 - A_{cap} \quad \text{and} \tag{4.131}$$

$$\Delta A_{WA} = -A_{base}. \tag{4.132}$$

Recalling the earlier description of the geometry of a spherical segment, Eqs. (3.30) and (3.32) one has:

$$A_{cap} = 2\pi R^2 (1 - \cos\theta_{AWP}) \quad \text{and} \tag{4.133}$$

$$A_{base} = \pi r^2 = \pi R^2 \sin^2\theta_{AWP} = \pi R^2 (1 - \cos^2\theta_{AWP}), \tag{4.134}$$

where r is the base radius of the spherical segment. Combining Eqs. (4.130) – (4.131) with Eq. (4.127) gives:

$$\Delta F_{W/A} = -2\pi R^2 (1 + \cos^2\theta_{AWP})(\sigma_{PA} - \sigma_{PW}) - \pi R^2 (1 - \cos^2\theta_{AWP})\sigma_{WA}. \tag{4.135}$$

Finally, substitution of Young's Equation in the form:

$$(\sigma_{PA} - \sigma_{PW}) = \sigma_{WA}\cos\theta_{AWP}, \tag{4.136}$$

and some simplification gives

$$\Delta F_{W/A} = -\pi R^2 \sigma_{WA}(1 + \cos\theta_{AWP})^2. \tag{4.137}$$

If $\Delta F_{W/A}$ is negative, which will always be the case for $\theta_{AWP} < 180°$, the particles should spontaneously enter the water/air interface, unless an even larger negative free energy change accompanies one or more of the other scenarios. It would appear that results similar to Eq. (4.137) would be obtained for the entry of the particle into the oil/air or oil/water interfaces, but such entry would be accompanied by an extension of the lens, *i.e.*, an increase in the oil/air and oil/water interfaces by an amount equal to the base area of the particle, and a corresponding decrease in the water/air interface. The free energy of such extension is seen to be minus the work of spreading for the oil on the water (*i.e.*, minus the spreading coefficient, $S_{O/W}$) times the base area. Thus for the entry of a particle into the oil/air interface or the oil/water interface, we obtain:

$$\Delta F_{O/A} = -\pi R^2 \sigma_{OA}(1 + \cos\theta_{AOP})^2 - A_{base}S_{O/W}, \text{ or} \tag{4.138}$$

$$\Delta F_{O/W} = -\pi R^2 \sigma_{OW}(1 + \cos\theta_{OWP})^2 - A_{base}S_{O/W}. \qquad (4.139)$$

A_{base} can be evaluated using the appropriate form of Eq. (4.134). Since the oil lens does not spread in the absence of the particles, the spreading coefficient $S_{O/W}$ is negative, and the final term of Eq. (4.138) or (4.139) is thus positive. If in either case, the first term in $\Delta F_{O/A}$ or $\Delta F_{O/W}$ is larger in magnitude than the spreading term, the process should occur spontaneously, and we have particle-assisted spreading.

Various pairs of scenarios may coexist if their free energies are equated. For example, particles immersed in the oil phase may coexist with particles in the water/air interface, if $\Delta F_O = \Delta F_{W/A}$, or

$$4\sigma_{OA}\cos\theta_{AOP} = \sigma_{WA}(1 + \cos\theta_{AWP})^2. \qquad (4.140)$$

This and other coexistence curves have been computed by Goedel and coworkers and are presented in "phase" diagrams of θ_{AOP} vs. θ_{OWP} for fixed values of σ_{OA}/σ_{OW} and $S_{O/W}/\sigma_{WA}$. An example is shown in Fig. 4-78, in which the parameters correspond to tetradecane on water. Data for various systems have since been obtained which support the validity of these phase

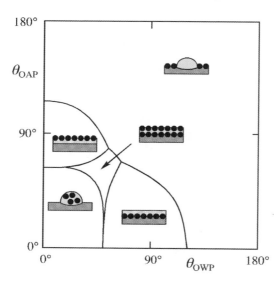

Fig. 4-78: Computed phase diagram showing various states for particle-oil-water systems. Results are shown for the case of tetradecane on water, for which $\sigma_{OA}/\sigma_{OW} = 0.5$, and $S_{O/W}/\sigma_{WA} = -0.1$. The three intermediate regions between the lower left and upper right hand portions of the diagram show particle assisted spreading by virtue of the presence of the particles at either the oil/air or oil/water interfaces, or both. Diagram is reproduced after [Ding, A., and Goedel, W. A., *J. Amer. Chem. Soc.*, **128**, 4930 (2006).]

diagrams.[122] Particles at fluid interfaces are generally unstable with respect to two-dimensional aggregation, as discussed further in Chap. 7, either by capillary forces (if they are large enough) or van der Waals forces if they are small (as assumed here), so that they form coherent rafts. Oil layers that are spread with the help of these rafts of particles may be transferred to solid substrates for examination or processing. Examples of such layers deposited on mica are shown in Fig. 4-79.

---

[122] Ding. A., Binks, B. P., and Goedel. W. A., *Langmuir*, **21**, 1371 (2005).

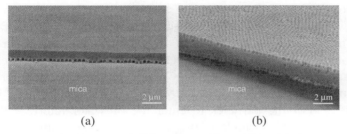

(a)   (b)

Fig. 4-79: SEM images of oil layers spread on water with the assistance of fine particles and transferred to a mica substrate. (a) methacrylate-coated silica particles and the oil trimethylolpropane trimethacrylate (TMPTMA), (b) methacrylate-coated silica particles and the oil pentaerythritol tetra-acrylate (PETA). From [Xu, H., and Goedel, W. A., *Langmuir*, **19**, 4950 (2003).]

The technology of particle assisted spreading has been used to form uniformly porous polymer membranes.[123] A thin layer of a nonvolatile, polymerizable oil, trimethylolpropane trimethacrylate (TMPTMA) was spread on water with the assistance of uniform sized silica particles. The particles were hydrophobized using silanes with hydrophobic organo-functional groups, as described earlier. The particles packed into the oil surface, and as the oil spread, eventually the particles bridged the oil layer between the air and the water. The oil was subsequently photopolymerized, and the particles removed (by etching with hydrofluoric acid), leaving a uniformly porous membrane. An example is shown in Fig. 4-80. Three-dimensional porous membranes could also be produced by simply spreading larger amounts of particle-monomer mixture on the water surface.[124]

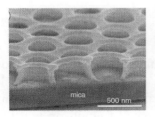

Fig. 4-80: A porous membrane produced by photo cross-linking a TMPTMA layer initially filled with surface-treated silica particles that were subsequently removed. From [Xu, H., and Goedel, W. A., *Angew. Chem. Int. Ed.*, **42**, 4694 (2003).]

4. Pickering emulsions

Emulsions are defined as systems of liquid droplets dispersed in liquids, and unless they are *micro*emulsions (swollen micelles), as discussed briefly in Chap. 3, they are inherently unstable toward coalescence. They are generally either oil droplets dispersed in water (O/W emulsions) or water droplets dispersed in oil (W/O emulsions). A more thorough description of emulsions is given in Chap. 9, but it can be stated here that their "stability," or more properly, their longevity or "shelf life," is determined by the time

[123] Yan, F., and Goedel, W. A., *Chem. Mater.*, **16**, 1622 (2004).
[124] Yan, F., and Goedel, W. A., *Adv. Mater.*, **16**, 911 (2004).

required for the liquid film separating a pair of droplets to thin to a critical value, at which point it ruptures, and the droplets merge. One of the most effective means of slowing or even halting this process is through the presence of particles at the droplet interfaces. The particles are presumed to hold the approaching liquid interfaces apart by a distance greater than the critical film thickness. Systems stabilized in this way are termed "Pickering emulsions."[125] For the particles to enter the oil/water interface they must produce a finite contact angle drawn in either phase against the other, as pictured in the right hand image of Fig. 4-77. A general rule of thumb about systems of this type is that the phase against which the particle is more wettable will be the continuous phase of the emulsion.[126] Thus, with reference to Fig. 4-77, if $\theta_{OWP} > 90°$, the particle will be preferentially wet by the oil, the larger portion of the particle will reside in the oil, and oil will be the continuous phase, leading to a W/O emulsion. This is sometimes referred to as the "oriented wedge rule." Other rules of thumb about Pickering emulsions are that larger particles lead to larger droplets, and that a sufficient number of particles must be provided to fully coat the droplet interfaces.[127] The latter rule means that for a given system, if the volume of the dispersed phase is continually increased, a point is often reached at which the emulsion breaks or even inverts. Finally, it is generally assumed that the particles at the interface must be much smaller that the droplets they are stabilizing against coalescence.[128] Thus for the stabilization of fine emulsions, the use of nanoparticles is required.

The degree of stability of a Pickering emulsion is related to the strength with which the particles are tethered to the interface, and this in turn may be estimated by the work required to detach the particles, as expressed by Eq. (4.137) applied to the O/W interface, i.e.,

$$W_{detach} = -\Delta F_{O/W} = \pi R^2 \sigma_{OW}(1 + \cos\theta_{OWP})^2. \qquad (4.141)$$

This shows W_{detach} to have a sharp maximum at $\theta_{OWP} = 90°$, and that it is directly proportional to the magnitude of the interfacial tension. Such observations have been validated by experiment.[129] The condition of $\theta_{OWP} = 90°$ is ambiguous with respect to which type of emulsion will be obtained (O/W or W/O), and indeed spontaneous inversion is often observed under these conditions. Thus a useful rule of thumb is to design the particles to have at least a slight preference for the phase that one wishes to be the continuous phase. The relative wettability of the particles can be tuned by treating their surfaces with silanes or functionalizing them in other ways. Although the use of Pickering emulsions dates back more than a century,

---

[125] Pickering, S. U., *J. Chem. Soc., 91*, 2001 (1907).
[126] Schulman, J. K., and Leja, J., *Trans. Faraday Soc., 50*, 598 (1954).
[127] Simovic, S., and Prestidge, C. A., *Langmuir, 20*, 8357 (2004).
[128] Levine, S., Bowen, B. D., and Partridge, S. J., *Colloids Surfaces, 38*, 325 (1989).
[129] Binks, B. P., and Lumsdon, S. O., *Langmuir, 16*, 3742 and 8622 (2000); *17*, 4540 (2001).

there is currently renewed interest in them that will lead to better understanding and new applications.[130]

5. *Armored bubbles and liquid marbles*

Just as small particles adsorbed to water/oil interfaces are able to stabilize emulsions, their presence under the right circumstances at air/liquid interfaces has been shown to stabilize foams or individual bubbles,[131-132] or on the other hand, liquid droplets in air.[133] Individual gas bubbles stabilized in this way have come to be known as "armored bubbles,"[134] and liquid droplets coated with hydrophobic particles have been called "liquid marbles." The criterion for adsorption of the particles at the gas-liquid interface is the existence of a finite contact angle of the liquid against the particle surface, and the tenacity of their adherence to the surface is maximized when this angle is near 90°. Under these conditions, it is almost impossible to force the particles out of the surface. Armored bubbles take on distinctly non-spherical (often polyhedral) shapes as the gas they contain dissolves or as they are pushed against a surface, as by gravity. Critical to this ability is the adhesion or friction between the particles that are jammed in the interface. The dissolution and disproportionation of small armored bubbles under the influence of the Kelvin effect is brought to a halt, presumably as the shape of the liquid surfaces between the particles becomes flat or even inverted relative to the curvature of the bubble itself. The addition of surfactant to a system of armored bubbles appears to lubricate the particles allowing them to shed from the surface as the bubbles dissolve, and permit the bubbles to maintain spherical shapes.[135]

Liquid marbles, *i.e.*, particle-coated droplets in air, are found to behave as soft solids that show little or no tendency to stick to other solid surfaces. When at rest on a horizontal surface, they assume the shape of sessile drops, as shown in Fig. 2-32, whose surface tension and density are essentially the same as that of the drop liquid. Their effective contact angle is near 180°, and they readily roll over solid substrates in response to gravity or other forces acting upon them in a manner similar to the motion of liquids on super-hydrophobic surfaces (Lotus effect). They may even be placed atop water surfaces, upon which they are observed to float. Their movement on tilted surfaces has been investigated and is found to exhibit some very interesting properties.[158] Large liquid marbles formed flat, pancake-like droplets whose descent velocity U_0 at steady state, which was reached almost immediately, was independent of pancake size and given by:

[130] Binks, B. P., *Curr. Opin Colloid Interface Sci.*, **7**, 21 (2002).
[131] Du, Z., Bilbao-Montoya, M. P., Binks, B. P., Dickinson, E., Ettelaie, R., and Murray, B. S., *Langmuir*, **19**, 3106 (2003).
[132] Binks, B. P., and Horozov, T. S., *Angew. Chem. Int. Ed.*, **44**, 3722 (2005).
[133] Aussillious, P., and Quéré, D., *Nature*, **411**, 924 (2001).
[134] Kam, S. I., and Rossen, W. R., *J. Colloid Interface Sci.*, **213**, 329 (1999).
[135] Subramanian, A. B., Mejean, C., Abkarian, M., and Stone, H. A., *Langmuir*, **22**, 5986 (2006).

$$U_0 = \frac{\sigma}{\mu} \sin\beta, \tag{4.142}$$

where σ and μ are the surface tension and viscosity of the liquid, respectively, and β is the angle of inclination, confined to small angles (< 10°). Smaller drops moved faster as the area of contact, for a given drop mass, decreased, and was given by

$$U = U_0 \frac{a}{\sqrt{2}R}, \tag{4.143}$$

where a is the capillary length: $a = \sqrt{2\sigma/\rho g}$, and R is the radius of the un-deformed drop. This result is consistent with that obtained earlier for drops rolling on super-hydrophobic surfaces,[136] and is exactly opposite to that for drops having a finite contact angle with the substrate, for which the smaller drops move more slowly, if at all. These are held back by contact angle hysteresis, as described by Eq. (4.4). Liquid marbles appear to provide an answer to the problem often encountered in microfluidics of moving small amounts of liquid about on a solid substrate. Electric or magnetic forces may be employed for controlling such motions.

When the angle of inclination β was made larger (24°), the drops rolled at speeds of the order of 1 m/s and deformed themselves into biconcave shapes rolling on their rims, and when they rolled off the edge of the plate and fell freely, they deformed themselves into peanut shapes.

6. Janus particles and nanoparticles at fluid interfaces

Another situation regarding particles at fluid interfaces exists if the particles have composite surfaces of different wettabilities, perhaps water wettable over half their surface and oil wettable over the other half. These "two-faced" or "Janus" particles, named after the Roman god of doorways, are currently generating considerable interest because of the many applications to which it appears they may be put. Computations suggest, for example, that they should be much more effective stabilizers of emulsions than homogeneous particles.[137] The energy required to detach a particle which is fully wet by water over half its surface and fully wet by oil over the remaining half is three times the energy needed to detach an equivalent sphere of contact angle 90°. Furthermore, Janus particles of average contact angle approaching 0° or 180° are still strongly attached, in contrast to homogeneous particles that are easily detached under such conditions. Both effects are the result of pinning at the interline, assumed to be sharp, between the hydrophobic and hydrophilic areas of the particle. Janus particles might also be used in directed drug delivery if part of their surface binds to certain tissues (such as tumor cells) one might wish to target, while

[136] Mahadevan, L., and Pomeau, Y., *Phys. Fluids*, **11**, 2449 (1999).
[137] Binks, B. P., and Fletcher, P. D. I., *Langmuir*, **17**, 4708 (2001).

the remaining part is used to bind the drug. If the Janus particles are functionalized to have positive charge on one side and a negative charge on the other they would constitute a large dipole that would be orientable in an electric filed. If they were simultaneously bichromal, they might then constitute the basis for electronic displays, such as "electronic paper." One might even envision particles with three or four different types of surface regions, with complementary colors leading to color electronic paper. Janus particles with opposite electron donor-acceptor properties on their two sides might lead to more efficient solar power systems, which require interfaces between electron donors and electron acceptors. They may also provide a platform for new microsensors.

Much of the attention given to Janus particles to date concerns methods for their synthesis. They may be prepared in a one-step process directly from different polymers either by jetting the molten materials side by side to produce what are called biphasic particles when the dual jet breaks up under capillary forces.[138] Such particles may also be produced by introducing the two different molten polymers to the opposite sides of a spinning disc, so that the two sheets of liquids join and subsequently break up into biphasic droplets as they are flung from the edge of the disc. More commonly, Janus particles are produced by topo-selectively treating uniform particles trapped at a solid surface or in a fluid interface. Particles at a solid surface may be treated on the exposed side by sputter coating or other types of vapor deposition, while particles at a water/oil interface may be treated by adsorbing or reacting chemicals selectively from either the water or the oil side. Many of these processes are shown in schematically in Fig. 4-81.

Janus particles, particularly nanoparticles, mimic surfactants in many ways, and as such provide building blocks for supra-particulate assemblies. Even uniform nanoparticles have been reported to reduce the surface tension or interfacial tension of fluid interfaces in which they are present. For example, Glaser et al.[139] used pendant drop tensiometry to determine the effect of Janus nanoparticles on the tension of the water-hexane interface. Janus particles of approximate diameter 14 nm were synthesized by nucleating Fe_3O_4 patches onto initially uniform gold nanoparticles and subsequently functionalizing the iron oxide with oleic acid or with oleylamine to render those portions hydrophobic. Homogeneous iron oxide particles and gold particles of the same size as the Janus particles were also produced. The iron oxide was surface functionalized in the same way as the Janus particles, while the gold particles were functionalized with SAM's of dodecyl thiol. While decreases of the order of 10 mN/m in interfacial tension were produced with the uniform particles, decreases of more than twice this

[138] Roh, K.-H., Martin, D. C., and Lahmann, J., *Nature Mater..* **4**, 759 (2005).
[139] Glaser, N., Adams, D. J., Böker, A., and Krausch, G., *Langmuir*, **22**, 5227 (2006).

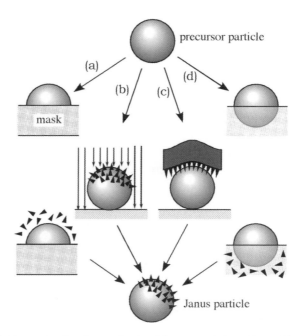

Fig. 4-81: Strategies for the synthesis of Janus particles. (a) masking/ unmasking techniques, (b) use of reactive directional fluxes, vapor deposition, (c) micro-contact printing, (d) reaction on one side of particles trapped at fluid interfaces. After [Perro, A., Reculusa, S., Ravaine, S., Bourget-Lami, E., and Duguet, E., *J. Matls. Chem.,***15**, 3745 (2005).]

were produced by the Janus particles. Miller *et al.*[140] point out such reductions should be regarded as changes in the "effective interfacial tension," since it is distinct from the "true" interfacial tension existing at the interface between the particles. The effective or apparent interfacial tension is identified with the creation of a unit area of new interface, including the particles from the bulk phase. A drop in tension is detectable by a Wilhelmy slide or a Langmuir barrier just as it is with a pendant drop. In many cases, as mentioned earlier, particles at fluid interfaces simply aggregate in two dimensions to form rafts, and in these cases one would not expect a reduction in effective interfacial tension, but if there is lateral repulsion (electrostatic, steric, *etc.*) between the particles and/or the particles are small enough to produce a two-dimensional osmotic pressure owing to their lateral Brownian motion, one can measure a reduction in boundary tension as a function of the interfacial particle density. Thus one may produce particle surface-pressure - area (π-A) isotherms in analogy with the isotherms of Langmuir surfactant monolayers discussed in Chap. 3. There are reports of the formation of micelle-like structures in dispersions of Janus nanoparticles.[141] Also, when fluid is leached from the interior of particle-

[140] Miller, R., Fainermann, V. B., Kovalchuk, V. I., Grigoriev, D. O., Leser, M. E., and Michel. M., *Adv. Colloid Interface Sci.*, **128**, 17 (2006).

[141] Pradhan, S., Xu, L.-P., and Chen, S., *Adv. Funct. Mater.*, **17**, 2385 (2007).

covered fluid spheres, hollow shells of nanoparticles, termed "colloidosomes"[142] (mimicking liposomes) are produced. Their potential use as porous microcapsules for controlled release applications is evident, since the permeability of the shell of solid particles can be controlled for a particular ingredient delivery by adjusting the size of the interstices between the particles.

K. The description of solid surfaces

In contrast to fluid interfaces, solid interfaces have *topography*, *i.e.*, roughness and texture, for extended surfaces, and pore size distribution and/or specific surface area, in the case of powders or porous media. A second difference is that solid surfaces are usually not chemically homogeneous. The consequences of solid surface roughness and spatial chemical heterogeneity on wetting behavior have been central to much of the foregoing discussion. While wettability measurements themselves permit one to infer something about these properties of solids, it is important to explore *independent* means for their experimental determination. In addition, one is often interested in imaging structures (usually three-dimensional) immobilized on solid surfaces. These would include enzymes and other biological entities, printed electronic circuits, *etc*. In this section we seek to provide descriptions of solid surface topography, in particular roughness and texture, and solid surface energetics and chemistry.

1. Solid surface roughness

An example of a rough surface is shown in Fig. 4-82. To describe the roughness, assume that a smooth datum surface, as outlined by the dashed lines in the figure, has been located such that the volume of material above it is equal to the deficit of material below it. Next imagine an array of equally spaced points (x_i, y_j) on the datum surface, and that the elevation, z_{ij}, of the rough surface at each of these points relative to the datum is determined. The surface roughness is then characterized by various statistical descriptors of such a data set. For example, the mean height, or average roughness, is computed as

$$\bar{z} = \frac{1}{MN} \sum_{i=1}^{M} \sum_{j=1}^{N} \left| z_{ij} \right|, \tag{4.144}$$

where M and N represent the number of data points within the selected area. The absolute value of z_{ij} is required, since otherwise $\bar{z}$ is trivially zero.

---

[142] Velev, O. D., Furusawa, K., and Nagayama, K., *Langmuir*, **12**, 2374 (1996).

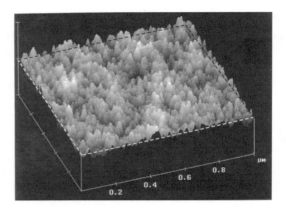

Fig. 4-82: A characteristic rough surface, as imaged by an AFM
scan. The datum plane is shown as the surface bordered by the
dashed lines.

Another common measure is the root-mean-square (rms) roughness:

$$S_q = \left[\frac{1}{MN} \sum_{i=1}^{M} \sum_{j=1}^{N} z_{ij}^2 \right]^{1/2} . \tag{4.145}$$

The skewness of the height distribution may be computed to give a measure
of any asymmetry in the height distribution:

$$S_{sk} = \frac{1}{MNS_q^3} \sum_{i=1}^{M} \sum_{j=1}^{N} z_{ij}^3 . \tag{4.146}$$

Sometimes the information needed is the shape of the profile along a
particular cross-section cut normal to the datum surface. Other measures of
interest are the maximum peak height, R_p, or valley depth, R_v, or peak-to-
valley distance, R_{max}.

2. Fractal surfaces

A powerful tool for characterizing the structure of many surfaces is
fractal geometry.[143] The concept of fractals arises in many situations of
interest in the study of interfacial, colloid and nano science, so it is useful at
this point to comment on fractals in a more general way. Fractal entities are
those that possess irregularities that are *self-similar* in structure when viewed
using different scales of measurement, *i.e.*, the shape is made up of smaller
copies of itself. Following Mandelbrot's classic paper,[144] the coast of Great
Britain is one of the most often-described fractal structures, and serves as a

---

[143] Several texts on fractals have become available. For example:
Birdi, K. S., **Fractals in Chemistry, Geochemistry, and Biophysics: An Introduction**,
Kluwer Pub., New York, 1993, and
Russ, J. C., **Fractal Surfaces**, Plenum Press, New York, 1994.
[144] Mandelbrot, B. B., *Science*, **156**, 636 (1967).

good example. When viewed in an atlas of maps, the west coast from Land's End at the tip of Cornwall in the south to Cape Wrath, the northwest tip of Scotland, presents a jagged structure similar in appearance to that which might be viewed from an airplane above Liverpool (perhaps a fifty mile distance on a clear day), or even to that of a five mile-long stretch of waterfront anywhere along the coast. To the extent that this is true, the coastline is self-similar, or *fractal*. One may inquire as to the length of this coastline between two points, say between the first two points mentioned above. If one simply draws a straight line ("as the crow flies") between the two points, an answer of about 600 miles is obtained. But if instead, a fifty-mile long yardstick is used, walking it along at water's edge between the set endpoints, will sample more of the bays and inlets along the way and get a larger number, about 800 miles in this case. If one uses a one-mile yardstick, the coastline will assume an even greater length. The answer for the length thus depends on the size of our yardstick, and for fractal lines does so in accord with a power law. The number of straight-line segments $N(\varepsilon)$ of length ε needed to walk the distance is given by

$$N(\varepsilon) = F\varepsilon^{-d_f}, \tag{4.147}$$

where d_f is termed the *fractal dimension*. The apparent length of the coastline obtained using a yardstick of length ε is thus

$$L(\varepsilon) = \varepsilon N(\varepsilon) = F\varepsilon^{1-d_f}. \tag{4.148}$$

F is evidently equal to D^{d_f}, where D is the straight-line distance (as the crow flies) between the two points, measured in the same units as ε. Richardson,[145] whose work inspired Mandelbrot's article, showed that the west coast of Britain was at least roughly a fractal line and obtained a fractal dimension for it of about 1.25.

A number of features of the above result are evident. First it is seen that for a fractal line, as the size of the yardstick decreases, the apparent L becomes larger without limit. For practical reasons, one must stop somewhere. Should it be an arbitrary length of one mile?, or the diameter of an average rock? or a grain of sand? This result stands in contrast to that for a smooth (Euclidiean) or differentiable line, which would produce a finite value for the length as $\varepsilon \to 0$. This is the result one obtains when the fractal dimension is 1, its minimum value. The maximum value of d_f is 2, for which the line approaches that of a two-dimensional (or cross-section of a) sponge, as shown in Fig. 4-83. The "length" of the line is then proportional to the "area" of the region it invades. An example of a fractal line of $d_f = 2$ is the two-dimensional projection of the random flight of a molecule in a gas, with the segment ε being the mean free path between collisions. The mean free

---

[145] As reported in Mandelbrot, B. B., **The Fractal Geometry of Nature**, p. 29, W. H. Freeman & Co., New York, 2000.

Fig. 4-83: A fractal line with $d_f \rightarrow 2$.

path may be varied by varying the temperature or pressure of the system. Another example is the two-dimensional trace resulting from the Brownian motion of colloid particles, in which ε is determined by the time interval between observations of the particle position. It should be noted that no curve in the real world is a true fractal in that they exhibit self-similarity only over some finite range of scales, reflective of the processes that produce them, and often they exhibit different values of d_f over different ranges of scale. An example of a fractal line that is of practical interest is the boundary between masked and unmasked regions in photolithography.

Of more interest in the present context are *surfaces* of fractal topography. A fractal surface is the two-dimensional analog of the fractal line described above. They may be created through simple fracture, grit blasting, acid or other chemical etching, or deposition of precipitates, among other processes. To describe the structure of the surface, one may envision it to be probed by a covering layer of spheres of a given radius, a_n, as shown in cross-section in Fig. 4-84(a). If the surface is flat, its area can be expressed in terms of the number $N(a_n)$ of such close-packed spheres required to cover it, *viz.*

$$A_{flat} = k\pi a_n^2 N(a_n), \text{ or } N(a_n) = (A_{flat}/k\pi)a_n^{-2} \qquad (4.149)$$

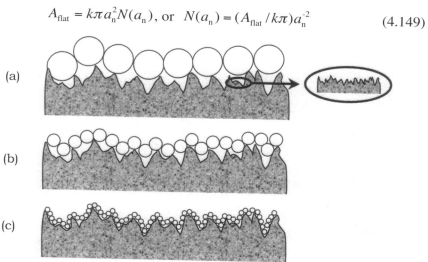

(a)

(b)

(c)

Fig. 4-84: Cross-section of a fractal surface. The nominal surface area depends on the size of the probe spheres used to measure it. Self-similarity is suggested by the enlargement of a small portion of the surface.

where the constant k accounts for the packing of the spheres on the surface (hexagonal, cubic, random). If the surface is rough (fractal), the number of spheres needed to cover it will increase as the sphere size decreases, as is evident by comparing Fig. 4-84(a)-(c). In fact we may write

$$N(a_n) = Ca_n^{-d_f}, \tag{4.150}$$

where d_f is the fractal dimension of the surface, and C is a constant with units of (area)^{d_f}. The area fractal dimension has a minimum value of 2 (the dimension of a Euclidean surface) and a maximum value of 3. A value of 3 could refer to a spongy surface of some depth on a substrate, and the "area" of such a "surface" would be proportional to the "volume" of the spongy region. The apparent area of a fractal surface, as obtained using probe spheres of radius a_n is given by

$$A(a_n) = k\pi a_n^2 N(a_n) = k\pi Ca_n^{2-d_f}, \tag{4.151}$$

One method for accessing the fractal dimension of a surface at the nano or sub-nano scale would be to measure the extent of adsorption of a series of molecules of different sizes.

The rugosity, r, of fractal surfaces may be expressed as

$$r = \frac{A(a_n)}{A_{flat}} = C'a_n^{1-d_f/2}. \tag{4.152}$$

It is evident that both the nominal area and the rugosity of a fractal surface approach infinity as the probe size approaches zero, but again, real surfaces are fractal only within certain ranges of scale. At the extreme, the lower limit of a_n to be used to compute r for Wenzel's Equation for the contact angle of a liquid on a fractal surface would be the effective molecular radius of the wetting liquid.

3. Surface texture

Roughness descriptors, such as those given by Eqs. (4.144)-(4.146), or by the rugosity r or the fractal dimension do not describe the surface texture. If the surface elevation variations are randomly periodic, we need some measure of this periodicity, such a correlation length, which may be thought of as a wavelength of the periodicity. A more detailed description may be given by treating a cross-section of the rough surface, which is a plot of surface elevation z against a straight-line distance, x, "parallel" to the surface. It would appear like the profile shown in Fig. 4-84, and could be subjected to a power spectral analysis. The Fourier transform of the oscillating profile $z(x)$ would reveal any patterns of underlying periodicity in the structure, i.e., it would show the characteristic wavelength(s) associated with the pattern. For example, underlying a low degree of roughness as characterized by the average or rms roughness may be a waviness of much greater wavelength. Power spectral density analysis of surface optical images, i.e., Fourier transform profilometry (FTP), is an often-used method for characterizing the spatial frequencies of the texture of a surface.[146-147] Other specific textural descriptors include the maximum peak height, valley depth and peak-to-valley distance in a given specimen area. For crystalline surfaces one is often interested in grain boundary and defect sizes (step heights, *etc.*) areal densities, and distributions.

4. *Measurement of surface roughness and texture by stylus profilometry*

Many techniques are available for measuring the topography (roughness, texture) of solid surfaces. The most straightforward of these is stylus profilometry, *i.e.* the use of a mechanical stylus, which is dragged across the surface while its elevation is monitored using a transducer, such as an optical interferometer, a variable capacitor or a linear variable differential transformer (LVDT). These transducers have a maximum vertical range between 0.5 and 1 μm, and a vertical sensitivity as great as 0.01 μm. The horizontal resolution depends on the dimensions of the stylus tip that cause the radius of curvature of the peaks to be exaggerated and the width of the troughs to be reduced. Modern instruments using piezo-electric translators produce vertical resolutions of 1 nm and lateral resolutions of 50 nm. Stylus instruments produce a one-dimensional profile of the surface as the tip is moved (or the sample is moved beneath it) in a given direction, suggested by Fig. 4-85, but may be made to yield areal information by rastering back and forth over the surface to give a family of such profiles, as shown in Fig. 4-86. Areas from 50 μm to 1 cm square are typically scanned. The instruments have the disadvantage of requiring direct contact with and possible damage to the surface studied. For untextured surfaces, roughness

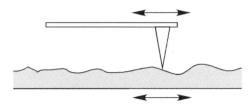

Fig. 4-85: Schematic of stylus profilometry.

is usually characterized in terms of the mean height, the average roughness or the rms roughness, as defined by Eqs. (4.144) – (4.146). These parameters are generally calculated automatically by the instrument used. For textured surfaces or patterned surfaces, such as that of the printed circuit shown in Fig. 4-86, the specific features are observed. Sub-nanometer vertical and side-to-side resolutions are obtainable using AFM profilometry techniques described later, but only rather small surface areas can be scanned, usually less than 100 x 100 μm.

[146] Takeda, M., and Mutoh, K., *Appl. Opt.*, **22**, 3977(1983).

[147] Bone, D. J., Bachor, H. A., and Sanderman, J., *Appl. Opt.*, **25**, 1653 (1986).

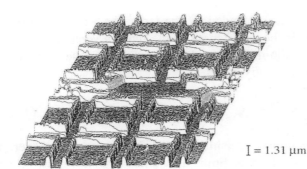

Fig. 4-86: Raster scan on an integrated circuit using a stylus profilometer.

$\underline{I} = 1.31\ \mu m$

L. Optical techniques for surface characterization

Geometrical and sometimes other features of solid surfaces may be observed directly using a variety of optical techniques, some of which are described below. Optical data are usually in the form of an instantaneously obtained two-dimensional map of a patch of the surface, although a number of the techniques used and described below involve lateral scanning. The key distinction from stylus profilometry is that no contact of a probe with the surface is required.

1. *Optical microscopy*

The most basic method for surface examination is that of reflected-light optical microscopy. The feature sizes that are resolvable by this method are limited, however, by the wavelength of the light. When light from the various feature points in the object are reconstituted in the image, the image points are in fact not points but small disks (called Airy disks) containing diffraction patterns. When the diffraction patterns surrounding neighboring image feature points start to overlap, these points are not resolvable. The minimum object distance d between resolvable points depends on the size of the Airy disks, and this in turn depends on the wavelength of light in the medium between the object and the objective lens, λ, and the fraction of the light emanating from an object point captured by the objective lens of the microscope. With reference to Fig. 4-87, this depends on the ratio of the lens diameter to the distance between the object and the lens, as expressed by

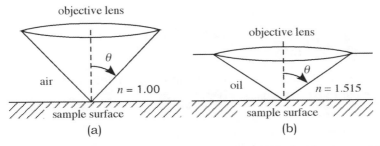

Fig. 4-87: Numerical aperture, $NA = n\sin\theta$ for (a) "dry" objective, and (b) oil-immersion objective.

$\sin\theta$, θ being the polar angle subtended by the lens. The wavelength of the light, $\lambda = \lambda_0/n$, where λ_0 is the wavelength of the light *in vacuo* and n is the refractive index of the medium. The resolution d, for the case of reflected light (non-luminous objects) is given by:

$$d = \frac{\lambda_0}{2n\sin\theta} = \frac{\lambda_0}{2NA}.$$
(4.153)

The quantity in the denominator, $n\sin\theta$, is called the numerical aperture (NA). Since θ has a practical upper limit just under 90°, $\sin\theta$ must be less than one, and with the refractive index of air $n = 1$, the upper limit of the NA of a "dry" objective is not more than 0.95. If the gap between the object (the cover slip) and the lens is filled with an oil, whose refractive index is $n > 1$, the NA can be increased. The usual oil for an "oil immersion" objective has $n = 1.515$, rendering an upper limit for NA of about 1.4. The resolution (minimum value of d) of an optical microscope is thus roughly half the wavelength of the light for a dry objective and one-third the wavelength of the light for an oil immersion objective.

The numerical aperture for a given microscope system depends on the NA of the condenser (for the light source) as well as the objective, such that:

$$NA_{syst} = \frac{1}{2}(NA_{obj} + NA_{cond}).$$
(4.154)

This shows that for optimum use of the objective aperture, the NA of the condenser should match, but not exceed) that of the objective lens. In many cases, the illuminating light comes down through the objective lens, so it serves simultaneously as the condenser and the objective. Another method of illumination passes light obliquely over the sample so that undeflected rays do not pass through the objective. Rays that are deflected off or scattered from the surfaces, however, appear as bright objects on a dark field. The method is called "darkfield" microscopy, and is discussed more fully in Chap. 5 in the section on Light Scattering. A combination of brightfield and darkfield illumination (called "epi-illumination") is especially effective in producing three-dimensional looking images. For self-luminous objects (as obtained when using fluorescent markers), the Rayleigh formula relating NA to the maximum resolution is

$$d = 0.61\frac{\lambda_0}{NA_{obj}}.$$
(4.155)

The depth of field, *i.e.*, the thickness of sample through which all features appear to be in focus, is given by:

$$\text{Depth of field} = \frac{n\lambda_0}{NA_{obj}^2}.$$
(4.156)

High *NA* objectives thus produce a depth of field of the order of one µm. For lower *NA* objectives, the depth of field is greater.

For flatness testing of relatively smooth surfaces, an interference setup is used. A half-silvered flat surface is put over the sample surface and illuminated from above. The light reflected from the test surface interferes with the light reflected from the half-silvered mirror to produce a pattern of fringes that are essentially contour lines.

Various methods of surface staining or marking may be used to enhance contrast between the features of the surface, and one of the most powerful of these is the use of fluorescent markers. Fluorescence microscopy is especially powerful for the examination biological surfaces and specimens since fluorophors can be designed to attach to certain surface chemical functionalities and not others. In the process, the fluorophor absorbs photon energy of a particular wavelength and very quickly (nanoseconds) re-emits light of a slightly longer (less energetic) wavelength. The difference between the excitation and the re-emission wavelengths, referred to as the Stokes shift, is the basis for spectrally separating the incoming and outgoing radiation. A strong illumination source is required because the relatively small number of fluorophors present and their low quantum efficiency. As shown in Fig. 4-88, the illuminating light is first passed through a band-pass filter that rejects all but the wavelength that excites the particular fluorophor present. Just before it passes through the objective (serving as a condenser) it is reflected off what is called a dichroic mirror. A dichroic mirror is a flat element that reflects light of short wavelength, but allows long-wavelength light to pass through. The excitation radiation then reaches the specimen, generates fluorescent

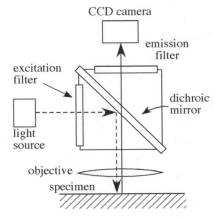

Fig. 4-88: Schematic diagram of a fluorescence microscope.

radiation which passes back up through the objective and through the dichroic mirror (because it is of longer wavelength), then through a second filter that removes all wavelengths of light except that emitted by the fluorophors, and then finally on to the imager (CCD camera, photodetector, *etc.*) The combination of excitation filter, the dichroic mirror and the emission filter is called a "filter set," and is designed for a particular fluorophor.

2. *Optical profilometry*

Optical profilometry is the general name for a variety of non-contact measurement technologies, most based on different types of interferometry. Resolution is strongly dependent on the technology used, but often is sub-nanometer in the vertical and 0.5 micron in the lateral. A common method of this type is scanning white light interferometry (SWLI). A beam of white light is divided by a beam splitter in a Michelson type of configuration with one half brought to focus upon and reflected from a point on the object surface and the other reflected from a reference mirror. The beams are recombined to produce an interference image. The position of either the sample, the lens or the reference mirror is translated, using a piezo translator, to produce the position of maximum constructive interference which is detected on each pixel of a CCD camera as the object surface is scanned. Software then produces a 3-dimensional profile of the surface. The use of white (mixed wavelength) light makes it somewhat easier to find interference fringes, but its validity requires that the zero order interference fringe be independent of wavelength. It is also common to use monochromatic laser illumination.

3. *Confocal microscopy*

An important variation of fluorescence microscopy that permits high resolution optical sectioning through thick specimens is that of confocal microscopy, or more precisely, laser scanning confocal microscopy (LSCM).[148] The setup is similar to that of ordinary fluorescence microscopy, with some very important differences, as shown in Fig. 4-89. The light source is a high intensity laser point source. The light is reflected off the dichroic mirror and then, by means of the objective lens, brought to focus on a sharp (diffraction limited) spot in the specimen. The illuminated spot is at a particular vertical as well as horizontal location. This spot then emits fluorescence that passes back up through the objective lens and an emission filter and is brought to focus at a particular location where an image of the original laser point source is formed. This is the *confocal plane*. It then moves on to the photodetector, where an image of the illuminated spot is formed. The image will be formed by fluorescent light emanating not only from the focal plane of the specimen, but also by light emanating from planes above and below it, causing a blurring of the image even when the depth of field is very small. This problem is solved by placing a pinhole aperture in the confocal plane. The fact is that the source image formed by light coming from planes above and below the specimen focal plane will be formed below and above, respectively, the confocal plane. The pinhole

---

[148] Sheppard, C. J. R., and Shotton, D. M., **Confocal Laser Scanning Microscopy**, Springer-Verlag, New York, 1997.

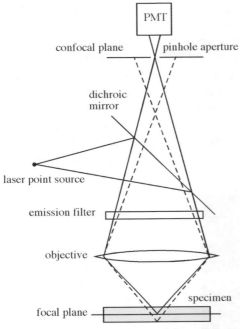

Fig. 4-89: Schematic of a laser scanning confocal microscope (LSCM). Light emerging from the spot on the focal plane of the specimen converges to an image of the laser point source at the confocal plane, where a pinhole aperture is inserted. This is shown blocking nearly all of the light that emerges from a plane below the focal plane (dashed lines), which comes to a focus ahead of the confocal plane.

aperture thus blocks essentially all light coming from above or below the focal place of the specimen spot. Since only a single spot is being observed, a simple photomultiplier tube can be used in synchrony with the laser to record the photon count from the illuminated spot, and then the spot can be scanned (rastered) horizontally (either the sample or the optics can be moved) to build up the image of the focal plane of the specimen. Then the vertical location of the specimen may be changed, the scanning repeated, and an image of another specimen plane obtained. Hundreds of vertical sections may be obtained in this way to build up a high-resolution, three-dimensional image of the specimen, which may be as thick as 200 μm.

4. *Electron microscopy*

The fact that electrons have wave-like properties under some conditions with effective wavelengths in the nanometer range provides the motivation for electron microscopy as an alternative means of measuring topography with resolutions over the full range of interest for colloid and nanoscience. In transmission electron microscopy (TEM) the image is produced by focusing a beam of electrons which have passed through and been scattered by a *thin* (< 1 μm) specimen of the solid (or a thin solid replica of it). While able to resolve horizontal features down even to the Ångström level, it does not readily reveal vertical information. Scanning electron microscopy (SEM) analyzes electrons scattered back from the surface of the specimen (removing the requirement that the specimens be thin) as the impinging beam of electrons is moved across the surface in a raster pattern. The image produced by a beam hitting the surface at an oblique angle produces a 3-d type of image, and true three-dimensionality may be produced from a pair of images obtained at different inclinations. SEM can produce a resolution of ≈ 10 nm. Both methods of electron microscopy suffer the limitations of requiring conductive samples under high vacuum conditions, and it is often difficult to separate artifacts from true sample features. A recent development is that of *environmental* scanning electron microscopy (E-SEM), which, due to the use of water vapor as a gas ionization detector, some of the limitations of conventional electron microscopy are overcome, *viz.*, non-conductive specimens may be examined under conditions of only moderate vacuum (> 20 Torr).[149] An example of an E-SEM image of a labile specimen, a sunflower petal, is shown in Fig. 4-90.

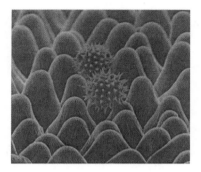

Fig. 4-90: E-SEM image of a sunflower petal with pollen particles.

5. Near-field scanning optical microscopy (NSOM)

Another important recent development is that of *near-field scanning optical microscopy* (NSOM).[150] (The acronym SNOM is also in common use.) In this method the resolution limits of ordinary (far-field) optical microscopy ($\approx$ half the wavelength of the light used ≈ 0.3 µm) are overcome by scanning the sample surface with a sub-wavelength-sized light source (the probe tip) from which the evanescent field of internally reflected light emerges, as shown in Fig. 4-91. This evanescent field falls off exponentially in intensity with distance away from the probe surface, extending out about 100 nm. The probe tip is thus held very close (≈ 5 nm) to the surface by means of a force-feedback and piezo positioning device similar to that used in atomic force microscopy (AFM), and in recent instrumentation, the techniques are combined. As the sample is scanned, an optical image of it is built up pixel by pixel. The probe is the apex of an aluminized optical fiber with a ≈ 25 nm diameter aperture through which the light passes. The portion of the surface illuminated is approximately of this size, so that any collected optical contrast originates from this region only, a region whose size (and not the wavelength of the light) limits the resolution. NSOM is very versatile in that it can operate in any ambient conditions and is completely non-invasive. The high resolution (< 1 nm) makes it an

[149] Baumgarten, N., *Nature*, **341**, 8-1 (1989).

[150] Moyer, P., Marchese-Ragona, S. P., and Christie, B., *Amer. Laboratory*, **26**(6), 30 (1994).

important tool of nanoscience. In even more recent developments, Raman and infrared spectroscopy are combined with NSOM to generate images with chemically specific information.[151]

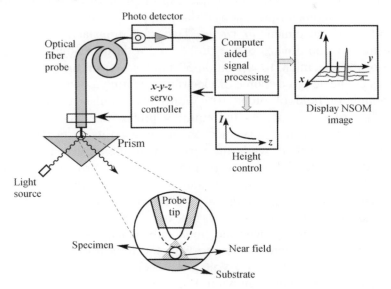

Fig. 4-91: Schematic of a typical setup of a near-field scanning optical microscope (NSOM). Optical fiber probe and collection-mode operation are employed. After [Ohtsu, M., and Hori, H., **Near-Field Nano-Optics**, Kluwer Academic, new York (1999).], p. 43.

M. Scanning probe microscopy (SPM)

One of the most important recent developments for probing surface structure and for nanoscience in general is that of *scanning probe microscopy* (SPM,[152,153,154] in which a sharp-tipped probe is moved with atomic precision over or near to a surface allowing surface topography and/or surface forces to be mapped. There are two principal manifestations of SPM. The first to be developed was *scanning tunneling microscopy* (STM), in which the basis for measurement is a tunneling electron current passing between a conducting tip and a conducting or semi-conducting solid surface. The second to be developed was *atomic force microscopy* (AFM), in which the essential quantity is a force exerted on or by a flexible cantilever, to which the sharp tip is attached, when the latter is put in contact or in the proximity of the surface to be studied.

[151] Henry, C. M., *Chem. & Engr. News*, **79** [18], 37 (2001).

[152] Strausser, Y. E., and Heaton, M. G., *Amer. Laboratory*, **26**(6), 20 (1994).

[153] DiNardo, N. J., **Nanoscale Characterization of Surfaces and Interfaces**, VCH, Weinheim, Germany, 1994.

[154] Wiesendanger, R., **Scanning Probe Microscopy and Spectroscopy**, Cambridge Univ. Press, Cambridge, UK, 1994.

1. *Scanning Tunneling Microscopy (STM)*

An early version of the idea[155] leading to STM used the concept of "field emission microscopy" and made use of piezo-electric translators to position (with nanometer precision) a sharp-tipped probe over the surface of the (conductive) specimen, *in vacuo*. Two such translators were used to position the *x-y* coordinates of the probe, and a third to adjust the vertical position of the specimen, to a position 50 nm away from the probe tip. The probe was connected to a source of electricity creating a potential difference great enough (a field of order 10^9 V/cm) to establish a current ("field emission") across the gap. As the emitting probe was rastered across the surface of the specimen, a servomechanism adjusted its vertical position to keep the current constant, while the variations in the vertical position were recorded. It was a breakthrough variation of this idea, proposed in 1981 by Binnig and Rohrer,[156] that led to Scanning Tunneling Microscopy (STM) (and the 1986 Nobel Prize for Physics). Their device moved an atomically sharp metal probe tip to within 1 nm of a hard conducting or semi-conducting specimen surface and took advantage of the spontaneous tunneling of electrons across this gap with the imposition of a small bias potential (usually of order 1 - 3 volts). A schematic of the more common configuration in use today is shown in Fig. 4-92. Tunneling currents, usually of the order of 0.01–10 nanoAmps, occur when the highest occupied

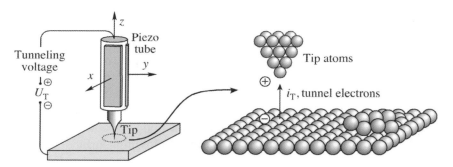

Fig. 4-92: Schematic of a scanning tunneling microscope (STM).

molecular orbital (HOMO) of the material on one side of the gap overlaps with the lowest unoccupied orbital (LUMO) of the material on the other side of the gap. The tunneling current, I_{tun}, for a given system decreases exponentially with the gap width in accord with

$$I_{tun} = K_1 \left(\frac{V}{d} \right) \exp(-K_2 d),$$

(4.157)

---

[155] Young, R. D., *Rev. Sci. Inst.*, **37**(3), 275 (1966).
[156] Binnig, G., Rohrer, H., Gerber, C., and Weibel, E., *Phys. Rev. Lett.*, **49**, 57 (1982); Binnig, G., and Rohrer, H., *Surface Sci.*, **126**, 236 (1983).

where d is the gap width, V is the imposed bias voltage, and K_1 and K_2 are materials constants. The decay constant K_2 depends on the tip and sample work functions, and is related to the energy difference between the HOMO and LUMO states of the opposing materials.

Tips for STM scanning are typically of tungsten wire electrochemically etched into sharp uniform tips. Often a subsequent chemical etching with hydrofluoric acid is used to remove or clean up oxide layers that quickly form. While tip curvature radii are typically ≈ 100 nm, the electron flow emerges from its outermost atom.

Positioning and monitoring of the tip may be accomplished using a piezo tripod, of the type used by Binnig and Rohrer, or a quartered piezo tube, as shown in Fig. 4-92, either of which can be moved with Ångström precision. Motion of the piezo elements is controlled by applying a voltage across them. A bias of ≈ 1 volt across such an element causes an expansion or contraction of ≈ 10 Å, so that sub-atomic movement in all three directions can be effected. X-y movements by a sufficient amount that square regions up to microns on a side may be imaged without the need to move the tip housing manually. In some cases, particularly with AFM to be described below, it is the sample rather than the tip housing that is moved using such piezo elements.

A scan is usually accomplished as follows. First the z-position is zeroed at the point of direct electrical contact. Then the probe tip is receded until the desired tunneling current is reached for the imposed bias. For metal samples, the sign of the bias is immaterial, but for semi-conductors, it is critical because the electron current in one direction is greatly resisted compared to the other. As the probe tip is moved across the surface, a feedback loop continuously adjusts the vertical position of the tip to maintain a constant current, corresponding to a constant gap width. The potential applied to the z-component of the piezotube is monitored as a function of the x,y positioning to produce an image of the topography. The extraordinary sensitivity of the tunneling current to the gap width, expressed in Eq. (4.157), permits atomic z-resolution. While this *constant current mode* of scanning is most common, STM can also be operated in a *constant height mode*, in which the tunneling current is monitored as one rasters over the surface at a constant z. This mode permits much faster scanning, but can be used only for ultra-smooth surfaces, such as freshly cleaved mica or highly-ordered pyrolytic graphite (HOPG), because only in these cases does one not run the risk of crashing the tip into larger surface features.

As stated above, STM is limited to conducting or semi-conducting samples. For such cases, its extraordinary resolution has permitted the investigation of the atomic structure of the surfaces of such materials and the manner in which such structure differs from that in bulk as a result of the asymmetry of inter-atomic forces acting upon them. One such achievement was the determination of the surface structure of Si(111)-7×7, a subject of controversy for many years. The bulk Si crystal has four nearest neighbors (coordination number = 4), but the surface atoms were found to be rearranged into a much larger repeating pattern.[157] STM has also been applied to adsorbed or Langmuir-Blodgett layers on conductive surfaces to reveal details of molecular structure. While generally applied in ambient air, STM can also be carried out *in vacuo* or under liquids. STM can also be used to manipulate atoms on surfaces, as demonstrated in the famous image of the logo: "IBM," produced by the arrangement of 35 xenon atoms on a Ni(110) substrate, mentioned in Chap. 1.

2. Atomic Force Microscopy (AFM)

A second more versatile form of scanning probe microscopy was introduced within a few years of the appearance of SPM[158] and is pictured schematically in Fig. 4-93. In this device, it is not the electric tunneling current which is measured, but instead the *force* with which the sample pushes against the tip attached to a long cantilever with a very low spring constant ($\approx$ 1 N/m). Called Atomic Force Microscopy (AFM), in its first manifestation, the tip-to-sample spacing was maintained constant by holding the force exerted by the cantilever at a constant low value ($\approx 10^{-9}$ N) while the tip was rastered over the sample, or more commonly, the sample was rastered beneath the tip over the area to be mapped. AFM in this mode resembles a stylus profilometer, but due to its use of the same type of piezo-electric positioners and feedback schemes as used in STM, is capable of sub-Ångström vertical sensitivity. Since the technique does not depend on the tunneling current, the sample surface need not be electrically conducting,

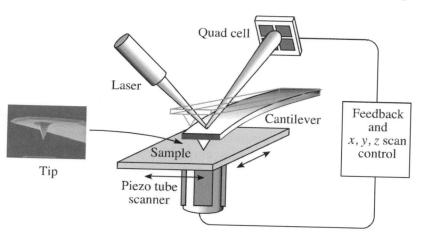

Fig. 4-93: Schematic of Atomic Force Microscopy (AFM), showing the beam deflection system used for vertical motion detection.

---

[157] DiNardo, N. J., **Nanoscale Characterization of Surfaces and Interfaces**, p. 12, VCH, Weinheim, Germany, 1994.
[158] Binnig, G., Quate, C. F., and Gerber, C., *Phys. Rev. Lett.*, **56**, 930 (1986).

and AFM scans may be carried out in ambient air, *in vacuo* or under liquid. An early example of the type of atomic profilometry that is possible with AFM is shown in Fig. 4-94, in which the tops of *individual*, ordered surfactant molecules on a mica surface under water are imaged. A different technique than that used in STM is required to detect the motion of the AFM tip as it moves over the sample surface. The most common device is the beam deflector shown in Fig. 4-93. A laser shines onto and reflects off the back of the cantilever and onto a position-sensitive photo-diode (PSPD) which detects beam motion and provides a correction signal to the z-piezo to keep the force (spacing) constant. The photo-diode is segmented either into halves (bi-cell), to detect up-and-down motion, or into quadrants (quad-cell),

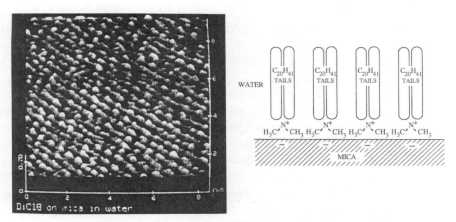

Fig. 4-94: AFM scan of closed-packed monolayer of surfactant on mica. Scanned image from [Tsao, Y. -H., Yang, S. X., Evans, D. F., and Wennerström, H., *Langmuir*, **7**, 3154 (1991).]

to detect both up-and-down and side-to-side motion. The equilibrium force applied by the tip to the sample (loading) is the product of the cantilever force constant (supplied by the manufacturer, but usually requiring calibration by the user) and its deflection from the zero point (zero deflection). The net force exerted at a given spacing is the result of the generally attractive van der Waals or other molecular or atomic forces (adhesion forces), possible electrostatic or magnetic forces, and Born repulsion. If scanning is carried out in ambient air, there is usually sufficient humidity to form a liquid bridge between the tip and the surface, if the latter is wetted. The resulting capillary force (often 10 -100 nN) may be a significant portion of the total force acting on the tip. It can be eliminated only if the surface is thoroughly cleaned, and the scanning is performed under high vacuum. Another way to eliminate capillary forces is to change the probing medium to a liquid, such as ethanol. Adhesive and capillary forces may maintain the tip in "contact" with the surface even under conditions of negative applied force (unloading), in a situation analogous to that obtained in a JKR experiment, as described earlier. The configuration during unloading may be unstable to vibrations and result in a sudden detachment of the tip from the surface.

Cantilevers may be very flexible, with force constants as low as 0.01 N/m, or rather stiff, with force constants as high as 100 N/m. The softness/hardness of the sample dictates the cantilever stiffness to be used. For very soft samples, even the most flexible cantilevers may produce damage (scratches) as the surface is moved beneath the tip. This problem may be addressed by oscillating the cantilever so that it touches the surface only at the bottom of these oscillations, usually carried out at a kHz or more, much faster than the scanning rate. This intermittent contact mode, or TappingMode™ (Digital Instruments Corp.) is believed to faithfully reproduce the topography of even the softest, liquid-like materials.

In another mode, termed lateral force microscopy (LFM), the twisting of the probe tip may be detected and recorded as the sample surface is moved beneath it, giving additional information. The tip and hence the cantilever may experience a change in torsion from either a change in surface slope or a change in surface lubricity or friction (nano-tribology), and hence local rheology. Using a quad-cell PSPD, both vertical topological data and LFM data may be obtained simultaneously during a scan. In a related technique, shear modulation force microscopy (SM-FM),[159] the end of the cantilever is driven side-to-side, usually with the tip at a given location on the surface of the sample. The specimen may then be heated to reveal changes in the friction related to changes in, for example, the local glass transition temperature, T_g, of the material.

In all of the above types of AFM, the probe tip is maintained in either continuous or intermittent physical contact with the sample surface. It is also possible to probe the force field emanating from the surface by oscillating the cantilever above it without contact. These non-contact modes are useful in particular for probing long-range interactions such as those of electrostatic or magnetic forces leading to electrostatic force microscopy (EFM) or magnetic force microscopy (MFM), respectively. In the former case, the tip is electrically charged, and in the latter, it is magnetically coated. MFM is an important tool for inspection of various magnetic data storage media, such as audio or video tapes, computer hard drives, *etc.* The smallness of the forces ($\approx 10^{-13}$ N) for probe-sample separations in the range of 10–100 nm requires the use of resonance enhancement techniques. The cantilever is driven (oscillated) above a given spot on the sample surface, using piezo-electric elements, at or near its mechanical resonance frequency, ω_0, given by

$$\omega_0 = \left(\frac{k}{m_{\text{eff}}} \right)^{1/2},$$
(4.158)

---

[159] Sills, S. E., Overney, R. M., Chau, W., Lee, V. Y., Miller, R. D., and Frommer, J., *J. Chem. Phys.*, **120**, 5334 (2004).

where k is the force constant of the cantilever, and m_{eff} is its effective mass, dependent on the mass distribution and geometry of the tip/cantilever assembly. ω_0-values are typically in the range of 1–100 kHz. Deviations from the resonance condition due to the immersion of the probe in the force field, f, are detected,[160] as the position of the cantilever is monitored using a PSPD (or by other means, such as tunneling or capacitance). When immersed in the force field at a position z above the surface, the effective force constant of the cantilever becomes

$$k' = k - \frac{df}{dz},$$ (4.159)

with the negative sign indicating an attractive force. The force gradient thus reduces the effective force constant of the cantilever, and the resonance frequency is correspondingly reduced in accord with

$$\omega_0' = \left(\frac{k'}{m_{eff}} \right)^{1/2} < \omega_0.$$ (4.160)

By monitoring the change in frequency as a function of z, one thus obtains

$$\frac{df}{dz} = k - k' = m_{eff}(\omega_0^2 - \omega_0'^2)$$ (4.161)

as a function of z, which can be integrated to give $f(z)$. One may alternatively monitor oscillation amplitude attenuation or phase shift to provide such information.

Another important AFM technique is force-distance profiling, or force spectroscopy (FS), wherein the force acting on a non-oscillating probe tip is monitored at a given x-y location on the surface as a function of probe-sample separation or displacement, from direct contact out to a distance where the force of interaction is zero. In practice, the sample surface is made to slowly approach the probe tip, and the force on the tip can be computed at any point from its location relative to its zero-deflection position and the force constant k. Thus a deflection force vs. scanner displacement curve may be generated, as shown schematically in Fig. 4-95. Starting from a large scanner displacement, at the right in the figure, the deflection force is zero. As the sample moves toward the tip, no force change is noted until the tip comes "in range" of the attractive force field. The tip is in a constant natural state of vibration, thus probing positions slightly nearer to the surface than indicated by the average scanner displacement. Eventually the vibrating tip probes a position where the attractive force is larger than can be resisted by the stiffness of the cantilever, and it "snaps" in to the surface. At this point, the cantilever is deflected downward, indicating an attractive (negative)

---

[160] Umeda, N., Ishizaki, S., and Uwai, H., *J. Vac. Sci. Tech.*, **B9**, 1318 (1991).

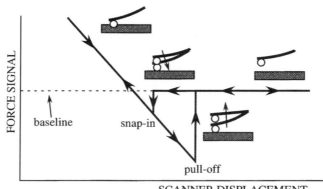

Fig. 4-95: Force profiling (z-probing) at a given surface location.

force. As the sample surface continues to be moved upward, the cantilever flattens out, and a point of zero deflection (zero force) is reached. This position may be identified with the topological location of the sample surface. As one continues to push the sample against the probe, causing it to indent the surface, a region of constant compliance is generally observed. When this linear region ends, it is common to reverse the direction of the sample movement, *i.e.*, to retract it. At first the deflection force retraces its original path (although sometimes hysteresis is observed), but then continues to exhibit attraction, or adhesion, as the scanner displacement moves past the position where snap-in had been observed. Eventually the force of the cantilever is sufficient to detach the tip from the surface, and the zero deflection state is recovered. The negative force at the point of detachment is referred to as the "pull-off" force, and is a measure of the tip-substrate adhesion. If the forces of interaction are repulsive at all separations, no snap-in is observed, and the hysteresis loop between snap-in and pull-off is missing from the force trace.

The insets in Fig. 4-95 show a sphere attached at the end of the cantilever rather than a sharp tip. Ducker *et al.* in 1991 were first to glue a colloid-sized particle (a 3.5 μm radius silica sphere) to an AFM cantilever and used the method described above to probe the interaction between this colloid particle and a planar surface.[161] The method is thus referred to as "colloidal probing." It hence provides the means, with due cautions being observed, of carrying out a JKR adhesion experiment leading to the work of adhesion. It has been used to probe not only the interaction between colloid particles and hard surfaces in various environments, but also to examine the interaction between colloid particles and fluid interfaces,[162-163] providing an

---

[161] Ducker, W. A., Senden, T. J., and Pashley, R. M., *Nature*, **353**, 239 (1991).

[162] Mulvaney, P., Perera, J. M., Biggs, S., Grieser, F., and Stevens, G. W., *J. Colloid Interface Sci.*, **183**, 614 (1996).

[163] Snyder, B. A., Aston, D. E., and Berg, J. C., *Langmuir,* **13**, 590 (1997).

important method for determination of disjoining pressure isotherms.[164] From the force trace of Fig. 4-95 it is evident that colloidal probing cannot access the details of the *attractive* portion of $f(z)$, giving only the distance at which snap-in occurs and the force corresponding to pull-off. One way to obtain more information would be to use a series of cantilevers of increasing stiffness, or a cantilever whose stiffness may be controlled, for example by using a magnetic force transducer.[165] An alternative approach is to follow the actual dynamics of the snap-in event and to extract information from the dynamic force trace.[166]

An important development that combines several of the techniques mentioned above is Pulsed-Force Mode (PFM) AFM. In it, the cantilever is oscillated at a frequency well below its resonance frequency (usually 0.1–5 kHz) and makes contact with the sample surface at the bottom of each oscillation cycle. Thus each cycle produces a force-distance profile, including snap-in, penetration and pull-off, as pictured schematically in Fig. 4-96. In this figure, the surface is approaching the probe as one moves from

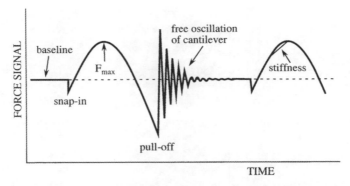

Fig. 4-96: Force trace from one cycle of a pulsed force mode (PFM) contact-detachment event.

left to right (opposite to the direction in Fig. 4-95). After snap-in, the probe tip is pushed into the sample until a controlled maximum force is reached. The distance corresponding to the location of F_{max} provides topographical information. The slope of the curve as the maximum force is approached gives a measure of the stiffness of the sample. Reversal of the sample direction leads to pull-off, and the difference between the zero force at the baseline and the pull-off force is a measure of the local adhesive strength. The oscillations following the pull-off event are superimposed on the forced oscillation of the cantilever. The sample is rastered beneath the tip of the oscillating cantilever providing a relative value of surface elevation, stiffness and adhesion at each sample point and hence maps of each of these features. The forces probed are of course not equilibrium forces, and the properties

[164] Basu, S., and Sharma, M. M., *J. Colloid Interface Sci.*, **181**, 443 (1996).
[165] Parker, J. L., aand Claesson, P. M., *Langmuir*, **8**, 757 (1992).
[166] Butt, H.-J., *J. Colloid Interface Sci.*, **166**, 109 (1994).

measured are relative rather than absolute, but the scan yields three very different ways to display contrast. Its use is exemplified in Fig. 4-97,[167] which shows topographical, adhesion force and stiffness maps of 5x5 μm silica specimens with patches of octadecyltrimethoxysilane (ODTS) primer (top row) and patches of ODTS with the intervening area back-filled with an aminopropylsilane (APS) primer. For the ODTS islands on bare silica, all three scans give good contrast. The elevation of the islands (a) is clearly greater than that of the silica substrate, while the adhesion of the SiN tip to the bare silica surface (b) is greater than to the waxy ODTS surface, and the stiffness of the hard silica substrate is clearly greater than that of the rather mushy ODTS material, as shown in panel (c). When the bare silica surface is back-filled with the adhesion-promoting primer APS, little contrast is exhibited in either the topographical image (d), because the layers are about equally thick, or the stiffness image (f), because they are equally mushy, but good contrast is exhibited between the adhesiveness of the two primers, as shown in panel (e).

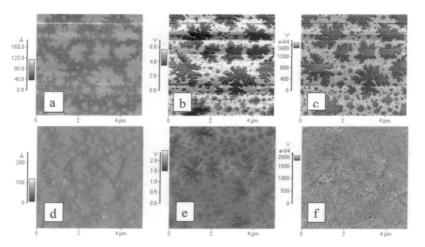

Fig. 4-97: Pulsed force mode AFM scans of silane primers on a silica substrate. (a) – (c) show topographical, adhesion force and stiffness scans, respectively, of ODTS islands on a silica substrate. (d) – (f) show topographical, adhesion force and stiffness scans, respectively, of ODTS islands backfilled with APS primer on a silica substrate. From [Buseman-Williams, J., and Berg, J. C., *Langmuir*, **20**, 2026 (2004).]

A very large number of other types of AFM probing have been developed, examining different types of properties, and new ones are being proposed at a rapid rate. For example, through the use of a heated tip one may measure and map thermal expansivity, thermal conductivity and phase transitions temperatures through dc and ac temperature changes (micro thermal analysis - μTA, *etc.*). Contact electrical potential differences, such

[167] Buseman-Williams, J., and Berg, J. C., *Langmuir*, **20**, 2026 (2004).

as between different metals, may be mapped using Kelvin probe force microscopy. Rheological properties may be mapped by tracking the amplitude and phase lag when in contact with a modulated sample (force modulation microscopy – FFM). A partial list is given in Table 4-11, together with the associated acronym for each. A practical guide to many of these techniques is given by Maganov and Whangbo,[168] but for more recent developments, one must consult the current instrumentation trade literature.[169]

Table 4-11: Various modes of AFM

Mode	Acronym	Quantity measured
Contact AFM imaging	AFM/C-AFM	Nanoscale profilometry
Lateral or Friction Force Microscopy	LFM/FFM	In-plane lubricity/friction
Kelvin Probe Force Microscopy	KPM	Contact potential difference
Micro Thermal Imaging Microscopy	μTIM	Thermal property mapping
Micro Thermomechanical Analysis	μTMA	Thermal expansions/softening
Scanning Force Microscopy	FS/SFS	Force-distance profiling
Chemical Force Microscopy	CFM	Chemical; bond force mapping
Pulsed-Force Mode	PFM	Topography, adhesion, stiffness mapping
Tapping or Intermittent Contact Mode	IC-AFM	Topography of soft materials
Phase Detection Imaging	PDI	Material properties by phase lagging
Scanning Electrochemical Microscopy	ECM/SECM	Current cyclic voltammograms
Non-Contact Mode	NCM	Near-field van der Waals forces
Magnetic Force Microscopy	MFM	Magnetic domain mapping
Electric Force Microscopy	EFM	Coulombic charge domain mapping
Scanning Near-Field Acoustic Microscopy	SNAM	Dynamic fluid damping
Scanning Near-Field Thermal Microscopy	SNTM	Heated thermocouple tip; thermal property mapping
Ultrasonic Force Microscopy	UFM	Material properties by Mhz to sub-GHz sample vibration

N. Surface area of powders, pore size distribution

Topographical information as it applies to porous or powder media generally refers to pore size distributions and specific area (surface area per unit mass). Many minerals, oxides, clays, *etc.*, are micro- or even nano-porous, having specific areas of 10 – 100 m^2/g. Activated charcoals are

168 Maganov, S. N., and Whangbo, M.-H., **Surface Analysis with STM and AFM: Experimental and Theoretical Aspects of Image Analysis**, VCH, New York, 1996.
169 Maganov, S. N., and Heaton, M. G., *Amer. Laboratory,* **30** [10], 9 (1998); Leckenby, J., Faddis, D., and Hsiao, G., *Amer. Laboratory*, **31** [31], 26 (1999).

readily available with specific areas of 1000 – 2000 m^2/g, and newly synthesized metal-organic cage-like materials, "metal organic frameworks" (MOF), have reached the 4500 m^2/g mark.[170] Highly porous materials are important as catalysts, adsorbents, and media for gas storage and separation. We have seen earlier how mercury porosimitry is used to determine pore size distributions. Specific areas are generally determined using gas adsorption onto finely divided and/or porous solids. Recall that adsorption refers to the concentration of one or more species at an interface, which in this case is the interface between the solid and a gas. If there is no specific chemical interaction between the gas species and the solid surface, it will adsorb in proportion to its partial saturation, $\phi = (p/p^s)$, where p is the partial pressure of the gas, and p^s is its equilibrium (saturation) vapor pressure at the temperature of the system. The relationship is given quite generally by the Brunauer-Emmett-Teller (BET) adsorption isotherm, Eq. (4.162):[171]

$$\frac{V}{V_m} = \frac{C\phi}{(1-\phi)[1+(C-1)\phi]},$$
(4.162)

which reflects the assumption that adsorption may occur in multilayers, even before the first monolayer is complete, as pictured schematically in Fig. 4-98, and that it becomes infinite as saturation conditions are approached. The heat of adsorption for the first monolayer is considered to be generally greater than the heat of condensation of the vapor, which characterizes the energetics of all layers subsequent to the first.

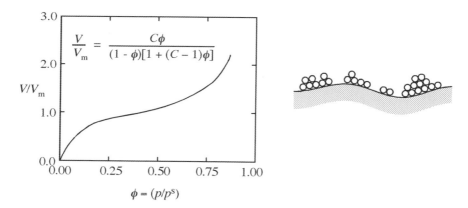

Fig. 4-98: BET isotherm for the non-specific adsorption of a gas onto a solid substrate, and qualitative sketch of adsorbed layer configuration.

In the equation for the isotherm, V is the volume of gas adsorbed per unit mass of solid, and V_m is the amount needed to cover the area of the solid with a complete single monolayer of molecules. C is a constant

[170] Chae, H. K., Siberio-Pérez, D. Y., Kim, J., Go, Y., Eddaoudi, M.. Matzger, A. J., O'Keeffe, M., and Yaghi, O. M., *Nature*, **427**, 523 (2004).

[171] Gregg, S.J., and Sing, K.S.W., **Adsorption, Surface Area and Porosity**, 2nd Ed., Academic Press, London, 1982.

characteristic of the particular solid and gas, and temperature, and may be given the interpretation:

$$C = \exp\frac{(\lambda_1 - \lambda)}{RT},$$
(4.163)

where λ represents the molar heat of condensation, and λ_1 represents the molar heat of adsorption of the first monolayer of adsorbate. It may be noted that $C \rightarrow 1$ as T is increased. The general form of the isotherm is shown in Fig. 4-98. Adsorption may be measured directly as the increase in weight of a solid sample or the decrease in volume of the gas upon exposure to the gas at given T and partial saturation, or by permitting adsorption equilibrium to be achieved at a given (low) T and then measuring the volume of the desorbed gas as T is raised to a much higher value (out-gassing). The measurement of the specific area of the solid is made by determining V_m, i.e., the volume of the adsorbing gas of known molecular size (area) required to form a complete monolayer on the solid surface. This is the *BET method*, and is usually effected by adsorbing N_2 or some other simple gas (such as Krypton) at cryogenic temperatures and out-gassing at room temperature. Adsorption data are plotted in accord with the linear form of the BET isotherm:

$$\frac{\phi}{V(1-\phi)} = \frac{1}{CV_m} + \frac{(C-1)\phi}{CV_m},$$
(4.164)

and the parameters C and V_m are determined from the least-squares slope and intercept of a plot of Eq. (4.164). Once V_m is known, the specific area of the solid, Σ, may be computed using Eq. (4.165) as:

$$\Sigma = \frac{V_m^{STP} N_{Av}}{22.414} A_{sc},$$
(4.165)

where V_m^{STP} is the volume V_m (in liters) corrected for non-ideality (usually negligible) and to conditions of standard temperature and pressure (0°C, 1 atm); N_{Av} is Avogadro's Number, and A_{sc} is the cross-sectional area of the adsorbed molecule. For N_2, $A_{sc} = 16.2 \times 10^{-20}$ m^2/molecule.

Modern commercial instrumentation for carrying out BET surface area measurements can also be used for porosimitry, i.e., for determination of pore size distributions. This is accomplished using capillary condensation, as described in Chap. 2.

O. Energetic characterization of solid surfaces:
Inverse Gas Chromatography (IGC)

Chemical characterization of solid surfaces may be approached in many ways. An impressive array of high-vacuum, destructive and non-destructive techniques is available, and most yield information on the atomic and chemical composition of the surface and layers just beneath it. These are reviewed elsewhere[172] and are beyond the scope of the present work. From the standpoint of their effect on wettability, adhesion and wicking, the properties of greatest importance appear to be (1) the dispersion (or Lifshitz-van der Waals) surface energy, σ^d, and (2) the propensity of the solid surface to enter into acid-base interactions (quantified in some way). Both may be measured by the simple but elegant technique of *inverse gas chromatography* (IGC),[173] which will be briefly described here. It is conventional gas chromatography with respect to the equipment and techniques involved, but is called "inverse" because it is the solid stationary phase, rather than the gas phase, whose properties are to be investigated. The stationary phase may be in the form of a powder, fiber mass, or thin coating on the wall of the column. The technique measures the adsorption of probe gases of various kinds on the test solid, and the characteristics of the resulting adsorption isotherms are related to the surface energetics and chemistry of the solid. Probe gases, such as the normal alkanes, known to adsorb only through dispersion interactions, may be used to determine the dispersion force component of the solid surface energy, while acidic and basic probe gases may be used to probe the acid and base properties of the solid.

The simplest mode of IGC is the "infinite dilution mode," effected when the adsorbing species is present at very low concentration in a non-adsorbing carrier gas. Under such conditions, the adsorption may be assumed to be sub-monolayer, and if one assumes in addition that the surface is energetically homogeneous with respect to the adsorption (often a reasonable assumption for dispersion-force-only adsorbates), the isotherm will be linear (Henry's Law), *i.e.*, the amount adsorbed will be linearly dependent on the partial saturation of the gas. The proportionality factor is K_{eq}, the adsorption equilibrium constant, which is the ratio of the volume of gas adsorbed per unit area of solid to its relative saturation in the carrier. The quantity measured experimentally is the *relative retention volume*, V_N, for a gas sample injected into the column. It is the volume of carrier gas required to completely elute the sample, relative to the amount required to elute a non-adsorbing probe, *i.e.*,

$$V_N = jF_{col}(t_R - t_{ref}), \tag{4.166}$$

where F_{col} is the volumetric flow rate of the carrier gas, t_R is the retention time (time required to elute the sample, which for dilute sample is taken as

[172] Walls, J. M., (Ed.), **Methods of Surface Analysis**, Cambridge Univ. Press, Cambridge, UK, 1989.

[173] Lloyd, D. R., Ward, T. C., and Schreiber, H. P., Eds., **Inverse Gas Chromatography**, ACS Symposium Series 391, Washington, DC (1989);
Brendlé, E., and Papirer, E., **Powders and Fibers: Interfacial Science and Applications**, M. Nardin and E. Papirer (Eds.), pp. 47-122, CRC Press, Boca Raton, 2007;
Wang, M. J., **Powders and Fibers: Interfacial Science and Applications**, M. Nardin and E. Papirer (Eds.), pp. 122-170, CRC Press, Boca Raton, 2007.

the time to reach the peak of the elution curve), and t_{ref} is the time required to elute the non-adsorbing reference gas (often taken as methane). j is the "James-Martin correction factor," used to correct for any pressure drop across the column, and given by:

$$j = \frac{3}{2} \frac{(p_i/p_o)^2 - 1}{(p_i/p_o)^3 - 1},$$
(4.167)

where p_i and p_o are the inlet and outlet pressures, respectively.

The relative retention volume is directly related to the slope of the adsorption isotherm:[174]

$$V_N = K_{eq} A_{tot},$$
(4.168)

where A_{tot} is the total area of the solid in the column.

The standard molar free energy change upon adsorption of the probe gas is thus given by

$$\Delta G^o_{ads} = -RT \ln K_{eq} + C_1 = RT \ln V_N + C_2,$$
(4.169)

where C_1 and C_2 are constants dependent upon the standard states chosen and the adsorbent surface area. Finally, the standard free energy of adsorption is related to the work of adhesion by

$$W_A = \frac{-\Delta G^o_{ads}}{a_{mol}} + C_3,$$
(4.170)

where a_{mol} is the molar area of the adsorbate gas, and the constant accommodates the choice of standard states. If the only interaction between the adsorbate and the adsorbent is of the Lifshitz-van der Waals (*i.e.*, dispersion force) type, the geometric mean mixing rule applies, and

$$W_A = 2\sqrt{\sigma_S^d \sigma_L^d},$$
(4.171)

where σ_S^d is the dispersion surface energy component of the solid we seek, and σ_L^d refers to the dispersion portion of the surface tension of the probe (in liquid form). For apolar probes, the latter is simply the surface tension itself. Combining the above four equations leads to:

$$RT \ln V_N = 2 a_{mol} \sqrt{\sigma_S^d \sigma_L^d} + C_4,$$
(4.172)

from which σ_S^d could be evaluated if one had knowledge of the trailing constant. The need for it is avoided, however, by using a series of chromatograms of different probes and extracting σ_S^d from the slope of a plot of $RT \ln V_N$ vs. $a_{mol} \sqrt{\sigma_L^d}$, as shown in Fig. 4-99. A homologous series of

---

[174] Conder, J. R., and Young, C. L., **Physicochemical Measurements by Gas Chromatography**, Wiley, New York, 1979.

alkane probes is generally used because good values for their adsorbed molar areas are available, *viz.* [175]

$$a_{mol} = [(n-2)(0.06) + 2(0.08)](N_{av}/10^{14}) \ cm^2/mol, \qquad (4.173)$$

where n (> 2) is the number of carbon atoms in the chain. Excellent linearity of the plot (the "alkane line") is generally found and lends credence to the procedure.

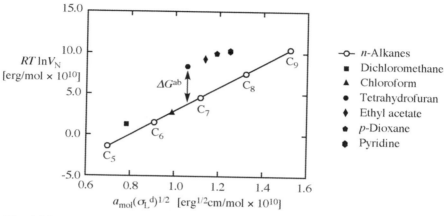

Fig. 4-99: An IGC plot for Whatman CF1 cellulose powder. Data at 63.3°C.

Information regarding the acid-base character of the solid surface is obtained by using acid and base probe gases on a solid for which the alkane line has already been obtained, as suggested by Schultz and Lavielle.[176] If acid-base interaction is involved in the adsorption, the retention volume should be greater than that corresponding to the dispersion force interaction, which should be the same as that of the "equivalent alkane," *i.e.*, the hypothetical alkane for which the value of $a_{mol}\sqrt{\sigma_L^d}$ is the same. The acid-base component of the work of adsorption is given directly by the difference in the $RT\ln V_N$ values:

$$W_{ads}^{ab} = \frac{\Delta G_{ads}^{ab}}{a_{mol}} = \frac{RT}{a_{mol}} \ln \frac{V_N}{(V_N)_{alkane}}, \qquad (4.174)$$

as shown for several probes in Fig. 4-99. The cellulose powder for which data are shown exhibits a dispersion surface energy of approximately 42 erg/cm² (mJ/m²) (at 63.3°C) and a predominantly acid character (as evidenced by the strong adsorption of the basic probes). When such data are obtained over a range of temperatures and differentiated in accord with the Gibbs-Helmholtz Equation, one obtains values of $(-\Delta H_{ads}^{ab})$:

[175] Dorris, G. M., and Gray, D. G., *J. Colloid Interface Sci.*, **71**, 93 (1979).

[176] Shultz, J., and Lavielle, L., **Inverse Gas Chromatography**, D. R. Lloyd, T. C. Ward, and H. P. Schreiber (Eds.), ACS Symposium Ser., No. 391, pp. 185-202, Washington, DC, 1989.

$$\frac{\partial \dfrac{\Delta G_{\text{ads}}^{\text{ab}}}{T}}{\partial T} = R \frac{\partial \ln \dfrac{V_N}{(V_N)_{\text{alkane}}}}{\partial T} = \frac{(-\Delta H_{\text{ads}}^{\text{ab}})}{T^2}, \text{ or}$$

$$(-\Delta H_{\text{ads}}^{\text{ab}}) = RT^2 \frac{\partial \ln \dfrac{V_N}{(V_N)_{\text{alkane}}}}{\partial T}. \tag{4.175}$$

Attempts to use the latter to evaluate such parameters as those of Drago or Gutmann have met with only marginal success.

When $(-\Delta H_{\text{ads}}^{\text{ab}})$-values are obtained using Eq. (4.175) for a number of acid-base probe vapors, it has been suggested that for a given solid one may write:[177-178]

$$(-\Delta H_{\text{ads}}^{\text{ab}}) = K_A \cdot DN + K_B \cdot AN, \tag{4.176}$$

where DN and AN are the tabulated Gutmann Donor and Acceptor Numbers, respectively, for the probes, and K_A and K_B are numbers describing the acid and base characters of the solid. Schultz et al.[114] put Eq. (4.176) in the form:

$$\frac{(-\Delta H_{\text{ads}}^{\text{ab}})}{AN} = K_A \frac{DN}{AN} + K_B, \tag{4.177}$$

and plotted $\dfrac{(-\Delta H_{\text{ads}}^{\text{ab}})}{AN}$ vs. $\dfrac{DN}{AN}$ for three different carbon fiber types tested by a variety of acid-base probes. Straight lines resulted, from which K_A and K_B values could be determined. Since then, other investigators have used this method for solid-surface acid-base characterization, but caution should be exercised in interpreting the values of such highly "derived" parameters.

Values of σ_s^d obtained by IGC can be compared with those obtained by wetting measurements, and are usually found to be significantly larger. They may be brought largely into agreement, however, when proper account is taken of the equilibrium spreading pressure.[179] Finally, it has been shown how IGC may be used to obtain complete adsorption isotherms by working beyond the infinite dilution regime, and from them to derive site energy distribution functions for (energetically heterogeneous) solids.[180] The various techniques for determining the acid-base properties of solid surfaces has recently been reviewed elsewhere.[181]

---

[177] Saint Flour, C., and Papirer, E., Ind. Engr. Chem., Prod. Res. Dev., 21, 666 (1982).

[178] Schultz, J., Lavielle, L., and Martin, C., J. Adhesion, 23, 45 (1987).

[179] Jacob, P. N., and Berg, J. C., J. Adhesion, 54, 115 (1995).

[180] Jacob, P. N., and Berg, J. C., Langmuir., 10, 3089 (1994).

[181] Sun, C., and Berg, J. C., Adv. Colloid Interface Sci., 105, 151 (2003).

Some fun things to do:
Experiments and demonstrations for Chapter 4

1. *Contact angles*

Contact angles made by sessile drops on horizontal flat surfaces may be formed, displayed and evaluated by projecting the image of the drop profile onto a screen.

Materials:

- Two Plexiglas® coupons, ≈ 2×2 in.
- Teflon tape
- 1-mL polyethylene disposable pipettes
- Video camera
- S-video cable
- Video projector
- Unipod or tripod (for mounting camera)
- Ring stand and clamps
- Large plastic protractor
- Fine grit sand paper

Procedure:

1) Mount the video camera on the tripod. Set up the ring stand and put the Plexiglas® coupon directly in front of the video camera and place a small drop of water on it about 2 cm in front of the lens. Connect the video camera with the projector using the S-video cable. Turn on the projector to see the live feed from the video camera and focus, as shown in Fig. E4.1.

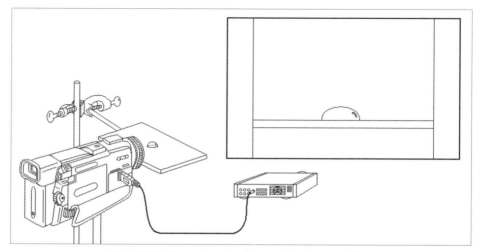

Fig. E4.1: Set-up for viewing sessile drop profiles.

2) Measure the contact angle of the water drop against the Plexiglas® on the screen using the protractor. It should be in the range of 50-70°. Then roughen a small patch of the surface with sand paper and determine the contact angle of a water drop on the roughened surface. It should be reduced, in accord with the Wenzel Equation, Eq. (4.7). Wrap some Teflon tape around one of the plastic coupons, and place a drop of water on it. The contact angle should be ≈100°. *Gently* roughen the Teflon with the sandpaper, and observe the resulting contact angle. In this case, it should be *increased.*

3) With the water drop on either the Plexiglas® or the Teflon, gradually tilt the plate, to observe contact angle hysteresis, until the drop eventually slides off the plate.

2. *Making self-assembled monolayers (SAM's), and measuring their wettability*

As described in Section L of Chap. 3, SAM's refer to highly ordered monolayers of alkane thiols on smooth surfaces of gold, silver or other "coinage metals." They are often terminally functionalized to produce a desired surface chemistry. One of their properties that can readily be measured is their wettability. In this experiment, five different SAM's are prepared (self-assembled) on gold surfaces. They will consist of varying area fractions of hydroxy-functionalized thiols (higher surface energy) and alkyl-funtionalized thiols (lower surface energy) to produce surfaces of varying water wettability. The resulting contact angles can be measured and then analyzed using the Cassie-Baxter Equation, Eq. (4.8).

Materials:
- Five 1×1 cm gold-coated silicon chips (SPI Supplies, West Chester, PA)
- 200 mL absolute ethanol
- < 1 g Dodecanethiol (HS-$C_{12}H_{25}$), CAS# 112-55-0
- < 1 g 11-Mercapto-1-undecanol (HS-$(CH_2)_{11}OH$), CAS# 73768-94-2
- Plastic forceps
- 5-mL syringe
- 1-mL polyethylene disposable pipettes

Procedure:

1) Prepare 50 mL of 1 mM dodecanethiol solution in ethanol (Stock A) and 50 mL of 1 mM 11-mercapto-1-undecanol solution in ethanol (Stock B).

2) Prepare five 10-mL samples by mixing Stock A and Stock B at different volume ratios—100%, 75%, 50%, 25%, and 0% of Stock

A. Place the gold-coated chips in each sample solution for 24 hours, to allow for annealing. Rinse the chips with DI water. The gold-coated chips should be handled from the edge with plastic forceps to avoid scratching their surfaces.

3) View and measure the water contact angles of the five samples as described in Expt. 1. Analyze the results using the Cassie-Baxter Equation, assuming the area fractions of the thiols on the gold surfaces are the same as their volume fractions in the bulk ethanol solutions.

3. *Use of a super-spreader*

Super-spreaders are trisiloxane ethoxylates that can be used to cause water to wet a variety of hydrophobic surfaces, as discussed in Section E.4. This can be demonstrated using the set-up of Expt. 1.

Materials:

- < 1 mL Silwet® Hydrostable® 611 Superspreader (Momentive Performance Materials, Inc., Albany, NY)
- Plexiglas® coupons, ≈ 2×2 in.

Procedure:

Place a drop of water on the Plexiglas® coupon as described in Expt. 1, and gently add the smallest drop possible to the drop as it is observed on the screen. This can be done by dipping the end of a straightened paperclip into the super-spreader and applying it to the drop.

4. *Contact adhesion*

The JKR experiment described in Section F.7, can be demonstrated using gelatin spheres that are easily prepared, and compared to the behavior of comparably-sized silica gel spheres (which are used as desiccants) or glass spheres.

Materials:

- Two 100-mL beakers
- 1 g anhydrous $CaCl_2$
- 1 g alginic acid sodium salt (Fisher Sci., CAS: 9005-38-3)
- Silica gel beads or glass spheres, 3-4 mm diameter
- Six 1-mL polyethylene disposable pipettes
- Six glass microscope slides
- Scoop, made by cutting half the bulb from one of the above pipettes
- Forceps

Procedure:

1) Prepare 50 mL each of solutions of 20 mg/mL of alginic acid and of $CaCl_2$ in de-ionized water.

2) Slowly drop droplets of the alginic acid solution into the $CaCl_2$ solution using a polyethylene pipette. The added drops will form gel spheres of 3-4 mm in diameter.

3) Using the scoop, carefully remove the gel spheres from the $CaCl_2$ solution and allow them to dry for 10 min.

4) Place a couple of the gel spheres and a silica gel bead or glass sphere of comparable size onto a horizontal microscope side and wait for 10 min. Then tilt the glass slide to observe the sticking of the gel spheres and the slipping of the hard spheres. The gel spheres have been able to deform themselves slightly and to establish adhesive molecular contact with the glass slide. It should remain attached even when the slide is inverted.

5) Place a gel sphere and a glass sphere on a microscope slide which is placed on an overhead projector. Observe the contact spot of the gel sphere, and the much smaller one for the glass sphere.

6) Use a second glass slide to press down on the gel sphere and observe the increase in the size of the contact spot with applied pressure.

5. *Playing with Magic Sand®*

Magic Sand® is sand with a proprietary hydrophobic coating that gives it some amazing properties. It is available from Steve Spangler Science (http://www.stevespanglerscience.com/product/1331). The website demonstrates many fun ideas for experiments.

Procedure:

1) Simply pour Magic Sand® into a beaker of water. Notice how the sand sticks together in large chunks. The sand is coated with air bubbles giving it a bright, shiny look. (Many hydrophobic bugs survive under water for long periods of time in the same way, using their entrained air as a diving suit.) Try to break up the chunks with a spatula. Then pour the water out of the beaker. All of the water can be recovered, and the Magic Sand® remains completely dry and free flowing.

2) Magic Sand® can be used to make a large "armored bubble." Gradually sprinkle more and more of the sand on the surface of water in a beaker. When the weight of the patch is great enough to overcome the surface tension force, the patch collapses on itself forming a macroscopic armored bubble that sinks.

3) Magic Sand® can be wetted with a soap solution. Repeat Step (1) above using a soap solution instead of water. It will now be found to act like ordinary "un-magic" sand.

6. *Liquid marbles*

As described in Section J.5, liquid marbles are liquid spheres (usually of water and a few mm in size) coated with particles and having the properties of soft solids. They can be rolled around, manipulated and even floated on water without losing their integrity.

Materials:

- 10 mL glycerol
- 10 g hydrophobic precipitated silica (0.5 – 3 μm) (*e.g.*, Dow Corning VM-2270 Aerogel fine particles)
- Three Petri dishes (10 cm)
- Plastic spoon
- 1-mL disposable pipettes

Procedure:

Spread about 5 g of the hydrophobic silica in one of the Petri dishes and deposit a drop of glycerol on top of it. To make a liquid marble, coat the drop with the silica particles using the plastic spoon. If the marble is too big, cut it with the spoon to make it smaller and then coat the new marbles with the silica particles. The size of the liquid marbles should be around 2-3 mm. After the marble is perfectly coated with the silica, pick it up with the plastic spoon and put into a clean Petri dish and roll it around. Fill another Petri dish with water, pick the marble with the spoon and gently put it on the surface of the water. The drop will float on the water, usually for at least 15 seconds.

7. *Wicking of water into filter paper*

Wicking of a liquid into a porous medium is usually well described by the Washburn Equation, Eq. (4.111), which for the case in which the liquid fully wets the medium ($\theta = 0°$) is:

$$x = \left[\frac{r_e \sigma}{2\mu}\right]^{1/2} t^{1/2} = k t^{1/2},$$

where x is the wicking distance, t is time, σ is surface tension, μ is viscosity, and r_e is the "wicking-equivalent" pore size in the medium. In this experiment the wicking of water into filter paper is monitored, a system in which full wetting is expected. If the format of Eq. (4.111) is obeyed, for a liquid of known surface tension and viscosity, the Washburn slope k can be used to evaluate the wicking-equivalent radius. This will be found to be usually an order of magnitude smaller than the actual pore openings in the medium as a result of the tortuosity of the flow, as the actual path traveled by the liquid is much greater than its projection along the direction of the movement of the front, x.

Materials:

- Rectangular strips of filter paper (*e.g.*, Whatman 41) at least 10 cm long and 1-2 cm wide, cut from filter paper circles using a paper cutter
- Large, flat-bottomed glass tube or jar with diameter about 10 cm and standing at least 15 cm high
- Thin wire
- Stop watch

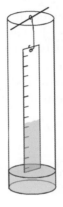

Fig. E4-2: Wicking of water into a filter paper strip.

Procedure:

1) Using a pencil, pre-mark the filter paper strip starting about 2mm from one end at 0.5 cm intervals to the opposite end. Punch a small hole near the end opposite to that where the marking started.

2) Suspend the filter paper inside the tube using a hang-down wire attached to the paper through the small hole, as shown in Fig. E4-2. The tube should have at least a 1-2 cm depth of water in the bottom, and hang-down wire length pre-adjusted so the paper strip dips about 2 mm into the water.

3) As soon as the end of the paper strip is put in the water, observe the position of the rising liquid. Note and write down the time at which the front passes each of the hash marks. This is best done with the help of an assistant, as the time to reach the first mark is only about 1 s. It slows down after the first few marks are passed. If Whatman 41 filter paper is used, the front will reach 10 cm in about 10 minutes.

4) When the data are collected, plot the wicking distance against the square root of time, and see if it is a straight line as predicted by the Washburn Equation. If so, evaluate the Washburn

slope k and the wicking-equivalent radius of the pores, r_e. For this you will need to look up the surface tension and viscosity of water at the temperature of your experiment.

8. *Capillary rise in a narrow groove*

A wetting liquid will spontaneously move into a narrow horizontal groove if the groove angle $\phi < (180° - 2\theta)$, where θ is the contact angle, as described in Section I.7. In a zero-gravity environment, the groove need not be horizontal, and this was the basis of the "zero-gravity cup" made from a folded plastic sheet used by astronaut Don Petit aboard the space station and sketched schematically in Fig. E4.3. The groove angle ϕ could be changed at will by squeezing the plastic sheet until the water flowed upward. It was featured recently in the *National Geographic* magazine.[182] If the groove is oriented vertically in the Earth's gravity, however, it will rise only to a certain distance, as given by the profile of Eq. (4.122) and shown in Fig. 4-74. This situation is easily reproduced using the glass plates described in Expt. 3 after Chap. 2. Simply hold the two glass plates together with a pair of rubber bands, separate them at one edge using one or two paper clips, and dip into a Petri dish of water.

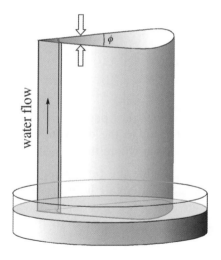

Fig. E4-3: Astronaut's zero-gravity cup.

[182] *National Geographic*, May, 2009, p. 29.

Chapter 5

COLLOIDAL SYSTEMS: PHENOMENOLOGY AND CHARACTERIZATION

A. Preliminaries

1. *Definition and classification of colloids*

Colloids refer to mixtures in which one material is dispersed in another, as shown in Fig. 5-1, with the dispersed material (the "dispersoid") subdivided into independent units in the size range between approximately 1 nanometer and 10 micrometers (or microns, μm) in linear dimensions. The lower end of this size range is that of nano-particles. The size limits are not hard-and-fast: some molecular clusters smaller than 1 nm (10Å) may behave as colloids, and many investigators put the upper size limit at one micrometer. Often, however, emulsions and suspensions with particle diameters up to even several tens of μm act like colloids in many respects. In any case, the term "colloid" excludes dispersions of "ordinary-sized" molecules at the low end and macroscopic phases visible to the naked eye at the high end, *i.e.*, both the regions with which the chemist or engineer feels familiar. Thus a colloid is a system "in between," in Ostwald's *"Welt der vernachlässigten Dimensionen,"* *i.e.*, "world of neglected dimensions."

The dispersion medium, *i.e.*, the continuous portion of the system, is assumed to be composed of small-to-moderately-sized molecules. One may therefore model (and hence define) colloids as dispersions of distinguishable particles (kinetic units) in the size range of 1 nm to 10 μm in a medium regarded as a *structureless continuum*, as suggested in Fig. 5-1.

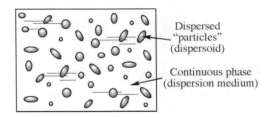

Dispersed "particles" (dispersoid)

Continuous phase (dispersion medium)

Fig. 5-1: Definition of a colloid: a dispersion of kinetic units in the size range from 1 nm to 10 μm in a structureless continuum.

When the dispersion medium is either a liquid or a solid, a colloid may be either *lyophobic* or *lyophilic*. Lyophobic (*i.e.*, "solvent hating") colloids are those in which the dispersoid constitutes a distinct phase, while lyophilic colloids refer to single-phase *solutions* of macromolecules or polymers. In lyophilic as opposed to lyophobic systems, there is no true "interface" between the "particles" and the medium. Solutions of micelles, vesicles and other self-assembled structures of amphiphiles are examples of lyophilic systems called *association colloids*. While lyophobic and lyophilic colloids show some of the same behavior, in other important ways they are fundamentally different. The primary focus in what follows is on lyophobic colloids.

Lyophobic colloids may be classified in terms of the types of phases which compose them, as is shown in Table 5-1, which helps to introduce some of the terminology used to name the various types of colloids and to

Table 5-1: Types of lyophobic colloid systems.

Dispersed phase	Medium	Name(s)	Examples
Liquid	Gas	Aerosol	Fog, fine sprays
Solid	Gas	Aerosol	Smoke
Gas	Liquid	Microfoam, "aphron"	Shaving cream
Liquid	Liquid	Emulsion	Milk, mayonnaise
Solid	Liquid	Sol	Paint, ink
Gas	Solid	Solid micro/nano-foam	Some foamed plastics
Liquid	Solid	Solid emulsion	Opal, pearl
Solid	Solid	Micro/nano composites	Pigmented plastics

give a framework that allows one to limit the scope of the subject. The coverage in this text is given primarily to systems in which the dispersion medium is a liquid, either aqueous or organic. Metal sols in water, pigment particles in paint, emulsions of oil-in-water, or *vice versa*, and foams all fit this classification.

The size ranges of various examples of colloid particles are shown in Fig. 5-2, comparing them to wavelengths in the electromagnetic spectrum and the resolvability of several methods for their examination. The narrow visible spectrum is seen to fall approximately in the middle of the colloid size range, suggesting the importance but also limitations, of optical (wavelength-dependent) techniques in their investigation. Optical microscopy, for example, resolves details only in the upper part of the colloid size range. A study of Fig. 5-2 will help familiarize one with colloidal dimensions.

2. General properties of colloidal dispersions

The major categories of differences between the properties of colloids on one hand and solutions of small-to-moderate molecular weight substances on the other hand are listed and discussed briefly below. All were sources of considerable puzzlement to early investigators. They are discussed in more detail later in this and subsequent chapters, and it will be evident that they constitute the basis for the study of colloidal systems.

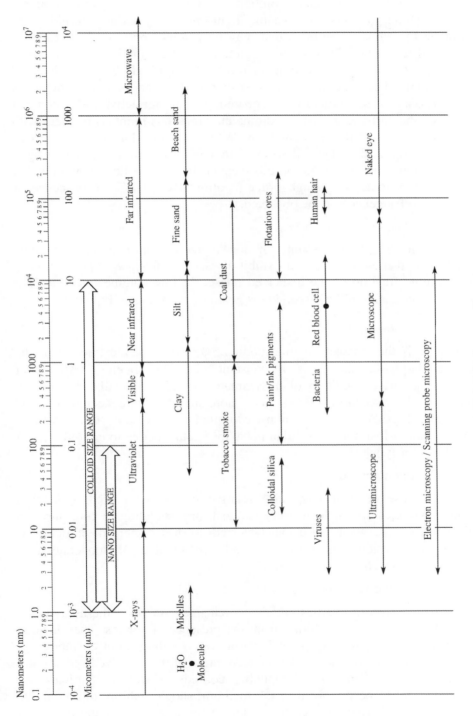

Fig. 5-2: Colloidal system dimensions.

1) Optical properties:

In contrast to ordinary solutions, colloidal systems, both lyophobic and lyophilic, often scatter visible light, *i.e.*, they exhibit turbidity, in contrast to ordinary solutions that may be colored, but are clear. Light is scattered in all directions when it encounters optical inhomogeneities (*e.g.*, particles) of dimensions within an order of magnitude or so of its own wavelength. The amount of light scattered, for very small particles, is proportional to the particle mass squared, and the refractive index contrast between the particles and the medium, and variations in intensity occur with the wavelength of the incident light. With incident white light, ranges of colors known as the "Tyndall spectra" may be observed. Light scattering is discussed in some detail later in this chapter. In dense colloidal systems that can achieve some long-range order ("colloidal crystals"), opalescence or optical birefringence may be observed.

2) Rheological properties:

Dilute lyophobic systems of non-interacting particles in Newtonian media are also Newtonian and exhibit a viscosity increase in proportion to the volume fraction of the dispersed phase. At low volume fractions ($\phi \leq 0.02$) the viscosity of the dispersion is given by the Einstein Equation:

$$\mu = \mu_0(1 + k\phi), \tag{5.1}$$

where μ_0 is the viscosity of the medium, ϕ is the volume fraction of the dispersed phase, and k is a constant (2.5 for spheres). At higher concentrations, particularly of unsymmetrical and/or flocculated particles, however, the viscosity increases much more sharply with concentration, and the colloid usually becomes non-Newtonian and in some circumstances viscoelastic. Most non-Newtonian behavior is attributable to colloids, either lyophobic or lyophilic.

3) Kinetic properties:

Because of their large size relative to moderate-sized molecules, colloid particles appear to be very sluggish in responding to concentration differences, *i.e.*, their diffusion (referred to for dispersed particles as "Brownian motion") is very slow relative to that of moderate-sized molecules in solution.

4) Colligative properties:

Colligative properties refer to changes in the thermodynamic properties of a solvent resulting from the presence of solutes. They include osmotic pressure, freezing point lowering, and boiling point elevation, all directly dependent on the number concentration of the dissolved species and independent of their size. Lyophobic colloids, since the particles are presumed not to be dissolved in the medium, should exhibit no colligative properties at all. Lyophilic colloids, due to the large size of the dissolved entities, exhibit these properties to only a very small extent. Relatively large amounts, in terms of mass, of dissolved macromolecules are needed to register measurable colligative properties.

5) Instability of lyophobic colloids:

The most important general distinction between lyophobic colloids and ordinary solutions is their inherent instability relative to a number of processes that act to destroy them. Thus any apparent stability ("shelf life") exhibited by lyophobic colloids is kinetic in nature, *i.e.*, it is the result of a low rate of self-destruction. Lyophilic colloids, on the other hand, are thermodynamically stable in the same sense that ordinary solutions are. When it is said that a system is "unstable," it is necessary to ask "with respect to what process?" The most important of these processes are: phase segregation, aggregation, coalescence, and size disproportionation. A brief description of each process, together with an expression for its rate (under certain ideal conditions), is given below, followed by more detailed examination later.

3. Dense vs. dilute dispersions

In most textbooks and in most basic research laboratories dealing with colloids, the assumption is that the dispersions are *dilute*, *i.e.*, the particle number density or volume fraction is sufficiently low that the particles (aside from their possible aggregation) act individually to affect the properties of the system. Ironically, perhaps the majority of the colloidal dispersions encountered in practice do not conform to this assumption. In rough terms, multi-particle effects begin to arise when the particle volume fraction exceeds approximately 0.02. Table 5-2 lists this and other benchmark volume fractions for the state of density of dispersions of spherical particles.

Table 5-2: Benchmark volume fractions for monodisperse dispersions of spheres.

$\phi \geq 0.02$	Multi-particle effects begin to influence dispersion properties	
$\phi = 0.024$	Average particle separation $\langle S_0 \rangle$ equals particle diameter	
$\phi_{rcp} \approx 0.64$	Random close-packed limit	
$\phi = 0.74$	Hexagonal close-packed limit	

A useful formula for computing the average particle separation distance between spheres, $\langle S_0 \rangle$, defined as the distance of closest approach of their surfaces, in terms of the dispersion volume fraction and particle radius, a, is:

$$\langle S_0 \rangle = a \left[\left(\frac{\phi_{rcp}}{\phi} \right)^{1/3} - 1 \right]. \tag{5.2}$$

It is perhaps surprising to realize the close proximity of particles that exists on average, even in dispersions of rather low volume fraction. Paint chemists and other formulators refer to the volume fraction as the "pigment volume concentration," *PVC*, defined as the volume fraction of particles in the particle-medium mixture, *i.e.*,

$$PVC = \frac{\text{vol. particles}}{\text{vol. particles + vol. medium}}. \tag{5.3}$$

The liquid medium is often referred to as a "binder." The *critical* pigment volume concentration, or *CPVC*, is defined as that value of the *PVC* corresponding to the situation where there is just enough binder present to fill all the interstitial regions between the particles, as shown in Fig. 5-3. The *CPVC* usually has a value between 0.5 and 0.6. Above this value, the mixture must contain some trapped air, while below it, there is no air, and the particles have room to move about in the binder liquid.

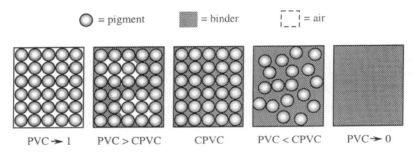

Fig. 5-3: Pigment volume concentration, *PVC*, in slurries.

Dense slurries in the vicinity of their *CPVC* are referred to more often as sludges or pastes, and their properties generally undergo drastic changes as the *CPVC* is traversed. It is common to see properties of such systems plotted as a function of their relative *PVC*, *i.e.*, $\Lambda = PVC/CPVC$. While all properties of these systems undergo changes with concentration, it is the rheological and optical behavior of dense dispersions that get the most attention. The viscosity of a slurry undergoes a drastic rate of increase as one moves from $\Lambda > 1$ to $\Lambda < 1$, and more importantly is the onset of viscoelasticity. The optical properties are of special importance to formulators of paints and coatings. Gloss paints have a low *PVC* (0.15–0.25), while primers, mid-sheen wall finishes, undercoats and flat paints are progressively higher.

While the "structure" of dilute dispersions is usually just assumed to be a uniform distribution of the widely spaced particles, that of dense dispersions is of considerable interest. If the particles are "non-interacting," *i.e.*, they do not aggregate or stick together, they may be arranged either randomly or in an ordered state referred to as a "colloidal crystal," exhibiting such optical properties as birefringence. If the particles are aggregated, they may be either *percolated*, in which case the entire volume of the dispersion may be viewed as containing a single aggregate, or they may be un-percolated, in which case the system is a collection of separate clusters of aggregated particles, none spanning the entire volume. These structures are typically found in sediment cakes, whose density may vary considerably from case to case, often determining such properties as filterability.

While most of what follows in the present chapter deals with the phenomenology of dilute colloids, comments are generally made topic by topic as to how the descriptions must be modified to describe the behavior of denser dispersions.

B. Mechanisms of lyophobic colloid instability

1. *Phase segregation: the "phoretic processes"*

Lyophobic colloids may be unstable with respect to the spatial distribution of the dispersed phase particles, since the systems are generally in the gravitational field, and the density of the dispersed phase is usually different than that of the dispersion medium. If the dispersed phase particles are denser than the medium, they will tend to settle toward the bottom of the container (sedimentation), and if they are lighter, they will tend to rise toward the top (creaming), as shown in Fig. 5-4, leading ultimately to the division of the colloid into two more or less well-defined regions, one dense

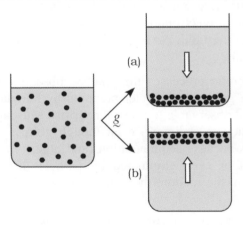

Fig. 5-4: Phase segregation by (a) sedimentation or (b) creaming.

in colloid particles and the other sparse. Sedimentation (or creaming) refers only to the process of segregation and not to any sticking together or merging of the particles. Other external force fields may also cause phase segregation. For example, an electric field will cause the electrophoresis of charged colloids, and the separation process of electro-decantation is based on this phenomenon. A number of analogous "phoretic" phenomena may occur when the colloid finds itself in other types of fields. Examples include magnetophoresis, thermophoresis, and diffusiophoresis, which refer respectively to particle motion in a magnetic field, a temperature gradient, and a concentration gradient. (In the same sense, sedimentation or creaming might be thought of as "graviphoresis.")

The rate equation for the sedimentation of an isolated spherical colloid particle is Stokes' Law:

$$v_{sed} = \frac{2a^2(\Delta\rho)g}{9\mu},$$

(5.4)

where a is the particle radius, $(\Delta\rho)$ is the density difference between the particle and the medium, g is the gravitational constant and μ is the viscosity.

The electrophoretic velocity of a charged particle in an electric field is given by:

$$v_{elect} = \frac{\varepsilon E_x \zeta}{4\pi\mu},$$

(5.5)

where ε is the dielectric constant of the medium, E_x is the electric field strength, and ζ (the "zeta potential," to be discussed in detail later) is the effective electrical potential at the particle surface relative to the interior of the medium. The structure of the rate equations for phoretic processes, as exemplified by Eqs. (5.4) and (5.5), is a ratio of the driving force for the process to the viscous forces resisting particle motion. The driving force is the product of the strength of the relevant external field: g in the case of sedimentation, and E_x in the case of electrophoresis, and the property of the colloid particle that couples with the field: $\Delta\rho$ in the case of sedimentation, and ζ in the case of electrophoresis.

Phase segregation by phoretic processes may not be observed even when it might be expected. The gold sols in water prepared by Faraday in the mid-1800's[1] for example, and observable today in the British Museum in London, have shown no tendency to sediment, even though gold is 19.3 times as dense as water. This is because sedimentation (and other types of phase segregation) is mitigated or even eliminated by particle diffusion (Brownian motion) resulting from the random thermal motion of all kinetic units in the system.

The linear rate (Stefan velocity) of Brownian motion is given by Fick's Law of diffusion, and the expression (to be derived later) for the diffusivity, applicable to spherical particles is:

$$v_{Brown} = -\frac{kT}{6\pi\mu a}\frac{d\ln n}{dx},$$

(5.6)

where k is the Boltzmann constant, T is absolute temperature, n is the local number concentration of the particles, and x is the distance of travel. It is the

---

[1] Faraday, M., *Phil. Trans.*, **147**, 145 (1857).

small size of the gold colloid particles (a few tens of nanometers) that renders Brownian motion so effective against gravitational settling. In addition to Brownian motion, another factor important in impeding phoretic processes is the existence of small but macroscopic thermal convection currents in the fluid medium, due to ever-present temperature fluctuations.

2. Thermodynamic criteria for stability

Phase segregation by phoretic processes does not in itself destroy the colloidal dispersion, and in principle at least, the system could be restored to its initial uniform condition by mixing. The remaining processes with respect to which a lyophobic colloidal system may be unstable, however, are not reversible and can be conveniently cast in thermodynamic terms.

The equilibrium and stability characteristics of a system at a constant intensive state, *i.e.*, constant T, p and composition, with respect to a given process are expressible in terms of the system Gibbs free energy, G. What is required is knowledge of how G depends on the "process coordinate," ξ, as is represented schematically in Fig. 5-5. Equilibrium with respect to the process represented by ξ, requires (for an incremental advancement of that process) that $dG = 0$, *i.e.*, G must be at a maximum, minimum, or point of inflection in the $G(\xi)$ curve. *Stability* requires, in addition, that the free energy be a local *minimum*, *i.e.*, $d^2G > 0$. Local maxima in G refer to unstable equilibrium states, while local minima in the vicinity of still lower minima in G correspond to *metastable* equilibria. At a point of inflection, *i.e.*, a point where $d^2G = 0$, stability is determined by the sign of the third (or

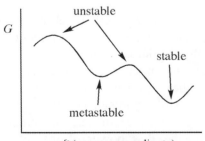

Fig. 5-5: General conditions for thermodynamic equilibrium: ($dG = 0$) and stability ($dG = 0$ *and* $d^2G > 0$).

the first non-vanishing) differential. Changes in G are effected in general by changes in several variables (sufficient in number to fix the state of the system), but at a fixed intensive state, only the "process variable," ξ may vary. Thus when considering a certain process, one need consider only changes in that particular term. With this in mind, one may consider the remaining processes pertinent to colloid stability.

3. Aggregation

Aggregation refers to the clumping or sticking together of the dispersed colloid particles to form aggregates or flocks. These structures, made up of so-called *primary particles*, may either eventually become cemented together or may remain loosely tied together so that upon restoration of conditions of "stability," gentle stirring can re-disperse them (a process referred to as "repeptization"). Consider the coming together and sticking of just two primary particles to form a doublet, as shown in Fig. 5-6.

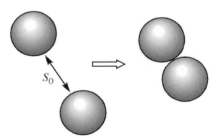

Fig. 5-6: Aggregation of two primary particles to form a doublet.

When two particles are close enough together, there are significant attractive forces between them, just as there are between molecules. The independent process variable (ξ) for aggregation is "S_0," the distance of separation between the particle surfaces. The process is one in which the separation between the particles, within the range of attraction, goes to zero (aggregation), and for the system:

$$G = G_0 + \Phi(\xi) = G_0 + \Phi(S_0). \tag{5.7}$$

The origin of $\Phi(S_0)$ may be no more than the collective van der Waals forces between the molecules that make up the particles, acting through the medium in which they are immersed and arising principally from molecular oscillations and electric fields which these oscillations establish (*i.e.*, London or "dispersion" forces). Recall that these inter*molecular* forces produce a pair potential energy, which in the case of individual molecules decreases as the sixth power of the intermolecular separation. The consequent interaction between the *macroscopic* colloid particles (containing perhaps $\geq 10^{18}$ molecules each), upon pairwise integration of the molecular interactions as first suggested by Hamaker,[2] is much longer-ranged, falling off approximately as the *first* power of the inter-particle separation, as shown in Fig. 5-7, and reaching a minimum (maximum negative value) when the particles touch ($S_0 \rightarrow 0$). At the point of contact, the interaction potential curve turns steeply upward, due to Born repulsion. In typical lyophobic colloids, the depth of the potential well is much greater than kT, so that the potential energy drop upon aggregation is essentially the same as the drop in system free energy. Thus, whenever lyophobic colloid particles get within range of one another's inter-particle attractive forces, they should spontaneously seek to clump together, as shown in Fig. 5-6, because in so doing they reduce the free energy of the system.

---

[2] Hamaker, H. C., *Physica*, **4**, 1058 (1937).

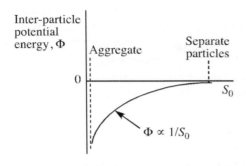

Fig. 5-7: Schematic of potential energy of attraction between a pair of colloidal particles.

The resulting aggregation, at least in its early stages, is a second-order process whose rate is given by:

$$\frac{dn}{dt} = -k_r n^2, \quad \text{with} \quad k_r = \frac{4kT}{3\mu}, \tag{5.8}$$

where n is the particle number density and k_r is the rate constant for aggregation. The expression for the rate constant corresponds to the case first described by Smoluchowski,[3] in which the particles reach each other through Brownian motion, and each collision results in the particles sticking together. The aggregation of lyophobic colloids will be examined in greater detail later in Chap. 7, and it will be seen that some colloids aggregate much more slowly. Observation shows many lyophobic colloid systems, unstable with respect to aggregation, nonetheless may exist in an un-aggregated state almost indefinitely.

4. *Coalescence*

Coalescence is a process that may occur when the dispersed phase is a fluid. Consider an emulsion, as shown in Fig. 5-8. The liquid droplets join together to form a larger droplet, and in so doing, destroy interfacial area. When two spheres of radius a merge to form a single sphere, its radius is $\sqrt[3]{2}a \approx 1.26a$, and the total area decreases as:

$$2(4\pi a^2) = 8\pi a^2 \xrightarrow{\text{coal.}} 4\pi(1.26a)^2 = 6.35\pi a^2. \tag{5.9}$$

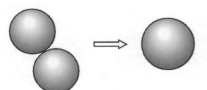

Fig. 5-8: Coalescence.

Of course the minimum area dividing a given volume of one phase from another is that corresponding to a single sphere. Coalescence is the major process that defines the stability of emulsions and foams.

---

[3] von Smoluchowski, M., *Physik. Z.*, **17**, 557, 585 (1916); *Z. Phys. Chem.*, **92**, 129 (1917).

Instability with respect to coalescence depends on the fact that there is a true interface between the particle material and the medium. True interfaces are repositories of free energy, and for interfaces between fluid phases, the free energy is identifiable with the interfacial tension, σ. Once again one may think of the system's free energy as split into two parts, one remaining constant during the process and the other representing that associated with the interface:

$$G = G_0 + \sigma A, \tag{5.10}$$

where σ is the interfacial free energy/area, and A the area. Any process reducing A (at constant T, p, and composition) reduces G, and hence should occur spontaneously.

In order to coalesce, droplets must approach one another closely enough, often assisted through sedimentation or creaming, to come under the influence of long-range van der Waals forces. As they near one another, they flatten to form a thin liquid film across a circular area, as shown in Fig. 5-9. In the simplest case, the rate of coalescence is governed by the time required

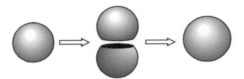

Fig. 5-9: Coalescence occurring through the formation, drainage and breakage of a thin liquid film between the approaching droplets or bubbles.

to drain this film to a critical thinness, h_c (usually $\approx$ 100 nm), at which point it becomes hydrodynamically unstable and ruptures. The Reynolds[4] expression based on lubrication theory, for the drainage time between two flat, circular plates is:

$$t_{\text{drainage}} = \frac{3\mu A_{\text{con}}}{4\pi p}\left(\frac{1}{h_c^2} - \frac{1}{h_0^2}\right) \approx \frac{3\mu A_{\text{con}}}{4\pi p h_c^2}, \tag{5.11}$$

where μ is the viscosity of the medium, A_{con} is the area of the circle of contact, p is the effective pressure pushing (or drawing) the surfaces together, and h_0 is the distance of separation which exists when the first flattening occurs. One often sees cases where the drops push against one another showing little or no tendency to coalesce, existing for days or even weeks. Examples of such stability were discussed in Chap. 2 as cases where the disjoining pressure in the in the film, Π, goes to zero, and its derivative with respect to film thickness is negative, *cf.* Eq. (2.123).

5. Particle size disproportionation

Particle size disproportionation, also referred to as "Ostwald ripening," or simply "aging" is the process whereby large particles grow at

---

[4] Reynolds, O., *Phil. Trans. Roy. Soc. (London)*, **A177**, 157 (1886).

the expense of smaller ones, as shown in Fig. 5-10. The driving force for size disproportionation is also that of minimizing interfacial area. The

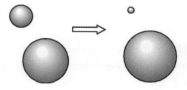

Fig. 5-10: Particle size disproportionation.

process requires that material making up the dispersed phase have a finite molecular solubility in the medium so that molecules or ions from smaller particles dissolve into the medium, while at the same time dissolved molecules incorporate themselves into the larger particles. This occurs because the solubility of the material from the smaller particles is greater than that of the material from the larger particles as a consequence of the Kelvin effect discussed in Chap. 2. The earlier discussion of the effect of curvature on equilibrium properties of materials focused on liquids, but the analogous situation with respect to the solubility of solid particles in the dispersion medium exists, as was first described by Ostwald.[5] Considering a pure particle of radius a into which only negligible amounts of the dispersing medium may dissolve, he obtained the expression for the solubility, $(x_2)_a$:

$$(x_2)_a = (x_2)_\infty \exp\left(\frac{2v_2\sigma}{aRT}\right), \tag{5.12}$$

where $(x_2)_\infty$ is the handbook solubility. For solids, σ refers to the interfacial free energy. For crystalline solids, there can be different types of crystal faces, each with a characteristic surface energy. So-called "Wulff crystals" are those for which, effectively, the average interfacial energy is taken over all the crystal faces of the particle.[6]

Usually the dispersed phase material is only very sparingly soluble in the dispersion medium, so disproportionation is slow. An expression for the instantaneous rate at which the smaller particles shrink in a bi-disperse colloid with particles of radii a_1 and a_2,[7] provides insight into the factors upon which the rate of size disproportionation depends.

$$\frac{da_2}{dt} = \frac{2Dv\sigma(x^s)_\infty}{RTa_2^2} \frac{n_2(a_2 - a_1)}{n_1(n_1a_1 - n_2a_2)}. \tag{5.13}$$

D is the diffusivity of the dissolved molecules, v is the molar volume of the dispersed phase, and n_1 and n_2 are the number densities of the larger and smaller particles, respectively. When $a_2 \ll a_1$, Eq. (5.13) takes the form:

[5] Ostwald, W., *Zeit. Phys. Chem. (Leipzig)*, **34**, 295 (1907).

[6] Defay, R., Prigogine, I., Bellemans, A., and Everett, D. H., **Surface Tension and Adsorption**, pp. 292-301, Longmans, London, 1966.

[7] Higuchi, W. I., and Misra, J., *J. Pharm. Sci.*, **51** [5], 459 (1962).

$$\frac{da_2}{dt} \propto -a_2^{-2},\tag{5.14}$$

which shows an acceleration in the reduction of the small particle sizes with time.

Size disproportionation results in the formation of rain from water aerosols, and its existence has profound consequences for the behavior of colloidal systems in general because the colloid particle size range (and particularly the nano-particle range) is in the region where the Kelvin effect on solubility is significant. As discussed earlier, the process is the basis for present understanding of homogeneous nucleation, through which macroscopic phase changes occur. The process of homogeneous nucleation always takes the system through a colloidal state.

C. Preparation of colloid particles and colloidal dispersions

1. *Classification of preparation strategies for lyophobic colloids*

Strategies for the preparation of a lyophobic colloidal dispersion may be broadly classified in accord with what the desired final system is and what one has to start with.

The first situation to discuss is one in which the colloid particles already exist as a starting material, either in dry form, as an aerosol or in some medium other than that which is desired. In such a case, "preparation" refers simply to the dispersion or re-dispersion of the given material. A number of lyophobic colloids do occur in Nature. Glacial silts and many naturally occurring clays consist of particles in the colloidal domain. Dispersions of biological origin, such as milk, cream, blood and liquid media containing bacteria, viruses and protozoa in water are also colloidal. In addition, a number of naturally occurring aerosols (pollen, spores, sea mists) exist. If the desired end product is a dispersion of colloidal particulates in a given liquid medium, and the dispersoid particles are available in the form of a dry powder, it is necessary that the desired liquid medium completely wet the powder and displace the air from the particle surfaces and the internal spaces between the powder particles.[8] Spontaneous wicking of the liquid into the powder mass requires an advancing contact angle < 90°, but full displacement of air pockets also requires minimal contact angle hysteresis, as discussed in Chap. 4.I.6. The latter condition is generally fulfilled only when the contact angle is 0°. If it is desired to transfer a given dispersion from one liquid to another, the receiving liquid must displace the original liquid from the surface of the particles. This generally requires that the desired liquid medium form a 0° contact angle

---

[8] Parfitt, G. D. (Ed.), **Dispersion of Powders in Liquids: with Special Reference to Pigments**, Appl. Sci. Publ., London, 1981.

against the particle surface in the presence of the original medium, analogous to the roll-up mechanism described in the context of detergency in Chap. 4.H.1. To accomplish this, appropriate wetting agents are often required.

It may also be necessary to break up clusters of powder particles that are present, exemplifying what is called a "top-down" process, often requiring considerable mechanical energy input. If the primary particles are in the nano size range ($d < 100$ nm), breakup of their clusters can be extremely difficult, depending on the nature of the inter-particle bonds. In some cases, the breakup of particles to form smaller ones can be induced by chemical changes in the medium, usually in combination with mechanical energy input. An example is the exfoliation of clays, described in more detail below.

Sometimes one wishes to produce a dispersion of particles *larger* than those present initially, and this may be accomplished by inducing controlled aggregation, described in Chap. 6, or possibly by "spherical agglomeration," described in Chap. 4.H.3. This is an example of a "bottom-up" method, as described below.

If the desired lyophobic colloid is not available in Nature, or the dispersoid is not available as a dry powder or aerosol, the two general approaches available for its preparation are *top-down* and *bottom-up* methods. Top-down methods produce initially, or start with, macroscopic chunks of material or agglomerates and employ methods to break them down to the desired colloidal dimensions. Bottom-up methods start with a solution or gas mixture containing species, often referred to as "precursors," that are induced to nucleate, react, self-assemble or precipitate to form the desired colloidal or nano-particles or structures.

Once prepared, by whatever means, colloidal dispersions are immediately vulnerable to self-destruction by the processes described earlier. Thus strategies for their preparation must also include consideration for their protection, in particular against aggregation. As described in Chap. 7, aqueous colloids are most often stabilized electrostatically, while either aqueous or non-aqueous dispersions may be stabilized sterically. The electrostatic mechanism requires the existence of a layer of charge on the particle surface, surrounded by a diffuse cloud of ions of opposite charge (counter-ions). It is the overlap of the like charges in the counter-ion clouds that may produce sufficient repulsion to drastically slow the approach of the particles required for their aggregation. The double-layer cloud shrinks and ultimately collapses as the electrolyte content of the medium is increased. Steric stabilization occurs when the particles are coated with a layer of solvent-compatible polymer. The overlap of these layers upon close approach of the particles sets up an osmotic repulsion that keeps them apart. Thus colloid preparation schemes include the presence of "stabilizers," or "capping agents" for their protection. Stabilizers, in aqueous media, are often ionic surfactants that confer charge to the surface, while capping agents are often diblock copolymers with the larger portion of their molar mass being soluble in the medium. In aqueous media, excess or "parasitical" electrolyte may be removed by dialysis, electro-dialysis or filtration and back dilution. Conventional filter paper retains only particles larger than about one micron and hence is generally not suitable for separating colloidal particles from dispersion media. There are membranes, however, which can be designed to retain particles of nearly any desired size range while passing smaller ones: Collodion and other cellulosic materials (cast in various ways),[9] polyamide and other polymeric membranes, and "track-etch" membranes[10] and various ceramic membranes[11] have been designed to retain particles as small as 20 nm in diameter. The process whereby the medium is separated from the sol is called *dialysis*. The transfer process can be assisted when the solutes are electrolytes by imposing an electric field, leading to *electrodialysis*. Another process for assisting this is that of using pressure, and this process is termed *serum replacement*,[12] or when higher pressures are used, *ultrafiltration*. This can simultaneously force solvent away from the sol by the process of reverse osmosis.

2. Top-down strategies

The most straightforward top-down method for producing colloidal particles is that of mechanical comminution ("crushing and grinding," *etc.*) of macroscopic chunks of the dispersoid material. Some general examples of this approach are listed in Table 5-3.

Table 5-3: Top-down methods for the preparation of colloids.
• Crushing and grinding (milling, "micronizing")
• Electrical disintegration (for metals)
• Laser ablation (SWNTs and MWCTs)
• Jet breakup
• Normal boundary motion (shaking, sonication)
• Tangential boundary motion (shearing)
• Electrical emulsification or atomization
• Lithography (photo-, E-beam, soft, *etc.*)

Comminution refers to the mechanical disintegration of macroscopic phases or agglomerates. Both dry and wet processes and equipment for carrying them out are available. In dry processes, mechanical energy is commonly applied to the macro-particles or agglomerates by grinding or impact to produce fracture. This is often accomplished by smashing the

[9] Nunes, S. P., and Peinemann, K.-V. (Eds.), **Membrane Technology**, 2nd Ed., Wiley-VCH, Weinheim, 2006.

[10] Tangirala, R., Revanur, R., Russell, T.P., and Emerick, T., *Langmuir*, **23**, 965 (2007).

[11] http://www.lenntech.com/ceramic-membranes-research.htm

[12] Ahmed, S. M., El-Aasser, M. S., Pauli, G. H., Poehlein, G. W., and Vanderhoff, J. W., *J. Colloid Interface Sci.*, **73**, 388 (1980).

particles together or against other hard objects in a rotating mill. Grinding media such as hard balls (ball mill), rods (rod mill) or other objects are often used. The fineness of the grind depends on the size of the grinding media. Impact mills, such as the Sturdevan, Inc. Micronizer® are also very effective at size reduction. In this instrument, high-speed gas jets of the fluidized particles are directed at one another causing the particles to fracture on impact. The particles are subsequently separated in the same device by a cyclone. Dry grinding or impact generally produces particles only in the upper part of the colloid size range or in the "sub-sieve" or "supra-colloidal" domain, and effective comminution often requires the use of "grinding aids," *e.g.*, surfactants that adsorb to the particle surfaces preventing re-fusion of cracks as they are generated.

Comminution of macro-particles or aggregates in liquid media is generally effected by inducing extremely high shear, sonication (or ultrasonication), cavitation or a combination of these. High shear may be created by the forcing of the liquid through small stationary orifices or gaps or into the narrow gap between a stationary component (stator) and a rapidly turning rotor, as pictured in Fig. 5-11. In more complex devices, such as the Microfluidics, Inc. Microfluidizer® or the Gaulin Corp. Homogenizer, liquid streams are made to collide at high velocity and pressure within an interaction chamber, making use of shear, turbulence, impact and cavitation.

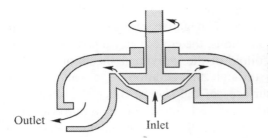

Fig. 5-11: Comminution in a colloid mill or homogenizer, where high shear is imparted to liquid passing between the rotor and stator.

Outlet ←   Inlet

Another important technique of comminution is laser ablation, in which a high-powered laser is directed at the surface of a bulk material producing *ejecta* that are subsequently collected, sorted and analyzed. It is an effective means for producing high quality single- or multi-walled carbon nanotubes (SWNTs or MWCTs),[13] and other types of nanopowders.

A specialized way of preparing metallic sols by comminution is that of electrical disintegration (Bredig method).[14] An electric arc is maintained under water between two metal electrodes producing a cloud of metal particles in the colloidal size range, as shown in Fig. 5-12. Often, traces of electrolyte must be added. Using high frequency alternating current, the

[13] Puretzky, A. A., Schittenhelm, H., Fan, X., Lance, M. J., Allard Jr., L. F., and Geohegan, D. B., *Phys. Rev. B.*, **65** (24), 245425 (2002).

[14] Overbeek, J. Th. G., **Colloid Science**, Vol. I, H. R. Kruyt (Ed.), p. 61, Elsevier, Amsterdam, 1952.

method can be used in non-aqueous media, and only slight decomposition of the medium occurs. Sols of almost any metal can be produced this way.

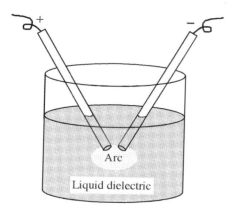

Fig. 5-12: Electrical disintegration of metals (Bredig method).

An example of comminution by "persuasion" rather than "compulsion" is afforded by most clays. Common clays, both naturally occurring and synthetic, are in the form of multilayered particles, as pictured schematically in Fig. 5-13. Chemically, many of them are aluminosilicates consisting of silica (SiO_4) tetrahedra bonded to alumina (AlO_6) octahedra in various ways. The individual sheets have a thickness of the order of 1 nm and widths varying from 100 nm to a few μm, and a single clay particle may have several hundred sheets. As described more fully in Chap. 6, the surfaces of the sheets generally carry a negative charge in contact with water, while under the right pH conditions, the edges carry a positive charge. It may be desired to produce a colloid of the individual sheets of the clay, and this may be accomplished through the addition of a solute which amphiphilically adsorbs to the surfaces of the sheets. Since the sheets are negatively charged, cationic surfactants are ideal for this purpose. The organophilic surfaces thus produced invite the subsequent intercalation of adsorbing polymers that may ultimately be capable of fully separating (exfoliating) the clay, as pictured in Fig. 5-13.

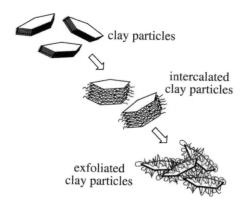

clay particles

intercalated clay particles

Fig. 5-13: Formation of intercalated and exfoliated clay through the use of adsorbing surfactant and polymer.

exfoliated clay particles

Dispersion of a liquid into a gas or another liquid can be accomplished by jetting the liquid through an orifice, shaking two liquids together (as in a paint mixer), sonicating or ultrasonicating the mixture of macro-phases, forcing the two liquids in parallel through a capillary slit, or allowing them to pass between rapidly rotating discs.

If it is desired to disperse a liquid in a gas to form a spray or aerosol, or into a second, immiscible liquid, to form an emulsion, a straightforward procedure is to force the liquid to be dispersed through a small circular orifice (or nozzle) into the desired fluid medium, as in Fig. 5-14. The breakup of a water capillary jet is shown in Fig. 1-6. The column of liquid produced becomes hydrodynamically unstable under the influence of interfacial tension forces when its length exceeds a value approximately equal to its circumference, after which small axisymmetric disturbances grow in amplitude along the jet, eventually pinching it off into droplets. In his classic study of laminar jet breakup, Lord Rayleigh[15] showed that

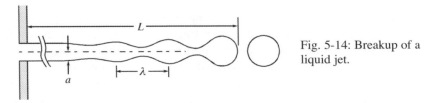

Fig. 5-14: Breakup of a liquid jet.

disturbances of a particular wavelength λ would be amplified more quickly than all others. For effectively inviscid jets breaking up in a vacuum, drops of diameter approximately twice the diameter of the orifice were predicted to form, and the unbroken jet length L was predicted to vary as

$$L \propto U \left(\frac{\rho a^3}{\sigma} \right)^{1/2} , \tag{5.15}$$

where U is the velocity of the liquid in the jet, ρ is its density, a is its radius, and σ its surface tension. Rayleigh's analysis is given in detail in Chap. 10, and all of his predictions, under the appropriate conditions, have been confirmed by experiment. In most cases, the drop size produced is comparable to the orifice dimensions, and for practical reasons, these can seldom be made less than 25 μm or so. Spray coating using droplets of this size or larger is an important process, and the use of spraying for the application of herbicides and pesticides is common practice. The method is also the basis for ink-jet printers, although most of these now produce drops one at a time ("on demand") by mechanical or thermal impulses imparted to the liquid reservoir upstream of the orifice. Ink-jet printing is increasing in importance as a method for computer-controlled patterned deposition of materials of a wide variety of types on a wide variety of substrates.

[15] Lord Rayleigh (J. W. Strutt), *Proc. Roy. Soc.*, **29**, 17 (1879).

While the drops produced by simple jet breakup are generally larger than colloidal, smaller drops may be produced upon evaporation or solvent leaching. Colloid-size droplets may also be produced by immersing the jet in an electric field co-linear with the jet axis. The interface becomes charged, and under the right conditions, the liquid emerging from the orifice will be pulled into a cone[16] (called a "Taylor cone") from the tip of which is drawn a jet about an order of magnitude smaller than that of the orifice. The process is the basis of electrospraying or electro-emulsification.[17]

Shaking liquids together or sonicating them produces essentially the same results as Rayleigh jetting. When a fluid-fluid mass is accelerated normal to its interface in the direction from the less dense toward the more-dense phase, it becomes unstable with respect to wave formation as shown in Fig. 5-15.[18] This is the well-known Rayleigh-Taylor type of instability.[19] The wavy disturbances are amplified until jet-like spikes are formed, which subsequently disintegrate. This is part of what happens during sonication, which may also introduce cavitation.

Other methods are based on shearing action. A common type of liquid colloid mill, or homogenizer, is shown in Fig. 5-11. When a fluid interface is sheared in this manner, it is also potentially unstable (termed Kelvin-Helmholtz instability[20]) to the growth of waves that eventually crest over and break up into drops as shown in Fig. 5-16. These can be further torn apart when caught in an intense shear field. This is how milk is homogenized, for example.

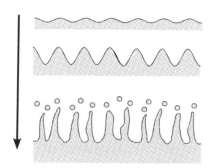

Fig. 5-15: Comminution by normal boundary motion, "Rayleigh-Taylor instability." System becomes unstable with acceleration toward the more dense phase.

Another important class of top-down synthesis procedures starts with the formation of a thin film of material on a substrate by spin-coating,

[16] Taylor, G. I., *Proc. Roy. Soc.*, **280A**, 383 (1964);
   Taylor, G. I., *Proc. Roy. Soc.*, **313A**, 453 (1969).
[17] Watanabe, A., Higashitsuji, K., and Nishizawa, K., *J. Colloid Interface Sci.*, **64**, 278 (1978).
[18] Rajagopal, E. S., in **Rheology of Emulsions**, P. Sherman (Ed.), p. 22, Pergamon, New York, 1963.
[19] Chandrasekhar, S., **Hydrodynamic and Hydromagnetic Stability**, pp. 428ff, Oxford, Clarendon Press, 1961.
[20] Chandrasekhar, S., **Hydrodynamic and Hydromagnetic Stability**, pp. 481ff, Oxford, Clarendon Press, 1961.

electroplating, chemical vapor deposition (CVD)[21] or physical vapor deposition (PVD),[22] including sputter coating and pulsed laser deposition (PLD). In the latter processes, the desired material is vaporized (often under vacuum) using high temperature and/or plasma arcing. The film, often

Fig. 5-16: The mechanism of "Kelvin-Helmholtz instability."

referred to as a "resist," is then etched into patterns or templates using one or another of various types of *lithography*. This may be done using a light beam (photolithography), an electron beam (E-beam lithography), X-rays (X-ray lithography), a focused ion beam, *etc*. Either the radiation covers a macroscopic area, and a mask is used to produce the desired pattern or a focused beam is used to etch the pattern directly. The patterned films may be broken up into particulates in the etching process, or they may be used as templates in bottom-up procedures, in which case the templates are filled electrolytically, or by subsequent CVD, PVD or other processes.

3. *Bottom-up strategies*

A wide and growing variety of ways exist for the production of colloidal dispersions from solutions of so-called precursor molecules by condensation methods, some of which are listed in Table 5-4. These may be

---

Table 5-4: Bottom-up methods for the preparation of colloids.

- Controlled precipitation (nucleation and growth from supersaturated solutions)
- Reduction of metal complexes in dilute solution (metallic particles)
- Sol-gel methods: hydrolysis/condensation of metal-organic precursors (metal oxide particles)
- Pyrolysis of metal-organic precursors in anhydrous solvents (semiconductor particles)
- Emulsion, dispersion, suspension polymerization (polymer colloids)
- Amphiphile self-assembly in solution (micelles, vesicles, *etc*.)
- Synthesis and growth within confined templates
- Vapor phase redox and other reactions
- Physical or chemical vapor deposition (PVD or CVD)
- Electro-deposition in templates

---

[21] Seshan, K. (Ed.), **Handbook of Thin-Film Deposition Processes and Techniques – Principles, Methods, Equipment and Applications**, 2nd Ed., Noyes, New York, 2002.

[22] Mahan, J. E., **Physical Vapor Deposition of Thin Films,** Vol. 1, Wiley, New York, 2000.

simply controlled precipitation from a super-saturated solution in which the process is stopped before the particles grow to supra-colloidal sizes,[23] or the inducement of a chemical reaction that produces the desired dispersoid as a product. Precipitation can be induced by bringing solutions together containing ions of an insoluble salt, or by cooling a saturated solution. One generally seeks to have a short burst of nucleation, followed by controlled growth of the nuclei. These are then allowed to grow under conditions that are below critical super-saturation where no new nuclei form. The rate of nucleation depends very strongly on the degree of supersaturation, as discussed in Chap. 2, while the rate of particle growth is controlled by the solution concentration and other factors. Often a very small amount of one of the reagents is used and is consumed before much growth can occur. Many sols of sparingly soluble salts, such as the silver halides, have been prepared this way. For example, silver iodide may be formed according to:

$$AgNO_3 + KI \rightarrow KNO_3 + AgI\downarrow,$$

and sulfur colloids may be produced by acidification of solutions of sodium thiosulfate[24]:

$$Na_2S_2O_3 + 2HCl \rightarrow 2NaCl + H_2O + SO_2 + S\downarrow$$

Colloids formed in this way in aqueous media are generally electrostatically stabilized *in situ* against aggregation. This sometimes requires the presence of an additional component to adsorb to the surfaces of the growing nuclei to confer additional charge to them and act as a stabilizer against aggregation. Typical stabilizers might be ionic surfactants or polyvalent anions, such as pyrophosphate ions. A detailed review on the preparation and properties of uniform-size mineral colloids by precipitation has been given by Matijevic.[25]

Sols of metallic particles are most often prepared by the chemical reduction of metal ion-containing precursors. Gold nanoparticles, as a specific case, are now commonly produced by the Turkevich recipe,[26] in which chlorauric acid in dilute ($\approx 2.5 \times 10^{-4}M$) boiling aqueous solution is reduced by citric acid (furnished by the addition of a 0.5% sodium citrate solution in the ratio of 1/20 to the acid solution). The reaction may be expressed as

$$HAuCl_4 + 3\text{"H"} \rightarrow 4HCl + Au\downarrow$$

where "H" represents the reducing agent. A wide variety of other metal sols are commonly produced using other precursors and reducing agents. Platinum colloids are produced from chloroplatinic acid (H_2PtCl_6), palladium from palladium chloride ($PdCl_2$), silver from silver perchlorate

[23] Heicklen, J., **Colloid Formation and Growth: A Chemical Kinetics Approach**, Academic Press, Plenum, New York, 1976.

[24] La Mer, V. K., and Dinegar, R. H., *J. Amer. Chem. Soc.*, **72**, 4847 (1950).

[25] Matijevic, E., *Chem. Mater.*, **5**, 412 (1993).

[26] Turkevich, J., Hillier, J., and Stevenson, P. C., *Disc. Faraday Soc.*, **11**, 55 (1951).

(AgClO$_4$), *etc.*, and reducing agents such as hydrogen, carbon monoxide, formaldehyde, various alcohols, ammonium ions, *etc.*, as summarized by Cao.[27] Again, a third component is often also present in these syntheses. It is generally a soluble polymer or a surfactant that adsorbs to the surfaces of the particles as they are formed and generally plays two roles. First, its presence on the particle surfaces provides steric stabilization against aggregation, as described in Chap. 7, and secondly, it provides a barrier to diffusion of dissolved material to the surface of the particles and thereby slows their growth. These materials are referred to as both stabilizers and "capping agents." For a given metal precursor, variations in the amounts and type of reducing agent(s) and stabilizer(s), as well as reaction conditions, dictate the properties of the colloid obtained.

An interesting special example of the production of a metal colloid is the basis for the fast-disappearing wet photographic process. A silver bromide colloid is dispersed uniformly in a gel medium that forms the "emulsion" coating of photographic film. When light or radiation of appropriate wavelength strikes one of the silver halide crystals, a series of reactions begins that produces a small amount of free silver in the grain. Initially, a free bromine atom is produced when the bromide ion absorbs a photon of light, in accord with:

$$AgBr + h\nu \rightarrow Ag^+ + Br + e\text{-}$$

The silver ion then combines with the electron to produce a silver atom. The free silver produced in the exposed silver halide grains constitutes what is referred to as the "latent image," which is later rendered visible by the development process.

An important variant of the nucleation-growth-protection scheme for producing metal and metal alloy colloids is the polyol process.[28] An inorganic salt, *e.g.*, AgNO$_3$ for the production of silver colloids, serves as the precursor that is reduced by a polyhydric alcohol, often ethylene glycol. Poly(vinyl pyrrolidone) (PVP) is a commonly-used stabilizer or capping agent, and the process is carried out at elevated temperature (160°C) due to the increased reducing power of the alcohol under these conditions. Judicious choice of the PVP/AgNO$_3$ ratio and the use of other capping agents has permitted control of the *shape* of the resulting particles, leading to nanospheres, nanocubes, nanowires and other possibilities.[29]

Colloids of metal oxides are produced by another important condensation method: the "sol-gel process."[30,31,32] The technique is

---

[27] Cao, G., **Nanostructures & Nanomaterials**, pp. 63ff, Imperial College Press, London, 2004.

[28] Fiévet, F., Lagier, J. P., and Figlarz, M., *Mater. Res. Soc. Bull.*, **14**, 29 (1989).

[29] Sun, Y., and Xia, Y., *Science*, **298**, 2176 (2002);
Wiley, B., Sun, Y., Mayers, B., and Xia, Y., *Chem. Eur. J.*, **11**, 454 (2005).

[30] Brinker, C. J., and Scherer, G. W., **Sol-Gel Science: The Physics and Chemistry of Sol-Gel Processing**, Academic Press, Boston, 1990.

exemplified by the Stöber process for producing monodisperse silica.[33] It starts with alkoxide precursors, $Si(OR)_4$, usually tetramethoxysilane (TMOS) or tetraethoxysilane (TEOS). These are hydrolyzed and subsequently condensed to yield the oxide, *e.g.*, SiO_2, ideally (using TEOS) in accord with:

$$Si(OC_2H_5)_4 + 4H_2O \rightarrow Si(OH)_4 + 4\,C_2H_5OH$$

followed by

$$Si(OH)_4 \rightarrow 2\,H_2O + SiO_2\downarrow$$

The reactions are carried out in water-alcohol mixtures, with ammonia as a catalyst. Depending on the precursor and alcohol used, the water/alcohol ratio and the amount of ammonia, spherical silica particles ranging in size from 50 nm to 2 μm are formed. The resulting colloid is generally stabilized electrostatically against aggregation, as described in Chap. 6, by adjusting the system *pH* to an appropriate level. When the particulate gels are dehydrated rapidly, one obtains open sponge-like monolithic structures termed "aerogels." These light-weight materials, when dried (usually by first immersing them in supercritical CO_2 that is subsequently vaporized) contain large amounts of internal trapped air and thus serve as effective thermal insulators. Slower dehydration leads to consolidated particulate structures termed "xerogels."

More generally, sol-gel synthesis may be used to produce a variety of metal oxides, and, depending on conditions, neither the hydrolysis nor the condensation reactions are complete. Thus more generally the hydrolysis and condensation reactions are:

$$M(OR)_4 + x\,H_2O \rightarrow M(OR)_{4-x}(OH)_x + x\,ROH$$

followed by

$$2\,M(OR)_{4-x}(OH)_x \rightarrow (OR)_{4-x}(OH)_{x-1}MOM(OR)_{4-x}(OH)_{x-1} + H_2O$$

etc., occurring in sequence and in parallel. Halogenated compounds, such as chlorosilanes, may also be used as precursors, and organic-inorganic colloids or monoliths may be formed by starting with precursors such as the organo-functional silanes of the type discussed in Chap. 4 and used for surface chemical modification of mineral oxides to improve adhesion.

Another important sol-gel route is through the use of the many different self-assembled surfactant structures formed in solution (vesicles, normal hexagonal tube structures, *etc.*, described in Chap. 3) as templates for mineralization. The inorganic precursors invade the aqueous portions of

[31] Pierre, A. C., **Introduction to Sol-Gel Processing**, Kluwer, Boston, 1998.

[32] Wright, J. D., and Sommerdijk, N. A. J. M., **Sol-Gel Materials: Chemistry and Applications**, Gordon and Breach, Amsterdam, 2001.

[33] Stöber, W., Fink, A., and Bohn, E., *J. Colloid Interface Sci.*, **26**, 62 (1968).

these structures, and when they are dried out, the surfactants may be removed by calcination, leaving behind a micro- or nano-structured monolith.

Semiconductor nanoparticles, *i.e.*, quantum dots, are gaining increasing importance in a number of applications, such as components in transistors, solar cells, light-emitting diodes (LEDs), and diode lasers as well as agents for medical imaging. They often (but not exclusively) consist of sulfides, selenides, tellurides, phosphides or arsenides of cadmium, zinc or indium, or combinations of these compounds. They are commonly produced from metal-organic precursors in anhydrous solvents at elevated temperatures.[34,35] As an example,[36] cadmium selenide may be produced by reacting dimethyl cadmium with tri-*n*-octylphosphine selenide at $\approx$ 300°C in an anhydrous solution of tri-*n*-octylphosphine (TOP) and tri-*n*-octylphosphine oxide (TOPO) which act not only as solvents, but also as capping agents. Schematically, the reaction is:

$$Cd(CH_3)_2 + (C_8H_{17})_3PSe \xrightarrow{\text{300°C, TOP, TOPO}} CdSe \text{ nanocrystals}$$

Dispersions of colloid-size polymer particles may be prepared by *emulsion polymerization*, as shown in Fig. 5-17, and the resulting system is termed a *polymer colloid* or a *latex*.[37] This is an extremely important and

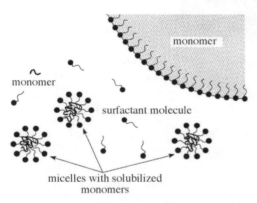

monomer

monomer

surfactant molecule

Fig. 5-17: Emulsion polymerization process.

micelles with solubilized monomers

highly varied process. It is often made to occur in the following way. Sparingly soluble monomer molecules are dispersed in water by means of an emulsifier-stabilized emulsion. The emulsifier is present at sufficient concentration to form micelles, and its structure is chosen such that the

[34] Cao, G., **Nanostructures and Nanomaterials**, pp. 74-82, Imperial College Press, London, 2004.

[35] Rogach, A. L., Talapin, D. V., and Weller, H., in **Colloids and Colloid Assemblies**, F. Caruso (Ed.), pp. 52-95, Wiley-VCH, Weinheim, Germany, 2004.

[36] Murray, C. B., Norris, D. J., and Bawendi, M. G., *J. Amer. Chem. Soc.*, **115**, 8706 (1993).

[37] Fitch, R. M. (Ed.), **Polymer Colloids**, Plenum, New York, 1970.
Fitch, R. M. (Ed.), **Polymer Colloids II**, Plenum, New York, 1980.
Buscall, R., Corner, T., and Stageman, J. F. (Eds.), **Polymer Colloids**, Elsevier, London, 1985.

micelles are spherical, *i.e.*, the critical packing parameter (*CPP*) is approximately 1/3, as discussed in Chap. 3. Monomer then finds itself in three locations in the system: a very small amount is dissolved directly in the water; a more significant amount is solubilized in the micelles, and most is in the emulsion droplets. To begin the process, a small amount of water-soluble initiator (such as a persulfate) is added to the system. Polymerization then begins primarily in the micelles. The bulk concentration of monomer remains constant as the latter are supplied by the monomer emulsion droplets. The process continues until monomer, initiator or emulsifier is exhausted, and may lead to a dispersion of latex particles of highly uniform size, an example of which is shown in Fig. 5-18. Their near-perfect sphericity and monodispersity leads to their use as calibration standards for electron microscopy. The polymerization process produces polymers and oligomers terminated with sulfate groups that end up on the surface of the

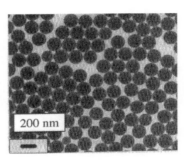

Fig. 5-18: Polystyrene latex prepared by emulsion polymerization.

200 nm

particles. These ionize to produce a negative charge over most of the practical *pH* range and are generally present at sufficient density to electrostatically stabilize the latex against aggregation, although the original emulsifier and additional ionic surfactant may also play a role.

Closely related processes for producing polymer colloids are *dispersion polymerization* and *suspension polymerization*. In dispersion polymerization, the medium is chosen to be a good solvent for the monomer. When an appropriate initiator is added, chain growth of the dissolved monomers proceeds until the polymers grow to a size that is no longer soluble, and they precipitate. A stabilizer is also present initially to adsorb to and protect the particles against aggregation. Colloids formed in aqueous media, *i.e.*, *hydrosols*, are often stabilized electrostatically, while those formed in non-aqueous media, *i.e.*, *organosols*, are stabilized sterically by the adsorption of soluble polymers. Both stabilizing mechanisms are described in Chap. 7. Suspension polymerization (also called *bead polymerization*) refers to the polymerization of monomer dispersed in an aqueous medium as an emulsion. The initial condition is similar to that for emulsion polymerization, but micelles are not present. The reaction in this case is effected using an *oil*-soluble initiator.

Vapor condensation or aerosol synthesis processes produce colloid-sized particles by chemical reactions in or precipitation from the vapor phase, and are thus closely related to the CVD (chemical vapor deposition) or PVD (physical vapor deposition) processes, which are considered to be top-down. The particulates formed in the gas phase are then collected on some convenient surface.

Template processes for producing colloid particles use patterns produced by the various types of lithography. These include, for example, electron-beam lithography, chemical etching, soft lithography, ink-jet printing or dip-pen nanolithography. Templates may also be produced by bombarding a polymer film with alpha particles (which punch holes in the film), possibly followed by chemical etching to produce holes of desired size ("track-etch" membranes). Etched holes may be filled by vapor deposition, or if the bottoms of the holes are electrically conductive, may be filled electrolytically.[38] Another example of the use of templates is the seeded growth of particles, particle chains, pillars or filaments, at a pattern of locations on a desired surface through the use of the patterned deposition of catalyst particles. For example, a "forest" of carbon nanotubes may be grown onto a smooth silicon surface by the selective deposition of carbon atoms onto sites that favor their growth along an axis normal to the surface. As another example, silicon oxide nano-wires may be grown on a silica substrate on which a pattern of catalytic gold nano-dots has been deposited. CVD is an especially powerful technique for producing desired micro- or nano-particles by deposition onto pre-patterned surfaces or templates.

D. Morphology of colloids: particle size, size distribution, and particle shape

1. Description of particle size distributions

In idealized colloids, the particles are spherical and all of the same size, *i.e.*, *monodisperse*. Unfortunately this happenstance is rare. Most colloids are not of spherical particles and have a range of particle sizes, *i.e.*, are *polydisperse*. Monodispersed, spherical colloids (such as the polymer latices described earlier) *can* be prepared, but in general, very special recipes and procedures must be used for doing so. Given that most colloids are polydisperse, it is important to see how to describe this distribution. An example of a polydisperse colloid is pictured in Fig. 5-19, which shows an optical micrograph of a sample of a Min-U-Sil® 15[39] dispersion. It is also to be noted that the particles are not spherical. Given some specific data on a system, as those from a micrograph, to describe the distribution of particle sizes, one must first decide on some measure of particle size that is accessible to the type of data available. Here what may be used is some effective particle diameter. This is unambiguous in general only when the

---

[38] Lai, M., and Riley, D. J., *J. Colloid Interface Sci.*, **323**, 203 (2008).

[39] Min-U-Sil® is a high purity natural silica produced by U. S. Silica Co., Berkeley Springs, WV, from ground silica sand.

projections of the particles are circular (corresponding, to spheres or circular disks). In the usual case, the shape is irregular. A common way to quantify the "effective" diameter is to choose some axis onto which the particles widths are projected. These widths are called "Feret diameters," and if the shapes and orientations are random, provide a good linear measure of size.[40] This "diameter" then becomes, in the language of statistics, the "random variable" whose distribution is sought. Alternatively, image analysis may be used to determine the projected area of each particle and compute from it the effective circular diameter. The pool of data is the "sample," and one hopes it to be representative of the total population from which it was withdrawn. One may capture and process the data contained in the image using image analysis software, such as *Image J*, a public domain Java image-processing program available from NIH.[41] Such software reveals that the sample in Fig. 5-19 contains 580 particles ranging in size (in terms of their Feret diameters)

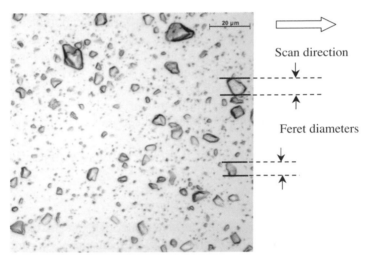

Fig. 5-19: Optical micrograph of Min-U-Sil® 15 dispersion, showing Feret diameters.

between 0.5 and 7.0 μm, in addition to smaller particles and a small number of larger ones. One way of representing or describing the distribution would be to list, in order of size, the measured diameter of every single particle in the sample. But most statistically meaningful samples contain many hundreds (perhaps hundreds of thousands) of entries, and since the random variable is continuous (as opposed to discrete), presumably no two particles have *exactly* the same size. Such a list would thus not be very efficient in telling at a glance the character of the polydispersity. It is convenient with a large sample (and particularly when the random variable is continuous) to

---

[40] "Martin's diameter," defined as the line parallel to the direction of scan that divides the particle profile into two equal areas, is also commonly used.

[41] http://rsb.info.nih.gov/ij/docs/

treat the data by discretizing or classifying them into equal-sized ranges. For very wide distributions, it is sometimes convenient to discretize the distribution *geometrically* rather than arithmetically, with each class twice, or ten times, *etc.*, the size of the preceding one. Table 5-5 shows data obtained from Fig. 5-19 for particle sizes (Feret diameters) in the range

Table 5-5: Size distribution of Min-U-Sil® 15 sample.

Class boundaries (μm)	Class mark, d_i (μm)	#Particles in class	Fraction of particles in class $= f_i$	#Particles of $d \leq d_i$	Fraction of particles of $d \leq d_i = q_i$
0.5 – 1.0	0.75	91	0.157	91	0.157
1.0 – 1.5	1.25	148	0.255	239	0.412
1.5 – 2.0	1.75	115	0.198	354	0.610
2.0 – 2.5	2.25	73	0.126	427	0.736
2.5 – 3.0	2.75	46	0.079	473	0.816
3.0 – 3.5	3.25	20	0.034	493	0.850
3.5 – 4.0	3.75	26	0.045	519	0.895
4.0 – 4.5	4.25	18	0.031	537	0.926
4.5 – 5.0	4.75	8	0.014	545	0.940
5.0 – 5.5	5.25	10	0.017	555	0.957
5.5 – 6.0	5.75	13	0.022	568	0.979
6.0 – 6.5	6.25	6	0.010	574	0.990
6.5 – 7.0	6.75	6	0.010	580	1.000

between 0.5 and 7.0 μm, split into 14 equal-sized classes. The usual practice is to have 10 – 20 classes. The "class-mark" is the arithmetic average of the random variable in the class. The first and second columns in Table 5.5 show the class boundaries and the class mark for each class. The third column counts the number of particles found in each class, and the fourth column gives the fraction f_i of the particles in each class i. The fifth column gives the cumulative number of particles in the given class i plus all smaller classes, and the final column gives the cumulative fraction, q_i, of particles in class i plus all smaller classes. One may prepare histograms from Column 4 and Column 6, showing the "frequency distribution," and the "cumulative distribution," respectively, as in Figs. 5-20 and 5-21. The frequency distribution is typically bell-shaped, while the cumulative distribution function is sigmoidal in shape.

While histograms of the type of Fig. 5-20 are very useful for expressing the particle size distribution, it is very useful to define various numerical descriptors for this purpose. The first of these is the *mean value* of the variable, in this case the mean (Feret) diameter $\bar{d}$, defined as

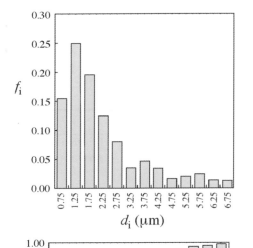

Fig. 5-20: Histogram of the frequency distribution of Feret particle diameters in the Min-U-Sil sample of Table 5-5.

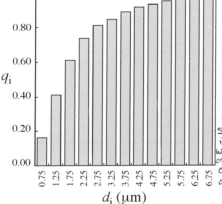

Fig. 5-21: Histogram of the cumulative distribution of Feret particle diameters in the Min-U-Sil sample of Table 5-5.

[handwritten annotation: Equation 5.17 assumes that the sample size n is sufficiently large that it represents the whole population from which the sample is withdrawn. Otherwise one needs the sample variance, which is obtained by multiplying the right hand side of Eq. (5.17) by the factor n/(n-1)]

$$\overline{d} = \frac{\Sigma n_i d_i}{\Sigma n_i} \equiv \sum \left(\frac{n_i}{\Sigma n_i} \right) d_i = \Sigma f_i d_i. \tag{5.16}$$

$\overline{d}$ is thus the arithmetic mean diameter, the most frequently used average to describe a distribution. For the distribution of Table 5-5, $\overline{d} = 2.08$ μm. It alone does not reveal anything about the "spread" in the distribution of diameters. This may be given by the *variance*:

$$\sigma^2 = \Sigma f_i \left(d_i - \overline{d} \right)^2, \tag{5.17}$$

or its positive square root, the *standard deviation*, σ. The larger the standard deviation, the wider the spread. For the sample distribution, $\sigma = 2.45$ μm.

The variance, σ^2, is an example of a "moment" of the distribution. In general, the kth moment of the distribution about a value d_0, is defined as:

$$m_k = \sum f_i (d_i - d_0)^k. \tag{5.18}$$

It is seen that the variance is the second moment of the distribution about the mean, [m_2] while the mean itself is the *first* moment about the *origin*:

$$\bar{d} = \Sigma f_i(d_i - 0)^1 = \Sigma f_i d_i. \tag{5.19}$$

Two higher moments are often used to further characterize distributions. The third moment about the mean, m_3, is ~~the skewness of the distribution~~:

a measure of the asymmetry of the distribution, and from it may be compared a dimensionless descriptor termed the skewness, sk,

$$m_3 = \Sigma f_i(d_i - \bar{d})^3. \tag{5.20}$$

$$sk = \frac{m_3}{m_2^{3/2}} = \frac{\Sigma f_i(d_i - \bar{d})^3}{(\sigma^2)^{3/2}}$$

It shows the asymmetry of the distribution shape. The third moment of the example distribution, normalized by $\bar{d}^3$, is $m_3/\bar{d}^3 = 2.39$. The rather large positive value of the skewness is indicative of a distribution tailing off toward the larger sizes. A fourth moment about the mean, m_4, is called the *kurtosis* ("peakedness"). For the example distribution, the normalized value is: $m_4/\bar{d}^4 = 5.32$. These four moments permit a quite complete reconstruction of the distribution function for particle diameters.

✳ see paper in front of book

2. Distributions based on different size variables and weighting factors

Information of the type shown in the micrograph of Fig. 5-19 provides, in principle, the size and shape of every particle in a sample of the colloid. Most instrumentation used to characterize colloidal media does not give such detail. Some techniques, such as the electro-zone (or Coulter) method and photo-zone methods, described later in this chapter, measure and record the volumes of individual particles as they are made to flow single-file though an aperture. The data obtained with these instruments are generally classified and presented to the user as discrete histograms or continuous distribution curves. Similar outputs are obtained for size distributions from sieving, differential sedimentation or centrifugation techniques, as well as the various chromatographic methods. Other instrumental techniques, such as those based on light scattering, surface area measurements (by adsorption), or viscosity yield only certain types of averages and sometimes information equivalent to one or two additional moments of the size distribution. It is thus important to be able to interpret and interconvert between distributions and averages based on different variables and different weighting factors.

One may be interested in the mean of some property other than the particle diameter, such as the particle surface area or volume (or mass). The number average particle surface area is given by

$$\bar{A} = \frac{\Sigma n_i A_i}{\Sigma n_i} \equiv \Sigma \left(\frac{n_i}{\Sigma n_i}\right) A_i = \Sigma f_i A_i, \tag{5.21}$$

and the number average particle volume by

$$\overline{V} = \frac{\sum n_i V_i}{\sum n_i} \equiv \sum \left(\frac{n_i}{\sum n_i} \right) V_i = \sum f_i V_i, \tag{5.22}$$

where A_i and V_i refer to the average particle area and volume, respectively, of the particles in class i. If the particles can be regarded as spheres, we may write

$$A_i = \pi d_i^2 \quad \text{and} \quad V_i = \frac{1}{6} \pi d_i^3, \tag{5.23}$$

leading to

$$\overline{A} = \pi \Sigma f_i d_i^2 = \pi \overline{d^2} \quad \text{and} \quad \overline{V} = \frac{1}{6} \pi \Sigma f_i d_i^3 = \frac{1}{6} \pi \overline{d^3}. \tag{5.24}$$

Thus one may compute the diameter of the particle of mean surface area, $\overline{d_s}$, as

$$\overline{d_s} = \left[\overline{d^2} \right]^{1/2} = \left[\overline{A} / \pi \right]^{1/2}, \tag{5.25}$$

and the diameter of the particle of mean volume, $\overline{d_v}$, as

$$\overline{d_v} = \left[\overline{d^3} \right]^{1/3} = \left[6\overline{V} / \pi \right]^{1/3} \tag{5.26}$$

from the original diameter distribution, as long as the particles are spherical. On the other hand, if the total area of a sample of a known total number of spherical particles is measured (as by an adsorption method), $\overline{d_s}$ may be determined directly. If the total mass (hence volume, if density is known) of a sample of a known total number of spherical particles is measured, $\overline{d_v}$ may be determined directly. It is evident that for any polydisperse sample, these diameters are different, and

$$\overline{d_n} < \overline{d_s} < \overline{d_v}, \tag{5.27}$$

where $\overline{d_n}$ is the original number average particle diameter. For the sample distribution of Table 5-5, these values are: 2.08; 2.48; and 2.89 μm, respectively.

Imagine next that one has experimental access not to particle diameters but instead to particle areas, which may then be discretized into equal-sized *area* classes, with class marks A_i. The distribution function would be given by values of f_i^s, *i.e.*,

$$f_i^s = \frac{n_i^s A_i}{\sum n_i^s A_i} = \frac{A_{it}}{A}, \tag{5.28}$$

where the superscript "s" is used as a reminder that the numbers are different than those corresponding to the diameter distribution. A_{it} is the total particle surface area in class i, and A is the grand total surface area in the sample (as before). One may convert at least approximately between a particle surface

*in Eqns. 5.29 – 5.34, f_i in the last step of these equations is not the same as that designed in 5.16. Instead of the number fraction in class, it is the "diameter fraction in class" and should be designated f_i^d

area distribution and a particle diameter distribution if the particles are spheres. Starting with the original diameter-delineated classes,

$$f_i^s = \frac{n_i^s A_i}{\sum n_i^s A_i} \approx \frac{n_i A_i}{\sum n_i A_i} = \frac{n_i \pi d_i^2}{\sum n_i \pi d_i^2} = \frac{n_i d_i}{\sum n_i d_i} \cdot d_i \cdot \frac{\sum n_i d_i}{\sum n_i d_i^2} = f_i d_i \frac{\sum n_i d_i}{\sum n_i d_i^2}. \quad (5.29)$$

The approximation is that the transformed classes, defined in terms of the new variable, are no longer equally spaced. If possible one should return to the original individual particle data, evaluate d_i^2 or d_i^3 for each datum, and then reclassify them.

The surface area distribution function may be used to obtain *area-averaged* properties. For example, the surface area-average particle diameter of the sample is

$$\overline{d}_{sv} = \sum f_i^s d_i. \quad (5.30)$$

This is clearly a different average diameter than $\overline{d}_s$ of Eq. (5.25), hence the different notation. What is different is the *weighting factor* used in the computation of the average. In general, to compute any average for a polydisperse system one must specify the variable to be averaged (particle diameter in the above case) and the weighting factor (particle surface area in the above case). Thus one obtains (for the above case) the *surface area average particle diameter*. This contrasts with the *number average particle diameter*, $\overline{d}_n$. The diameter averages we have defined as $\overline{d}_s$ and $\overline{d}_v$ do not fit this classification and are a source of some confusion. The latter are computed from the number average particle surface area and volume, respectively, assuming spherical particles.

The surface area-average particle diameter of Eq. (5.30), $\overline{d}_{sv}$, is also called the Sauter mean diameter, SMD, (after the German scientist J. Sauter). If the particles are spheres, it may be computed as

$$\overline{d}_{sv} = \sum f_i^s d_i = \sum \frac{n_i^s A_i}{\sum n_i^s A_i} d_i \approx \sum \frac{n_i \pi d_i^2}{\sum n_i \pi d_i^2} d_i = \frac{\sum n_i d_i^3}{\sum n_i d_i^2}. \quad (5.31)$$

The SMD has the physical significance of being the diameter of a particle having the same volume-to-surface-area ratio as the entire sample. This is the average diameter accessible (for smooth surface particles) from BET or other adsorption measurements, which yield the *specific area, i.e.*, the area per unit mass of the sample, Σ. The Sauter mean diameter is then given by

$$\overline{d}_{sv} = \left(\frac{\text{volume}}{\text{mass}}\right)\left(\frac{\text{mass}}{\text{area}}\right) = \frac{1}{\rho \Sigma}. \quad (5.32)$$

There is also instrumentation that provides measurement of the individual particle volumes in a sample. An example is the electro-zone method or Coulter counter described at the end of this chapter. Such data can be discretized to produce a set of classes equally spaced in terms of volume and characterized by class marks V_i. The distribution function would be then given by values of f_i^v:

$$f_i^v = \frac{n_i^v V_i}{\sum_i n_i^v V_i} = \frac{V_{it}}{V},$$
(5.33)

where V_{it} is the total particle volume in class i, and V is the grand total particle volume in the sample. Here again it is possible to interconvert between, for example, the volume distribution and the number distribution of particle diameters and to compute a particle *volume* average particle diameter, $\overline{d}_{vv}$. Using an analysis the same as that used in Eq. (5.29) one obtains:

$$f_i^v = \frac{n_i^v V_i}{\sum_i n_i^v V_i} \approx \frac{n_i V_i}{\sum_i n_i V_i} = \frac{n_i \frac{1}{6}\pi d_i^3}{\sum_i n_i \frac{1}{6}\pi d_i^3} = \frac{n_i d_i}{\sum_i n_i d_i} \cdot d_i^2 \cdot \frac{\sum_i n_i d_i}{\sum_i n_i d_i^3} = f_i d_i^2 \frac{\sum_i n_i d_i}{\sum_i n_i d_i^3}.$$
(5.34)

and the volume-average particle diameter in the sample is given by an analysis the same as that leading to Eq. (5.31), *viz.*

$$\overline{d}_{vv} = \sum_i f_i^v d_i = \frac{\sum_i n_i d_i^4}{\sum_i n_i d_i^3}.$$
(5.35)

This is sometimes referred to as the De Broucker mean diameter. It is evident that the volume average is the same as the mass (or weight) average, since the mass distribution is given by simply multiplying both the numerator and denominator in Eq. (5.35) by the density ρ (or ρg).

The means defined in Eqs. (5.31) and (5.35) are examples of what are called "moment means," in that they are ratios of different moments of the distribution about the origin. In shorthand notation, the Sauter mean diameter of Eq. (5.31) is D[3,2], the ratio of the third to the second moments of the diameter distribution, and the De Broucker mean diameter of Eq. (5.35) is D[4,3], the ratio of the fourth to the third moments. In the same way, the number mean diameter of Eq. (5.16) is D[1,0].

Rayleigh light scattering or turbidity measurements of a polydisperse colloid, as described later in this chapter, produce *volume average particle volumes*, i.e.

$$\overline{V}_v = \frac{\sum_i n_i V_i^2}{\sum_i n_i V_i}.$$
(5.36)

This is equivalent to the moment mean V[2,1]. The indicated particle diameter, let's call it $\overline{d}_{ls}$, would then be (for spherical particles):

$$\overline{d}_{ls} = \left[6\overline{V}_v / \pi\right]^{1/3},$$
(5.37)

another example of an equivalent mean-sphere diameter, such as $\overline{d}_s$ and $\overline{d}_v$.

Another type of average, the "Z-average," is one encountered in some light scattering and centrifugation measurements. The "Z" stands for the German *Zentrifuge*. It weights the averaged quantity, for example the particle diameter, with the *square* of the particle mass (or volume), *viz.*

$$\text{Z-average particle diameter} = \frac{\sum n_i m_i^2 d_i}{\sum n_i m_i^2}, \tag{5.38}$$

where m_i is the average particle mass in class i.

It is evident that when some average particle property in a colloid sample is given, both the particular property *and* the weighting factor must be clearly understood. The distinction between the weighted average properties and the equivalent mean-sphere properties that can be derived from the averages must also be understood. It is clear finally, that the various "average" particle diameters (for example) will differ from one another for all but monodisperse distributions, and that the extent of the difference between the various averages will depend on the actual particle distribution in any case. Since different experimental methods lead to different averages, a comparison of results from these different techniques will provide insight into the nature of the distribution. For example, the ratio of a mass average property to the number average of the same property is often used as a measure of the polydispersity in the sample, *i.e.*, the greater such a ratio differs from unity, the greater the "polydispersity." Some commonly encountered "averages" are summarized in Table 5-6.

3. Normal (Gaussian) and log-normal distributions

Two theoretical distribution functions are often useful in describing particle size distributions. The first is the normal (Gaussian) distribution:

$$f(x) = \frac{1}{\sigma\sqrt{2\pi}} \exp\left[-\frac{1}{2}\left(\frac{x - \bar{x}}{\sigma}\right)^2\right], \tag{5.39}$$

Table 5-6: Various averages for polydisperse systems.

Average	Averaged variable	Weighting factor
Number average diameter $= \bar{d}_n = \dfrac{\sum n_i d_i}{\sum n_i}$	Particle diameter, d_i	Number of particles in class, n_i
Area average diameter $= \bar{d}_{sv} = \dfrac{\sum n_i A_i d_i}{\sum n_i A_i}$	Particle diameter, d_i	Particle surface area in class, $n_i A$
Volume average volume $= \bar{V}_v = \dfrac{\sum n_i V_i^2}{\sum n_i V_i}$	Particle volume, V_i	Particle volume in class, $n_i V_i$
Z-average diameter $= \bar{d}_z = \dfrac{\sum n_i m_i^2 d_i}{\sum n_i m_i^2}$	Particle diameter, d_i	Particle mass squared in class, $n_i m_i^2$

where x is the random variable, and σ is the standard deviation. This distribution is shown in Fig. 5-22. The fraction of the particles whose size lies between x and $x+dx$ is $f(x)dx$, or alternatively, $f(x)$ is the probability that any particle withdrawn from the population will have a size in this range. When x represents any event whose outcome depends solely on chance, any sufficiently large sample of such outcomes will have the above distribution

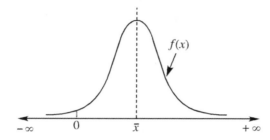

Fig. 5-22: Normal (Gaussian) distribution. In use for particle size distributions, it must be constrained to non-negative values.

(the "central limit theorem"). The curve is symmetrical about $\overline{x}$, which is the mean, and σ is the standard deviation in the same sense as defined earlier, with the summation replaced by an integral:

$$\overline{x} = \frac{\int_{-\infty}^{\infty} x\, f(x)dx}{\int_{-\infty}^{\infty} f(x)dx}, \quad \text{and} \quad \sigma^2 = \frac{\int_{-\infty}^{\infty} (x - \overline{x})^2 f(x)dx}{\int_{-\infty}^{\infty} f(x)dx} \qquad (5.40)$$

Guassian distributions are symmetrical, *i.e.*, all odd moments of the distribution about the mean are zero. Real distributions of particle size cannot, of course, conform exactly to the Gaussian distribution because the random variable must extend to negative numbers. Many actual cases, however, do have the form of a truncated Gaussian distribution. It sometimes approximates quite closely the distribution of particle sizes produced by condensation methods.

The cumulative Gaussian distribution function, *i.e.*, the fraction of the particles greater than x_i in size, is equal to

$$q_i = \int_{x_i}^{\infty} f(x)dx. \qquad (5.41)$$

It is of interest to note that if one introduces the auxiliary variable, $t = \dfrac{(x - \overline{x})}{\sigma}$, the cumulative distribution function corresponding to the Gaussian frequency distribution takes the form:

$$q(t_i) = \frac{1}{\sqrt{2\pi}} \int_{t_i}^{\infty} \exp\left(-\tfrac{1}{2}t^2\right) dt, \qquad (5.42)$$

which is closely related to the *error function*, erf(t_i):

$$\mathrm{erf}(t_i) = \frac{1}{\sqrt{2\pi}} \int_0^{t_i} \exp\left(-\tfrac{1}{2}t^2\right)dt .$$

(5.43)

Equation (5.41) may then be written:

$$q(t_i) = \frac{1}{2} \pm \mathrm{erf}(t_i) .$$

(5.44)

It is evident that at the mean, $x = \bar{x}$, $t = 0$, and $q = \frac{1}{2}$. It can also be shown from computed values of the error function, that when x is at a value one standard deviation above its mean, $q = 0.8143...$, and when it is one standard deviation below the mean, $q = 0.1587...$

If one plots an experimentally observed cumulative distribution function on probability coordinates, as is shown in Fig. 5-23(a) for the

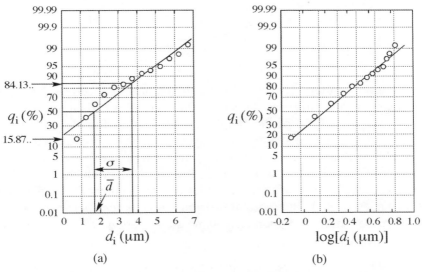

Fig. 5-23: Cumulative distribution for sample of Table 5-5 plotted on (a) normal, and (b) log-normal coordinates.

sample distribution of Table 5-5, a straight line will result if the distribution is in fact Gaussian. Deviations from straight-line behavior, as are clearly evident in this case, indicate deviations from Gaussian behavior. (The correlation coefficient for the best-fitting straight line is $R^2 = 0.936$.) If an acceptable straight line *could* be drawn, one could read off at once the values of $\bar{x}$ and σ. $\bar{x}$ is the value of the abscissa corresponding to $q = \frac{1}{2}$, and σ is the distance away from $\bar{x}$ where $q = 0.84134...$ or $0.15866...$ The values read from this construction are approximately $\bar{d} \approx 1.7$ and $\sigma \approx 1.9$ μm, in (not surprisingly) poor agreement with the values of 2.08 and 2.45 μm computed directly from the data.

Another useful distribution is the *log normal* distribution, in which it is the *logarithm* of the particle size that is normally distributed, so that

$$f(x) = \frac{1}{x\,\sigma_{\log x}\sqrt{2\pi}}\exp\left[-\frac{1}{2}\left(\frac{\log x - \overline{\log x}}{\sigma_{\log x}}\right)^2\right]. \tag{5.45}$$

Such a distribution is skewed to the smaller particle sizes, as shown qualitatively in Fig. 5-24, when plotted on Cartesian coordinates. Such a

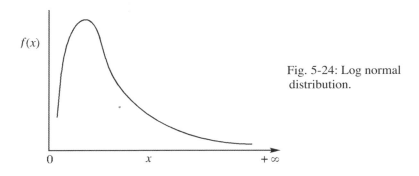

Fig. 5-24: Log normal distribution.

distribution is often observed when colloids are produced by comminution methods, particularly by crushing and grinding. Examination of the histogram, Fig. 5-20, of the frequency distribution for the sample data of Table 5-5 suggests that the data may be closely approximated by a log-normal distribution. Thus the plot of the log of the particle diameter on probability coordinates, as shown in Fig. 5-23(b) produces a much better straight line ($R^2 = 0.985$) than the plot of the diameter itself. The average that characterizes the log normal distribution is the *geometric* mean, and the standard deviation is a geometric standard deviation. The geometric mean, $\overline{d}_g$, of n diameters, for example, is the nth root of the product of the n diameters. It is computed for a classified data set as

$$\overline{d}_g = \text{antilog}\,\frac{\sum n_i (\log d_i)}{\sum n_i}. \tag{5.46}$$

For the sample distribution of Table 5-5, the geometric mean diameter is 1.53 μm, and it is in fact always the case that $\overline{d}_g < \overline{d}_n$. The geometric standard deviation, σ_g, is given by

$$\sigma_g = \text{antilog}\left\{\frac{\sum\left[n_i(\log d_i - \log \overline{d}_g)^2\right]}{\sum n_i}\right\}^{1/2}. \tag{5.47}$$

4. Particle shape

Colloid particles exist in a very wide variety of different shapes, although many are in fact spherical. Emulsions, for example, have spherical droplets, the configuration which minimizes the total surface free energy under normal droplet number density conditions (although in very dense emulsions, the droplets assume polyhedral shapes as they are pressed together). A number of bottom-up syntheses are designed to produce spherical particles. Examples include the perfect spheres of various polymers created by emulsion polymerization, or those of silica and other mineral oxides that can be made by sol-gel methods. Most particulates created by comminution methods are polyhedral or quasi-spherical. Solid sol particles that have measurable solubility also tend toward the spherical shape by the same mechanism as that of particle size disproportionation, although this process is often very slow. Many colloids, however, consist of particles that cannot be adequately described as spheres.

Particle shape is important to many colloidal properties and to many of the applications of colloids. Opalescence or gloss in coatings is often achieved through the flow-induced orientation of non-spherical pigment particles, and optical birefringence may be the result of the alignment of needle-like particles. The ability of such particles to orient themselves in a flow field also contributes to the rheology of colloids, as discussed in Chap. 8. Non-spherical particles are also subject to rotational Brownian motion[42], a phenomenon not evident for spherical particles. (Ordinary, or translational Brownian motion is discussed later in this chapter.) Rotational Brownian motion is accessible using depolarized light scattering or nuclear resonance methods, and has been used as a basis for determining particle size and shape.[43] Shape is also important to many of the interesting thermoelectric, piezoelectric, opto-electronic, catalytic and other properties of particles. Shape-controlled synthesis of colloid particles, particularly nanoparticles, is currently a thrust of much research, as it is appreciated that so many of the important properties of colloids are shape-dependent. Some of the strategies for the synthesis of particles of different shapes are discussed later in this section.

The various types of microscopy provide direct means of studying particle geometry. Optical microscopy, with a resolvability of approximately 0.5 μm, covers only the upper part of the colloid size range, so that electron microscopy is the most common method for obtaining shape information. For particles deposited on surfaces, the methods of scanning probe microscopy are finding increased use. Two examples of scanning electron micrographs of particles are shown in Fig. 5-25. The virus particles are seen to be cylindrical in shape, with an aspect ratio of approximately 150. The hair-like colloids formed by vanadium pentoxide are typical of what are now more popularly referred to as "nanowires."[44] In contrast to elongated

[42] Takeo, M., **Disperse Systems**, pp. 52-60, Wiley-VCH, Weinheim, Germany, 1999.

[43] Bauer, D. R., Opella, S. J., Nelson, D. J., and Pecora, R., *J. Amer. Chem. Soc.*, **97**, 2580 (1975).

[44] Gigargizov, E. I., **Highly Anisotropic Crystals**, Reidel, Dortrecht, The Netherlands, 1987.

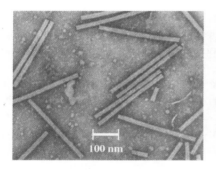
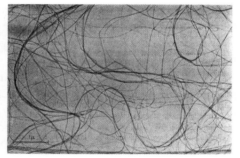

Fig. 5-25: Electron micrographs of (a) tobacco mozaic virus (From American Phytopathological Soc.) particles, and (b) a filamentous vanadium pentoxide V_2O_5 sol. From [Hall, C. E., **Introduction to Electron Microscopy**, McGraw-Hill, New York (1953).]

shapes, as those shown in Fig. 5-25, most naturally occurring clays consist of platelet-shaped particles, as exemplified by the kaolinite shown in Fig. 5-26. The figure shows the faces of the platelets, some of which are as thin as

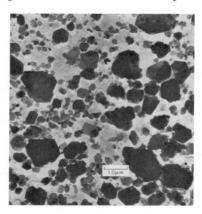

Fig. 5-26: Scanning electron micrograph of sodium kaolinite clay. From [Luh, M.-D., and Baker, R. A., *J. Colloid Interface Sci., 33*, 539 (1970).]

only 50 nm. To facilitate the description of the various shape possibilities for colloid particles, it is convenient to attempt to portray as many of them as possible, at least approximately, in terms of ellipsoids of revolution, as shown in Fig. 5-27. An ellipsoid is a volume obtained by rotating an ellipse about either its major or its minor axis. When the major axis is the axis of symmetry, one obtains a *prolate* ellipsoid (an egg), and when the minor axis is the axis of symmetry, an *oblate* ellipsoid (a curling stone, or an M&M's candy) results. In either event, one takes:

a = 1/2 length of axis of symmetry, and

b = 1/2 diameter of the circle normal to the axis of symmetry.

Thus one has:

$a = b$:   sphere

$a > b$:   prolate $a \gg b$: rod, ribbon or thread

$a < b$:   oblate $a \ll b$: disk or plate

In this context, either *a* or *b* may be used to specify particle size, while the aspect ratio *a/b* describes the shape. The particles of Fig. 5-25 are clearly prolate, while oblate particles are exemplified by the disk-shaped clay particles shown in Fig. 5-26.

Not all particle shapes can be adequately described in terms of simple ellipsoids of revolution. For example, cubical particles, as the silver particles shown in Fig. 5-28,[45] or the spiny or star-shaped gold particles shown in

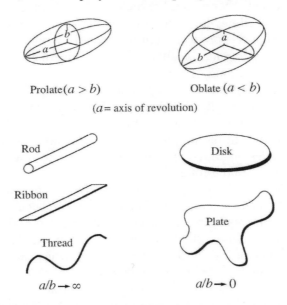

Prolate($a > b$)   Oblate ($a < b$)

(a = axis of revolution)

Rod

Disk

Ribbon

Plate

Thread

$a/b \rightarrow \infty$ $a/b \rightarrow 0$

Fig. 5-27: Ellipsoids of revolution as models for the shape of non-spherical particles.

Fig. 5-28: TEM image of silver nanocubes synthesized by the polyol process. From [Wiley, B., Sun, Y., Mayers, B., and Xia, Y., *Chem. Eur. J.*, **11**, 454 (2005).]

Fig. 5-29, can only be described as equivalent spheres, with their special features requiring additional descriptors. Also, neither particle roughness nor porosity is taken into account in the ellipsoid model. A generic type of

Fig. 5-29: SEM images of (a) spiny structures and (b) star-shaped particles of gold produced using an EDTA-controlled synthesis. From [Wang., W., Liu, Y., Zhou, X., Sun, J. and You, T., *Chem. Lett.*, **36**, 924 (2007).]

---

[45] Sun, Y., and Xia, Y., *Science*, **298**, 2179 (2002).

particle that is often encountered is one formed by the aggregation and sintering of smaller, primary particles. These may form globular structures, approximated as porous spheres, but often they form more ramified structures that are best described in terms of a fractal dimension. Aggregate structures are described in more detail in Chap. 7.

A wide variety of methods for particle shape controlled syntheses exist. Bottom-up precipitation processes can be made to produce a wide variety of oxide and other mineral particles of a wide range of shapes simply by carefully controlling the conditions of the precipitation. For example, Matijevic and Steiner[46] produced uniform, nearly monodisperse colloids of cubic, ellipsoidal, pyramidal, rodlike as well as spherical iron oxide particles from solutions containing ferric ions and nitrate, perchlorate or chloride ions. The exact compositions, temperature and aging time were varied to produce the different results. Another way to produce anisotropic particles is through the use of capping agents that block growth from solution to specific facets of the crystal. In the polymer-mediated polyol process, for example, silver nanoparticles in the shape of cubes, spheres, rods, wires or other shapes could be effected through control of the relative concentrations of the precursor ($AgNO_3$) and the capping agent [poly(vinylpyrrolidone), PVP] in the solvent (ethylene glycol). Another example of such a process is the growth of the spiny or star-shaped gold particles from $HAuCl_4$ solutions containing the sodium salt of eththylenediamine tetra-acetic acid (Na_2EDTA), shown in Fig. 5-29. Growth of rods, whiskers or nanowires can be induced to occur from precursors either in a vapor phase or a solution at specific sites on a surface where "catalyst" material, often in the form of nanodroplets of liquid metal, has been deposited. As an example of growth from the vapor phase, *i.e.*, the VLS (= vapor-liquid-solid) process, silicon and germanium nanowires are grown from SiH_4 and GeH_4 precursors onto gold or other metal nano-droplet catalyst sites on silicon or other substrates.[47] Semiconductor nanowires of various types have also been formed by the VLS process.[48] Templates, usually produced by some lithographic process, are also commonly used for producing particles of various shapes. The templates may be filled by condensation from a vapor phase, precipitation from a liquid phase or electrolytically. Wang and coworkers[49] produced nanobelts of various oxides by vapor condensation directly onto smooth, defect-free alumina substrates.

An interesting process for the production of prolate polymeric particles of systematically controlled aspect ratio was proposed by Keville *et al.*[50] Microspheres of poly(methylmethacrylate), PMMA, produced by

[46] Matijevic, E., and Steiner, P., *J. Colloid Interface Sci.*, **63**, 509 (1978).
[47] Bootsma, G. A., and Gassen, H. J., *J. Cryst. Growth*, **10**, 223 (1971).
[48] Duan, X., and Lieber, C. M., *Adv. Mater.*, **12**, 298 (2000).
[49] Pan, Z. W., Dai, Z. R., and Wang, Z. L., *Science*, **291**, 1947 (2001);
   Wang, Z. L., *Adv. Mater.*, **15**, 432 (2003).
[50] Keville, K. M., Frances, E. I., and Caruthers, J. M., *J. Colloid Interface Sci.*, **144**, 103 (1991).

dispersion polymerization, were dispersed in hexane containing poly(dimethylsiloxane) (PDMS), prepolymers, polymerization catalysts and cross-linking agents. The mixture was poured into a Teflon mold and cured to produce an elastomeric PDMS gel with 1-20% PMMA particles by volume. The composite elastomer block was cut into strips, heated to a temperature just above the T_g of the PMMA beads, stretched, and finally cooled so that they contained PMMA particles of an aspect ratio determined by the extent of stretching. The particles were subsequently recovered by degradation/dissolution of the matrix material. Figure 5-30 shows PMMA

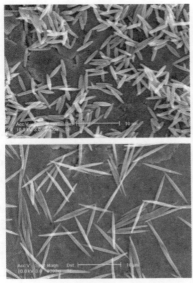

Fig. 5-30: Scanning electron micrographs, SEM's, of PMMA ellipsoids. Particles in the upper figure have an aspect ratio $a/b =$ 5.2, and the lower, 12.8. From [Mohraz, A., and Solomon, M. J., *Langmuir*, **21**, 5298 (2005).]

particles produced by this method by Mohraz and Soloman, who focused on obtaining 3-dimensional images of the structure of dense dispersions of these particles by confocal scanning laser microscopy, CLSM, and on the mathematical description of their assembly.

E. Sedimentation and centrifugation

1. *Individual particle settling: Stokes' Law*

When a colloidal dispersion finds itself in a gravitational field, its particles move downward or upward depending on whether they are denser or less dense than the medium, leading to sedimentation or creaming, respectively. The phenomenon is important in such industrial processes as wastewater clarification, and is an important basis for the separation of the particles in a dispersion based on their size. Finally, it is the basis of important analytical techniques for determining particle size and size distribution in a colloid. Consider first the sedimentation of a single isolated spherical particle. If at $t = 0$, the system is at rest, the particle must accelerate to its terminal velocity. The time this takes is very small for small particles, so they are effectively instantly at their terminal velocity. A 10-μm diameter sphere of density 2.0 g/cm³ in water at 20°C will travel a distance of only $\approx$ 40 nm before it reaches 99% of its terminal velocity. The calculated distance is only $\approx$ 1 Å for a 1-μm diameter sphere under the same conditions. The terminal velocity of settling for a small individual solid spherical particle, shown in Fig. 5-31, is given by balancing the driving force, F_{driv}, *i.e.*, the net gravitational force:

$$F_{\text{driv}} = \frac{1}{6}\pi d^3\left(\rho_p - \rho\right)g,$$ (5.48)

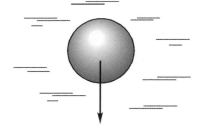

Fig. 5-31: A settling sphere.

where d is the particle diameter, ρ_p is the particle density, ρ the medium density and g the gravitational acceleration, against the resisting force, which is the sum of viscous and form drag:

$$F_{\text{res}} = fv_\infty = 3\pi\mu dv_\infty,$$ (5.49)

where μ is the viscosity of the medium (assumed Newtonian), and v_∞ is the terminal velocity. f is termed the "friction factor." The result is Stokes' Law:

$$v_\infty = \frac{d^2\left(\rho_p - \rho\right)g}{18\mu}.$$ (5.50)

In applying Stokes' Law it is important that ρ_p be the *effective* particle density. For solid particles this may differ substantially from the nominal density of the solid because they may possess pores or roughness concavities that are filled with the fluid medium or even with vapor pockets (if the medium is a liquid).

For particles of more general shape, the driving force may be written:

$$F_{\text{driv}} = m_p g - mg = m_p\left(1 - \frac{\rho}{\rho_p}\right)g,$$ (5.51)

where m_p is the particle mass, and m is the mass of the displaced fluid. The resisting force is always proportional to the terminal velocity and takes the form:

$$F_{\text{res}} = fv_\infty,$$ (5.52)

where the friction factor f depends on the size and shape of the particle and has dimensions [=] mass/time. For a sphere, recalling Eq. (5.49), one has

$$f_{sph} = 3\pi\mu d. \tag{5.53}$$

"Stokes-like" laminar flow holds for small non-spherical particles, with f always $> f_{sph}$, but seldom is f /f_{sph} greater than about 2.[51] For particles of arbitrary shape, a generalized Stokes Law then takes the form:

$$v_\infty = \frac{m_p}{f}\left(1 - \frac{\rho}{\rho_p}\right)g. \tag{5.54}$$

It is evident that a measurement of v_∞ gives only m_p /f, or it provides the radius of a "hydrodynamically equivalent" sphere.

Stokes' Law is strictly valid only for solid spheres and for Reynolds Number, $Re \le 0.1$ (the creeping flow regime), where

$$Re = \frac{dv_\infty\rho}{\mu}. \tag{5.55}$$

In the colloid size range, Stokes' Law is always valid. As an example, consider 1-μm diameter spherical gold particles ($\rho_p = 19.3$ g/cm^3), dispersed in water at 20°C ($\rho = 1.0$ g/cm^3; $\mu = 0.01$ g/cm·s). This yields $v_\infty \approx 10$ μm/s, and $Re \approx 10^{-5}$. Stokes' Law is also usually valid for droplets or bubbles in the colloid size range, which might be subject to internal circulation, if one of the phases is water. Traces of ever-present surfactant contamination generally prevent such circulation, however, in spheres of under 1 mm diameter, as discussed further in Chap. 10.

2. Multi-particle, wall and charge effects on sedimentation

A dispersion is a swarm of particles rather than a single one, but for sufficiently dilute dispersions, they sediment independently. This requires that the particles be non-aggregating and that the dispersion be dilute enough that particle-particle hydrodynamic interactions are negligible. The "hydrodynamic influence" of a particle moving through a medium extends a sufficient distance from its surface that the degree of dilution required for independent settling is a volume fraction $\phi \le 0.02$. For denser non-aggregating dispersions, the rate of sedimentation is reduced from that given by Stokes' Law, and often assumes an expression of the form:

$$v_{sed} = v_{Stokes}(1 + a\phi + b\phi^2 + ...), \tag{5.56}$$

where a and b are constants, evaluated empirically. For moderately dense ($\phi \le 0.1$) dispersions of spheres, the series may be truncated after the first term,

---

[51] Happel, J., and Brenner, H., **Low Reynolds Number Hydrodynamics, with Special Applications to Particulate Media**, Noordhoff Intl. Pubs., Leiden, 1973.

and Batchelor[52] has obtained $a = -6.55$. For dispersions up to higher volume fractions, an expression of the form

$$V_{sed} = V_{Stokes}\left[1 - \frac{\phi}{p}\right]^{k_1 p},$$

(5.57)

has been proposed[53], where k_1 and p are empirical constants. Wall effects may be important even in dilute dispersions if sedimentation is occurring in a tightly confined region.[54] For a spherical particle of diameter d settling in creeping flow parallel to and with its center a distance l from a vertical wall, the terminal velocity is decreased by a factor K:

$$K = \left[1 - \frac{9}{16}\lambda + O(\lambda^3)\right],$$

(5.58)

where $\lambda = d/l$. The implication is that particles more than 50 diameters away from the wall will have their terminal velocities reduced by less than 1%. Wall effects are thus negligible for sedimentation in ordinary-sized vessels, but will be important in microfluidics applications.

If the particles are aggregating during sedimentation, the effective kinetic units are increasing in size during the process, and the sedimentation rate may be greatly *increased*. In fact, voluminous flocks may be produced having a sweeping effect as they settle, collecting un-aggregated particles and small flocks. Inducing aggregation, as discussed in more detail in Chap. 7, is central to wastewater clarification by sedimentation.

It must be noted finally that colloidal particle surfaces, particularly in aqueous media, often carry an electric charge, as acquired by one of several mechanisms described in the next chapter. There will then be small ions of opposite charge (counter-ions) in the immediate neighborhood of the surface to maintain macroscopic electro-neutrality. When the particles start to sediment due to gravity (or in a centrifugal field), they would appear to leave their counter-ions behind, the latter being kept in suspension by diffusion (since they are so small). But as the charged particles move away from their counter-ions, an electric field is set up which pulls them along behind. The net effect is the establishment of a sedimentation potential, with counter ions moving along behind the particles, as suggested in Fig. 5-32. Since all are moving together, the total drag is greater than it would be if the colloid particle surfaces were not charged. The force resisting sedimentation becomes:

$$F_{res} = v_\infty\left(f_p + z_p f_{ci}\right),$$

(5.59)

[52] Batchelor, G. K., *J. Fluid Mech.*, **52**, 245 (1972).
[53] Ekdawi, N., and Hunter, R. J., *Colloids Surfaces*, **15**, 147 (1985).
[54] Clift, R., Grace, J. R., and Weber, M. E., **Bubbles, Drops and Particles**, pp. 221ff, Academic Press, New York, 1978.

where f_p is the friction factor of the particle; z_p is the "valence" number of the particle, *i.e.*, number of electronic charges/particle; and f_{ci} is the friction factor for the hydrated counter-ion. While the friction factor for the ion is usually small compared to that of the particle, the "valence" of the particle is

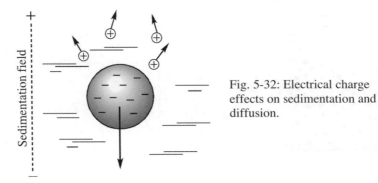

Fig. 5-32: Electrical charge effects on sedimentation and diffusion.

often very large, so that the term $z_p f_{ci}$ is comparable to f_p, and the additional drag may be very significant. On the other hand, the driving force for sedimentation:

$$F_{driv} = m_p\left(1 - \frac{\rho}{\rho_p}\right) + z_p m_{ci}\left(1 - \frac{\rho}{\rho_{ci}}\right) \approx m_p\left(1 - \frac{\rho}{\rho_p}\right). \qquad (5.60)$$

will not be affected much, because the mass of the counterions, m_{ci}, is so small. Equating the two, $F_{driv} = F_{res}$, at steady state gives:

$$v_\infty = \frac{m_p}{\left(f_p + z_p f_{ci}\right)}\left(1 - \frac{\rho}{\rho_p}\right). \qquad (5.61)$$

The effect often decreases v_∞ by as much as a factor of two or more. As described in more detail in Chap. 6, electrical charge separation at particle interfaces may be reduced or effectively eliminated by adding sufficient electrolyte to the solution, screening out the charge and causing the particles to sediment as though they were electrically neutral.

3. Differential sedimentation; particle size analysis

Sedimentation provides a powerful means for measuring colloid particle size, based on Stokes Law analysis, and a number of commercially available instruments have been designed to exploit it. If the dispersion is monodisperse (a rare circumstance), all the particles will settle at the same rate, and there will be a clean line of demarcation between the particle-free supernatant and the dispersion, moving downward as shown in Fig. 5-33(a). The dispersion will be of uniform concentration in particles, and the front will move like a curtain falling at the rate v_∞. The motion of the front can be followed optically. Alternatively, one could weigh the sediment continuously as it collects at the bottom of the vessel. This gives a read-out, as shown in Fig. 5-33(b). If the weight of sediment per unit area in the pan is W, the flux of the descending particles is $J = dW/dt$ and is equal to the number concentration of the particles below the "curtain," n, times the sedimentation velocity, v_∞:

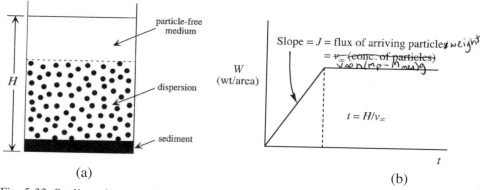

Fig. 5-33: Settling of a monodisperse suspension: (a) the "falling curtain" of particles all settling at the same rate, and (b) the measured sediment weight/area, W, as a function of time.

$$J = nv_\infty = \frac{dW}{dt} = nV_\infty(m_p - m_{med})g \qquad (5.62)$$

The collected mass in the weighing pan stops increasing at a time t_s, the total settling time, equal to the original height of the dispersion, H, divided by the settling velocity. Thus measurement of t_s gives at once the value of v_∞, and using Stokes Law, Eq. (5.50) for spherical particles, the particle size. Measurement of the rate of settling, dW/dt, then gives the particle number density, n.

In a differential sedimentation of a polydisperse dispersion one measures the weight/area of particles collected in a weighing pan as a function of time, $W(t)$, to obtain a curve as shown in Fig. 5-34. At any given time, t, $W(t)$ will consist of two parts: 1) the weight of all the particles of the original dispersion that are of a given size (m_i) and larger, viz., $W_{m>m_i}$, and

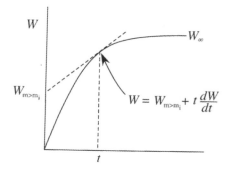

Fig. 5-34: Measurement of weight/area, W, of particles gathered in a weighing pan as a function of time during a differential sedimentation experiment with a polydisperse dispersion.

2) the partial amounts of all the smaller particles that are still settling, given by $t(dW/dt)$. (dW/dt) is the sum of the sedimentation fluxes of all the particles of mass $m < m_i$. Thus

$$W(t) = W_{m>m_i} + t\frac{dW}{dt}. \tag{5.63}$$

The particles that are completely settled out at time t and constitute $W_{m>m_i}$ are those whose hydrodynamically equivalent diameters are greater than dictated by Eq. (5.50), i.e.,

$$d_i > \sqrt{\frac{18\mu H}{(\rho_p - \rho)gt}}. \tag{5.64}$$

As $t \to \infty$, all the particles will have settled, and $W \to W_\infty$. The values of $W_{m>m_i}$ as a function of time are recovered from the measured $W(t)$ using Eq. (5.63), and the values of time are converted to diameter using Eq. (5.64). Then the final cumulative weight (or mass) distribution of the particle diameter is given by

$$q_w(d \leq d_i) = 1 - \frac{W_{m>m_i}}{W_\infty}. \tag{5.65}$$

The technique of "differential sedimentation" described above is limited to rather large-particle dispersions (because of excessive sedimentation times required for colloid-size particles) and is awkward because $W(t)$ must be differentiated to obtain the particle size distribution. An alternative approach is to measure the concentration of particles remaining in suspension at some location (e.g., at a position H measured from the top surface of the dispersion) as a function of time. At any instant of time, t_i, the dispersion at location H will contain particles of all sizes in the dispersion, in their original proportions, except those which have sedimented more than a distance H in the time t_i. These will not be present at level H at time t_i (or $t > t_i$) as they have irreplaceably settled "out of the picture." The missing particles are thus those for which $v_\infty t_i > H$, i.e., those for whom the particle diameter is given by Eq. (5.64). Thus, using an optical device that senses the mass concentration of particles at level H as a function of time $C_H(t_i)$, yields the mass concentration of particles finer than d_i given by Eq. (5.64). The total mass fraction of particles at level H as a function of time is given by

$$w_H(t_i) = \frac{C_H(t_i)}{C_0}, \tag{5.66}$$

where C_0 is the total mass concentration of particles in the original dispersion. t_i may be eliminated parametrically from Eq. (5.66) using Eq. (5.64), so that one has finally

$$q_w(d < d_i) = \frac{C_H(d_i)}{C_0}, \tag{5.67}$$

where $q_w(d < d_i)$ is again the mass fraction of particles smaller that d_i, *i.e.*, the *cumulative* mass distribution of particle size.

A simple but rather crude device for making the above measurements is the *Andreason pipette*, a graduated cylinder containing the sedimenting dispersion, with a thin sampling pipette immersed from the top to a level H below the surface. Small samples are withdrawn at a series of time intervals and analyzed offline for the concentration of the particles they contain.

A more sophisticated technique is that provided by the SediGraph® Particle Size Analyzer (Micromeritics Corp., Norcross, GA), a widely used commercial instrument for carrying out the above type of measurements. A finely collimated, low-energy X-ray beam is used to measure the local particle mass concentration, and the measured transmittance of the suspension to the X-ray beam, relative to that of the suspending liquid, is directly proportional to the mass concentration of the suspended solids. The transmitted radiation is detected by means of a scintillation counter. The use of an X-ray source limits the instrument to suspensions of particles that absorb X-rays, such as metals, oxides, silicates and other minerals. Other materials, such as carbon, sulfur, organic pigments, polymer latices, *etc.*, are transparent to the beam. The instrument is calibrated by adjusting "100% transmittance" to correspond to the particle-free suspending liquid, and "0% transmittance" to correspond to the initial, unsedimented dispersion. In principle, the X-ray beam could then be located at any position H, and the relative particle mass concentration monitored as a function of time. In practice, however, the cell is moved downward in accord with a prescribed schedule, resulting in a large reduction in the time required for measurement. As an example, consider an aqueous suspension at 20°C of silica particles ($\rho_p = 2.6$ g/cm³) whose diameters range from 50 to 0.2 μm. In accord with Eq. (5.51), the largest particles settle at a rate of approximately 2 mm/s. To obtain accurate measurements at the course end of the size distribution, the location of the X-ray beam should be at least 2.5 cm below the top of the cell (*i.e.*, H should be > 2.5 cm). In order to extend the size analysis down to 0.2 μm, using the sample depth of $H = 2.5$ cm, would require over 200 hours! Moving the cell downward (decreasing H) during the course of the run, the run time is reduced to less than one hour in this case. The schedule governing the movement of the cell is dictated by the system properties in each case. In operation, H is made to vary inversely with $t^{1/2}$. Under these circumstances, in view of Eq. (5.64), d_i varies inversely with $t_i^{1/2}$:

$$d_i = K/t_i^{1/2}, \tag{5.68}$$

where K is a constant. H decreases at a rate proportional to $\sqrt{(\rho - \rho_0)/\mu}$. Using such a schedule for the movement of the cell allows a wide particle size range to be covered, and depending on the density difference and viscosity, particles as small as 0.2 µm can generally be accommodated.

4. *Centrifugation*

If the particles involved are very small and/or the density difference is slight, simple sedimentation may be impractical as an analytical tool. In the nano-colloid size range, Brownian motion also begins to interfere with the process, as described later in this chapter. Such limitations may be overcome using centrifugation. The dispersion may be put in a tube and spun around in a plane parallel with the tube axis, as shown in Fig. 5-35. Letting R be the

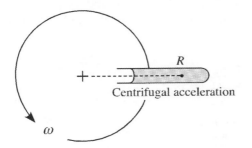

Fig. 5-35: Centrifugation.

distance measured from the center of rotation, the *local* centrifugal acceleration is given by $\omega^2 R$ (where ω = angular velocity). Using the centrifugal acceleration in place of g in Stokes' Law (for spheres or particles of general shape), one obtains expressions for the rate of centrifugation:

$$v_\infty = \frac{dR}{dt} = \frac{d^2(\rho_p - \rho)\omega^2 R}{18\mu} \tag{5.69}$$

or

$$v_\infty = \frac{dR}{dt} = \frac{m_p}{f}\left(1 - \frac{\rho}{\rho_p}\right)\omega^2 R. \tag{5.70}$$

Since $\omega^2 R$ is usually made to be many hundreds or even thousands times g, ordinary gravity is of no significance in the centrifuge. $\omega^2 R$ values up to $10^5 g$ have been employed in the *ultra*centrifuge, a device first used by T. Svedberg,[55] (who received the Nobel Prize for Chemistry in 1926 for this work). Since the local acceleration depends on R, the steady state rate of sedimentation increases with R (unlike the case of gravity settling). It is therefore useful to define a quantity that is constant during centrifugation, *viz.*, the "sedimentation coefficient," S, defined as

$$S = \frac{1}{\omega^2 R}\frac{dR}{dt} = \frac{\ln(R_2/R_1)}{\omega^2(t_2 - t_1)}, \tag{5.71}$$

[55] Svedberg, T. **The Ultra-Centrifuge**, Oxford Univ. Press, Oxford, 1940.

where the last form relates to experiments in which the movement of a sedimenting front from position R_1 to position R_2 over the time interval ($t_2 - t_1$) is observed. S has units of time and is generally of the order of a few 10^{-13} s. For monodisperse particles, the front can be observed optically without ambiguity. Ultracentrifugation is an especially powerful technique for studying lyophilic colloids, *e.g.*, dissolved proteins. Such solutions usually contain only a few different dissolved compounds, and the centrifuge solves the problem of the very slow gravitational settling of such solutions.

A common optical method for determining the movement of the fronts is the schlieren method,[56] as shown in Fig. 5-36, although a number of other optical methods are also used. The schlieren method passes a beam of collimated light through the medium, and the rays of this light are bent away

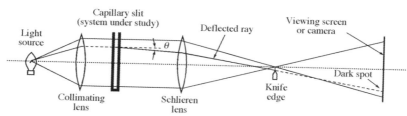

Fig. 5-36: Schematic of schlieren system.

from the optical axis at an angle proportional to the local refractive index gradient. The light is then sent through a lens that brings it to a point at the focal point of the lens, where deflected beams are gathered at a point off the optical axis. These can be blocked or transmitted by means of an optical stop (such as a sharp knife edge.) The intensity of light I at any point in the image is given by

$$I \propto \frac{dn}{dR} \propto \frac{dC}{dR},$$

(5.72)

where n is the refractive index, C is the concentration of particles and R is

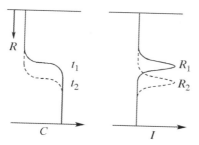

Fig. 5-37: Interpretation of schlieren trace. C is the particle concentration, and I is the optical intensity.

the distance (with reference to Fig. 5-35). Thus the motion of the concentration front can be followed with time, as shown in Fig. 5-37.

[56] Holder, D. W., and North, R. J., **Schlieren Methods**, Notes on Applied Science, No. 31, HMSO, London, 1963.

One can interpret the result of a sedimentation rate experiment with Eq. (5.69) or (5.70). For spheres, the value of the particle diameter can be obtained as:

$$d = \sqrt{\frac{18\mu S}{(\rho_p - \rho)}}.$$
(5.73)

For non-spherical particles, the best one can do is get the *equivalent* sphere radius, or evaluate: m_p / f:

$$\frac{m_p}{f} = S\left(1 - \frac{\rho}{\rho_p}\right)^{-1}.$$
(5.74)

Centrifugal analyses for obtaining particle size distributions in polydisperse systems parallel the analyses used for gravitational sedimentation, and commercial instrumentation is available. For example, the Horiba CAPA-700 Particle Size Distribution Analyzer (Horiba Instruments, Inc., Irvine, CA) is a tube-type centrifuge of the type suggested by Fig. 5-35 that spins up to 10,000 rpm, sufficient to handle many colloids of particles as small as 10 nm. The optical density at a particular location along the tube axis is monitored as the tube (and a twin tube containing only the dispersion medium, located on the opposite side of the axis of rotation, and used as a reference) is spun about the axis. A green LED, 560 nm light source is used, and the detector is a silicon photo-diode. The instrument outputs the absorbance *vs.* time curve, the cumulative and the frequency mass distributions and the attendant statistics. Another configuration available is the disk centrifuge. In this system the medium (serum) of the colloid to be analyzed is first placed in a cell consisting of the space between two parallel transparent discs. When the cell is spinning a small sample of the colloid to be analyzed is injected onto the surface of a liquid (spin fluid) previously injected into the disk cavity. The particles sediment radially through the liquid in the centrifugal field. Light from a narrow-band, LED is passed through the disc at a particular location. The transmitted light is measured by a photodiode and recorded by the computer as a function of time. Hydrodynamically stable sedimentation is ensured by using a gradient method, *i.e.*, an overcoat of a thin layer of increasing density and viscosity is injected before the sample. The gradient cushions the hydrodynamic shock experienced by the particles as they enter the spin fluid so that a smooth transition and subsequent laminar flow is ensured. The same type of information as that obtained from the tube centrifuge is provided.

F. Brownian motion; sedimentation-diffusion equilibrium

1. *Kinetic theory and diffusion*

Sedimentation or centrifugation set up concentration gradients that are opposed by *diffusion*, the macroscopic manifestation of random thermal motion of molecules or individual particles. According to classical kinetic theory, all kinetic units, regardless of size, possess on average, translational kinetic energy in the amount of $3/2kT$. There is assumed to be no preference amongst the three directions, so kinetic units move in any given direction (say x) with a velocity that depends on their mass in accord with:

$$\frac{1}{2}mv_x^2 = \frac{1}{2}kT.$$

(5.75)

This kinetic energy, whose magnitude depends on T, leads to straight-line, i.e., ballistic, motion. The computed x-velocity of a hydrogen molecule at room temperature is approximately 1100 m/s, whereas that of an isolated 1-picogram colloid particle is 2 mm/s. Actually, the molecules experience many collisions with other molecules and with the walls leading to direction changes and zig-zag paths. The average distance between collisions is the mean free path, λ, which for molecules in an ideal gas is given by

$$\lambda = \frac{1}{\sqrt{2}\pi n d^2},$$

(5.76)

where n is the molecular number density in #/cm^3, and d is the molecular diameter. For gas molecules at room temperature and atmospheric pressure, λ is of the order of 10–100 nm. For liquids of ordinary-sized molecules, λ is of the order of one Å, and the frequency of collisions 10^{12} - 10^{13} s^{-1}. All that can generally be observed in the laboratory is the macroscopic consequences of such motions, i.e., the tendency of media of non-uniform composition to move toward uniform composition with the passage of time, as described phenomenologically by Fick's Law:

$$J_A = -D_{AB}\left(\frac{\partial C_A}{\partial x}\right),$$

(5.77)

where J_A is the flux of moles of A across a surface normal to x, and C_A is its concentration. Equation (5.77) defines D_{AB}, the diffusivity of A in B. The corresponding conservation equation for species A is:

$$\frac{\partial C_A}{\partial t} = -D_{AB}\left(\frac{\partial^2 C_A}{\partial x^2}\right).$$

(5.78)

One can obtain values for D_{AB} in the laboratory by measuring concentration as a function of time and position and comparing results to solutions of the above differential equations. Typically for liquids of ordinary-sized molecules, D_{AB} is of the order of 10^{-5} cm^2/s, which for the rather steep steady concentration gradient of 0.01 M/cm ($= 10^{-5}$ mol/cm^4) leads to a flux of: $J_A = (10^{-5})$ cm^2/s(10^{-5}) mol/cm^4 = 10^{-10} mole/cm^2·s. This diffusive flux can be expressed in terms of an equivalent linear velocity, viz.

$$v_{\text{diff}} = \frac{J_A}{C_A},\qquad\qquad (5.79)$$

termed the "Stefan velocity." Assuming at the position of interest $C_A = 10^{-3}$ mol/cm^3:

$$v_{\text{diff}} = \frac{J_A}{C_A} = \frac{10^{-10} \text{ mol cm}^3}{\text{cm}^2 \text{ s } 10^{-3} \text{ mol}} = 10^{-7} \text{ cm/s}.$$

This would be the rate of convective flow required to produce the same flux.

2. *Brownian motion*

When the kinetic units reach colloidal dimensions or larger, the D-values become so low that ordinary measurements of concentration changes with time are not feasible. The diffusion of colloid particles, however, may be tracked individually since their motion can be observed by optical microscopy (at least when they are in the upper part of the colloid size range). One may attempt an analogy to the random thermal motion of molecules. If one could somehow tag a single molecule and track its motion, it might appear in two-dimensional projection as shown in Fig. 5-38, with the straight line segments being the "free paths" between collisions, and the average value of the length, the mean free path. Interpreting the "mean free path" as the "distance between collisions," however, doesn't make sense for colloid particles. Instead, the situation is more like that of a blimp (the colloid particle) being bombarded by billions of basketballs (the molecules

Fig. 5-38: Colloid particle path due to uneven bombardment by molecules of the dispersion medium. Positions shown are recorded at equal time intervals.

of the medium). Spatial fluctuations in the density of "basketballs" leads to a slow, lumbering motion of the "blimp." One may spot and record its location at time intervals, as shown Fig. 5-39, retaining the idea of the mean free path by logging the "blimp's" position at specified time intervals, t, and constructing the motion as straight steps. Figure 5-39(a) actually shows such a record for the motion of a gamboge (a resin obtained from any of several trees of the genus *Garcinia*, of south- central Asia, and yielding a golden-yellow pigment) particle obtained by Perrin. Note that the general features of the motion are retained if only every fifth or every tenth position is recorded, as shown in Fig. 5-39(b) and (c). There is no loss of statistical generality if a

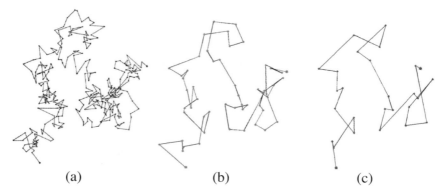

(a)   (b)   (c)

Fig. 5-39: (a) An observed Brownian motion path of a gamboges particle.
[Perrin, J., **Atoms**, 2nd English Ed., p.116 Constable & Co., Ltd, London,
1923,] (b) The same path, but with only every fifth position shown and
connected by straight lines and (c) every tenth position shown and connected
by straight lines. The similar shapes of the traces reveal their fractal nature.

sufficient number of such positions is tracked, *i.e.*, the diffusion path is a
fractal structure. Furthermore, one can idealize the steps as of equal length,
ℓ, equal to their mean value characteristic of the time interval between
which particle positions are recorded. It is such movement that is known as
Brownian motion (after Robert Brown, the botanist, who in 1828 reported
such movements of pollen particles in water). Consider next how to obtain
D-values from such observations.

Figures 5-38 and 5-39 represent two-dimensional projections of a
three-dimensional "random flight" of a colloid particle. Consider a further
simplification in which all movements are projected onto a single line. *i.e.*,
along some particular axis, say the x-axis. Regarding each step as the same
length, ℓ, and assuming the particle has only the equally-probable options of
moving forward or backward, produces a one-dimensional "random walk."
Consider a plane of area A oriented normal to a one-dimensional
concentration gradient, as shown in Fig. 5-40. Consider a time interval t

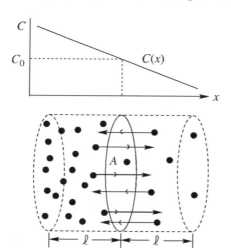

Fig. 5-40: Diagram used in
the derivation of the Einstein-
Smoluchowski Equation.

during which all particles take one step in the x-direction. Half go forward ($+x$ direction), and half go backward. Since the average concentration on the left is greater than on the right, the collective result of all of the particle jumps will be a net transfer of particles in the $+x$ direction. In words, during the time interval t we have

$$\begin{bmatrix} net \text{ transfer from} \\ \text{left to right} \end{bmatrix} = \begin{bmatrix} \text{transfer from left} \\ \text{chamber to right} \end{bmatrix} - \begin{bmatrix} \text{transfer from right} \\ \text{chamber to left} \end{bmatrix}$$

$$= \begin{bmatrix} 1/2 \text{ of particles in} \\ \text{left chamber at } t = 0 \end{bmatrix} - \begin{bmatrix} 1/2 \text{ of particles in} \\ \text{right chamber at } t = 0 \end{bmatrix}$$

$$= \frac{1}{2} \begin{bmatrix} \text{vol. of} \\ \text{chamber} \end{bmatrix} \begin{bmatrix} \text{ave. conc. in} \\ \text{left chamber} \end{bmatrix} - \frac{1}{2} \begin{bmatrix} \text{vol. of} \\ \text{chamber} \end{bmatrix} \begin{bmatrix} \text{ave. conc. in} \\ \text{right chamber} \end{bmatrix}.$$

The factor 1/2 is used since half the particles move *away* from A in both chambers, assuming the concentration of particles C varies linearly with x. The above statement becomes:

$$JAt = \frac{1}{2} A\ell \left[C_0 - \frac{\ell}{2} \frac{dC}{dx} \right] - \frac{1}{2} A\ell \left[C_0 + \frac{\ell}{2} \frac{dC}{dx} \right]. \qquad (5.80)$$

Solving for the flux and comparing with Fick's Law:

$$J = -\frac{\ell^2}{2t} \left(\frac{dC}{dx} \right) = -D \left(\frac{dC}{dx} \right), \qquad (5.81)$$

leads to:

$$D = \overline{\ell^2}/2t, \qquad (5.82)$$

termed the Einstein-Smoluchowski Equation or the Einstein diffusivity equation. The above is a simplified version of the derivation that was obtained by Einstein (1906)[57] in terms of fluctuation theory, von Smoluchowski (1906)[58] in terms of the statistics of the random walk, and Langevin (1908)[59] in terms of the classical mechanics of the movement of a single particle subject to the stochastic forces of molecular bomabardment.

This extremely important result permits the computation of D-values from direct observations of Brownian motion. t is the time interval between which particle positions are observed, and $\overline{\ell^2}$ is the average square of the distance between positions (the *mean square displacement*). The computed value of D has meaning when a sufficiently large number of steps have been averaged. Each single "step" observed, of course, is the net effect of many smaller steps, *i.e.*, the "degree of fineness" of the observation does not affect the computed D, since $\overline{\ell^2}$ and t vary together. Using this procedure, D's for colloid particles of various kinds have been obtained experimentally. The

[57] Einstein, A., **Investigations in the Theory of Brownian Movement**, Dover, New York, 1956.
[58] von Smoluchowski, M., *Ann. der Phys.*, **21**, 756 (1906).
[59] Langevin, P., *Compt. rend.*, **146**, 530 (1908).

modern technique for the determination of particle diffusivity is that of photon correlation spectroscopy (PCS), described later in this chapter.

Einstein's derivation of Eq. (5.82) yielded another very important result, *viz.*,

$$D = \frac{kT}{f},\qquad(5.83)$$

where f is the usual friction factor associated with the movement of the particle through the medium. Using diffusivity measurements, as analyzed using Eqs. (5.82) and (5.83), together with sedimentation or centrifugation measurements, as analyzed using Stokes Law or its equivalent, several important results can be obtained as follows.

1) A measurement of D gives f. If no other information is available, this permits computation of the hydrodynamically equivalent spherical diameter of the particle using Eq. (5.53) so that

$$d_{\text{sph}} = \frac{f}{3\pi\mu},\qquad(5.84)$$

2) If both the diffusion coefficient, D, *and* the sedimentation coefficient, S, are measured, one can obtain the particle mass, m_p, regardless of particle shape. Combining Eqs. (5.74) and (5.83) leads to:

$$m_\text{p} = \frac{kTS}{D(1 - \rho/\rho_\text{p})}.\qquad(5.85)$$

3) The value of m_p together with f can be used to obtain information concerning the particle shape. Taking f_0 as the friction factor for a sphere of equivalent mass, the ratio f/f_0 can be related analytically to the aspect ratio $\beta = b/a$ for both prolate and oblate ellipsoids of revolution (*cf.* Fig. 5-27), as shown in Fig. 5-41. For prolate ellipsoids of revolution:

$$\frac{f}{f_0} = \frac{(1-\beta^2)^{1/2}}{\beta^{2/3}\ln\left[\dfrac{1+(1-\beta^2)^{1/2}}{\beta}\right]},\qquad(5.86)$$

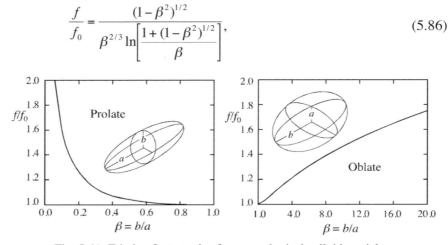

Fig. 5-41: Friction factor ratios for non-spherical colloid particles.

and for oblate ellipsoids of revolution:

$$\frac{f}{f_0} = \frac{(\beta^2 - 1)^{1/2}}{\beta^{2/3} \tan^{-1}(\beta^2 - 1)^{1/2}} \cdot \tag{5.87}$$

The friction factor for the desired ellipsoid of revolution may then be calculated from f_0, computed as:

$$f_0 = 3\pi \mu d_{eq} = 3\pi \mu \left(\frac{6m_p}{\pi \rho_p}\right)^{1/3} \approx 11.69 \mu \left(\frac{m_p}{\rho_p}\right)^{1/3} \tag{5.88}$$

where d_{eq} is the diameter of the sphere of equivalent mass.

4) An early value of Avogadro's Number was determined by simultaneous diffusion and sedimentation measurements on spherical particles of known size. Jean Perrin (1910)[60] studied Brownian motion of particles of gamboge (and mastic) tree resins, whose particles were well characterized as monodisperse spheres of known density. Perrin used the Einstein and Einstein-Smoluchowski Equations to obtain:

$$D = \frac{kT}{f} \equiv \frac{RT}{N_{Av} f}, \tag{5.89}$$

where R is the gas constant, so upon substitution of Eqs. (5.53) and (5.82):

$$N_{Av} = \frac{RT}{fD} = \frac{2RTt}{3\pi\mu d\overline{\ell^2}}. \tag{5.90}$$

This led to somewhat high values of N_{Av}, but in reasonable agreement with results of the Millikan oil drop experiment and the distribution of gas density in the gravitational field, as well as those obtained by X-ray diffraction. Perrin received the Nobel Prize in Physics in 1926 for his work on Brownian motion.

3. Sedimentation (centrifugation) – diffusion equilibrium

The steady state distribution of particle number density is obtained by comparing the rate of particle sedimentation, v_∞, as given by Stokes' Law, to the rate of back diffusion, as expressed by the Stefan velocity, v_{diff}. The latter is reckoned as

$$v_{diff} = \frac{J_A}{C_A} = -\frac{D}{C_A}\left(\frac{dC_A}{dz}\right) \equiv -D\frac{d\ln n}{dz}, \tag{5.91}$$

where z is the direction of travel (assumed parallel to g), and the concentration C_A has been replaced with the number concentration, n, of the particles. The numerical values of v_∞ and v_{diff} may be compared for a

[60] Perrin, J., **Brownian Movement and Molecular Reality** (F. Soddy, Trans.), Dover Publ., New York, 2005.

representative system as a function of particle size. Taking the particles to be spheres, the density difference between the particle and medium to be $\Delta\rho = 1$ g/cm³, the viscosity of the medium to be $\mu = 0.01$ g/cm·s (that of water at 20°C), and the particle concentration gradient to be such there is a 1% decrease in n per cm of distance z (so that $\dfrac{d\ln n}{dz} = -0.01$ cm⁻¹) produces the results shown in Table 5-7. It is seen that sedimentation rates are far greater for larger particles, but for sufficiently small particles, the rates of sedimentation and back-diffusion become comparable.

Table 5-7: Comparison of sedimentation velocities and diffusion velocities as a function of particle size for case when: $T = 20°C$, $\Delta\rho = 1$ g/cm³, $\mu = 0.01$ g/cm·s, and $d\ln n/dz = 0.01$ cm⁻.

$d(\mu m)$	v_∞ (μm/s)	v_{diff} (μm/s)
1.0	1.09	4.29×10^{-7}
0.1	1.09×10^{-2}	4.29×10^{-6}
0.01	1.09×10^{-4}	4.29×10^{-5}
0.005	2.72×10^{-5}	8.58×10^{-5}

At steady state, there is a balance between sedimentation and diffusion described by equating v_∞ to v_{diff}, as given by Eqs. (5.50) and (5.91), respectively, and leading to (for spherical particles):

$$n(z) = n_0 \exp\left[\frac{-\pi d^3 (\rho_p - \rho)g}{6kT} z\right],$$
(5.92)

an exponential decrease in particle density with elevation, as shown in Fig. 5-42.

Fig. 5-42: Sedimentation-diffusion equilibrium.

Concentration profiles for poly(styrene) latex particles in water at 25°C are shown in Fig. 5-43. Faraday's gold sol, shown in Fig. 1-5, may be analyzed using Eq. (5.92). Assuming that the particles are of $d \approx 10$ nm, their number density at an elevation of 10 cm above the bottom of the vessel (*i.e.*, $z = 10$

cm) is predicted to be approximately 90% of the density of the particles at the bottom of the vessel.

For particles of general shape, the resulting distribution is:

$$n(z) = n_0 \exp\left[-\frac{m_p\left(1 - \rho/\rho_p\right)gz}{fD}\right].$$

(5.93)

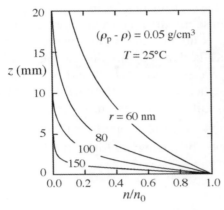

Fig. 5-43: Particle number-density profiles for sedimentation-diffusion equilibrium for polystyrene particles in water.

Such profiles can be analyzed for gravity settling (for which the above results are computed) or in a centrifugal field, to yield the particle mass, m_p, (or equivalent), or its molecular weight. For lyophilic colloids (such as dissolved proteins), the schlieren method can be used to yield the concentration gradient.

Number density profiles obtained in a centrifuge under conditions of equilibrium between centrifugal settling and diffusion are readily derived: Again: $v_\infty = v_{\text{diff}}$, so

$$\left(\frac{m_p}{f}\right)\left(1 - \frac{\rho}{\rho_p}\right)\omega^2 R = -D\frac{d\ln n}{dR},$$

(5.94)

and integrating yields:

$$n(R) = n_0 \exp\left[-\left(\frac{m_p}{2fD}\right)\left(1 - \frac{\rho}{\rho_p}\right)\omega^2\left(R^2 - R_0^2\right)\right].$$

(5.95)

From this steady-state experiment, one can evaluate m_p/fD; then by spinning the centrifuge at a greater ω, one can achieve sedimentation by overriding diffusion, and by examining its rate, get m_p/f. Thus one may obtain both m_p and D separately. The sorting out of shape effects from hydration (for lyophilic colloids) remains ambiguous until some other type of measurement can be found to separate them. Also, when the distribution is polydisperse, the results may be difficult to interpret. Each particle size will reach its own distribution equilibrium, and the observed turbidity will be the result of the sum of the contributions of all particle sizes at that level.

Consider next the effect that particle surface electric charge separation may have on particle diffusion and the sedimentation-diffusion equilibrium. Its influence in reducing particle sedimentation rates has already been described by Eq. (5.61), so it remains to describe its effect on particle *diffusion*. The process is influenced by the presence of the counter-ions that surround the particles, as suggested schematically in Fig. 5-32. These counter-ions also diffuse in response to their concentration gradient by diffusing upward, and in so doing, set up an electric field that tends to drag the particles upward. The net effect is that the particles appear to diffuse upward faster than if their surfaces were not charged. The driving force for diffusion of the particle plus its counter-ions is proportional to the thermal energy of all the moving kinetic units in the structure, *i.e.*, $(1 + z_p)kT$, whereas the resistance to movement is again proportional to: $f_p + z_p f_{ci}$, giving the effective particle diffusivity, by analogy to Einstein's Law:

$$D_p = \frac{(1 + z_p)kT}{(f_p + z_p f_{ci})}.$$
(5.96)

Comparing this result to the expression for diffusivity in the uncharged case shows:

$$\frac{D_{charged}}{D_{uncharged}} = \frac{(1 + z_p)}{(f_p + z_p f_{ci})} \frac{f_p}{1} \approx \frac{1 + z_p}{2} \gg 1.$$
(5.97)

Surface charge thus causes sedimentation itself to be slowed, and the counter process of diffusion to be enhanced. The net effect leads to an altered steady state distribution given by:

$$n(z) = n_0 \exp\left[\frac{-m_p(1 - \rho/\rho_p)g\,z}{(1 + z_p)kT}\right],$$
(5.98)

which could easily reflect a change in n of a factor of 100 or so at any given elevation.

It is worthwhile to look at the problem of sedimentation-diffusion equilibrium from a somewhat different point of view, *i.e.*, in terms of the Boltzmann distribution. Any collection of particles in a force field, in accord with the postulates of kinetic theory, will distribute itself such that the ratio of the probability $\mathcal{P}$ of finding a particle at some location in the field, z, relative to the probability of finding it at some reference location, z_0, is:

$$\frac{\mathcal{P}(z)}{\mathcal{P}(z_0)} = \frac{n(z)}{n(z_0)} = \exp\left(\frac{-w}{kT}\right),$$
(5.99)

where w is the reversible work required to move the particle from z_0 to z.[61] The ratio of probabilities is also the ratio of the number concentrations of the particles, $n(z)/n(z_0)$. With the force field of gravity, and z the distance measured above some datum plane normal to g, the work required to raise a particle of mass, m_p, density, ρ_p, in a medium of density ρ is:

$$w = m_p\left(1 - \rho/\rho_p\right)g,$$ (5.100)

so that

$$n(z) = n_0\exp\left[\frac{-m_p\left(1 - \rho/\rho_p\right)gz}{kT}\right],$$ (5.101)

known as the "barometric formula." This must equal the result of Eq. (5.93), so that:

$$\exp\left[\frac{-m_p\left(1 - \rho/\rho_p\right)gz}{kT}\right] = \exp\left[\frac{-m_p\left(1 - \rho/\rho_p\right)gz}{f D}\right],$$ (5.102)

from which it is seen that $kT = fD$, providing a proof of the Einstein diffusivity equation, Eq. (5.83).

4. Practical retrospective regarding sedimentation and other phoretic processes

It is clear that sedimentation, centrifugation and the analogous phoretic processes are among the most powerful for the separation of a colloid into its parts, i.e., the fractionation of the dispersoid particles as well as their separation from the medium. This applies to wastewater clarification in settling ponds, to the recovery of products produced in many types of liquid phase chemical reactors, particularly bioreactors, and to the preparative fractionation of particles by size. It is also seen to be the basis of important methods for particle size distribution measurement. From a practical point of view, examination of Stokes' Law, Eq. (5.50), reveals strategies for either enhancing or retarding sedimentation:

$$v_\infty = \frac{d^2\left(\rho_p - \rho\right)g}{18\mu}.$$ (5.50)

The rate of the process can be enhanced significantly by increasing the *size* of the sedimenting particles, as it is seen to depend to the second power on their diameter, d. The most effective way to increase d is to induce particle aggregation. While this is examined in detail in Chap. 7, it is evident from what has been discussed thus far that it may often be accomplished in aqueous media by the addition of sufficient electrolyte to reduce electrostatic repulsion between the particles. As will be seen later, this is done most

---

[61] The usual assumption of mean field theory is being used here, *i.e.*, the potential energy at a particular location is that of the field of mean force at that location.

efficiently using polyvalent ions of charge opposite to that of the charge of the particle surfaces. Since most surfaces in water carry a negative charge, this means polyvalent cations. The use of aluminum or ferric salts or cationic polyelectrolytes to induce settling is thus understood. They also produce voluminous flocks, which as they settle, sweep or scavenge smaller particles in their sedimentation path. The addition of electrolyte also reduces or eliminates the drag due to the electric field of the non-uniformly-distributed counterions around the particles during settling.

It has also been shown how particle separation is enhanced by replacing gravity with a much greater centrifugal acceleration. Another way to enhance the sedimentation rate is to subject the system to a hydrodynamic downflow, v, in which the effective velocity of settling becomes $v_\infty + v$. This is the basis of filtration processes and sieving, as well as the inertial impaction and the use of cyclones for separating smaller from larger particles. Sedimentation may be retarded, conversely, by immersing the system in a hydrodynamic *up*flow, with the imposed fluid velocity opposite in sign to the settling velocity. This is the basis for the process of *elutriation*, wherein smaller, lighter particles are separated from larger, heavier ones by the upflowing liquid.

Sedimentation may also be retarded by increasing the viscosity of the medium, μ, a practice often used in the determination of the size distribution of large, dense particles. To provide significant shelf life against sedimentation, however, one must modify the rheology of the medium more significantly by providing an adequate level of viscoelasticity to suspend the particles indefinitely. It has been found recently[62] that the "gel-trapping" of colloids against sedimentation requires that the dimensionless yield stress of the medium, Y, defined as

$$Y = \frac{\tau_y}{d(\rho_p - \rho)g},$$ (5.103)

where τ_y is the yield stress of the medium, and the other parameters are the same as in Eq. (5.50). Y must exceed a critical value of approximately 0.06, and the "loss tangent" of the medium, $\tan\delta$, defined as:

$$\tan\delta = \frac{G''}{G'}$$ (5.104)

where G'' and G' are the "loss modulus" and the "storage modulus" of the medium, respectively, must be less than 1.0. The rheology of colloidal systems is discussed in more detail in Chap. 8.

It should be appreciated, finally, that much of what has been described regarding sedimentation is applicable by analogy to the other phoretic processes.

---

[62] Laxton, P. B., and Berg, J. C., *J. Colloid Interface Sci.*, **285**, 152 (2005).

G. Measurement of particle size and size distribution: overview

1. *Classification of methods*

An enormous number of different techniques are available for providing information on particle size and size distribution in colloidal dispersions. Only a few of the most important of these are discussed in detail in this text, and usually in the context of the phenomenology being described. Comprehensive monographs on particle sizing are available, for example the two-volume series by Allen.[63] It is useful at this point to separate the methods by category, and a listing is given in Table 5-8.

Table 5-8: General methods of particle sizing.

1. Direct optical methods (microscopy)
   - Light microscopy ($d > 0.5$ μm)
   - Confocal microscopy ($d > 0.5$ μm)
   - Electron microscopy (1 nm $< d <$ 5 μm)
   - Nearfield scanning optical microscopy (NSOM) (1 Å $< d <$ 5 μm)
2. Indirect optical methods (light scattering)
   - Darkfield (ultra) microscopy ($d > 5$ nm)
   - Tubidimetry, nephelometry (10 nm $< d <$ 100 μm)
   - Classical light scattering (1 nm $< d <$ 10 μm)
        Rayleigh scattering
        Rayleigh-Gans-Debye (RGD) scattering
        Mie scattering
        Fraunhofer diffraction; laser diffraction
   - Dynamic light scattering (1 nm $< d <$ 3 μm)
3. Sedimentation methods
   - Natural sedimentation (creaming) ($d > 0.5$ μm)
   - Centrifugation ($d > 1$ nm)
4. Aperture (one-at-a-time) methods
   - Electro-zone (Coulter) method ($d > 0.5$ μm)
   - Photo-zone method ($d > 0.5$ μm)
   - Time-of-transition method ($d > 0.5$ μm)
5. Chromatographic methods
   - Hydrodynamic (size exclusion) chromatography ($d > 50$ nm)
   - Capillary hydrodynamic chromatography (CHDF) ($d > 50$ nm)
   - Field flow fractionation (FFF) ($d > 50$ nm)
6. Acoustic methods
   - Ultrasound amplitude attenuation (10 nm $< d <$ 1 mm)

---

[63] Allen, T., **Particle Size Measurement,** Vol. 1: Powder Sampling and Particle Size Measurement; Vol. 2: Surface Area and Pore Size Determination, Kluwer Publ., Norwell, MA, 1997.

2. Microscopy

Direct optical methods refer to microscopy, which has been discussed in Chap. 4 in the context of methods for examining solid surfaces and structures supported on solid surfaces, but will be reiterated and amplified briefly here as it applies to colloidal dispersions. What is generally desired for individual particles or aggregates is knowledge of their size, shape and structures, while for dispersions one seeks in addition, the volume (or mass) fraction of the dispersed phase, the particle size and spatial distribution and the state of aggregation. Microscopy examines directly the light reflected, refracted or fluorescing from the specimen, in contrast to the "scattering" methods, which examine the light that is re-emitted from them. Ordinary light microscopy accesses only the upper end of the colloid size range as its resolving power is between 0.3 and 0.5 μm in typical cases. For dispersions in liquids of particles in this size range, samples of the dispersion confined between a glass microscope slide and an overlying cover slip may be observed conveniently using an inverted microscope in which the sample is generally illuminated from above and observed with an upward-facing objective from below. Since the depth of field is generally of the order of one μm, crude optical sectioning may be carried out recording images at various sections through the sample.

Low optical contrast between the dispersed phase particles and the medium (as often occurs with emulsions, for which both the particles and the background are transparent) may be overcome using fluorescence markers, as mentioned in Chap. 4, but two additional methods for enhancing contrast not discussed there are interference contrast microscopy and differential interference contrast microscopy (DIC). (The latter is also termed Nomarski microscopy, after its inventor.) Simple interference contrast microscopy makes use of the fact that light passing through an object will have a different optical path length from that of the surrounding medium, due its differing refractive index. This difference is made visible by placing an annular-shaped optical stop in the illumination path ahead of the condenser and an annular "phase plate" at the conjugate plane behind the objective (actually built into the objective). The phase plate is thinner in the annular region where an image of the condenser annulus is formed. The un-phase-shifted light will pass through this thinner annular window, but where the light of differing optical path lengths is recombined, destructive interference occurs, and the corresponding image area appears dark. A special phase contrast objective must be used to implement phase contrast microscopy, but without the condenser annulus, the system behaves the same as in conventional microscopy.

Differential interference contrast (DIC) (or Nomarski) microscopy also makes use of optical elements placed ahead of the condenser and behind the objective, but employs polarization. The illuminating beam first passes through a plate that linearly polarizes it. Next the light passes through a modified Wollaski prism, which splits the beam into two parallel beams separated by a distance of a few tens of nanometers (well below the distance of optical resolvability) that sample optically adjacent points of the specimen. When the rays from the adjacent specimen points are recombined by a second prism located behind the objective they will produce an equivalent ray with the same linear polarization only if the refractive index is the same at the two adjacent specimen points. If the refractive index was even slightly different at the two points, the combined ray will be elliptically polarized. This difference is detected by a second polarizer, located behind the second prism and oriented 90° to the polarizer in the illuminating beam. The background, where no refractive index gradients exist, will appear dark (actually gray, because the second polarizer is slightly offset from 90° with the first), and the places where the particle edges exist will show gradients in brightness that are both sharp and give the final image a three-dimensional appearance.

Laser scanning confocal microscopy (a special type of fluorescence microscopy), discussed in Chap. 4, is a powerful means for providing precise optical sectioning so that the three-dimensional structure of either a dispersion or a microstructure may be obtained.[64] Its application to the study of dense colloids of non-spherical particles is especially compelling.[65]

The resolvability limitations (due to the wavelength of light in ordinary farfield illumination) of ordinary microscopy, which preclude its use for examining nano-particles, are overcome using the techniques of nearfield scanning optical microscopy (NSOM) or electron microscopy, both discussed in Chap. 4. Neither is conducive, however, to obtaining real-time observations of dispersions of particles that are labile, moving about or undergoing aggregation processes. A technique that may in some cases overcome such problems is that of *cryo*-electron microscopy, in which a labile sample is flash frozen by indirect contact with liquefied gases, then fractured and sometimes "etched" through sublimation of the frozen dispersion. The technique is thus also known as "freeze-etch" electron microscopy.[66] The etched sample is coated at cryogenic temperatures and the replica viewed in the microscope under ordinary conditions. The presumption is that the labile structures have been frozen in place, and that the freezing rate is sufficiently rapid that nucleation of ice crystals (or other solvent crystals) does not take place. Figure 5-44 shows cryo-electron micrographs of a dispersion of vesicles, as described in Chap. 3.

A method that is referred to as "microscopy," but is not one of direct observation, but instead based on the fact that colloid particles *scatter* light

[64] Sheppard, C. J. R., Hotton, D. M., and Shotten, D., **Confocal Scanning Laser Microscopy (Microscopy Handbooks)**, Bios Sci. Publ., Oxford, 1997.

[65] Mohraz, A., and Solomon, M. J., *Langmuir*, **21**, 5298 (2005).

[66] Severs, N. J., and Shotten, D. M., Eds., **Rapid Freezing, Freeze Fracture, and Deep Etching**, Wiley-Liss, New York, 1995.

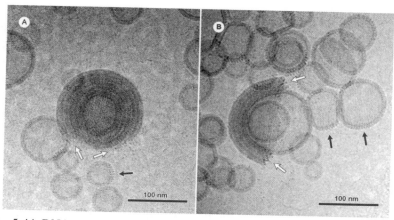

Fig. 5-44: DNA-coated unilamellar and multilamellar vesicles. Some of the structures display unclosed outer layers, indicated by white arrows. Black arrows indicate locations of parallel DNA helices. From [Huebner, S., Battersby, B. J., Grimm, R., and Cevc, G., *Biophys. J,*, **76**, 3158 (1999).]

(as described in the next section) is *darkfield* or *ultra*-microscopy.[67] Its inventor, Zsigmondy,[68] used it for tracking the motion of colloid particles an order of magnitude smaller than resolvable in an ordinary optical microscope, work for which he received the 1925 Nobel Prize in Chemistry. An ordinary optical microscope is used, but the sample is illuminated obliquely by means of a darkfield condenser, as shown in Fig. 5-45, so that the field of view in the microscope is dark, unless particles are present which scatter light up into the objective. These are visible as bright spots on the dark field. It is not possible to view details of the particle structure or sometimes even to be quantitative about its size, but the presence of particles as small as 5 nm can be detected. A flow version of the ultra-microscope has been used to count particles passing through a narrow beam of light and therefore to determine particle number densities in colloids.[69] In addition to providing a static or "snapshot" picture of the state of a colloidal dispersion, the various optical techniques which do not require "off-line" sample preparation can also be adapted for examining the dynamics of colloids undergoing phase segregation (sedimentation, electrophoresis, *etc.*), aggregation, coalescence, or size disproportionation.

A common feature of all of the direct methods (except darkfield microscopy) is that they have the advantage of revealing information on particle size, shape and structure as well as number. Modern image analysis capabilities make the handling of such data quite painless. The major difficulty with the direct methods is that they always involve a very small

[67] Zocher, H., and Stiebel, F., *Z. Phys. Chem.*, **147**, 401 (1930).

[68] Zsigmondy, R., **Colloids and the Ultramicroscope** (J. Alexander, Trans.), Wiley, New York, 1909.

[69] Davidson, J. A., Collins, E. A., and Haller, H. S., *J. Polym. Sci.*, Part C, No. **35**, 235 (1971).

sample size (even though they may contain several thousand particles), and statistical inferences made from such data may not be valid.

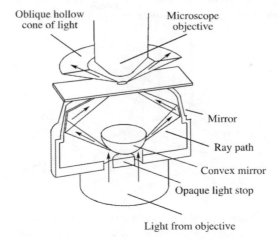

Fig. 5-45: Cardiod condenser for darkfield ultramicroscopy.

H. Light scattering

When particles are illuminated with electromagnetic radiation of wavelength comparable to or larger than the size of the particles themselves they scatter the radiation, *i.e.*, part of the incident photons are re-radiated ("scattered") in all directions, as suggested in Fig. 5-46. The scattering intensity I_θ measured at a given angle θ and distance from the scattering center is sensitive in general to the particle size, shape, refractive index, *etc.*, and is therefore rich in information regarding the dispersion. The phenomenon of light scattering in its many variations for use in measuring particle size, size distribution and other properties is discussed below.

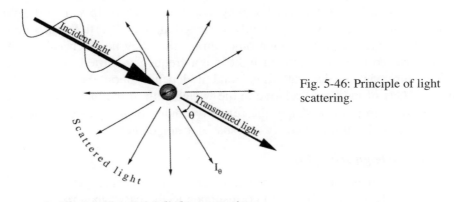

Fig. 5-46: Principle of light scattering.

1. *Classical (static) light scattering*

Recall first that light refers to that small part of the electromagnetic spectrum that is visible to the human eye, encompassing wavelengths of approximately 400–700 nm, and that electromagnetic radiation of all wavelengths consists of an oscillating electric field, **E**, together with the orthogonal magnetic field, **B**, it induces, traveling at the "speed of light," 3×10^8 m/s in a perfect vacuum, as pictured in Fig. 5-47. When light passes

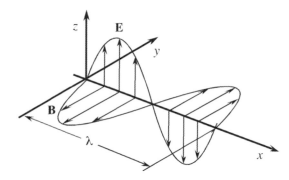

Fig. 5-47: Schematic of a monochromatic, vertically polarized light beam of wave length λ, consisting of an oscillating electric field component **E** and induced magnetic field **B**.

through a transparent medium (other than a perfect vacuum), its oscillating electric field induces synchronous oscillating dipoles in the molecules of the medium, which re-emit the radiation. As long as the medium is homogeneous, this effect is self-canceling, and the light is simply refracted. However, if the medium contains discrete variations in refractive index, as caused by the presence of particles, or of microscopic density or composition fluctuations, some of the incident energy is re-emitted from these inhomogeneities and detectable as "scattering" in all directions, as pictured schematically in Fig. 5-46. It may be manifest as visible opalescence or cloudiness (called the Tyndall effect), or it may require instrumentation for its detection.

Light incident on colloidal (*i.e.*, meso-heterogeneous) media may be either transmitted or scattered, but may also be *absorbed*. In standard light scattering techniques, it is generally assumed that the particles are non-absorbing (which requires among other things that they be electrically non-conductive) so that the scattered light is not frequency-shifted in any way. This is termed "elastic scattering," and it depends upon the size of the particles (say of radius, a) relative to the wavelength of the incident light, λ, as well as upon particle shape, the refractive index of the particles and the medium, particle number density and spatial arrangement.

2. Rayleigh scattering

The simplest situation is known as *Rayleigh scattering* (first described by Lord Rayleigh in 1871)[70] and corresponds to the case when $a \ll \lambda$, *i.e.*, for nanoparticles. The entire particle volume is in an effectively uniform

---

[70] Lord Rayleigh (J. W. Strutt), *Phil. Mag.*, **41**, 107; 274 (1871).

electric field at any instant (as suggested in Fig. 5-46). Our interest is in determining the intensity of scattered light resulting from the impingement of an incident light beam of intensity I_0 on a single spherical particle of radius a (for which $a << \lambda$) detected at some position away from the particle. Ordinary incident light consists of a mixture of different wavelengths ("white light") with the electric field and orthogonal magnetic field oscillations occurring in all directions perpendicular to the direction of propagation ("unpolarized light"), and the components oscillating in different directions out of phase with one another ("incoherent light"). It *is* possible to produce a beam of light of a single wavelength (using a monochromatic source or a polarizing filter), which oscillates in only a single direction, and in which all components are in phase (using a laser). Most lasers produce beams of monochromatic, polarized, coherent light.

Consider the scattering center (the particle) to be located at the origin of the coordinate system pictured in Fig. 5-48. It is illuminated by an incident beam, traveling in the $+x$ direction. The x-axis and the y-axis lie in a horizontal plane, called the "scattering plane," and the z-axis points vertically upward. There are a number of angles and directions for us to keep straight. The point of observation lies in the scattering plane at an angle θ with respect to the direction of propagation. This is the "scattering angle," and it may vary from 0° (forward scattering) to 180° (back scattering). More

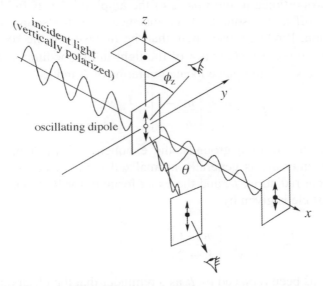

Fig. 5-48: Geometry used for the description of Rayleigh scattering. Incident light, plane polarized in the z-direction travels along the horizontal $+x$ axis and encounters the scattering center located at the origin. The horizontal x-y plane is the scattering plane, in which the scattered light is observed at some distance from the scattering center, and an angle θ (scattering angle) measured away from the line of incidence. An observation point *off* the scattering plane is also shown, located some distance from the scattering center and at an angle ϕ_z measured away from the z-axis.

generally one refers to $\theta < 90°$ as forward scattering and $\theta > 90°$ as back scattering. The point of observation (*i.e.*, the location of the photo-detector) is placed a distance r from the scattering center, with $r \gg a$.

For now, consider that the incident light is plane polarized in the z-direction, *i.e.*, the electric field vector **E** oscillates up and down, inducing a dipole in the particle which oscillates up and down, producing radiation of its own, in all directions, which is the scattered light. When the particle size is much smaller than λ, the entire particle at any instant finds itself in a uniform electric field, and the dipole (and the field) that is induced in the particle at any instant as a result of the instantaneous incident field is directly proportional to the polarizability of the particle. The latter is a product of the molecular polarizability α, and the number of molecules in the particle. The number of molecules in the particle is proportional to its volume, *i.e.*, to $(4/3)\pi a^3$, and it is known from the continuum theory of dielectrics[71] that the molecular polarizability α is proportional to $[(m^2 -1)/(m^2+2)]$ (the Clausius-Mossotti factor), where m is the refractive index of the particle relative to that of the medium, *i.e.*, $m = n_p/n_m$. As long as attention is confined to elastic scattering, these refractive indices are real numbers, independent of the wavelength of the irradiation. Some numerical values for it, and the corresponding Clausius-Mossotti factors for dispersions in water, are listed in Table 5-9 later in this chapter. The *intensity* (energy flux) of the scattered light is proportional to the *square* of the amplitude of its field. The intensity also falls off as the square of the distance from the source to the point of observation. Rayleigh also knew that the intensity of the scattered light, I_s, depended on the wavelength λ of the light in the medium so that at a given point in the scattering plane located in terms of θ and r:

$$I_s \propto |E|^2 \propto \frac{a^6 \lambda^b}{r^2} \left[\frac{m^2 -1}{m^2 + 2} \right]^2, \qquad (5.105)$$

where on dimensional grounds, the exponent "b" must be set to -4. The scattering intensity is generally normalized with respect to the intensity of the incident radiation, I_0, and the exact formula for Rayleigh scattering by a single particle is given by

$$\frac{I_\theta}{I_0} = \frac{16\pi^4 a^6}{r^2 \lambda^4} \left[\frac{m^2 -1}{m^2 + 2} \right]^2, \qquad (5.106)$$

where I_s has been replaced by I_θ as a reminder that the observation is made at the scattering angle θ. As an aside, the dependence of scattering intensity on λ explains why the sky is blue at midday and red at sunset. What is seen at midday is light scattered from particles (molecules) in the atmosphere; this is rich in the shorter wavelengths (blue). At sunset one sees mostly transmitted

---

[71] *cf.* Atkins, P. W., **Physical Chemistry**, 3rd Ed., p. 773, Oxford University Press, Oxford, 1982.

light, which has had the short wavelength portion scattered out, and is therefore enriched in the longer wavelength (red) portion of the spectrum.

The point of observation of scattered light has so far been limited to the scattering plane. A point of observation *off* the plane (say above it), can be designated in terms of the polar angle ϕ_z, as shown in Fig. 5-48. At this point, the dipole is observed to be oscillating in the $\pm z$ direction projected onto a surface perpendicular to the line joining the scattering center to the point of observation. The electric field induced in this plane will be the projection of the electric field vector onto this surface, *viz.* $E\sin\phi_z$, and the intensity becomes $I_\theta\sin^2\phi_z$. Thus if one looks straight down on the oscillating dipole ($\phi_z = 0°$), the scattered field intensity is zero, whereas if the viewing surface is parallel to the oscillating dipole ($\phi_z = 90°$), as is the case for any point of observation in the scattering plane, the scattered intensity is maximum ($\sin\phi_z = \sin 90° = 1$) at the value given by the Eq. (5.106).

The effect of having a point of observation of a vertically oscillating dipole off the horizontal scattering plane is geometrically identical to having a *horizontally* plane polarized incident beam (and hence a horizontally oscillating induced dipole: assume it is oscillating back and forth along the $\pm y$-axis) viewed at various angles in the horizontal scattering plane. For such a case, the scattering intensity is given by $I_\theta\sin^2\phi_y = I_\theta\cos^2\theta$. Un-polarized incident light propagating in the x-direction may be resolved into equal-amplitude components (producing two components, each of half-unit intensity) oscillating along the $\pm z$-axis and the $\pm y$-axis, so that the total intensity of the scattered light is given by the sum of that due to the two components, giving the Rayleigh Equation *viz.*

$$\frac{I_\theta}{I_0} = \frac{8\pi^4 a^6}{r^2\lambda^4}\left[\frac{m^2-1}{m^2+2}\right]^2 (1+\cos^2\theta). \qquad (5.107)$$

When all of the light from the source is vertically polarized, as from the laser sources in modern instruments, $(1 + \cos^2\theta)$ is simply replaced by 2.

Equation (5.107) can be generalized to correspond to the total intensity of scattered light not just from a single scattering center, but from *all* the particles in a given sample, provided the concentration of particles is sufficiently low that light scattered from a given particle passes directly to the detector without first encountering any other particles, *i.e.*, there is *single scattering*. This usually requires a particle volume fraction ≤ 0.01. Otherwise, one observes *multiple scattering*. One must also assume that the particles are randomly ordered in space so that regular patterns of constructive or destructive interference do not occur. This latter condition is called *independent* scattering. It is such interference (non-independence) in large specimens of pure liquids or crystalline solids that prevent net scattering by the sum of individual atoms or molecules of such media from being observed. If the conditions of single, independent scattering prevail, the scattering intensity I_θ observed from a volume V_s of a dispersion of particles, each of volume v_p, at a number density of ρ_N particles/volume, is just the sum of the scattering from each particle. Thus:

$$\frac{I_\theta}{I_0} = \frac{9\pi^2 \rho_N v_p^2}{2r^2 \lambda^4} \left[\frac{m^2 - 1}{m^2 + 2}\right]^2 V_s (1 + \cos^2 \theta). \tag{5.108}$$

A useful quantity independent of the sample volume, the distance of observation and the scattering angle is the *Rayleigh ratio*, R_θ:

$$R_\theta = \left(\frac{I_\theta}{I_0}\right) \frac{r^2}{V_s (1 + \cos^2 \theta)} = \frac{9\pi^2 \rho_N v_p^2}{2\lambda^4} \left[\frac{m^2 - 1}{m^2 + 2}\right]^2. \tag{5.109}$$

The intensity of scattered light can be measured in a simple device such as that shown schematically in Fig. 5-49. For colloids conforming to the requirements of the Rayleigh analysis, its value can be used to determine the particle size v_p, if ρ_N or the volume fraction $\rho_N v_p = \phi$ is known, or ρ_N, if v_p is known.

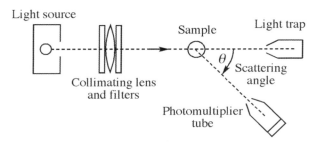

Fig. 5-49: Schematic of classical light scattering apparatus.

3. *Turbidity*

A useful (and simpler) measurement is that of the intensity of light *transmitted* through a dispersion of optical path length ℓ, *viz.*, I_t. Recall that the optical energy incident upon a sample is the sum of the energy transmitted plus that which is scattered out and that absorbed. (In the present case, we are assuming that the particles of interest are non-absorbing.) Assume a light beam of radius R illuminates a cylindrical volume of specimen: $\pi R^2 \ell$, where ℓ is the depth of the cuvette, as shown in Fig. 5-50. An energy balance on the illuminated volume gives:

$$\begin{array}{c} \text{incident} \\ \text{energy flow} \end{array} = \begin{array}{c} \text{transmitted} \\ \text{energy flow} \end{array} + \begin{array}{c} \text{rate of energy} \\ \text{scattered} \end{array}, \text{ or}$$

$$I_0 \pi R^2 = I_t \pi R^2 + \left(\frac{\text{scattered energy}}{\text{volume}}\right) \pi R^2 \ell. \tag{5.110}$$

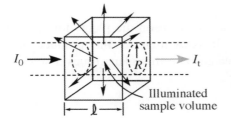

$I_0 \longrightarrow$ $\longrightarrow I_t$

Illuminated
sample volume

$\longmapsto \ell \longmapsto$

Fig. 5-50: Sample volume for
transmitted light and turbidity
measurement.

The total scattered energy/volume per unit incident intensity is defined as the turbidity, τ, of the sample. The above equation then gives (upon cancellation of πR^2 and division by I_0)

$$1 = \frac{I_t}{I_0} + \tau\ell, \text{ or } \frac{I_t}{I_0} = 1 - \tau\ell. \tag{5.111}$$

Since the fraction of the energy that is scattered out of the dilute specimen is very small, one may write:

$$\ln\frac{I_t}{I_0} = \ln(1 - \tau\ell) \approx -\tau\ell, \text{ or } \frac{I_t}{I_0} \approx \exp(-\tau\ell). \tag{5.112}$$

This last equation is the *Beer-Lambert Law*. The experimentally determined turbidity, using an ordinary spectrophotometer, is seen to be:

$$\tau = -\frac{1}{\ell}\ln\frac{I_t}{I_0}, \tag{5.113}$$

and its units are seen to be length^{-1}.

The turbidity of a sample of Rayleigh scatterers can be obtained by integrating the equation for I_θ, from Eq. (5.108), over all scattering angles, *i.e.*, over the area of a reference sphere of radius r:

$$\tau = \frac{1}{V_s} \int \left(\frac{I_\theta}{I_0}\right) dA_r, \tag{5.114}$$

where $dA_r = 2\pi r^2 \sin\theta d\theta$. The integration leads to:

$$\tau = \frac{24\pi^3 \rho_N v_p^2}{\lambda^4} \left[\frac{m^2 - 1}{m^2 + 2}\right]^2. \tag{5.115}$$

When particle size is sought, data should be obtained at several low concentrations and extrapolated to zero concentration. It is seen that the turbidity of a sample of Rayleigh scatterers leads to the same information as the Rayleigh ratio.

Another common method for expressing the turbidity (or by analogy the Rayleigh ratio) is in terms of particle *scattering cross sections*, C_s, *i.e.*,

$$\tau = \rho_N C_s, \tag{5.116}$$

where C_s is conceptualized as an area projected on the plane normal to the direction of the light propagation. It is written as the product of the geometrical area of the particle projected onto this plane times a *scattering efficiency factor*, Q_s. Thus for spheres,

$$C_s = \pi a^2 Q_s \tag{5.117}$$

For spherical Rayleigh scatterers, the scattering efficiency is

$$Q_s = \frac{8}{3}\left(\frac{2\pi a}{\lambda}\right)^4 \left[\frac{m^2 - 1}{m^2 + 2}\right]^2. \tag{5.118}$$

When the particles are not spherical, but otherwise satisfy the Rayleigh conditions, the intensity of scattering is greater at all angles, and the angular dependence for unpolarized light is no longer $(1 + \cos^2\theta)$, but the dependence on $\rho_N v_p^2 / \lambda^4$, is unchanged. The intensity, for dispersions of uniform ellipsoids, is obtained by multiplying by a correction factor, $C(\theta)$, called the "Cabannes factor," which depends on the particle shape and θ.[72] The appropriate correction for turbidity is denoted C_τ. A check of any angular dependence of scattering intensity will reveal nonsphericity.

When the dispersion is polydisperse, but otherwise satisfies the Rayleigh assumptions, the measurement of scattered intensity or turbidity yields a volume (or mass, or weight) average volume of the form:

$$(v_p)_v = \frac{\sum \rho_{Ni} v_{pi}^2}{\sum \rho_{Ni} v_{pi}}. \tag{5.119}$$

Rayleigh was also able to recast his scattering formula into terms of the refractive index of the dispersion, n_{disp}, rather than that of the particles and medium separately.[73] In terms of turbidity the result is

$$\tau_E = \tau - \tau_0 = C_\tau \frac{32\pi^2 n_{disp}^2}{3 N_{Av} \lambda_o^4} \left(\frac{dn_{disp}}{dc_2}\right)^2 c_2 M_2, \tag{5.120}$$

where τ_E is the excess turbidity of the dispersion over that of the medium, τ_0; λ_0 is the wavelength *in vacuo*, c_2 is the mass concentration of particles (or dissolved macromolecules) in g/cm^3 of medium, N_{Av} is Avogadro's Number, and M_2 is the molecular weight of the particle or macromolecule.

Dropping the Cabannes factor (*i.e.*, assuming particle sphericity), and neglecting the background turbidity of the medium, one may write the above equation in the form:

---

[72] Kerker, M., *Ind. & Eng. Chem.*, **60**, 31 (1968).
[73] For more details, see Hiemenz, P. C., and Rajagopalan, R., **Principles of Colloid and Surface Chemistry**, 3rd Ed., pp. 204-207, Marcel Dekker, NY, 1997.

$$\tau = Hc_2 M_2 ,\tag{5.121}$$

where H is the cluster of constants indicated by Eq. (5.120). One might also express this formulation in terms of the Rayleigh ratio, *viz.*:

$$R_\theta = K_\theta c_2 M,\tag{5.122}$$

where

$$K_\theta = \frac{3H}{16\pi}(1+\cos^2\theta) = \frac{2\pi^2 n_{disp}^2}{N_{Av}\lambda_o^4}\left(\frac{dn_{disp}}{dc_2}\right)^2(1+\cos^2\theta).\tag{5.123}$$

When the dispersion is truly a solution, the equivalent expression for the turbidity is[71]

$$\tau = -Hc_2\left(\frac{\bar{v}_1}{RT}\right)\left(\frac{\partial\mu_1}{\partial c_2}\right)_{T,p}^{-1},\tag{5.124}$$

where $\bar{v}_1$ is the partial molar volume of the solvent, and μ_1 is its chemical potential. Expanding for μ_1 in terms of the virial equation gives:

$$\tau = \frac{Hc_2}{\left(1/M_2 + 2Bc_2 + ...\right)},\tag{5.125}$$

where B is the second virial coefficient. The above equation can be rearranged and expressed in terms of the Rayleigh ratio as:

$$\frac{Hc_2}{\tau} = \frac{K_\theta c_2}{R_\theta} = \frac{1}{M_2} + 2Bc_2.\tag{5.126}$$

Quantities on the left hand side may be measured experimentally, so that a plot of Hc_2/τ_E or $K_\theta c_2/R_\theta$ *vs.* c_2 will yield a straight line of slope $2B$ and intercept $1/M_2$. One thus has a means of determining both the molecular weight and the second virial coefficient for a solution of macromolecules, as shown in Fig. 5-51.

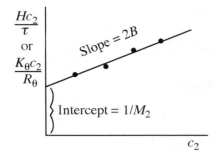

Fig. 5-51: Determination of the molecular weight and second virial coefficient for dissolved macro-molecules treated as Rayleigh scatterers.

4. Rayleigh-Gans-Debye (RGD) scattering

Rayleigh generalized his analysis to particles of arbitrary size and shape, provided the difference of refractive index between the particle and medium was not too large. Further contributions were later made by Gans and Debye, and the theory is now termed Rayleigh-Gans-Debye (RGD) scattering. The basis of the theory is shown in Fig. 5-52. The general criterion for the validity of the RGD approximation is:

$$\frac{2\pi d}{\lambda}|m-1| << 1,$$

(5.127)

where d is the largest linear dimension of the particle. The particle is thought

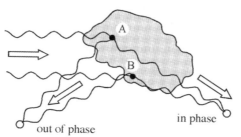

Fig. 5-52: Mutual interference between wavelets emanating from points A and B.

of as being subdivided into small volume elements, such as A and B, each of which is a Rayleigh scatterer. Their contributions to the scattering can be summed to give the net scattering intensity at any point of observation. The closeness of the refractive indices permits this summing because there is then little distinction between light traveling through the particle or the medium. It is evident that there will be mutual interference between the scattered rays emerging from the different volume elements. The net result is a reduction of the scattered intensity that may be expressed as:

$$\left(\frac{I_\theta}{I_0}\right)_{RGD} = \left(\frac{I_\theta}{I_0}\right)_{Rayleigh} P(\theta),$$

(5.128)

where $P(\theta) \leq 1$, is a "shape factor" or "form factor," which depends upon the size and shape of the particles and the wavelength of the light, as well as the scattering angle, θ. A useful quantity in expressing these relationships is the magnitude of the "scattering wave vector,"[74]

$$Q = \frac{4\pi}{\lambda}\sin\frac{\theta}{2}.$$

(5.129)

A general expression for the form factor is:

---

[74] The electrical field portion of an electromagnetic wave propagating in the direction $\mathbf{r}$ is given by

$$E(t,\mathbf{r}) = A\cos(\varphi + \mathbf{k}\cdot\mathbf{r} + \omega t),$$

where A is the amplitude, φ is the starting phase of the wave, ω is the angular frequency, and $\mathbf{k}$ is known as the *wave vector*. Its magnitude is the angular wave number $2\pi/\lambda$, and its direction is that of $\mathbf{r}$. If we designate $\mathbf{k}_0$ as wave vector of the incident radiation, and $\mathbf{k}_S$ as the wave vector of the scattered radiation (at the scattering angle θ), the scatter*ing* wave vector is defined as: $\mathbf{Q} = \mathbf{k}_0 - \mathbf{k}_S$. Its magnitude can be shown (as an exercise) to be given by Eq. (5.129).

$$P(Q) = \frac{1}{N^2} \sum_i \sum_j \frac{\sin(Qr_{ij})}{Qr_{ij}}, \qquad (5.130)$$

in which the scattering object has been subdivided into N differential elements with r_{ij} being the separation distance between the ith and j^{th} elements,[75] as suggested in Fig. 5-53. $P(Q)$ has been worked out for spheres, random coils, rods and other shapes, as shown. The analytical result for spheres of radius a is:

$$P(Q) = \left| \frac{3[\sin(Qa) - (Qa)\cos(Qa)]}{(Qa)^3} \right|^2. \qquad (5.131)$$

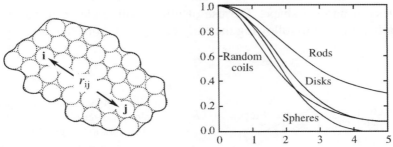

Fig. 5-53: Form factors for particles or macromolecules of various shapes. The abscissa is Qd/n_m, where d is the diameter for the sphere or disk, the length for the rod, and the radius of gyration for the random coil; n_m is the refractive index of the solvent. After [Marshall, A.G., **Biophysical Chemistry: Principles, Techniques and Applications**, Wiley, New York, 1978.]

It can be shown that for particles of various shapes, $P(Q)$ can be expressed as a power series in (QR_g), where R_g is the radius of gyration of the scatterer. The leading terms are:

$$P(Q) = 1 - \frac{1}{3}(QR_g)^2 + \dots O(QR_g)^4. \qquad (5.132)$$

For spheres of radius a: $\quad R_g = \sqrt{3/5}\ a$

For rods of length L: $\quad R_g = \frac{1}{2\sqrt{3}} L$

For dissolved macromolecules, R_g is the radius of gyration of the coil. Only these leading terms need be considered when Q (or θ) $\rightarrow$ 0. Taking logarithms of both sides of the scattering intensity equation yields

$$\ln I(Q) = \ln\left[1 - \frac{1}{3}(QR_g)^2 + \dots \right] + \text{const.} \approx -\frac{1}{3}(QR_g)^2 + \text{const.}, \qquad (5.133)$$

---

[75] More details are given in Hiemenz, P., and Rajagopolan, R., *loc. cit.*, pp. 216ff.

so that plots of log$I$ vs. Q^2 at low scattering angles have initial slopes of -1/3 R_g^2, and are called *Guinier plots*, as pictured schematically in Fig. 5-54. The region in which this analysis is valid (the Guinier region) is considered to be $(QR_g) < 1$. The technique when applied with a laser light source, is referred to as low-angle laser light scattering (LALLS). The use of Guinier plots to determine particle size and shape information through the form factor tacitly assumes monodispersity. When the system is polydisperse in particle size and/or shape, the de-convolution of the angular dependence of the scattering intensity is difficult.

One situation in which the effects of shape and polydispersity are avoided is at *extremely* small scattering angles (< 5°), or more generally, extremely small Q-values, because under these circumstances, $P(Q) \to 1$, regardless of particle shape. The use of ultra-small-angle scattering is thus especially powerful in following the early course of particle aggregation,

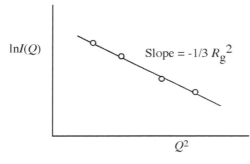

Fig. 5-54: Schematic of Guinier plot for determining particle size from low angle (low Q) laser light scattering (LALLS) measurements.

because the scattering intensity yields information on the time change of average particle size (volume) without the need to know the aggregate shape.[76]

It is evident that in the RGD region, static light scattering probes the structural details of scatterers of the order of Q^{-1} ([=] length) in size. Thus when Q^{-1} is significantly larger than the characteristic length dimensions of the particle (R_g in the above examples), a Guinier plot may be employed, and when $Q^{-1} >> R_g$, $P(Q) \to 1$, so that no structural detail is revealed. This is analogous to the general inability of electromagnetic radiation to probe structural details of particles significantly smaller than the wavelength of the radiation. At the opposite extreme, *i.e.* if Q^{-1} is significantly *smaller* than the characteristic length of the scatters, it will probe only small parts of the scattering structure. Ultimately, when Q^{-1} becomes so small that the only optical inhomogeneity of comparable dimensions is that associated with the surfaces (interfacial *layers*) of the particles, whose "thickness" is normally a few Å, one enters what is called the *Porod region*. Under these conditions, the form factor $P(Q)$ is proportional to Q^{-4}, and the measured intensity of

[76] Young, W. D., and Prieve, D. C., *Langmuir*, **7**, 2887 (1991).

scattered light is proportional to the total surface area in the sample (assuming the surfaces to be smooth). Thus in the Porod region:

$$I(Q) \propto (\text{total surface area})Q^{-4}. \tag{5.134}$$

Scattering measurements in the Porod region may thus be used to determine the surface area. If the surfaces are rough, the surface fractal dimension, d_{sf}, (see Chap. 4) may also be determined. This fractal dimension varies between 2 (for completely smooth surfaces) and 3 for a completely porous or spongy surface. The more general result for rough surfaces becomes

$$I(Q) \propto (\text{total surface area})Q^{-(6-d_{sf})}. \tag{5.135}$$

An important intermediate case exists in which Q^{-1} is between approximately 0.1 and 1.0 times the size of the scatterer. Analysis of the scattering intensity as a function of Q^{-1} in this range in principle reveals the internal structure of the scatterers. In particular, one may probe the internal structure of growing aggregates of primary particles falling in this size range, as discussed in Chap. 7. The truncated expansion for $P(Q)$ used in the Guinier region is no longer useful, but a simple alternative formulation of the expression for $I(Q)$ can be obtained for aggregates of fractal structure, viz.

$$I(Q) = \frac{I_{pp}}{(Qa_1)^{d_f}}, \tag{5.136}$$

where I_{pp} is the scattering intensity from the primary particle dispersion, and a_1 is the primary particle radius. Thus the slope of the plot of $\log I$ vs. log (Qa_1) yields, if it is a straight line (indicative of fractal structure), the (negative of the) fractal dimension of the aggregates, as shown in Fig. 5-55.

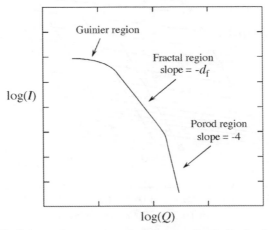

Fig. 5-55: Schematic representation of scattering behavior in the RGD region.

For the case of dilute solutions of dissolved macromolecules in the RGD range, one may modify Eq. (5.126) by dividing by the shape factor $P(\theta)$ to get:

$$\frac{K_\theta c_2}{R_\theta} = \left[\frac{1}{M_2} + 2Bc_2\right]\frac{1}{P(\theta)} \approx \left[\frac{1}{M_2} + 2Bc_2\right]\left[1 + \frac{16\pi^2}{3\lambda^2}R_g^2\sin^2\left(\frac{\theta}{2}\right)\right]. \quad (5.137)$$

Plots of the above equation may be prepared for various constant values of c_2 (mass concentration) at several scattering angles θ, permitting a double extrapolation to be made to $\theta \to 0$ and $c_2 \to 0$. Such a construction is known as a *Zimm plot*, and the classic example of it shown in many textbooks is reproduced in Fig. 5-56. (They seldom look this nice.) The constant (2000) in the abscissa is chosen arbitrarily to spread out the data. Results from the Zimm plot are obtained as follows:

$$(\text{slope})_{\theta=0} = 2B$$

$$(\text{intercept})_{\theta=0,\, c_2=0} = 1/M_2 \quad\quad\quad (5.138)$$

$$(\text{slope})_{c_2=0} = 16\pi^2 R_g^2/3\lambda^2 M_2$$

RGD scattering is, unlike Rayleigh scattering, not symmetrical about 90°. A convenient way to extract information from this fact is to determine the ratio of the scattering intensity at two angles symmetric about 90°, *viz.*

$$D_\theta = \frac{R_\theta}{R_{(180-\theta)}} = \frac{P(\theta)}{P(180-\theta)}. \quad\quad\quad (5.139)$$

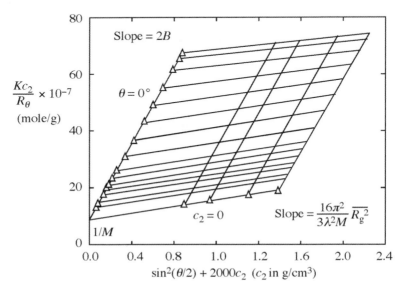

Fig. 5-56: Zimm plot for cellulose nitrate in acetone. After [Benoit, H., Holtzer, A.M., and Doty, P., *J. Phys. Chem.*, **58**, 635 (1954).]

This is the *dissymmetry* at θ. It takes on characteristic values for particles of different size and geometry and may be used to extract such information.

5. *Mie scattering*

For particles outside the RGD range, a general theory is required, involving the full solution of the Maxwell Equations of electrodynamics. The analytical solution for *spherical* scatterers of radii up to about the wavelength of the incident light is credited to Mie in 1908[77] (although Lorenz is generally recognized as having done it first). For spheres larger than those in the Rayleigh range, it yields an increasingly complex angular dependence of scattering intensity as the particle radii approach the wavelength of the light. It provides, when applicable, an exquisitely sensitive measure of particle size. A rough qualitative representation of the angular dependence of scattering intensity as it depends on the ratio of particle size to incident wavelength is shown in Fig. 5-57. As this ratio increases, forward scattering (low angle) is more and more favored over side and back scattering. In the Mie region, the pattern is often more complex than shown, containing many more lobes. A more accurate representation of this increasing complexity is shown in Fig. 5-58. To give an idea of the

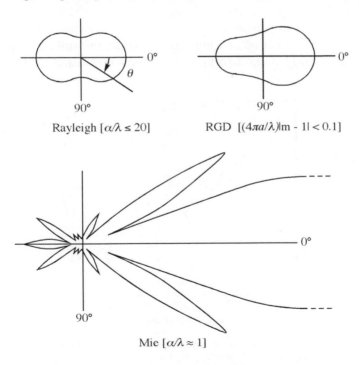

Rayleigh $[\alpha/\lambda \leq 20]$ RGD $[(4\pi a/\lambda)|m - 1| < 0.1]$

Mie $[\alpha/\lambda \approx 1]$

Fig. 5-57: Dependence of scattering intensity for unpolarized light as a function of scattering angle for different (spherical) particle sizes relative to the wavelength of the light.

[77] Mie, G., *Ann. Phys.*, **25**, 377 (1908).

structure of so-called Mie scattering, we may look at the result for the Rayleigh ratio for a vertically polarized incident beam and a point of observation in the horizontal plane, *viz.*:

$$R_\theta = \frac{\rho_N \lambda^2}{4\pi^2} \left| \sum_{n=1}^{\infty} \frac{2n+1}{n(n+1)} (a_n \pi_n + b_n \tau_n) \right|^2, \qquad (5.140)$$

where a_n and b_n correspond to the electric and magnetic multipole moments, respectively, and are composed of Ricatti-Bessel and Hankel functions and their derivatives. The only parameters the arguments of these functions contain are: $\alpha = 2\pi a/\lambda$ and the relative refractive index. The functions π_n and τ_n reduce to first-order Legendre polynomials but depend ultimately only on the scattering angle θ. A qualitative implication of Eq. (5.140) is that for (spherical) particles of a given size, the observed scattering intensity depends on both the scattering angle θ and the wavelength of the incident light, λ. It turns out that for a given θ, light of a given wavelength produces the most intense scattering. Thus if a monodisperse (or nearly monodisperse) colloid in the Mie range is illuminated with white light, the observed color of the scattered light will depend on the scattering angle. The phenomenon is referred to as a *higher order Tyndall spectrum* (HOTS). It was observed notably for the nearly monodisperse sulfur sols produced by La Mer (*cf.* Section C) by acidification of solutions of sodium thiosulfate. In principle, if a sol is sufficiently monodisperse, HOTS provides a means for sizing the particles.

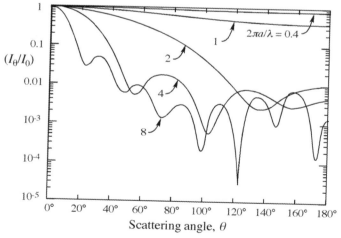

Fig. 5-58: Mie scattering intensity for vertically polarized incident and detected light, with $2\pi a/\lambda = 0.4$, 1, 2, 4 and 8.

The result corresponding to Eq. (5.140) for the turbidity of a dispersion containing ρ_N identical spheres per unit volume is

$$\tau = \frac{\rho_N \lambda^2}{2\pi} \sum_{n=1}^{\infty} (2n+1)\left\{|a_n|^2 + |b_n|^2\right\}. \tag{5.141}$$

Equation (5.141) can be used to show that the specific turbidity (τ/ρ_N) of a dispersion is maximum for a particular value of $\alpha = 2\pi a/\lambda$, depending on m. For $m = 1.1$ (*e.g.*, paraffin oil droplets in water), maximum specific turbidity occurs at $\alpha \approx 4.5$; for $m = 1.2$ at $\alpha \approx 7.5$, and for $m = 1.3$ at $\alpha \approx 14$. This corresponds to particles of diameter 0.7, 1.2 and 2.2 µm, respectively. It can be shown further that for sufficiently large particles (generally for $a \geq 2.5$ µm), turbidity is given by the formula:

$$\tau_{\alpha \to \infty} = 2\pi\rho_N a^2, \tag{5.142}$$

i.e., the scattering efficiency takes on a constant value of 2. Analysis of turbidity results for particles in this size range is thus quite simple. For further details, see *e.g.*, the classical texts by Kerker,[78] van de Hulst, and Bohren and Huffman.[79]

6. Fraunhofer diffraction; laser diffraction

When $a \gg \lambda$, *i.e.*, for very course dispersions, non-absorbed light is refracted through the particles, reflected from their surfaces or diffracted around their edges (Fraunhofer diffraction). The latter is an important basis for determining the size distribution in dispersions of particles larger than $\approx$ 3 µm,[80] but is different from the classical light scattering phenomena described above. Fraunhofer diffraction occurs when a collimated beam of light (generally produced by a laser point source located at the focal point of a collimating lens) passes through a field of particles in the appropriate size range, as shown in Fig. 5-59. The light is then brought to a focus by a second lens on a detector surface located at the rear focal point of the second lens. If a single quasi-spherical particle is located in the field, the collimated light will be diffracted about its edges, and a circular diffraction pattern will be produced on the detector surface centered at the point where the optical axis intersects that surface, regardless of the position of the particle along or off the optical axis. The diffraction pattern of concentric rings produced will be the inverse of that which would have been produced by a pinhole in a mask located in the particle field. The latter is given by the Airy function (named for the British astronomer George Airy, who first derived it):

$$I = I_0 \left[\frac{2J_1(x)}{x}\right]^2, \tag{5.143}$$

[78] Kerker, M., **The Scattering of Light**, Academic Press, New York, 1969.

[79] Bohren, C. F., and Huffman, D. R., **Absorption and Scattering of Light by Small Particles**, Wiley, New York, 1983.

[80] Weiner, B. B., "Particle and Droplet Sizing using Fraunhofer Diffraction" in **Modern Methods of Particle Size Analysis**, Chemical Analysis, Vol. 73, H. G. Barth (Ed.), p. 135, Wiley-Interscience, New York (1984).

where I is the local intensity, I_0 the incident intensity, J_1 is the first-order spherical Bessel function, and x is given by

$$x = \frac{2\pi a s}{\lambda f_F},$$ (5.144)

where a is the particle radius, s is the radial distance from the optical axis, λ is the wavelength of the light and f_F is the focal length of the focusing lens. The pattern of concentric rings is dependent uniquely upon the size of the diffracting particle, being broader for smaller particles. The actual pattern recorded by a set of concentric photosensitive rings will be the superposition of the patterns produced by all of the particles in the scattering volume, and for polydisperse samples, must be de-convoluted to yield a particle size (volume) distribution. The measurement is effected essentially instantaneously and so may be used to study aggregating and/or flowing

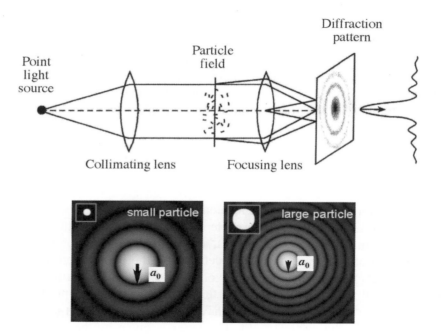

Fig. 5-59: Schematic of setup for observation of Fraunhofer diffraction, showing diffraction patterns characteristic of small and large particles.

dispersions. It is to be noted that Fraunhofer diffraction produces a different type of scattering data. Rather than a single measure of scattering intensity acquired at a particular point corresponding to a specific scattering angle or Q, a diffraction pattern, as pictured in Fig. 5-59, is acquired by a CCD device and analyzed pixel by pixel in the radial direction.

The generalization of Fraunhofer diffraction, known as *laser diffraction*, is applicable to the study of particles as small as 1 μm or less.

The sample distribution is illuminated by a broadened laser beam that scatters light into what is termed a Fourier lens. This lens focuses the scattered light onto a detector array and, using appropriate inversion algorithms (more complex than that of the Fraunhofer diffraction analysis), a particle size distribution is inferred from the diffracted light data. The analysis is limited to quasi-spherical particle geometries, and returns a volume-weighted distribution of equivalent sphere diameters. With refinements in both hardware and analytical methods (and the software to implement them) over the past two decades, laser diffraction has become the most widely used technique for particle size analysis. One of the most serious limitations of standard laser diffraction is its inability to accurately characterize sub-micron particles. This has been largely overcome using various strategies, one of which is termed *polarization intensity differential scattering* (PIDS). This method uses three different wavelengths of light (450, 600 and 900 nm) in vertical and horizontal planes of polarization to irradiate the sample. The PIDS detectors collect back-scattered light to provide what are believed to be unambiguous scatter patterns for different sizes of submicron particles. Enhanced laser scattering is easy to use, and in a number of commercial instruments has been combined with other (light scattering or microscopy) techniques to cover a size range from nanoparticles to those as large as several millimeters.

7. Inelastic scattering: absorbance; the Raman effect

If the particles of the dispersion absorb a fraction of the light incident upon them, the intensity of the transmitted light is reduced by both this absorption as well as scattering. The Beer-Lambert Law takes the form:

$$\frac{I_t}{I_0} = \exp(-\alpha_{ext}\ell), \tag{5.145}$$

where α_{ext}, the extinction coefficient is now the sum of both the turbidity and the absorbed fraction:

$$\alpha_{ext} = \rho_N(C_{sca} + C_{abs}), \tag{5.146}$$

where C_{sca} is the scattering cross-section (defined earlier), and C_{abs} is the absorption cross-section. The latter may be written in the usual way as

$$C_{abs} = \pi a^2 Q_{abs}, \tag{5.147}$$

with Q_{abs} = the absorption efficiency factor. Absorption occurs when incident radiation bumps the electrons in the medium (*i.e.*, the particles) to higher energy states. This always occurs in metals, as the conductance electrons are easily moved to higher energy states. In other cases, energy may be absorbed, for example, providing vibrational excitation of specific chemical bonds.

Absorption is frequency (or wavelength) dependent. For example, glass and water are opaque to ultraviolet light but transparent to visible light, and color is imparted to a material through the selective removal of certain wavelengths from an incident beam of white light. In order to account for absorption one must recognize that the refractive indices of materials are in general complex numbers, *i.e.*

$$n = n_1 + i\kappa, \tag{5.148}$$

where n_1 is the real part, $i = \sqrt{-1}$, and κ is the absorption coefficient. Both are frequency-dependent. Standard refractive index measurements are taken at the yellow sodium D line, 589 nm. For a number of dielectric materials, the absorption coefficient may be regarded as negligible over much of the visible spectrum, but for metals it is always of significant magnitude. Refractive indices are also weakly dependent on temperature, and the reference temperature is usually 20°C. Some values for the static (real) value of the refractive index of a number of dielectric materials at the reference wavelength and temperature are given in Table 5-9. The real and imaginary

Table 5-9: Some static refractive index values for dielectric materials at $\lambda = 589$ nm, $T = 20°C$.

Material	n_1	m (in water)	$2\pi/\lambda\|m-1\|$ (nm^{-1})	$(m^2-1)/(m^2+2)$
Vacuum	1 (exactly)	0.750	0.004152	-0.1706
Air (STP)	1.00003	0.750	0.004152	-0.1706
Water (liquid at 20°C)	1.333	1.000	0.000000	0.0000
Water (ice)	1.31	0.983	0.000287	-0.0115
Acetone	1.36	1.020	0.000337	0.0135
Ethanol	1.36	1.020	0.000337	0.0135
Teflon	1.35 – 1.38	1.013-1.035	0.000212-0.000586	0.0085-0.0234
Crown glass (pure)	1.50 – 1.54	1.125-1.155	0.002082-0.002581	0.0815-0.1004
Flint glass (pure)	1.60 – 1.62	1.200-1.215	0.003329-0.003579	0.1281-0.1372
Silica (natural quartz)	1.55	1.163	0.002706	0.1050
PMMA latex	1.49	1.118	0.001958	0.0768
PS latex	1.59	1.193	0.003205	0.1235
Paraffin oil	1.48	1.110	0.001833	0.0720
Diamond	2.419	1.815	0.013542	0.4332
Titanium dioxide	2.6 – 2.9	1.950-2.176	0.015798-0.019539	0.4832-0.5544
Silicon	4.01	3.008	0.033380	0.7285

parts of the complex refractive index of gold are shown in Fig. 5-60 as a function of wavelength. The rate at which electromagnetic energy is removed from the wave as it traverses the medium is dependent on the imaginary part of the refractive index, κ. The functional dependence of κ on the frequency depends on the material. Since the refractive indices of both the particles and the medium are complex numbers, the Clausius-Mossotti factor, $\left[(m^2 - 1)/(m^2 + 2)\right]$, is also a complex number. For nanoparticles, satisfying the Rayleigh criterion, it can be shown that the absorption efficiency factor is given by

$$Q_{abs} = \frac{8\pi a}{\lambda} \text{Im}\left\{\frac{m^2 - 1}{m^2 + 2}\right\},$$
(5.149)

where "Im" represents the imaginary part of the Clausius-Massotti function. The scattering efficiency factor is given by

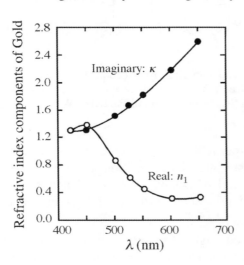

Fig. 5-60: The real and imaginary parts of the refractive index of gold. Data from [van de Hulst, H. C., **Light Scattering by Small Particles**, Wiley, New York, 1957.]

$$Q_{sca} = \frac{8}{3}\left(\frac{2\pi a}{\lambda}\right)^4 \text{Re}\left\{\frac{m^2 - 1}{m^2 + 2}\right\}^2,$$
(5.150)

where "Re" represents the real part of the square of the Clausius-Mossotti function. This result is identical to that given earlier for elastic scattering. The resulting scattering cross-sections for gold nanoparticles of three different sizes, as computed from the above data, data for water and formulas for Q_{ext}, Q_{abs} and Q_{sca} above are shown in Fig. 5-61. Note that as the particles become larger, the wavelengths of light selectively removed gradually increases. Thus transmitted light from the smaller-particle dispersions appears reddish (as is the case for Faraday's dispersions discussed in the opening chapter), while those of the larger particles appear more bluish.

Even under conditions when it can be said that elastic scattering prevails, a small fraction (about 1 in 10^7) of the light is scattered inelastically, so that this scattered fraction has a lower frequency (larger wavelength) than the incident light. This phenomenon is termed the *Raman effect*. The energy loss is generally due to changes in the vibrational energy of the molecules (although rotational, electronic or other energy modes may be involved as well) due, in turn, to changes in the polarizability of these

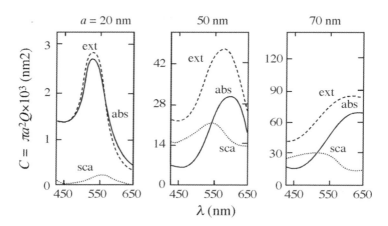

Fig. 5-61: Extinction, absorption and scattering cross-sections for spherical gold nanoparticles of different sizes as a function of wavelength (in air). Data from [van de Hulst, H. C., **Light Scattering by Small Particles**, Wiley, New York, 1957.]

vibrational modes. Thus bonds particularly susceptible to exhibiting the Raman effect are those with distributed electron clouds, such as the pi-electrons in carbon-carbon or other double bonds. Changes in the polarizability of strongly polar bonds, such as those of O-H in water are essentially negligible.[81]

The shift of the scattered light to lower energy and longer wavelengths reflects (in the language of quantum mechanics) an instantaneous shift to a virtual higher energy state followed by a nearly coincident de-excitation ("Raman scattering"), back not to the undisturbed state, but to one of higher energy. This difference between the incident energy and the scattered energy, in terms of wave number, is called the Stokes shift. The net energy absorbed is dissipated as heat, but this is immeasurably small. If a solution or dispersion of Raman scattering molecules or particles in a solvent medium of non-Raman active molecules is examined as a function of wave number, the spectrum obtained is a powerful tool for chemical analysis. Since the intensity of Raman scattering is in general but a small fraction of the elastic scattering signal, instrumentation for Raman spectroscopy requires a powerful source and an effective means for filtering out the elastic scattering signal.

A small fraction of the entities subject to Raman scattering are already in a higher state of excitation, and this fraction increases as the temperature is increased. These scatterers may be bumped to an even higher virtual energy state, followed by de-excitation to the initial state. The result is a

---

[81] The selection rules for Raman absorption are complementary to those for infrared absorption. In the latter case, absorption is governed by a change in dipole moment rather than polarizability. Bonds that are Raman active are generally IR-inactive, and vice versa. Thus the combination of the two spectroscopies provides a powerful analytical arsenal.

shift to *higher* wave numbers of this Raman scattered light, called the anti-Stokes shift. Since this is a tiny fraction of an already tiny fraction of the incident radiation, it is generally negligible. Instrumentation for its use as a non-invasive measurement of temperature, however, has been developed.

Some special situations exist that lead to significantly enhanced Raman signals. Raman spectroscopy is generally performed using lasers in the visible spectrum, where the wave numbers are below the first electronic transitions of most molecules. However, for a number of molecules, such as porphyrins, fullerenes (Buckyballs, *etc.*), polydiacetylenes, carotenoids and other biologically important species, the wavelength of the exciting laser is within the electronic spectrum of the molecule, leading to *resonance enhanced Raman scattering*, and signals 10^2-10^4 times larger than those of conventional Raman scattering. It is thus an especially powerful tool for such materials.

A second situation that is leading to increased research and instrument development is *surface-enhanced Raman scattering* (SERS). SERS is relevant to molecules (or entities) that find themselves adsorbed to or in the close vicinity (a few Å) of the surfaces of particular metals, specifically, silver, gold or copper. The "plasma wavelengths" of these metals are in the visible spectrum, and when the wavelength of the incident radiation is close to the appropriate value for the metal, the conductance electrons of the metal oscillate in synchrony with the light, leading to what is called *surface plasmon resonance*. A "plasmon" may be thought of as a swarm of these electrons sloshing back and forth with the incident oscillating electric field. The result of this situation is an enormous enhancement of the local field (typically by 10^3-10^6), with consequent enhancement of the Raman scattering from the molecules near these surfaces. This effect is non-specific, but is strongly dependent on the size and shape of the metal particles. It is greatest when the particles are in the nano size range (< 100 nm) and presenting sharp edges and corners. A second more specific SERS occurs when a charge transfer complex is established between the adsorbate molecule and the metal surface. The classic illustration of the phenomenon is for pyridine adsorbed to silver, but other aromatic oxygen or nitrogen-containing compounds, such as amines or phenols, are strongly SERS active. More detail regarding Raman scattering can be found in many sources.[82]

8. *Scattering from denser dispersions*

An underlying assumption in all of the foregoing is that the dispersions under investigation are sufficiently dilute that one has *single* scattering, *i.e.*, light scattered by any particle passes directly to the detector without being scattered by any other particles. Of course as the particle volume fraction increases, multiple scattering will occur, and this may be

---

[82] Schrader, B., **Infrared and Raman Spectroscopy**, B. Schrader (Ed.), Chap. 4, VCR Publ., New York, 1995.

436 INTERFACES & COLLOIDS

accounted for in a formal way with the use of an additional corrective factor, $S(\theta)$, or $S(Q)$, called the *structure factor*,[83] so that:

$$\left(\frac{I_\theta}{I_0}\right)_{RGD} = \left(\frac{I_\theta}{I_0}\right)_{Rayleigh} P(\theta)S(\theta). \qquad (5.151)$$

It is evident that the structure factor should depend upon the "structure" of the dispersion. Under ideal circumstances, *e.g.*, monodisperse, spherical particles, the angular dependence of the structure factor can be quite revealing. For "colloidal crystals," *i.e.* particles dispersed in a regular array (face-centered cubic, hexagonal close-packed, *etc.*), the appearance of the structure factor plotted as a function of the scattering angle magnitude Q has the damped-oscillation appearance of a radial pair distribution function. It can be shown in fact, that the Fourier transform of $S(Q)$ *is* the radial pair distribution function under these ideal circumstances. For the messier real situations in which there is size and shape polydispersity, it is extremely difficult to interpret structure factors, and for randomly structured dispersions, $S(\theta) \rightarrow 1$. Their use is thus primarily for getting insight into the structure of model dispersions and colloidal crystals.

In recent years, many efforts have been made to use static light scattering for the investigation of dense, randomly dispersed colloids. One set of devices, marketed under the name Turbiscan®[84], employs an illuminator-detector system in the form of a collar surrounding a cylindrical vessel. It moves up and down following the time course of aggregation, coalescence and sedimentation or creaming in the sample dispersion. Transmission (forward scattering) is automatically engaged for dilute dispersions, while back scattering (at 135°) is used for dense dispersions. Analysis of the latter depends on proprietary algorithms based on Mie theory but calibrated against experimental data.

9. Dynamic light scattering (photon correlation spectroscopy)

Another technique based on light scattering was developed following the invention of the laser in 1961. Termed "dynamic light scattering" (DLS), it has become one of the routine methods for particle size measurement, generally applicable to sub-micron particle dispersions, and many instruments using it are commercially available. It also goes by other essentially equivalent names such as "quasi-elastic light scattering" (QELS), and "photon correlation spectroscopy" (PCS), reflecting the common way the technique is implemented.[85] In classical (static) light scattering, it is the time-averaged scattering intensity that is observed, and the magnitude is just

[83] The form factor, defined earlier, is sometimes called an "internal structure factor," often leading to confusion.
[84] http://www.turbiscan.com/home/lab_present1.htm
[85] Weiner, B. B., "Particle Sizing using Photon Correlation Spectroscopy," in **Modern Methods of Particle Size Analysis**, Chemical Analysis, Vol. 73, H. G. Barth (Ed.), p. 93, Wiley-Interscience, New York (1984).

the algebraic summation over the intensity due to each particle. This simple summation occurs because the individual particles are randomly arrayed at any instant, and the lack of coherence in the incident light means that phase relationships are maintained only over a few intermolecular spacings (<< scattering volume dimensions). Thus all interference effects are "averaged out." In dynamic light scattering, measurement of the particle movement is carried out. There are two modes for making such measurements.

One of the modes is based on the Doppler effect. Since the particles in the sample are moving about (by Brownian motion), the scattered light may appear to have a *slightly* different frequency than the incident light, higher if the particle is moving toward the detector, and lower if moving away. Thus the scattering appears not to be precisely elastic, and is termed "quasi-elastic." As a result of the Doppler effect, the scattered light intensity exhibits a frequency dependence given by the Lorentzian distribution function:

$$I_s(\omega) = A \frac{DQ^2}{(\omega - \omega_0)^2 + (DQ^2)^2},$$ (5.152)

where ω is the frequency of the scattered light, ω_0 is the frequency of the incident light, Q is the magnitude of the scattering vector, D is the Brownian diffusivity and A is an instrument constant. It is depicted in Fig. 5-62, in which it is shown that the half-width of the distribution at the point where the scattered light intensity is half its maximum corresponds to DQ^2. Thus in principle one could use a power spectrum analyzer to determine the diffusivity of the particles, and through the Stokes-Einstein Equation, the particle size. In practice, however, this is impossible because the Doppler broadening amounts to only about 10^3 Hz, in comparison with the frequency of the incident light of approximately 10^{16} Hz. The problem is solved by mixing the scattered light with a portion of the unscattered incident beam, which amounts to the superposition of two beams of *nearly* the same frequency. Thus one may write for the mixed beam:

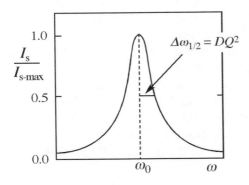

Fig. 5-62: Lorentzian distribution of scattered light intensity resulting from the Doppler effect.

$$I_{\text{s-mix}} = \tfrac{1}{2}BI_s + \tfrac{1}{2}BI_0 = \tfrac{1}{2}B\cos\omega t + \tfrac{1}{2}B\cos\omega_0 t, \qquad (5.153)$$

where B is a constant. This expression may be recast in terms of the sum and difference of these frequencies, *i.e.*, the average frequency: $\omega_{\text{av}} = \tfrac{1}{2}(\omega + \omega_0)$, and the modulation frequency: $\omega_{\text{mod}} = \tfrac{1}{2}(\omega - \omega_0)$, in terms of which one may write $\omega = \omega_{\text{av}} + \omega_{\text{mod}}$ and $\omega_0 = \omega_{\text{av}} - \omega_{\text{mod}}$, so that

$$I_{\text{s-mix}} = \tfrac{1}{2}B\cos(\omega_{\text{av}}t + \omega_{\text{mod}}t) + \tfrac{1}{2}B\cos(\omega_{\text{av}}t - \omega_{\text{mod}}t), \qquad (5.154)$$

and recalling a trig identity, and noting that $\omega_{\text{av}} \approx \omega_0$:

$$I_{\text{s-mix}} = [B\cos(\omega_{\text{mod}}t)]\cos(\omega_{\text{av}}t) \approx B_{\text{mod}}(t)\cos(\omega_0 t). \qquad (5.155)$$

Thus the mixed beam oscillates with essentially the same frequency as the incident beam, but with an amplitude which oscillates at the much lower frequency of ω_{mod}, as shown schematically in Fig. 5-63. These oscillations are referred to as *beats*, and they are readily measured in the technique known as *heterodyning*. The phenomenon is analogous to that which is audible (in a Freshman physics demonstration) when two tuning forks of nearly the same characteristic frequency are sounded together. While used in a number of research laboratories, heterodyning is not available in commercial instrumentation for a variety of practical reasons.[86]

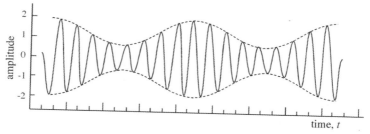

Fig. 5-63: Modulated wave produced by the mixing of two waves of nearly the same frequency.

The more commonly used mode of dynamic light scattering monitors the fluctuations of scattered light intensity due to time-varying interference between the light scattered from the various particles in the sample volume as they move about by Brownian motion. With a narrow-beam laser light source and a small scattering volume, an observer would see an intensity resulting from the vectorial sum of the scattering from each of the particles in the scattering volume, as pictured in Fig. 5-64. The intensity observed would depend on the nature of the interference (constructive or destructive) between the scattered rays emerging from the individual particles, and this in turn depends on the mutual positions of the particles. Because the particles

[86] Lebedev, A. D, Ivanova, M. A., Lomakin, A. V., and Noskin, V. A., *Appl. Optics*, **36**, 7518 (1997).

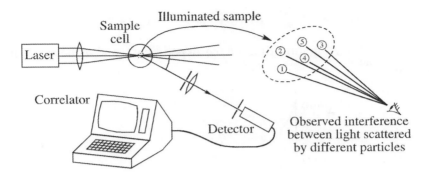

Fig. 5-64: Principle of photon correlation spectroscopy (PCS).

are moving about by Brownian motion, their mutual positions are changing, and the observed intensity exhibits fluctuations on a time scale related to the diffusion rate, examples of which are shown in oscilloscope traces of Fig. 5-65 for samples containing particles of three different sizes. It is evident that the smaller particles produce fluctuations of higher frequency. Over a time interval, τ, beginning at time t, and short relative to the changes in particle positions, two values of the measured intensity would "correlate" strongly, i.e., $I(t)$ and $I(t + \tau)$ would be nearly the same. If many pairs of intensity values were measured over this short time interval, the average value of the intensity product, $I(t)I(t+ \tau)$, would be very nearly equal to $\langle I^2(t)\rangle$, the average of the *square* of the instantaneous intensity. On the other hand, if the

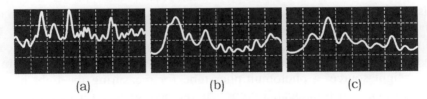

(a)   (b)   (c)

Fig. 5-65: Effect of the size of polystyrene latex spheres on the time fluctuations of scattered light intensity. Diameters of spheres: (a) 0.085 mm; (b) 0.220 mm; (c) 1.011 mm.

intensities are measured at the beginning and end of a time interval *large* compared to the time required for the particle positions to change significantly, then the intensity-values would be essentially "un-correlated," and the average value of the intensity product, $I(t)I(t+ \tau)$, would be just the square of the long-time averaged intensity, *viz.*, $\langle I(t)\rangle^2$. At intermediate time intervals, the intensity-values would be *partially* correlated.

The extent of such correlation may be quantified by the "auto-correlation function" for the intensity of scattered light, as pictured in Fig. 5-66, and defined as:

$$G(\tau) = \lim_{T \to \infty} \frac{1}{T} \int_0^T I(t)I(t + \tau)dt = \langle I_s(t)I_s(t + \tau) \rangle, \qquad (5.156)$$

where T is the integration time of the experiment, and τ the time shift or delay time.

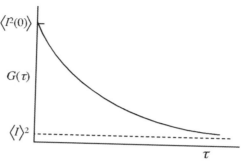

Fig. 5-66: Schematic of an intensity autocorrelation function.

The autocorrelation function for a given dispersion may be obtained in a dynamic light scattering apparatus, as shown schematically in Fig. 5-64, with analog processing by multiplying the intensity signal $I(t)$ with a delayed version of the signal $I(t+\tau)$ and averaging the product over the time T. As the experimental time T is extended, $G(\tau)$ evolves into an unchanging function for a given dispersion. The method is known as *homodyning*, to contrast it with heterodyning, in which the scattered light is mixed with incident light. For particles in the size range between nanometers and microns in media of viscosity near that of water, the relevant time range is between microseconds and milliseconds. Thus dynamic events, such as aggregation, whose time constants are large relative to this acquisition time, may be tracked by dynamic light scattering. The essential features of the equipment are the laser light source, a photomultiplier tube for measuring the light intensity fluctuations and a dedicated computer (autocorrelator) for determining the intensity correlation over various time intervals. The photo-multiplier is essentially counting photons, and the technique is referred to as "photon correlation spectroscopy" (PCS). The resulting auto-correlation function, as it develops over time, appears as shown in Fig. 5-66.

The usefulness of $G(\tau)$ is that it can be directly related to the particle diffusivity as follows for a dispersion of monodisperse particles:

$$G(\tau) = A_0 + A\exp(-\Gamma\tau), \qquad (5.157)$$

where A_0 is the background signal level, A is an instrument constant, and Γ is the decay constant. What is important is that the measured decay constant gives the particle diffusion coefficient in accord with:

$$\Gamma = Q^2 D, \qquad (5.158)$$

where Q is the magnitude of the scattering wave vector defined earlier, *i.e.*, $Q = (4\pi/\lambda)\sin(\theta/2)$.

When D is evaluated, we can use the Stokes-Einstein Equation to compute the particle radius, a, assuming spherical shape:

$$a = \frac{kT}{6\pi\mu D}. \tag{5.159}$$

It is seen that dynamic light scattering provides the modern, more powerful means for carrying out Perrin's experiments yielding D-values from Brownian motion.

The effective particle radius so obtained is a *hydrodynamic radius* and is highly sensitive to the presence of adsorbed polymers or oligomers at the particle surfaces or to the presence of a diffuse electrical double layer, as discussed in Chap. 6. PCS is limited to dilute dispersions ($\phi \le 0.001$), and to obtain the infinite dilution value for D generally requires a lower value of ϕ to assure independence between the particles.

When the sample is polydisperse, the auto-correlation function is given by

$$G(\tau) = \int_0^\infty g(\Gamma)e^{-\Gamma\tau}d\Gamma , \tag{5.160}$$

which represents the superposition of the exponentials for the particle sizes in the distribution and in which $g(\Gamma)$ is the normalized distribution of the decay constants, Γ. It is not easy to unambiguously extract the particle size distribution function from $G(\tau)$, but software is available for fitting non-negatively constrained normal, log-normal and some other distributions. For fairly narrow, mono-modal distributions, one may expand the logarithm of the auto-correlation function in terms of the moments of the distribution in what is termed the *method of "cumulants."*

$$\ln[G(\tau)] = \left[-\overline{\Gamma}\tau + \frac{\mu_2\tau^2}{2!} - \frac{\mu_3\tau^3}{3!} + \ldots\right], \tag{5.161}$$

where $\overline{\Gamma}$ is the mean decay constant, and μ_2, μ_3, *etc.* represent the second, third and higher moments of the decay constant distribution about its mean. These quantities are related in somewhat complicated ways to the mean, standard deviation, *etc.* of the particle size distribution. PCS yields an unusual average radius (the inverse Z-average), and one that is quite highly sensitive to the presence of outsized particles. A discussion of the various averages involved in dynamic light scattering is given by Thomas,[87] and further general discussion of dynamic light scattering is given by Berne and Pecora,[88] and a review by Finsy.[89]

[87] Thomas, J. C., *J. Colloid Interface Sci.*, **117**, 187 (1987).
[88] Berne, B. J., and Pecora, R., **Dynamic Light Scattering**, Krieger, Malibar, FL, 1990.
[89] Finsy, R., *Adv. Colloid Interface Sci.*, **52**, 79 (1994).

One of the important applications of photon correlation spectroscopy derives from its ability to follow time changes occurring in a dispersion, such as those resulting from aggregation or coalescence, as discussed in Chap. 7, or to the accumulation of polymeric adsorbate. PCS measurements taken as a function of time give the effective average hydrodynamic radius of the particles as a function of time.

In addition to the measurement of temporal intensity fluctuations due to the light scattered from a given dispersion sample volume, a dynamic version of laser diffraction may also be employed.[90] The illumination of the particles by laser light results in a diffraction pattern showing a ring structure created by diffraction on the particles, as shown in Fig. 5-67, but also shows a fine structure from the diffraction between the particles, *i.e.* its near-order. As the particles move about, their mutual positions change with time, resulting in changes of the phases and thus in fluctuations in the fine structure of the diffraction pattern, *i.e.*, the intensity at any given point in the diffraction pattern fluctuates with time. For static laser diffraction, these fluctuations are averaged out. The fluctuations can be analyzed in the time domain by a correlation function analysis or in the frequency domain by frequency analysis, and the two methods are linked by Fourier transformation.

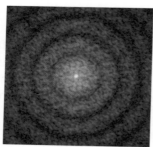

Fig. 5-67: Speckle pattern produced in dynamic laser diffraction.

10. *Dynamic light scattering from denser dispersions*

Returning to the question of the degree of dilution of the dispersion, a high degree of dilution in PCS measurements is required for two reasons: (1) to avoid multiple scattering, and (2) to avoid the influence of neighboring particles on particle diffusion. It is not always clear which of these demands is the more severe, but it is usually a matter of particle size. The first consideration usually requires $\phi < 0.001$, for particles smaller than 1 μm, but less when larger particles are present. The second factor depends on the particular system. When particle-particle interactions can be neglected (as assumed, except when they aggregate), one gets the so-called *self*-diffusion coefficient (diffusion of a single particle in an infinite medium), which is described by the Stokes-Einstein Equation. When such interactions do exist, they have two origins: (1) long-range van der Waals, electrostatic or other

[90] Asakura, T., and Takai, N., *Appl. Phys. A, Matls. Sci. & Proc.*, **25**, 179 (2004).

interactions described in Chap. 7, or (2) hydrodynamic effects. When either or both factors are important, one obtains a *mutual* or *collective* diffusion coefficient, D_m, which is dependent on particle concentration. Hydrodynamic effects (as given by the effective friction factor) and attractive inter-particle forces act to retard motion, and hence to cause D_m to be less than D, whereas repulsive inter-particle interactions cause D_m to increase with concentration. When multiple scattering can be neglected (as for very small particles), data over a range of particle concentrations may be extrapolated to zero to yield the self-diffusion coefficient, *i.e.*, $D_m(\phi) \to D$, as $\phi \to 0$.

Instrumentation has been developed[91] for applying PCS to dense dispersions. Referred to as *photon cross correlation spectroscopy* (PCCS), it uses a double set of incident beams, detectors and correlators to produce two identical-Q correlation functions, permitting the determination of a 3D *cross* correlation that is used to eliminate the effects of multiple scattering.[92-93] Called "two-color dynamic light scattering" (TCDLS), it employs two laser light sources of different wavelengths (green and blue). Scattered light from the sample is viewed simultaneously at two different angles such that the scattering angle is the same for light scattered from each source, and it is shown that the signal that is the cross-correlation of the two is that due only to single scattering. The ratio of the single-scattered light to the total scattered light, as given by their respective autocorrelation functions, produces a *dynamic structure factor* $S_{dyn}(Q)$, and hence information on the mutual diffusivity of the particles. The challenge, of course, is to determine the relationship between $S_{dyn}(Q)$ and the dispersion structure. Another technique that has been proposed is that of collecting PCS data from the tip of a fiber optic probe immersed in the dispersion. Only particles near the probe tip face are observed, in back-scattering mode. The method has been called fiber optical quasi-elastic light scattering (FOQELS).[94]

A relatively recent development is that of *diffusing wave spectroscopy* (DWS).[95,96] In this technique, multiple scattering is exploited rather than avoided. The autocorrelation function for light that is presumed to have undergone a statistically large number of scattering events in traversing the sample is obtained. The path of the light is considered analogous to a diffusion path (hence the name), as pictured schematically in Fig. 5-68. The attenuation of the time-dependent correlation function due the Brownian

[91] http://www.sympatec.com/PCCS/PCCS.html

[92] Schäzel, K., Drewel, M., and Ahrens, J., *J. Phys.: Cond. Matter*, **2**, SA393 (1990).

[93] Stieber, F., and Richtering, W., *Langmuir*, **11**, 4724 (1995).

[94] Tanaka, T., and Benedek, G. B., *Appl. Optics*, **14**, 185 (1975).

[95] Pine, D. J., Weitz, D. J., Chaikin, P.M., and Herbolzheimer, E., *Phys. Rev. Lett.,* **60**, 1134 (1988).

[96] David A. Weitz, D. A., and Pine, D., in **Dynamic Light Scattering**, W. Brown (Ed.), pp. 652-720, Oxford Univ. Press, New York, 1992.

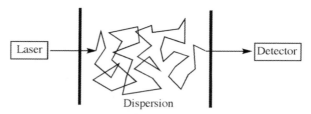

Fig. 5-68: Schematic of diffusing wave spectroscopy (DWS).

motion of the particles is determined for each ray and the total effect summed up. DWS may be operated in either transmission or backscattering modes, and the equipment is essentially the same as that for conventional PCS. So far the method has been applied to concentrated, random dispersions,[97] colloidal crystals and foams.[98]

I. Aperture, chromatographic and acoustic methods for particle sizing

Last on the list of methods for particle sizing given in Table 5-8 are a few that don't fit into the categories of sedimentation/centrifugation or optical techniques discussed thus far. They offer a number of specific advantages in many cases, and most of them are available as commercial instruments.

1. Aperture (one-at-a-time) methods

Aperture methods refer to situations where the particles of a small sample of a dispersion are made to flow single file through a small aperture where they are counted and sized one at a time. The most venerable among these is the "electro-zone" method, known originally as the Coulter method, and the instrument the Coulter Counter, after the earliest commercial device for its implementation. It finds particular use for the counting blood cells in medical laboratories. The particles to be analyzed are suspended in a dilute saline solution. As shown in Fig. 5-69, the instrument contains the dispersion to be analyzed both in an external chamber and in a tube connected to it by a small aperture near the bottom of the tube. In operation, a measured volume (usually 1 cm³) of the dispersion is pulled through the aperture between the two chambers by retracting the piston at the top of the tube. During the flow of the sample through the aperture, the electrical resistance between the two chambers is monitored. Each time a particle passes through the opening, a brief spike in the resistance is recorded. The

---

[97] Maret, G., and Wolf, P. E., *Z. f. Physik B (Cond. Matter)* **65** (4), 409 (1987).

[98] Duran, D. J., Weitz, D. A., and Pine, D. J., *Science,* **252**, 686 (1991).

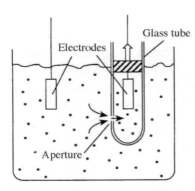

Fig. 5-69: Electrozone (or Coulter) method.

larger the particle volume, the larger will be the spike. These electrical pulses are counted and sized to yield the number density and the volume-weighted size distribution, as well as the number concentration of the particles in the dispersion. Tubes with apertures of various sizes are available, so that the technique may used for particles between approximately 0.5 μm and 1 mm. Key to the success of the method is the avoidance of aggregates or multiple numbers of primary particles passing through the aperture at the same time. In the newer instruments, the shape of the electrical pulse is analyzed to allow rejection of signals from multiple particles passing at the same time. The electrozone method has the advantage of not requiring optical transparency of the dispersion, and may therefore be used for particle sizing in opaque media, such as inks. It has the disadvantage of being limited to aqueous media of appropriate conductivity.

A closely related aperture technique uses optical sensing rather than electrical resistance to monitor and size particles flowing through a narrow aperture. The detector may consist of a narrow beam of collimated light passing through the sensing zone where the passage of particles interrupts the beam, causing a reduction in the amount of light reaching a photo-detector on the other side. The amount of light blockage is proportional to the cross-sectional area of the particle passing through. In another variation, the light source is a laser, and the quantity measured is the amount of low-angle forward-scattered light occurring as the particle traverses the beam. Realistically, these systems can size particles down to ≈ 0.5 μm at best, and it is critical to work with very dilute systems to avoid the simultaneous passage of multiple particles (coincidence error). A common technique is to use a flow system in which the serum is circulated through the sensor, with a filter in the line, until the background count is sufficiently low. Then a small amount of the sample dispersion is injected upstream of the sensor, and its contents analyzed as it passes through. In newer instruments, this process is carried out automatically. An advantage of the photozone in contrast to the electrozone methods is that it may be used in non-aqueous, *i.e.*, non-electrolyte systems. An advantage of both types of aperture methods is that they can provide information for complex, *i.e.*, multimodal types of distributions.

Another one-particle-at-a-time method uses a laser beam to rapidly scan a dilute dispersion or an aerosol, while a photodiode placed directly behind the dispersion detects a reduction in the beam intensity as the beam traverses a particle. Using the speed with which the beam is scanned across the dispersion and the time that its intensity is reduced, the size of the particle can be estimated. The technique is thus referred to as the *time-of-transition* method.[99] Many thousands of particles may be scanned per second, and an algorithm accounting for the probability that the beam is traversing the full breadth of the particle or just a portion of it is applied to produce a particle size distribution. Current instruments are capable of analyzing particles as small as 0.5 μm.

2. *Chromatographic methods*

A method analogous to size exclusion or gel permeation high performance liquid chromatography (HPLC) for the separation and analysis of polymer solutions by molecular weight can be applied to the separation and analysis of polydisperse colloids. Called "hydrodynamic chromatography[100]" (HDC), it is pictured schematically in Fig. 5-70. As the dispersion passes through a packed bed of particles, the smaller ones are

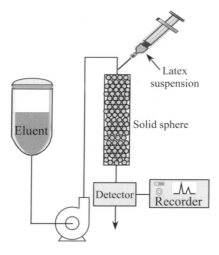

Fig. 5-70: Schematic of setup for hydrodynamic chromatography of polydisperse colloidal dispersions.

retained for longer periods of time in the void spaces. The particles are thus eluted according to size, with the larger particles exiting first. A detector is required to sense the concentration of particles as they are leaving, and the common ones used are spectrophotometers (measuring optical extinction at a particular wavelength) or refractive index detectors. Both detect the volume

[99] Weiner, B. B., Tscharnuter, W. W., and Karasikov, N., **ACS Symposium Ser.**, No. 693, p. 88, 1998.

[100] Small, H. J., *J. Colloid Interface Sci.*, **48**, 147 (1974).

concentration of the particles as they pass through the unit, and the detector output is plotted against the volume of eluent passing through the column to produce an elution curve as shown in Fig. 5-70 or 5-72(b). The total retention volume, V_R, i.e., the volume required to reach the peak of the elution curve, for particles of a given size is determined by calibration. Problems with the method include its relatively low resolution, the necessity for calibration standards, and most seriously, the fact that not all the sample injected may exit from the packed column, i.e., some the particles may stick to or become entrapped in the packing. A similar method that appears to overcome some of the limitations of HDC is "capillary hydrodynamic fractionation" (CHDF),[101] available commercially. In this technique, analogous to capillary chromatography, a sample of the dispersion is injected into the carrier liquid flowing through an empty small-bore capillary tube. Brownian particles suspended in Hagen-Poiseuille (parabolic) flow in the capillary, as pictured in Fig. 5-71, sample all radial positions accessible to them. Considering the location of the particle as characterized by its center of gravity, however, a particle is excluded from a zone next to the capillary wall equal to the particle radius. Thus on average, larger particles spend a greater proportion of their time near the center of the capillary, where the velocity is highest, and are eluted first.

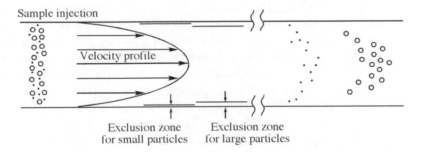

Fig. 5-71: Capillary hydrodynamic fractionation (CHDF).

A variation on capillary chromatography is "field flow fractionation" (FFF), invented by Giddings.[102] It employs sedimentation or centrifugation in combination with a flow that is normal to the settling direction,[103] and is pictured in Fig. 5-72(a). The sample of dilute dispersion is injected into a spinning annular channel. The centrifugal field sets up a gradient of particles in the channel with the larger (heavier) particles near the bottom (outside) surface and the smaller ones layered on top. The particles near the center of the channel (the smaller ones) are swept out first by virtue of the greater velocity at the center of channel, and the sample is thereby fractionated with

[101] DosRamos, J. G., and Silebi, C. A., J. Colloid Interface Sci., **135**, 165 (1990).

[102] Giddings, J. C., Karaisakis, G., Caldwell, K. D., and Meyers, M. N., J. Colloid Interface Sci., **92**, 81 (1983).

[103] Caldwell, K. D., in **Modern Methods of Particle Size Analysis**, H. G. Barth (Ed.), pp. 211-250, Wiley, New York, 1984.

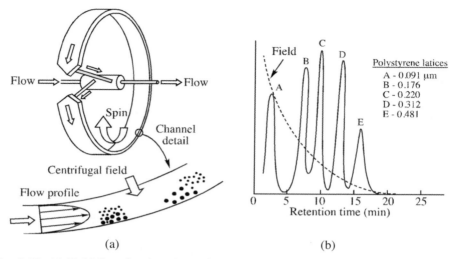

Fig. 5-72: (a) Field flow fractionation unit. Smaller particles are eluted ahead of larger particles. (b) Analysis of a sample of mixed polystyrene latices.

respect to the position of the particles in the field. Separation is enhanced by use of a programmed variation in the centrifugation rate during the run. The eluent shows the particles separated into peaks representing different sizes, as shown in Fig. 5-72(b). The sedimentation (centrifugation) FFF described is only one of the possibilities. Other types of fields can be used to produce various types of phoretic motion that establish a steady-state particle profile across the flow channel when balanced against Brownian motion or diffusion. In thermal FFF a temperature gradient is established between the walls of the flow channel, and the particles take up positions across the channel based on their thermophoresis. Brownian motion tends to move particles from the hot toward the cold regions, based on the increased intensity of Brownian motion with temperature, but thermophoresis depends also on the chemical composition of the particle surfaces,[104] some compositions being more effective thermal absorbers than others. Electrical FFF employs the profile established by the competing effects of electrophoresis and Brownian motion, and magnetic FFF would use the profile of magnetophoresis against Brownian motion, *etc*. The various phoretic process may also be put in competition with sedimentation/centrifugation to establish profiles across the flow channel to effect separation.

3. Acoustic methods

Finally, ultrasound attenuation techniques are being used to yield information about size distribution in dispersions.[105] The principles of the

[104] Ratanathanawongs, S. K., Shiundu, P. M., and Giddings, J. C., *Colloids Surfaces A*, **105**, 243 (1995).
[105] Povey, M. J. W., **Ultrasonic Techniques for Fluids Characterization**, Academic Press, New York, 1997.

method are based on the measurement of the loss of amplitude of a monochromatic, longitudinal acoustic pressure wave generated using a quartz crystal activator and sensed with a quartz crystal transducer. Attenuation results principally from, in addition to intrinsic energy absorption by the system, viscous losses caused by relative movement of the medium with respect to the particle caused in turn by the density difference between them. A second mechanism, termed one of "thermal losses," is the result of the temperature response to the pressure waves at the particle interfaces. This is especially important for "soft" particles, such as emulsion droplets or latices. In terms of the applied frequency, ω, three ranges of behavior can be distinguished. At intermediate frequencies the attenuation, α, is greatest when the shear wavelength (relaxation distance of the velocity profile at the particle surface, or "viscous depth") is comparable to the particle radius, a. At lower frequencies, both the particles and the medium can follow the sound wave with little relative motion, while at higher frequencies, there is little motion of either the particles or the medium. The typical measurement provides the attenuation $vs.$ frequency, a trace that can be fit to model responses to yield particle size. As an example, for submicron monodisperse particles at low to intermediate frequencies:

$$\frac{\alpha}{\phi} = \frac{1}{9}\omega^2 a^2 \frac{\rho}{c\mu}\left(\frac{\rho_p}{\rho}-1\right)^2,$$

(5.162)

where ϕ is the volume fraction of particles, c is the sound velocity, μ is the medium viscosity, and ρ and ρ_p are the medium and particle densities, resp.

It is apparent that acoustic methods have the advantage of not requiring transparency of the medium, and it is also true that they can be applied to dense as well as dilute ($\geq$ 1 wt%) dispersions, and that with a single instrument, particle sizes from 10 nm to 1 mm can be accommodated. They have the disadvantage of requiring input data on the heat capacities, thermal conductivities, compressibilities, acoustic absorption of the materials, and the presence of air bubbles in the dispersion renders the technique inoperable because of their ability to absorb acoustic energy. Nonetheless, these techniques are often used in conjunction with the measurements of the electrokinetic properties of the dispersion, as discussed in Chap. 6.

Some fun things to do:
Experiments and demonstrations for Chapter 5

1. Synthesis of a colloid of sulfur particles in water and its interaction with light

A nearly monodisperse colloid of elemental sulfur in water may easily be prepared by one of the bottom-up methods of La Mer.[106] A solution of sodium thiosulfate ($Na_2S_2O_3$) is acidified with sulfuric acid (H_2SO_4) (other acids may also be used), leading to the precipitation of elemental sulfur in accord with:

$$Na_2S_2O_3 + H_2SO_4 \rightarrow Na_2SO_4 + H_2O + SO_2\uparrow + S\downarrow$$

Colloidal particles slowly grow from nuclei of only a few atoms each to diameters of a few hundred nm over a period an hour or two. Eventually the particles aggregate into clusters (as described in Chap. 7) and sediment to the bottom of the vessel in a day or so, as sulfur (sp. gr. ≈ 2) is denser than water.

The model colloid produced above can be used to illustrate some basic aspects of light scattering. In the early stages (up to 30 min after initiation), the particles are only a few nanometers in diameter, and are therefore in the Rayleigh region. The scattering intensity, I, of unpolarized white light is given by

$$I \propto \frac{\rho_N v_p^2}{\lambda^4}(1 + \cos^2\theta),$$

where ρ_N is the particle number density, v_p is the particle volume, λ is the wavelength of the incident light and θ is the scattering angle (see Fig. 5-46). The light observed at a scattering angle of 90° will be essentially all scattered light, while that observed at 180° will be largely transmitted light. In the early stages of colloid particle growth, the colloid illuminated from the side with white light will take on a bluish appearance, since, in accord with the above equation, the shorter wavelength (blue) part of the spectrum is preferentially scattered. When observed head-on ($\theta \approx 180°$), the colloid will appear yellowish to reddish, since the blue portion the spectrum has been scattered out. This is the same as the explanation for a blue sky at midday and a yellow-red sunset.

Various methods may be used to illuminate the colloid for observation. The growing colloid in a beaker can be placed on an opaque mask with a 1-cm wide slit on an overhead projector, as shown in Fig. E5-1. As the particles nucleate and grow, the turbidity and its bluish hue will become evident in a few minutes when viewed from the side. It will

---

[106] La Mer, V. K., and Barnes, M. D., *J. Colloid Sci.*, **1**, 71 (1946).

intensify with time as both the particle number density and the particle volume grow. The projected image of the slit will begin to take on a yellowish-to-orange hue, as this light is transmitted ($\theta \approx 180°$), and the shorter wavelength light has been scattered out. Another way to detect the presence of the colloid, even before the turbidity is evident to the naked eye, is to pass a laser beam through it and observe the turbidity in the beam, known as the Tyndall cone. Finally, a narrow collimated beam of white light, as produced by a fiber optic illuminator, may be passed through the colloid which can be viewed at various angles, primarily in the back-scattering range ($90° < \theta < 180°$) to observe changes in the color of the scattered light, known as higher order Tyndall spectra (HOTS). This phenomenon is generally observable only when the colloid consists of highly monodisperse spherical particles.

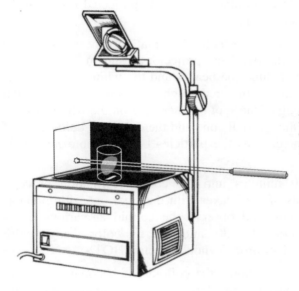

Fig. E5-1: Set-up for observing colloid formation and optical properties.

Materials:

- Laser pointer
- Overhead projector and screen
- 250-mL beaker
- Sodium thiosulfate (3M solution) (10 mL)
- Sulfuric acid (3M solution) (10 mL)
- Cardboard sheet large enough to cover the overhead projector, with window measuring 1× 5 cm at its center.
- Fiber Optic Illuminator producing a 1-2cm diameter beam (*e.g.*, Model 190, Dolan-Jenner Ind., Inc., Woburn, MA). This produces a narrow beam of collimated white light.
- Two pieces of cardboard (≈6×6 in.), one white, one black, joined at the edge.

Procedure:

1) Prepare the sodium thiosulfate and sulfuric acid solutions as specified above, in small beakers.

2) Cover the surface of the overhead projector with the cardboard with the window described above, and place the 250-mL beaker over the hole in the cardboard.

3) Add 200 mL of water to the beaker. Add 0.5 mL of 3M sulfuric acid to the water in the beaker. Gently stir. Then add 0.5 mL of 3M sodium thiosulfate. Gently stir.

4) Turn on the overhead projector and turn off the room lights. Over the next 10 minutes, notice the gradual appearance of bluish turbidity in the beaker viewed from the front and the gradual development of a yellow-to-orange hue in the projected image of the 1×5 cm window on the screen.

5) Periodically during the colloid formation process, stand the cardboard sections on the overhead projector as shown in Fig. E5-1, with the black surface behind the beaker and the white surface to the side. Then shine the laser pointer through it to observe the Tyndall cone and the defocusing of the spot of laser light on the white surface. Both the intensity of the Tyndall cone and the degree of defocusing of the laser spot will increase as the particles become more numerous and grow in size.

6) At about 10 minutes into the colloid formation process, illuminate the contents of the beaker with the fiber optic illuminator, turn the room light out, and observe the colloid at various back-scattering angles from 90° to 180° and note the color changes with angle. These are the higher order Tyndall spectra (HOTS).

7) Note that after 2-3 hours the colloid has become whitish in color (as the particles have become larger and begun to aggregate) and that after 24 hr, has sedimented to the bottom of the beaker.

2. Synthesis of monodisperse silica particles via the Stöber process

Monodisperse, spherical silica is an ideal model system for investigation of various colloidal phenomena. The particles can be synthesized by the well-known Stöber method[107] which uses an alkoxide precursor, most often tetraethoxy orthosilicate (TEOS), reacting with water in an alcohol (usually ethanol) medium with a basic catalyst (usually ammonia). The process begins with hydrolysis of TEOS in a basic solution of water and alcohol:

---

[107] Stöber, W., Fink, A., and Bohn, E., *J. Colloid Interface Sci.*, **26**, 62 (1968).

$$Si(OC_2H_5)_4 + 4H_2O \xrightarrow{\text{NH}_3} Si(OH)_4 + 4C_2H_5OH$$

Condensation of the tetrahydroxide occurs following and concurrent with its formation in accord with

$$Si(OH)_4 \rightarrow SiO_2 \downarrow + H_2O$$

to produce colloidal silica particles.

In this experiment, silica particles with a diameter of ≈ 600 nm are synthesized. By changing the reactant concentrations, the type of alcohol and silicon precursor used, a wide range of particle sizes ranging from a few nanometers to a few microns in diameter can be obtained.

Materials:

- 6 mL of tetraethoxy orthosilicate (TEOS)
- 10 mL of ammonium hydroxide solution (29%)
- 12 mL of de-ionized water
- 75 mL of absolute ethanol
- 200-mL Erlenmeyer flask with a stopper
- Magnetic stirrer and stir bar
- Bench top centrifuge
- Fume hood

Procedure:

1) In a fume hood, add water, ethanol and ammonium hydroxide solution to the flask in the amounts specified above. Start the magnetic stirrer.

2) Add TEOS to the flask and cap it with a stopper. The system should become cloudy and opaque within 10 minutes.

3) After 3 hours, separate the particles by centrifugation. The particles can then be re-dispersed in water or solvent. The diameter of the particles will be approximately 600 nm.

4) If a microscope is available, view a sample of the colloid under darkfield illumination. The particles will be visibly exhibiting Brownian motion. If a microscope is not available, view the Tyndall cone formed by the colloid in a test tube, as shown in Fig. E5-2, in a darkened room. The light in the cone is seen to be flickering as a result of Brownian motion.

3. Sedimentation of a monodisperse colloid

The particles of a monodisperse colloid will all settle at the same rate, observable as the rate at which the particle front descends, $-(dh/dt)$ (the "falling curtain"). A variety of suitable latices can be synthesized by dispersion polymerization or purchased for demonstration of this phenomenon, and their size can be estimated using Stokes' Law, Eq. (5.50) in the form:

$$d = \sqrt{\frac{18\mu(-dh/dt)}{(\rho_p - \rho)g}} \quad .$$

Materials:

- Monodisperse latex in the size range of 5-15 μm. An example is poly(methyl methacrylate-*co*-ethylene glycol dimethacrylate) of diameter 8 μm, available from Sigma Aldrich (Product #463167). These particles have a density of 1.19 g/mL.
- 25-mL graduated cylinder, marked in 0.2-mL delineations.
- Kitchen timer

Procedure:

1) Calibrate the graduated cylinder. For the usual 25-mL cylinder, the distance between the 0.2-mL hash marks is 0.8 mm.

2) Place about 1 g of dry latex in the cylinder, fill with water to exactly the 0-line, shake thoroughly and let stand.

3) Set the timer for 20 minutes (depending the size and density of latex used). When the timer sounds, measure distance the particle front has descended. Calculate the rate of descent and hence the particle size.

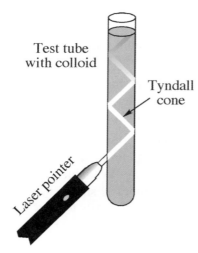

Fig. E5-2: Viewing a Tyndall cone with a laser pointer. Close examination of the cone reveals flickering that results from Brownian motion of the particles.

Chapter 6

ELECTRICAL PROPERTIES OF INTERFACES

A. Origin of charge separation at interfaces

1. *Overview*

One of the most important aspects of interfaces is the electrical charge separation that exists there. It usually plays a decisive role in the aggregation stability of aqueous colloids, in which it is evident that the approach of two particles in Fig. 1-3 would be resisted by electrostatic forces. Interfaces rarely carry a *net* charge when the whole region of inhomogeneity that constitutes the interfacial layer is taken into account. For example, when a particle surface acquires a charge, an excess of ions of opposite charge (counterions) gather in the vicinity of the surface and neutralize it, as shown in Fig. 1-3. In aqueous media, this is usually accomplished within a few tens of nanometers of the surface. In media of lower dielectric constant, the distance may be much greater. The structure formed in the vicinity of the interface is termed an "electrical double layer," and is described in terms of the surface charge density, $\overline{\sigma}_o$, the surface potential, ψ_0, relative to the bulk solution, and the potential and ion concentration profiles, $\psi(x)$ and $C_i(x)$, respectively, where x is the distance measured away from the interfacial plane. There are three major issues: 1) the origin(s) of electrical charge separation at interfaces, 2) the structure of the resulting double layer, and 3) the electrostatic forces between surfaces bearing such double layers.

A number of different mechanisms may give rise charge separation at aqueous interfaces, as listed in Table 6-1. These are described below, followed by a brief discussion of charging at non-aqueous interfaces.

Table 6-1: Origins of charge separation at aqueous interfaces.

---

1. Preferential adsorption/desorption of lattice ions
2. Specific adsorption of ions
3. Ionization of surface functional groups
4. Isomorphic substitution
5. Accumulation/depletion of electrons

---

2. Preferential adsorption/desorption of lattice ions

A mechanism relevant to surfaces of sparingly soluble crystalline materials, such as the silver halides, is that of preferential adsorption (or desorption) of ions from the surface lattice of those ions, as shown in Fig. 6-1. AgI has been much studied[1] and can serve as an example. If AgI crystals

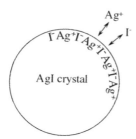

Fig. 6-1: Example of interface charging through preferential adsorption or desorption of lattice ions.

are slurried into pure water, I^- ions adsorb onto the surface of the lattice more strongly than Ag^+, and the surface acquires a negative charge given by:

$$\overline{\sigma}_0 = F(\Gamma_{Ag^+} - \Gamma_{I^-}),\qquad(6.1)$$

where $\overline{\sigma}_0$ is the surface charge density, and F is Faraday's constant, and the Γ's are the molar adsorptions of the ions. This situation is manifest by the existence of a surface potential, ψ_0, (in this case, a negative one) relative to the bulk solution far from the surface. Of course an equivalent and indistinguishable result could be attributed to preferential *de*-sorption (or escape) of Ag^+ ions into solution. From either point of view, the situation is described thermodynamically by the requirement that the effective chemical potential of the ions (Ag^+ or I^-) be the same in the crystal lattice as in the solution far from the surface. When equilibrium is established, there will in general be an imbalance in the ion population in the crystal surface creating an electric field normal to the surface. The ions (like Ag^+) possess potential energy (hence, free energy) in part by virtue of their position in this field. The effective chemical potential of the silver ions is thus the "total" or *electro*-chemical potential of the ions, depending on both the electrical potential where it is located and on the local chemical environment. Guggenheim suggested splitting the total chemical potential up into its "chemical" and "electrical" parts as:

$$\tilde{\mu}_i = \mu_i + z_i F\varphi,\qquad(6.2)$$

where z_i is the valence of the ion, φ is the local electrical potential, and F is, again, the Faraday constant (the magnitude of charge associated with one

[1] Lyklema, J., in **Colloidal Dispersions**, J. W. Goodwin (Ed.), Roy. Soc. of Chem., Spec. Pub. 43, pp. 47-70, London, 1982.

mole of electrons, *viz.*, 96,519 Coul). For the ion in the surface of the solid, one may write

$$\tilde{\mu}_i^S = \mu_i^{\ominus S} + z_i F \varphi_s , \tag{6.3}$$

where $\mu_i^{\ominus S}$ is the standard state *chemical* potential in the solid.[2] For the ion out in the bulk solution

$$\tilde{\mu}_i^L = \mu_i^{\ominus L} + RT \ln C_i + z_i F \varphi(x), \tag{6.4}$$

where $\mu_i^{\ominus L}$ is the standard state *chemical* potential in the liquid and φ is the electrical potential out in the solution. This potential is not measurable in any absolute sense. Therefore it is set equal to zero far from the surface ($x \rightarrow \infty$). Relative to this reference, the potential is designated as ψ, and its value at the surface is designated as ψ_0. Thus the equations (6.2) – (6.4) take the form:

$$\tilde{\mu}_i = \mu_i + z_i F \psi, \tag{6.2a}$$

$$\tilde{\mu}_i^S = \mu_i^{\ominus S} + z_i F \psi_0, \tag{6.3a}$$

and deep into the solution ($x \rightarrow \infty$),

$$\tilde{\mu}_i^L = \mu_i^{\ominus L} + RT \ln C_i . \tag{6.4a}$$

Equating the total electro-chemical potentials, one obtains:

$$\psi_0 = \frac{(\mu_i^{\ominus L} - \mu_i^{\ominus S})}{z_i F} + \frac{RT}{z_i F} \ln C_i, \tag{6.5}$$

which shows how the surface potential depends on the concentration of either of the lattice ions (called the *potential-determining ions,* or *pdi*'s) in the solution. These may be adjusted by the addition to the solution of electrolytes that contain them, *e.g.*, in this case $AgNO_3$ or KI. Equation (6.5) is the *Nernst Equation*,[3] and it establishes the general link between chemical change and electrical potential change. The constant it contains (*i.e.*, the first term on the right hand side) may be evaluated experimentally by noting that there should be *some* bulk concentration of ion i which will make the surface potential, ψ_0, equal to zero. This concentration is termed the *point of zero charge*, C_{pzc}, or *PZC*. Colloid titrations (described later in this chapter) may be used to determine the conditions under which ψ_0 is zero and thence to the C_{pzc} for either of the potential-determining ions. The Nernst Equation thus takes the form:

---

[2] This assumes that the potential in the interior of the solid is the same as its value at the surface, ψ_0. The true situation is a bit more complicated: ψ_0 generally differs from the interior potential by an amount χ, which depends, *inter alia*, on the orientation of water (or other solvent) dipoles at the surface.

[3] Nernst, W., Z. *Physik. Chem.*, **4**, 129 (1889).

$$\psi_0 = \frac{RT}{z_i F} \ln \frac{C_i}{(C_i)_{PZC}}. \tag{6.6}$$

For the silver ions of AgI, the *PZC* is found to be 3×10^{-6} M (or in terms of pAg_{pzc}, 5.52). For the iodide ions, the *PZC* is 2.5×10^{-11} M. The two concentrations are related by the solubility product: $C_{Ag}C_I = K_{sp} = 7.5 \times 10^{-17}$ M^2. For the practical use of the Nernst Equation in the form of Eq. (6.6), it is useful to note that

$$\frac{RT}{F} \equiv \frac{kT}{e} = (\text{at } T = 298°K) \ 25.7 \text{ mV}, \tag{6.7}$$

where e is the protonic charge.

Once the concentration of the potential-determining ions is set, the surface potential is fixed and constant with respect to the solution composition of other ions (called *indifferent electrolyte*). The silver halides are thus examples of *constant-potential* surfaces, and the surfaces of other sparingly soluble salts behave similarly.

Oxides are often treated as constant potential surfaces with OH⁻ and H⁺ ions regarded as potential-determining. The *PZC* is thus conveniently specified as a *pH* value, and the surface potential is given by

$$\psi_0 = \frac{RT}{z_{H^+} F} \ln \frac{[H^+]}{[H^+]_{pzc}} \equiv \frac{2.303..RT}{(1) F} \left(pH_{pzc} - pH \right)$$

$$\xrightarrow{T=298K} 59.2 \left(pH_{pzc} - pH \right) \text{ mV} \tag{6.8}$$

This is not strictly correct, as OH⁻ and H⁺ are not lattice ions, but the Nernst Law usually holds for them over at least one or two *pH* units. This behavior is due to the presence of surface hydroxyl groups that are capable of acquiring a proton at low *pH*'s and losing a proton at high *pH*'s (called "surface regulation").

$$M - OH_2^+ \underset{K_+}{\overset{H^+}{\rightleftarrows}} M - OH \xrightarrow[K_-]{OH^-} M - O^- \tag{6.9}$$

The surface regulation model described later produces Nernstian behavior over certain ranges of *pH*.[4] An extensive compilation of *PZC*-values for mineral oxides has been given by Kosmulski, and a brief subset of these values is given in Table 6-2. It is convenient to refer to surfaces or colloids whose particles have surfaces that obey Eq. (6.6) as "Nernstian." They exemplify surfaces that are electrochemically completely reversible, or "non-polarizable" in that any attempt to change the potential (as if the surface were that of an electrode) results in an immediate adjustment of the chemical makeup of the surface, or conversely, any chemical adjustment of

---

[4] Russel, W. B., Saville, D. A., and Schowalter, W. R., **Colloidal Dispersions**, pp. 98-99, Cambridge Univ. Press, Cambridge, UK, 1989.

the surface produces an immediate change in the potential. Such a surface is contrasted with its opposite, the completely "polarizable" surface, exemplified by the mercury/solution interface discussed later in this section, for which the potential may be changed without any charge passing the boundary.

3. Specific adsorption of charged species

A second important mode of surface charging is through the *specific* adsorption of ions, shown in Fig. 6-2. These are often ionic surfactants (*e.g.*,

Table 6-2: Ranges of *PZC*-values generally found for representative mineral oxides (in terms of *pH*)*

Oxide	PZC range
α-Al_2O_3	8 – 9
Fe_2O_3	6 – 8
Fe_3O_4	6 – 8
MgO	10 – 12
MnO_2	2 - 4
SiO_2	2 – 3
SnO_2	3 – 4
TiO_2	6 – 7
ZnO	9 – 10
ZrO_2	5 – 7

* Approximate averages of values taken from [Kosmulski, M., **Chemical Properties of Material Surfaces**, Surf. Sci. Ser. 102, Marcel Dekker, New York (2001).]

sodium dodecylsulfate, SDS) that adsorb primarily through hydrophobic bonding, as pictured in Fig. 6.2(a) for the case in which initial surface charge is either zero or is of the same sign as the surfactant ion. The surface charge density $\bar{\sigma}$ is given by

$$\bar{\sigma} = \bar{\sigma}_0 + z_i e \Gamma_i, \tag{6.10}$$

where $\bar{\sigma}_0$ is its initial value, z_i is the valence of the adsorbing surfactant ion, Γ_i is its adsorption in *ions* per unit area. The amount of adsorption is given by the adsorption isotherm, and hence the change in surface charge density is governed by the bulk concentration of the adsorbing ion(s) (*charge-determining ions*, or *cdi*'s), and is indifferent to the presence of small, non-specifically adsorbing ions. This is thus an example of a *constant charge density* surface.

The specifically adsorbing charge carriers may also be polyvalent ions of either the same or opposite charge to that the original surface. In the latter case, sufficient adsorption may occur to neutralize or reverse the original charge. Figure 6.2(b) provides a schematic example (the tetravalent pyrophosphate ion) of polyvalent ion adsorption in that such species are invariably both hydrolyzed and hydrated to various extents. They may adsorb by ion pairing, ion exchange, ion complexing reactions, acid-base reactions or other mechanisms. Figure 6.2(c) shows polyelectrolytes, which may adsorb by any of the above mechanisms. They are often "hydrophobically modified" with un-ionized groups or blocks that adsorb through hydrophobic bonding.

Whether the adsorbing ions are surfactants, polyvalent ions or polyelectrolytes, they are often used to disperse colloid particles due to the electrostatic repulsive forces they establish, and are referred to collectively as ionic dispersants.

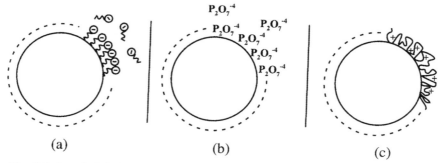

Fig. 6-2: Interfacial charging by specific adsorption of ions. (a) ionic surfactants, (b) polyvalent anions (or sometimes cations), (c) polyelectrolytes.

4. *Ionization of surface functional groups*

The third mechanism of charging in aqueous media involves ionization of chemical functional groups on the surface, as shown in Fig. 6-3. An example of this situation is that of polymer latices with imbedded, ionized functional groups, most often sulfates, whose origin is the persulfate initiator used in the emulsion polymerization process. There is a fixed population of chemical groups on the surface, and hence the surface charge density is constant. The charge density is often specified in terms of the area per unit charge on the surface, termed the *parking area* (nm^2/charge). Sulfate groups are fully ionized over nearly the entire *pH* range, but for weak acid groups, such as -COOH, or weak basic groups such as $-NH_2$, the degree of ionization (hence surface charge density) is a function of *pH*, producing a situation somewhat analogous to that of the oxides described earlier. In such cases, the ionization of the surface groups imposes a relationship between the surface charge density and potential, *i.e.*, *surface regulation*. The mass action expression for the ionization equilibrium can be converted into such a relationship.[5,6] The surface regulation boundary condition is intermediate in

[5] Chan, D. Y. C., in **ACS Symposium Ser.**, Vol. 323, J. A. Davis and K. F. Hayes (Eds.), p. 99, ACS, Washington, DC, 1986.

[6] James, R. O., and Parks, G.A., in **Surface and Colloid Science**, Vol. 12, E. Matijevic (Ed.), pp. 119-216, Plenum, New York, 1982.

its behavior to the constant charge density and constant potential boundary conditions. As another example, shown in Fig. 6-3(b), many naturally-occurring (protein) materials have surfaces that are populated with both amino ($-NH_2$) *and* carboxyl (-COOH) groups which ionize in the presence of water to an extent dependent on the *pH* of the medium. At low *pH*, amino groups pick up a proton to become ($-NH_3^+$), giving the surface a positive surface charge, while at high *pH*, the carboxyl groups ionize to give it a negative charge. At some intermediate *pH*, the surface has an equal population of anionic and cationic groups and is effectively neutral, giving a second interpretation of the *PZC*.

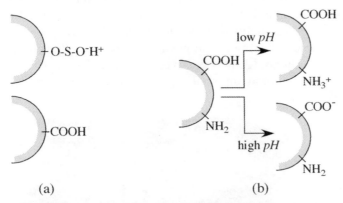

(a)   (b)

Fig. 6-3: Interfacial charging by ionization of surface functional groups. (a) strong acid and weak acid sites often found on latexes, (b) carboxyl and amino groups commonly found on protein surfaces.

5. *Isomorphic substitution*

The fourth mechanism, shown in Fig. 6-4, concerns certain minerals that have ions of different valence (but about the same size) than the main constituents incorporated into their lattices. For example, in many clays, Al^{3+} ions may be substituted for Si^{4+} on tetrahedral silicon sites in the crystal lattice, yielding a net negative charge. This is an ion exchange process referred to as *isomorphic substitution* (also called "isomorphous" substitution). For clay materials it is quantified by titration measurements in terms of the "*cation exchange capacity*," (*CEC*) defined as the number of

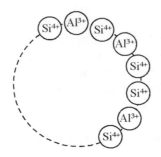

Fig. 6-4: Interfacial charging by isomorphic substitution.

milliequivalents of monovalent cation required to neutralize 100 g of dry clay.[7] As examples, most illite clays have a *CEC* of 20–40, while most kaolinite clays have a *CEC* of 1–10. The resulting surface charge density is fixed and does not adjust itself in response to changes in indifferent electrolyte composition in the bulk.

Recall that clay particles are typically plate-like in their structure, as pictured in Fig. 5-26. It is generally accepted that the face charges of the particles are negative and of constant charge density due to isomorphic substitution, but that the edges or rims are of constant potential dictated by the system *pH*. Thus under moderate-to-low *pH* conditions, the rim charge may be positive. This circumstance is illustrated by the well-known 1942 micrograph, Fig. 6-5, published by van Olphen. It shows negatively charged gold nanoparticles clinging to the positively charged edges of kaolinite clay particles while shunning the negatively charged faces. The negative face charges may be predominant, so that "from a distance" such particles appear to carry a negative charge.[8]

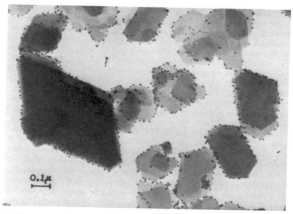

Fig. 6-5: Electron micrograph of a mixture of a kaolinite and a negative gold sol. From [Van Olphen, H., **An Introduction to Colloid Clay Chemistry**, 2nd Ed., p. 95, John Wiley & Sons, New York, 1977.]

6. Accumulation/depletion of electrons

A fifth mechanism, which appears to be important in at least two situations, is that of the direct electron transfer from one phase to another. This is thought to be one of the charging mechanisms at interfaces between un-oxidized metals (mercury, gold, platinum, *etc.*) and solutions, and amounts to spontaneous polarization. Conductance electrons of the metal move spontaneously to the lowest unoccupied orbital (LUMO) of the adjacent material. A similar process has been proposed for latices and oils in

---

[7] van Olphen, H., **An Introduction to Clay Colloid Chemistry**, 2nd Ed., pp. 280ff, Wiley, New York, 1977.

[8] Secor, R. B., and Radke, C. J., *J. Colloid Interface Sci.*, **103**, 237 (1985).

contact with water and has been called "electron injection."[9] It is thought to occur by way of direct tunneling of electrons when the highest occupied molecular orbital (HOMO) of the water molecules overlaps with the LUMO of the hydrocarbon molecules, analogous to the phenomenon occurring at the metal-solution interface. It has been observed that polymeric latex particles and oil droplets in water both often possess significant negative surface potentials. It is difficult experimentally, however, to distinguish this mechanism from that of anionic (usually OH⁻) adsorption, which is often assumed to be the origin of negative charge at such surfaces.[10]

7. Interface charging in non-aqueous systems

In solvents with moderate dielectric constants (> 10, in comparison with ≈ 80 for water), at least some degree of ionization and ion solvation of added electrolytes is possible, and charging mechanisms parallel to those that occur in water may occur. Referred to as "leaky dielectrics," these include low molecular weight alcohols, amines, aldehydes, ketones, carboxylic acids and their aqueous solutions. Ionic surfactants and even simple ionic salts in such media may dissociate to some extent, with one ionic species adsorbing to the surface, balanced by the ions of opposite charge to form a double layer. For example, Fu and Prieve[11] investigated solutions of KOH in a solvent of acetone with 10-20% v/v water and found the KOH to act as a weak electrolyte, and used electrical conductivity measurements to obtain the dissociation constant. It was also determined using the technique of total internal reflection microscopy (TIRM), to be described later, that a sufficient number of ions adsorbed to the surface of model polystyrene latex particles to confer a surface potential of a few millivolts in magnitude, relative to the interior of the solution. Labib and Williams[12] studied the electrophoresis of a series of different oxide particles in a series of leaky dielectric solvents of varying Gutmann Donor Number (recall Chap. 4, Table 4-7). For a given oxide, as the Donor Number of the solvent was increased, a point was reached where the direction of electrophoretic motion reversed. The Donor Number where this occurred was identified as the donicity of the solid surface. Such a formulation suggests that the charging mechanism is an acid-base interaction between the solvent and the particle surface followed by ionic dissociation at the surface. This picture would then appear to require the formation of ion-solvent complexes of sufficient size to provide the charge separation needed to confer electrophoretic mobility.

---

[9] Fowkes, F. M., and Hielscher, F. H., "Electron Injection from Water in Hydrocarbons and Polymers," ACS Org. Coat. Plast. Chem. Preprint **42**, 169 (1980).

[10] Pashley, R. M., Francis, M. J., and Rzechowicz, M., *Curr. Opin. Colloid Interface Sci.*, **13**, 236 (2008).

[11] Fu, R., and Prieve, D. C., *Langmuir*, **23**, 8048 (2007).

[12] Labib, M. E., and Williams, R., *J. Colloid Interface Sci.*, **97**, 356 (1984); Labib, M. E., and Williams, R., *Colloid Polym. Sci.*, **264**, 533 (1986).

It is difficult to imagine the processes described above occurring in fully nonpolar media, such as alkanes, with a dielectric constants of ≈ 2. The observation of electrophoresis of particles in these media, however, suggests that surface charging does occur. For example, Sanfeld et al.,[13] studied theoretically the system of 1-µm radius silica particles dispersed and evidently stabilized in benzene solutions of tetraisoamylammonium picrate (TIAP). It is likely that traces of polar impurities, particularly water, play a key role. Not only the magnitude, but also the direction of electrophoresis (hence the sign of the surface charge) has been observed to depend on the presence and amount of traces of water.[14] It is supposed that the water is adsorbed to particle surfaces, where traces of salt or other ionizable solutes, in addition to the water itself, may reside. Ions in solution are likely to exist primarily in water solubilized in the interior of inverse micelles, in this case presumably produced by the TIAP.

In another example, Pugh et al. studied the system of 75-nm diameter carbon black particles dispersed in dodecane (with a water content of 25 mg/L and a measured dielectric constant of 5). A commercial dispersant OLOA 1200 (from Chevron Chemicals) was added at various levels up to 4 wt%. OLOA 1200 is a polyisobutylene succinimide (PIBS), with the isobutylene chain having about 70 carbons, and polyamines (such as diethylene triamine) attached to the chain by a succinimide or succinamide group. The amine groups confer a basic functionality to the compound. Electrophoresis measurements of the carbon particles suggested large effective negative surface potentials (actually, zeta potentials, to be discussed later in this chapter) as the OLOA 1200 content approached 4 wt%. The mechanism for the surface charging was assumed to be as pictured in Fig. 6-6.

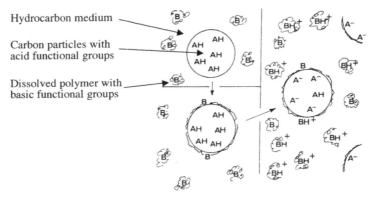

Fig. 6-6: Mechanism for surface charging of carbon particles in dodecane by the dispersant OLEA 1200, as proposed by Pugh et al.. From [Pugh, R. J., Matsunaga, T., and Fowkes, F. M., *Colloids Surfaces*, **7**, 183 (1983).]

[13] Sanfeld, A., Devillez, C., and Wahrmann, S., *J. Colloid Interface Sci.*, **36**. 359 (1971).

[14] Mysko, D. D., and Berg, J. C., *Ind. Eng. Chem. Res.*, **32**, 854 (1993).

The surfaces of the carbon black particles are known to contain acidic oxygen-containing groups such as phenols and carboxylic acids. These groups attract the amphiphilic adsorption of the polyamines of the dispersant, where proton transfer from the acidic groups to the amines occurs. The final step is the desorption of the proton carrying dispersant into the surrounding medium, leaving a negative charge on the particle surface. It is possible, although not mentioned by the authors, that the charged sites on the carbon surface were solvated by small amounts of water also adsorbed to the surface, and the positive charges in the solution may have resided in water solubilized in inverse micelles of the dispersant. The larger issue is the presumed (and necessary) existence of the stabilized BH^+ ions in the solution.

As indicated above, preceding the issue of surface charging in nonpolar media is the issue of the origin and maintenance of charge in the bulk solution. Ions exist as separate entities in aqueous or leaky dielectric media because they are hydrated (or solvated), and their solvation sheaths protect them from recombining. A pair of adjacent ions has an amount of thermal energy of $\approx kT$ available to it to break it apart. The electrostatic energy holding the pair of (monovalent) ions together is given by Coulomb's Law:

$$\Phi_{att} = \frac{1}{4\pi\varepsilon\varepsilon_0} \frac{e^2}{r},$$

(6.11)

where e is the protonic charge, ε is the dielectric constant, ε_0 the permittivity of free space, and r the distance of separation between the ions. If the attractive energy is comparable to or less than the disruptive energy kT, the ion pair would be expected to remain dissociated. Setting $\Phi_{att} \approx kT$ and solving for the indicated distance of separation defines the *Bjerrum length*:

$$\lambda_B = \frac{1}{4\pi\varepsilon\varepsilon_0} \frac{e^2}{kT},$$

(6.12)

the distance between the ions required for stable dissociation. For water at room temperature, the Bjerrum length is 0.7 nm, comparable to twice the thickness of the typical hydration sheath. For an alkane, however, with a dielectric constant of only 2, one has $\lambda_B \approx 28$ nm. Thus for stable dissociation to occur in a nonpolar medium, the ions would need to "hide" in a structure providing a sheath of substantial thickness. These structures are thought to be reverse micelles, or polymers or copolymers that are capable of sequestering an ion. In the study of Sanfeld *et al.*, it was evidently reverse micelles of TIAP that stabilized the charges, while in the study of Pugh *et al.*, the OLOA molecules they used formed the needed micelles. These have later been shown to have a diameter of $\approx$ 20nm.[15] Many more examples are

[15] Kim, J., Anderson, J. L., Garoff, S,, and Schlangen, L. J. M., *Langmuir*, **21**, 8620 (2005).

given in the extensive review of electrical charges and surface charging in nonaqueous media by Morrison.[16]

The question remains as to how the micelles or other structures acquire their charges. They enter or are formed, often if not always assisted by small amounts of water, in the solution as undissociated neutral entities. It is now widely believed that as a result of non-specific, ever-present thermal interactions, these entities undergo a charge disproportionation reaction:

$$2\,M \;=\; M^+ + M^-,$$

so that at any time there will a small (perhaps *very* small) fraction of the micelles or other entities that carry charge. If the size of the entities, M, are comparable to the Bjerrum length, they can be maintained in solution on a dynamic basis.

The next question is that of how such charges are transferred to particle or other surfaces, but the most likely answer appears to be through donor-acceptor interactions as described above. It is clear that more research is needed in this area.

B. Electric double layer formation and structure

1. *The Helmholtz model; electrostatic units*

Regardless of how charge separation is generated, a structure will be developed such that the surface charge is neutralized by an adjacent layer in the solution containing an excess of ions of charge opposite to that of the surface, *i.e.*, *counterions*. Ions of the same charge as the surface are termed *coions*. The simplest picture was proposed by Helmholtz (1879)[17] and consisted of opposed monolayers of charge (a "molecular capacitor") pictured in Fig. 6-7. Thus it was termed an *electrical double layer*. The

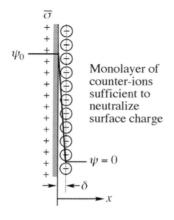

Monolayer of counter-ions sufficient to neutralize surface charge

Fig. 6-7: Helmholtz double layer model.

[16] Morrison, I. D., *Colloids Surfaces A*, **71**, 1 (1993).
[17] Helmholtz, H., *Ann. d. Physik u. Chemie (Wiedemann's Ann.)*, N. F., **7**, 337 (1879).

separation distance between the charged layers, δ, was presumably the hydrated radius of the counterions, and all of the potential drop occurred across this distance. The parallel-plate capacitor equation relating the surface charge density, $\overline{\sigma}$, to the potential drop is:

$$\overline{\sigma} = \frac{\varepsilon\varepsilon_0}{\delta}\psi_0[\text{SI units}] = \frac{\varepsilon}{4\pi\delta}\psi_0[\text{cgs units}] \tag{6.13}$$

The equations used hereinafter are written assuming the use of SI (or MKS, or "rationalized") units. Charges are in units of Coulombs, and potentials are in Volts. These equations often require the use of the electric permittivity of free space, $\varepsilon_0 = 8.854 \times 10^{-12}$ Coul^2/J-m [=] Coul/Volt-m. The dielectric constant of the medium, ε, is equal to the permittivity of the medium ε_m divided by the permittivity of free space, ε_0. The cgs system is used in many textbooks and in particular, the older literature. In this system, the unit of charge (the esu) is defined in the context of Coulomb's Law with forces in dynes and distances in centimeters. An equation appropriate to cgs units may be converted to one appropriate to SI units by replacing 4π with $1/\varepsilon_0$. A more detailed discussion of the problems associated with electrostatic units is given by Hiemenz and Rajagopalan.[18]

2. The Gouy-Chapman Model; Poisson-Boltzmann Equation

It was soon realized that the Helmholtz picture was too simple to explain many of the observations concerning such systems. It was proposed later by Gouy (1910)[19] and Chapman (1913)[20] that the layer of counterions should be *diffuse*, as shown in Fig. 6-8.

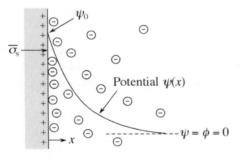

Fig. 6-8: The Gouy-Chapman double layer model.

[18] Hiemenz, P. C., and Rajagopalan, R., **Principles of Colloid and Surface Chemistry**, 3rd Ed., App. B., Marcel Dekker, New York, 1997.
[19] Gouy, G., *J. Phys.*, **9** (4), 457 (1910);
   Gouy, G., *Ann. Phys.*, **7** (9), 129 (1917).
[20] Chapman, D. L., *Phil. Mag.*, **25** (6), 475 (1913).

The abrupt concentration change required in the Helmholtz model should be eroded by diffusion (thermal motion) that would seek to make the concentration uniform. At equilibrium, a balance between the orienting effects of the electric field set up by the surface charge and the randomizing effects of back diffusion would be established, yielding a high concentration of counterions right next to the surface, and a diminishing concentration of them moving away from it. This, together with the surface charge, would produce a variation of the potential in the solution, $\psi(x)$, from its surface value, ψ_0 (at $x = 0$) to zero far from the surface. Gouy and Chapman modeled the electrical double layer in accord with four critical descriptive assumptions listed in Table 6-3. The first two assumptions treat the ions as structureless, achemical point charges that take up residence solely on the basis of their charge. In particular, the model admits of no adsorption beyond residence next to a surface dictated by considerations of charge.

Table 6-3: Defining assumptions of the Guoy-Chapman model of the double layer.

1. The ions are point charges (*i.e.*, they have no volume)
2. There is no *specific* adsorption of ions
3. The dielectric constant of the medium is constant within the double laye
4. The surface charge is uniformly distributed over the surface

---

The third assumption has dielectric properties of water near the surface the same as those of bulk water in the absence of an electric field. Actually, the fields developed in the inner portion of the double layer are often large enough to expect dielectric breakdown and a dielectric constant as much as an order of magnitude smaller than the handbook value. The final assumption neglects the fact that charges on the surface are in reality located at discrete points, and treats the charge as though it were uniformly smeared over the surface. Based on these assumptions, Gouy and Chapman modeled the double layer in terms of the potential, $\psi(x)$, and ion concentration profiles, $C_i(x)$, the surface charge density, $\overline{\sigma}_0$, *etc.*

To develop an equation giving the potential profile in the diffuse part of the double layer, consider first a collection of ions in space, as pictured in Fig. 6-9. One can compute the force on (and hence, by integration, the

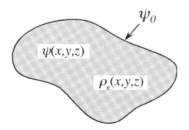

Fig. 6-9: Ionic cloud in space; basis for derivation of Poisson's Equation.

potential at the location of) any charge q_i in this region from Coulomb's law by vectorially summing all of the contributions of the various ions present to the potential at the point (unit charge) of interest:

$$\underline{F}_i = q_i \sum_{j \neq i} \frac{q_j}{4\pi\varepsilon\varepsilon_o} \frac{\underline{r}_{ij}}{r_{ij}^3}, \tag{6.14}$$

where $\underline{F}_i$ is the electrostatic force on the charge q_i exerted by the other charges, q_j. $\underline{r}_{ij}$ is the spatial displacement vector from q_i to q_j, and r_{ij} is its magnitude.[21] The distribution of the ions in space can be well described by "smearing" them over the volume yielding a space charge distribution, $\rho_e(x,y,z)$ [=] charge/volume. The result of the vectorial summation (integration) in a medium of dielectric constant ε is Poisson's Equation,

$$\nabla^2\psi = -\frac{1}{\varepsilon\varepsilon_o}\rho_e, \tag{6.15}$$

where ψ is the local potential. The space charge density is given by

$$\rho_e = \rho_e(x,y,z) = \sum_i z_i e n_i(x,y,z), \tag{6.16}$$

where n_i is the number concentration, [=] #/vol., of ions i in the solution at the point of interest, and z_i their respective valence.

The one-dimensional version (x-direction only, Cartesian coordinates) pertinent to the present purposes of describing the diffuse part of the double layer adjacent to an infinite flat surface is:

$$\frac{d^2\psi}{dx^2} = -\frac{1}{\varepsilon\varepsilon_0}\sum_i z_i e n_i(x). \tag{6.17}$$

This is a mixed differential equation in that it contains both the potential, $\psi(x)$, and the ionic concentrations, $n_i(x)$, as dependent variables. Without additional information, it cannot be integrated to solve for either. This is not surprising because we have yet to account for the diffusion (randomizing) effect. This is accomplished through the use of a Boltzmann factor, which expresses the ratio of the probability of finding a given ion at position x to that of finding it out in the bulk solution ($x = \infty$), which will be the relevant ratio of observed concentrations:

$$\frac{n_i(x)}{n_i(\infty)} = \exp\left(\frac{-w_i}{kT}\right), \tag{6.18}$$

where w_i is the reversible work required to bring the ion from $x = \infty$ to $x = x$, viz.:

---

[21] See, e.g., Reitz, J. R., Milford, F. J., and Christy, R. W., **Foundations of Electromagnetic Theory**, pp. 24ff, Addison-Wesley, Reading, MA, 1980.

$$w_i = z_i e \psi(x).$$
(6.19)

When the Boltzmann expressions for the ionic concentrations are substituted into the Poisson Equation, one obtains the *Poisson-Boltzmann* (PB) Equation, in terms of the potential alone:[22]

$$\frac{d^2\psi}{dx^2} = -\frac{1}{\varepsilon\varepsilon_0} \sum_i z_i e n_{i,\infty} \exp\left(\frac{-z_i e \psi}{kT}\right).$$
(6.20)

Equation (6.20) can in principle be integrated to give $\psi(x)$, but it is useful at first to consider the case in which two simplifications are made. First, assume a single symmetrical $(z\text{-}z)$ electrolyte. The right hand side of Eq. (6.20) then contains just two terms, such that:

$$\frac{d^2\psi}{dx^2} = -\frac{zen_{i,\infty}}{\varepsilon\varepsilon_0}\left[\exp\left(\frac{ze\psi}{kT}\right) - \exp\left(\frac{-ze\psi}{kT}\right)\right] = \frac{2zen_{i,\infty}}{\varepsilon\varepsilon_0}\sinh\left(\frac{ze\psi}{kT}\right).\text{[23]}$$
(6.21)

While this equation may be solved analytically, consider first an additional simplification, *viz.*, that surface potential is relatively low ($|z\psi_0| \le kT/e \approx 25$ mV). This is referred to as the Debye-Hückel (DH) approximation. Under such conditions, $\left(\dfrac{ze\psi}{kT}\right)$ is sufficiently small that $\sinh\left(\dfrac{ze\psi}{kT}\right) \approx \left(\dfrac{ze\psi}{kT}\right)$, so that

$$\frac{d^2\psi}{dx^2} = \frac{2z^2 e^2 n_\infty}{\varepsilon\varepsilon_0 kT}\psi = \kappa^2\psi,$$
(6.22)

called the *linearized* PB Equation. For many situations encountered it is unrealistic, but its investigation produces an excellent qualitative picture of the Gouy-Chapman diffuse double layer. For a boundary condition of specified ψ_0, it results in the simple exponential decay for $\psi(x)$, *viz.*

$$\psi(x) = \psi_0 \exp(-\kappa x).$$
(6.23)

Substitution of Eq. (6.23) into the Boltzmann Equation, Eq. (6.18) yields a roughly exponential decay function for counterion concentration (and corresponding increase in coion concentration):

$$C_i(x) = C_{i,\infty} \exp\left[-\frac{z_i e \psi}{kT}\right] = C_{i,\infty} \exp\left[-\frac{z_i e}{kT}\psi_0 \exp(-\kappa x)\right].$$
(6.24)

---

[22] The Poisson equation is based on the additive contribution of charge to the potential at any location, whereas the Boltzmann equation suggests an exponential relationship between charge and potential. At lower potentials, these requirements are self-consistent, but at higher potentials there is at least a formal inconsistency, which is generally overlooked.

[23] Recall the definition of the hyperbolic sine function: $\sinh X = \frac{1}{2}(e^X - e^{-X})$, and the related hyperbolic cosine function: $\cosh X = \frac{1}{2}(e^X + e^{-X})$, and the hyperbolic tangent: $\tanh X = \sinh X / \cosh X$.

DH ion concentration profiles in a 0.01N solution of NaCl against a surface of potential $\psi_0 = 25$ mV, are plotted in accord with Eq. (6.24) in Fig. 6-10. A key parameter in these profiles is the decay constant κ:

$$\kappa = \sqrt{\frac{2e^2z^2n_\infty}{\varepsilon\varepsilon_0 kT}} = \sqrt{\frac{2000\,e^2z^2N_{Av}^2 C}{\varepsilon\varepsilon_0 RT}}, \qquad (6.25)$$

where C is the z-z salt concentration in mol/L. It is the so-called "*Debye parameter*" and has units [=] 1/length. κ^{-1}, called the *Debye length* is a measure of the "thickness of the double layer." It is also appropriately termed the electrostatic "screening length." For a single symmetrical electrolyte in water at 25°C, it can be readily computed in the form

$$\kappa^{-1} = \frac{0.304}{|z|\sqrt{C}} \quad [=] \text{ nm}. \qquad (6.26)$$

κ^{-1} is typically in the range of a few to a few 10's of nm for aqueous electrolyte solutions.

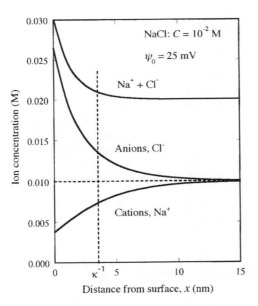

Fig. 6-10: Ion concentration profiles near a charged surface.

Figure 6-10 shows the important result that the diffuse part of the double layer is enriched in counterions and depleted in coions, but also that the *sum* of the ion concentrations in the double layer exceeds their total concentration in bulk. This might appear to violate species conservation, but the effect is generally negligible because the total ion inventory in the double layer is negligible relative to that in the bulk in all but highly concentrated colloidal dispersions or in highly confined spaces.[24] The excess

---

[24] In more concentrated colloidal dispersions, it is important to compute the relevant salt concentrations on the basis of the volume of solvent, not the volume of the dispersion, a

ion concentration in the double layer relative to the bulk will have important consequences for understanding electrostatic repulsion between approaching colloid particles, as discussed in Chap. 7.

Because of its central importance, some further comments regarding the Debye parameter κ should be made. While it may appear that it came out of the DH approximation, neither its definition nor the formula for its computation depends on that assumption. It is simply a property of the electrolyte solution and a measure of its electrostatic screening power. For the case of a single symmetric electrolyte, examination of Eq. (6.24) shows that it increases significantly (hence the double layer *thickness*, κ^{-1} decreases) with ion valence and concentration, as shown in Fig. 6-11. For mixed electrolytes, the Debye parameter is expressed in terms of the *ionic strength, I*:

$$I = \frac{1}{2}\sum z_i^2 C_i,$$ (6.27)

where the C_i are the molarities of all the ions in solution.[25] In terms of ionic strength,

$$\kappa = \sqrt{\frac{2000e^2 N_{Av}^2 I}{\varepsilon\varepsilon_0 RT}}.$$ (6.28)

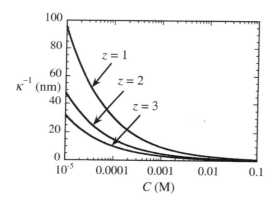

Fig. 6-11: The "Debye length," or "double layer thickness" as it depends on salt concentration for different counterion valences, in water at 25°C.

For non-aqueous media at 25°C, one obtains:

$$\kappa^{-1} = \frac{0.0343\sqrt{\varepsilon}}{\sqrt{I}} \xrightarrow{z-z} \frac{0.0343\sqrt{\varepsilon}}{|z|\sqrt{C}} \ [=] \ \text{nm},$$ (6.29)

with ε = the dielectric constant of the solvent. The lower dielectric constants of solvents, as compared to water (ε = 78.5), would appear to produce

---

significant proportion of which may be occupied by the particles. Also in denser colloids, it may be necessary to account for electrolyte that is removed from the bulk solution by adsorption to the particle surfaces.

[25] Ionic strength is more formally defined using the ionic *molalities, m_i* (moles/kg of solvent) of the ions rather than the molarities, but for the low concentrations generally involved, the two are numerically essentially the same.

thinner double layers, but the *much* lower ion concentrations in solvents generally assures that they are more than an order of magnitude thicker.

Among the other double layer properties that are of interest is the surface charge density on the solid, $\overline{\sigma}_{\text{solid}} \equiv \overline{\sigma}_0$. Because the double layer taken as a whole is electrically neutral, the surface charge density on the solid may be equated to the (negative) integral of the space charge density, ρ_e, in the solution from the surface outward. The space charge density is expressible in terms of the potential profile using the Poisson Equation. Thus

$$\overline{\sigma}_0 = -\overline{\sigma}_{\text{soln}} = -\int_0^\infty \rho_e dx = \int_0^\infty \varepsilon\varepsilon_0 \frac{d^2\psi}{dx^2}. \tag{6.30}$$

Integrating gives:

$$\overline{\sigma}_0 = \varepsilon\varepsilon_0 \left[\left(\frac{d\psi}{dx} \right)_\infty^{\;\;0} - \left(\frac{d\psi}{dx} \right)_0 \right] = -\varepsilon\varepsilon_0 \left(\frac{d\psi}{dx} \right)_0. \tag{6.31}$$

Equation (6.31) shows that the surface charge density is proportional to the negative of the slope of the potential function at the surface. For a single z-z electrolyte at low potentials ($z\psi_0 \leq 25$ mV), *i.e.*, the DH approximation, one obtains:

$$\overline{\sigma}_0 = \varepsilon\varepsilon_0 \psi_0 \kappa. \tag{6.32}$$

Thus a DH double layer may be described as a simple capacitor with a "plate spacing" of κ^{-1}, as suggested in Fig. 6-12. Increasing z or C thus effectively

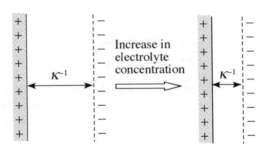

Fig. 6-12: Comparison of Debye-Hückel double layer to a parallel plate capacitor.

decreases the capacitor plate spacing, or "compresses" the double layer. The effect of a ten-fold increase in the concentration of a 1-1 electrolyte for a DH system is shown in Fig. 6-13, in which the double layer thickness is seen to decrease by factor of about three, while the surface charge density *increases* by the same factor.

Relaxing the DH approximation for a single z-z electrolyte against a flat surface requires the general solution to Eq. (6.21). For the case of a constant surface potential ψ_0 one obtains:

$$\tanh\left(\frac{ze\psi}{4kT}\right) = \tanh\left(\frac{ze\psi_0}{4kT}\right)\exp(-\kappa x). \qquad (6.33)$$

Figure 6-14(a) shows the potential profile in a 0.01 M solution of NaCl against a surface of potential $\psi_0 = 80$ mV in comparison with the profile that would have been obtained using the DH approximation. It is of approximately the same shape, but varies more steeply near the wall, but at larger distances from the wall, the curves merge. The ion concentration profiles for this case are shown in Fig. 6-14(b), together with the result of the DH approximation for the counterion. Using the general solution for a symmetrical electrolyte, the surface charge density is computed as:

$$\overline{\sigma}_0 = \frac{2kT\varepsilon\varepsilon_0\kappa}{ze}\sinh\frac{ze\psi_0}{2kT} = \left[8kT\varepsilon\varepsilon_0 n_\infty\right]^{1/2}\sinh\frac{ze\psi_0}{2kT}. \qquad (6.34)$$

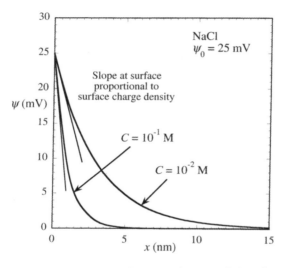

Fig. 6-13: Potential profiles in a 1-1 electrolyte solution against a surface of a constant potential of 25 mV at concentrations of 0.01 and 0.1 M.

To recap briefly, the picture of the electrical double layer that emerges from Gouy-Chapman theory is that of a layer of charge uniformly distributed over a surface, against which resides an electrolyte solution containing a more or less diffuse layer of a sufficient excess of counterions to effect electrical neutrality. The source of the counterions may be (in addition to the water itself) simply the potential determining ions opposite in sign to those that dominate the surface, or in the case of constant charge density surfaces, the other ions from the salt yielding the adsorbate ions, or those associated with the ionizing surface functional groups. More often, however, the major source of the counterions is a "background" or "supporting" electrolyte that is present, constituting the bulk of the electrolyte content of the system. The key solution parameter describing the decay of the potential and the ion concentration profiles to their respective bulk values moving away from the surface is the Debye parameter, κ, or its reciprocal, the Debye length, regarded as the "thickness" of the double layer. For a given bulk solution,

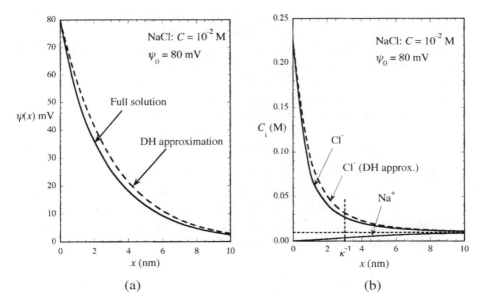

Fig. 6-14: Potential (a) and ion concentration (b) profiles in a 1-1 electrolyte when the surface potential is 80 mV. Debye-Hückel profiles are shown for comparison.

the model also provides the surface charge density, if the surface potential is specified, or the surface potential, if the surface charge density is specified.

3. *Boundary conditions to the Poisson-Boltzmann Equation*

The potential and ion concentration profiles obtained above are for the case of specified surface potential, ψ_0. An equally important type of boundary condition corresponds to that of constant surface charge density. A Debye-Hückel double layer, such as that described by Eq. (6.23) may be obtained under the constant surface charge density condition if the charge density corresponds to a sufficiently low surface potential. In view of Eq. (6.32), the boundary condition at the surface ($x = 0$) for Eq. (6.22) becomes:

$$\left(\frac{d\psi}{dx}\right)_{x=0} = -\frac{\overline{\sigma}_0}{\varepsilon\varepsilon_0} = \text{constant}, \tag{6.35}$$

and the resulting potential distribution is:

$$\psi(x) = \frac{\overline{\sigma}_0}{\varepsilon\varepsilon_0\kappa}\exp(-\kappa x), \tag{6.36}$$

as plotted in Fig. 6-15 for monovalent electrolyte concentrations of 0.01 and 0.1 M. In this case, while the surface charge density remains constant at 5.71 mCoul/m^2 (the value corresponding to a surface potential of 25 mV at C = 0.01 M), the surface potential drops from 25 mV to 7.91 mV.

The cases of constant surface potential and constant charge density bracket the range of electrostatic boundary conditions encountered, but intermediate cases arise when one or more functional groups on the surface ionize to extents dependent upon the bulk solution composition, in particular, the *pH*. The simplest case occurs when there is only a single type of ionizing site on the surface, *e.g.*, -COOH, subject to only one type of association/dissociation reaction. In this case it would be:

$$-\text{COOH} \xrightleftharpoons{K} -\text{COO}^- + \text{H}_s^+ , \tag{6.37}$$

where H_s^+ refers to hydrogen ions in the solution adjacent to the surface. The effective concentration of hydrogen ions at the surface is in general different from that in the bulk solution, since the former are present at an electric field of potential ψ_0 relative to that in the bulk. Thus the two effective concentrations are related by a Boltzmann Equation:

$$\left[\text{H}_s^+\right] = \left[\text{H}^+\right] \exp\left(-\frac{e\psi_0}{kT}\right). \tag{6.38}$$

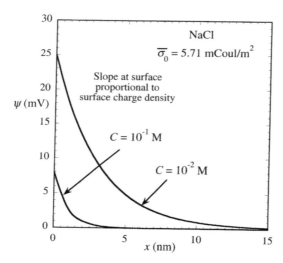

Fig. 6-15: Potential profiles in a 1-1 electrolyte solution against a surface of a constant charge density of 5.71 mCoul/m² at concentrations of 0.01 and 0.1 M.

The mass-action law for the site dissociation thus gives

$$K = \frac{\{-\text{COO}^-\}\left[\text{H}_s^+\right]}{\{-\text{COOH}\}} = \frac{\{-\text{COO}^-\}}{\{-\text{COOH}\}}\left[\text{H}^+\right]\exp\left(-\frac{e\psi_0}{kT}\right), \tag{6.39}$$

where curly brackets {} are used to denote surface concentrations (sites/area). The concentrations of dissociated and undissociated sites are related by the fact that the total site density, N_s, is constant, *i.e.*

$$N_s = \{-\text{COOH}\} + \{-\text{COO}^-\}. \tag{6.40}$$

The surface charge density is given by

$$\overline{\sigma}_0 = -e\{-COO^-\},$$

and substituting from Eqs. (6.39) and (6.40):

$$\overline{\sigma}_0 = \frac{-eN_s}{1 + \dfrac{[H^+]}{K}\exp\left(-\dfrac{e\psi_0}{kT}\right)}. \tag{6.41}$$

One thus has a relationship between $\overline{\sigma}_0$ and ψ_0 dependent upon pH. It is evident that as pH goes down ($[H^+]$ goes up), $\overline{\sigma}_0 \to 0$, but at high pH, $\overline{\sigma}_0 \to -eN_s$ (*i.e.*, all the sites are ionized). Again, the surface charge density decreases monotonically with pH, and no PZC is observed.

Two-site dissociation models are used to describe either mixed populations of two different sites or single amphoteric sites. If one has two different acidic sites (*e.g.*, -COOH and $-SO_4H$), then behavior qualitatively similar to the single-site model is observed, *viz.*

$$\overline{\sigma}_0 = \frac{-eN_{s1}}{1 + \dfrac{[H^+]}{K_1}\exp\left(-\dfrac{e\psi_0}{kT}\right)} + \frac{-eN_{s2}}{1 + \dfrac{[H^+]}{K_2}\exp\left(-\dfrac{e\psi_0}{kT}\right)}. \tag{6.42}$$

If the surface has both acidic *and* basic sites, however, one can identify a pH at which the positive and negative site densities are equal (the PZC). Single amphoteric sites, as exemplified by the -OH groups on metal oxides, also yield a PZC. These two-site dissociation models are described in more detail elsewhere,[26,27] where it is shown that if the majority of sites are ionized at the PZC, Nernstian behavior should be expected. This is often not the case, however, for many oxides.

4. Double layers at spherical and cylindrical surfaces

The one-dimensional Poisson-Boltzmann Equation can also be cast in spherical or in cylindrical coordinates to describe double layers surrounding spheres or infinite cylinders, respectively. For spheres:

$$\frac{1}{r^2}\frac{d}{dr}\left(r^2\frac{d\psi}{dr}\right) = \frac{\kappa^2 kT}{ze}\sinh\left[\frac{ze\psi}{kT}\right], \tag{6.43}$$

and for cylinders:

$$\frac{1}{r}\frac{d}{dr}\left(r\frac{d\psi}{dr}\right) = \frac{\kappa^2 kT}{ze}\sinh\left[\frac{ze\psi}{kT}\right], \tag{6.44}$$

[26] Hunter, R. J., **Foundations of Colloid Science**, Vol. I, Chap. 6; Vol. II, Chap. 12, Clarendon Press, Oxford, 1987.

[27] James, R. O., and Parks, G. A., in **Surface and Colloid Science**, Vol. 12, E. Matijevic (Ed.), pp. 119-216, Plenum, New York, 1982.

where r is the radial coordinate in each case. Analytical solutions are not available for either equation except in the limit of the Debye-Hückel approximation, but for this case, one obtains, for spheres and cylinders of radius "a" and constant surface potential, ψ_0:

$$\text{spheres: } \quad \psi(r) = \psi_0 \frac{a}{r} \exp\left[-\kappa(r-a)\right], \quad \text{and} \tag{6.45}$$

$$\text{cylinders: } \quad \psi(r) = \psi_0 \frac{K_0(\kappa r)}{K_0(\kappa a)}, \tag{6.46}$$

where K_0 is the zeroeth-order Bessel function of the second kind. In both cases, the potential falls off more steeply the higher the curvature of the surface, as shown in Fig. 6-16. This fact has consequences for the electrostatic stability of nano-colloidal particles.

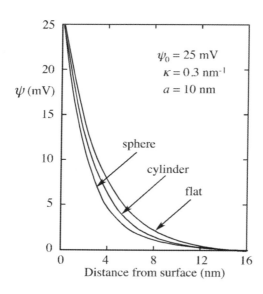

Fig. 6-16: Comparison of potential profiles (Debye-Hückel case) for flat, cylindrical and spherical surfaces.

5. The free energy of double layer formation

It is instructive to consider the spontaneous formation of an electric double layer from a thermodynamic point of view, and it will be useful later in developing expressions for the energy associated with the mutual approach of surfaces bearing double layers. The hypothetical process of forming a double layer (at constant T and p) starts with an uncharged surface in contact with a large reservoir containing a uniform distribution of all the relevant ions, and ends with the surface charged to the appropriate potential ψ_0 (by means of the net adsorption of one of the lattice ions and/or specifically-adsorbing ions) and the adjacent solution exhibiting the appropriate spatial distribution of all the ions. Since the formation of the double layer is spontaneous, $\Delta G < 0$ for the process. In accord with

Guggenheim's decomposition of the electro-chemical potentials into chemical end electrostatic components, we may write

$$\Delta G = \Delta G_{elect} + \Delta G_{chem}, \tag{6.47}$$

on a per unit area basis. The process may be followed in terms of Γ_i, the adsorption of (for example) a potential-determining ion. As the adsorption proceeds, both the surface potential ψ_0' and the surface charge density $\bar{\sigma}'$ increase from zero. At any point in the double layer formation process, the increment of electrical work required is given by

$$\delta W_{elect} = d\Delta G_{elect} = \psi_0' d\bar{\sigma}'. \tag{6.48}$$

The change of the electric component of the free energy of the system is positive right after adsorption of the first ion and grows larger as adsorption proceeds, owing to the electrostatic repulsion that must be overcome for the ions to adsorb to a surface of the same charge. The magnitude of the coefficient at each stage accounts not only for the ionic adsorption itself, but also for the concomitant arrangement of ions in the adjacent portion of the double layer.

The change of the chemical component of the free energy of the system is regarded as a constant (negative) quantity as Γ_i increases (neglecting any lateral non-electrostatic interactions between the adsorbing ions). The adsorption stops when

$$d\Delta G_{total} = d\Delta G_{elect} + d\Delta G_{chem} = 0. \tag{6.49}$$

Thus the constant increment of chemical work done in forming the double layer is equal (and of opposite sign) to the final increment of the electrical work, *viz.*

$$d\Delta G_{chem} = -(d\Delta G_{elect})_{final} = -\psi_0 d\bar{\sigma}' \tag{6.50}$$

where ψ_0 is the final surface potential of the double layer. Thus

$$\Delta G_{chem} = -\psi_0 \int_0^{\bar{\sigma}} d\bar{\sigma}' = -\psi_0 \bar{\sigma}. \tag{6.51}$$

Therefore

$$\Delta G_{total} = \Delta G_{chem} + \Delta G_{elect} = -\psi_0 \bar{\sigma} + \int_0^{\bar{\sigma}} \psi_0' d\bar{\sigma}' = -\int_0^{\psi_0} \bar{\sigma}' d\psi_0' \tag{6.52}$$

To compute the free energy of double layer formation, we thus require the dependence of the surface charge density on the surface potential $\bar{\sigma}(\psi_0)$. This may be given in terms of a double layer model, such as that of Guoy and Chapman. As seen earlier, for a single z-z electrolyte, we have

$$\bar{\sigma} = \frac{2kT\varepsilon\varepsilon_0\kappa}{ze}\sinh\frac{ze\psi_0}{2kT}. \tag{6.53}$$

Upon substituting into Eq. (6.52), the free energy of double layer formation is then given by

$$\Delta G = -\left(\frac{2kT}{ze}\right)^2 \varepsilon\varepsilon_0 \kappa \left[\cosh\left(\frac{ze\psi_0}{2kT}\right) - 1\right].$$ (6.54)

For small values of surface potential, *i.e.*, $z\psi_0 < 25$ mV, the expression reduces to

$$\Delta G = -\frac{1}{2}\varepsilon\varepsilon_0 \kappa\psi_0^2 = -\frac{1}{2}\overline{\sigma}\psi_0.$$ (6.55)

6. The Stern Model; structure of the inner part of the double layer

The Gouy-Chapman model provides a good first picture of double layer structure, but its predictions are unacceptable under many circumstances. For example, a quick calculation of the predicted surface concentration of monovalent cations in a 1.0 M solution adjoining a surface for which ψ_0 has the reasonable value of -100 mV is over 50 M! This results from the assumption that ions are regarded as point charges (*i.e.*, they have no volume). A second major problem with the model is that it treats all ions of the same valence as being identical with respect to their presence adjacent to the surface, *i.e.*, their adsorption. Ions may adsorb for reasons other than their charge, however, *e.g.*, they may exhibit *specific* adsorption. Both of these problems are addressed in an improved model proposed by Stern (1924),[28] pictured in Fig. 6-17. In this model, the double layer consists of an inner and an outer portion. The inner portion is modeled as a monolayer of counterions, generally not exactly equal to the amount required to achieve neutralization of the surface charge. The number of such ions in this layer (termed the *Stern layer*) was assumed to be given by the Langmuir

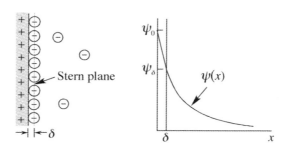

Fig. 6-17: Stern model of the electric double layer.

adsorption isotherm. The closeness of their approach to the surface and to each other is limited by the magnitude of their hydrated radii. The maximum packing in the layer corresponds to the saturation adsorption Γ_∞ as expressed by the isotherm. The plane through the centers of the first layer of ad-ions, taken to be located at $x = \delta$, is termed the *Stern plane*, and all of the charge in the Stern layer is taken to reside in this plane. If the charge in the Stern layer is less than that needed for complete neutralization of the surface

[28] Stern, O., *Z. Elektrochem.*, **30**, 508 (1924).

charge, the balance of the neutralization is presumed to occur in a diffuse, Gouy-Chapman layer outside the Stern layer. The potential drops (or rises) linearly within the Stern layer from its surface value ψ_0 to its value at the Stern plane, ψ_δ. For the diffuse (*i.e.*, Gouy-Chapman) layer, the same equations hold as before, but the potential ψ_δ, replaces ψ_0, and the coordinate x is replaced by $(x - \delta)$. Thus the potential profile in the diffuse layer is:

$$\tanh\left(\frac{ze\psi}{4kT}\right) = \tanh\left(\frac{ze\psi_\delta}{4kT}\right)\exp\left[-\kappa(x-\delta)\right]. \tag{6.56}$$

Some features of Eq. (6.56) should be noted. First, since $\tanh X \approx 1$ for $X \geq 1$, it is evident that for high values of the Stern potential, specifically, for

$$\psi_\delta \geq \frac{4kT}{ze} \approx \frac{102.8}{z} \text{ mV (for water at 25°C)}, \tag{6.57}$$

the prefatory constant on the right hand side of Eq. (6.56), sometimes abbreviated as γ_0, is $\approx$ unity. This means that once the Stern potential exceeds the value given by Eq. (6.57), further increases will not affect the potential profile. The second observation is based on the fact that $\tanh X \rightarrow X$ as $X \rightarrow 0$. In practical terms, X is within 0.1% of $\tanh X$ when $X \leq 0.1$. Thus, far enough out in the solution away from the surface,

$$\psi(x-\delta) \approx \frac{4kT}{ze}\tanh\left(\frac{ze\psi_\delta}{4kT}\right)\exp\left[-\kappa(x-\delta)\right]. \tag{6.58}$$

Note finally, that the surface charge density at the Stern plane is given by

$$\overline{\sigma}_\delta = -\varepsilon\varepsilon_0\left(\frac{d\psi}{dx}\right)_\delta = \frac{2kT\varepsilon\varepsilon_0\kappa}{ze}\sinh\frac{ze\psi_\delta}{2kT}. \tag{6.59}$$

Because the Stern model can accommodate specific adsorption, it is possible to exhibit charge reversal (as might be expected in the case of surfactant counterions) or charge intensification (as might be expected in the case of surfactant coions) in the Stern layer, as shown in Fig. 6-18. Ions in the Stern layer are sometimes referred to as "bound."

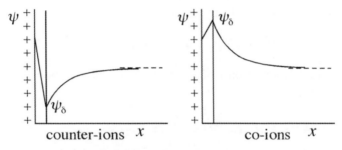

Fig. 6-18: Potential profiles in the case of specific adsorption of counterions (left) and coions (right), in accord with the Stern model.

Some insight into the Stern model can be obtained by implementing the Langmuir adsorption isotherm to compute the charge in the Stern plane, $\overline{\sigma}_\delta$. The Stern layer is usually occupied principally by one type of ion, $e.g.$, the adsorbate or dispersant ion for a double layer formed by the specific adsorption of such an ion, or simply a counterion from the supporting electrolyte. Assuming a positively charged surface, as shown in Fig. 6-18, this would be an anion. Considering its adsorption, as described by the Langmuir adsorption equation, Eq. (3.185)

$$K_- = \frac{\Gamma_- C_B^0}{C_-(\Gamma_\infty - \Gamma_-)}, \tag{6.60}$$

where Γ_- is the adsorption of the anion, and Γ_∞ is the saturation adsorption; C_- is the bulk concentration of the ion, and $C_B^{\;0}$ is the concentration of the solvent, water. Thus

$$C_- / C_B^0 = x_-, \tag{6.61}$$

the mole fraction of the anion in solution, and

$$\Gamma_- / \Gamma_\infty = \theta_- = N_- / N_s, \tag{6.62}$$

is the "fractional coverage" of the surface by the anions, where N_- is number of adsorbed anions per unit area, and N_∞ is the total number of adsorption sites per unit area. Equation (6.60) may then be written as

$$K_- = \frac{\theta_-}{x_-(1 - \theta_-)}, \tag{6.63}$$

with the equilibrium constant for the adsorption of the anions, K_-, given by (cf. Eq. (3.186))

$$K_- = \exp\left(-\frac{\Delta\tilde{\mu}_{ads}^\Theta}{RT}\right), \tag{6.64}$$

where the "~" is needed as a reminder that the standard free energy of adsorption must be formulated in terms of the $electro$-chemical potentials to account for the effect of the electric field. Specifically

$$\Delta\tilde{\mu}_{ads}^\Theta = \Delta\mu_{ads}^\Theta - zF\psi_0 = \Phi_{ads} - zF\psi_0, \tag{6.65}$$

where Φ_{ads} is the $chemical$ part of adsorption potential. Then the charge in the Stern plane is given by

$$\overline{\sigma}_{St} = -zeN_s\theta_- = -zeN_s\frac{K_- x_-}{(1 + K_- x_-)}$$

$$= -zeN_s \frac{x_- \exp(-\Phi_{ads}/RT)\exp(ze\psi_\delta/kT)}{[1 + x_- \exp(-\Phi_{ads}/RT)\exp(ze\psi_\delta/kT)]}. \tag{6.66}$$

Unfortunately, the chemical adsorption potential, Φ_{ads}, is not known *a priori*, nor is there a convenient way to measure it, but it is clear that large (negative) values for it will lead to a high charge density in the Stern layer. Electrokinetic measurements, described later in this chapter, do provide reasonable access to the Stern potential, ψ_δ.

7. *The mercury solution interface; electrocapillarity and refinements to the double layer model*

Much of what is known about the structure of the electric double layer comes from studies of the mercury/solution interface because such systems permit the control and measurement of a number of parameters inaccessible (at least directly) in other systems. The apparatus is an electrolytic cell containing an aqueous solution, a reference electrode and an electrode consisting of a mercury drop whose surface is in contact with the solution. The electrodes are connected externally to a potentiometer that is used to control the potential drop across the cell and hence the potential difference, E, across the Hg/solution interface, which is assumed to be "perfectly polarized," *i.e.*, no electric charge crosses the interface as E is varied. The mercury is in a reservoir connected to a tube from which drops are exuded (called a "dropping mercury electrode"), with their weight giving the interfacial tension by the drop weight method (see Chap. 2.H.7). In another configuration, the mercury is in a downward tapering capillary tube, so that its position in the tube is indicative of the interfacial tension.

Re-derivation of the Gibbs adsorption equation, Eq. (3.83), applied to the system at constant temperature, noting that electrochemical potentials, $\tilde{\mu}_i$, are needed for all ionic species involved, leads to

$$-d\sigma = \overline{\sigma}_0 dE + \Gamma_- d\mu_{salt} = \overline{\sigma}_0 dE + 2RT\Gamma_- d\ln a_\pm, \tag{6.67}$$

when using a reference electrode reversible to the cation, such as an H_2 electrode. Γ_- is the adsorption of the anion, and $a_\pm$ is the mean ion activity of the salt, often simply replaced with some measure of the electrolyte concentration. A similar equation involving Γ_+ could be obtained when using a reference electrode reversible to the anion, but in any case the adsorptions are related through

$$\overline{\sigma}_0 = F(\Gamma_+ - \Gamma_-). \tag{6.68}$$

Varying the potential E across the interface at constant electrolyte composition (and T, p) is described by Eq. (6.67) in the form

$$\left(\frac{\partial \sigma}{\partial E}\right)_{T,p,\mu_{salt}} = -\overline{\sigma}_0. \tag{6.69}$$

This important result is the *electrocapillary equation*, or the Lippmann Equation.[29] Its full derivation and many other details concerning electrocapillary experiments can be found in many references, *e.g.*, Bockris and Reddy.[30] Plots of the interfacial tension versus applied potential, $\sigma(E)$, are known as electrocapillary curves. They are very close to perfect concave downward parabolas whose maximum for each case is termed the "electrocapillary maximum" (*e.c.m.*), indicative of the potential where the surface charge is zero. The *e.c.m.* is a direct analogy to the *PZC* for Nernstian surfaces. To either side of the *e.c.m.*, the interfacial tension σ falls as the charge density on the interface increases, either positive or negative. The lateral repulsions between the like charges on the interface act like a surfactant in reducing the interfacial tension. A wealth of information may be derived from such curves obtained for different electrolytes at different concentrations.

One useful parameter obtainable from an electrocapillary curve is the differential capacitance of the double layer, $\overline{C}$, defined as[31]

$$\overline{C} = \left(\frac{\partial \overline{\sigma}_0}{\partial E}\right)_{T,p,\mu_{salt}} = \left(\frac{\partial \overline{\sigma}_0}{\partial \psi_0}\right)_{T,p,\mu_{salt}}. \tag{6.70}$$

This property measures the amount of charge that can be accommodated for a given potential drop across the double layer, and since that depends on the amount and spatial distribution of the charges, is an important indicator of structure. Differentiation of the Lippmann Equation and comparison with Eq. (6.70) leads to

$$\overline{C} = -\left(\frac{\partial^2 \sigma}{\partial E^2}\right). \tag{6.71}$$

For a Stern double layer, which consists of an inner and diffuse layer in series, Eq. (6.70) may be written as

$$\frac{1}{\overline{C}} = \left(\frac{\partial \psi_0}{\partial \overline{\sigma}_0}\right)_{T,\mu_{salt}} = \left(\frac{\partial(\psi_0 - \psi_\delta)}{\partial \overline{\sigma}_0}\right)_{T,\mu_{salt}} + \left(\frac{\partial \psi_\delta}{\partial \overline{\sigma}_0}\right)_{T,\mu_{salt}} = \frac{1}{\overline{C}_{St}} + \frac{1}{\overline{C}_{diff}}. \tag{6.72}$$

The reciprocals of the capacitances of the layers in series are additive. The capacitance of the diffuse part of the double layer may be computed by differentiating Eq. (6.34) to obtain:

[29] Lippmann, G., *Ann. Chim. Phys.*, **5**, 494 (1875);
Lippmann, G., *J. Phys. Radium*, **2**, 116 (1883).
[30] Bockris, J. O'M., and Reddy, A. K. N., **Modern Electrochemistry II**, pp. 698ff, Plenum Press, New York, 1970.
[31] The differential capacitance is to be contrasted with the integral capacitance, $(\overline{\sigma}_0/\psi_0)$, and is a more useful quantity, since ψ_0 values themselves are generally not accessible, while changes in ψ_0 are.

$$\overline{C}_{\text{diff}} = \left(\frac{\partial \overline{\sigma}_0}{\partial \psi_\delta}\right) = \varepsilon\varepsilon_0\kappa \cosh\frac{ze\psi_\delta}{2kT}, \tag{6.73}$$

which permits the capacitance of the inner double layer to be isolated. The charge in the Stern layer can be obtained from

$$\overline{\sigma}_{\text{St}} = \overline{\sigma}_0 - \overline{\sigma}_\delta. \tag{6.74}$$

$\overline{\sigma}_0$ and $\overline{\sigma}_\delta$ are obtained using Eq. (6.34) and (6.59). This can lead ultimately to an evaluation of the specific adsorption potential in the Stern Equation, Eq. (6.66).

Electrocapillary studies have led to further refinements to the Gouy-Chapman-Stern model. Grahame (1947)[32] suggested, for example, that in the inner part of the double layer, a distinction should be made between ions that are *specifically* adsorbed and generally considered to be dehydrated on the side adjacent to the surface and those that are non-specifically adsorbed within the Stern layer, and considered to retain their full hydration sheath. This conclusion was based primarily on the observation that as one changed salts with a given cation from one anion to the next, the *e.c.m.* was shifted (in terms of E), and each salt yielded a different cathodic (ascending) branch to the curve. Changing cations in salts of a given anion, however, often yielded coincident curves. The shift in the *e.c.m.* and the anion-to-anion variation in the electrocapillary curves is a telltale indication of their specific adsorption. Since the dehydrated ions (anions) are closer to the surface, the plane through their centers is termed the "inner Helmholtz plane," while the surface drawn through the centers of the hydrated ions in the Stern layer is designated the "outer Helmholtz plane." Grahame also noted that anions are generally less and more weakly hydrated than cations and hence are more likely to populate the inner Helmholtz plane than cations. The Stern layer may thus consist of two layers at different degrees of closeness to the surface, so that the Stern plus Guoy-Chapman structure is sometimes called a "triple layer." In further developments, Levine and coworkers[33] have begun to work out the problems associated with the discreteness-of-charge in the surface, *i.e.*, to relax the assumption of charge smeared over the plane of the surface.

8. Oriented dipoles at the interface: the χ-potential

A "dirty little secret" not in the description of double layer structure so far is the effect that solvent (water) dipoles may have on the interface. While these molecules are randomly ordered in the bulk of the solution, their dipoles may be more or less oriented at the interface, as suggested in Fig. 6-19. It is believed that water molecules are on average oriented at a surface for which $\psi_0 = 0$ (at the *PZC* or the *e.c.m.*) with their negative portions

---

[32] Grahame, D. C., *Chem. Rev.*, **41**, 441 (1947).
[33] Levine, S., Mingins, J., and Bell, G. M., *J. Electroanal. Chem.*, **13**, 280 (1967).

directed toward the surface. This leads to a jump in the *true* surface potential ψ_0' of an amount χ, *i.e.*,

$$\psi_0' = \psi_0 + \chi.$$ (6.75)

This so-called χ-potential can presumably change with a change in surface potential ψ_0 (as by changing the concentration of the potential determining ions in a Nernstian colloid, or changing E in an electrocapillary experiment with the Hg/solution system) or more importantly with the presence of even small amounts of polar organic solutes in the electrolyte. Thus Mackor[34] reported a dramatic shift in the measured Stern potential, ψ_δ, (using micro-electrophoresis, to be described later in this chapter) of AgI colloid particles when small amounts (3-4 mol%) of acetone were added to the solution. The *PZC* itself was shifted from $pAg = -5.52$ to -2.70, suggesting that acetone molecules orient their dipoles with their positive portions directed toward the surface. As reported by Sparnaay, this amounted to a χ-potential shift of

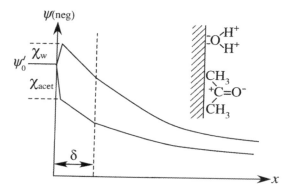

Fig. 6-19: Possible potential distributions within the Stern layer (δ) at the interface between AgI and water (top curve) and a water solution with acetone (4 mol%). The dipole orientations are shown to the right. After [Sparnaay, M. J., "The Electrical Double Layer," Vol. 4, Topic 14: Properties of Interfaces, (D. H. Everett, Ed.) in **The international Encyclopedia of Physical Chemistry and Chemical Physics**, pp.136-137, Pergamon Press, Oxford, 1972.]

-199 mV, as oriented acetone molecules (with their positive portions directed toward the surface) displaced water molecules. The upshot of this is that one should be using true surface potentials, which include the effect of the χ-potential, in the Nernst Equation and elsewhere. In principle, the χ-potential could be computed as:

---

[34] Mackor, E. L., *Rec. Trav. Chim. Pays-Bas*, **70**, 747 (1951).

$$\chi = \frac{\sum_i N_i \delta_{\perp i}}{\varepsilon_0}, \tag{6.76}$$

where N_i is the number density of dipoles of component i on the surface, and $\delta_{\perp i}$ is the normal component of its dipole moment at the surface. These quantities are not known *a priori,* however, so the χ-potential is in general unknown, and is generally swept under the rug. This is apparently tolerable as long as one does not perform experiments in which it is likely to change, but experiments involving changes in solvent composition must take it into account.

C. Electrostatic characterization of colloids by titration methods

1. *Colloid titrations*

There are three principal titration techniques that are used to characterize the various properties of electrical double layers that surround colloid particles, *viz.*: *potentiometric, conductometric* and *electrokinetic* titrations. All are referred to as *colloid titrations*. Some of the properties that can be obtained are as follows. For a Nernstian colloid, one seeks the point of zero charge (*PZC*). For a colloid with ionizable functional groups, the surface density of those groups (or their "parking area"), as well as their acidity or basicity (and strength or weakness) is the desired information. For clays, one seeks the cation exchange capacity (*CEC*). It is also possible to monitor the surface charge density as a function of the nominal surface potential as one moves away from the *PZC*, and therefore to calculate the double layer capacitance. Finally, under most circumstances it is possible to measure the Stern potential. In potentiometric titrations, one adds potential-determining ion to a colloidal dispersion while monitoring the change in the bulk concentration of that ion by means of an ion-specific electrode. Stoichiometric accounting allows one to determine the amount of the ion associated with the surface of the particles. In conductometric titrations, one adds ions of the type associated with the functional groups on the surface while monitoring the conductance of the dispersion. The location and the sharpness of the "endpoint(s)" provide the desired information. The cation exchange capacity of a clay colloid may be obtained using either type of titration. An electrokinetic titration is one in which the *electrophoretic mobility,* u_E, is measured as a function of electrolyte composition (potential-determining ions, charge-determining ions, supporting electrolyte, *etc.*). U_E is defined as the velocity of particles undergoing electrophoresis, divided by the strength of the electric field inducing it. The so-called *zeta potential,* which is derived from it, may often be identified with the Stern potential, ψ_δ. The concentration of a pdi or cdi at which the electrophoretic mobility (or zeta potential) is zero is denoted the *isoelectric point* (*IEP*). Electrokinetics will be discussed in the final sections of this chapter. While all types of titrations may be applied to all types of colloids, potentiometric titrations are especially convenient for the determination of the *PZC*, while both potentiometric and conductometric titrations are convenient for the study of clays or of colloids with ionizable functional groups at their surfaces.

2. Potentiometric titrations

A potentiometric titration is illustrated for a colloid with a surface of constant potential, determined by H^+/OH^- ion concentrations (*i.e.*, *pH*), as is typical of most oxides. An electrolytic cell is constructed, as shown in Fig. 6-20, consisting of the colloidal dispersion (containing a known weight of dry colloid of known specific area dispersed in a known volume of water) in a "background electrolyte" of say 10^{-3} M KNO_3. The background electrolyte

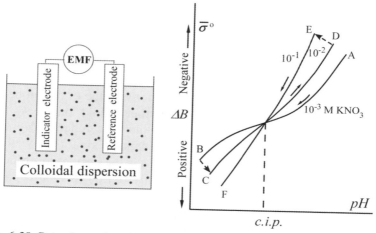

Fig. 6-20: Potentiometric colloid titration. After [Lyklema, J., **Fundamentals of Interface and Colloid Science, Vol. I**, pp. 5.101ff, Academic Press, London (1995).]

should consist of ions that are known not to specifically adsorb to the surface of the particles, and its concentration must be sufficiently high that the concentration of potential-determining ions (pdi) is always negligible in comparison to it. The cell has an indicator electrode, whose EMF depends directly on the concentration of one of the potential-determining ions in the solution. In the present case, the indicator electrode measures *pH* (typically a glass electrode), which is be pre-calibrated. The other is a reference electrode, such as a calomel electrode, whose EMF is fixed. The *pH* of the dispersion is noted, and titration proceeds as follows. In the example shown in Fig. 6-20, the titration starts at Point A, on the basic side. One then adds acid (a standard solution of say HNO_3), and monitors the *pH* as a function of the amount of added acid. One must keep track of the net amount of titrant (as acid or base) added and plot this as the ordinate against the measured *pH*, as the abscissa. The titration should be stopped when the *pH* reaches as low a value as possible while satisfying the requirement that the pdi concentration stays at least an order of magnitude lower than that of the background electrolyte, Point B in the figure. In this way, an inflected curve of the type AB in Fig. 6-20 is generated. It is shown with the usual convention of plotting negative charges upward (hence plotting the net amount of *base* added, *i.e.* ΔB = base added - acid added) and negative potentials (increasing *pH*'s) to the right. The background electrolyte is then increased (say to 10^{-2} M KNO$_3$), which produces an increase in the (positive) surface charge density (due the increased screening that exists at the higher electrolyte concentration), and the protons needed for that are withdrawn from solution, slightly increasing the *pH*. Curve CD is then generated by titrating with a standard base (say KOH). The background electrolyte concentration is then once again increased, and curve EF is generated. Again, the extremes of *pH* must be limited by the requirement that the H$^+$ or OH$^-$ concentrations must not exceed that of the background electrolyte. If ideal conditions prevail (*i.e.*, no *specific* adsorption of any of the ions present, no chemical alterations of the solid, no impurities, *etc.*), a common intersection point (*c.i.p.*) between the curves obtained at different background electrolyte concentrations is found. The *c.i.p.* gives the *pH* at which the surface charge density is 0, assumed to be the *PZC*, because this would be the only point where differences in screening due to differences in background electrolyte concentration would have no effect. The origin of the coordinates is then shifted to this point. The values of the ordinate relative to this origin are directly related to the surface charge density, $-\overline{\sigma}_{\text{solid}} \equiv -\overline{\sigma}_0$, and some insight into the measured curves is obtained by examining the calculated relationship between $-\overline{\sigma}_0$ and the surface potential (which in turn is directly related to the *pH* through the Nernst Equation). For a single symmetrical electrolyte, in accord with Eq. (6.34):

$$-\overline{\sigma}_0 = \left[2kT\varepsilon\varepsilon_0 n_\infty \right]^{1/2} \sinh \frac{ze\psi_0}{2kT}. \tag{6.77}$$

For a 1-1 electrolyte in water at 25°C:

$$-\overline{\sigma}_0 = -117.4\sqrt{C(\text{M})} \sinh \frac{\psi_0(\text{mV})}{51.4} \ [=] \ \text{mC/m}^2. \tag{6.78}$$

C(M) refers to the total electrolyte concentration, which for all practical purposes is that of the background electrolyte alone. It is evident from the above equation that the increased screening due to the increase in background electrolyte concentration allows the surface charge density to be larger for a given surface potential; specifically, $-\overline{\sigma}_0$ is proportional to $\sqrt{C}$. At least three such titrations are needed to verify the existence of a *c.i.p.*, which if found, validates the assumption of no specific adsorption of either the potential-determining or indifferent ions, as well as the use of the Nernst Equation in the vicinity of the *PZC*.

One may determine the appropriate value for the surface charge density at any point from the *adjusted* value of the net amount of base added by subtracting from it the *computed* amount of base that would be required to effect the same *pH* change in an equal volume of colloid-free solution. Alternatively, one may perform an identical titration against the colloid-free solution (the "blank"), producing a curve which must be adjusted vertically so that its adjusted $\Delta B_{blank} = 0$ at the *pH* corresponding to the *PZC*. Running a blank is often desirable because it automatically subtracts out the effects of dissolved CO_2. The difference between the adjusted values of ΔB and ΔB_{blank} (*i.e.*, ΔB_{net}) is converted to $-\overline{\sigma}_0$ with knowledge of the total dry mass of colloid present, m_{coll}, and its specific area Σ:

$$-\overline{\sigma}_0 = -F\left(\Gamma_H - \Gamma_{OH}\right) = F\frac{\Delta B_{net}}{(\text{Area})_{total}} = F\frac{\Delta B_{net}}{m_{col}\Sigma},\qquad(6.79)$$

and may thus produce a curve of surface charge density *vs. pH* (or, with the help of the Nernst Equation, *vs.* ψ_0). Figure 6-21 shows the classical results

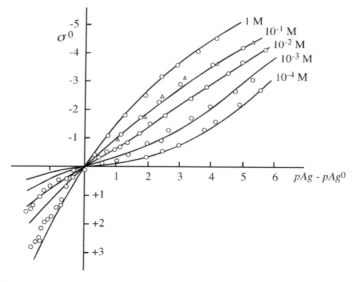

Fig. 6-21: Potentiometric titration of a AgI colloid by Verwey and de Bruyn. The drawn curves correspond to a background electrolyte of 7:1 KNO_3:$NaNO_3$. (o) refers to $NaClO_4$, and (Δ) to $NaNO_3$. The surface charge values are only relative, since the specific area Σ was not known. After [Overbeek, J. Th. G., **Colloid Science**, Vol. 1, p.162, Elsevier, Amsterdam, 1952.]

reported by Overbeek obtained for silver iodide colloids, where Ag^+ and I^- are the potential-determining ions (rather than H^+ and OH^-, as in the previous illustration). The common intersection point is taken as the point where $\overline{\sigma}_0 = 0$, and the *pAg*-value at which this occurs was found to be $pAg^0 = 5.52$. Surface charge densities could not be determined because the specific area was not accurately known. pAg and ψ_0 values are convertible because Nernst's Law, Eq. (6.6), applies.

In Fig. 6-21, only the curves for supporting electrolyte concentrations ≤ 10^{-3} M, in the vicinity of the *PZC* (around the inflection point) show exactly the shape predicted by Eq. (6.78). At higher salt concentrations, the Stern layer starts to come into play. This is also true for oxide titration curves shown schematically in Fig. 6-20. Moving away from the *PZC*, differences between the titration curves for the silver halides and for mineral oxides become evident. The silver iodide curves (for low-to-moderate electrolyte concentrations) are concave to the *pAg* axis, while the oxide curves are convex to the *pH* axis. When numerical values for the charge densities, $\bar{\sigma}_0$, obtained for the oxides are compared with those for AgI (or for the Hg/solution interface), they are found to be many times larger, and to be inconsistent with Eq. (6.1). This is attributed to the spongy, porous nature of oxide surfaces, allowing them to accommodate more charge than their nominal surface area would suggest.

It should be noted that differentiation of titration curves of the type shown in Figs. 6-20 or 6-21 leads to the differential capacitance of the double layer, in accord with Eq. (6.70), so it is a derived property for Nernstian surfaces, while it is directly measurable for the Hg/solution interface.

Potentiometric titrations may also be carried out on colloids with fixed populations of ionizable functional groups on their surfaces (such as latices). A schematic example of the measured *pH vs.* the amount of base added is shown in Fig. 6-22 for a latex with sulfonic acid groups. The curve

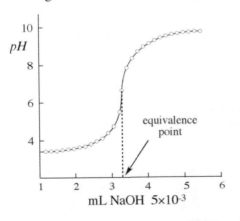

Fig. 6-22: Potentiometric titration of a latex colloid with sulfonic acid groups.

has a knee at the equivalence point (taken as the point of maximum slope), which becomes sharper when the groups are stronger. The sharpness (for weak groups) may also be increased by adding indifferent electrolyte to screen the resulting surface charge. The net amount of base added to reach the equivalence point may be used to compute the total number of ionized groups in the sample, and with knowledge of the surface area of the sample, the number of functional groups per unit area and the surface charge density, $\bar{\sigma}_0$, which for strong (fully-dissociated) groups is constant over the entire *pH* range. For weak acid groups, the *pH* at the equivalence point may be identified as the effective pK_a value for the groups, and the Henderson-Hasselbalch Equation (the titration equation for a weak acid) may be used to determine the degree of dissociation α as a function of *pH*:

$$pH = pK_a + \log\frac{\alpha}{1-\alpha}. \tag{6.80}$$

3. Conductometric titrations

Conductometric titrations are more commonly used for colloids with ionizable functional groups, such as latices.[35] The groups are usually acidic, and the common titrant is NaOH. As titrant is added to the colloidal dispersion, the conductance in Siemens (or Mho/m) is monitored. Results of the type shown in Fig. 6-23 may be obtained. The measured conductance is assumed to reflect the presence of ions in the bulk solution only, *i.e.*, the contribution of ions in the double layer itself (surface conductivity) is ignored.[36] The leftmost curve shows the situation encountered when the acid groups are strong, such as sulfuric or sulfonic acid groups. At first, as NaOH is added, the conductance decreases because the more conductive (mobile) H^+ ions are neutralized with OH^- ions and are replaced with the less mobile

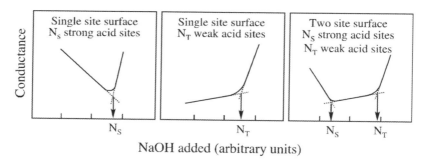

Fig. 6-23: Schematic of conductometric titrations, showing cases in which strong acid sites, weak acids and a mixed population of sites are present.

Na^+ ions. Once all the H^+ ions associated with the surface groups are neutralized, further addition of base just increases the total electrolyte content of the solution, and its conductivity rises. The extrapolated breakpoint is identified as the equivalence point. As a practical matter, the level of background electrolyte should be kept as low as possible in conductometric titrations because the effect of the NaOH addition would otherwise be swamped out. Furthermore, the solution must be thoroughly

[35] Bangs, L. B., **Uniform Latex Particles**, Form No. 1661-84, Seragen, Inc., Indianapolis, IN (1984).

[36] Overbeek, J. Th. G., in **Colloid Science**, Vol. 1, Irreversible Systems, H. R. Kruyt (Ed.), pp. 115-190, Elsevier, New York, 1952, for further discussion of this assumption.

de-gassed and the titration carried out under nitrogen to avoid the effects of dissolved CO_2. In the case of weak acid groups, such as those of carboxylic acid, a result of the type shown in the middle part of Fig. 6-23 would be encountered. As NaOH is added, only the few H^+ ions from the partially dissociated groups are neutralized by OH^- ions, so the resulting decrease in conductance may be overshadowed by the increase due to added electrolyte. When the equivalence point is reached, no further neutralization occurs, and the conductance increases more rapidly. The equivalence point may be identified by extrapolation as in the previous case. The rightmost curve shows the case when the surface contains a mixed population of strong and weak acidic groups. The equivalence points for both may be identified if they are sufficiently far apart. In all the above cases, the number of ionizable groups per unit area, or more commonly, the *parking area* per group (in nm^2/group) may be computed if the total particle surface area is known.

Either potentiometric or conductometric titrations may be used to determine the cation exchange capacity of clays.[37] A convenient procedure is to first convert the clay into the barium form by soaking the clay powder in $BaCl_2$ solution to exchange all of exchangeable cations in the clay with Ba^{++}. The clay is then titrated with a solution of a soluble sulfate ($MgSO_4$, $ZnSO_4$, *etc.*), which produces a reaction of the type:

$$Ba\text{-clay} + Mg^{++} + SO_4^{=} \rightarrow BaSO_4 + Mg\text{-clay}. \qquad (6.81)$$

In the potentiometric titration using magnesium sulfate, the activity of Mg^{++} in the dispersion is monitored by means of a Mg^{++}-specific electrode, *i.e.*, the potential of the electrode is proportional to the concentration of Mg^{++} ions in solution. With the complete insolubility of $BaSO_4$, addition of the $MgSO_4$ at first causes little change in either the electrode potential (because the added Mg^{++} ions are all tied up with the clay) or the solution conductance (because the Mg^{++} ions are simply traded for Ba^{++} ions). At the equivalence point, the potential of the Mg^{++}-specific electrode begins to rise sharply and then levels off. The amount of Mg^{++} added at the point of inflection determines the cation exchange capacity of the clay. In the conductometric titration, when the cation exchange capacity is reached, the conductance begins to rise sharply.

4. *Donnan equilibrium and the suspension effect.*

An interesting situation arises if one constructs a cell containing a colloidal suspension with a clear supernatant above it, with one electrode in the suspension and the other in the supernatant. Often a substantial difference in potential is observed, even though the supernatant and the medium surrounding the suspended particles are identical. This was

---

[37] Chiu, Y. C., Huang, L. N., Uang, C. M., and Huang, J. F., *Colloids Surfaces*, **46**, 327 (1990).

evidently first encountered by soil scientists, who noted an apparent difference between the *pH* of water in soil suspensions and the filtrate from the same suspensions.[38] Often the cells used for potentiometric titrations make use of a liquid junction, which produces the same effect. Now known as the "suspension effect" (or Pallman effect), it is traceable to the phenomenon of Donnan equilibrium.[39]

Donnan equilibrium refers to the equilibrium distribution of salts across a membrane that is permeable to some, but not all, of the ions in the system. It produces a situation identical to that of the suspension effect, and arises in particular when the system contains a macro-ion, such as a polyelectrolyte (*e.g.*, a protein), or charged colloidal particles. Consider a solution or suspension of a macro-cation M of valence $+z$ in Chamber II of the osmotic equilibrium system shown in Fig. 6-24(a). It will be accompanied by a sufficient number of (assume monovalent) counterions

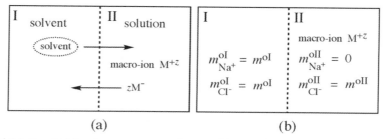

Fig. 6-24: Ion distribution across a semipermeable membrane. (a) Chamber I is pure solvent, and Chamber II contains macro-ions and diffusible counterions. (b) Initial state, with Chamber I containing solvent and a supporting electrolyte, NaCl, of diffusible ions, and Chamber II contains macro-ions and diffusible counterions, Cl⁻.

X^- to assure electroneutrality. Chamber I is initially pure water. Solvent (water) will flow from side I to side II until the equilibrium osmotic pressure is established, but in addition, the counterions will want to cross over to side I in an attempt to equalize the concentration (activity) of this "diffusible" ion between the chambers. A few of them will do so, but this will set up, across the membrane, a potential difference of sufficient magnitude to restrain others from crossing. A more realistic situation is that exemplified in Fig. 6-24(b), in which, in addition to the macro-ions and their counterions, there is a supporting electrolyte of which both anions and cations are diffusible. Often the supporting electrolyte has one of its ions the same as the counterion to the macro-ion. The objective is to calculate the equilibrium distribution of the ions across the membrane.

To fix ideas, assume that the counterion to the macro-ion M^{+z} is Cl⁻, and the supporting electrolyte is NaCl. The initial state, as shown in Fig. 6-24(b), is as follows: 1) both chambers contain 1 kg of water; 2) the amount

[38] Gillespie, L. J., and Hurst, L., *Soil Sci.*, **4**, 313 (1917).

[39] Chernoberezhski, Yu. M., "The Suspension Effect," in **Surface and Colloid Science**, Vol. 12, E. Matijevic (Ed.), pp. 359-453, Plenum Press, New York, 1982.

of M^{+z} ions in Chamber II is such that the required (for electro-neutrality) of counterions is $m_{Cl^-}^{oII} = m^{oII}$; 3) Chamber I contains NaCl at a molality, $m^{oI} = m_{Na^+}^{oI} = m_{Cl^-}^{oI}$. Several things then happen as the system moves toward equilibrium. Na^+ ions move from I to II, and Cl^- ions will accompany them in an attempt to maintain electro-neutrality as the system seeks to equalize the activity of the diffusible salt on both sides of the membrane. Solvent (water) at any instant senses which chamber has the higher osmolality (total molality of dissolved species) and flows across the membrane in the appropriate direction until the equilibrium osmotic pressure is established. At equilibrium, the chemical or electro-chemical potentials of all species that can cross the membrane must be equal in the two chambers, *i.e.*

$$\text{i) } \mu_{H_2O}^I = \mu_{H_2O}^{II}$$

$$\text{ii) } \tilde{\mu}_{Na^+}^I = \tilde{\mu}_{Na^+}^{II} \tag{6.82}$$

$$\text{iii) } \tilde{\mu}_{Cl^-}^I = \tilde{\mu}_{Cl^-}^{II}$$

It turns out that the concentrations of all the species are different in I and II at equilibrium, so that other factors affecting the chemical potential must come into play. Equilibrium with respect to the solvent (water) is achieved through the development of the osmotic pressure difference. The inequality of the ion concentrations in I and II is accompanied by an electrical potential difference across the membrane (the membrane potential), $\psi^{II} - \psi^I$. The "chemical potentials" expressed in (ii) and (iii) above are in fact *electro*-chemical potentials introduced in Eq. (6.2). Thus

$$\tilde{\mu}_{Na^+}^I = \mu_{Na^+}^\ominus + zF\psi^I + RT\ln m_{Na^+}^I, \tag{6.83}$$

where z is the valence of the ion (+1, in this case) and F is Faraday's constant. It has been assumed for now that the solution may be regarded as ideal-dilute, so that the ion activity is simply given as its molality. The effects of non-ideality approximately cancel out in the final result, so the assumption is tenable. Substitution of expressions like those of Eq. (6.83) into (ii) and (ii) above gives

$$\psi^{II} - \psi^I = -\frac{RT}{F}\ln\left[\frac{m_{Na^+}^{II}}{m_{Na^+}^I}\right] = \frac{RT}{F}\ln\left[\frac{m_{Cl^-}^{II}}{m_{Cl^-}^I}\right], \tag{6.84}$$

the last of which leads directly to the equilibrium requirement that

$$m_{Na^+}^I \cdot m_{Cl^-}^I = m_{Na^+}^{II} \cdot m_{Cl^-}^{II}. \tag{6.85}$$

While the anion and cation concentrations (molalities) are the same in Chamber I, they are unequal in the chamber containing the macro-ion. Equation (6.84) shows the ion molalities appearing in ratios, suggesting that the activity coefficients that should accompany them to account for solution non-ideality will indeed tend to cancel out.

The solution for the ion concentrations is completed with a material balance. If in the attainment of equilibrium, Δm moles of NaCl transfers from I to II, the equilibrium molalities are given by

$$m_{Na^+}^{I} = m_{Cl^-}^{I} = m^{oI} - \Delta m,$$

$$m_{Na^+}^{II} = \Delta m, \text{ and} \tag{6.86}$$

$$m_{Cl^-}^{II} = m^{oII} + \Delta m.$$

Substitution of Eqs. (6.86) into Eq. (6.85) gives

$$\Delta m = \frac{\left(m^{oI}\right)^2}{\left(m^{oII} + 2m^{oI}\right)}. \tag{6.87}$$

in terms of which all ion molalities are computed using Eq. (6.86).

Once the equilibrium ion molalities are computed, one may return to Eq. (6.84) to evaluate the membrane potential $\psi^{II} - \psi^{I}$. In the case of a colloidal suspension with a supernatant serum above it, there is no physical membrane, but gravity prevents the denser particles, which are the macro-ions, from invading the overlayer. The effect is thus the same as a membrane, and the potential difference noted in the suspension effect is the same as the membrane potential.

D. Electrokinetics

1. *The electrokinetic phenomena*

The existence of a diffuse electric double layer in a fluid phase adjacent to a charged solid surface makes possible a number of phenomena associated with the relative movement of the fluid containing the diffuse charge with respect to the surface. The study of such phenomena leads to the *zeta potential*, which is often close in value to the Stern potential and one of the most important properties of the electrical double layer which can be "measured." The relative motion is parallel to the surface (shearing). Consider a flat solid surface bearing a negative charge, with a diffuse layer of cations in the adjoining electrolyte solution, as shown in Fig. 6-25. Next imagine that an electric field is set up parallel to this surface. The excess positive ions in the diffuse layer will migrate toward the cathode ("down the field"), and their motion, through the action of viscosity, will move the liquid along the surface. Fluid motion generated in this manner is called *electro-osmosis*. It may be made to occur in a capillary tube, producing a plug flow velocity profile, as

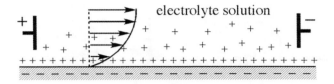

Fig. 6-25: Electro-osmosis.

shown in Fig. 6-26, in contrast to the very different parabolic profile resulting from an imposed pressure drop. Electro-osmosis is most often studied as flow through a porous plug across which a field is imposed and monitored with a bubble flow meter, as shown in Fig. 6-27.

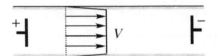

Fig. 6-26: Electro-osmosis inside a capillary,

An apparently different electrokinetic phenomenon occurs if the solid surface along which the field is imposed is that of a particle suspended in a liquid medium, as shown in Fig. 6-28. The imposition of the field will then set the solid particle in motion, with the liquid medium at rest, *i.e.*, *electrophoresis*, as described in Chap. 5. If the particle is large relative to the double layer thickness, it is to be noted that electro-osmosis and electrophoresis are phenomenologically identical processes, and if the coordinate system is made to move with the solid, the flow fields expressing the motion of the liquid relative to the solid are identical.[40]

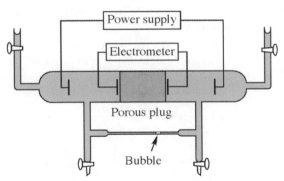

Fig. 6-27: Electro-osmosis measuring system.

Other electrokinetic phenomena may be conceived by considering additional means of inducing relative motion between a charged surface and its adjoining diffuse double layer. They are summarized in Table 6-4.

---

[40] This is not precisely true in all situations, as discussed later.

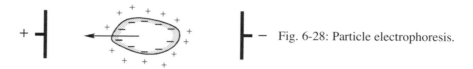

Fig. 6-28: Particle electrophoresis.

If, by means of a pressure gradient, liquid is forced through a capillary whose walls bear an electric charge and against which there is a diffuse double layer, there will be a net convection of counterions in the direction of the flow. If it is tap water being forced through a glass capillary, as shown in Fig. 6-29, the capillary wall will have a negative charge, and cations will be "pumped" to the right, setting up an electric field. The potential difference between the electrodes in Fig. 6-29 becomes greater as time passes, until a steady state is achieved in which the back conduction of electricity comes

Table 6-4: The electrokinetic effects.

Externally imposed EMF produces motion.
- Liquid moves, solid is stationary: *Electro-osmosis*
- Solid moves, liquid is stationary: *Electrophoresis*

Externally imposed motion produces EMF (and back current)
- Liquid forced to move past solid: *Streaming potential (current)*
- Solids forced to move through liquid: *Sedimentation potential*

into balance with its convection from left to right. The potential drop at steady state, $E(+) - E(-)$, is the *streaming potential*, and the back current, i_s, is the *streaming current*. If the electrical resistance for the streaming current is very large (as is the case for poorly-conducting non-aqueous media), the potential drops may become very large. Hence when pumping hydrocarbon fuels, unless the pump line is grounded, sparks can be produced, perhaps leading to what is known as an "electrokinetic explosion." It was largely the

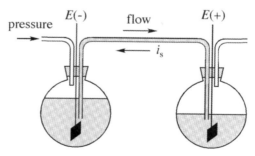

Fig. 6-29: Apparatus for creating streaming potential.

attempt to understand such events that motivated early research into the existence of electric charges in apolar media, such as hydrocarbon fuels.[41] It

---

[41] Klinkenberg, A., and van der Minne, J. L., **Electrostatics in the Petroleum Industry**, Elsevier, New York, 1958.

is now believed, as discussed at the beginning of this chapter, that such charges can be stabilized in nonpolar media only when sequestered within rather large structures such as reverse micelles, polymers, *etc*. Early work attributed the charge stabilization to such impurities as metal soaps produced over time by corrosion reactions with the container walls abetted by traces of water.[42] One (difficult) way to avoid the buildup of charges would be to keep the medium absolutely dry, while an alternative (more successful) approach is to intentionally add compounds that assist in the stabilization of charges to the point that the conductivity is high enough to retard the buildup of charge. It is believed that 1 μMho/m (*i.e.* μSiemen/m) is adequate, and compounds used for this purpose include tetraisoamylammonium picrate (TIAP) (at 0.1mM in benzene) and the cerium salts of mono- and di-alkyl (C_{14}–C_{18}) salicylic acid (at 0.0025 mM in gasoline).

In a final permutation, one can induce relative motion between a stationary liquid and a moving solid, as occurs during sedimentation or centrifugation of charged colloids. This leads to a *sedimentation potential*, (or centrifugation potential), as shown in Fig. 6-30, and the corresponding sedimentation current, *etc*. These phenomena, discussed earlier in connection with sedimentation rates, are known collectively as the "Dorn effect."

The four "electrokinetic effects" listed in Table 6-4 are seen to be different manifestations of the same general phenomenon. Additional electrokinetic effects may be conceived by using either alternating potentials or alternating pressure waves of sufficient frequency. The AC equivalent of the DC effects of electro-osmosis and electrophoresis are pictured schematically in Fig. 6-31, which shows a colloidal dispersion subjected

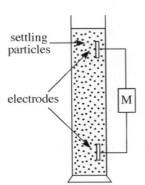

settling particles

electrodes

M

Fig. 6-30: Sedimentation potential (Dorn effect).

either to an alternating electric field or to an ultrasound wave. An imposed ultrasonic pressure wave will tend to displace the counterion clouds from around the colloid particles, whose greater inertia will tend to keep them in place. On the down- sloping side of the pressure wave, the counterion clouds will be displaced from the particle in the opposite sense. The effect is an

---

[42] Piper, J. D., Fleiger, A. G., Smith, C. C., and Kerstein, N. A., *Ind. Eng. Chem.*, **31**, 307 (1939).

alternating electric field set up in the dispersion, with the maximum polarity observed using inert electrodes placed apart a distance equal to one-half the wavelength of the ultrasound wave. The measured amplitude is the "ultrasonic vibration potential" (UVP), or for present purposes, the *colloid vibration potential* (CVP). Its existence was first shown by Debye and is sometimes called the Debye effect. It may also be observed in ionic solutions in the absence of colloids, due then to differences in inertia that exist between anions and cations of different mass. Present-day instrumentation also monitors the same phenomenon in terms of the alternating *current* produced, yielding the colloidal vibration current (CVI).

The opposite effect is also measurable. When the dispersion is subjected to an alternating potential, an acoustic wave is generated which, by means of a piezoelectric detector, may be transduced into an alternating potential of a certain amplitude, depending on the dispersion and the double layer properties. The amplitude in this case is called the electrokinetic sonic amplitude (ESA). These phenomena are now known collectively as *electro-acoustic effects*.

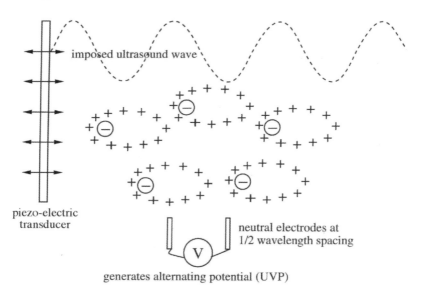

Fig. 6-31: Electro-acoustic effects: ultrasonic vibration potential (UVP) or electrokinetic sonic amplitude (ESA).

2. *The zeta potential and its interpretation*

Since all the DC electrokinetic effects are essentially the same phenomenon, only one description is needed, and electro-osmosis will serve that purpose. Consider the liquid to be in a relatively large capillary tube. The surface is charged negatively, and a diffuse layer of cations adjoins it. The double layer thickness, κ^{-1}, is thus small relative to the capillary radius, so one may treat the capillary wall as though it were flat, as implied by Fig.

6-32. The position measured away from the wall is y, and E_x is the strength of the electric field, imposed in the $+x$ direction. The objective is to derive an expression for the electro-osmotic velocity profile, $v_x(y)$, and in particular the velocity far from the surface, V_e (the electro-osmotic velocity), in terms of the electrical double layer potential profile and the fluid properties. A force balance is drawn on a thin control volume in the fluid (called a "shell") as shown. Without any pressure forces in the x direction, there are two forces which in the steady state are in balance: the Maxwell (or electrical) stresses, F_{el}, and viscous resistance, F_{res}. The electrical term accounts for the force exerted by the electric field on the net positive charge, q, within the control volume acting in the $+x$ direction, viz.

$$F_{el} = qE_x = \rho_e dV\, E_x,\qquad(6.88)$$

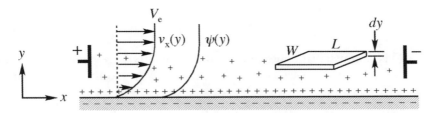

Fig. 6-32: Diagram for derivation of electro-osmotic velocity profile.

where $dV = LWdy$ is the volume of the shell. Using the one-dimensional form of the Poisson Equation, Eq. (6.15), the space charge density, ρ_e, is expressed in terms of the field due to the double layer. There should be no distortions of this field, $d\psi/dy$, due to the imposed field, E_x, which is orthogonal to it. Then

$$\rho_e = -\varepsilon\varepsilon_0 \frac{d^2\psi}{dy^2},\qquad(6.89)$$

and

$$F_{el} = -\varepsilon\varepsilon_0 \frac{d^2\psi}{dy^2} E_x LWdy.\qquad(6.90)$$

The net resisting force in the $-x$ direction is the difference between the larger viscous shear stress τ_{yx} acting of the bottom surface of the shell, $y = y$ (where the velocity gradient is steeper) and the smaller viscous shear stress at the top surface, $y = y + dy$, exerted by the fluid in the shell on the fluid above it: Then

$$F_{res} = LW\, d\tau_{yx} = LW \frac{d\tau_{yx}}{dy} dy.\qquad(6.91)$$

The viscous shear stress, for a Newtonian fluid, is given by Newton's Law of viscosity:

$$\tau_{yx} = -\mu \frac{dv_x}{dy},$$
(6.92)

where μ is the viscosity. Substituting Eq. (6.92) into (6.91) gives:

$$F_{res} = -LW\mu \frac{d^2v_x}{dy^2} dy.$$
(6.93)

Equating the forces at steady state:

$$\varepsilon\varepsilon_0 \frac{d^2\psi}{dy^2} E_x LW dy = LW\mu \frac{d^2v_x}{dy^2} dy.$$
(6.94)

Simplifying, and integrating once with respect to y yields:

$$\varepsilon\varepsilon_0 E_x \frac{d\psi}{dy} = \mu \frac{dv_x}{dy} + C,$$
(6.95)

where C is a constant of integration. It can readily be evaluated by noting that far away from the surface ($y \to \infty$), both $d\psi/dy$ and $dv_x/dy \to 0$, so $C = 0$.

The second integration is then taken from some position very near to the surface, $y \approx 0$, where $v_x = 0$ (the "slip plane") out to some indefinite y. This gives:

$$\varepsilon\varepsilon_0 E_x (\psi - \zeta) = \mu(v_x - 0),$$
(6.96)

where ζ is taken as the value of the potential *at the slip plane*. Since the capillary tube is large in radius relative to κ^{-1}, one expects to have $\psi \to 0$ long before the centerline is reached. The desired electro-osmotic velocity profile is thus:

$$v_x(y) = \frac{\varepsilon\varepsilon_0 E_x}{\mu}[\psi(y) - \zeta],$$
(6.97)

and the electro-osmotic velocity, V_e, is

$$V_e = -\frac{\varepsilon\varepsilon_0 E_x \zeta}{\mu}.$$
(6.98)

There will be "plug flow" at this velocity across most of the cross-section of the tube, as shown in Fig. 6-26.

To evaluate V_x, or the entire velocity profile, one needs the value of the potential at the slip plane, ζ, termed the *zeta potential*. Note the sign: if ζ is negative (positive ions in the double layer), then V_x is positive for positive E_x; *i.e.*, fluid moves toward the cathode. For water at 20°C, the above formula gives:

$$V_e = 0.690(-E_x)\zeta \quad [=] \ \mu m/s,$$
(6.99)

where E_x [=] Volt/cm, and ζ [=] mV. Thus for fields of a few Volts/cm and a typical zeta potential of 50 mV, the electro-osmotic velocity is seen to be a few tens of μm/s.

It is critical to inquire into the meaning of the zeta potential. The exact location of the "slip plane" is not known independently, but it seems reasonable to assume that it is not far away from the Stern plane, as depicted in Fig. 6-33. In fact, it is often assumed that they coincide, so that

$$\psi_\delta \approx \zeta. \tag{6.100}$$

The slip plane may be just *slightly* further out into the solution than the outer edge of the Stern layer, but Eq. (6.100) is generally considered adequate. When polymers or other macromolecules are adsorbed, however, the hydrodynamic slip plane will be moved outward, perhaps beyond the range of the diffuse double layer, so that ζ may approach 0, despite the existence of a large Stern potential.[43,44] The zeta potential is seen to depend on all those factors that determine the structure of the diffuse double layer, including the presence of neutral species. In fact, the measured ζ is one of the few things that can be obtained that directly describes the double layer, and it plays a vital role in that regard. The condition under which the zeta potential is zero is especially important, and is referred to as the *isoelectric point (IEP)*. The *IEP* is generally taken as the point where $\psi_\delta \approx \zeta = 0$.

3. *Electrokinetic measurements; micro-electrophoresis*

The foregoing discussion suggests that the zeta potential might be determined using any of the DC electrokinetic effects. The zeta potential itself is not measured but must be derived from measured quantities with the help of an appropriate model. For systems with extended surfaces, such as flat specimens or large and/or irregular particles, the most common method is based on streaming potential (or streaming current) measurements. Commercial instrumentation is available in which the serum is made to flow through a packed bed of the particles or fibers for whose surface the zeta potential is sought. The measured streaming potential or current is related to the zeta potential through an analysis similar to that detailed for electro-osmosis. Instruments designed to determine the zeta potential for planar solids use a parallel plate capillary, *i.e.*, narrow flow channel bound by the specimen or a device in which a low cell is clamped to the specimen surface. An interesting new device (called the "Zetaspin®" - www.zetaspin.com) uses a spinning disk electrode. The specimen of interest in the form of a disk or wafer of diameter from a few mm to a few hundred mm, is affixed to the bottom of the rotating disk. The resulting analytically accessible radial flow

[43] Maier, H., Baker, J. A., and Berg, J. C., *J. Colloid Interface Sci.*, **119**, 512 (1987).
[44] Miller, N. P., and Berg, J. C., *Colloids Surfaces*, **59**, 119 (1991).

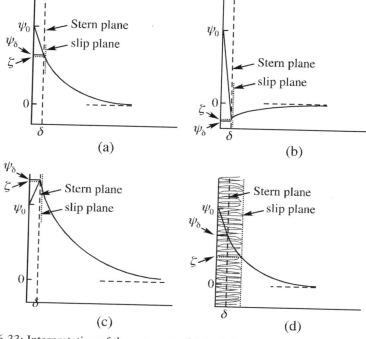

Fig. 6-33: Interpretation of the zeta potential in different situations. (a) ordinary Stern adsorption, (b) specific adsorption of counterions, (c) specific adsorption of coions, (d) macromolecule adsorption.

of the diffuse double layer generates a measurable potential that can be related directly to the streaming potential.[45]

For colloidal systems, the most common method used is based on electrophoresis, and in this application, referred to as *micro-electrophoresis*. One measures the velocity of particles (in the simplest instruments, by means of a measuring darkfield microscope) in an imposed electric field, as indicated schematically in Fig. 6-34. In this configuration, patterned after the Rank Brothers Mark II Apparatus (Rank Brothers, Cambridge, UK), the cell

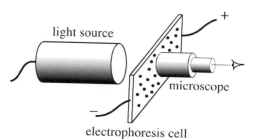

Fig. 6-34: Schematic of device for measuring zeta potential by micro-electrophoresis.

[45] Sides, P. J., and Hoggard, J. D., *Langmuir*, **20**, 11493 (2004);
    Hoggard, J. D., Sides. P. J., and Prieve, D. C., *Langmuir*, **21**, 7433 (2005);
    Sides, P. J., Newman, J., Hoggard, J. D., and Prieve, D. C., *Langmuir*, **22**, 9765 (2008).

is rectangular and oriented on its edge. Other instruments use cells of different shapes and/or orientations. The electric field is established, and the velocity of the particles is measured with a stopwatch as they move past a set of grid lines visible in the eyepiece of a darkfield microscope. One obtains the *electrophoretic mobility*, u_E, defined as the particle velocity divided by the imposed field strength, *i.e.*,

$$u_E = V_p / E_x,$$ (6.101)

usually given in units of $(\mu m/s)/(Volt/cm)$. If it is assumed that the particles have smooth surfaces and are large relative to the thickness of the double layer that surrounds them, then the geometry is effectively that of a diffuse double adjacent to a flat plate. Then setting up the axes on the particle itself, the velocity field in the vicinity of the solid surface is identical to that obtained for electro-osmosis, so that $V_p = -V_e$. Then using Eq. (6.98):

$$u_E = \frac{V_p}{E_x} = \frac{\varepsilon \varepsilon_0 \zeta}{\mu}.$$ (6.102)

This is the Helmholtz-Smoluchowski (HS) Equation, and it provides a direct means for evaluating the zeta potential from a measurement of u_E valid for large, smooth, non-conducting particles. The usual criterion is $\kappa a > 200$, where a is the particle radius. (More conservatively, especially when polyvalent ions are present, the criterion should be set at $\kappa a > 300$.) This turns out to limit the use of Eq. (6.102) to colloids of particle diameter ≥ 1 μm) in aqueous media of ionic strength $I \geq 10^{-3}$ M. The particles must also be non-conducting so that the electric field lines are not diverted through them, as discussed below. The general relationship between measured electrophoretic mobility and zeta potential is deferred to the next section.

Commercial instrumentation is available for making measurements of electrophoretic mobility, as exemplified by the Rank Brothers system mentioned above, and the PenKem Lazer Zee Meter™, no longer manufactured, but present in many industrial laboratories. In the latter instrument, an entire swarm of particles may be viewed at one time. A rotating prism device is used, and the rate of its rotation is adjusted by the user to compensate for the particle motion, until the swarm appears to be stationary. The prism rotation rate is then translated internally into the electrophoretic velocity, and using the HS theory (even when it may *not* be valid) into a directly read zeta potential.

It is important to realize that V_p is *not* the velocity observed in such instruments unless careful precautions are taken. The walls of the glass or quartz cell generally bear a negative charge, so that a diffuse layer of cations exists adjacent to them. This sets up an *electro-osmotic* flow through the cell (with or without particles), as shown in Fig. 6-35. But since there can be no net flow through the cell, a back-pressure is built up, sending fluid in the reverse direction in the core of the cell, as shown. At steady state, these

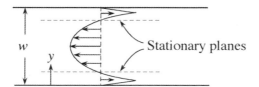

Fig. 6-35: Velocity profile in an electrophoresis cell, viewed from the top. The cell width w is typically 1-2 mm.

flows balance, giving a velocity profile like that shown in Fig. 6-35. To obtain an expression for this velocity profile, consider fully developed flow between parallel plates, with no edge effects. The profile is obtained by adding the electro-osmotic flow and the back-flow due to the generated pressure gradient:

$$v_x(y) = v_{eo} + v_{press} = -\frac{\varepsilon\varepsilon_0 E_x \zeta}{\mu} + \frac{w^2}{2\mu}\left(\frac{dp}{dx}\right)\left[\left(\frac{y}{w}\right)^2 - \left(\frac{y}{w}\right)\right].$$ (6.103)

The pressure gradient (dp/dx) is evaluated by noting that the net volumetric flow, Q_{flow}, is zero:

$$Q_{flow} = \int_0^w v_x(y)dy = 0 = -\frac{\varepsilon\varepsilon_0 E_x \zeta w}{\mu} + \frac{w^3}{12\mu}\left(\frac{dp}{dx}\right),$$ (6.104)

so that:

$$v_x = -\frac{\varepsilon\varepsilon_0 E_x \zeta}{\mu}\left[1 + 6\left(\frac{y^2}{w^2} - \frac{y}{w}\right)\right].$$ (6.105)

The observed motion of a particle in such a cell is the sum of its electrokinetic velocity and its convective velocity, dependent on its location in the cell. Only at the two locations where $v_x = 0$ will the observed velocity be the true electrophoretic velocity. These "stationary planes," are located at $y = 0.2113w$, and $0.7887w$. (When the finite height and length of the cell are accounted for, the stationary planes are somewhat nearer the wall.) Particles are viewed in these planes using a low depth-of-field objective such that out-of-plane particles are not in focus. Finding a stationary plane is thus a critical point in the use of direct electrophoretic mobility measurements. Some of the newer instruments for measuring electrophoretic mobility, *e.g.*, the Brookhaven Instruments ZetaPlus® or ZetaPALS® use cells in which the electrode pair is far enough removed from the cell walls that electro-osmosis is negligible, and the problem is avoided.

A number of more sophisticated techniques and commercial instruments are available for the measurement of electrophoretic mobility, most of which are based on one or another of the modes of dynamic light scattering described in Chap. 5. Using photon correlation spectroscopy (PCS) in the ordinary mode, the decay constant Γ of the intensity autocorrelation function gives directly the diffusion constant, *i.e.*, $\Gamma = Q^2 D$, where Q is the magnitude of the scattering wave vector. When the sample chamber is an electrophoresis cell, there is a *drift* of particle motion

superimposed upon the Brownian motion, and the autocorrelation function is
shifted with the decay constant now given by:

$$\Gamma = Q^2 D \pm iQV_{\text{p}}. \tag{6.106}$$

Thus electrophoretic PCS yields directly the average particle electrophoretic
velocity, and hence the electrophoretic mobility.

The most common device now used to measure electrophoretic
mobility in colloids is that of laser-doppler electrophoresis (LDE), as
exemplified by the Malvern Instruments ZetaSizer® unit for LDE shown
schematically in Fig. 6-36. A beam splitter is placed in the path of the
incident laser light to produce a pair of beams that are made to cross at one
of the stationary planes in the electrophoresis cell, creating a pattern of
interference fringes. The fringe spacing, Δ, is given by

$$\Delta = \frac{\lambda}{2n\sin(\theta/2)}, \tag{6.107}$$

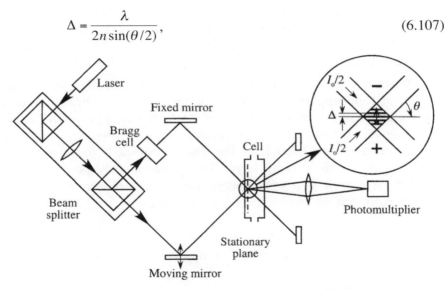

Fig. 6-36: Schematic of apparatus for laser Doppler electrophoresis (LDE). After [Hunter,
R. J., **Foundations of Colloid Science**, 2nd ed., p. 392, Oxford Univ. Press, Oxford, UK,
2001.]

where λ is the wavelength of the incident light, n is the refractive index of
the medium, and θ is the scattering angle. The electric field in the sample
cell is oriented normal to the fringes so that when a particle drifts across
them, a flickering in the scattering intensity is detected. Since the distance
between the fringes is known, and the flickering frequency is measured
(again, using a measured autocorrelation function), one obtains the average
particle velocity, as $V_{\text{p}} = \tau_{\Delta}\Delta$, where τ_{Δ} is the time required for a particle to
traverse one fringe spacing. The sign of the particle charge is determined by
setting the mirror directing one of the crossing beams into oscillatory
motion. Determining whether the particles appear to move faster or slower
as the mirror is moving in the direction of the field allows determination of

the direction of particle movement relative to the field. The LDE technique can be made up to 1000 times more sensitive (*i.e.*, able to detect very low mobilities, as might occur in non-aqueous or high-salt media) by phase modulation of one of the crossed beams (using a Bragg cell), so that the fringes are made to move in the direction opposite to that of the particle drift. This technology is termed phase angle light scattering (PALS). Commercial LDE instruments generally yield a histogram of the electrophoretic mobilities of the particles in the sample dispersion, as well as the particle size distribution.

4. Relationship of zeta potential to electrophoretic mobility

The HS relationship of the measured electrophoretic mobility to the zeta potential given by Eq. (6.98) is valid, as mentioned earlier, only for very large values of κa, restricting its use to rather large particles under relatively high salt conditions. Nonetheless, its comparative simplicity often causes it to be used for electrophoresis in aqueous media even when it is not warranted, and some commercial instruments, as the Lazer Zee Meter™ mentioned earlier, are "programmed" to apply it under all conditions.

When the HS conditions do not hold, the relationship between zeta potential and electrophoretic mobility may be more complicated. Another simple situation exists, however, for very *small* κa-values ($\kappa a < 0.1$), as may occur when the particles are in the nano size region, or when the solution has a low concentration of electrolyte. The latter is especially the case for non-aqueous media, with κ^{-1} very large. This situation may be analyzed by assuming the particle is a small sphere moving in response to the electric field in accord with the motive force (Maxwell stress):

$$F_{\mathrm{mot}} = qE_x,$$
(6.108)

where q is total charge on the surface of the sphere, given by

$$q = 4\pi a^2 \overline{\sigma}_0.$$
(6.109)

The surface charge density on the sphere, $\overline{\sigma}_0$, is obtained using Eq. (6.27):

$$\overline{\sigma}_0 = -\varepsilon\varepsilon_0 \left(\frac{d\psi}{dr}\right)_{r=a}.$$
(6.110)

The potential surrounding a small sphere is given by Eq. (6.45):

$$\psi(r) = \psi_0 \frac{a}{r}\exp\left[-\kappa(r - a)\right] \approx \zeta \frac{a}{r}\exp\left[-\kappa(r - a)\right],$$
(6.111)

leading to, upon combining Eqs. (6.109) – (6.111):

$$q = 4\pi\varepsilon\varepsilon_0 a(1 + \kappa a)\zeta.$$
(6.112)

The motive force is balanced by the hydrodynamic resistance of the sphere in moving through the liquid:

$$F_{res} = 6\pi\mu a V_p.$$ (6.113)

Equating $F_{mot} = F_{res}$ gives:

$$u_E = \frac{2\varepsilon\varepsilon_0\zeta}{3\mu}.$$ (6.114)

This is the Hückel Equation, and it is valid only in the limit of $\kappa a <$ 0.1. Thus the two extremes, pictured in Fig. 6-37, have formulas for the electrophoretic mobility that differ only by a constant multiplicative factor. For intermediate conditions, this factor is presumably a function (at least) of κa, so that:

$$u_E = \frac{2}{3}C(\kappa a)\frac{\varepsilon\varepsilon_0\zeta}{\mu}.$$ (6.115)

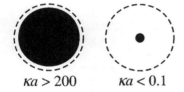

Fig. 6-37: Limiting cases for the relationship of zeta potential to measured electrophoretic mobility.

$\kappa a > 200$ $\kappa a < 0.1$

The coefficient $C(\kappa a)$, defined by the above equation, thus varies between 1 and 3/2 as κa varies between < 0.1 and > 200. It is clear that many colloidal suspensions have properties falling between the limiting cases of HS and Hückel. The difference between the HS and the Hückel results comes from the different nature of the resistance to particle movement. When the double layer is thin (κa large), the primary resistance is *retardation*, *i.e.*, the tendency of the counterions to move in the direction opposite to that of the particle. The HS analysis in effect considers *only* retardation by equating the particle movement in one direction exactly equal to the virtual movement of the surrounding liquid in the opposite direction. Retardation in the case of small κa is negligible (although the counterions seek to migrate, this motion is not transmitted to the particle), and the resistance to particle movement is just that of its own hydrodynamic friction, $f = 6\pi\mu a$, for a sphere. It has been shown[46] that the ratio of the retardation force to the hydrodynamic friction force is proportional to κa.

The situation corresponding to the intermediate range was first considered by Henry (1931)[47], subject to the assumptions that 1) the particles are spheres, 2) the diffuse double layer is undistorted by the externally applied field, *i.e.*, the double layer field and the imposed field are

[46] Dukhin, S. S., and Deryaguin, B. V., in **Surface and Colloid Science**, Vol. 7, E. Matijevic (Ed.), pp. 1-335, Wiley-Interscience, New York, 1974.

[47] Henry, D. C., *Proc. Roy. Soc.*, **133A**, 106 (1931).

superposed (Helmholtz and Smoluchowski had assumed that these fields were simply orthogonal.), and 3) potentials are low, *i.e.*, $ze\psi_s/kT < 1$ (the Debye-Hückel approximation). His analysis accounted for both retardation and hydrodynamic friction, and his results give the factor, C, in the form of two rather awkward power series expressions in κa. A more convenient expression has been provided by Ohshima, *viz.*:[48]

$$C(\kappa a) = 1 + \frac{1}{2}\left[1 + \frac{2.5}{\kappa a(1 + 2e^{-\kappa a})}\right]^{-3}. \tag{6.116}$$

C varies smoothly from 1 to 3/2, as shown in the topmost curve in Fig. 6-38. It is evident that the use of Henry's result requires knowledge of the particle size, a, in contrast to the limiting HS or Hückel results that are independent of particle size.

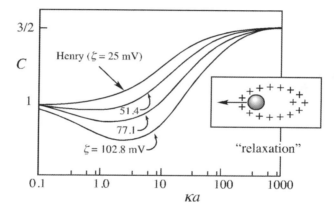

Fig. 6-38: Variation of the "constant" C with κR for various effective zeta potentials for a 1-1 electrolyte. After [Wiersema, P. H., Loeb, A. L., and Overbeek, J. Th. G., *J. Colloid Interface Sci.*, **22**, 78 (1966).]

Henry also allowed for particle conductivity, which could be accounted for by expressing the prefatory factor in the form:

$$C'(\kappa a, \lambda) = 1 + 2\lambda[C(\kappa a) - 1], \tag{6.117}$$

where

$$\lambda = \frac{(k_m - k_p)}{(2k_m + k_p)}, \tag{6.118}$$

and k_m and k_p are the conductivities of the bulk electrolyte and the particles, respectively. The function C' approaches 1 for small κa regardless of conductivity, and approaches 0 for highly conductive particles, when κa is

---

[48] Ohshima, H., *J. Colloid Interface Sci.*, **168**, 269 (1994).

large. For the interesting special case of $k_p = k_m$, $\lambda = 0$, the Hückel result holds for all values of κa. It has been pointed out,[49] however, that in most practical cases, "conducting" particles are rapidly polarized in the applied electric field and behave more nearly as non-conductors.

The most restrictive of Henry's assumptions is that of the low potential approximation. In the more general case of higher potentials, the diffuse double layer shape is distorted by the electrophoretic motion itself, as shown in Fig. 6-38, and the additional drag is termed the *relaxation effect*.

Numerical solutions for the general case, but still limited to zeta potentials less than about 100 mV, were first published by Wiersema *et al.* [50], but the complete solution (for spherical particles) was given finally by O'Brien and White.[51] Software (Mobility™) implementing their treatment is available. Dukhin and Derjaguin (*loc. cit.*) present approximate solutions that match those of O'Brien and White quite closely over significant ranges of κa. Their result was simplified further by O'Brien and Hunter[52] into an especially convenient form which reproduces the O'Brien and White computations to an accuracy of order $(\kappa a)^{-1}$. It expresses the dimensionless electrophoretic mobility:

$$\tilde{U}_E = \frac{3}{2} \frac{\mu}{\varepsilon \varepsilon_0} \left(\frac{e}{kT} \right) \cdot u_E \tag{6.119}$$

as a function of the dimensionless zeta potential:

$$\tilde{\xi} = \frac{\zeta}{\left(kT / e \right)}, \tag{6.120}$$

in the form, for a z-z electrolyte:

$$\tilde{U}_E = \frac{3}{2} \tilde{\xi} - \frac{6 \left[\dfrac{\tilde{\xi}}{2} - \dfrac{\ln 2}{z} \left(1 - \exp\left(-z\tilde{\xi} \right) \right) \right]}{2 + \dfrac{\kappa a}{\left(1 + 3\tilde{m}_{\pm}/z^2 \right)} \exp\left(\dfrac{-z\tilde{\xi}}{2} \right)}, \tag{6.121}$$

where $\tilde{m}_{\pm}$ is the average dimensionless mobility of the counterions and coions, defined as:

$$\tilde{m}_{\pm} = \frac{2\varepsilon\varepsilon_0 RT}{3\mu} \frac{z_{\pm}}{\Lambda^0_{\pm}}, \tag{6.122}$$

[49] Shaw, D. J., **Introduction to Colloid and Surface Chemistry**, 4th Ed., p. 202, Butterworth-Heineman, Oxford, 1992.

[50] Wiersema, P. H., Loeb, A. L., and Overbeek, J. Th. G., *J. Coll. Interface Sci.*, **22**, 78 (1966).

[51] O'Brien, R. W., and White, L. R., *J. Chem. Soc. Faraday II*, **74**, 1607 (1978).

[52] O'Brien, R. W., and Hunter, R. J., *Can. J. Chem.*, **59**, 1878 (1981).

with Λ^0 the limiting (infinite dilution) molar conductance of the ion. In water at 25°C, with Λ^0 [=] ohm^{-1} cm^2 equiv^{-1}: $\tilde{m}_\pm = 12.86\left(z_\pm / \Lambda^0_\pm\right)$. The accounting for the relaxation effect requires more input information than needed in the more simplified analyses, specifically, the limiting molar conductances of the ions present in the system.

A feature automatically taken into account when relaxation effects are described explicitly is "surface conductivity" associated with the diffuse part of the double layer. Since the ion concentration in the double layer around the particle is higher than in the surrounding medium, there will be an excess "surface conductivity" that affects the distribution of the electric field near the particle surface. This effect is negligible for cases of small κa, where the field is relatively unaffected by the presence of the particle, but may be important in the HS region. Booth[53] and Henry[54] suggest that the zeta potential computed by the HS Equation, ζ_{HS}, should be corrected in accord with

$$\zeta_{HS} = \zeta\left(1 + \frac{k_s}{k_m a}\right), \tag{6.123}$$

where, again, k_m is the bulk electrolyte conductivity ([=] Cs^{-1}V^{-1}m^{-1}) and k_s ([=] Cs^{-1}V^{-1}) is the "surface conductivity." There may also be a surface conductivity associated with the inner part of the double layer, which is taken into account by k_s, if there are data for it. A more detailed discussion of surface conductivity is given by Lyklema.[55]

Another complicating factor is that of particle porosity (as might be encountered in the electrophoresis of aggregates). Both computations[56] and experiments[57] indicate that there may be significant contributions due to the effective particle conductivity due in turn to the electrolyte-filled pores and to the electro-osmotic "jetting" of liquid through the pores.

A comprehensive monograph on zeta potential and related phenomena has been prepared by Hunter.[58]

5. Electrokinetic titrations

When the electrophoretic mobility (or zeta potential) of a colloid is monitored during a titration against potential-determining or charge-determining ions, or even indifferent electrolyte, one is performing an *electrokinetic titration*, as mentioned earlier. Such a procedure when using potential-determining or charge-determining electrolyte as titrant, locates

[53] Booth, F., *Trans. Faraday Soc.,* **44**, 955 (1948).

[54] Henry, D. C., *Trans. Faraday Soc.,* **44**, 1021 (1948).

[55] Lyklema, J., **Fundamentals of Interace and Colloid Science**, Vol. II, pp. 4.31ff, Academic Press, London, 1995.

[56] Miller, N. P., O'Brien, R. W., and Berg, J. C., *J. Coll. Interface Sci.,* **153**, 237 (1992).

[57] Miller, N. P., and Berg, J. C., *J. Coll. Interface Sci.,* **159**, 253 (1993).

[58] Hunter, R. J., **Zeta Potential in Colloid Science**, Academic Press, London, 1981.

what is called the *isoelectric point* (*IEP*) of the colloid, defined as the point at which the zeta potential is zero. It is effectively the point at which the Stern potential is zero. If the background electrolyte concentration is varied, and a common intersection point (*c.i.p.*) is found, one may infer that the ions of the background electrolyte are not specifically adsorbed. Comparison of the isoelectric point with the point of zero charge (obtained by a potentiometric titration) yields information concerning specific adsorption of the titrant ions in the Stern layer. If the two points are coincident, one may assume that specific adsorption of pdi's has not occurred. If specific adsorption of cations has occurred, the *IEP* will be shifted to a higher *pH* (or *pAg*, *etc.*) while the *PZC* (as determined from a potentiometric titration) will be shifted to a lower *pH*.[59] Specific adsorption of anions will shift the *IEP* and the *PZC* in the opposite directions. Thus if one finds, for example, pH_{IEP} < pH_{PZC}, one may infer specific adsorption of anions (OH^-), *etc.*

The result of a typical electrokinetic titration of an oxide in the vicinity of its *PZC* is shown schematically in Fig. 6-39, where the measured value of the zeta potential is compared with the surface potential, ψ_0,

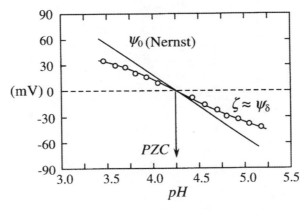

Fig. 6-39: Electrokinetic titration of an oxide whose *PZC* = 4.25, shown in comparison with the surface potential, ψ_0, computed using the Nernst Equation.

computed using the Nernst Equation, Eq. (6.8). If it is assumed that the zeta potential is equivalent to the Stern potential, the difference between it and ψ_0 is indicative of the contents of the Stern layer. The difference must also account for changes in the χ potential that may occur with changes in the polarity of the surface.

Aside from its value in checking models for surface electrical structure, the zeta potential is useful as a monitoring variable in dealing with the often very complex suspensions encountered in practice. For electrically stabilized colloids, it is a good measure of the stability to aggregation, *viz.*:

---

[59] Lyklema, J., **Fundamentals of Interface and Colloid Science, Vol. II**, p. 3.107ff, Academic Press, London, 1995.

1) the larger $|\zeta|$, the more stable will be the dispersion, and

2) as $|\zeta| \to 0$, the dispersion is less stable and will tend to aggregate.

An example is shown in Fig. 6-40, in which the commonly used coagulant $FeCl_3$ (providing trivalent cations under appropriate *pH* conditions) is being used to treat a wastewater sample. The plot of zeta potential *vs.* dosage shows that the condition of $\zeta = 0$, correlates with a minimum in turbidity and chemical oxygen demand (*COD*), as well as the ill effects associated with coagulant over-dosing. The latter is due to charge reversal of the colloid caused by specific adsorption of the counterions. The reason that repeptization occurs is that when the charge of the particle has been reversed, the counterion changes, generally to a monovalent species, for which the critical coagulation concentration (*CCC*) is much higher. This is discussed in more detail in the next chapter.

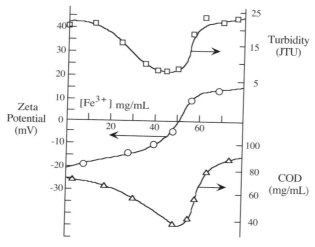

Fig. 6-40: QC tests showing correlation of the zero point of zeta potential with the measured turbidity and chemical oxygen demand (*COD*) of a $FeCl_3$-treated waste water sample.

6. Electro-acoustic measurements

One limitation that the different experimental methods discussed thus far have in common is their need to detect the motion of individual particles or swarms of individual particles. This requires that the dispersions be dilute (usually less than 2 vol%), non-aggregating and transparent. Many of the dispersions of practical interest are more concentrated, requiring dilution before measurements can be made. It is important in that case to dilute the dispersion with the serum of the original colloid (obtained as filtrate), rather than water, so that the state of the double layer remains unchanged. In some cases, however, it is not practical, or even possible, to do this. An approach well suited for the study of the electrokinetic properties of dense and/or opaque dispersions is electro-acoustics, *i.e.*, the *UVP* (*CVP* or *CVI*) or *ESA* phenomena described earlier. Commercial instruments from a number sources are available. The *CVP* and *ESA* potentials (or the *CVI*) are related to the zeta potential (but not identical to it). The first theoretical relationship between the *UVP* and ζ was developed by Enderby and co-workers,[60,61] but later investigated in detail in a series of papers by O'Brien.[62] O'Brien's result for (relatively) dense suspensions of spherical particles is summarized below, revealing the rather complex dependence of this relationship on other dispersion properties. The quantity obtained is the *dynamic* electrophoretic mobility, $u_{\text{E-dyn}}$, which for large κa, takes the form:

$$u_{\text{E-dyn}} = \frac{2}{3}\varepsilon\varepsilon_0\zeta(1+f)\frac{G}{\mu}, \tag{6.124}$$

where: ε and ε_0 are the dielectric constant and permittivity of free space, respectively; ζ is the zeta potential; f is a correction term for the "electric backfield"; G is a correction term for particle inertia, and μ is the viscosity of the medium. The *ESA* and the *UVP* are related to the dynamic electrophoretic mobility with the formulas:

$$ESA = \frac{P}{E_{\text{applied}}} = iC\phi(\Delta\rho)u_{\text{E-dyn}}F, \text{ and} \tag{6.125}$$

$$UVP = \frac{\Psi}{P_{\text{applied}}} = iC\phi u_{\text{E-dyn}}\frac{F}{K}, \tag{6.126}$$

where
$$P, P_{\text{applied}} = \text{pressure amplitude of sonic wave}$$
$$E_{\text{applied}} = \text{electric field amplitude}$$
$$i = \sqrt{-1}$$
$$C = \text{instrument constant}$$
$$\Psi = \text{potential difference at electrode}$$
$$\Delta\rho = \text{density difference between particle and fluid}$$
$$\phi = \text{volume fraction of particles}$$
$$F = \text{correction factor; a function of the electrode}$$
geometry, frequency, and speed of sound in the medium
$$K = \text{high frequency complex conductivity}$$

A similar formulation has been developed by Dukhin and Goetz[63] relating the dynamic mobility to the colloid vibration current (*CVI*).

[60] Enderby, J. A., *Proc. Roy. Soc.*, **207A**, 329 (1951).

[61] Booth, F., and Enderby, J. A., *Proc. Phys. Soc.*, **65**, 321 (1952).

[62] O'Brien, R. W., *J. Fluid Mech.*, **91**, 17 (1979);
O'Brien, R. W., *Adv. Colloid Interface Sci.*, **16**, 281 (1982);
O'Brien, R. W., *J. Fluid Mech.*, **190**, 71 (1988);
O'Brien, R. W., *J. Fluid Mech.*, **212**, 81 (1990).

[63] Dukhin, A. S., and Goetz, P. J., **Ultrasound for Characterizing Colloids**, Vol. 15, Elsevier, Amsterdam, 2002.

The evaluation of zeta potentials from measured electroacoustic potentials (or current) is clearly challenging, not just because of the complexity of the analysis, but also the requirement for additional properties and parameters. This might not be necessary, however, because despite the complexity, there is a close relationship between the *ESA* or *UVP* and ζ. When one goes up, so does the other, and they zero out under the same conditions, enabling identification of the *IEP*. An example of an electrokinetic titration of a titanium dioxide colloid in the presence of varying amounts adsorbed poly (vinyl alcohol) is shown in Fig. 6-41, which shows the effect of the adsorbed polymer in moving the slip plane outward from the particle surface without changing the *IEP*.

The especially important aspect of electro-acoustic measurements (if one can tolerate not knowing the zeta potential itself) is that they can be applied directly to *dense* (and/or opaque) dispersions (unlike those of electrophoresis) and to particles of arbitrary size. Instrumentation is available for both aqueous and non-aqueous systems. Reviews have been given by Marlowe, *et al.,* [64] Hunter,[65] and Dukhin and Goetz (*loc. cit.*)

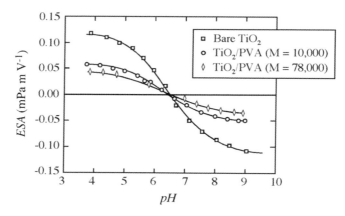

Fig. 6-41: Effect of PVA adlayer on electrokinetic sonic amplitude (ESA) signal for titania. [KCl] = $1.3 \cdot 10^{-3}$M. *IEP* unaffected by presence of adlayer. Adlayer appears to be non-draining. After [Miller, N. P., and Berg, J. C., *Colloids Surfaces*, **59**, 119 (1991).]

E. Dielectrophoresis and optical trapping

1. *Dielectrophoresis*

There is another important type of electrokinetic phenomenon that does not depend on the presence of a diffuse electrical double layer. Instead, it relies on the interaction of induced dipoles with electric fields, and can result in particle translation, rotation or trapping. When particle translation is

[64] Marlowe, B. J., Fairhurst, D., and Pendse, H. P., *Langmuir*, **4**, 611 (1988).
[65] Hunter, R. J., *Colloids Surfaces A*, **141**, 37 (1998).

the result, it is referred to as *dielectrophoresis*.[66] The principle is as follows. When a particle is immersed in a dielectric medium, and an electric field is imposed on the system, a dipole is induced in the particle whenever there is a mismatch between the dielectric constants (which scale with the polarizabilities) of the particle and the medium. If the field is uniform, the electric forces are balanced, and the net force on the particle is zero. If there is a field *gradient*, however, *i.e.*, the field is non-uniform, there will be a force on the particle, as pictured schematically in Fig. 6-42. If the dielectric constant (or the conductivity) of the particle is higher than that of the medium, as shown in the figure, a dipole will be induced that aligns with the field, and the particle will experience a net force directed toward the region of highest field intensity. The particle thus migrates up the field gradient in what is called *positive dielectrophoresis*. If the particle has a lower dielectric constant (or conductivity) than the medium, it will be polarized in the opposite sense, and migrate down the field gradient, exhibiting *negative dielectrophoresis*. Since the force acting on the particle is sensitive only to the field *gradient*, rather than to the field polarity, the direction of migration of the particle is unaffected by reversing the field. Thus the dielectrophoresis will occur in the same way in an AC field as in a DC field, and AC has the advantage of reducing or eliminating the effect of any electrophoretic force

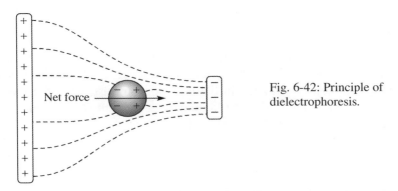

Fig. 6-42: Principle of dielectrophoresis.

due to the presence of an electric double layer, if there is one. Since dielectrophoresis is generally applied in the AC mode, it is sometimes referred to as *AC electrokinetics*. This is unfortunate, because it leads to confusion with the electro-acoustic phenomena described earlier.

The dielectrophoretic force, F_{DEP}, acting on a homogeneous, isotropic dielectric sphere of radius a, is given by:

$$F_{DEP} = \frac{1}{2}\alpha_p \nabla E^2 = 2\pi a^3 \varepsilon_m \, Re[K(\omega)]\nabla E^2, \qquad (6.127)$$

where E is the electric field strength and α_p is the polarizability of the particle. ε_m is the real part of the permittivity of the medium, and Re

[66] Jones, T. B., **Electromechanics of Particles**, Cambridge Univ. Press, New York, 1995

designates the real part of $K(\omega)$, the frequency-dependent Clausius-Mossotti factor in the form:

$$K(\omega) = \frac{\varepsilon_p^* - \varepsilon_m^*}{\varepsilon_p^* + 2\varepsilon_m^*},$$ (6.128)

where ε_p^* and ε_m^* are the complex permittivities of the particle and the medium, given for a material of finite conductivity by

$$\varepsilon_i^* = \varepsilon_i' + i\varepsilon_i'' + \frac{k_i}{i\omega},$$ (6.129)

where ε_i' and ε_i'' are the real and imaginary parts of the dielectric permittivity, respectively, and k_i is the electrical conductivity. ε_i'' is related to the dissipation or loss of energy within the system associated with the passage of current through a medium of finite conductivity. ω is the (angular) frequency of the applied field, and $i = \sqrt{-1}$. In many cases of practical interest, dielectric losses may be neglected, so that $\varepsilon_p'' = \varepsilon_m'' = 0$. Under these circumstances, the Clausius-Mossotti factor takes the form

$$K(\omega) = \frac{\varepsilon_p - \varepsilon_m - i\dfrac{(k_p - k_m)}{\omega}}{\varepsilon_p - 2\varepsilon_m - i\dfrac{(k_p + 2k_m)}{\omega}},$$ (6.130)

where the primes have been dropped from notation for the real parts of the permittivity coefficients. The frequency dependence of $K(\omega)$ derives not only from its explicit appearance in the equation but also from the fact the permittivities and conductivities are frequency dependent. The simplest case occurs for fully non-conductive materials in a DC field, in which case:

$$K(\omega) = K = \frac{\varepsilon_p - \varepsilon_m}{\varepsilon_p - 2\varepsilon_m}.$$ (6.131)

Several facts emerge from the equation for F_{DEP}. First, the force is independent of the sign of E (since it is squared). Second, the force varies directly with particle volume and thus becomes very small for small particles. It is generally applicable to particles in the micron size range, but with increasing technological developments has been applied increasingly to smaller particles, where the competition with the randomizing effects of Brownian motion is critical. The size dependence also suggests that particle separation based on size can be effected using dielectrophoresis. Lastly, for the more general case, the force is seen to be dependent on the angular frequency of the applied field. Even the sign of $K(\omega)$ may change with frequency. Thus particle separations may be effected by properly tuning the frequency of the applied field.

2. Electrorotation and traveling wave dielectrophoresis

A related phenomenon, and one dependent on the frequency dependence of $K(\omega)$ occurs if the particle is suspended in a rotating electric field. A dipole M will be established in the particle that rotates in synchrony with the rotating field. The time required for the dipole to form (relaxation), however, will cause the induced dipole to lag the rotation of the field E by a particular amount. The resulting interaction between the field and the lagging particle dipole induces a torque, T_{elect}, on the particle, causing it to rotate, a phenomenon called *electrorotation*,[67] An example of a physical setup that might be used to induce such rotation is shown in Fig. 6-43.

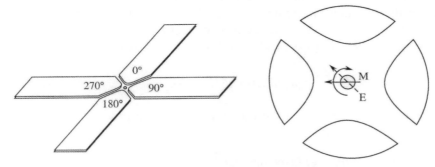

Fig. 6-43: Schematic of quadrupole electrode arrangement for investigating particle electrorotation.

Known as a quadrupole arrangement, four electrodes pointing toward a central region are made of thin films of gold or other conductor deposited on a glass or silica slide. A second glass or silica slide is placed on top, enclosing the liquid medium between the electrodes and the glass surfaces. The gap between the electrode tips is typically 50–1000 μm. The electrodes are given polarities of 90° advancing phase of their neighbor to the left and alternated such as to produce a field rotating at an angular frequency ω. An enlargement of the central region with the particle is shown on the right hand side. The cell is then placed in an optical microscope, and the rotation of a suitably tagged particle may be recorded photographically.

The torque on the uniform particle of radius a is given by

$$T_{elect} = -4\pi\varepsilon_m a^3 \, \text{Im}[K(\omega)]E^2, \tag{6.132}$$

where Im refers to the imaginary part of the Clausius-Mossotti factor. The negative sign indicates that the particle rotation lags that of the electrical field, and is proportional to the field strength squared and the cube of the

67 Zimmermann, U., and Neil, G. A., **Electromanipulation of Cells**, CRC Press, Cleveland, OH, 1996.

particle radius. The angular rate of rotation of the particle, Ω, is affected by the viscous drag of the medium, and taking this into account is given by[68]

$$\Omega = -\frac{\varepsilon_m \text{Im}[K(\omega)]E^2}{2\mu}, \tag{6.133}$$

where μ is the viscosity of the medium. Measurement of the rotation of the particle as a function of the frequency of rotation of the electric field provides a convenient means for determining the complex permittivity (or dielectric constant) of the particle.

A linear analogue to electrorotation occurs when the electrode components are arranged not in a circle but in a row, producing a traveling electric field. When the electric field moves rapidly enough, the induced dipole in the particle will again lag it, and a force will be induced on the particle causing it to move along the row of electrodes in a process known as *traveling-wave dielectrophoresis.*[69] The force on the particle is given by[70]

$$F_{\text{TWD}} = -\frac{4\pi}{\lambda}\varepsilon_m a^3 \text{Im}[K(\omega)]E^2, \tag{6.134}$$

where λ is the wavelength of the traveling wave. The process may be used for electrostatic pumping.[71]

3. Optical trapping; laser tweezers

The electrodes shown in Fig. 6-43 need not be "rotated" but may be stationary, with opposing components of opposite sign. This produces a field that is minimum at the center of the cell and maximum at the electrode tips. Particles experiencing positive dielectrophoresis would thus be drawn to these electrode tips. Particles experiencing *negative* dielectrophoresis, however, would be drawn to the center and trapped. Such non-contact *trapping* is a significant tool for controlling a particle's location.

When the alternating electric field is a wave of laser light (with a frequency of $\approx 10^{15}$ Hz), steep gradients can be produced as shown in Fig. 6-44 using a lens, and a particle near the focal point of the beam (called the *beam waist*) experiences a trapping force, sometimes called a *gradient force*, as given by Eq. (6.127). In addition to the gradient force, the trapped particle will experience a net *scattering force* in the direction of light propagation. One can envision the scattering force as the result of photons thought of as particles bombarding the trapped particle on its upbeam side. For trapped

[68] Arnold, W. M., and Zimmermann, U., *J. Electrostatics*, **21**, 151 (1988).

[69] Batchelder, J. S., *Rev. Sci. Instrum.*, **54**, 300 (1983);
   Hughes, M. P., Pethig, R., and Wang, X.-B., *J. Phys. D: Appl. Phys.*, **29**, 474 (1996).

[70] Huang, Y., Wang, X.-B., Tame, J., and Pethig, R., *J. Phys. D: Appl. Phys.*, **26**, 312 (1993).

[71] A number of interesting dielectrophoretic phenomena may be viewed at:
   http://www.youtube.com/watch?v=WBj3sBHumGw&feature=player_embedded

particles comparable in size to the wavelength of the laser light (the usual situation), exact computation of the scattering force requires full solution to

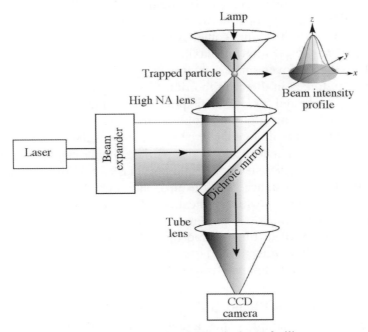

Fig. 6-44: Schematic of an optical trap facility.

the Mie scattering problem. (For Rayleigh scatterers, the computation would be straightforward.) This is the basis for the powerful technique of *optical trapping* for the positioning and manipulation of particles whose refractive index differs from the medium in which it is immersed. The detection of gradient and scattering forces caused by the interaction of focused laser light on particles was first reported by Ashkin[72] at Bell Labs in 1970. Then in 1986, Ashkin and coworkers[73] reported construction of an optical trap, now commonly referred to as *optical tweezers* or *laser tweezers*, for the positioning and manipulation of micron-sized particles, including latex or metal particles, viruses, bacteria, living cells, *etc.*[74] The instruments are most often custom-built, starting with the microscope and laser, but commercial instrumentation is also available.

The essential feature of the optical trap is that it is capable of holding the object particle in position in three dimensions. For stability in the *z*-direction, *i.e.*, along the beam as shown in Fig. 6-44, it is essential that the *z*-gradient forces exceed the scattering forces, so that the particle is not pushed forward out of position. This requires a very steep gradient in the light intensity, which is produced using an objective of high numerical aperture

[72] Ashkin, A., *Phys. Rev. Lett.*, **24**, 156 (1970).
[73] Ashkin, A., Dziedzic, J. M., Bjorkholm, J. E., and Chu, S., *Opt, Lett.*, **11**, 288 (1986).
[74] Neuman, K. C., and Block, S, M., *Rev. Sci. Insts.*, **75**, 2787 (2004).

(NA), usually between 1.2 and 1.4. The particle then positions itself just slightly down-beam from the exact center of the trap. Stability against movement in the x-y plane depends on the intensity gradient *across* the laser beam. Most optical traps operate with a Gaussian beam intensity profile, so that movement away from the center is resisted as the particle always seeks to move toward the position of maximum field intensity. The end result is that as the particle moves small distances away from its equilibrium position it experiences forces in direct proportion to its displacement, *i.e.*, it acts as though it were tethered by Hookean springs. Thus the optical tweezers may be used not only to position and hold a particle at a desired position, but once its "stiffness" is calibrated, can be used to assess the strength of various external forces brought to bear on the particle, including those needed in micro-rheological studies. It in principle provides an additional technique to probe colloid force-distance relationships, although the achievable spatial resolution puts limits on this application. The technique can be used to assemble micro-structures one particle at a time, or several at a time using independently controlled, time shared optical traps or "holographic optical tweezers" in which a propagating laser wave front is modified by passing it through a pattern of interference fringes.[75]

The major components of an optical trap system are the laser (usually Nd:YAG 1064 nm wavelength), a microscope with a high NA objective to create the trap in the sample plane, a dichroic mirror to direct the expanded laser beam up through the microscope objective while allowing the particle to be imaged by a CCD camera on the opposite side of the objective. A precise record of the particle position can be kept by reflecting some of the laser light from the particle to a quadratic photodiode cell (not shown in Fig. 6-44) the same as that used for monitoring the position of the cantilever tip of an atomic force microscope (AFM) described in Chap. 4.M.2.

The range of dielectrophoretic phenomena offer many possibilities for the manipulation of particles, including particle separations based on size and/or electrostatic/conductive properties, their transport to various locations on surfaces for assembly, or their contactless trapping in potential energy wells or levitated against gravity for assembly or further manipulation or examination. Through the use of such technologies as electron beam lithography, a virtually infinite variety of electrode configurations may be created for these purposes. Finally, it should be mentioned that an analogous array of phenomena of *magnetophoresis* exists for magnetizable particles in non-uniform magnetic fields.

[75] Patina, J. P., and Furst, E. M., *Langmuir*, **20**, 3940 (2004).

Some fun things to do:
Experiments and demonstrations for Chapter 6

1. *Electrophoresis of an electrocratic colloid*

When an electric field is imposed on an electrocratic colloid, the particles migrate to the electrode opposite to their surface charge. For an oxide, the surface charge depends on the *pH* relative to the *PZC* of the oxide. For titanium dioxide, for example, the *PZC* is usually approximately 6-7, so that at low *pH* (say 3), the particles should be positively charged and migrate toward the anode, while at *pH* 9, they should migrate toward the cathode. This can be demonstrated with a home-built cell, as described below.

Materials:

- Electrophoresis cell constructed and dimensioned in accord with Fig. E6-1. Exact dimensions are not critical. Use metal rods and copper plates as electrodes, held in a wood cover.
- 250-mL beaker filled with a dispersion (≈5 wt%) of TiO_2 of particle size approximately one μm
- 0.01N standard solutions of HCl and NaOH, and hydrion paper

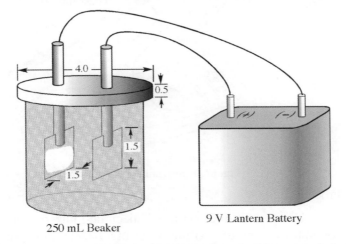

250 mL Beaker   9 V Lantern Battery

Fig. E6-1: Apparatus for monitoring electrophoresis. Dimensions in inches.

Procedure:

Adjust the *pH* of the above dispersion to 9 (check with hydrion paper), connect the electrodes to the terminals of the lantern cell as shown, place the electrodes in the dispersion and wait ≈ 15 min. Withdraw the electrode system from the dispersion and note that the TiO_2 particles have coated the electrode connected to the positive terminal. Adjust the *pH* to 3 and repeat the above, noting that under these conditions, the particles coat the negative terminal.

2. *Streaming potential*

When a liquid supporting a diffuse electrical double layer against a solid surface is made to flow along the surface, electric charge is convected downstream, establishing a potential difference between electrodes located upstream and downstream in the flow. This can be demonstrated with a simple home-built device.

Materials:

- Streaming potential apparatus constructed and dimensioned in accord with Fig. E6-2. Pack an 8-inch length of 0.25-in. diameter glass tube with $1/16^{th}$ inch glass beads. Use small plugs of glass wool to retain the beads in the tube, and fit each end with tight-fitting ≈ 2 inch lengths of Tygon® tubing. Insert pieces of copper wire in the tubing at the ends of the glass column to serve as electrodes, and connect these (using alligator clips and lead wires) to the terminals of a digital voltmeter with a minimum internal resistance of 10 MΩ.

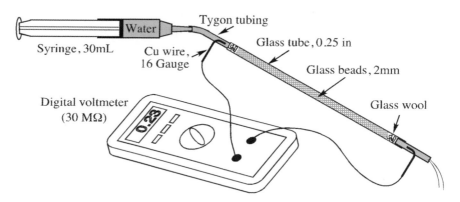

Fig. E6-2: Streaming potential apparatus.

Procedure:

Using a 30-mL syringe, inject de-ionized water through the column, catching the exiting water in a beaker. Note the potential and its polarity recorded on the digital voltmeter. It will be in the range of 100 – 150 mV, depending the speed of the water flow, with the positive potential at the downstream end. The potential recorded is closely related to the streaming potential. Next try the same experiment using tap water, and 0.005, 0.01, 0.05, and 0.10N NaCl solutions. Explain the differences observed. Also, vary the flow rates and note any differences in voltage developed. Finally, draw liquid *from* the beaker *into* the syringe and note the change in polarity.

Chapter 7

INTERACTION BETWEEN COLLOID PARTICLES

A. Overview and rationale

When sufficiently close, colloid particles exert forces on one another that determine their fate with respect to aggregation. These forces also play a central role in the rate of the process, the structure of the aggregates formed and the ultimate structure of the colloidal system. Particles of any specific type in a fluid medium are universally subject to attractive forces, called long-range van der Waals forces, that are the integrated result of the attraction between the molecules in the particles, with the mitigating influence of the molecules in the medium. The assumption of pairwise additivity of forces, credited to Hamaker, has been introduced in Chap. 2 in the context of describing the forces acting across a fluid film. Another approach to the description of long-range van der Waals forces, credited to Lifshitz, views these forces as the mutual electromagnetic interactions between macroscopic bodies. Electromagnetic waves emanate from any body as a result of the collective oscillations of the electrons they contain and interact with any nearby body as dictated by its dielectric response function, *i.e.*, its dielectric "constant" expressed as function of frequency. While the Lifshitz (quantum electrodynamic) theory is daunting, its results are tractable, and its predictions are considered reliable. As noted briefly in Chap. 5, if no interactions other than van der Waals attractions are present, aggregation of the particles in all but the most dilute colloidal dispersions will proceed quite rapidly. However, in many cases, inter-particle repulsive forces are also present. In particular, in aqueous media of low-to-moderate electrolyte content, the overlap of the diffuse electric double layers upon mutual particle approach results in significant repulsive forces. The interplay between van der Waals attractive forces and repulsive electrostatic forces is described by what is called "DLVO theory," which gives the most basic account of colloid stability with respect to aggregation, at least in aqueous media. Systems that can be adequately described in this way are termed *electrocratic colloids*. DLVO theory, even in aqueous media however, is often only a starting point to the description colloid behavior. Other types of interaction forces have come into focus in recent years, and their influence on aggregation phenomena is becoming more clearly understood.

The objectives of the current chapter are to first examine the van der Waals forces between approaching particles, followed by the development of expressions for the electrostatic repulsive forces and DLVO theory. The kinetics of aggregation and aggregate structure in electrocratic systems are examined next. Methods for the measurement of the interaction forces and the resulting aggregation kinetics are considered, followed by examination of stabilization, termed *steric stabilization*, resulting from the repulsive interactions between particles covered with adsorbed polymers. Finally, other types of interaction forces, falling into none of the above categories, are briefly described.

B. Long-range van der Waals interactions

1. *The Hamaker (microscopic) approach*

Van der Waals forces include electrostatic interactions between permanent charge distributions in the molecules (dipoles, quadrupoles, *etc.*), and induced charge distributions, as well as the ever-present London (or dispersion) forces due to the oscillations of the orbital electrons and the consequent interactions with the synchronous induced dipoles in the neighboring molecules. It is sometimes the case that the dispersion forces are all that need be considered. In accord with Eq. (2.7), the dispersion interaction potential between a pair of molecules depends on the distance of molecular separation, r, in accord with

$$\Phi_{molec} = -\frac{B}{r^6},$$

$$(7.1)$$

where B (in the simplest representation) depends only on the molecular polarizability and the ground state energy. It was Hamaker (1937),[1] after de Boer (1936),[2] who suggested that the van der Waals interaction between a pair of *macroscopic* objects (such as colloid particles) could be obtained by the pairwise summation of the energies acting between all the molecules in one body with all those in the other body. This six-fold integration was carried out in Chap. 2, Eqs. (2.12) and (2.13) for a pair of semi-infinite half-spaces interacting across a distance, D_0, with the result

$$\Phi_{macro}(\text{half spaces}) = -\frac{A}{12\pi D_0^2},$$

$$(7.2)$$

where A, designated the *Hamaker constant*, was given by:

$$A = \pi^2 \rho_N^2 B,$$

$$(7.3)$$

where ρ_N is the molecular density of the half spaces. What was done for semi-infinite half spaces can also be done for other geometries. Three of the

---

[1] Hamaker, H. C., *Physica*, **4**, 1058 (1937).
[2] De Boer, J. H., *Trans. Faraday Soc.*, **32**, 118 (1936).

other results that have been obtained for the interaction potential are summarized below, where D_0 is the distance between flat surfaces, and for curved surfaces, S_0 is the distance of closest approach.

- Plates of thickness δ:

$$\Phi = -\frac{A}{12\pi}\left[\frac{1}{D_0^2} + \frac{1}{(D_0 + 2\delta)^2} + \frac{2}{(D_0 + \delta)^2}\right]. \tag{7.4}$$

- Sphere of radius a and a flat-surface half space (or crossed cylinders of radius a):

$$\Phi = -\frac{A}{6}\left[\frac{a}{S_0} + \frac{a}{2a + S_0} + \ln\left(\frac{S_0}{2a + S_0}\right)\right]. \tag{7.5}$$

- Spheres of radii a_1 and a_2:

$$\Phi = -\frac{A}{6}\left[\frac{2a_1a_2}{S_0^2 + 2a_1S_0} + \frac{2a_1a_2}{S_0^2 + 2a_1S_0 + 2a_2S_0 + 4a_1a_2}\right.$$

$$\left. + \ln\left(\frac{S_0^2 + 2a_1S_0 + 2a_2S_0}{S_0^2 + 2a_1S_0 + 2a_2S_0 + 4a_1a_2}\right)\right]. \tag{7.6}$$

For the important special case of two spheres of equal radii, with $S_0 \ll a$:

$$\Phi = -\frac{Aa}{12S_0}. \tag{7.7}$$

It is clear that the integration has changed the dependence of the potential on the distance of separation from that of the inverse sixth power for individual molecules to the inverse *first* power for macroscopic particles, hence the name *long-range* van der Waals interactions.

Results for the interaction between macroscopic bodies always take the form:

$$\Phi = A \cdot f(\text{geometry}), \tag{7.8}$$

the product of the Hamaker constant, A, and a function of the geometry of the interacting systems. The geometric factor has been worked out for other shapes, and for layered and coated materials. An important result was obtained by Vold[3] for prolate ellipsoids of revolution, showing that the strongest interaction was obtained for an end-to-end configuration, explaining the tendency of particles to form end-to-end string aggregates.

Equation (7.3) suggests that A may be computed *ab initio* from molecular parameters, *i.e.*, from knowledge of B_{vdW} (which will hereafter designated simply as B). B may be represented as the sum of Keesom (dipole-dipole), Debye (dipole-induced dipole) and London (dispersion)

[3] Vold, M. J., *Proc. Indian Acad. Sci.*, **46**, 152 (1957).

contributions. London theory assumes that molecules have only a single absorption frequency (the "ground state" or ionization frequency, v_I) with an oscillator strength given closely by the first ionization potential. The data needed for the evaluation of B are the molecular dipole moments, u_i, polarizabilities, α_{0i}, and first ionization potentials, I_i. No account of the effect of the intervening medium between the molecules was taken, nor was the possibility of more than a single resonance frequency considered. For this, a more general approach is required, such as that of McLachlan[4] below, and described in detail by Israelachvili,[5] and Mahanty and Ninham.[6] McLachlan's result for the interaction between molecules of type 2 and 3 in a medium of molecules of type 1 is:

$$B = \frac{6kT}{(4\pi\varepsilon_0)^2} \sum_{m=0,1,...}^{\infty} {}' \frac{\alpha_2(iv_m)\,\alpha_3(iv_m)}{\varepsilon_1^2(iv_m)}, \qquad (7.9)$$

where ε_0 is the permittivity of free space, $\alpha_j(iv_m)$ is the polarizability of molecules j (2 or 3), and $\varepsilon_1(iv_m)$ is the dielectric constant of the medium 1, all at *imaginary* frequencies iv_m, where the real frequencies are given by

$$v_m = \left(\frac{2\pi kT}{h}\right)m \approx 4\times 10^{13}\,m\ \text{s}^{-1}\text{(at 300K)}, \qquad (7.10)$$

where h is Planck's constant. The prime next to the summation sign in Eq. (7.9) means the $m = 0$ term in the summation is multiplied by 1/2. The expression of polarizabilities and dielectric constants as functions of the imaginary part of complex frequencies is a mathematical device that appears strange, but one may immediately express these as functions of real quantities. For the total polarizabilities (dropping the subscript):

$$\alpha(iv_m) = \frac{u^2}{3kT(1 + v_m/v_{rot})} + \frac{\alpha_0}{1 + (v_m/v_I)^2}, \qquad (7.11)$$

where u is the dipole moment of the molecule (if any), α_0 is the zero-frequency, or static polarizability, v_{rot} is the average rotational relaxation frequency of the molecule (usually $\approx 10^{11}$ s^{-1}), and v_I is the first (and assumed only) ionization absorption resonance frequency (usually $\approx 3 \times 10^{15}$ s^{-1}). For apolar molecules, the first term on the right of Eq. (7.11) is zero.

     Consider the interaction of molecules 2 and 3 in a vacuum, so that $\varepsilon_1 = 1$. Then the first term on right hand side of Eq. (7.9), *i.e.*, for m = 0, gives the zero-frequency component of B, and is seen to give the Keesom and Debye contributions for molecules with permanent dipoles. Specifically,

[4] McLachlan, A. D., *Proc. Roy. Soc.*, **202A**, 224;
     McLachlan, A. D., *Mol. Phys.*, **6**, 423 (1963);
     McLachlan, A. D., *Discuss. Faraday Soc.*, **40**, 239 (1965).
[5] Israelachvili, J. N., **Intermolecular & Surface Forces**, 2nd Ed., Chaps. 5 and 6, Academic Press, London, 1991.
[6] Mahanty, J., and Nihham, B. W., **Dispersion Forces**, Academic Press, London, 1976.

$$B_{v=0} = -\frac{1}{(4\pi\varepsilon_0)^2}\left[\frac{u_2^2 u_3^2}{3kT} + \alpha_{02}u_3^2 + \alpha_{03}u_2^2\right].\tag{7.12}$$

The non-zero frequencies start with $v_1 \approx 4 \times 10^{13}$ s^{-1} >> v_{rot}, so the dipole contribution can be neglected for finite frequencies. It can be shown that for many substances, the summation of terms for non-zero frequencies can be reduced to a single term:

$$B_{v>0} = -\frac{3\alpha_{02}\alpha_{03}}{4(4\pi\varepsilon_0)^2}\frac{hv_{12}v_{13}}{(v_{12}+v_{13})} \approx -\frac{3\alpha_{02}\alpha_{03}}{4(4\pi\varepsilon_0)^2}\frac{I_2 I_3}{(I_2+I_3)},\tag{7.13}$$

the same as London's original result, in which the ground state energies, hv_j have been replaced by the first ionization potentials, I_j. It is the cases for which there are multiple resonance frequencies (specific absorbance frequencies) that the full Eq. (7.9) must be used. The non-zero frequency terms of course require knowledge of any specific absorption frequencies for the materials. The B required for the evaluation of the Hamaker constant is

$$B = B_{v=0} + B_{v>0}.\tag{7.14}$$

This representation is quite good for many substances, and good results are obtained using its calculated values of B, in Eq. (2.7), for predicting the volumetric behavior of gases. Calculating Hamaker constants A from Eq. (7.3) using such B-values leads to $A \approx 0.5 \times 10^{-19}$ J for alkanes and for CCl_4, and $\approx 1.5 \times 10^{-19}$ J for water, in reasonable agreement with currently accepted values for these materials. The full range of Hamaker constants for all condensed phase materials interacting across a vacuum (or gas at low-to-moderate pressure) runs from about 1.0×10^{-20} J to as high as about 50×10^{-20} J for some metals. A number of values of Hamaker constants for different materials obtained in different ways have been tabulated, are a few are collected later in Table 7-1.

The values of A computed using Eqs. (7.12) - (7.14) take no account of the dispersion medium. The McLachlan approach, using the Hamaker expression for A in terms of B given by Eq. (7.9) *does* take the effect of the medium into account, but its results are not necessarily easy to implement. There is a simpler approximate way to account for the effect of the medium, however, as discussed earlier in Chap. 2. The effective Hamaker constant between a pair of particles of substance 2 in a dispersion medium 1, for example, may be estimated by regarding the process whereby the two particles come together as a displacement "reaction," as pictured below.

This gives the effective Hamaker constant as

$$A_{212} = A_{11} + A_{22} - 2A_{12}.\tag{7.15}$$

If the *cross*-Hamaker constant, A_{12}, can be taken as the geometric mean of the corresponding pure component constants,[7] one obtains

$$A_{212} = A_{11} + A_{22} - 2\sqrt{A_{11}A_{22}} = \left[\sqrt{A_{11}} - \sqrt{A_{22}}\right]^2. \qquad (7.16)$$

A_{212} is thus always positive regardless of the relative magnitudes of A_{11} and A_{22}. The greater the difference between Hamaker constants of the particles and the medium, the greater will be the effective attractive interaction potential energy between particles. For unlike particles, it is possible to analyze the process as analogous to that shown above Eq. (7.15), yielding

$$A_{213} = \left(\sqrt{A_{22}} - \sqrt{A_{11}}\right)\left(\sqrt{A_{33}} - \sqrt{A_{11}}\right). \qquad (7.17)$$

It can thereby be seen that the effective Hamaker constant A_{213} will be negative (suggesting van der Waals *repulsion*) if A_{11} is intermediate in value between A_{22} and A_{33}. Otherwise it is positive.

Equations such as (7.15) – (7.17) have been a recurring theme in this text. They arose first in Chap. 2 in the context of the discussion of the disjoining pressure in thin films, and again in Chap. 4 in the context of the discussion of adhesion.

2. Retardation

The London component of the van der Waals interaction between molecules has been computed with the assumption that the electron cloud oscillations in one molecule are synchronous with the oscillations in its neighbor with which it is interacting. As the molecules get farther apart, however, the finite time required for the propagation of the electromagnetic radiation between them begins to erode this assumption, producing the effect termed *retardation*. For the interaction between molecules in a gas phase the effect is generally small because at the distances for which it begins to appear (generally > 5 nm), the interaction has dropped to near zero in any case. But for the collective molecular interactions that produce long rang van der Waals forces between macroscopic bodies, they may be very important. The key parameter for estimating the importance of retardation, at least for interactions between molecules, is the ratio

$$y = \frac{2\pi v r}{c}, \qquad (7.18)$$

where v is the frequency of oscillation of the electron cloud about the molecule ($v = v_1 \approx 3 \times 10^{15}$ Hz) , and c is the speed of light. For large distances, corresponding to $y > 10$, Casimer and Polder[8] found that the dispersion force interaction decreased substantially, and that its dependence on r went from $1/r^6$ to $1/r^7$. Thus B for dispersion force interactions changed

---

[7] This is valid only to the extent that dispersion forces dominate the interaction.
[8] Casimer, H. B. G., and Polder, D., *Phys. Rev.*, **73**, 360 (1948).

from its unretarded value (valid when $y < 1$), given by Eq. (7.13) written for a pair of like molecules, to

$$B(y > 10) = -\frac{23}{4\pi} \frac{\alpha_0^2}{(4\pi\varepsilon_0)^2} \frac{hc}{r} . \tag{7.19}$$

For intermediate separations, the Casimer-Polder results were expressed using a correction factor to the unretarded result, so that:

$$B(y) = -\frac{3\alpha_0^2 I}{8(4\pi\varepsilon_0)^2} f(y), \tag{7.20}$$

where $f(y)$ was approximated by the formulae:[9]

$$(y < 3) \quad f(y) = 1.01 - 0.14y$$
$$\tag{7.21}$$
$$(y > 3) \quad f(y) = 2.45/y - 2.04/y^2$$

The expressions of Eq. (7.21) have been the starting point for the computation of the effect of retardation on the interactions between semi-finite half spaces, spheres, and spheres and plates, *etc.* For example, Clayfield *et al.*[10] integrated the pairwise molecular interactions to produce "exact" results for the sphere-sphere and the sphere-thick plate cases in the form of extremely elaborate correction factors to the unretarded dispersion interactions. Gregory[11] put these results in more tractable form. For the interaction between semi-infinite half spaces, the correction factor for Eq. (7.20) was given as

$$f(D_0) = \left(\frac{1}{1 + 0.0532 D_0}\right), \tag{7.22}$$

with D_0 [=] nm. For the interaction between a sphere and a flat-surface half space, Eq. (7.5), the correction factor was

$$f(S_0) = \left(\frac{1}{1 + 0.14 S_0}\right), \tag{7.23}$$

and finally, for the interaction between spheres, the correction factor for Eq. (7.7) was

$$f(S_0) = \left[1 - 0.0532 S_0 \ln\left(1 + 18.80/S_0\right)\right], \tag{7.24}$$

both with S_0 [=] nm.

It should be recalled that these computations are based on the assumption of only a single resonance frequency, whereas in the case of multiple frequencies, each would have its own retardation characteristics.

---

[9] Overbeek, J. Th. G., in **Colloid Science**, Vol. I, H. R. Kruyt (Ed.), p. 266, Elsevier, Amsterdam, 1948.

[10] Clayfield, E. J., Lumb, E. C., and Mackey, P. H., *J. Colloid Interface Sci.*, **37**, 382 (1971).

[11] Gregory, J., *J. Colloid Interface Sci.*, **83**, 138 (1981).

Nonetheless, the above corrections suggest that long-range van der Waals forces between spheres drop by 35% at a separation of only 5 nm, and by 70% at a separation of 20 nm, suggesting that failure to account for retardation may lead to serious over-estimation of these interactions.

3. *The Lifshitz (macroscopic) approach*

The assumption of molecular pairwise additivity in the Hamaker approach amounts to the neglect of all multi-body effects. This assumption is a good approximation in systems all of whose components are in the gaseous state and often even for interactions between condensed phases with a vacuum or a dilute gas as the intervening medium. It can be quite poor, however, in dealing with condensed phase systems interacting in a condensed phase medium. The pairwise additivity assumption is avoided in the completely different approach to the problem proposed by Lifshitz,[12] in which the interacting bodies are treated as continua. In this theory, interaction between *macroscopic* bodies is attributed to the fluctuating electromagnetic field in the gap between them. The material property that reflects these effects, *i.e.*, both the propagation and the reception of electromagnetic energy, is the *dielectric response function*, $\varepsilon(\omega)$:

$$\varepsilon(\omega) = \varepsilon'(\omega) + i\varepsilon''(\omega), \tag{7.25}$$

a complex number with both real and imaginary parts expressed as functions of the *angular* frequency of the electromagnetic radiation,

$$\omega = 2\pi\nu \ [=] \ \text{radians/s}. \tag{7.26}$$

The real part, $\varepsilon'(\omega)$, corresponds to the transmission, and the imaginary part, $\varepsilon''(\omega)$, to the absorption of radiation by the material. These parameters may also be put into terms of the perhaps more familiar complex refractive index, mentioned briefly in the context of describing inelastic light scattering (absorption in the visible portion of the spectrum) in Chap. 5. The refractive index was written as

$$n(\omega) = n_1(\omega) + i\kappa(\omega), \tag{7.27}$$

where the real part, n_1, is what is generally thought of as the refractive index, and $\kappa(\omega)$ is the absorption coefficient. The dielectric response function is given by

$$\varepsilon(\omega) = n^2(\omega) = \left[n_1(\omega) + i\kappa(\omega)\right]^2, \text{ so that} \tag{7.28}$$

$$\varepsilon'(\omega) = n_1^2(\omega) - \kappa^2(\omega), \text{ and} \tag{7.29}$$

$$\varepsilon''(\omega) = 2n_1(\omega)\kappa(\omega). \tag{7.30}$$

---

[12] Lifshitz, E. M., *J. Exp. Theoret. Phys.(USSR)*, **29**, 94 (1955);
Lifshitz, E. M., *Sov. Phys. JETP*, **2**, 73 (1956).

There are a number of methods by which the required spectroscopic data may be obtained in the laboratory, and in the following, it is assumed such data are available.

The Lifshitz theory is first used to compute the energy of interaction between half spaces of materials 2 and 3, separated by a medium of material 1. The resulting expression is then templated against Eq. (7.2) to give an expression for the Hamaker constant in the form:[13]

$$A_{213} = -\frac{3}{2}kT\sum_{m=0}^{\infty}{}'\int_{r_m}^{\infty} x\left\{\ln\left[1 - \Delta_{21}\Delta_{31}e^{-x}\right] + \ln\left[1 - \overline{\Delta}_{21}\overline{\Delta}_{31}e^{-x}\right]\right\}dx, \qquad (7.31)$$

where:
$$\Delta_{jk} = \frac{\varepsilon_j s_k - \varepsilon_k s_j}{\varepsilon_j s_k + \varepsilon_k s_j} \quad ; \quad \overline{\Delta}_{jk} = \frac{s_k - s_j}{s_k + s_j}$$

$$s_k^2 = x^2 + \left(\frac{2\xi_m D_0}{c}\right)^2 (\varepsilon_k - \varepsilon_1) \quad ; \quad r_m = \frac{2D_0\xi_m\sqrt{\varepsilon_1}}{c}$$

where c is the speed of light. The dielectric response functions have been expressed as functions of the imaginary part of a complex frequency,

$$\varepsilon_k = \varepsilon_k(i\xi_m), \quad \text{with} \quad \xi_m = \left(\frac{2\pi kT}{h}\right)m \quad . \qquad (7.32)$$

The dependence on D_0 provides the description of retardation effects, so that $A_{213} = A_{213}(D_0)$. Putting $D_0 \rightarrow 0$ recovers the non-retarded result, and $D_0 \rightarrow \infty$ gives the fully retarded result. Both the complexity of the theoretical expression and the data it appears to require are challenging.

For the non-retarded case, the second term in the integrand of Eq. (7.31) vanishes, and upon expanding the remaining logarithm and integrating term-by-term gives:

$$A_{213} = \frac{3}{2}kT\sum_{m=0}^{\infty}{}'\sum_{s=1}^{\infty}\frac{(\Delta_{23}\Delta_{13})^s}{s^3}, \qquad (7.33)$$

where
$$\Delta_{jk} = \frac{\varepsilon_j(i\xi_m) - \varepsilon_k(i\xi_m)}{\varepsilon_j(i\xi_m) + \varepsilon_k(i\xi_m)}. \qquad (7.34)$$

Ninham and Parsegian[14] showed that the functions $\varepsilon(i\xi_m)$ could be represented for many substances by

$$\varepsilon(i\xi_m) = 1 + \sum_{j=1}^{N}\frac{C_j}{1 + (\xi_m/\omega_j)^2}, \qquad (7.35)$$

where the ω_j are the characteristic relaxation (angular) frequencies, and

---

[13] Prieve, D. C., and Russel, W. B., *J. Colloid Interface Sci.*, **125**, 1(1988).
[14] Parsegian, V. A., and Ninham, B. W., *Nature*, **224**, 1197 (1969);
Ninham, B. W., Parsegian, V. A., *Biophys. J.*, **10**, 646 (1970).

$$C_j = \frac{2}{\pi} \frac{f_j}{\omega_j},$$ (7.36)

with f_j being the strength of the oscillator j for each absorption peak in the spectrum. Hough and White[15] showed further that most materials could be represented by a single effective absorption peak in the infrared (IR) part of the spectrum and a single effective peak in the ultraviolet (UV) part of the spectrum, and for a substantial subset, only the UV term is needed. Thus Eq. (7.35) becomes

$$\varepsilon(i\xi_m) = 1 + \frac{C_{IR}}{1 + (\xi_m/\omega_{IR})^2} + \frac{C_{UV}}{1 + (\xi_m/\omega_{UV})^2}.$$ (7.37)

The frequencies ω_{IR} and ω_{UV} are taken as the frequencies of the major absorption peaks in the IR and UV regions, respectively, and C_{UV} and C_{IR} can be evaluated as

$$C_{UV} = n_{UV\text{-vis}}^2 - 1, \text{ and } C_{IR} = \varepsilon(0) - C_{UV} - 1,$$ (7.38)

where $n_{UV\text{-vis}}$ is the refractive index in the UV-vis range, and $\varepsilon(0)$ is the static dielectric constant. Thus the Hamaker constant may be evaluated for many substances using Eq. (7.33). For the case of slabs of material 2 and material 3 interacting across a medium 1, and assuming that the UV absorption frequencies are approximately the same for the slabs and the medium, *viz.* v_1 (again, typically $\approx 3 \times 10^{15}$ s^{-1}), Eq. (7.33) becomes

$$A_{213} = \frac{3}{4}kT\left(\frac{\varepsilon_1 - \varepsilon_3}{\varepsilon_1 + \varepsilon_3}\right)\left(\frac{\varepsilon_1 - \varepsilon_2}{\varepsilon_1 + \varepsilon_2}\right) + \frac{3hv_1}{8\sqrt{2}} \frac{(n_3^2 - n_1^2)(n_2^2 - n_1^2)}{(n_1^2 + n_2^2)^{1/2}(n_1^2 + n_3^2)^{1/2}\left[(n_1^2 + n_3^2)^{1/2} + (n_1^2 + n_2^2)^{1/2}\right]}$$ (7.39)

where the ε's are the static dielectric constants, and the n's are the refractive indices in the visible range.

Water is somewhat more complex in that it exhibits five distinct absorption frequencies in the IR spectrum and six in the UV spectrum, which must be accounted for as separate terms in Eq. (7.35). The summation must also include the term $d/(1 + \xi\tau)$, where d and τ relate to the strength and relaxation times of the oscillator in the microwave region, due to the large dipole moment for water. Also, for media containing electrolyte, the interaction between permanent charge distributions may be screened, eliminating the zero-frequency contribution and leading to lower values of the Hamaker constant in such cases.

The computations from Lifshitz theory above have all been based on the interactions between semi-infinite half spaces, and this theory does not yield the clean factoring of dielectric and geometric contributions suggested by Eq. (7.8). Nonetheless it is common practice to use the half-space values so computed together with the geometric factors that emerged from the original Hamaker calculations and exemplified by Eqs. (7.2) – (7.7). Smith,

---

[15] Hough, D. B., and White, L. R., *Adv. Colloid Interface Sci.*, **14**, 3 (1980).

et al.[16] consider cases where this practice might not be appropriate. In particular, it may be problematic if there is a large mismatch between the dielectric permittivities of the two media, as between polymeric particles and water, and the surface separations become large, leading to retardation.

The general use of Lifshitz theory to predict Hamaker constant depends on the availability of the needed spectral data, but in recent years this database has become essentially complete. To the extent that good direct experimental data exist for the Hamaker constant, excellent agreement is found with the calculations based on Lifshitz theory. As mentioned earlier, the older pairwise additivity theory gives close to the same results for gases and often for media interacting across a vacuum or dilute gas, but it often fails badly for predicting interactions across condensed phase media. In this case, the Lifshitz theory often predicts values of the Hamaker constant significantly higher than the simple mixing rules would suggest. It appears that expressions for effective Hamaker constants in condensed media, such as those of Eqs. (7.15) and (7.16) need to be multiplied by factors ranging from about 1.5 to 2.5 to agree with Lifshitz calculations.[17] Hough and White (*loc. cit.*) suggest that the Lifshitz theory calculations are on sufficiently solid ground that Hamaker constants may regarded as known quantities in experiments seeking information about other phenomena convoluted with long range van der Waals forces.

Table 7-1: Hamaker constants ($\times 10^{20}$ J) calculated from Lifshitz theory. From [Hough, D. B., and White, L. R., *Adv. Colloid Interface Sci.*, **14**, 3 (1980).]

Material(M)	M|air|M	M|water|M	M|water|air	M|air|water	water|M|air
Pentane	3.75	0.336	0.153	3.63	0.108
Octane	4.50	0.410	-0.200	3.97	0.527
Dodecane	5.04	0.502	-0.436	4.20	0.848
Quartz (fused)	6.50	0.833	-1.01	4.81	---
Water	3.70	0	0	3.70	0
Silica (fused)	6.55	0.849	-1.03	4.83	---
PMMA	7.11	1.05	-1.25	5.03	---
PS	6.58	0.950	-1.06	4.81	---
PTFE	3.80	0.333	0.128	3.67	---
Ag, Au, Cu	---	30-40[18]	---	---	---

A few Hamaker constants as computed by Lifshitz theory are given in Table 7-1. Larger compilations are given in Hough and White (*loc. cit.*), Bergström,[19] Hunter,[20] Visser (*loc. cit.*), Lyklema,[21] Morrison and Ross,[22] and elsewhere.

---

[16] Smith, E. R., Mitchell, D. J., and Ninham, B. W., *J. Colloid Interface Sci.*, **45**, 55 (1973).

[17] Visser, J., *Adv. Colloid Interface Sci.*, **3**, 331 (1972).

[18] Parsegian, V. A., and Weiss, G. H., *J. Colloid Interface Sci.*, **81**, 285 (1981).

[19] Bergström, L., *Adv. Colloid Interface Sci.*, **70**, 125 (1997).

4. *Measurement of Hamaker constants*

Since the effect of long-range van der Waals forces is so pervasive in so many different situations of interest, it is not surprising that there are many routes to its experimental determination. Most important are the methods that permit the *direct* measurement of colloid forces. These methods have not only corroborated the Lifshitz results for Hamaker constants, but have also characterized and quantified other important "colloid forces" between surfaces, not yet discussed.

The surface forces apparatus (SFA), first developed by Tabor and Winterton in 1969[23] and taken to the next level by Israelachvilli and coworkers[24,25] is shown in Fig. 7-1. A pair of atomically-smooth, freshly-cleaved mica surfaces (with or without adsorbed layers of various kinds) are brought together as crossed cylindrical surfaces of radii of a few mm to within distances controllable to within one or two Å using a piezo-electric tube similar to that described in Chap. 4 as part of the device for executing

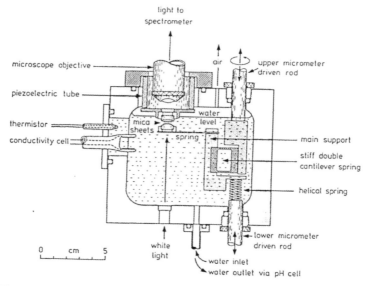

Fig. 7-1: The surface forces apparatus of Israelachvili and coworkers. From [Israelachvilli, J.N., **Intermolecular and Surface Forces**, 2nd. Ed., p. 170, Academic Press, London, 1991.]

[20] Hunter, R. J., **Foundations of Colloid Science**, 2nd Ed., p. 572, Oxford Univ. Press, Oxford, 2001.

[21] Lyklema, J., **Fundamentals of Interface and Colloid Science**, Vol. 1, App. 9, Academic Press, London, 1991.

[22] Morrison, I. D., and Ross, S., **Colloidal Dispersions**, Wiley-Interscience, New York, 2002.

[23] Tabor, D., and Winterton, R. H. S., *Proc. Roy. Soc.*, **312A**, 435 (1969).

[24] Israelachvilli, J. N., and Tabor, D., *Proc. Roy. Soc. Lond. A*, **331**, 19 (1972); Israelachvilli, J. N., and Tabor, D., *Prog. Surf. Membr. Sci.*, **7**, 1 (1973).

[25] Israelachvilli, J. N., and Adams, G. E., *Nature*, **262**, 774 (1976); Israelachvilli, J. N., and Adams, G. E., *J. Chem. Soc. Faraday Trans. I*, **74**, 975 (1978).

scanning tunnel microscopy (STM) or atomic force microscopy (AFM) measurements. The upper cylindrical surface is attached to the piezoelectric tube, while the bottom cylinder is attached to a calibrated cantilever spring. The separation distances are determined using interferometry or capacitance measurements. The force is measured by moving the upper cylinder up or down by a known amount (by expanding or contracting the piezoelectric crystal) and comparing this to the actual displacement observed. The difference, when multiplied by the spring constant of the cantilever (which can be varied by a factor up to 1000 during a run by clamping it at various positions) gives the force acting between the surfaces. Both attractive and repulsive forces can be measured, to $\pm 10^{-8}$ N.

An alternative type of actuator in the form of a bimorph strip has also been used to sense the force between the approaching surfaces leading to an instrument termed the MASIF (Measurement and Analysis of Surface Interaction Forces).[26] A bimorph strip is composed of two thin panels of ceramic elements bonded together with a flexible metallic panel as its central electrode. By wiring these two elements in such a way as to make one elongate and the other contract by applying voltage, inflection deviation occurs conforming to the waveform of the applied voltage.

The attractive force, F, between two crossed cylinders of radius a that is measured is geometrically identical to that between a sphere of radius a and a planar half space. The expression for this force can thus be obtained by differentiating the potential function given by Eq. (7.5), yielding, for the case of $S_0 \ll a$:

$$F_{\text{crsd cyl}} = \frac{d\Phi_{\text{sph-half sp}}}{dS_0} = -\frac{Aa}{6S_0^2}. \tag{7.40}$$

The general validity of Eq. (7.40) does not depend on the factorability of energy and geometry implicit in Eq. (7.5).[27] Comparing Eq. (7.40) with Eq. (7.2) for the *energy* of interaction between planar half spaces at the same spacing, it is noted that

$$F_{\text{crsd cyl}}(D_0) = 2\pi a \Phi_{\text{half sp}}(D_0). \tag{7.41}$$

One thus plots the results of an SFA experiment as F/a or $F/2\pi a$ vs. the spacing D_0 because of the direct relationship to the energy of interaction between planar half spaces.

SFA measurements have provided quantitative confirmation of Hamaker constants for a wide variety of different systems in aqueous and non-aqueous media as computed by Lifshitz theory, and the predictions of retardation effects have also been replicated. The technique is not limited to mica, of course, as that may serve as a substrate for the deposition (by a

[26] Parker, J. L., *Prog. Surface Sci.*, **47**, 205 (1994).

[27] Equation (7.40) may also be derived on purely geometric grounds using the *Derjaguin approximation* to be discussed in the next section.

variety of means) of virtually anything from sputter-coated metals, oxides, carbon, *etc.*, to adsorbed polymers, proteins or lipid bilayers. Other atomically smooth surfaces have also been developed as substrates. In addition to the validation of Lifshitz theory, the SFA technique has provided measurement (and confirmation of theory) for electrostatic interactions between surfaces with double layers and steric interactions between surfaces with polymer adlayers, both about to be discussed. Beyond this, SFA measurements have swung open the door on Nature to reveal a variety of other interaction forces to be discussed briefly at the end of this chapter. One example is shown in Fig. 7-2. It shows results obtained for the forces measured between mica surfaces in octamethylcyclotetrasiloxane, a liquid made up of large ($d \approx 9$Å) spherical molecules. The dashed line is that given by the expected Lifshitz-Hamaker interaction,

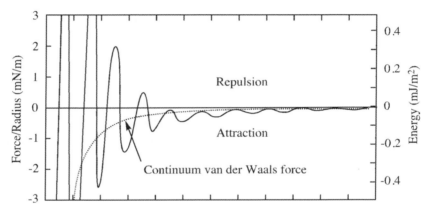

Fig. 7-2: Measured force between mica surfaces in Octamethylcyclotetrasiloxane ($d \approx 9$Å). After [Horn, R. G., and Israelachvili, J. N., *J. Chem. Phys.*, **75**, 1400 (1981).]

$$\frac{F}{a} = -\frac{A}{6D_0^2}, \qquad (7.42)$$

with a Hamaker constant, A, of 1.35×10^{-20} J. The ordinate on the right is the computed energy of interaction between flat surfaces (per unit area) in accord with $\Phi = F/2\pi a$. But the more interesting result is the actual measurement shown as the oscillatory solid line. The oscillations are the result of the solvent structuring forces, as successive layers of molecules are squeezed out of the liquid between the approaching mica surfaces, and for this case as well as that of other spherical, fairly rigid molecules like CCl_4, benzene, toluene, cyclohexane, *etc.*, the periodicity matches fairly closely with the mean molecular diameter. A similar result is obtained for water when the surfaces are within 2 nm of each other.[28] This result parallels the results for disjoining pressure in thin films pictured in Chap. 2, Fig. 2-67.

[28] Israelachvili, J. N., and Pashley, R. M., *Nature*, **306**, 249 (1983).

Other methods have also been developed for the direct measurement of colloid-type forces.[29] These include atomic force microscopy (AFM), which is adapted for making inter-surface force measurements by attaching a small sphere to the probe tip, as first done by Ducker, *et al.*[30] and Butt,[31] and reviewed by Gan.[32] The sphere is then brought into controlled distances of a test surface and the forces measured.

Another method, which was first used to determine the repulsive interaction between a colloid particle, denser than the medium in which it is dispersed, and a horizontal flat plate by tracking the average displacement over time of the particle from the flat surface. The method of detection is that of *total internal reflection microscopy* (TIRM), introduced by Prieve and coworkers,[33] in which a light beam is passed through the material of the flat plate, acting as a wave guide, and producing an evanescent beam projecting above the surface and decaying exponentially in intensity moving away from the surface. The vertical displacement of the particle from the surface is then determined (with an optical microscope) from the brightness of its image in the evanescent field. The Prieve method is applicable in its original form only to repulsive forces, but with the use of optical trapping techniques to hold a particle within any desired distance of the surface, attractive forces are also accessible.[34] The various techniques for the direct measurement of surface forces are reviewed by Claesson *et al.*[28]

Many other approaches lead to the indirect determination of long-range van der Waals forces, quantified by the Hamaker constant. Consider the close relationship between the Hamaker constant for liquids and the surface tension that were touched upon in Chap. 2, Eq. (2.14):

$$\sigma = \frac{A}{24\pi D_0^2}. \tag{7.43}$$

This may be regarded as the result of applying Eq. (7.2) to the *attractive* interaction ($-\Phi$) between a pair of planar liquid half spaces interacting at a distance D_0, in this case the distance between adjacent layers of molecules in a liquid, the accepted value for which is 0.165 nm. Consistent with arguments leading to Fowkes' combining rule, Eq. (2.27), only dispersion forces should be accounted for in Eq. (7.43), leading to

$$A \approx 2.1 \times 10^{-21} \sigma^d, \tag{7.44}$$

[29] Claesson, P. M., Ederth, T., Bergeron, V., and Rutland, M. W., *Adv. Colloid Interface Sci.*, **67**, 119 (1996).

[30] Ducker, W. A., Senden, T. J., and Pashley, R. M., *Nature*, **353**, 239 (1991).

[31] Butt, H.-J., *Biophys. J.*, **60**, 1438 (1991).

[32] Gan. Y., *Rev. Sci. Inst.*, **78**, 081101 (2007).

[33] Prieve, D. C., *Adv. Colloid Interface Sci.*, **82**, 93 (1999).

[34] Walz, J. Y., and Prieve, D. C., *Langmuir*, **8**, 3043 (1992).

where A [=] J, and σ^d [=] mN/m. Equation (7.44) leads to surprisingly good agreement with Hamaker constants computed for many liquids using Lifshitz theory.

The measurement of the disjoining pressure isotherm for thin liquid films, as effected for example, using the biconcave meniscus arrangement of Sheludko described in Chap. 2 (Fig. 2-64) affords another route to the Hamaker constant. In accord with Eq. (2.109), the disjoining pressure isotherm for van der Waals liquids is given by

$$\Pi(h) = -\frac{A}{6\pi h^3}.\qquad(7.45)$$

This is compared with results such as those shown in Fig. 2-65 to yield Hamaker constants. Disjoining pressure isotherms may also be measured using atomic force microscopy (AFM).[35-36]

Hamaker constants may also be extracted from aggregation rate experiments to be described later in this chapter, but that process is convoluted with a number of additional approximations.

C. Electrostatic interactions; DLVO theory

1. *Electrostatic repulsion between charged flat plates*

The next objective is to describe quantitatively the repulsive forces that arise when surfaces bearing electric double layers approach one another. To this end, we consider first flat plates of indeterminate thickness whose surfaces carry an effective surface potential, *i.e.*, a Stern potential, ψ_δ, which is considered to remain constant as the surfaces approach. Figure 7-3 shows the potential profile between the surfaces when they are a distance D apart, close enough that the diffuse portions of their separate double layers have begun to overlap, as well as the isolated potential profiles against the external plate surfaces. The distance measured away from Stern plane of the inner surface of the left plate is x. To obtain the appropriate expression for the force (per unit area) tending to push the plates apart, consider the forces on the thin volume (shell) of fluid shown in Fig. 7-3. The shell has dimensions $L \times W \times \Delta x$, and we seek to reckon the force on it in the $-x$ direction. At equilibrium, this force must be constant across the gap, *i.e.*, with x, and hence will be equal to the force pushing the left plate away from the right plate. The *net* force on the left plate is obtained by subtracting the force on the outer surface of the left plate. The total force on the shell, whose left surface is located at $x = x$, consists of two terms. The first is the net pressure force exerted on the surfaces at $x = x$ and $x = x + \Delta x$, and the second is the body force due to the interaction of the net charge enclosed in

[35] Basu, S., and Sharma, M. M., *J. Colloid Interface Sci.*, **181**, 443 (1996).
[36] Bowles, A. P., Hsia, Y.-T., Jones, P. M., Schneider, J. W., and White, L. R., *Langmuir*, **22**, 11436 (2006).

the shell, $\rho_e LW\Delta x$, where ρ_e is the net charge/volume in the shell, and $LW\Delta x$ is its volume, with the mean electric field strength within the shell, $\overline{(d\psi/dx)}$. The latter is the Maxwell stress, and is the same force considered in the

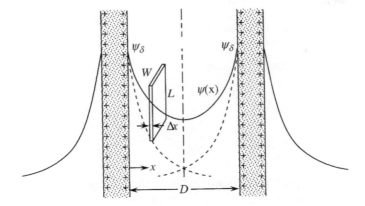

Fig. 7-3: Electrostatic repulsion between flat plates from overlap of diffuse double layers. Dashed lines show isolated potential profiles.

derivation of the velocity profile for electro-osmotic flow in Chap. 6, Eqs. (6.76) - (6.78).) The sum of the x-forces on the shell is thus:

$$\Delta F_{\text{shell}} = \left[p\big|_x - p\big|_{x+\Delta x} \right] LW + \rho_e \overline{\left(\frac{d\psi}{dx} \right)} LW\Delta x. \tag{7.46}$$

In the usual way of performing a "shell balance," divide through by the volume of the shell and take the limit as its thickness goes to zero:

$$\lim_{\Delta x \to 0} \left\{ \frac{\Delta F_{\text{shell}}}{LW\Delta x} = \frac{\left[p\big|_x - p\big|_{x+\Delta x} \right]\cancel{LW}}{\cancel{LW}\Delta x} + \rho_e \overline{\left(\frac{d\psi}{dx} \right)} \frac{\cancel{LW}\cancel{\Delta x}}{\cancel{LW}\cancel{\Delta x}} \right\}, \text{ or} \tag{7.47}$$

$$\frac{d(F_a)}{dx} = -\frac{dp}{dx} + \rho_e \frac{d\psi}{dx} = 0, \tag{7.48}$$

where, since the force/area on the shell is constant with x, its derivative may be set equal to zero. The constant (F/area) sought, which has now been written as F_a for brevity, can be obtained by integrating Eq. (7.48). Before doing that, however, the Poisson Equation, Eq. (6.15), is used to express the space charge density in terms of the potential profile:

$$\rho_e = -\varepsilon\varepsilon_0 \frac{d^2\psi}{dx^2}, \tag{7.49}$$

so that the second term on the right hand side of Eq. (7.48) becomes

$$\rho_e \frac{d\psi}{dx} = -\varepsilon\varepsilon_0 \frac{d^2\psi}{dx^2} \frac{d\psi}{dx}. \tag{7.50}$$

Then substitution of the mathematical identity:

$$\frac{d^2\psi}{dx^2}\frac{d\psi}{dx} \equiv \frac{1}{2}\frac{d}{dx}\left(\frac{d\psi}{dx}\right)^2,$$
(7.51)

allows Eq. (7.48) to be put in the form

$$\frac{d}{dx}\left[p - \frac{\varepsilon\varepsilon_0}{2}\left(\frac{d\psi}{dx}\right)^2\right] = 0,$$
(7.52)

or integrating:

$$F_a = p - \frac{\varepsilon\varepsilon_0}{2}\left(\frac{d\psi}{dx}\right)^2 = \text{const}$$
(7.53)

The pressure p may be identified with the excess osmotic pressure due to the dissolved ions in the double layer region relative to its value in the outer solution far from the plate surfaces. Thus

$$p(x) = \Pi(x) - \Pi(\infty).$$
(7.54)

Since this excess osmotic pressure gets higher (due to the increased ion concentration, as noted in Fig. 6-10) as one gets closer to the plate surface, it is clear that there is a trade-off between the osmotic and Maxwell stresses in the space between the plates. Taking advantage of the symmetry of the potential profile, it is noted that at the mid-plane, $x = D/2$, $(d\psi/dx) = 0$, so that the force/area that is sought is

$$F_a = p\big|_{x=D/2} = \Pi(x = D/2) - \Pi(\infty).$$
(7.55)

To evaluate the excess osmotic pressure at any location the total ion concentration is needed, in accord with the van't Hoff Equation:

$$\Pi(D/2) - \Pi(\infty) = \left[n(+) + n(-)\right]_{D/2} kT - 2n_\infty kT,$$
(7.56)

where $\left[n(+) + n(-)\right]_{D/2}$ is the sum of the ion concentrations at the mid-plane, and n_∞ is the total ion concentration out in the solution away from the plates. The ion concentrations are related to the local potential using the Boltzmann Equation, Eqs. (6.16) and (6.17), so that for a single z-z electrolyte

$$F_a = n_\infty kT\left[\exp\left(\frac{ze\psi(D/2)}{kT}\right) + \exp\left(\frac{-ze\psi(D/2)}{kT}\right) - 2\right]$$

$$= 2n_\infty kT\left[\cosh\left(\frac{ze\psi(D/2)}{kT}\right) - 1\right].$$
(7.57)

Thus to evaluate the repulsive force/area, what is needed is the potential at the mid-plane, $\psi(D/2)$. This requires solution to Eq. (6.21) with the boundary conditions

$$\psi = \psi_\delta \quad \text{at} \quad x = 0 \text{ and } x = D. \tag{7.58}$$

The result was obtained in the classical work of Verwey and Overbeek,[37] in the form of an integral, *viz.*

$$-\kappa D = \frac{1}{\sqrt{2}} \int_{\psi_\delta}^{\psi(D/2)} \frac{(ze/kT)d\psi}{\left[\cosh(ze\psi/kT) - \cosh(ze\psi(D/2)/kT)\right]^{1/2}}, \tag{7.59}$$

for which they provide tabulated numerical results for $\psi(D)$. (The result for more general boundary conditions is given by Chan *et al.*[38]) An acceptable approximation, if the plate spacing is not too close, is that the potential at the mid-plane is just the sum of the potentials that would be exhibited by isolated double layers at a distance of $D/2$ from the left and right surfaces, *i.e.*,

$$\psi(D/2) \approx 2\psi_{\text{iso}}(D/2). \tag{7.60}$$

Consistent with the assumption that the plates are not too closely spaced, ψ_{iso} may be estimated using Eq. (6.58), so

$$\psi(D/2) \approx \frac{8kT\gamma_\delta}{ze}\exp(-\kappa D/2), \tag{7.61}$$

where γ_δ is shorthand for $\tanh(ze\psi_\delta/4kT)$. One more approximation consistent with the assumption of a rather low mid-plane potential resulting from the plate spacing being not too close is that the argument of the cosh function in Eq. (7.57) is small. Under these conditions,

$$\cosh X \approx 1 + \frac{1}{2}X^2 + \ldots, \tag{7.62}$$

so that Eq. (7.57) becomes

$$F_a = n_\infty \frac{z^2 e^2}{kT}\psi^2(D/2). \tag{7.63}$$

Finally, substitution of Eq. (7.61) gives

$$F_a \approx 64 n_\infty kT\gamma_\delta^2 \exp(-\kappa D). \tag{7.64}$$

Recall that for large $|z\psi_\delta|$ (≥ 100 mV), $\gamma_\delta \approx 1$.

In order to get the potential function, Φ_R, one must integrate Eq. (7.64) with respect to the distance of separation. Taking advantage of symmetry

$$\Phi_R = -2\int_\infty^{D/2} F_a\, dD = -2\int_\infty^{D/2} 64 n_\infty kT\gamma_\delta^2 \exp(-\kappa D) dD$$

[37] Verwey, E. J. W., and Overbeek, J. Th. G., **Theory of the Stability of Lyophobic Colloids**, Chap. IV, Elsevier, Amsterdam, 1948.
[38] Chan, D. Y. C., Pashley, R. M., and White, L. R., *J. Colloid Interface Sci.*, **77**, 283 (1980).

$$= \frac{64 n_\infty kT\gamma_\delta^2}{\kappa} \exp(-\kappa D). \tag{7.65}$$

Comparison of the result of Eq. (7.65) with the exact result deduced from the numerical solution of Eq. (7.59), as carried out by Verwey and Overbeek,[39] shows it to be quite good when $\kappa D \geq 1$, regardless of γ_δ, despite the several approximations inherent in it. Lastly, it was derived for the boundary condition of constant effective surface potential, but is thought to be valid for the constant charge density condition as well, since, subject to its assumption of small double layer overlap, little discharging would occur.

Recall, parenthetically, that the (repulsive) force per unit area acting between a pair of opposed surfaces across a thin fluid film is the *disjoining pressure*, discussed in Chap. 2. Equation (7.65) then gives the electrostatic component to the disjoining pressure isotherm.

2. Electrostatic interactions between curved surfaces; the Derjaguin approximation

What is more often needed than the interaction between flat surfaces is Φ_R for that between *spherical* colloid particles. Under the right conditions, such a function can be derived from the above equation by splitting the spherical surfaces into parallel-facing annular disks, as proposed by Derjaguin[40] and shown in Fig. 7-4. For the approximation to be valid, the radius of the approaching spheres (more generally, the radii of curvature of the approaching surfaces[41]) must be large relative to the distance of

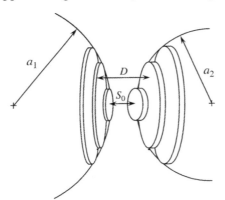

Fig. 7-4: Schematic for the derivation of the Derjaguin approximation.

separation of the surfaces, *i.e.*, $S_0 \ll a_1$ or a_2. Consider the interaction between opposing annular disks of radii x and a distance D from each other. The contribution to the repulsive force due to the interaction between these annular disks is

---

[39] Verwey, E. J. W., and Overbeek, J. Th. G., **Theory of the Stability of Lyophobic Colloids**, p. 85, Elsevier, Amsterdam, 1948.

[40] Derjaguin, B. V., *Kolloid Z.*, **69**, 155 (1934).

[41] White, L. R., *J. Colloid Interface Sci.*, **95**, 286 (1983).

$$dF_R = F_a(D)2\pi x \, dx \tag{7.66}$$

From the geometry shown in Fig. 7-5, it is evident that

$$D = S_0 + z_1 + z_2 \tag{7.67}$$

and from the Pythagorean theorem

$$a_1^2 = (a_1 - z_1)^2 + x^2 = a_1^2 - 2a_1z_1 + z_1^2 + x^2 \approx a_1^2 - 2a_1z_1 + x^2, \tag{7.68}$$

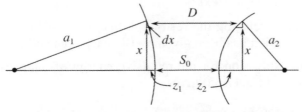

Fig. 7-5: Geometric construction for derivation of Derjaguin approximation.

so that

$$x^2 \approx 2a_1z_1 = 2a_2z_2, \tag{7.69}$$

and

$$D = S_0 + z_1 + z_2 = S_0 + \frac{x^2}{2}\left(\frac{1}{a_1} + \frac{1}{a_2}\right). \tag{7.70}$$

Then

$$dD = \left(\frac{1}{a_1} + \frac{1}{a_2}\right)x \, dx, \tag{7.71}$$

and Eq. (7.66) becomes

$$dF_R = F_a(D)2\pi\left(\frac{1}{a_1} + \frac{1}{a_2}\right)^{-1} dD, \tag{7.72}$$

which upon integration between $D = S_0$ and ∞ gives the total repulsive force between the spherical surfaces:

$$F_R = 2\pi\left(\frac{a_1a_2}{a_1 + a_2}\right)\int_{S_0}^{\infty} F_a(D)dD = 2\pi\left(\frac{a_1a_2}{a_1 + a_2}\right)\Phi_{R\text{-flat plate}}(S_0). \tag{7.73}$$

This result, known as the *Derjaguin approximation*, gives the total *force* of interaction between a pair of spherical surfaces as proportional to the repulsion *energy* of interaction between a pair of flat surfaces. It must be recalled that Eq. (7.73) is restricted to the case where the distance of particle approach is much less that the particle radii, *i.e.*, $S_0 \ll a$. Since distances of approach of interest for the interaction of overlapping double layers is $S_0 <$

2κ, the Derjaguin approximation as applied in this case is for reasonably large values of κa (> 10). Verwey and Overbeek[42] addressed themselves to the case of small κa (< 5), which would be the case relevant to nanoparticles or to particles in organic media with very thick double layers. While the exact results are complex, an approximate expression valid for either constant potential or constant surface charge density condition (with error up to 40%) was given as

$$\Phi_R \approx 2\pi\varepsilon\varepsilon_0 a\psi_\delta^2 \exp(-\kappa S_0).$$ (7.74)

A result completely relaxing the restriction that $S_0 \ll a$ for the case of the sphere-plate interaction has been given by Bhattacharjee and Elimelich.[43]

Using the expression for the repulsive energy of interaction between flat surfaces given in Eq. (7.65) gives

$$F_R = \frac{128\pi n_\infty kT\gamma_\delta^2}{\kappa}\left(\frac{a_1 a_2}{a_1 + a_2}\right)\exp(-\kappa S_0).$$ (7.75)

The case of a sphere of radius a and a flat plate, equivalent to that of crossed cylinders of radius a, is recovered by setting $a_1 \rightarrow a$ and $a_2 \rightarrow \infty$:

$$(F_R)_{\text{crsd cyl}} = \frac{128\pi n_\infty kT\gamma_\delta^2}{\kappa} a\exp(-\kappa S_0).$$ (7.76)

Hunter[44] reports quantitative corroboration of Eq. (7.76) using the surface forces apparatus to investigate the interaction between mica surfaces in a series of aqueous KCl solutions.

For the approach of two spheres of radius a, Eq. (7.75) gives

$$(F_R)_{\text{sph-sph}} = \frac{64\pi n_\infty kT\gamma_\delta^2}{\kappa} a\exp(-\kappa S_0).$$ (7.77)

The potential energy of electrostatic repulsion between equal-sized spheres, $(\Phi_R)_{\text{sph-sph}}$ is obtained by recalling the relationship between force and potential: $F_R = -(d\Phi_R/dS_0)$. Thus

$$(\Phi_R)_{\text{sph-sph}} = \int_{S_0}^\infty (F_R)_{\text{sph-sph}} dS_0 = \frac{64\pi a n_\infty kT\gamma_\delta^2}{\kappa^2} \exp(-\kappa S_0).$$ (7.78)

It appears that the integration toward ∞ would violate the requirements of the Derjaguin approximation, but the fact is that the potential function drops off exponentially with separation so that the interactions at large separations make only a negligible contribution to the total.

[42] Verwey, E. J. W., and Overbeek, J. Th. G., **Theory of the Stability of Lyophobic Colloids**, p. 260, Elsevier, Amsterdam, 1948.

[43] Bhattacharjee, S., and Elimelich, M., *J. Colloid Interface Sci.*, **193**, 273 (1997).

[44] Hunter, R. J., **Foundations of Colloid Science**, 2nd Ed., pp. 609-610, Oxford Univ. Press, Oxford, UK, 2001.

3. *DLVO theory: electrocratic dispersions*

A number of aqueous colloidal dispersions are described by the interplay between van der Waals attractions, which should lead to their rapid aggregation, and electrostatic repulsion upon close approach and overlap of their double layers. The stability of the colloid to aggregation is determined by the *net* interaction potential, *i.e.*,

$$\Phi_{net} = \Phi_A + \Phi_R. \qquad (7.79)$$

Expressions for the attractive and repulsive components of the net interaction potential between a pair of spherical particles of radius a at a distance of closest approach S_0 have been developed above as Eqs. (7.7) and (7.78), respectively. They are represented schematically in Fig. 7-6, together with their summation. A third curve labeled Φ_{SR} is also added, representing a short-range, very steep repulsion. The summing of the van der Waals attraction and electrostatic repulsion in accord with Eq. (7.78) is commonly referred to as *DLVO Theory* - after **D**erjaguin and **L**andau[45] (in the Soviet Union) and **V**erwey and **O**verbeek[46] (in the Netherlands) who pioneered this approach in the 1930's and 1940's. The resultant "DLVO curve" has a number of important features. First, it often displays an intermediate maximum in potential, Φ_m, that represents a "potential energy barrier" to aggregation, once being overcome, allowing aggregation to proceed. If the particles thus approach to within distances to the left of the location of the local maximum in Φ_{net}, they will be drawn toward one another into a "primary minimum" in Φ_{net} to form an aggregate. The means of overcoming such a barrier is provided by the energy of Brownian motion, whose average

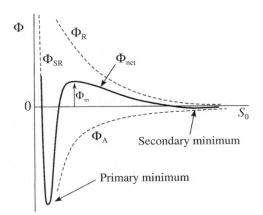

Fig. 7-6: Interaction potential for a pair of spheres as a function of their distance of closest approach, S_0. The "DLVO curve."

[45] Derjaguin, B. V., and Landau, L., *Acta Physico-chim. USSR*, **14**, 633 (1941).
[46] Verwey, E. J. W., and Overbeek, J. Th. G., **Theory of the Stability of Lyophobic Colloids**, Elsevier, Amsterdam, 1948.

intensity approximates kT. It is thus useful to prepare plots of Φ_{net} in units of kT. It is generally believed that if this barrier is less than a few kT, aggregation will occur rapidly. A pair of particles approaching one another will thus have a probability of aggregating that is determined by the height of Φ_m in kT units, and it is this feature that determines a colloid's "stability" toward aggregation. The depth of the primary minimum is determined by the existence of short-range repulsion which itself is not a part of DLVO theory. This may be due to a tightly bound hydration layer. Its presence and its "thickness" are critical to the "repeptization" or restoration of a colloid that has been aggregated, as will be discussed below. There may also be a "secondary minimum," as shown in Fig. 7-6, which may be responsible for the formation of loose flocks, possibly disrupted by stirring. Aggregation in the primary minimum is commonly termed *coagulation*, while aggregation into the secondary minimum is termed *flocculation*.

A colloid whose stability behavior is described by DLVO theory is termed an *electrocratic colloid* to distinguish it from those systems in which interactions other than van der Waals attraction and electrostatic repulsion are important. Many colloids are in fact *electrocratic*, and DLVO theory in any case is the starting point for the description of other and more complex systems. Since we have workable analytical expressions for both Φ_A, from Eq. (7.7), and Φ_R, from Eq. (7.78), we have an analytical expression for Φ_{net} that permits exploration of the stability behavior of electrocratic colloids with respect to the system parameters on which it depends. For particles of a material 2, of radius a, in a medium of material 1 (presumably water):

$$\Phi_{net} = -\frac{A_{212}a}{12S_0} + \frac{64\pi a n_\infty kT}{\kappa^2}\tanh^2\left(\frac{ze\psi_\delta}{4kT}\right)\exp(-\kappa S_0). \tag{7.80}$$

Taking the medium as water at $T = 25°C$ containing a single z-z electrolyte, and particles of radius $a = 100$ nm, Eq. (7.80) reduces to:

$$\frac{\Phi_{net}}{kT} = -2.025\cdot10^{21}\frac{A_{212}}{S_0} + \frac{1119.8}{z^2}\tanh^2\left(\frac{z\psi_\delta}{102.8}\right)\exp(-3.288z\sqrt{C}S_0), \tag{7.81}$$

where A_{212} [=] J; S_0 [=] nm; ψ_δ [=] mV, and C [=] M. Equation (7.81) permits investigation of the importance of the Hamaker constant, A_{212}, the Stern potential, ψ_δ, the salt concentration, C, and the valence z on the DLVO curve, and in particular on the magnitude of the barrier it presents to aggregation.

It is clear that the magnitude of the Hamaker constant plays a significant role in determining the DLVO curve, in that larger values of A_{212} yield lower barriers, other factors being equal. One generally has little control over it, however. It may be thought that it might be changed by coating the particles with a layer of material of different Hamaker constant, such as a polymer or surfactant and an expression for the effective Hamaker interaction between two spherical core-shell particles has been derived by

Vold.[47] The result derived later by Israelachvili[48] for two infinite flat plates of material 2 covered by adlayers of material 3 facing each other across medium 1 is simpler and more instructive for present purposes:

$$\Phi_A = \frac{1}{6\pi}\left[\frac{A_{313}}{D_0^3} - \frac{2A_{231}}{\left(D_0 + \delta\right)^3} + \frac{A_{232}}{\left(D_0 + 2\delta\right)^3}\right], \qquad (7.82)$$

where D_0 is the distance between the adlayer surfaces, δ is the adlayer thickness, and A_{ijk} values are computed using the combining rules shown earlier. Two limiting cases are revealing. When $D_0 \gg \delta$ (large separations relative to adlayer thickness):

$$\Phi_A \approx \frac{1}{6\pi}\left[\frac{A_{212}}{D_0^3}\right], \qquad (7.83)$$

essentially the same as the result without the adlayer present. On the other hand, when $D_0 \ll \delta$ (small separations relative to the adlayer thickness):

$$\Phi_A \approx \frac{1}{6\pi}\left[\frac{A_{313}}{D^3}\right]. \qquad (7.84)$$

Thus for large plate separations (relative to the adlayer thickness), the presence of the adlayer has little effect on the interaction, while at close spacings, the properties of the adlayer dominate. In particular, for direct contact, i.e., adhesion, the adlayer is decisive even when it is as thin as a monolayer of small molecules. The general effect of an adsorbed polymer layer on colloid stability, however, is usually better understood in terms of "steric stabilization," to be discussed in more detail later. It involves more than simply altering the effective Hamaker constant.

The effect of the Stern potential, ψ_δ, can be very great. For Nernstian colloids, its magnitude can be controlled by varying the surface potential, ψ_0, which in turn is determined by the relative concentration of the potential determining ions (for oxides, the pH), although it is to be recalled that the Stern potential is often considerably less than ψ_0. While ψ_0-values may exceed 250 mV, such high values strongly attract counterion binding in the Stern layer, and one seldom encounters Stern potentials in excess of 100 mV. In other situations, it is $specific$ adsorption of ionic components (particularly surface active agents) that determine the Stern potential. Even if the quantitative effect of such additions cannot be predicted on the basis of theory, there is a means of experimentally estimating ψ_δ as the zeta potential, determined from electrokinetic measurements. Figure 7-7 shows the effect of varying the Stern potential for a colloid of 100-nm radius silica

[47] Vold, M. J., *J. Colloid Sci.*, **16**, 1 (1961).
[48] Israelachvili, J. N., *Proc. Roy. Soc.*, **331A**, 39 (1972).

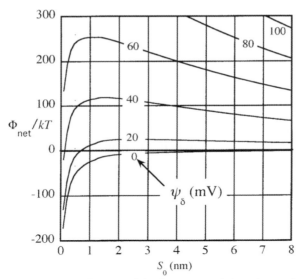

Fig. 7-7: The influence of the Stern potential on the total potential
energy of interaction of two spherical particles: $a = 100$ nm, $T = 298$ K,
$A_{212} = 0.849$ x 10^{-20} J, $z = 1$, $C = 1$ mM.

particles dispersed in an aqueous medium with a background electrolyte
concentration of 10^{-3}M univalent electrolyte. The relevant Hamaker
constant, from Table 7-1, is 0.849 x 10^{-20} J. The plot shows Φ_{net} normalized
by kT, computed in accord with Eq. (7.81). It is evident that potentials below
about 20 mV are insufficient to produce a barrier greater than a few kT,
thought to be required to assure reasonable stability.

The most generically controllable variables in an electrocratic
dispersion are the salt (indifferent electrolyte) concentration and its valence.
Figure 7-8 shows DLVO curves for the same colloid as depicted in Fig. 7-7,
i.e., 100-nm radius silica particles dispersed in water at a temperature of
25°C. The Stern potential has been set to 30 mV. It is evident that the
potential barrier decreases sharply with increase in electrolyte concentration
as the double layer is compressed and ultimately collapsed. It appears that
there is a critical concentration, somewhere between 200 and 400 mM in this
case, for which the barrier goes to zero and at which the colloid would be
expected to undergo rapid aggregation. This is termed the *critical
coagulation concentration* (*CCC*). It should correspond to the situation
pictured schematically in Fig. 7-9. If the criterion for coagulation is taken as
$\Phi_{max} = 0$, that condition may be located by requiring:

$$\Phi_{net} = 0, \quad \text{and} \quad \frac{d\Phi_{net}}{dS_0} = 0, \tag{7.85}$$

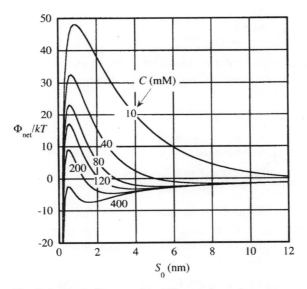

Fig. 7-8: The influence of indifferent electrolyte concentration on the total potential energy of interaction of two spherical particles: $a = 100$ nm, $T = 298$ K, $A_{212} = 0.849$ x 10^{-20} J, $z = 1$, $\psi_\delta = 30$ mV.

Application of the above criteria, *e.g.* to Eq. (7.81), gives the value of the *CCC* and the value of S_0 where the potential is zero. It leads to *CCC's* in range of 50 - 250 mM for monovalent electrolytes, depending on the other parameters describing the system, but the most interesting outcome is the dependence of the *CCC* on valence. At high values of ψ_δ (> 100 mV), the dependence takes the form:

$$CCC = (const)\frac{\varepsilon^3 \gamma_\delta^4}{A^2 z^6} \propto \frac{1}{z^6},\qquad (7.86)$$

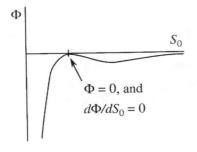

Fig7-9: Determination of the critical coagulation concentration (*CCC*).

where $\gamma_\delta = \tanh\left(\dfrac{ze\psi_\delta}{4kT}\right) \approx$ constant ≈ 1. Under these conditions, the *CCC* depends very steeply on the valence of the electrolyte, and is independent of the exact value of the Stern potential. At low values of ψ_δ (< 20 mV), one obtains

$$CCC = (\text{const})\frac{\varepsilon^3 \psi_\delta^4}{A^2 z^2} \propto \frac{1}{z^2}, \tag{7.87}$$

which shows a much weaker dependence on z and an explicit dependence on the Stern potential. At intermediate values of ψ_δ one has

$$CCC \propto \frac{1}{z^n}, \tag{7.88}$$

where, to reasonable approximation,

$$n \approx 0.51 |\psi_\delta|^{0.49}. \tag{7.89}$$

Brief comment should be made as to the possible effect of temperature and particle size on coagulation behavior of electrocratic sols. Equation (7.80) shows that the Hamaker attraction term is independent of temperature, while the electrostatic repulsion term varies as $T \exp(-\text{const}/\sqrt{T})$ for low potentials and $T^2 \exp(-\text{const}/\sqrt{T})$ for high potentials. This shows that increases in temperature should produce increases in the repulsive term, but since $T [=]$ K, the dependence is fairly small. With respect to the effect of particle size, it is to be noted that both terms in the expression for the interaction potential vary directly with particle radius a. Thus the barrier to coagulation, Φ_m, should decrease with decreases in a, while the energy of a particle moving by Brownian motion should be independent of particle size. This should explain in part why electrocratic nano-colloids are more difficult to stabilize than those consisting of larger particles, other factors being equal, although this result is not always observed.

4. Jar testing, the Schulze-Hardy Rule and agreement with theory

CCC's are most simply determined in the laboratory by "jar testing," in which the dispersion is contacted with electrolyte over a range of concentrations, as pictured in Fig. 7-10 for the As_2S_3 sol, and permitted to stand for some pre-determined time, often $\approx$ 30 minutes. When coagulation and sedimentation are evident, a new series of salt concentrations spanning the gap between the aggregated case and the next salt concentration below it is made, *etc.*, until an accurate value of the CCC is determined.[49] The fact that it is so sharp (usually to within $\pm$ 1 mM) makes this an accurate method. Other more sophisticated techniques are discussed following the treatment of aggregation kinetics.

---

[49] If a salt containing a potential determining ion is used, one may obtain a more complicated result. For example, when using NaOH as a coagulant for an iron oxide sol, one obtains precipitation at $C \approx 0.5$ mM, followed by re-dispersion at higher salt concentrations, and then finally coagulation at $C \approx 100$ mM, yielding what may be called an *irregular aggregation series*. The first event is due to reaching the *PZC* and consequent charge reversal, and the second event due to the non-specific collapse of the double layer.

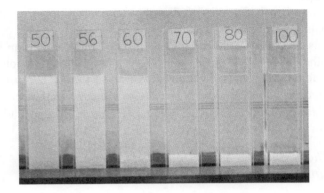

Fig. 7-10: Aggregation jar test series for As_2S_3 sol with 1-1 electrolyte concentrations in mM. The *CCC* appears to lie between 60 and 70 mM. From [Overbeek, J. Th. G., **Colloidal and Surface Chemistry, A Self-Study Subject, Part 2**, Lyophobic Colloids, p. 6.6, MIT, Cambridge, MA (1972).]

Some classical results of jar testing of As_2S_3 colloids are shown in Table 7-2. With a few exceptions, the *CCC* of a particular electrolyte depends only on the valency of the counterion. The *CCC* changes by a factor of $(z_i/z_j)^n$ when the counterion valency is changed from z_i to z_j. For the data shown, as z varies from 1/2/3, the *CCC* varies from $\approx$ 50/0.65/0.09, corresponding to $n \approx 6$, suggesting a high Stern potential.

Table 7-2: Critical coagulation concentrations (*CCC*) for an aqueous sol of As_2S_3 (a negative sol), in mM. From [Overbeek, J. Th. G., **Colloidal and Surface Chemistry, A Self-Study Subject, Part 2**, Lyophobic Colloids, p. 6.6, MIT, Cambridge, MA (1972).]

Monovalent cations		Divalent cations		Trivalent cations	
Salt	*CCC*	Salt	*CCC*	Salt	*CCC*
LiCl	58	$MgCl_2$	0.72	$AlCl_3$	0.093
NaCl	51	$MgSO_4$	0.81	$Al(NO_3)_3$	0.095
KCl	49.5	$CaCl_2$	0.65	1/2 $Al_2(SO_4)_3$	0.096
KNO_3	50	$SrCl_2$	0.63^5	$Ce(NO_3)_3$	0.080
1/2 K_2SO_4	65.5	$BaCl_2$	0.69		
HCl	31	$ZnCl_2$	0.685		
1/2 H_2SO_4	30	$(UO_2)(NO_3)_2$	0.64		
Morphine chloride	0.42	Quinine sulphate	0.24		
New Fuchsin	0.11	Benzidine nitrate	0.09		

The empirical observation that the concentration of a salt required to induce coagulation depended on the inverse sixth power of its cationic valence was recorded well before the turn of the century and came to be known as the *Schulze-Hardy Rule*.[50] Since most colloid particles in aqueous media are negatively charged, focus was put on the valence of the cation. It was a crowning achievement of the DLVO theory that it could provide a theoretical rationalization for the Schulze-Hardy Rule that had been known and used for so long.

5. *The Hofmeister series; ion speciation and ionic specific adsorption*

The experimental *CCC*-values shown in Table 7-2 reveal that not all counterions of the same valence have the same effect on coagulating the colloid. Different ions (of a given valence) have different specific tendencies to enter the Stern layer and hence to "bind" to the surface and effect more efficient charge neutralization. These tendencies are described by neither the generic Guoy-Chapman model nor the DLVO theory. To a rough approximation, it can be said that ions of a given valence have a tendency to adsorb in direct proportion to their *un-hydrated* size (which interestingly is often inversely proportional to their hydrated size). For monovalent cations, the tendency to bind often varies as:

$$Li^+ < Na^+ < K^+ < H_3O^+ < NH_4^+ < Rb^+ < Cs^+ \qquad (7.90)$$

A series of the above type is referred to as a *Hofmeister series*, or a *lyotropic series*. For divalent cations, one has:

$$Mg^{2+} < Ca^{2+} < Sr^{2+} < Ba^{2+}. \qquad (7.91)$$

These series are roughly consistent with what is observed in Table 7-2. The morphine chloride, the new Fuchsin dye, quinine sulfate and benzidine nitrate, on the other hand, provide large, strongly and specifically adsorbing cations which more effectively neutralize surface charge (reduce the Stern potential). For monovalent anions, the Hofmeister series goes as:

$$F^- < OH^- < Cl^- < ClO_4^-, NO_3^- < I^- < CNS^-. \qquad (7.92)$$

More recent reviews of specific ion effects suggest that different Hofmeister series' may apply under different application circumstances.[51]

Application of DLVO theory to coagulation phenomena involving polyvalent counter-ions, as described above, is vastly over-simplified. There are two major reasons for the additional complications, as pointed out by

[50] Schulze, H., *J. Prakt. Chem.*, **25**, 431 (1882);
   Schulze, H., *J. Prakt. Chem.*, **27**, 320 (1883);
   Hardy, W. B., *Proc. Roy. Soc.*, **66**, 110 (1900);
   Hardy, W. B., *Z. Phys. Chem.*, **33**, 385 (1900).
[51] López-León, T., Jódar-Reyes, A. B., Bastos-Gonzáles, D., and Ortega-Vinuesa, J. L., *J.Phys. Chem. B*, **107**, 5696 (2003);
   Lyklema, J., *Adv. Colloid Interface Sci.*, **100-102**, 1 (2003).

Matijevic in a paper[52] that should be required reading for anyone who seeks to apply the simple theory without further thought. These are 1) ion speciation in solution, and 2) specific ionic adsorption. First, when a salt (particularly one yielding polyvalent ions) is dissolved, the species are hydrolyzed and hydrated to different extents dependent largely on the system pH. If an additional salt is present, complexes may form between the ions of the different salts in what is called an "antagonistic effect." These processes produce a variety of ions of varying valence in the solution, the overall process being referred to as *speciation*. As an illustration, an aluminum salt such as $Al(NO_3)_3$ or $AlCl_3$ will produce at least the following ions in solution (neglecting the waters of hydration), dependent on pH: $Al_8(OH)_{20}^{4+}$, $Al_7(OH)_{17}^{4+}$, Al^{3+}, $Al(OH)^{2+}$, $Al(OH)_2^+$ and $Al(OH)_4^-$. If another salt such as K_2SO_4 is present, a variety of (Al^{3+})-(SO_4^{2})-(OH^-) complexes form, ranging in valence from +4 to -2. Thus to apply the Schulze-Hardy Rule, one must know the valence of the counterion species present. High cation valences are favored in general by low pH conditions. The second factor concerns the adsorption of the ionic species to the solid surface. In a certain way, this effect is taken care of by using the Stern potential in the formulation of the DLVO curve. Whatever the mechanism of adsorption (ion exchange, coordination, acid-base), the result should be reflected in ψ_δ, which is generally accessible as the zeta potential. But if one wishes to have an *a priori* idea of the effect of certain salts, concentrations, and pH's on the

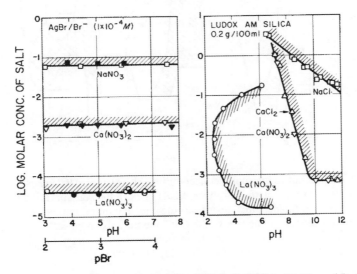

Fig. 7-11: Left: *CCC* values for $NaNO_3$ ($\square$, $\blacksquare$), $Ca(NO_3)_2$ ($\triangledown$,$\blacktriangledown$), and $La(NO_3)_3$ ($\bigcirc$,$\bullet$), respectively, as a function of pH ($\square$,$\triangledown$,$\bigcirc$) and pBr ($\blacksquare$,$\blacktriangledown$,$\bullet$) for a negatively charged AgBr sol. Right: *CCC*'s for NaCl ($\square$), $CaCl_2$ ($\triangledown$), and $La(NO_3)_3$ ($\bigcirc$), respectively, as a function of pH for a Ludox AM silica sol. From [Matijevic, E., *J. Colloid Interface Sci.*, **43**, 217 (1973).]

52 Matijevic, E., *J. Colloid Interface Sci.*, **43**, 217 (1973).

stability of a colloid, one needs to inquire further. It is generally the bulkier ions that adsorb more strongly, even if they are present in only small amounts. The particular colloid surface, of which DLVO theory takes no account, also comes into play. Figure 7-11 shows the sharp contrast in the behavior of two different colloids to the same set of coagulating electrolytes. The silver bromide colloid on the left shows full agreement with the Schulze-Hardy Rule, with *CCC*-values depending only on the counterion valence (as $1/z^6$), and independent of *pH*. For the silica colloid on the right there is a huge influence of *pH*, which clearly affects the adsorption of the ions into the Stern layer. In both diagrams, the shaded area refers to the region in which coagulation is observed.

When strong specific adsorption of counterions occurs, one observes first charge neutralization as salt concentration is increased, and then charge reversal of sufficient magnitude that the colloid is re-stabilized. This occurs at a critical *stabilization* concentration *(CSC)*. Figure 7-12 shows this phenomenon as observed for three different sols in $Al(NO_3)_3$ or $AlCl_3$ solutions as a function of *pH*.

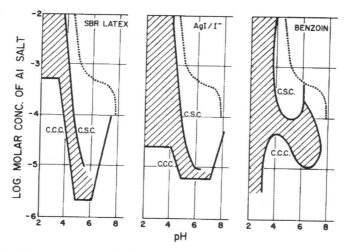

Fig. 7-12: The entire log molar concentration $Al(NO_3)_3$ or $AlCl_3$-*pH* stability domain for styrene-butadiene rubber (SBR) latex (left), AgI sol (middle), and benzoin (right). Shaded area designates the coagulation region; below the *CCC* line the sols remain stable, above the *CSC* line the sols are re-stabilized due to charge reversal. (…) indicates the formation of $Al(OH)_3$ precipitate. From [Matijevic, E., *J. Colloid Interface Sci.*, **43**, 217 (1973).]

6. Repeptization

Repeptization refers to the restoration of a colloid after it has been coagulated. It has generally been observed that once an electrocratic sol has been coagulated by dosing it with sufficient electrolyte, it will spontaneously repeptize when the original conditions of stability are restored by dilution, provided that the coagulating electrolyte is monovalent. For polyvalent coagulants, the issue is more complex, and repeptization may not occur, particularly if the colloid has been permitted to stand in the coagulated state for an extended time. This is only the first puzzling result. With reference to Fig.7-6, it would appear that once a pair of particles has aggregated into a deep primary minimum, it would never acquire sufficient energy to overcome the huge barrier to its separation. Frens and Overbeek[53] provided both a thermodynamic and a kinetic rationalization for spontaneous repeptization as follows. The thermodynamic argument rests on the presumption that a thin layer (of *something*) on the particle surfaces prevents their touching, or even their approach to within closer than a few tenths of a nanometer, say $2d$. The resulting steep (near infinite) repulsion Φ_{SR} was already introduced in Fig. 7-6. For hydrophilic particles, like oxides, this could be a thin layer of bound water. In any case, consider two possible situations pictured in Fig. 7-13, in which the steep repulsive branch of the interaction potential, Φ_{SR}, is represented as a vertical line at two possible locations, depending on the thickness of what may be called the bound solvent layer, $2d$. In case (b), that thickness is great enough that the primary minimum lies above the energy level associated with the separated particles, and the "activation barrier" to their separation is low enough to be easily overcome. Close examination of actual DLVO curves in Figs. 7-7 and 7-8 suggest that a thickness of only ≈ 0.2 nm, corresponding to little more than a monolayer of water, may be sufficient. Additional evidence for the existence of such a layer and the resulting steep, repulsive "hydration force" is discussed later in this chapter. The second explanation depends on rate phenomena, specifically the rate of double layer relaxation. When there is a sudden change in electrolyte concentration, time is required for the charge in the Stern layer to adjust to the new conditions. For changes that are rapid relative to this relaxation time, the double layer acts like one of constant change density, regardless of the equilibrium boundary condition. Figure 6-15 is helpful in visualizing this. The lower curve corresponds to the

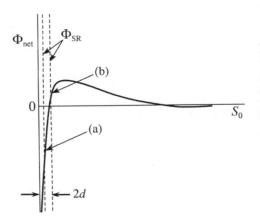

Fig. 7-13: DLVO curve showing the location of the primary minimum as it depends on the thickness of a bound solvent layer, $2d$. In case (a), the thickness is not great enough to permit repeptization, but in case (b) it is.

[53] Frens, G., and Overbeek, J. Th. G., *J. Colloid Interface Sci.*, **38**, 376 (1972); Overbeek, J. Th. G., *J. Colloid Interface Sci.*, **58**, 408 (1977).

coagulated condition. When the colloid is diluted, the surface charge at first remains constant and the potential shoots up to values higher than the equilibrium potential, as suggested by the upper curve. The DLVO curve moves upward, and the depth of the primary minimum, at least momentarily, is significantly reduced. Repeptization may be effectively complete before the potential comes back down to its equilibrium value. Again, a solvent layer thickness of only ≈ 0.2 nm is required for this to occur.

For the case of polyvalent counterions, as stated earlier, aggregation is often irreversible, particularly if the coagulated colloid is permitted to age by over 30 min or so. Evidently a cold sintering process may occur. The situation is more complex, as detailed in the previous section, but with simple un-complexed divalent ions, the shapes of the DLVO curves, even with the same assumed solvent layer thickness, do not yield a shallow enough primary minimum upon dilution to produce repeptization. The subject of repeptization of electrocratic dispersions destabilized by polyvalent counterions needs more study.

7. Interaction between dissimilar surfaces: hetero-aggregation

Consider next the electrostatic interaction between opposing flat plates bearing *different* electrical double layers. Various possibilities exist. The surfaces may be of constant potential, constant charge density or charge-regulated, and they may be of the same (but unequal) potential or charge density, or of opposite sign. The completely general problem, as treated by McCormack et al.[54] is quite complex, but a constant-potential model based on the Debye-Hückel approximation, as given by Hogg, Healy and Fuerstenau (HHF)[55] captures most of the physics of the situation and appears to be applicable to surface (Stern) potentials at least as high as 60 mV. Consider the situation pictured in Fig. 7-14 showing the opposing plates of unequal effective Stern potentials of ψ_1 and ψ_2, respectively, in contrast to the situation pictured in Fig. 7-3. The potential profile is given by the solution to the Debye-Hückel double layer equation (for a symmetrical supporting electrolyte)

$$\frac{d^2\psi}{dx^2} = \kappa^2\psi,$$ (7.93)

subject to the boundary conditions: $\psi = \psi_1$ at $x = 0$, and $\psi = \psi_2$ at $x = D$. The solution is:

$$\psi = \psi_1\cosh(\kappa x) + \left(\frac{\psi_2 - \psi_1\cosh(\kappa D)}{\sinh(\kappa D)}\right)\sinh(\kappa x).$$ (7.94)

[54] McCormack, D., Carnie, S. L., and Chan, D. C., *J. Colloid Interface Sci.*, **169**, 177 (1995).
[55] Hogg, R., Healy, T. W., and Fuerstenau, D. W., *Trans. Faraday Soc.*, **62**, 1638 (1966).

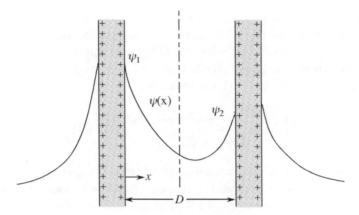

Fig. 7-14: Potential profile between plates of different effective surface potential.

It embraces the situations in which ψ_1 and ψ_2 are either the same or opposite signs.

The potential energy of interaction between the plates (per unit area) is obtained by recalling the expression developed earlier for the free energy of double layer formation:

$$\Phi_R = \Delta G_D - (\Delta G_{\infty 1} + \Delta G_{\infty 2}), \qquad (7.95)$$

where ΔG_D is the free energy of formation of the pair of partially overlapping double layers in the gap of width D, and $\Delta G_{\infty 1}$ and $\Delta G_{\infty 2}$ are the free energies of formation of the isolated double layers. For low surface potentials (DH approximation), ΔG_D is given by the sum:

$$\Delta G_D = -\frac{1}{2}(\overline{\sigma}_1 \psi_1 + \overline{\sigma}_2 \psi_2), \qquad (7.96)$$

where $\overline{\sigma}_1$ and $\overline{\sigma}_2$ are the charge densities on the surfaces of the overlapping double layers. These are given by:

$$\overline{\sigma}_1 = -\varepsilon\varepsilon_0 \left(\frac{d\psi}{dx}\right)_{x=0} = -\varepsilon\varepsilon_0 \kappa \left(\frac{\psi_2}{\sinh(\kappa D)} - \frac{\psi_1}{\tanh(\kappa D)}\right), \qquad (7.97)$$

and

$$\overline{\sigma}_2 = -\varepsilon\varepsilon_0 \left(\frac{d\psi}{dx}\right)_{x=D} = +\varepsilon\varepsilon_0 \kappa \left(\frac{\psi_2}{\tanh(\kappa D)} - \frac{\psi_1}{\sinh(\kappa D)}\right). \qquad (7.98)$$

For the isolated double layers, $\overline{\sigma}_1 = \varepsilon\varepsilon_0 \kappa \psi_1$, etc.

Combining the above equations gives:

$$\Phi_R^{\psi}\Big|_{\text{plates}} = \frac{1}{2}\varepsilon\varepsilon_0 \kappa \left[(\psi_1^2 + \psi_2^2)\left(1 - \frac{1}{\tanh(\kappa D)}\right) + \frac{2\psi_1\psi_2}{\cosh(\kappa D)}\right], \qquad (7.99)$$

where the superscript on Φ_R^ψ is a reminder that this is the expression corresponding to the use of the constant-potential boundary condition. When Φ_R is integrated in accord with the Derjaguin approximation to obtain the interaction between equal spheres of radius a with a minimum approach distance of S_0 (with $S_0 \ll a$), the result is

$$\Phi_R^\psi\Big|_{spheres} = \frac{1}{2}\varepsilon\varepsilon_0 a\left(\psi_1^2 + \psi_2^2\right)\left\{\frac{2\psi_1\psi_2}{\left(\psi_1^2 + \psi_2^2\right)}\ln\left[\frac{1+e^{-\kappa S_0}}{1-e^{-\kappa S_0}}\right] + \ln\left[1-e^{-2\kappa S_0}\right]\right\}. \quad (7.100)$$

Examination of the above equation reveals that Φ_R^ψ (now perhaps misnamed, with its subscript $_R$) is negative (hence *attractive*), not only whenever the signs of ψ_1 and ψ_2 are opposite, but also when ψ_1 and ψ_2 are of the *same* sign but different enough magnitudes, for close distances of approach (low values of S_0). It is always negative if one of the surface potentials is zero while the other is finite. A similar expression to that above was obtained by Wiese and Healy[56] for the case of constant-charge-density surfaces, *viz.*

$$\Phi_R^\sigma = \frac{1}{2}\varepsilon\varepsilon_0 a\left(\psi_1^2 + \psi_2^2\right)\left\{\frac{2\psi_1\psi_2}{\left(\psi_1^2 + \psi_2^2\right)}\ln\left[\frac{1+e^{-\kappa S_0}}{1-e^{-\kappa S_0}}\right] - \ln\left[1-e^{-2\kappa S_0}\right]\right\}. \quad (7.101)$$

For this case, the interaction is repulsive whenever the surface potentials are of the same sign or one of them is zero. It is attractive when the signs are opposite, except at very close distances of approach, where it will become repulsive.

D. Kinetics of aggregation

1. *Classification of aggregation rate processes and nomenclature*

Aggregation rate processes[57] may be classified in two ways: first with respect to the process by which the particles are brought together, and second by the probability of their sticking when they do make contact. The two main processes bringing particles together are Brownian motion, always present but not always significant, and convection, *i.e.*, transport by bulk fluid motion. Aggregation by Brownian motion alone is termed "*perikinetic*," while that which is dominated by shearing of the dispersion is termed "*orthokinetic*" aggregation. Their domains of importance are governed largely by the particle size. There is generally only a narrow range of particle size in a given case when both mechanisms are significant; *i.e.*, usually one or the other is dominant.

---

[56] Wiese, T. R., and Healy, T. W., *Trans. Faraday Soc.*, **66**, 490 (1970).
[57] When dispersions are said to "break down," the process for sols is aggregation, and for emulsions, it is the sequence of aggregation followed by coalescence. Only the aggregation step is considered here, in particular its application to sols (dispersions of solid particles). Coalescence is dealt with in Chap. 9 dealing with emulsions and foams.

If it is assumed that the particles in question have no interactions between them until there is a physical collision, the process is termed *diffusion-limited aggregation* (DLA), or more properly, *transport limited* aggregation. Under these conditions, every particle contact results in aggregation, *i.e.*, the sticking probability is unity. Furthermore, if it is assumed that as soon as an aggregate is formed, it becomes a "particle," or a "cluster," then cluster-cluster aggregation can occur. This is referred to as *diffusion-limited cluster-cluster aggregation* (DLCCA) to contrast it with the situation in which the aggregates grow solely by addition of monomers: *diffusion-limited monomer-cluster aggregation* (DLMCA).

Once particles get together, by whatever mechanism, they may or may not aggregate, depending on the sticking probability. This is controlled by the interaction profile, such as the DLVO curve, that the particles experience as they approach one another. In an electrocratic dispersion with a high electrostatic barrier, Φ_m, the probability of particles approaching to the point of contact is low, *i.e.*, aggregation into the primary minimum means overcoming a potential energy "barrier," somewhat analogous to the activation energy for a bimolecular chemical reaction as described by Eyring's absolute rate theory. Thus such a process is termed *reaction-limited aggregation* (RLA). A further distinction between reaction-limited cluster-cluster and reaction-limited *monomer*-cluster aggregation, *i.e.*, (RLCCA) and (RLMCA) can be made. In any event, if the system of two kinetic units possesses a potential energy that is a function of their mutual position, it must be taken into account in the evaluation of encounters between such particles. In all cases, the rate of aggregation will depend, other factors being equal, on the concentration of the particles in the dispersion.

2. Smoluchowski theory of diffusion-limited (rapid) aggregation

The process of aggregation involves, in the early stages, the collisions of pairs of *primary particles* to form doublets. It is assumed that every such contact during which the particles approach each other to a distance within the energy barrier (if there is one) is successful in creating such a doublet. Subsequent collisions enlarge the size of the aggregates, and in the usual case they eventually leave the main volume of the dispersion by sedimentation or creaming. Attention here is focused on the early stages of aggregation when such subsequent processes need not be considered. It may thus be assumed that each aggregation event results in the loss of a single particle. The number of collisions occurring per unit volume per unit time is proportional to the square of the number concentration of particles in the dispersion, n. Thus:

$$-\frac{dn}{dt} = k_r n^2,$$

(7.102)

where n = the total number of particles/volume, and k_r is a rate constant. This is the same as the rate equation for any bimolecular reaction. If n_0 is the initial particle concentration, it may be integrated to yield:

$$\frac{1}{n} - \frac{1}{n_0} = k_r t.$$
(7.103)

Equation (7.103) should describe the rate of aggregation during the early stages of the process, and it is usually found to hold approximately to the point where half the particles in the original dispersion have coagulated. Setting $n = 1/2n_0$, and solving for t, gives the "half-life" for an aggregating dispersion:

$$t_{1/2} = \frac{1}{k_r n_0}.$$
(7.104)

It can be seen that for any given rate constant, the lifetime of the dispersion depends inversely on the initial particle concentration, *i.e.,* concentrated dispersions subject to incipient instability may be much more difficult to maintain than dilute dispersions.

Although attention has been focused on early-stage aggregation, if one wishes to generalize beyond the early stages, a population balance that yields the concentration change of each of the k-fold particles with time is required. It is given by the net rate of the formation of k-fold particles from smaller particles and their disappearance by aggregation with other particles, *viz.*

$$\frac{dn_k}{dt} = \frac{1}{2} \sum_{i=1; i+j=k}^{i=k-1} k_{ij} n_i n_j - \sum_{i=1}^{i=\infty} k_{ik} n_i n_k,$$
(7.105)

where n_i is particle number concentration of i-fold particles and k_{ij} is the aggregation frequency function (rate constant), also often called the *aggregation kernel.* The leading terms of the population balance take the form:

$$\frac{dn_1}{dt} = -k_{11} n_1^2 - k_{12} n_1 n_2 - \ldots$$
(7.106)

$$\frac{dn_2}{dt} = \frac{1}{2} k_{11} n_1^2 - k_{12} n_1 n_2 - \ldots$$
(7.107)

$$\frac{dn_3}{dt} = k_{12} n_1 n_2 - \ldots, \text{ etc.}$$
(7.108)

Summing over all aggregate sizes, $n = \sum n_i$, gives:

$$\frac{dn}{dt} = -\frac{1}{2} k_{11} n_1^2 - k_{12} n_1 n_2 - \ldots,$$
(7.109)

which for early times, when $n \approx n_1$, gives

$$\frac{dn}{dt} = -\frac{1}{2} k_{11} n^2, \tag{7.110}$$

so that we identify k_r in the earlier rate law with $\frac{1}{2} k_{11}$.

Consider next the evaluation of the rate constant, k_r, under various conditions. The simplest situation is that in which the particles may be taken as hard, non-interacting spheres of uniform size. Collisions occur in such a system only when they make direct physical contact with one another, as considered by Smoluchowski (1916-1917)[58]. The rate process resulting from this model is termed "rapid flocculation," or more properly in terms of current usage, "rapid aggregation." The mechanism by which particles migrate toward one another is Brownian diffusion only. It is useful to fix attention on one single particle and to assume for the moment (it will be relaxed later) that the particle is fixed in position, as shown in Fig. 7-15. Over the course of time, other particles colliding with it become a part of it and thus "disappear." The fixed particle has thus become a "sink" for other particles that happen to come in contact with it. Eventually a quasi-steady

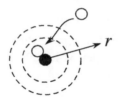

Fig. 7-15: A particle diffuses toward a "target" particle to aggregate. A quasi-steady state is soon achieved during which the flow across each concentric spherical surface is the same.

state is reached (a consequence of the spherical geometry) in which the total rate of transfer of particles across any spherical surface of radius r about the central particle approaches a constant equal to the rate of capture of particles by the central particle. Actually, this condition is achieved very rapidly. Brownian diffusion sets up and responds to the radial particle concentration profile that exists from the central particle surface outward to far away from the central particle.

The flux of particles across a surface toward the central particle (*i.e.*, in the *minus r*-direction) in accord with Fick's Law, is given by:

$$J = D\nabla n = D\frac{dn}{dr}, \tag{7.111}$$

where D is the particle diffusivity. The flow of particles across a spherical surface of any radius r is then

$$JA = 4\pi r^2 D\frac{dn}{dr} = \text{constant} = C. \tag{7.112}$$

[58] von Smoluchowski, M., *Physik. Z.*, **17**, 557, 585 (1916);
von Smoluchowski, M., *Z. Phys. Chem.*, **92**, 129 (1917).

The constant C may be evaluated by integrating Eq. (7.112) between the point of particle contact, $r = 2a$ (where a = the particle radius) and a distance far from the central particle, *i.e.*, $r \rightarrow \infty$. The boundary conditions are:

$$\text{at } r = 2a, \quad n = 0, \quad \text{and}$$

$$\text{at } r \rightarrow \infty, \quad n = n_0. \tag{7.113}$$

Thus:

$$\int_0^{n_0} dn = n_0 = \int_{2a}^{\infty} \frac{C}{4\pi D r^2} dr = \frac{C}{8\pi D a}, \tag{7.114}$$

so that the rate of particle capture is:

$$JA = C = 8\pi Dan_0. \tag{7.115}$$

Actually, the central particle is not fixed in space, but may itself diffuse at the same rate as the other particles. One must thus multiply C by a factor of two to get the total effective particle capture rate by the "central particle:"

$$C_{\text{eff}} = 16\pi Dan_0. \tag{7.116}$$

This same expression would hold if the point of observation were at some time slightly later than our original $t = 0$, *i.e.*, one can put $n \approx n_0$.

The total rate of particle capture per unit volume of the system would be (the capture rate per particle) × (particles per volume), but one must multiply by 1/2 to keep from counting each particle twice (*i.e.*, both as a captor and a captive). The total rate of particle loss per unit volume is thus (at time t):

$$8\pi Dan^2 = -\frac{dn}{dt}. \tag{7.117}$$

Comparison with the original rate equation for rapid coagulation gives:

$$k_r = 8\pi Da. \tag{7.118}$$

Since the particles are treated as hard spheres, one may express D in terms of the Einstein-Smoluchowski Equation as:

$$D = \frac{kT}{f} = \frac{kT}{6\pi \mu a}, \tag{7.119}$$

and

$$k_r = \frac{4kT}{3\mu}. \tag{7.120}$$

Equation (7.120) is the principal Smoluchowski result for rapid perikinetic aggregation. The half-life of the dispersion is then given by:

$$t_{1/2} = \frac{3\mu}{4kTn_0},$$ (7.121a)

and is seen to depend directly on the viscosity of the medium and inversely on temperature and initial particle concentration. Interestingly, it is independent of particle size at a given number concentration. If the volume fraction ϕ is held constant, however, and the particle size is varied, a very different picture emerges, as shown in Eq. (7.121b) for spherical particles of diameter d:

$$t_{1/2} = \frac{\pi\mu d^3}{8kT\phi}.$$ (7.121b)

For a given volume fraction, the half-life drops steeply with particle size, providing one of the explanations for the difficulty in stabilizing nano-particle dispersions. Table 7-3 shows numerical results for various cases. The numbers are in reasonable agreement with observations for aggregation in the absence of barriers. The existence of barriers will of course increase the half-life; in the limit of stabilized dispersions, $t_{1/2} \rightarrow \infty$. There may also be situations in which the observed rate of aggregation is *greater* than that predicted by "rapid aggregation" theory, as will be explained below. It is important to note that in systems subject to rapid aggregation, dispersions with a half-life of the order of one hour are *very* dilute.

Table 7-3: Half-lives for perikinetic aggregation of spherical particles in water at 20°C.

(a) For various number concentrations of 1-μm diameter particles.

n_0 (#/cm^3)	ϕ (for $d = 1$ μm)	$t_{1/2}$
10^{11}	0.052	1.85 s
10^9	0.00052	3.09 min
10^7	0.0000052	5.14 hr

(b) For various particle diameters at a constant volume fraction of 0.05.

d	n_0	$t_{1/2}$
1 μm	9.55×10^{10}	2.16 s
100 nm	9.55×10^{13}	2.16 ms
10 nm	9.55×10^{16}	2.16 μs

It is easy to consider the Smoluchowski treatment for the case of spherical particles of *different* size. Assuming the collisions involved are between particles of radii a_1 and a_2, respectively, the distance apart upon collision becomes $a_1 + a_2$ and the effective diffusivity becomes $1/2(D_1 + D_2)$. The equation for C_{eff} then takes the form:

$$C_{eff} = 4\pi(D_1 + D_2)(a_1 + a_2)n_0, \tag{7.122}$$

and upon substitution of the Einstein-Smoluchowski expressions for the diffusivities:

$$k_r = \frac{4kT}{3\mu}\left[\frac{(a_1 + a_2)}{2\sqrt{a_1 a_2}}\right]^2. \tag{7.123}$$

For differences in particle radii of a factor of two, the "correction factor" takes on a value of only 1.06. Thus the uncorrected rate constant equation can accommodate at least some polydispersity.

Smoluchowski used the above approximation, *i.e.*, he assumed that k_{ij} in the general population balance could be taken as a constant, to solve the general population balance for Brownian aggregation, beginning with singlets only. His result gives for aggregates consisting of k primary particles:

$$n_k(t) = n_0\frac{(t/t_{1/2})^{k-1}}{(1 + t/t_{1/2})^{k+1}}, \tag{7.124}$$

and for the total number of singlets plus aggregates:

$$n_{tot}(t) = \sum_{k=1}^{\infty} n_k(t) = \frac{n_0}{(1 + t/t_{1/2})}. \tag{7.125}$$

These results, assuming a constant aggregation kernel, have produced a remarkable description of experiments by Higashitani and Matsuno with polystyrene latices, shown in Fig. 7-16.

A more general treatment of this problem, with arbitrary initial conditions, has been given by Swift and Friedlander[59], who found that a "self-preserving" particle size distribution satisfied the Smoluchowski Equations. Experiments with heterogeneous hydrosols revealed that the size distributions *were* self-preserving and that the rate of aggregation was second-order in total particle concentration, consistent with Smoluchowski theory.

3. *The hydrodynamic drainage effect*

A potential difficulty with the Smoluchowski theory is its tacit assumption that the description of independent particle movement holds for the motion of the particles even as they approach very near to one another prior to aggregation. This cannot be the case. One would expect the flow field about each particle to be influenced by the proximity of its neighbor.

[59] Swift, D. L., and Friedlander, S., *J. Colloid Interface Sci.*, **19**, 621 (1964).

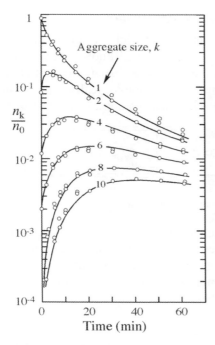

Fig. 7-16: Results of Higashitani and Matsuno showing aggregate size populations (up to $k = 10$) formed from an initial dispersion of polystyrene latex particles of diameter 0.974 μm in aqueous KCl solutions at 20°C. Size distributions were measured using the Coulter (electrozone) method. The solid lines are computed according to Eq. (7.124), all using a single aggregation rate constant. After [Higashitani, K., and Matsuno, Y., *J. Chem. Eng. Japan*, **12**, 460 (1979).]

Specifically, a non-negligible time would be required to drain the viscous film separating the particles during their final approach. Spielman[60] and Honig *et al.*[61] have examined this problem and found it to be especially severe when $\kappa a \gg 1$. They found a convenient way to take the effect into account by modifying the effective diffusivity of the particles approaching one another, *viz.*

$$\frac{D_{\text{eff}}(S_0)}{D(S_0 \to \infty)} = \frac{1 + 2a/3S_0}{1 + 13a/6S_0 + a^2/3S_0^2} \xrightarrow{S_0/a \to 0} 2\left(\frac{S_0}{a}\right). \qquad (7.126)$$

Since $D_{\text{eff}} \to 0$ as the particles near contact, actual encounters require an attractive force at least proportional to a/S_0, and van der Waals attractions do meet this requirement.

Reductions in the aggregation rate by as much as a factor of ten are predicted due to the viscous drainage effect. One might wonder that the viscosity does not appear explicitly in Eq. (7.126), but if the viscosity of the liquid in the thin film is the same as it is in the bulk, the effect cancels out. The way that the importance of the drainage effect can be verified experimentally is by the determination of the absolute rate constant for rapid aggregation in comparison to the prediction of Eq. (7.120). Typically reductions from a factor of two to five are found.

[60] Spielman, L. A., *J. Colloid Interface Sci.*, **33**, 562 (1970).
[61] Honig, E. P., Roebersen, G. J., and Wiersema, P. H., *J. Colloid Interface Sci.*, **36**, 97 (1971).

4. *Orthokinetic (shear flow induced) aggregation*

The situation considered thus far is one in which particle collisions are brought about solely by Brownian motion. It is well known, however, that stirring or other processes that cause shearing of a dispersion, can enhance the aggregation rate. Figure 7-17 shows phenomenologically why shearing action should enhance particle contact. The particles entrained in the faster

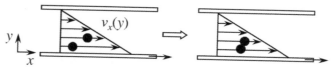

Fig. 7-17: Orthokinetic aggregation.

moving streamlines catch up with those in the slower moving stream lines. Smoluchowski also derived an expression for collision frequency, hence orthokinetic aggregation, in a laminar shear field. His expression for the number of collisions per unit volume per unit time by uniform shear flow in the absence of Brownian motion effects is:

$$-\frac{dn}{dt} = \frac{16}{3}\dot{\gamma}a^3 n^2,$$
(7.127)

where $\dot{\gamma}$ is the shear rate, *viz.*, $|dv_x/dy|$ (*cf* Fig. 7-17), with dimensions of time^{-1}. Note that in orthokinetic aggregation, the rate depends very strongly (to the third power!) on the particle radius, whereas Brownian coagulation is independent of it. The total rate of aggregation, considering both effects, may be taken to good approximation as the sum of the Brownian and the shear-rate effects:

$$-\frac{dn}{dt} = \left(\frac{4kT}{3\mu} + \frac{16\dot{\gamma}a^3}{3}\right)n^2.$$
(7.128)

Brownian coagulation is dominant in systems with sufficiently small particles and is favored by high temperature and low viscosity. Some numerical comparisons are shown in Fig. 7-18. Particle size is the key parameter. For 5-μm radius or more, even in an unstirred system the ever-present thermal convection currents ($\dot{\gamma} \approx 10^{-2}$ s^{-1}) will be dominant in inducing aggregation. Also since larger particles or aggregates settle more rapidly than the smaller particles, sedimentation may enhance the rate of aggregation in the absence of stirring. In a colloid system with $a \approx 1$ μm, orthokinetic aggregation can easily be made as much as 10^4 times greater than the perikinetic rate by intentional mixing. On the other hand, for the very smallest of colloid particles ($a < 0.1$ μm), even the most vigorous stirring will have little effect in assisting the initial stages of Brownian aggregation. While temperature and viscosity effects (and the effect of temperature on viscosity) are important in Brownian aggregation, they cannot compare to the sensitivity of orthokinetic aggregation rate to particle size. Swift and Friedlander[56] also extended the Smoluchowski treatment of orthokinetic aggregation to the case of polydisperse systems and found the same type of self-preserving distribution function exists for this case as for the perikinetic case.

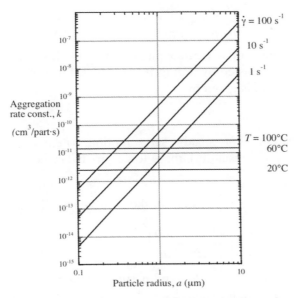

Fig. 7-18: Comparison of perikinetic and orthokinetic contributions to the aggregation rate constant. Note that stirring is unimportant when $a < 0.1$ μm, and Brownian motion is unimportant when $a > 5$ μm.

5. Reaction-limited (slow) aggregation; the stability ratio W

The rate of slow (reaction-limited) aggregation was first described by Fuchs (1934) as follows.[62] In Smoluchowski theory (excluding orthokinetic effects for the moment), particles move only under the influence of Brownian motion (diffusion), which can be thought of as net movement in response to a spatial variation in a total potential, V_{total}, that is the sum of the *chemical* potential plus the potential energy due to the proximity of a nearby particle.

$$V_{total} = \mu + \Phi, \tag{7.129}$$

where Φ is the particle pair interaction potential, *e.g.*, the DLVO curve of Fig. 7-6. The "force" which moves the particles under the influence of a spatially varying potential is the negative gradient of that potential function,

$$F_{motive} = -\nabla V_{total} = -\nabla \mu - \nabla \Phi. \tag{7.130}$$

The chemical potential for a dilute system of particles may be written as

$$\mu = \mu^{\ominus} + kT \ln n, \tag{7.131}$$

[62] Fuchs, N., Z. Phys., **89**, 736 (1934).

in which the standard chemical potential is a function of temperature only. The particle number concentration, n, is the appropriate measure of concentration. Considering also that we are only interested in movement in the r-direction (in the spherical coordinate system), Eq. (7.130) may be written as

$$F_{\text{motive}} = -\frac{kT}{n}\frac{dn}{dr} - \frac{d\Phi}{dr}.$$ (7.132)

The movement of a particle in response to this driving force in a viscous medium is resisted by a drag force, which for a spherical particle in a Newtonian medium is given by

$$F_{\text{res}} = 6\pi\mu a v_{\infty},$$ (7.133)

as discussed earlier. For steady state motion, (i.e., $v = v_{\infty}$, which is established effectively instantly), these forces are balanced, i.e.,

$$F_{\text{motive}} = F_{\text{res}}.$$ (7.134)

Substitution leads to the expression for the steady-state diffusion velocity:

$$v_{\infty} = -\frac{1}{6\pi\mu a}\left(\frac{kT}{n}\frac{dn}{dr} + \frac{d\Phi}{dr}\right),$$ (7.135)

and the steady-state rate of flow of particles toward a central reference particle (minus-r direction):

$$C = JA = -(nv_{\infty})(4\pi r^2) = \frac{2r^2}{3\mu a}\left(kT\frac{dn}{dr} + n\frac{d\Phi}{dr}\right).$$ (7.136)

Recall that in developing the expression for fast aggregation, one next integrated the equivalent equation between $r = 2a$ (where $n = 0$) and $r \to \infty$ (where $n = n_0$) to obtain the expression for C. To repeat that procedure here, Eq. (7.136) is put in the form:

$$\frac{dn}{dr} + \frac{n}{kT}\frac{d\Phi}{dr} = \frac{3\mu aC}{2r^2 kT},$$ (7.137)

and multiplying through by the integrating factor exp (Φ/kT) and integrating between the specified limits leads to:

$$C = \frac{2kTn_0}{3\mu a}\left[\int_{2a}^{\infty}\frac{1}{r^2}\exp\left(\frac{\Phi}{kT}\right)dr\right]^{-1},$$ (7.138)

producing an aggregation rate constant of

$$k_r = \frac{2kT}{3\mu}\left[\int_{2a}^{\infty}\frac{1}{r^2}\exp\left(\frac{\Phi}{kT}\right)dr\right]^{-1} = \frac{k_r(\text{fast agg.})}{W},$$ (7.139)

with

$$W = 2a \int_{2a}^{\infty} \frac{1}{r^2} \exp\left(\frac{\Phi}{kT}\right) dr .$$ (7.140)

It has been shown[63] that to good approximation when a potential barrier exists

$$W \approx \frac{1}{2\kappa a} \exp(\Phi_m / kT) .$$ (7.141)

It is seen that under the influence of a potential energy of interaction, the aggregation rate constant differs from that for fast aggregation by the factor $1/W$. W is called the "*stability ratio*," and it may be interpreted equivalently as the ratio of the rapid to the slow aggregation rate constant for doublet formation, the ratio of the half-life for slow aggregation to that for rapid aggregation, or to $1/f$, where f is the probability that a given particle collision will result in aggregation (*i.e.*, a "sticking probability").

In the usual case, there is a barrier to aggregation into the primary minimum of the potential energy curve, and the stability ratio takes on values substantially greater than unity. It should be noted that in the absence of a barrier to aggregation, but in the presence of inter-particle *attractive* forces, Φ will be negative, and W will take on values less than unity. This leads to a slight enhancement of aggregation rate over that predicted by the Smoluchowski Equation, but the effect is rarely very great ($\leq 10\%$).

Finally, it should be noted that the distinction between perikinetic and orthokinetic aggregation also exists for slow or reaction-limited aggregation, but it is evident that for systems whose barriers to entering the primary minimum are substantially in excess of $\approx 15kT$ may require a large shear rate to induce aggregation. On the other hand, intense shear may break apart larger aggregates so that eventually a steady-state flock size distribution is achieved.

6. *Secondary minimum effects*

Aggregation has so far been assumed to imply passage of the particle pair into the primary minimum, and, since the depth of the minimum is very large (*i.e.*, many times kT), the event is properly treated by Smoluchowski and Fuchs as irreversible, unless the original conditions of stability are restored. On the other hand, it may be possible to have aggregation into the relatively shallow secondary minimum. Such aggregation will often be reversible, particularly if the depth of the secondary minimum is less than a few kT, and the resulting flocks will be loose and easily broken by stirring. This is referred to as "weak aggregation," and, as stated earlier, the word "flocculation" has been associated with it, and the term "coagulation" used

---

63 Reerink, H., and Overbeek, J. Th. G., *Discuss. Faraday Soc.*, **18**, 74 (1954).

for irreversible aggregation, of the type we have been discussing. More general treatments, in which flocculation into the secondary minimum may be followed by coagulation into the primary minimum have been given by Ruckenstein[64], Marmur[65] and others.

In part due to its simplicity, Marmur's method for relating the stability ratio to the pair interaction potential when both secondary- and primary-minimum aggregation may occur is especially useful. He assumed the kinetic energy distribution amongst colloidal particles to mimic the Maxwell distribution in an ideal gas, and considered the fate of a pair of particles arriving at a distance of separation corresponding to the secondary minimum. The particles were assumed to aggregate if: 1) the sum of their kinetic energies was *less* than Φ_{min}, the depth of the secondary minimum (so that they would experience aggregation into that minimum), *or* 2) their total kinetic energy was *greater* than $\Delta\Phi$, the difference in height between the top of the energy barrier and the depth of the secondary minimum, in which case they would aggregate into the primary minimum. The total fraction of successful collisions was given by:

$$\frac{1}{W} = f_p + f_s,$$
(7.142)

where f_p is the fraction of collisions resulting in coagulation into the primary minimum, and f_s the fraction entering the secondary minimum. The expressions for f_p and f_s, dictated by the ideal gas form of the kinetic energy distribution are:

$$f_p = 1 - \frac{4}{\sqrt{\pi}} \int_0^{\sqrt{\Delta\Phi}} x_1^2 \exp(-x_1^2) \left[\text{erf}(x_p) - \frac{2}{\sqrt{\pi}} x_p \exp(-x_p^2) \right] dx_1,$$
(7.143)

where $x_p = \sqrt{\Delta\Phi - x_1^2}$, $x_1^2 \le \Delta\Phi$, and

$$f_s = 1 - \frac{4}{\sqrt{\pi}} \int_0^{\sqrt{-\Phi_{min}}} x_1^2 \exp(-x_1^2) \left[\text{erf}(x_s) - \frac{2}{\sqrt{\pi}} x_s \exp(-x_s^2) \right] dx_1,$$
(7.144)

where $x_s = \sqrt{-\Phi_{min} - x_1^2}$ and $x_1^2 \le -\Phi_{min}$. It is evident that all that is required for the evaluation of W is the depth of the energy minimum and the height of the energy barrier.

Marmur's model successfully simulated aggregation rate data, and in particular successfully described the observed weak dependence of the stability ratio on particle size. The situation in which a secondary minimum

[64] Ruckenstein, E., *J. Colloid Interface Sci.*, **66**, 531 (1978).
[65] Marmur, A., *J. Colloid Interface Sci.*, **72**, 41 (1979).

exists adjacent to an insurmountable barrier to primary-minimum aggregation produces flocculation, the kinetics of which is discussed following the description of steric stabilization below.

7. *Kinetics of hetero-aggregation*

The kinetics of aggregation between unlike particles, *i.e.*, hetero-coagulation, may be important in many applications. Of particular interest is the situation in which particles of opposite charge are mixed together in the same dispersion. The kinetics of hetero-coagulation is treated in a straightforward way using the method of Hogg, Healy and Fuerstenau (HHF)[66], who, as discussed earlier, obtained the expression for the electrostatic interaction between particles of opposite charge for colloids of constant surface potential in the Debye-Hückel range. The overall early-time rate of aggregation in a colloid consisting of particles of types 1 and 2 could be expressed as k_r/W_{total}, with the total stability ratio given in terms of the stability ratios for the three possible types of pair interactions as

$$\frac{1}{W_{total}} = \frac{x_1^2}{W_{11}} + \frac{x_1 x_2}{W_{12}} + \frac{x_2^2}{W_{22}}, \tag{7.145}$$

where x_1 and x_2 are the number fractions of particles of types 1 and 2, and

$$W_{ij} = 2a \int_{2a}^{\infty} \frac{1}{r^2} \exp\left(\frac{\Phi_{ij}}{kT}\right) dr. \tag{7.146}$$

Of particular interest is the case of *selective* hetero-aggregation, in which the effective surface potentials of the two species are both sufficiently high to prevent homo-aggregation, and

$$W_{total} \approx \frac{1}{x_1 x_2} W_{12}. \tag{7.147}$$

The full role of surface chemistry (mode of charge regulation during particle interaction) in primary- and secondary-minimum aggregation and in hetero-coagulation are discussed by Prieve and Ruckenstein.[67] As cited earlier, McCormack *et al.*[68] have provided an analysis which allows the determination of the pair interaction potential between parallel plates under completely general electrostatic boundary conditions, *i.e.*, all possible combinations of surface boundary conditions and all realizable levels of potential and charge density. A recent study of hetero-aggregation in aqueous sol-emulsion systems[69] implements the McCormack *et al.* double layer analysis for the stability of mixed colloids of spherical particles and shows the importance of hetero-aggregation phenomena in the effort to

[66] Hogg, R., Healy, T. W., and Fuerstenau, D. W., *Trans. Faraday Soc.*, **62**, 1638 (1966).
[67] Prieve, D. C., and Ruckenstein, E., *J. Colloid Interface Sci.*, **73**, 539 (1980).
[68] McCormack, D., Carnie, S. L., and Chan, D. C., *J. Colloid Interface Sci.*, **169**, 177 (1995).
[69] Sunkel, J. M., and Berg, J. C., *J. Colloid Interface Sci.*, **179**, 618 (1996).

replace solvents with water in coating dispersions. One outcome of that study was that if the experimentally measured zeta potentials are used in place of the surface potentials, reasonable agreement is found with the simpler HHF analysis, even in situations where the restrictive assumptions in that treatment were not valid.

8. *Measurement of early-stage aggregation kinetics (W)*

Experimental measurements of rates of aggregation in different systems, as might be accomplished by direct particle counting (as the Coulter method used for the results shown in Fig. 7-16, turbidimetry, photon correlation spectroscopy (PCS) or some other light scattering technique will provide values for the stability ratio and therefore information on the Φ-function itself. The deduction of such information, however, is not straightforward because Φ is buried in the integral expression for W. Nonetheless, the study of early-stage aggregation kinetics provides some important information concerning the interaction between colloidal particles.

The measurement of aggregation rate (*e.g.*, by light scattering) also provides an alternative means to the more traditional method of "jar testing" for locating the *CCC* (or *CFC*) in a given case. As an example of this, Ottewill and Shaw[70] reported *CCC*-values and additional information from studies of the aggregation of monodisperse polystyrene latex particle dispersions they investigated, *viz.*, $a \approx 30$ nm. For particles of this size, the turbidity of the dispersion is given by the Rayleigh expression, Eq. (5.115):

$$\tau = Anv^2 = A\phi v, \tag{7.148}$$

where A is a cluster of constants, $v = \bar{v}_m$ is the *mass*-average volume per particle at the instant of interest, and ϕ is the volume fraction of the particles (a constant). (Note: Slightly different notation than that of Chap. 5 is being used.) Differentiation with respect to time, t, gives:

$$\frac{d\tau}{dt} = A\phi \frac{d\bar{v}_m}{dt}. \tag{7.149}$$

When the aggregation has proceeded only to the point where the dispersion consists of primary particles and doublets:

$$\bar{v}_m = \frac{\sum n_i v_i^2}{\sum n_i v_i} = \frac{n_p v_0^2 + n_d (2v_0)^2}{n_p v_0 + n_d (2v_0)} = \left(\frac{n_p + 4n_d}{n_p + 2n_d}\right) v_0, \tag{7.150}$$

where v_0 is the volume of a primary particle, n_p = number density of primary particles and n_d = number density of doublets. From a material balance, $n_{p0} = n_p + 2n_d$, where n_{p0} = initial number density of primary particles. Substituting:

---

[70] Ottewill, R., and Shaw, D. J., *Disc. Faraday Soc.*, **42**,154 (1966).

$$\bar{v}_m = \left(\frac{2n_{p0} - n_p}{n_{p0}}\right)v_0,$$ (7.151)

and

$$\frac{d\tau}{dt} = \frac{A\phi v_o}{n_{p0}}\left(-\frac{dn_p}{dt}\right).$$ (7.152)

Thus as $t \rightarrow 0$

$$\left(\frac{d\tau}{dt}\right)_{t\rightarrow 0} = \frac{A\phi v_0}{n_{p0}}(k_r n_{p0}^2) = A\phi v_0 n_{p0} k_r = \tau_0 n_{p0} k_r = \tau_0 n_{p0}\frac{k_r}{W}.$$ (7.153)

Finally, then

$$\log\left(\frac{d\tau}{dt}\right)_{t\rightarrow 0} = (\text{const}) - \log W,$$ (7.154)

as shown in Fig. 7-19. The formula is easily modified to accommodate RGD scattering. For larger particles (Mie region), a more complicated expression can be developed, but becomes especially simple for particles of $a \geq 2.5$ μm, cf. Eq. (5.142). Aggregation may be followed until there is a 5% reduction in

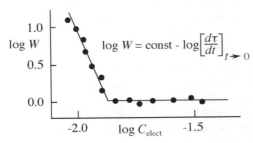

Fig. 7-19: Stopped-flow spectrophotometry measurement of aggregation rate.

the particle number concentration before significant amounts of higher order aggregates are present. In plotting the results of such experiments it is common to reckon W relative to the value obtained for rapid aggregation, therefore dividing out any effect of hydrodynamic drainage. Some calculated stability diagrams are shown in Fig. 7-20.

Using the expression for Φ_{tot} which holds for low and constant particle surface potentials in aqueous media with a symmetrical electrolyte and neglecting retardation effects in reckoning the long range van der Waals interactions, Reerink and Overbeek[71] noted that the computed expression for W should take the form:

$$\log W_{\text{theor}} = k_1 - k_2 \log C_e,$$ (7.155)

where C_e is the concentration of z-z electrolyte and k_1 and k_2 are constants. More specifically,

---

[71] Reerink, H., and Overbeek, J. Th. G., *Disc. Faraday Soc.*, **18**, 74 (1954).

$$\frac{d\log W_{theor}}{d\log C_e} = -k_2 = -2.06 \cdot 10^7 \frac{a\gamma_\delta^2}{z^2},$$ (7.156)

where recall that $\gamma_\delta \approx ze\psi_\delta / 4kT$. Thus, the slope of the curve of $\log W$ vs. $\log C_e$ should be:

$$\text{Slope} = -2.06 \cdot 10^7 \frac{ae^2\psi_\delta^2}{(kT)^2}$$ (7.157)

for "slow aggregation," and it should be zero for rapid aggregation.

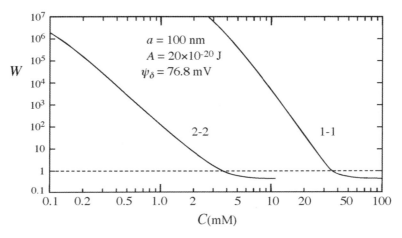

Fig. 7-20: Stability ratio W as a function of electrolyte concentration for 1-1 and 2-2 electrolytes, calculated from Fuchs theory. Note that the compute W-values for rapid aggregation are somewhat less than 1, due to the attractive van der Waals interactions. After [Overbeek, J. Th. G., **Colloidal and Surface Chemistry, A Self-Study Subject, Part 2**, Lyophobic Colloids, MIT, p. 15.10, Cambridge, MA (1972).]

The *CCC* is clearly identifiable as the concentration of electrolyte when the $\log W$ curve becomes horizontal. Using the slope of curves such as this, a value for ψ_δ may be inferred, and using that value for ψ_δ, one can examine the Φ-function and try different values of the Hamaker constant, A_{212} until the best fit is found. When this was done, Reerink and Overbeek (*loc. cit.*), Ottewill and Shaw (*loc. cit.*) and others have obtained values in reasonable agreement with values computed in other ways. It is also possible to determine that a significant slowing of aggregation exists only when the barrier exceeds approximately 15 kT.

Photon correlation spectroscopy (PCS) provides one of the most powerful means of following the course of aggregation.[72,73] One obtains (*cf.* Chap. 5) the effective average particle diffusivity (hence the hydrodynamic radius) as a function of time. As doublets, triplets, *etc.* form, the observed

[72] Barringer, E. A., Novich, B. E., and Ring, T. A., *J. Colloid Interface Sci.*, **100**, 584 (1984).
[73] Virden, J. W., and Berg, J. C., *J. Colloid Interface Sci.*, **149**, 528 (1992).

mean hydrodynamic radius increases. The rate constant for doublet formation can be obtained from the initial slope of the curve of the mean hydrodynamic radius of the particles versus time, as shown in Fig. 7-21, in accord with

$$k_{11} = \frac{1}{\alpha n_0 \bar{a}_h} \left(\frac{d\bar{a}_h}{dt} \right)_{t \to 0},$$
(7.158)

where $\bar{a}_h$ is the mean hydrodynamic particle radius, n_0 is the initial particle number density, and α is an optical constant dependent on the size of the particles and aggregates, their respective form factors, the scattering angle and the wavelength of the illumination. It is common practice to determine the slope corresponding to rapid aggregation (shown as $W = 1$ in Fig. 7-21) and to ratio the slopes obtained under other conditions to it. For very slow rates of aggregation, $W \to \infty$, one obtains a horizontal line.

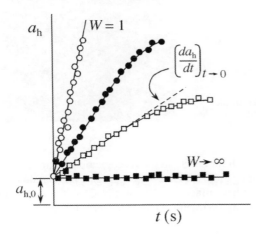

Fig. 7-21: Aggregation rates by photon correlation spectroscopy (PCS).

9. Surface aggregation

While the clumping together of larger particles ($d \geq 10$ μm) at fluid interfaces by meniscus (or capillary) forces has been discussed in Chap. 4, the aggregation of colloidal particles at interfaces, when these forces are negligible, is also an important phenomenon. Aggregation of particles at interfaces is important in flotation, preparation of anti-stick surfaces, and in fluid-phase contacting operations of all kinds, where particulate contamination often accumulates and aggregates at the interfaces. One of the most important consequences of interfacial aggregation is that it provides a pathway for the degradation of an otherwise stable dispersion in which particles migrate to the surface, are trapped because they are not fully wet out by either phase ($\theta \neq 0°$), and aggregate under the influence of weaker net inter-particle repulsions than exist in bulk. When a pair of particles at the air-water interface of an electrocratic dispersion interact, the van der Waals attraction is increased because the effective Hamaker constant for the particles in air is larger than for the particles in the solution, and at the same time, the electrostatic repulsion will be reduced because the double layer exists only in the solution. Early reports of unexpected coagulation behavior due to interfacial aggregation were given by Freundlich and coworkers[74] (although their initial interpretation was incorrect) and Heller and coworkers.[75]

The surface aggregation of an otherwise stable dispersion of 1-μm diameter polystyrene (PS) latex particles in water at various concentrations of NaCl was investigated using the setup shown in Fig. 7-22(a). The particles were stabilized with surface sulfate groups with a parking area of (2.15 nm^2/SO$_4^{2-}$). The contact angle of the water against the latex surface

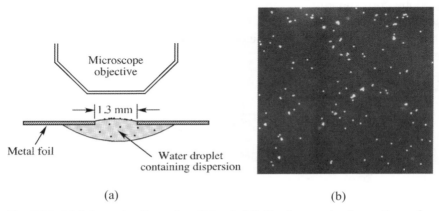

(a)   (b)

Fig. 7-22: (a) Schematic of sample cell for darkfield microscopic observations of particle adsorption and surface aggregation. (b) Observed aggregate population of 1-μm diameter PS primary particles at the surface of a 100 mM NaCl solution after 1.6 hr. Field of view shown: 160 μm. From [Williams, D. F., and Berg, J. C., *J. Colloid Interface Sci.*, **152**, 218 (1992).]

was measured at 102°, so that particles in the interface were approximately 40% submerged. Particles were observed to arrive at the surface as singlets and to form doublets, triplets, *etc.* over time in accord with a population balance:

$$\frac{d\Gamma_1}{dt} = J_{ads}(t) - k_{11}\Gamma_1^2 - k_{12}\Gamma_1\Gamma_2 - \ldots, \tag{7.159}$$

$$\frac{d\Gamma_2}{dt} = \frac{1}{2}k_{11}\Gamma_1^2 - k_{12}\Gamma_1\Gamma_2 - \ldots, \tag{7.160}$$

[74] Freundlich, H., and Basu, S. K., *Z. Phys. Chem. (Leipzig)*, **115**, 203 (1925); Freundlich, H., and Kroch, H., *Z. Phys. Chem. (Leipzig)*, **124**, 115 (1926); Freundlich, H., and Loebmann, S., *Kolloid Beih.*, **28**, 391(1931); Freundlich, H., and v. Rechlinghausen, R., *Z. Phys. Chem. Abt. A*, **157**, 325 (1931).
[75] Heller, W., and Peters, J., *J. Colloid Interface Sci.*, **32**, 592 (1970); Heller, W., and Peters, J., **33**, 578 (1970); Heller, W., and Peters, J., **35**, 300 (1971); DeLauder, W. B., and Heller, W., *J. Colloid Interface Sci.*, **35**, 308 (1971).

$$\frac{d\Gamma_3}{dt} = \frac{1}{2} k_{12} \Gamma_1 \Gamma_2 - \cdots, \tag{7.161}$$

where Γ_1, Γ_2, and Γ_3 are the surface number densities of singlets, doublets and triplets. Every effort was made to avoid convection so that both the rate of arrival of particles to the surface and their subsequent movement in the surface would be governed by diffusion. The arrival of particles and their aggregation in the surface was monitored, and Fig. 7-22(b) shows a typical image obtained, in this case for a salt concentration of 100 mM and a time of 1.6 hr. Attention was focused on doublet formation governed by Brownian motion as described by the radial component of Fick's Law in cylindrical coordinates. Two important distinctions from the corresponding process in bulk, as described by Smoluchowski, emerged from the analysis. First, in surface aggregation, no quasi-steady state (a consequence of spherical coordinates) occurred, so the rate "constant" for doublet formation continues to decrease with time during the course of the process. Secondly, the rate constant increases significantly with particle size, so that the constants appearing in the population balance cannot be treated as equal. Ultimately, semi-quantitative agreement was achieved between the model calculations and the observed aggregation. As salt concentration was increased, both the model calculations and the observed results showed that aggregation commenced at salt concentrations approximately two orders of magnitude less than those observed for bulk aggregation.

10. *Electrostatic stabilization and aggregation rates in apolar media*

It has been shown in Chap. 6.A.7 that mechanisms exist for the stabilization of ions in apolar media and for the surfaces of particles in such media to acquire charge. Although substantial zeta potentials (electrophoretic mobilities) can be measured, the ionic strengths and surface charge densities are almost immeasurably small, at least when the dielectric constant is very low (≈ 2). It is nonetheless beyond doubt that electrostatic forces can play a key role in colloid stabilization in such systems.[76] Furthermore, direct measurements of repulsive electrostatic forces between surfaces have been made.[77]

It has been argued that if only one knew the ionic strength so that the appropriate screening length, κ^{-1}, as computed using Eq. (6.25), classical DLVO theory could be applied to assess the electrocratic stabilization,[78] and a recent study of poly (methyl methacrylate) (PMMA) latex particles dispersed in dodecane doped with 12 mM Aerosol OT (known to produce inverse micelles in this medium) provides an example validating this

[76] Van der Minne, J. L., and Hermaie, P. H., *J. Colloid Sci.*, **8**, 38 (1952);
  Pugh, R. J., Matsunaga, T., and Fowkes, F. M., *Colloids Surfaces*, **7**, 183 (1983).
[77] Briscoe. W. H., and Horn, R. G., *Langmuir*, **18**, 3945 (2002);
  McNamee, C. E., Tsuji, Y., and Matsumoto, M., *Langmuir*, **20**, 1791 (2004);
  Prieve, D. C., Hoggard, J. D., Fu, R., Sides P. J., and Bethea, R., *Langmuir*, **24**, 1120 (2008).
[78] Lyklema, J., *Adv. Colloid Interface Sci.*, **2**, 65 (1968).

belief.[79] In this study the particle pair interaction potential was extracted from the quasi 2-dimensional structure of the dispersion as determined from a large ensemble of microscopic images. Analysis of the data yielded the pair distribution function, from which the pair interaction potential was derived using methods of statistical mechanics. The particle-particle repulsion closely matched the functional form of the screened Coulombic term in the DLVO Equation, Eq. (7.78), with the best-fitting surface potentials ranging from approximately 50-100 mV and Debye lengths from $0.2 - 1.4$ μm.

With screening lengths of the order of one μm, a picture much different from that envisioned for aqueous colloids emerges, particularly for small particles at anything much greater than infinitesimal number concentrations. As pointed out early on by Overbeek and coworkers[80] and others,[81] and brought to focus further by Morrison,[82] repulsion should not be expected to arise from the partial overlap of the diffuse double layers of approaching particles. Instead, the average particle separation is much less than the Debye length, so the particles are always swimming around *inside* each other's double layer. Overbeek pointed out that in ordinary electrocratic dispersions, the repulsive force generated is equal to the *gradient* in the overlap potential. In apolar media, since this potential varies so slowly over the distance of average particle separation, no such forces can develop. An analogous situation exists in even modestly concentrated dispersions in water when little or no background electrolyte is present. It was proposed by Osmond and others cited that a more realistic picture is one regarding the particles as charged entities in a pure dielectric medium, *i.e.*, one free of any double layer at all, so that $\kappa \rightarrow 0$. The repulsive electrostatic term in the DLVO Equation would then just be that derived from Coulomb's Law, which gives for the force between a pair of particles of charge q each:

$$F_R = \frac{q^2}{4\pi\varepsilon\varepsilon_0 r^2},$$
(7.162)

where r is the distance between particle centers. Assuming the particles are spheres of radius a, the charge is given by Eq. (6.109):

$$q = 4\pi a^2 \overline{\sigma}_0,$$
(6.109)

where $\overline{\sigma}_0$ is the surface charge density, related to the potential profile by Eq. (6.31):

[79] Hsu, M. F., Dufresne, E. R., and Weitz, D. A., *Langmuir*, **21**, 4881 (2005).
[80] Koelmans, H., and Overbeek, J. Th. G., *Discuss. Faraday Soc.*, **18**, 52 (1954);
    Albers, W., and Overbeek, J. Th. G., *J. Colloid Sci.*, **14**, 510 (1959).
[81] Osmond, D. W. J., *Discuss. Faraday Soc.*, **42**, 247 (1966);
    McGown, D. N. L., and Parfitt, G. D., *Kolloid-Z. Z. Polym.*, **219**, 51 (1967).
[82] Morrison, I. D., *Langmuir*, **7**, 1920 (1991);
    Morrison, I. D., *Colloids Surfaces A*, **71**, 1 (1993).

$$\bar{\sigma}_0 = -\varepsilon\varepsilon_0 \left(\frac{d\psi}{dr}\right)_{r=a}. \tag{6.31}$$

The potential profile surrounding a small sphere is given by Eq. (6.45), so that:

$$\psi(r) = \psi_0 \frac{a}{r} \exp[-\kappa(r-a)] \rightarrow \psi_0 \frac{a}{r}, \tag{7.163}$$

since we are putting $\kappa \rightarrow 0$.

Combining the above four equations gives

$$F_R = \frac{4\pi\varepsilon\varepsilon_0\psi_0^2 a^2}{r^2}. \tag{7.164}$$

The repulsive energy is obtained by integrating the force from an infinite separation to the distance of $r = S_0 + 2a$, where S_0 is the distance of closest approach of their surfaces:

$$\Phi_R = -\int_{\infty}^{S_0+2a} F_R\, dr = -4\pi\varepsilon\varepsilon_0\psi_0^2 a^2 \int_{\infty}^{S_0+2a} \frac{dr}{r^2} = \frac{4\pi\varepsilon\varepsilon_0\psi_0^2 a^2}{(S_0+2a)}. \tag{7.165}$$

It is this expression that should replace the second term in the DLVO Equation, Eq. (7.80). so that the total interaction is given by:

$$\Phi_{net} = \Phi_A + \Phi_R = -\frac{A_{212}a}{12S_0} + \frac{4\pi\varepsilon\varepsilon_0\psi_0^2 a^2}{(S_0+2a)}. \tag{7.166}$$

It is instructive to put in some numbers for a typical case. Consider PMMA particles of radius $a = 200$ nm in dodecane ($\varepsilon \approx 2$), with a surface potential of 50 mV. The appropriate Hamaker constant is $A_{212} = 0.178\times10^{-20}$ J. This gives

$$\Phi_{net} = -\frac{2.97\times10^{-20}}{S_0(nm)} + \frac{2.225\times10^{-17}}{(S_0+400)(nm)} \text{ J}. \tag{7.167}$$

It is clear that the repulsive term dominates until $S_0 \rightarrow 0$, *i.e*, they are repelled until they essentially touch, and then they stick.

To assess the stability in any situation, one needs to evaluate the stability ratio W as given by the Fuchs integral, Eq. (7.140):

$$W = 2a \int_{2a}^{\infty} \frac{1}{r^2} \exp\left(\frac{\Phi_{net}}{kT}\right) dr. \tag{7.140}$$

This can be evaluated analytically because, in view of the comment following Eq. (7.167) above, only the second term in Φ_{net} need be considered, and the result is

$$W = \frac{kT}{2\pi\varepsilon\varepsilon_0 a\psi_0^2}\left[\exp\left(\frac{2\pi\varepsilon\varepsilon_0 a\psi_0^2}{kT}\right) - 1\right].$$ (7.168)

Morrison[83] uses Eq. (7.168) and shows that (at $T = 298K$) in order to have robust stability (*i.e.*, $W > 10^5$), one must satisfy the criterion

$$\varepsilon a\psi_0^2 > 10^3 \text{ (with } a \text{ [=] } \mu\text{m and } \psi_0 \text{ [=] mV)}$$ (7.169)

and cites literature to show good agreement with experiment. The example above of 200 nm radius PMMA particles with a surface potential of 50 mV gives $\varepsilon a\psi_0^2$ a value of exactly 10^3 and would thus be considered a system of marginal electrocratic stability. It is evident that to stabilize small nano-particles in low dielectric constant media electrostatically requires very high surface potentials that may be difficult to achieve.

E. Steric stabilization and other colloid-polymer interactions

1. *Polymer adsorption and steric stabilization*

The almost exclusive focus thus far on electrocratic dispersions may leave one wondering how colloid stability toward aggregation might be achieved in non-aqueous media, or in aqueous media under high salt conditions. The most common way of obtaining this is by the adsorption of macromolecules to the particle surfaces, leading to what is termed "steric stabilization." The use of this strategy to stabilize dispersions has a respectable antiquity, dating back as far as the ancient Egyptians, who used it to stabilize pigment dispersions in fresco paints and in inks. It was also a key to the long-term stability of Faraday's celebrated gold sols, which required a bit of gelatin to guarantee their absence of aggregation. Quantitative understanding of the mechanism of steric stabilization has developed with increased knowledge of the nature of polymeric adsorption.[84]

It is useful to give a brief preview of the stabilization from the standpoints of both molecular models and classical thermodynamics. Start by considering the approach of two particles coated with adsorbed layers of thickness δ, as shown in Fig. 7-23. The coated particles are assumed to be electrically neutral, and δ is assumed to be at least comparable with the range of influence of van der Waals attractions. The potential energy of interaction between the particles (or equivalently, ΔG for the process of bringing them from infinite separation to a distance apart, S_0) is not appreciable until the adsorbed layers are brought in contact and start to

[83] Morrison, I. D., *Langmuir*, **7**, 1920 (1991).
[84] Fleer, G. J., Cohen Stuart, M. A., Scheutjens, J. M. H. M., Cosgrove, T., and Vincent, B., **Polymers at Interfaces**, Chapman & Hall, London, 1993.

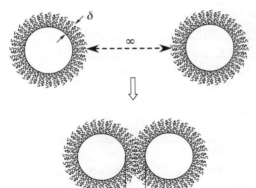

Fig. 7-23: Approach of particles with adsorbed macromolecules.

interpenetrate. The Gibbs free energy of close approach:

$$\Delta G_R = G_{S_0} - G_\infty, \qquad (7.170)$$

may be composed of: 1) changes in surface free energy in close approach (such as by desorption of adsorbate or its moving laterally away along the surface), 2) the excess chemical potential of adsorbate and solvent caused by over-lapping of layers ("mixing effect"), and 3) elastic energy arising from compression of the adsorbed layers ("volume restriction effect"). It is generally assumed that the adsorbed molecules are irreversibly and locally adsorbed ("anchored") so that factor (1) above need not be considered further. The nearness of approach of the particles depends also on their mutual kinetic energy upon collision. This may not bring particles into the reach of the region of the steep rise of the elastic repulsion, $S_0 < \delta$. (There are situations, however, where such interactions are of crucial importance.) The first concern is thus with penetrations associated with distances of approach: $\delta < S_0 < 2\delta$ and to factor (2) above, $viz.$, the changes in chemical potential of the adsorbate and solvent molecules during such approach.

Considerable experience with steric stabilization in practice reveals the reasonableness of the above narrowing of mechanistic forms at the outset. It is known that the best stabilizers are block or graft copolymers that consist of both $anchor$ $groups$ and stabilizing moieties ($buoy$ $groups$). The anchor groups should be both insoluble in the dispersion medium and attached either chemically or physically (by mechanisms described earlier) to the adsorbent particles to prevent their escape as the polymer-covered particles approach one another. The stabilizing moieties should be as fully solvent-compatible as possible, and of sufficient molar mass to provide the needed adlayer thickness. These conditions are achieved by suitable choice of the polymer for aqueous or non-aqueous dispersion media. Examples are shown in Table 7-4, where it is seen that the chemical groups are the same in both cases, with roles of "anchor" and "buoy" reversed.

It is useful at the outset to catalog some of the observations are commonly made concerning sterically stabilized dispersions.

Table 7-4: Typical stabilizing moieties (buoy groups) and anchor groups for stabilizing polymer adsorbates.

Dispersion	Anchor polymer	Stabilizing moieties
Aqueous	Polystyrene	Poly(oxyethylene)
	Poly(vinyl acetate)	Poly(vinyl alcohol)
	Poly(methyl methacrylate)	Poly(acrylic acid)
	Poly(dimethylsiloxane)	Poly(acrylamide)
	Poly(ethylene)	Poly(vinyl pyrrolidone)
Nonaqueous	Poly(acrylonitrile)	Polystyrene
	Poly(oxyethylene)	Poly(lauryl methacrylate)
	Poly(ethylene)	Poly(dimethylsiloxane)
	Poly(vinyl chloride)	Poly(vinyl acetate)
	Poly(acrylamide)	Poly(methyl methacrylate)

1) The stability of the dispersions is highly sensitive to changes in temperature, the transition from high stability to catastrophic destabilization often occurring over just a $1°$ - $2°C$ range. An example is shown in Fig. 7-24. The temperature at which flocculation occurs is termed the *critical*

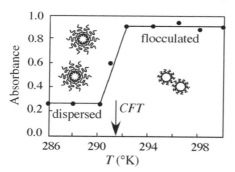

Fig. 7-24: Increase in turbidity of a poly(vinyl acetate) dispersion, stabilized by poly(ethylene oxide), on reaching the *CFT* in an aqueous electrolyte solution. After [Napper, D. H., *J. Colloid Interface Sci.*, **58**, 390 (1977).]

flocculation temperature (*CFT*). The cartoon in the figure suggests that the polymer layer undergoes an abrupt change in thickness at the *CFT*, as is discussed further below. *Aqueous* dispersions are most often (but not always) flocculated by *raising* the temperature, whereas *non-aqueous* dispersions are usually (but not always) flocculated by *decreasing* temperature. Some cases have been found which do not have an accessible *CFT* and one at least (polyacrylonitrile latex particles stabilized by polystyrene in methyl acetate) has been found to exhibit both an upper and a lower *CFT*. This temperature sensitivity stands in sharp contrast to the relative temperature *in*sensitivity of electrocratic dispersions.

2) Temperature-induced instability is almost always readily reversible (spontaneous) upon reheating or re-cooling the flocculated dispersion. This contrasts with the sometimes-observed irreversibility of aggregated electrocratic dispersions.

3) Catastrophic instability may also be observed at a sharply defined point upon addition of other solvents to the dispersion medium. An example is shown in Fig. 7-25. The amount required to induce flocculation is called the *critical flocculation volume* (*CFV*). The additives that have such an effect are ones that reduce the solvency of the dispersion medium for the stabilizing moiety of the adsorbate. Once again, the resulting flocculation is reversible when the original solvency conditions are restored.

4) Less commonly, changes in pressure can also induce sudden changes in stability, and a *critical flocculation pressure* (*CFP*) may be identified. Combinations of changes in the above-mentioned variables can be contrived to induce or to reverse flocculation of a sterically stabilized dispersion, the sharply defined conditions for which define a *critical flocculation point*, (*CFPT*).

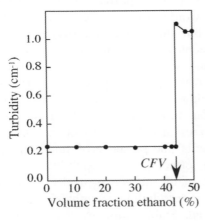

Fig. 7-25: Catastrophic onset of flocculation of a non-aqueous colloid upon addition of a non-solvent (ethanol) for the stabilizing moieties [poly (hydroxystearic acid)]. After [Napper, D. H., *Ind. Eng. Chem. Prod. Res. Develop.*, **9**, 467 (1970).]

5) Provided there is good anchoring and total coverage, stability conditions appear to depend only on the nature of the stabilizing moiety, essentially independent of the particles and the specific anchor group(s).

6) For large enough molecular weights of the stabilizing moiety (say > 10^4), stability conditions are essentially independent of molecular weight.

2. *Thermodynamic considerations: enthalpic vs. entropic effects*

Repulsion (and stability) exists only if the free energy change upon particle approach and adlayer overlap, ΔG_R is positive, and the transition from stability to instability occurs when the sign of ΔG_R changes from positive to negative. It is useful to decompose ΔG_R into enthalpic and entropic components:

$$\Delta G_R = \Delta H_R - T\Delta S_R. \tag{7.171}$$

Positive values of ΔG_R may be obtained in two ways:

1) There may be a large configurational entropy decrease due to the loss in volume accessible to the stabilizing polymer chains during

interpenetration, while at the same time ΔH_R is small. It would be zero if the mixing of polymer chains and solvent were completely athermal, but it is usually at least slightly negative, since the segment-solvent interactions are usually slightly weaker than the segment-segment and solvent-solvent interactions. Under such conditions, the entropy term dominates, and the result is termed entropic steric stabilization. This behavior is the more common in non-aqueous media.

2) Enthalpic stabilization occurs when ΔH_R is relatively large and positive, *i.e.*, when heat is absorbed as the inter-particle solution is made more concentrated in segments of the stabilizing chains at constant temperature. This implies that the segment-segment and solvent-solvent interactions are weaker than the segment-solvent interactions and is most commonly observed in aqueous dispersions stabilized by hydrated polymer chains. It is thought to be associated with the partial dehydration of those chains upon interpenetration. There would also be a decrease in segmental entropy in the interaction region, but this would be overridden by the entropy gain of the freed water molecules, making the total entropy change positive. The net positive ΔS_R would still, however, not be sufficient to overcome the positive ΔH_R in determining the sign of ΔG_R. The simple picture of dehydration of stabilizing chains cannot be the complete explanation of enthalpic stabilization, however, for at least two reasons. First, such stabilization can sometimes also be observed in non-aqueous media, and secondly, in aqueous media, the presence of small amounts of electrolyte may have a profound effect on the stability. The latter observation suggests the importance of the influence of electrolyte on the association properties of water. Full understanding of enthalpic stabilization seems not yet to have been achieved. Stabilization by a combination of entropic and enthalpic effects may occur, but appears to be rare.

The behavior of a dispersion upon raising or lowering the temperature is the key clue to the type of stabilization involved. For an entropically stabilized dispersion, the size of the $T\Delta S_R$ term dominates ΔG_R, but it becomes smaller as T is lowered. Eventually, $T\Delta S_R \approx \Delta H_R$ so that $\Delta G_R \approx 0$, and flocculation ensues. Thus entropically stabilized dispersions flocculate on cooling. For enthalpically stabilized dispersions, the situation is reversed. The $T\Delta S_R$ term is less than ΔH_R to start with, but becomes larger as temperature is increased. Enthalpically stabilized systems are thus subject to flocculation upon heating. (In aqueous systems of this type, however, it is often required that at least a small amount of electrolyte be present to assure such flocculation upon heating.) The simple temperature-change criterion for determining the type of stabilization is still not completely unambiguous since ΔH_R and ΔS_R may also be functions of temperature and may become either larger or smaller as temperature is increased. Some examples of the different types of sterically stabilized dispersions are shown in Table 7-5[85]. They are classified as to their behavior near their *CFT*-values, since the

[85] Napper, D. H., *J. Colloid Interface Sci.*, **58**, 390 (1977).

above criterion of flocculation on temperature increase or decrease samples the signs of ΔH_R and ΔS_R only near those points. To determine whether a dispersion is stabilized entropically or enthalpically at some point far removed from the *CFT* would require independent knowledge of the above signs at the conditions of interest.

Table 7-5: Classification of sterically stabilized dispersions near the *CFT*.

Stabilizer	Dispersion Medium		Classification
	Type	Example	
Poly(laurylmethacrylate)	Nonaqueous	*n*-Heptane	Entropic
Poly(12-hydroxystearic acid)	Nonaqueous	*n*-Heptane	Entropic
Polystyrene	Nonaqueous	Toluene	Entropic
Polystyrene	Nonaqueous	Methyl acetate	Enthalpic
Polyisobutylene	Nonaqueous	2-Methylbutane	Enthalpic
Poly(ethylene oxide	Aqueous	0.48 M MgSO$_4$	Enthalpic
Poly(vinyl alcohol)	Aqueous	2 M NaCl	Enthalpic
Poly(methacrylic acid)	Aqueous	0.02 M HCl	Enthalpic
Poly(acrylic acid)	Aqueous	0.2 M HCl	Entropic
Polyacrylamide	Aqueous	2.1 M (NH$_4$)$_2$SO$_4$	Entropic

3. *Fischer theory*

The basis for the model presently used to describe steric stabilization is due to Fischer[86] and is as follows. When the polymer sheathes approach and overlap, it is assumed that they first interpenetrate to some extent, as shown in Fig. 7-26. This causes the chemical potential of the *solvent* in the interaction zone (zone of overlap) to decrease, thereby establishing a

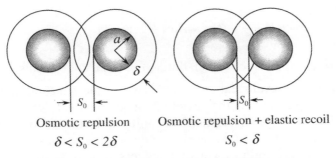

Osmotic repulsion   Osmotic repulsion + elastic recoil

$\delta < S_0 < 2\delta$ $S_0 < \delta$

Fig. 7-26: Mechanisms of steric repulsion.

difference in chemical potential, manifested as osmotic pressure, between solvent molecules in the overlap zone and those in the external dispersion

[86] Fischer, E.W., *Kolloid-Z.*, **160**, 120 (1958).

medium. As a result, solvent external to the interaction zone diffuses inward, forcing the stabilizing moieties, and the particles to which they are attached, apart. If the interpenetration is such that the distance of approach is less than the thickness of a single polymer sheath, there is also an elastic recoil effect (volume restriction). Considering just the osmotic effect, the repulsive potential energy which pushes the particles apart, *i.e.*, ΔG_R, was expressed by Fischer in terms of the excess osmotic pressure developed in the lens-shaped overlap zone, as shown in Fig. 7-27:

$$\Delta G_R = (\Pi_{\text{overlap}} - \Pi_{\text{ideal}})\Delta V_{\text{overlap}},\qquad(7.172)$$

where Π_{overlap} is the osmotic pressure of the solution in the overlap zone, Π_{ideal} is the osmotic pressure in an (assumed) ideal dilute solution, and $\Delta V_{\text{overlap}}$ is the volume of the overlap zone.

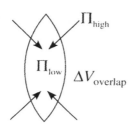

Fig. 7-27: Osmotic pressure effects in the lens-shaped adlayer overlap zone.

The osmotic pressure in an ideal dilute solution is given by van't Hoff's Equation:

$$\frac{\Pi}{c_2} = \frac{RT}{M_2},\qquad(7.173)$$

where c_2 is the mass concentration of solute in the solution (grams per unit volume) and M_2 is the solute molecular weight. For more concentrated solutions, the osmotic pressure is given using the virial expansion:

$$\frac{\Pi}{c_2} = RT\left[\frac{1}{M_2} + B_2 c_2 + B_3 c_2^2 + \ldots\right],\qquad(7.174)$$

where B_2, B_3, *etc.*, are the second, third, *etc.*, virial coefficients.

ΔG_R is then evaluated assuming that: 1) the concentration of solute (stabilizing polymer) in the external medium $c_2 \approx 0$; 2) the mass concentration of polymer segments is uniform in the adlayer at, $c_2 \approx 2c$, where c is the average mass concentration of segments of the stabilizing moieties in the adsorption layer of thickness δ, and 3) the non-ideality is sufficiently simple as to involve only pairwise interactions, so that the third and higher virial coefficients in Eq. (7.174) can be neglected. Under these conditions, substitution into Eq. (7.172) gives:

$$\Delta G_R = 2RTB_2 c^2 \Delta V_{\text{overlap}},\qquad(7.175)$$

which is Fischer's Equation. Although Fischer's simplifying assumptions can be criticized on a number of counts, the theory predicts essentially all of the qualitative features of incipient flocculation of sterically stabilized dispersions.

One must first note that B_2, the second virial coefficient, may be positive, negative or zero. If B_2 is positive, the medium is a good solvent, and if negative, it is a bad solvent for the solute 2. Positive B_2 values imply physically strong solvent-solute interactions (even solvation), whereas negative B_2's imply strong segment-segment interactions in the solvent. A solvent for which (under the temperature and pressure conditions of interest) there is an exact balance between segment-segment, solvent-solvent and segment-solvent interactions such that the solution acts as though it were ideal even at fairly high concentrations is one for which $B_2 = 0$ (and B_3, etc., also = 0). The situation represented by $B_2 = 0$ is analogous to the Boyle point (temperature) of a real gas, in which there is an exact compensation between the molecular excluded volume effect (which would by itself lead to a higher pressure) and intermolecular attractive forces (which would by themselves lead to a lower pressure), so that at that point, the ideal gas law is obeyed. This is essentially a canceling (at the Boyle point) between entropic and enthalpic effects, respectively. In a solvent for which $B_2 = 0$ the "volume restriction" entropy loss effect experienced by the polymer chains is just compensated for by adequate net chain-chain attraction, an enthalpy effect. The polymer chains easily telescope each other inasmuch as their volume fraction in the adsorption layers is seldom over 10%.

The second virial coefficient embodies both entropy and enthalpy effects and obviously depends on temperature. In accord with the Flory-Huggins theory of polymer solutions,[87] for which the reader is referred to Flory's original text or to other texts on polymer solutions, B_2 is proportional to $(\psi_1 - \kappa_1)$, where κ_1 and ψ_1 are enthalpy- and entropy-of-dilution parameters such that in the formation of a solution of polymer (2) in solvent (1):

$$\Delta \bar{h}_1 = RT\kappa_1\phi_2^2, \text{ and } \Delta \bar{s}_1 = RT\psi_1\phi_2^2, \tag{7.176}$$

where ϕ_2 is the volume fraction of polymer, and $\Delta \bar{h}_1$ and $\Delta \bar{s}_1$ are the partial molar quantities of mixing (referring to the solvent). These latter quantities may be derived from Flory's expression for the solvent chemical potential in a polymer solution:

$$(\mu_1 - \mu_1^\ominus) = RT\left[\ln(1-\phi_2) + \phi_2 + \chi_1\phi_2^2\right]. \tag{7.177}$$

The quantity $(\psi_1 - \kappa_1)$ is thus related to the Flory-Huggins interaction parameter χ_1, defined by Eq. (7.177). It is thus seen that

$$\psi_1 - \kappa_1 = 1/2 - \chi_1. \tag{7.178}$$

---

[87] Flory, P., **Principles of Polymer Chemistry**, p. 522, Cornell Univ. Press, 1953.

It is common to define a temperature, such that

$$\theta = \frac{\kappa_1 T}{\psi_1}, \text{ or } \kappa_1 = \psi_1 \frac{\theta}{T}. \tag{7.179}$$

Then finally:

$$B_2 \propto (\psi_1 - \kappa_1) = \psi_1 \left(1 - \frac{\theta}{T}\right). \tag{7.180}$$

It is seen that at $T = \theta$, the second virial coefficient (and all deviation from ideality) must vanish. This is termed the θ–*temperature*. B_2 can be positive in two different ways: 1) ψ_1 is positive with $T > \theta$ (entropic stabilization), or 2) ψ_1 is negative and $T < \theta$ (enthalpic stabilization). For entropically stabilized dispersions, the solutions of stabilizing moieties in the dispersion medium should be such that $\psi_1 > 0$ and B_2 is positive (and the dispersion stable) because $T > \theta$. As the dispersion is cooled, eventually $T = \theta$, $B_2 = 0$ and flocculation ensues. In contrast, for enthalpically stabilized dispersions, the polymer-medium solution is such that $\psi_1 < 0$ and B_2 is positive because $T < \theta$. Thus, as T is raised until $T = \theta$, B_2 becomes 0, and flocculation is observed.

These predictions of the Fischer theory are corroborated by experiment, as shown in Table 7-6, in which it is shown that incipient flocculation occurs very close to the θ-temperature, *i.e.*, the temperature at which $B_2 = 0$. Also in agreement with Fischer theory, there is no significant dependence on the molecular weight of the stabilizing moieties. As mentioned earlier, changes in system conditions other than temperature (such as solvent composition or pressure) may lead to $B_2 = 0$, and these may be referred to as θ-*conditions*.

Table 7-6: Comparison of *CFT*'s with θ-temperature for some aqueous dispersions. From [Napper, D.H., **Polymeric Stabilization of Colloidal Dispersions**, Academic Press, London, 1983,] p. 116.

Stabilizing moieties	MW	Dispersion medium	CFT (°K)	θ (°K)
Poly(oxyethylene)	1 400	0.39 M MgSO$_4$	317 ± 2	330 ± 10
	10 000		318	319 ± 3
	49 000		316	314 ± 3
	315 000		315	314 ± 3
	1 000 000		317	315 ± 3
Poly(acrylic acid)	9 800	0.2 M HCl	287 ± 2	287 ± 5
	19 300		289	289 ± 5
	89 700		281	287 ± 5
Poly(acrylic acid)	16 000	0.2 M HCl	286 ± 2	287 ± 5
Poly(vinyl alcohol)	26 000	2.0 M NaCl	320 ± 3	300 ± 3
	57 000		301	300 ± 3
	270 000		312	300 ± 3

4. *Steric repulsion plotted on DLVO coordinates*

Consider next putting the results of Fischer theory into a form that may be represented on DLVO coordinates. For a pair of spheres, each of radius a, with adlayers of thickness δ, the adlayer overlap volume, $\Delta V_{overlap}$, is given by:[88]

$$\Delta V_{overlap} = \frac{2}{3}\pi\left(\delta - \frac{S_0}{2}\right)^2\left(3a + 2\delta + \frac{S_0}{2}\right), \tag{7.181}$$

where S_0 is the distance of closest approach of the *bare* particle surfaces. The above equation is valid for $\delta \leq S_0 \leq 2\delta$. Substituting for ΔV into Fischer's Equation (which assumes uniform segment mass concentration c in the adlayers) gives:

$$\Delta G_R = \frac{4\pi}{3}RTB_2c^2\left(\delta - \frac{S_0}{2}\right)^2\left(3a + 2\delta + \frac{S_0}{2}\right), \tag{7.182}$$

This result applies only to the mixing effect, *i.e.*, it does not include elastic recoil, which would add its contribution to the repulsive energy when $S_0 < \delta$. If ΔG_R (a *free* energy) is identified with the potential energy of steric repulsion (*i.e.*, the change in the system entropy as the particles approach is neglected), the steric repulsion curve may be plotted on DLVO coordinates (on a per particle rather than per mole basis) as

$$\Phi_{steric} = \frac{4\pi}{3}kTB_2c^2\left(\delta - \frac{S_0}{2}\right)^2\left(3a + 2\delta + \frac{S_0}{2}\right), \tag{7.183}$$

shown qualitatively in Fig. 7-28. The steric repulsion is seen to yield a steep rather featureless curve, lifting from $\Phi = 0$ where $S_0 = 2\delta$, twice the adlayer

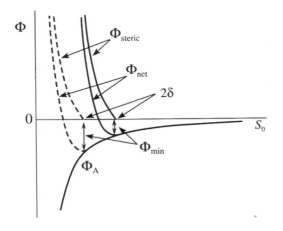

Fig. 7-28: DLVO-type curves for steric stabilization.

---

[88] This may readily be derived by noting that the overlap lens is just a pair of spherical lenses, whose geometry has been described in Chaps. 3 and 4.

thickness. When added to the curve for van der Waals attraction it yields the net interaction curve shown.

The net interaction curve is seen to lack a deep primary minimum, and therefore should show aggregation behavior similar to that of secondary-minimum aggregation encountered in some cases for particles with electric double layers. If the minimum that exists has a depth of less than approximately kT, the system should not exhibit significant flocculation. The curve shown in Fig. 7-28 corresponds to the situation when the system is "reasonably removed" from θ-conditions, and the adlayer thickness is roughly comparable to the range of the long-range van der Waals interactions. In general, the "secondary minimum" will have a depth depending on the thickness of the adlayer. A second set of curves (dashed) is shown for a thinner adlayer, which produces a deeper minimum in the net interaction curve. All of the qualitative observations regarding steric stabilization that were listed earlier can now be rationalized. The catastrophic flocculation with temperature or solvent composition changes are explained in terms of θ-conditions, and the absence or near-absence of a primary minimum in the total interaction curve explains the easy reversibility of flocculation of sterically stabilized dispersions. While the quantitative modeling of the steric interaction has been considerably refined beyond the treatment given above (*cf.* Napper[89]), the simplified treatment based on Fischer theory is adequate to represent its most important qualitative features.

The most important properties of the system upon which the steric repulsion depends are δ, the adlayer thickness, and some measure of the solvency of the (buoy blocks of the) polymer adlayer in the medium, such as B_2. These are not entirely independent quantities. While δ depends chiefly on the molar mass of the buoy groups of the adsorbed polymer and is roughly equivalent to the end-to-end distance of these parts of the polymer as free molecules in solution under the conditions of the system, it also depends on the solvency.

The relevant polymer adlayer thicknesses are those that are sensed hydrodynamically, and recent evidence confirms that even a very sparse population of solvent-compatible long tails is sufficient to yield a large hydrodynamic layer thickness and a high degree of stability.[90] This suggests that the optimum type of stabilizer is highly polydisperse, with at least a small proportion with a very high molecular weight. The hydrodynamic adlayer thickness is conveniently measured using dynamic light scattering (PCS) in which the hydrodynamic diameter of adlayer-covered particles is compared with that of bare particles.[91,92] It is the core diameter of adlayer-

[89] Napper, D. H., **Polymeric Stabilization of Colloidal Dispersions**, Academic Press, London, 1983.

[90] Stenkamp, V. S., and Berg, J. C., *Langmuir*, **13**, 3827 (1997).

[91] Cohen Stuart, M., Waajen, F. H. W. H., Cosgrove, T., Crowley, T. L., and Vincent, B., *Macromolecules*, **17**, 1825 (1984).

covered particles that is sensed by static light scattering. Viscometric methods (see Chap. 8) may also be used to determine layer thickness. Small-angle neutron scattering (SANS) techniques have been used to obtain the details of the polymer segment density distribution in the adlayer,[93] but are insensitive to the tails (which appear to be decisive in determining stability).

Information on solvency can be obtained by studying free solutions of the stabilizing polymer (the buoy groups) in the solvent by a variety of methods. Solvency may be characterized in terms of B_2, the second virial coefficient, χ_1, the Flory-Huggins interaction parameter, or alternatively, the θ-temperature. If the θ-condition has been determined for a given polymer-solvent system (e.g., by noting incipient flocculation brought about by changing solvency conditions), the additional information needed can be obtained by knowing the confirmation of the polymer molecule under the conditions of interest. When a polymer molecule (of sufficient length) is in dilute solution it assumes a random coil configuration, which under θ-conditions has an rms end-to-end distance of r_0. When in better-than-θ solvents, the molecule uncoils slightly and expands so that its rms end-to-end distance becomes $r = \alpha r_0$, where α is the *"intramolecular expansion factor."* For better-than-θ solvents, $\alpha > 1$, and for poorer-than-θ solvents, $\alpha < 1$. An experimental value for α can be secured in a number of ways, including light scattering, ultracentrifugation and viscosity measurements. Once determined, it can be related to the parameters needed to help predict steric stability.[94] The key fact is that as one traverses the boundary of the θ-condition from good solvency to poor solvency, the free polymer coil (and the corresponding adlayer, as pictured in Fig. 7-24) shrinks.

When solvency conditions are good for stabilization, but the adlayer thickness is small (perhaps that comparable to a simple surfactant adlayer), the depth of the potential energy minimum may be sufficiently great to lead to the formation of rather strong and large flocks. The flocculation is still reversible, however, since the particle surfaces never come in contact, and inter-particle sintering is prevented. While such a system is not "stable" in the sense of flocculation not occurring, it is described as *lubricated*. Lubrication may be adequate in many practical situations in which stability is sought, and may be the only "stability" accessible in dense colloids.

Another factor that may be important is "multipoint anchoring," under which conditions the free-solution thermodynamics of the buoy polymer may no longer adequate, and the correlation between the *CFPT* and θ-conditions may not hold. In a dispersion of polystyrene latex particles stabilized in water with polyethylene oxide under low *pH* conditions (in

---

[92] Baker, J. A., and Berg, J. C., *Langmuir*, **4**, 1055 (1988).

[93] Fleer, G. J., Cohen Stuart, M. A., Scheutjens, J.M.H.M., Cosgrove, T., and Vincent, B., **Polymers at Interfaces**, Chapman & Hall, London, 1993.

[94] Hiemenz, P. C., and Rajagopalan, R., **Principles of Colloid and Surface Chemistry**, 3rd Ed., pp. 616-618, Marcel Dekker, NY, 1997.

which extensive H-bonding between the undissociated carboxylic acid groups of the polystyrene latex and the ether oxygens of the stabilizing chains gave dense multipoint anchoring), Napper[95] found stability at much worse than θ-conditions and termed it "enhanced steric stabilization."

It is useful to summarize some of the practical observations concerning steric stabilization of colloids through polymer adsorption, as shown in Table 7-7.

Table 7-7: Some typical values of quantities associated with steric stabilization.

1. Anchor groups usually constitute 10-25% of the total adsorbing macromolecule; the remaining 75-90% are buoy groups.

2. Total amount adsorbed are typically 1-10 mg/m^2 of particle surface.

3. The MW of the stabilizing polymer is usually > 3,000.

4. The minimum bulk concentration required to achieve adequate coverage is 1000-10,000 ppm.

5. The thickness of the adlayer is usually $\geq$ 10 nm.

The practical implications of steric stabilization of dispersions are enormous. First of all, since electrostatic stabilization is often not an option for dispersions in apolar or high-salt media, steric stabilization of some kind must be used. Electrostatically stabilized dispersions in a medium such as *n*-heptane are stable only to about 1% solids, whereas sterically stabilized (or at least lubricated) dispersions of up to 60% solids are easily prepared. Even in aqueous media, steric stabilization has several distinct advantages over the more traditional electrostatic stabilization, *e.g.*,

1) Electrostatically stabilized dispersions can seldom tolerate ionic strengths of 0.01 M or more, whereas, *e.g.*, dispersions stabilized by poly(oxyethylene) polymer at room temperature can withstand ionic strengths of 1 to 10 M without flocculating.

2) Electrocratic dispersions are subject to electro-viscous effects, as discussed in Chap. 8. When such dispersions are made to flow, the charged particles must move relative to one another. The presence of the diffuse double requires the expenditure of extra energy to move the counterion atmospheres past one another. This shows up as extra viscosity, which depends on the solids content of the dispersions. It often puts an upper limit of about 20% solids on such a dispersion if it is to be pumped around. Sterically stabilized dispersions up to over 50% solids often present reasonable viscous properties.

---

[95] Napper, D. H., *J. Colloid Interface Sci.*, **58**, 390 (1977).

3) Sterically stabilized dispersions are usually insensitive to changes in *pH* or to the concentration of any potential determining ions.

4) The easy reversibility of flocculation of sterically stabilized or lubricated dispersions (in contrast to the sometimes-irreversible coagulation in the primary minimum of electrocratic dispersions) gives them excellent freeze-thaw stability.

5. *Electro-steric stabilization*

In aqueous media, electrostatic and steric stabilization may act in concert to produce *electro-steric stabilization*, as suggested by Fig. 7-29. This may be effected through the adsorption of a polyelectrolyte or through the adsorption of a neutral polymer to a surface supporting a diffuse double layer. Polyelectrolytes are the more common electro-steric stabilizers. In their usual application, they are anchored to particle surfaces of opposite charge, but bringing sufficient excess molar mass and charge to produce a thick charged layer at the surface.

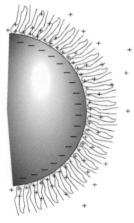

Fig. 7-29: Electro-steric stabilization.

In cases of adsorbed neutral polymers or oligomers where steric stabilization is weak (due, for example, to the adlayer being thin), stability may be achieved with the assistance of electrostatic repulsion.[96,97] Under higher salt conditions, stability may be broken down when the double layer is collapsed, or at higher concentrations, as the polymer adlayer loses solvency.[98] Electro-steric stabilization may also be employed in non-aqueous media.[99,100] When it exists in either medium, the colloid may exhibit extraordinarily robust stability. The electrostatic component is relatively insensitive to temperature fluctuations and small solvent composition fluctuations, while the steric component is largely insensitive to moderate swings in electrolyte concentration. Although one can find investigators

[96] Einarson, M. B., and Berg, J. C., *J. Colloid Interface Sci.*, **155**, 165 (1993).

[97] Virden, J. W., and Berg, J. C., *J. Colloid Interface Sci.*, **153**, 411 (1992).

[98] Einarson, M. B., and Berg, J. C., *Langmuir*, **8**, 2611 (1992).

[99] Pugh, R. J., and Fowkes, F. M., *Colloids Surfaces*, **9**, 33 (1984).

[100] Mysko, D. D., and Berg, J. C., *Ind. & Engr. Chem. Res.*, **32**, 854 (1993).

treating electro-steric DLVO-type pair interaction curves as the simple summation of the electrostatic and steric curves, this is probably an oversimplification, *i.e.*, it is more likely that the presence of the polymer adlayer will influence the electric double layer, and conversely, the presence of the ions in the double layer will influence the properties of the polymer adlayer.

6. Bridging flocculation

Polymers may act to influence colloid behavior in at least two other important ways besides steric stabilization due to adlayers. First, at very low concentrations (often < 50 ppm), particularly with very high molar mass polymers, the same polymer molecule may become adsorbed unto two or more particles, therefore linking them. This leads to de-stabilization by "sensitized" or *bridging* flocculation, extremely important in such processes as wastewater clarification and in papermaking, in which it is used to retain particles of filler within the paper structure ("retention aides"). Cationic polyelectrolytes are most commonly used flocculants of this type. This type of aggregation, including its kinetic aspects, has been described in detail by Gregory.[101] Figure 7-30 pictures the process schematically. The dispersion

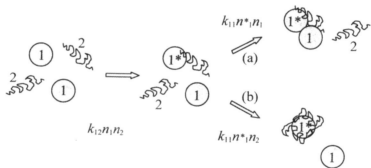

Fig. 7-30: Bridging flocculation. Polymer (2) adsorbs first to one particle (1) followed by adsorption to a second particle, as in route (a). In route (b), either the adsorbed polymer re-conforms, or a second polymer molecule adsorbs to the first particle.

contains particles (1) and dissolved polymer molecules (2). In the first step, a polymer molecule adsorbs to a particle in a process that may be represented by the rate function $k_{12}n_1n_2$, where n_1 and n_2 are the number concentrations of particles and polymer molecules. This process produces a particle with an adsorbed polymer, (1*), and may be followed by "reaction"

[101] Gregory, J., in **Chemistry and Technology of Water Soluble Polymers**, C. A. Finch (Ed.), pp. 307-320, Plenum, New York, 1983;
Gregory, J., in **The Effect of Polymers on Dispersion Properties**, Th. F. Tadros (Ed.), p. 301, Academic Press, London, 1982;
Gregory, J., **Particles in Water: Properties and Separation Methods**, CRC Press, Boca Raton, USA, 2005.

(a) in which a second particle is attached to (1*), causing bridging. The rate of this processes is $k_{11}n_1^*n_1$, where n_1^* is the concentration of particles with an adsorbed polymer. It competes with process (b), occurring at the rate: $k_{11}n_1^*n_2$ in which the first particle takes on a second polymer molecule, so that bridging does not occur. A second process that may tend to prevent bridging is that of reconfirmation of the adsorbed polymer into a thinner layer on the surface. While highly simplified, this model reveals that in order to favor bridging, the polymer concentration, n_2, must be kept as low as possible relative to that of the particles. It suggests that dosing should be done with dilute polymer solutions, or that the particulate dispersion be added to the flocculant solution. It is also evident that bridging is favored by high molecular weight of the flocculant, and in practice it is usually $> 10^6$ Daltons.

7. Depletion flocculation

A second important class of phenomena concerns the presence of polymers that are *not adsorbed*. The presence of free polymer may have significant consequences on the stability of the dispersion. There are a few situations when a dissolved polymer does not adsorb to the particle surfaces (*e.g.*, neutral polysaccharides do not adsorb to a number of different surfaces from water, or polyelectrolytes do not adsorb to surfaces of the same charge), but the main situation in which free polymer exists is the one in which the available surfaces have already been saturated with polymer adsorbate. In any event, when no further polymer attachment to a surface can occur, the centers of mass of the free polymer molecules are *excluded* from a zone near the (adlayer-covered) surface roughly equal in thickness to the radius of gyration of the polymer coil in solution. As shown in Fig. 7-31, when a pair of particles with depletion layers come sufficiently close to one

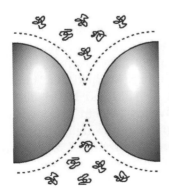

Fig. 7-31: Effect of free (unadsorbed) polymer: "depletion flocculation."

another that dissolved polymer is excluded from the region between them, an osmotic pressure difference is established which causes flow out of this region, leading to flocculation. This is the mechanism of *depletion flocculation*, as first described by Asakura and Oosawa[102] for the interaction

---

[102] Asakura, S., and Oosawa, F., *J. Chem. Phys.*, **22**, 1255 (1954);

between flat plates, and generalized to sphere-sphere interactions by Vrij.[103] Many developments and refinements have occurred since, and the subject is reviewed by Jenkins and Snowden[104] and by Piech and Walz.[105]

The simplest treatment considers the unadsorbed particles to act as hard spheres of radius equal to their radius of gyration, R_g, and that these molecules are completely excluded from the layer of this thickness (the depletion zone) surrounding the colloid particles, of radius a. When particles approach such that their depletion zones overlap, the energy of attraction is given by the product of the difference in the osmotic pressure between the polymer solution and the polymer-free overlap volume (where it is presumably zero) times the volume of the overlap lens:

$$\Delta G_{dep} = -(\Pi_{soln} - \Pi_{overlap})\Delta V_{overlap} = -\Pi_{soln}\Delta V_{overlap}, \qquad (7.184)$$

a direct analogy to Eq. (7.164). The osmotic pressure in the solution is given by the van't Hoff Equation, Eq. (7.183), written in the form

$$\Pi_{soln} = n_2 kT, \qquad (7.185)$$

where n_2 is the number concentration of the polymer molecules in the solution. The volume of the overlap lens is given by Eq. (7.181), written as:

$$\Delta V_{overlap} = \frac{2}{3}\pi\left(R_g - \frac{S_0}{2}\right)^2\left(3a + 2R_g + \frac{S_0}{2}\right), \qquad (7.186)$$

where, if a polymer adlayer is present, a includes its thickness, and S_0 is measured between the outer edges of that layer. The attractive depletion potential then becomes:

$$\Delta G_{dep} = \Phi_{dep} = -\frac{2}{3}\pi n_2 kT\left(R_g - \frac{S_0}{2}\right)^2\left(3a + 2R_g + \frac{S_0}{2}\right). \qquad (7.187)$$

Results are plotted in Fig. 7-32 for three different values of the ratio of the radius of gyration of the free polymer to the radius of the particle, R_g/a. Despite its simplicity, Eq. (7.179) reflects the features of depletion interactions in reasonable agreement with more sophisticated models[106] and with experimental data obtained using SFA, AFM and TIRM. The influence of temperature, molecular weight, solvency, electrolyte content,[107] are accounted for through R_g. At significantly higher free polymer concentrations, one may find "depletion stabilization." Depletion effects may also be attributed to the presence of polyelectrolytes and of surfactant

Asakura, S., and Oosawa, F., *J. Polym. Sci.*, **33**, 183 (1958).

[103] Vrij, A., *Pure Appl. Chem.*, **48**, 471 (1956).

[104] Jenkins, P., and Snowden, M., *Adv. Colloid Interface Sci.*, **68**, 57 (1996).

[105] Piech, M. and Walz, J. Y., *J. Colloid Interface Sci.*, **253**, 117 (2002).

[106] Joanny, J. F., Leibler, L., and de Gennes, P.-G., *J. Polym. Sci.:Polym. Phys. Ed.*, **17**, 1073 (1979).

[107] Seebergh, J. E., and Berg, J. C., *Langmuir*, **10**, 454 (1994).

micelles, ionic or nonionic, and to the presence of a second stable colloidal species (of smaller particles) that do not hetero-aggregate with the particles under consideration.

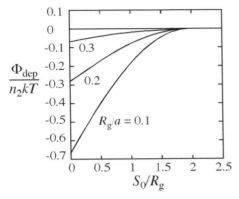

$$\frac{\Phi_{dep}}{n_2 kT}$$

Fig. 7-32: Normalized depletion interactions plotted for different ratios of the free polymer size to the particle size.

When higher order concentration effects are taken into account, *i.e.*, the ideal-dilute van't Hoff Equation is relaxed, and interactions between the free polymers are permitted, the depletion effect described by the simple analysis is seen to be just the short-range component of a more general structural interaction producing longer range oscillatory effects that have been both predicted and observed.[108] The repulsive component of such oscillations is thought to be the main cause of a "depletion stabilization" phenomenon that has been reported for more concentrated free polymer solutions.

In summary, a number of different interactions may occur in colloidal dispersions containing polymers, dependent upon different properties. It is possible, however, that this variety of behavior may be exhibited by a single colloid-polymer system, dependent only on the polymer concentration, as shown schematically in Fig. 7-33. At very low polymer concentrations (< 50 ppm) the dominant interaction may be bridging flocculation. As the polymer concentration is increased, and over a rather wide range of polymer concentration, the major effect will be steric stabilization. Then, as the particle surfaces can accommodate no further adsorption, further increases in polymer concentration lead to depletion flocculation, and possibly on to depletion stabilization.

8. *Electrophoretic displays; electronic paper*

An application involving many of the concepts that have been developed in Chaps. 6 and 7, and one illustrating the usual situation of technology perhaps outrunning science, is that of electrophoretic displays[109]

---

[108] Piech, M. and Walz, J. Y., *J. Colloid Interface Sci.*, **253**, 117 (2002).
[109] Comiskey, B., Albert, J. D., Yoshizawa, H., and Jacobsen, J., *Nature*, **394**, 253 (1998).

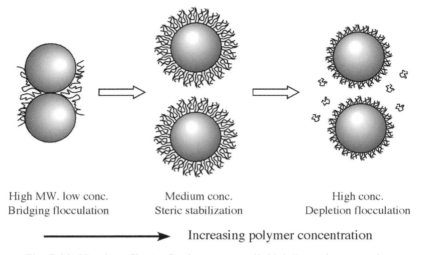

<table>
High MW. low conc. | Medium conc. | High conc.
Bridging flocculation | Steric stabilization | Depletion flocculation
</table>

High MW. low conc.
Bridging flocculation

Medium conc.
Steric stabilization

High conc.
Depletion flocculation

⟶ Increasing polymer concentration

Fig. 7-33: Varying effects of polymers on colloidal dispersions as polymer concentration is increased.

or "electronic paper." Devices currently using the technology are electronic books, such as Amazon.com's Kindle™ or Sony's Reader Digital Book, tablet PC's, phones, and other handheld devices. They are capable at present of producing high resolution black and white images by electrophoretically directing black or white charged colloidal particles to the back of a viewing screen as prescribed by a matrix-addressable electrode. The technology of E-Ink Corporation, Cambridge, MA, is illustrated in broad concept in Fig. 7-34; the details are proprietary. Colloids of sub-micron particles dispersed in a liquid medium contained within 50-100 μm diameter capsules. The

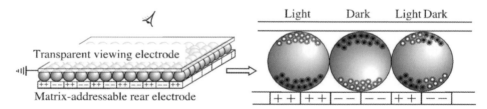

Fig. 7-34: Schematic of E-Ink Corporation technology for electrophoretic displays.

dispersoid is a mixture of white particles, such as titanium dioxide, and dark particles, such as carbon black. The dispersion medium is a low dielectric constant apolar liquid. (While the process might have been more simply constructed using an aqueous medium in the capsules, that would greatly increase the power consumption required: more charges to move around.) One of the particle types carries a positive charge and the other a negative charge, with particle surface charging is the result of the presence of inverse micelles or other dissolved structures in the apolar solution, as described in Chap. 6.A.7. The fact that one type of particle is positively charged and the other negatively charged is probably due to a difference in the acid-base chemistry of the particle surfaces and their interaction with the particular charge-stabilizing medium. This is likely to be the result of proprietary surface treatments. The particles, since they carry opposite charges, must be sterically stabilized against hetero-aggregation, and this requires an adlayer thickness at least comparable to the relevant Bjerrum length, given by Eq. (6.12).

The capsules are confined between two plates that serve as electrodes. The top plate is either glass or a transparent plastic coated with a conductive film, and grounded. The back electrode consists of a finely-segmented matrix-addressable electrode, each segment of which can be made positive or negative. This drives the particles either to the top or the bottom of the capsules by electrophoresis in accord with their respective charges. The segments of the electrode are small enough that the different particles may be acted upon differentially within a given capsule, making high resolution images possible. The technology of electrophoretic displays is in its infancy relative to the possibilities one might envision. It will probably not be long before high resolution color will be available.

F. The kinetics (and thermodynamics) of flocculation

Sterically stabilized dispersions have pair interaction potentials displaying a shallow minimum analogous to the "secondary minimum" sometimes encountered in electrocratic systems. Such shallow minima are also sometimes displayed in the presence of unadsorbed polymer (depletion effects) in either the presence or absence of a polymer adlayer. The kinetics of the reversible aggregation into such minima are the same as described earlier for the case of a secondary minimum with an infinite barrier. There are of course also systems that employ both steric and electrostatic modes of stabilization, and these might fall back on electrostatic barriers when θ-conditions for the polymer adsorbate chains in the dispersion medium prevail,[110] or on the steric barrier when the electric double layer is collapsed.[111]

For flocculation, i.e., reversible aggregation into a shallow minimum, a steady state condition may be achieved in which the rate at which particles flocculate is balanced by the rate at which the flocks decompose (as a result of collisions with other particles). A doublet may flocculate with an additional primary particle to form a triplet and so on, so that at steady state, there may be a population of primary particles plus aggregates of a range of sizes. In the early stages, aggregates grow primarily through the addition of single particles rather than through the collision of clusters. Thus, following Everett,[112] one may envision the process as occurring through a sequence of

[110] Einarson, M. B., and Berg, J. C., *Langmuir*, **8**, 2611(1992).
[111] Einarson, M. B., and Berg, J. C., *J. Colloid Interface Sci.*, **155**, 165 (1993).
[112] Everett, D. H., **Basic Principles of Colloid Science**, Royal Soc. of Chem., Letchworth, UK, 1988.

"reactions" in which a primary particle is added to an aggregate to form the next larger aggregate, *etc.*:

$$A + A = A_2$$
$$A_2 + A = A_3$$
$$\vdots$$
$$A_i + A = A_{i+1}$$
$$\vdots$$

$$(7.188)$$

The aggregate population at steady state is given by the Boltzmann distribution:

$$\frac{n_{i+1}}{n_i} = \exp\left[\frac{-w|_{i+1\rightarrow(i+1)}}{kT}\right], \tag{7.189}$$

where n_i is the number concentration (particles/vol) of aggregates of size i in the dispersion, and w is the work (free energy change) to transform a system consisting of an aggregate of size i plus a primary particle to an aggregate of size (i + 1), *i.e.*

$$w|_{i+1\rightarrow(i+1)} = \Delta w(i,\ i+1) = \Delta H(i,\ i+1) - T\Delta S(i,\ i+1) \tag{7.190}$$

The enthalpy contribution to Δw is minus the depth of the potential well ($-\Phi_{min}$), which may be assumed effectively independent of aggregate size. The principal contribution to $\Delta S(i, i+1)$ is the loss of entropy of the primary particle in joining the aggregate. This should be dependent on the volume fraction of the primary particles in the dispersion, $\phi_1 = v_1 n_1$:

$$\Delta S(i,\ i+1) \approx k\ln(v_1 n_1), \tag{7.191}$$

where v_1 is the volume of a primary particle. Thus:

$$w = \left(-\Phi_{min}\right) - kT\ln(v_1 n_1), \tag{7.192}$$

and substituting into Eq. (7.189):

$$\frac{n_{i+1}}{n_i} = v_1 n_1 \exp\left[\frac{\Phi_{min}}{kT}\right]. \tag{7.193}$$

Thus, if $\dfrac{\Phi_{min}}{kT}$ and $v_1 n_1$ are not sufficiently large, $\dfrac{n_{i+1}}{n_i} < 1$ (perhaps $\ll 1$), and the concentration of aggregates will decrease with size, so that the dispersion will consist of singlets and a few small aggregates. On the other hand, if $\dfrac{\Phi_{min}}{kT}$ and $v_1 n_1$ are jointly sufficiently large, larger aggregates will form in increasing numbers, and the dispersion will flocculate. It is important to realize that the fate of the dispersion is determined not only by the depth of the potential well, but also by the initial concentration of primary particles. When entropy effects are taken into account, it is seen that while a dilute dispersion of given particles may show little aggregation, a more concentrated one may yield a steady-state flock distribution.

There is a close analogy between flocculation (*i.e.*, weak aggregation) and the phase separation of a solution, so the process may be envisioned in terms of thermodynamics rather than kinetics. Phase diagrams of the type shown in Fig. 7-35 may be computed using statistical mechanical methods[113] and have been observed experimentally.[114] This type of "phase behavior" is

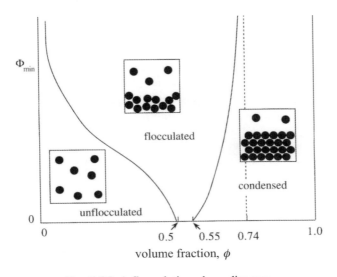

Fig. 7-35: A flocculation phase diagram.

characteristic of dispersions sterically stabilized through the adsorption of polymers to the particle surfaces or influenced by the presence of unadsorbed polymer. There is clearly a trade-off between particle volume fraction and the depth of the energy minimum, as suggested above. At high volume fractions, an order-disorder transformation (called a Kirkwood-Alder transformation[115]) may occur, leading to a solid-like phase (a "colloidal crystal") even in the absence of an energy minimum. For monodisperse spheres, this occurs at $\phi \approx 0.55$. At $\phi = 0.74$, one reaches the maximum packing density for equal-sized spheres.

G. Other non-DLVO interaction forces

The understanding of colloid stability with respect to aggregation is not complete with van der Waals and electrostatic (*i.e.*, "DLVO") interactions, or even with the addition of steric, lubrication and depletion

[113] Russel, W. B., Saville, D. A., and Schowalter, W. R., **Colloidal Dispersions**, pp. 332ff, Cambridge Univ. Press, Cambridge UK, 1989.
[114] Hachisu, S., and Kobayashi, Y., *J. Colloid Interface Sci.*, **46**, 470 (1974).
[115] Ziman, J. M., **Models of Disorder**, Cambridge Univ. Press, Cambridge, 1979.

effects. Some additional important interactions may arise under different circumstances, as listed in Fig. 7-36. These include the *structural forces* or "solvent structure forces" shown earlier in Fig. 7-2, as revealed by the surface forces apparatus, but also in the stepwise thinning often observed in thin liquid films, particularly those containing dissolved macromolecules or micelles, or monodisperse ultra-fine colloids.[116] They arise from packing constraints accompanying the ordering of solvent molecules or dispersed entities that occur as confining surfaces approach one another. These structural forces are oscillatory in nature and require that the approaching surfaces and the molecules (or other structures) in the medium be amenable to the establishment of at least medium-range order. Random micro-roughness in the approaching surfaces may be enough to substantially eliminate the oscillatory component of the structural interaction. Instead in such cases, one observes a monatonically decreasing repulsive force between the surfaces. When the medium is aqueous, this repulsive force is sometimes identified with another interaction called the *hydration force*. It

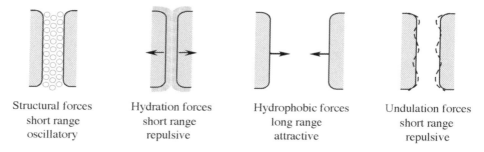

Structural forces	Hydration forces	Hydrophobic forces	Undulation forces
short range	short range	long range	short range
oscillatory	repulsive	attractive	repulsive

Fig. 7-36: Some additional ("non-DLVO") inter-particle interactions.

is (sometimes strongly) repulsive, short-ranged, and is found empirically to follow an exponential decay law:

$$\Phi_{\text{hydration}} = +A\exp(-S_0/\lambda_{\text{hyd}}), \tag{7.194}$$

where S_0 is the distance between the surfaces. The constant A depends on the degree of surface hydration, but is usually in the range of 3-30 mJ/m^2, and the decay length, λ_{hyd}, is of the order of 0.6-1.1 nm[117]. Repulsive hydration effects have been invoked earlier to explain the repeptization of electrocratic dispersions. They are also used to explain the rather high electrolyte concentrations (relative to DLVO predictions) often required to coagulate hydrophilic oxide dispersions, as well as dispersions of latices with high surface concentrations of ionized groups (usually sulfates), although "hairy

[116] Nikolov, A. D., and Wasan, D. T., *Langmuir*, **8**, 2985 (1992);
    Bergeron, V., and Radke, C. J., *Langmuir*, **8**, 3020 (1992).
[117] Israelachvili, J. N., **Intermolecular and Surface Forces**, 2nd Ed., pp. 276ff, Academic Press, London, 1991.

layers" thought to be present in such systems may be responsible.[118] The degree of hydration of the surface is dependent on the exchange of ions of different degrees of hydration between the solution and the surface. It is clear that solvent structural forces and hydration forces are closely related, and in some cases may be effectively one and the same.

Another important effect called the "hydrophobic interaction,"[119] arises when either or both of the approaching surfaces, in aqueous media, is hydrophobic, such that the water contact angle against the solid is above a certain value. At the molecular level, a hydrophobic surface is one incapable of binding with water molecules via ionic or acid-base (H-bonding) interactions. It is thus related to the "hydrophobic bonding" that exists between hydrophobic molecules (or portions of molecules) in water and derives from the entropically unfavorable structuring of water that exists adjacent to such molecules or surfaces. The effect is a potentially strong attractive force, and evidence is accumulating, particularly from the surface forces apparatus, of its importance for the interaction between extended surfaces. Its importance for colloid stability has not yet been so firmly established, as hydrophobic particles are difficult, if not impossible, to disperse in any case. Empirical observations for quasi-flat hydrophobic surfaces facing each other across an aqueous medium of thickness $S_0 \leq 10$ nm, the interaction energy per unit area is of the form:

$$\Phi_{\text{hydrophobic}} = -B\exp(-S_0/\lambda_{\text{phob}}), \qquad (7.195)$$

with the constant $B \approx 20\text{-}100$ mJ/m^2 (dependent upon the "degree of hydrophobicity"), and the decay length, $\lambda_{\text{phob}} \approx 1\text{-}2$ nm. The effect may thus be both large and long-ranged, at least for low-curvature surfaces.

The origin of these long-range attractions has been the subject of controversy, but the consensus emerging is that it is due most often to the presence of nano-bubbles populating hydrophobic surfaces in contact with water.[120] The detailed force traces, $F(z)$, obtained by the surface forces apparatus and AFM are consistent with this explanation, and tapping mode AFM[121] and ellipsometry[122] have been used to document the presence of nano-bubbles at such surfaces. A relatively simple experimental procedure using AFM has been proposed for determining the strength and range of hydrophobic interactions between a quartz sphere and a silane-hydrophobized glass surface by means of systematic electrolyte titration of the electrostatic repulsion.[123]

A key question for colloidal dispersions (or simple surfaces of any kind) in contact with water, is what should be taken as the demarcation

[118] Seebergh, J. E., and Berg, J. C., *Colloids Surfaces*, **100**, 139 (1995).

[119] Claesson, P. M., and Christenson, H. K., *J. Phys. Chem.*, **92**, 1650 (1988).

[120] Attard, P., *Adv. Colloid Interface Sci.,* **104**, 75 (2003).

[121] Simonsen, A. C., Hansen, P. L., and Klösgen, B., *J. Colloid Interface Sci.*, **273**, 291 (2004).

[122] Mao, M., Zhang, J., Yoon, R.-H., and Ducker, W. A., *Langmuir*, **20**, 1843 (2004).

[123] Aston, D. E., and Berg, J. C., *Colloids Surfaces*, **163**, 247 (2000).

between "hydrophilic" and "hydrophobic" behavior? The nano-bubble explanation above would suggest that "hydrophobicity" of a smooth surface requires a contact angle greater than 90°. On the other hand, evidence is accumulating that surfaces show hydrophilic behavior only when the advancing contact angle of water is less than about 66°,[124] or in terms of the adhesion tension, $W_w = \sigma_L \cos\theta > 30$ mN/m. Under these conditions, hydrophobic (amphipathic) adsorption would be minimized, and surfaces approaching one another would experience repulsion. On the other hand, surfaces for which $\theta_{adv} > 66°$, hydrophobic adsorption and attraction should be observed. The wide-ranging consequences of this generalization are yet to be fully explored.

Still other effects have been identified, in particular with deformable surfaces such as those of emulsion droplets or vesicles, due to possible undulations of the surface. These are repulsive, and the (disjoining) pressure between two surfaces subject to undulations takes the form:

$$\Pi(S_0) \approx \frac{(kT)^3}{2k_1 S_0^3},$$
                                                                      (7.196)

where k_1 is the first (Helfrich) bending modulus of the surface (cf. Chap. 2.G.4). The format of this relation has been verified experimentally, and it is also found that $k_1 \rightarrow \infty$ if the membrane surface bears any charge. These forces may be important for the interactions between colloidal particles with deformable surfaces such as micro bubbles or droplets, or between vesicles.

The structural, hydration, hydrophobic and undulation effects have been treated only briefly here. A more detailed discussion is to be found in Israelachvili.[125]

Finally, an important class of systems is the magnetic colloids from which many electronic data storage media are made. The particles of such dispersions may vary from the super-paramagnetic, which are ultra-fine nanoparticles ($d < 50$ nm) and possess thermally fluctuating magnetic dipoles, to the ferromagnetic, which are larger and possess permanent dipoles. The magnetic interaction between the particles differs from the interactions discussed so far in that it is orientation dependent. The mutual orientation of the particles, in turn, depends on the presence and strength of any external magnetic field. The field tends to increase the attractive interaction by aligning the dipoles. Scholten and Tjaden[126] derived an orientation-averaged (Boltzmann distribution of orientations) magnetic pair potential for super-paramagnetic particles in the absence of an applied field, while Chan and Henderson[127] obtained the potential for a pair of spherical

---

[124] Vogler, E., *Adv. Colloid Interface Sci.*, **74**, 69 (1998).
[125] Israelachvili, J. N., **Intermolecular & Surface Forces**, 2nd Ed, pp. 260-311, Academic Press, London, 1992.
[126] Scholten, P. C., and Tjaden, D. L. A., *J. Colloid Interface Sci.*, **73**, 254 (1980).
[127] Chan, D. Y. C., and Henderson, D., *J. Colloid Interface Sci.*, **101**, 419 (1984).

ferromagnetic particles in an external field. The magnetic interaction is found to be so strong that by comparison, the long-range van der Waals interaction is insignificant. Nonetheless, it is possible to stabilize such dispersions electrostatically. Young and Prieve[128] have used small-angle light scattering to monitor the initial flocculation rate of γ-Fe$_2$O$_3$ dispersions as a function of applied magnetic field and found a strong dependence of aggregation rate on the magnetic field strength, which reached a plateau when the field was strong enough to completely align the particles. They also observed the formation of the long, rope-like flocks known to occur.

H. Aggregate structure evolution; fractal aggregates

1. *Stages of the aggregation process*

While early-onset aggregation, *i.e.*, doublet formation, is extremely important for studying colloid particle interactions, one is usually interested for practical reasons in what happens after that. Aggregates usually continue to grow in size and take on different structural forms under different circumstances. It is helpful, if somewhat over-simplified, to think of three phases of the aggregation process: early-stage, middle-stage, and late-stage aggregation, as pictured in Fig. 7-37. Early-stage aggregation refers primarily to doublet formation, which dominates at first even in dense dispersions. In the middle stage, aggregates grow and take on structures that

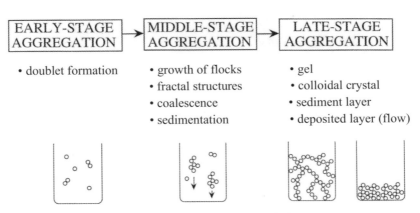

Fig. 7-37: Aggregate structure evolution.

may be described with the help of fractal geometry, as described below. In the late stage, these aggregates usually sediment (or cream) out of the dispersion into a layer at the bottom (or top) of the system, if it is quiescent, or deposit by inertial impaction onto baffles, walls, *etc.*, if the system is flowing. These sediments or deposits will have a structure reflecting that of the aggregates that formed them. Under some circumstances, the formed

---

[128] Young, W. D., and Prieve, D. C., *Ind. Eng. Chem. Res.*, **35**, 3186 (1996).

flocks are so voluminous that the entire system percolates or gels into a volume not much smaller than that of the original dispersion. The aggregates formed in the middle stage will generally be more open and voluminous, the more rapid the aggregation. The assumption is that the particles stick at the initial point of contact, and that the flocks are unable to densify themselves by sliding of their primary particles around one another. Slow aggregation leads to denser aggregates or flocks.

For aggregates formed out of electrocratic dispersions, this appears to be the case. Aggregates are sometimes found to sinter (form inter-particle solid bridges) over a period of time as brief as a few minutes. Flocks formed from initially sterically stabilized dispersions, however, are "lubricated," and individual particles thus can slide around one another and restructure the aggregate to a denser configuration.[129] This is also the case for flocks formed by depletion effects. Bridged flocks present yet another picture. These are formed quickly during the early stages after addition of the flocculating polymer, usually under conditions of high shear, and are initially voluminous. They then self-consolidate, as bridging ends, and the bridged flocks themselves begin to aggregate. The structures formed in the presence of polymers under different circumstances are often self-consolidating and more deformable than those formed from electrocratic dispersions.

If the colloid is stable to aggregation, the particles may sediment or cream into a dense regular array, termed a *colloidal crystal*. A polydispersity of only ± 10%, however, may be enough to prevent this from occurring. In that case, a dense, but irregularly packed sediment is produced.

2. Fractal aggregates

The present discussion concerns "sticky" aggregates. As mentioned above, these are often describable in terms of *fractals*, *i.e.*, self-similar structures that take on the same appearance at any degree of magnification, as shown in Fig. 7-38. The density of the aggregate (mass, or number of

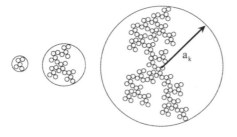

Fig. 7-38: Fractal aggregate structures.

primary particles, per unit volume) thus remains the same as it grows. If the aggregate at time $t = 0$ is just a single primary particle, of mass $= m_1$, and the mass at some later time is m_k (suggesting that it consists of k primary

[129] Huang, A. Y., and Berg, J. C., *J. Colloid Interface Sci.*, **279**, 440 (2004).

particles), the corresponding linear dimensions of the aggregate, vary in accord with

$$k = \left(\frac{m_k}{m_1}\right) = \left(\frac{a_k}{a_1}\right)^{d_f},$$ (7.197)

where "a" may represent, for example, the radius of the circumscribing sphere of the flock, and d_f is the *fractal dimension* of the aggregate. Aggregates described by Eq. (7.197) are *mass fractals*. Maximum aggregate density occurs if the primary particles are spheres which coalesce into larger spheres, and the fractal dimension is $d_f = 3$. The minimum fractal dimension of $d_f = 1$ occurs with a linear aggregate. Most aggregates in practice have $1.75 < d_f < 2.5$, with the lower number referring to rapid (diffusion-limited) aggregation, in which cluster-cluster interactions predominate, and the higher number referring to very slow (reaction-limited) aggregation ($W > 100$), in which primary particles add to the aggregates one at a time. These results also concur with computer simulations of aggregate growth.[130]

Light scattering measurements (*cf.* Chap. 5.H) provide the most convenient way of following aggregate structure evolution experimentally. Fractal structures in the appropriate size range scatter light in accord with RGD theory, and if in addition, $a_1 < 1/Q < a_k$, it has been shown that the scattering intensity from a fractal aggregate consisting of k primary particles depends on the magnitude of the scattering wave vector Q in accord with[131]

$$I_k(Q) = \frac{k^2 I_1}{(Qa_k)^{d_f}} = \frac{k I_1}{(Qa_1)^{d_f}},$$ (7.198)

where I_1 is the scattering intensity due to a single (spherical) primary particle. The total scattering intensity from a dispersion of such aggregates is then given by

$$I(Q) = \sum_{k=1}^{\infty} n_k I_k(Q) = \frac{I_0}{(Qa_1)^{d_f}},$$ (7.199)

where I_0 is the initial scattering intensity of the primary particle dispersion. The above relation should be valid as the fractal aggregates grow with time. Thus a plot of $\log I$ *vs.* $\log Q$ should yield a straight line of slope $-d_f$, as is shown in Fig. 7-39 for an aggregating aqueous gold sol. The fractal dimension in this case is $d_f = 1.8$, corresponding to open, highly branched aggregates. Higher values of the scattering vector were produced using neutron scattering, bringing one into the Porod region, as described in Chap. 5, Section H.4.

---

[130] Meakin, P., in **On Growth and Form**, H. E. Stanley, and N. Ostrowsky (Eds.), Martinus Nijhoff, Dortrecht, 1986.

[131] Schaefer, D. W., Martin, J. E., Wiltzius, P, and Cannell, D. S., *Phys. Rev. Lett.*, **52**, 2371 (1984).

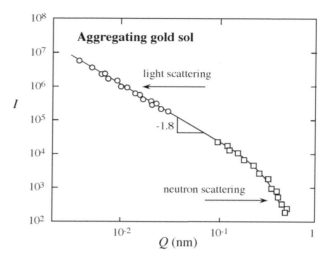

Fig. 7-39: Determination of aggregate fractal dimension by light scattering. After [Dimon, P., Sinha, S. K., Weitz, D. A., Safinya, C. R., Smith, G. S., Varaday, W. A., and Lindsay, H. H., *Phys. Rev. Lett.*, **57**, 595 (1986).]

Photon correlation spectroscopy (PCS) monitors the change in the aggregate diffusivity over time, and therefore follows the growth in the *hydrodynamic* dimensions with time. It thus reflects the total volume of the aggregate, including the trapped solvent. Thus the hydrodynamic dimensions of the aggregates approximate the dimensions of their circumscribing spheres. Combining the measured rate of growth of the hydrodynamic dimensions of the aggregates with the results of the population balance yields their fractal dimension.

It has been shown,[132] by combining static and dynamic light scattering techniques in a study of a charged polystyrene colloid that the fractal dimension of the aggregates scales directly with the normalized stability ratio, W, in accord with:

$$d_f = 1.69 + \log_{10} W ,$$ (7.200)

for $(1 < W < 100)$.

The use of controlled hetero-aggregation makes possible the production of aggregates of a wide variety of structures, including string-like aggregates consisting of alternating particle types, as pictured in Fig. 7-40.

The rate of sedimentation of an aggregating colloid, as pictured in Fig. 7-41, depends on the fractal dimension of the aggregates. Stokes' Law for the rate of settling of an aggregate of k primary particles $(v_\infty)_k$ of density ρ_p is:

---

[132] Kim, A. Y., and Berg, J. C., *Langmuir*, **16**, 2101 (1999).

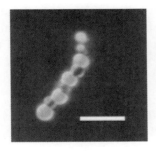

Fig. 7-40: Linear alternating aggregate formed by negatively charged (darker) and positively charged (lighter) polystyrene particles. Scale bar: 3 µm. From: [Kim, A. Y., Hauch, K. D., Berg, J. C., Martin, J. E., and Anderson, R. A., *J. Colloid Interface Sci.* **260**, 149 (2003).]

$$(v_\infty)_k = \frac{2a_k^2(\rho_k - \rho)g}{9\mu} = \frac{2k^{2/d_f}a_1^2(\rho_k - \rho)g}{9\mu}, \tag{7.201}$$

where ρ_k is the density of the aggregate, assuming it is the circumscribing sphere containing the aggregate itself plus the entrained solvent. It depends on the fractal dimension and the number of primary particles in the aggregate in accord with

$$\rho_k = k^{1-3/d_f}(\rho_p - \rho) + \rho. \tag{7.202}$$

Combining Eqs. (7.193) and (7.194) gives

$$(v_\infty)_k = \frac{2a_1^2 k^{1-1/d_f}(\rho_p - \rho)g}{9\mu}. \tag{7.203}$$

It is evident that the smaller d_f (*i.e.*, the more open the flock), the slower will be the sedimentation for an aggregate of a given number of primary particles. Measurement of aggregate sedimentation rates has been used a method for determining fractal dimension.[133]

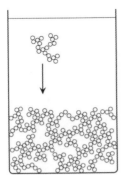

Fig. 7-41: Sediment structure obtained under conditions of rapid aggregation.

The late stage of aggregation, *i.e.*, the formation of sediments, deposits or gels, depends on the preceding events. If the aggregation is rapid so that open aggregates form, the resulting sediment will be voluminous. Thus for waste water clarification, for example, one seeks rapid aggregation,

[133] Tang, P., Greenwood, J., and Raper, J. A., *J. Colloid Interface Sci.*, **247**, 210 (2002).

which leads to rapid sedimentation, followed by *thickening* of the resulting voluminous sediment by mechanically breaking up or crushing it. One similarly seeks rapid aggregation and open sediment structures if the solids are to be removed by filtration, since the open structure will allow the liquid to be easily drained from the solid network.

3. *The effect of particle size on aggregation phenomena; coating by nanoparticles*

Aggregation phenomena become decreasingly important as particle size increases near the upper end of the colloid size range and beyond. One does not think of a pile of marbles, or buckshot or even dry beach sand particles as "aggregating." This is particularly true if the particles are hard as opposed to soft or deformable. For hard particles as large as 3–5 μm in diameter, aggregation may often be ignored, because the adhesive forces engaged upon particle-particle contact are small relative to other forces acting on the particles, such as gravity or hydrodynamic forces. This is because the areas of actual molecular contact between the particles are small even if their surfaces are smooth, and even smaller if they are rough. Dispersed particles in this size range are thus often considered to be "non-interacting." If one *does* seek to form aggregates of these larger particles, one of the most effective strategies is to engage capillary bridging ("spherical agglomeration") described earlier in Chap. 2. I.1.

The aggregation of small particles to large ones, or to large clusters or to walls remains very important. The coating of clean extended surfaces by nanoparticles, for example, is often so tenacious that it may be regarded as irreversible. Often, a layer-by-layer deposition technique is applied. Various post-treatments may be used to render the coatings even more permanent. This method has been used to render surfaces super-hydrophilic[134] or super-hydrophobic[135] or to impart other desired properties, such as color, anti-reflectivity,[136] lubricity, roughness, magnetic susceptibility,[137] *etc.*, and is emerging as one of the most important technologies for surface modification.[138-139]

[134] http://nanotechweb.org/cws/article/tech/23086

[135] Dambacher, G. T., *Kunstofffe*, **2**, 65 (2002).

[136] http://www.xerocoat.com/Anti-Reflective-Coatings.aspx

[137] LaConte, L. E. W., Nitin, N., Zurkiya, O., Caruntu, D., O'Conner, C. J., Hu, X., and Bao, G., *J. Magnetic Resonance Imaging*, **26**, 1634 (2007).

[138] Maenosono, S., Okubo, T., and Yamaguchi, Y., *J. Nanoparticle Res.*, **5**, 5 (2003).

[139] Gammel, F., in **NATO Science Series, II: Mathematics, Physics and Chemistry**, Vol. 155, Iss. Nanostructured, p. 261, 2004.

Some fun things to do:
Experiments and demonstrations for Chapter 7

1. *Determination of the Critical Coagulation Concentration (CCC) by jar testing*

The critical coagulation concentration (*CCC*) is the concentration of indifferent electrolyte required to coagulate an electrocratic colloid. Although it varies greatly from one colloid to the next, for a monovalent electrolyte it is often found to lie between 50 and 200 mM. For an electrolyte providing a counterion of higher valence, z, the empirical Schulze-Hardy Rule states and DLVO theory predicts that the *CCC* varies as z^{-6} when the effective surface potential (Stern potential) is high, *i.e.*, $|\psi_\delta| \geq 100$ mV. For lower Stern potentials, the exponent on z is lower, down to 2 for $|\psi_\delta| \leq 20$ mV. The *CCC* can usually be determined to within a few mM by jar testing, in which a series of "jars," *i.e.* test tubes, are arranged with a given concentration of the colloid and a systematically varying amount of electrolyte and observing the lowest concentration of salt required to induce aggregation and subsequent sedimentation over a given period of time, usually ≈ 30 min.

Materials:

- A convenient negatively-charged colloid is prepared as a 1 wt% dispersion of a processed montmorillonite clay (Nalco 650) in pure water.
- Two dozen 20-mL test tubes and three test tube racks.
- Stock solutions of 0.80M KCl, 0.05M $CaCl_2$ and 0.03M $AlCl_3$.
- Stock solutions of 0.1N HCl and 0.1N NaOH.
- Micropipette.
- Laser pointer.

Procedure:

1) Start with each test tube half filled (8 mL) with the stock colloid.

2) Add a sufficient amount of the stock KCl solution (using the micropipette) to each test tube, so that when water is added to bring the total volume in each tube to 12mL, the salt concentrations are 0, 50, 100, 150 and 200mM, shaking each thoroughly and placing in the rack.

3) Wait 30 min, and then in a darkened room and with a black surface behind the array of tubes, shine the laser pointer through the array of tubes from the side of the highest concentration of salt at a midpoint elevation, as shown in Fig. E7-1. Observe that tubes with 200, 150 and 100mM concentrations of KCl are clear, but a Tyndall

cone is observed in the tubes of lower concentration. This suggests that the *CCC* lies between 50 and 100mM. The value could be refined, if desired, by setting up a new array with concentrations of 60, 70, 80mM... KCl, *etc*.

4) Allow the array of test tubes to age overnight, which will give time enough for all of the colloids to sediment out, and observe the height of the resulting sediments. The un-aggregated colloid should have the smallest height, the next smallest height should be in the tube(s) that did not shown coagulation at 30 min, and the greatest sediment heights should be for those cases in which rapid aggregation was induced with an excess of salt. Greater sediment heights for the rapidly aggregated colloid results from the lower fractal dimension of the aggregates formed.

5) The above procedure may be repeated with $CaCl_2$, covering a range of concentrations from 0.2 to 10mM, and with $AlCl_3$, covering a range from 0.05 to 1 mM. In the case of the $AlCl_3$, the *pH* of the dispersion should be adjusted to $\approx$ 3 using the stock HCl solution. This produces Al^{+3} as the primary cation. If the *pH* is adjusted to $\approx$ 8, a very different result is obtained, suggesting the primary cationic species is perhaps monovalent.

6) For a classroom demonstration one may use two test tubes, each with the stock colloid. Using a salt shaker, add a copious amount of salt to one of the tubes, then shake both of them and let stand for 30 min. It will be obvious that the salt has coagulated the colloid.

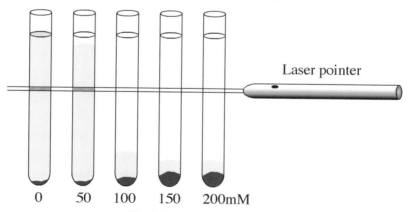

Fig. E7-1: Schematic of jar testing.

2. *Formation of a colloidal gel*

Under some circumstances, an electrocratic colloid will aggregate into a percolated structure, or a gel. A clay may aggregate into an edge-to-face structure, as pictured in Fig. E7-2, even at rather low concentrations due to the positive charges around the edges of the particles and the negative charges on the faces.

Materials:

- About 10 g of Laponite® RD, a synthetic clay (available from Southern Clay Products, Inc., Gonzales, TX)[140]
- 100-mL beaker

Procedure:

Add 7 g of Laponite RD clay to $\approx$ 100 mL of de-ionized water while stirring. Let stand for 30 min, or for a very stiff gel, overnight. The beaker may then be inverted without spilling any of the dispersion.

Fig. E7-2: A percolated suspension of Laponite clay particles.

3. Destruction of a colloidal gel

The gel prepared above may be destroyed and the stable colloid restored with the addition of a small amount of specifically adsorbing, high valence anions which neutralize the edge charges without collapsing the double layer at the particle faces.

Materials:

- About 10 g of gel prepared as above
- 100-mL beaker
- Solution of tetra-sodium pyrophosphate, TSPP, $Na_4PO_4 \cdot 10H_2O$ prepared by adding 50 mg to 10mL of water

Procedure:

Blend in 5 mL of the TSPP solution. The fluidity of the dispersion will be restored.

---

[140] Cummins, H. Z., *J. Non-Crystalline Solids*, **353**, 3891 (2007).

Chapter 8

RHEOLOGY OF DISPERSIONS

A. Rheology: scope and definitions

The term "rheology" refers to the deformation and flow of materials in response to applied stresses, a subject of considerable depth and breadth.[1] It is possible to classify and even define materials with respect to their rheology in various ways. A *fluid* (gas or liquid), for example, is defined as a material that is incapable of sustaining shear stress at rest. Even an infinitesimal shear force deforms a fluid without limit, and any work done in shear deforming it at a finite rate is irreversibly degraded to heat, *i.e.*, lost. In other materials, such as solids, even infinitesimal deformations are resisted, and when the causative stresses are removed, at least part of the work done in deforming the material may be recovered. Colloidal materials, both lyophilic and lyophobic, exhibit a broad spectrum of rheological behavior which is important, first because it governs how such materials are to be handled, *i.e.*, transported, pumped, coated, sprayed, injected, *etc.*, and second because the detailed nature of such behavior provides valuable clues concerning colloid structure. The focus in what follows is on lyophobic colloids, *i.e.*, dispersions of particles in a second phase.

To refine the rheological classification of materials further, one may define a solid as a material that resists deformation (shearing, stretching, *etc.*) in direct proportion to the *extent* of the deformation (strain), a property termed *elasticity*. An *ideal* or "Hookean" solid is one for which the proportionality between stress (force/area) and strain is linear, a statement known as *Hooke's Law*, and the proportionality factor it defines is the material's *modulus* (*e.g.*, Young's modulus). Real (as opposed to *Hookean*) solids have moduli that depend in general on the strain, as well as possibly strain history. On the other hand, a fluid's resistance is only to the *rate* of its deformation (*i.e.*, *rate of strain*) and the proportionality defines the property of *viscosity*. In the simplest case, the proportionality is linear, as expressed by *Newton's Law of viscosity*, and the coefficient it defines is the Newtonian *viscosity*. Real (as opposed to *Newtonian*) fluids have viscosity functions

---

[1] Goodwin, J. W., in **Colloidal Dispersions**, J. W. Goodwin (Ed.), pp. 165-195, Roy. Soc. of Chem., London, 1982;
Bird, R. B., Armstrong, R. C., and Hassager, O., **Dynamics of Polymeric Liquids**, Vols. I and II, Wiley, New York, 1987;
Barnes, H. A., **A Handbook of Elementary Rheology**, Univ. of Wales, Institute of Non-Newtonian Fluid Mechanics, Aberystwyth, 2000.

that depend in general on the strain rate, as well as possibly on strain rate history. In between solids and liquids are *viscoelastic* materials that exhibit resistance to both the extent *and* the rate of their deformation. Whatever its nature, the relationship between stress and strain and/or strain rate is termed a material's *constitutive equation*, the simplest examples of which are Hooke's Law of elasticity and Newton's Law of viscosity. Real materials span the range from those for infinitely rigid (Euclidean) solids to inviscid (Pascalian) fluids, as summarized in Table 8-1.

Table 8-1: Spectrum of rheological behavior of materials: $\tau =$ stress; $\gamma =$ strain; $\dot{\gamma} =$ strain rate; $G =$ modulus; $\eta =$ viscosity function; $t =$ time.

Material	Constitutive equation
Infinitely rigid (Euclidean) solid	$G \to \infty$
Linear elastic (Hookean) solid	$\tau = G\gamma$ ($G =$ const)
Nonlinear elastic solid	$\tau = G(\gamma)\gamma$
⋮	
Viscoelastic material	$\tau = \tau(\gamma, \dot{\gamma}, t \dots)$
⋮	
Nonlinear viscous (non-Newtonian) fluid	$\tau = \eta(\dot{\gamma})\dot{\gamma}$
Linear viscous (Newtonian) fluid	$\tau = \mu\dot{\gamma}$ ($\mu =$ const)
Inviscid (Pascalian) fluid	$\mu \to 0$

All gases and essentially all pure non-polymeric liquids and non-polymeric solutions are Newtonian. Dilute colloids of non-interacting particles are also usually Newtonian fluids, but more concentrated ones are typically non-Newtonian and quite often viscoelastic. Colloids of interacting (strongly repelling, aggregating or aggregated) particles often have rather complex rheology. One aspect of colloids, of course, is that they are in general changing with time, and that their deformation and deformation history affects their properties.

B. Viscometry

1. *Newton's Law of viscosity*

The simplest rheology to be exhibited by a colloidal dispersion is that of a Newtonian fluid. To describe this behavior, consider a fluid confined between parallel plates, as shown in Fig. 8-1. If the top plate is moved at a steady rate and the bottom held stationary, most pure fluids exhibit in the steady state, a linear velocity profile as shown. The force/area (F/A) which must be exerted by the top plate on the fluid is the shear stress, τ, and it is linearly proportional to the shear rate (the velocity gradient), $\dfrac{dv_x}{dy} \equiv \dot{\gamma}$, in

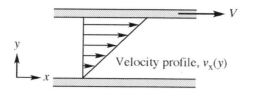

Fig. 8-1: Simple one-dimensional shear (Couette) flow.

Velocity profile, $v_x(y)$

accord with Newton's Law of viscosity:

$$\frac{F}{A} = \tau = -\mu\frac{dv_x}{dy} = -\mu\dot{\gamma}. \tag{8.1}$$

Equation (8.1) defines the viscosity of the fluid, μ, and fluids for which it is a constant (with shear rate, time, *etc.*) are *Newtonian* fluids. The viscosity so defined is the *dynamic* viscosity. It contrasts with the related quantity $\nu = \mu/\rho$, called the *kinematic* viscosity, where ρ = fluid density. The units of viscosity are summarized in Table 8-2.

The key structural requirements for a colloidal dispersion to be Newtonian are that the dispersion medium itself be Newtonian (which will

---

Table 8-2: Units of viscosity.

cgs system: μ [=] g/cm·s [=] Poise (P); 1 P = 100 centiPoise (cP)
                ν [=] cm²/s [=] 100 centiStokes (cS)

SI system: μ [=] N·s/m² [=] Pascal·sec (Pa·s)
                ν [=] m²/s

Conversion factor: 1 Pa·s = 10 P = 1000 cP

---

exclude most polymeric liquids and solutions of polymers), that the dispersion be stable with respect to aggregation, and dilute enough that particle-particle interactions are negligible. For particles bearing thick diffuse electrical double layers, the required degree of dilution may be quite great (< 1 vol %). For sterically stabilized dispersions, the allowable concentration may be an order of magnitude higher.

2. Measurement of viscosity

Viscosity can be measured by a variety of instruments in which a flow field is established and relevant forces and/or flow rates are determined, as shown in Figs. 8-2 to 8-5. One of the simplest devices is the capillary viscometer, shown in Fig. 8-2, in which steady flow is established in a round capillary tube by means of an imposed pressure drop along its length. The device is usually immersed in a constant temperature bath during measurement. The pressure drop is usually produced hydrostatically in a tube oriented vertically, with the liquid flowing from a reservoir at the top to one at the bottom. Steady state is quickly established, and end effects are either negligible or accounted for in calibration constants. The time required for a given volume of liquid to flow through the tube is measured using a stopwatch. The velocity profile in the tube is parabolic, and the resulting expression for the viscosity is:

$$\mu = \frac{\pi \Delta p R^4}{8QL},$$

(8.2)

where Δp is the imposed pressure drop, R is the tube radius, Q is the volumetric flow rate, and L is the tube length.

The most common methods for measuring viscosity (or other rheological parameters) use a device in which the material to be tested is confined between the surfaces of a pair of tools, one of which is made to rotate while the other is held stationary.[2] Modern instruments permit measurements in two modes. In the controlled strain or strain rate (also called "shear rate") mode, the rate of rotation is controlled, and the resulting stress (or torque) is measured. In the other mode, the stress or torque is

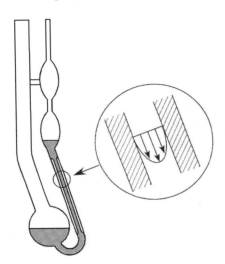

Fig. 8-2: Capillary viscometer (Cannon-Fenske type). Parabolic laminar flow is established in the capillary tube.

controlled (by means of a feedback mechanism) and the strain or strain rate is monitored. Temperature is also routinely controlled and monitored. One of the common configurations is that of the Couette, or cup-and-bob viscometer, shown in Fig. 8-3. The liquid is confined in the annular space between concentric cylinders, and one measures the torque $\mathcal{T}$ required to

---

[2] Mezger, T. G., **The Rheology Handbook**, 2nd Ed., Vincentz Network, Hannover, Germany, 2006.

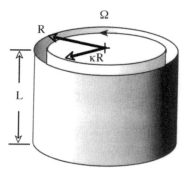

Fig. 8-3: Couette (cup-and-bob) viscometer.

maintain a steady rotation (angular velocity Ω) of one of the cylindrical surfaces (or alternatively, the rotational speed required to yield a given torque). The viscosity is given by:

$$\mu = \frac{\mathcal{T}}{4\pi L \Omega R^2 \left(\dfrac{\kappa^2}{1-\kappa^2} \right)}, \tag{8.3}$$

where κ is the ratio of the inner to the outer radius of the annulus, and R is the radius of the outer cylindrical surface. The gap width is very small relative to R, so that the velocity profile is essentially linear (shear rate constant) in conformance with Fig. 8-1.

A second common tool is the parallel plate device, shown in Fig. 8-4. In the controlled strain rate mode, one measures the torque for given plate

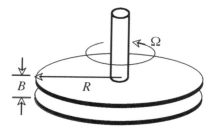

Fig. 8-4: Parallel plate viscometer.

dimensions and spacing, and a given rate of rotation. Liquid is held in place during the measurement by surface tension. The viscosity is given by:

$$\mu = \frac{2B\mathcal{T}}{\pi \Omega R^4}, \tag{8.4}$$

where B is the plate spacing. Unlike in the Couette viscometer, the shear rate is not uniform in this device.

The cone-and plate viscometer, shown in Fig. 8-5, has liquid confined in the narrow region between a shallow conical surface and a flat plate, and held there by the surface tension of the liquid. Either the cone or the plate is

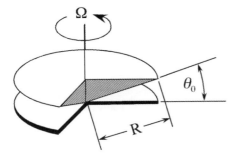

Fig. 8-5: The cone-and-plate viscometer.

made to rotate, while the opposite member is held stationary. The required torque is measured (in the controlled strain rate mode) and related to the viscosity by:

$$\mu = \frac{3\mathcal{T}\theta_0}{2\pi R^3 \Omega},$$ (8.5)

with the various quantities defined in Fig. 8-5. An important feature of the cone-and-plate arrangement is that in it the shear rate is effectively constant.

There are two common problems that occur during the rheological measurements of dispersions, particularly dense dispersions, as pictured in Fig. 8-6. Particle jamming is especially important in the cone-and-plate

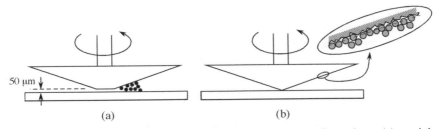

Fig. 8-6: Problems that occur during measurements with dispersions: (a) particle jamming, addressed by blunting the cone tip, and (b) slippage, addressed by roughening the tool surface.

configuration, and for this reason, cone tools used for such systems are purposely blunted. Typically a spacing of 50 μm is provided. The second problem is one of slippage at the tool surface. A thin film of the dispersion medium coats the tool surface as it moves, producing a falsely low measurement of the resistance. This problem can sometimes be addressed by roughening the tool surface, but in other cases must be addressed by using a different tool altogether, *e.g.*, the vane tool, shown in Fig. 8-7. It can be created with a variety of different vane numbers and aspect ratios. Exact theoretical analyses of the flow created with these devices can be complex,

Fig. 8-7: The vane tool.

but under optimum circumstances (thick enough media) can be well approximated by the cup-and-bob arrangement, with the outer edges of the vane defining the location of the inner "cylinder."[3]

Many other devices exist for the measurement of viscosity or other rheological properties; the reader is referred to Whorlow[4] for further elaboration.

C. The viscosity of colloidal dispersions

1. *Dilute dispersions; Einstein theory*

Even though dilute colloids may be Newtonian, their viscosity is affected by the presence of the colloidal particles. Einstein (in 1906, and in corrected form in 1911)[5] derived an expression for the effective viscosity of dilute dispersions of rigid, non-interacting, identical spheres in a Newtonian medium, *viz.*,

$$\mu = \mu_0 (1 + k\phi), \tag{8.6}$$

where μ_0, is the viscosity of the medium, ϕ is the volume fraction of the particles, and k is a constant equal to 2.5 for spheres. It comprises the leading terms of a power series in is ϕ and is valid only for $\phi \le 0.02$. The presence of the particles increases the effective viscosity due to the additional viscous drag around the particle surfaces and the energy expended in rotating the particles. For emulsions, k is reduced due to the ability of the droplets to exhibit internal circulation, reducing the distortion of the continuous phase flow around the drops. One proposed modification gives[6]

[3] Barnes, H., *J. Non-Newt. Fluid Mech.*, **81**, 133 (1999).

[4] Whorlow, R. W., **Rheological Techniques**, Ellis Horwood, Chichester, 1980.

[5] Einstein, A., *Ann. Physik*, **19**, 289 (1906);
Einstein, A., *Ann. Physik*, **34**, 591 (1911).

[6] Sherman, P. in **Encyclopedia of Emulsion Technology**, Vol. 1, P. Becher (Ed.), pp. 405-437, Marcel Dekker, New York, 1983.

$$k = 2.5\left[\frac{\mu_i + 0.4\mu_o}{\mu_i + \mu_o}\right], \tag{8.7}$$

where μ_i is the viscosity of the emulsion droplets.

The use of Einstein's Equation for dilute dispersions of spheres provides a convenient means for determining polymer adlayer (hydrodynamic) thicknesses, δ, in sterically stabilized systems or the thicknesses of large solvation layers. For spherical particles of known bare radius a and adlayer thickness δ, the effective (or inferred) volume fraction is given by:

$$\phi' = \phi(1 + \delta/a)^3. \tag{8.8}$$

This method parallels those of sedimentation, centrifugation and dynamic light scattering, which can be used for the same purpose. It is tempting to use the same procedure to estimate electrical double layer thickness as $\delta \approx \kappa^{-1}$. This is only *very* roughly the case, and will be considered in more detail below where electro-viscous effects are described.

2. Denser dispersions of non-interacting particles

As ϕ increases, Eq. (8.6) breaks down due to the importance of inter-particle crowding, although the dispersion may continue to be Newtonian (provided the particles are truly non-interacting and the shear rate is low, as described below). Measured viscosities are higher than predicted by Einstein's Law, which may be extended both theoretically and empirically by equations of the type:

$$\mu = \mu_0\left[1 + 2.5\phi + b\phi^2 + O(\phi^3)\right]. \tag{8.9}$$

Experimental values of b for spherical particles generally range from about 5 to 8. An example is shown schematically in Fig. 8-8. Theory has been applied to the case of low Peclet Number, *Pe, i.e.*, for low rates of shear.

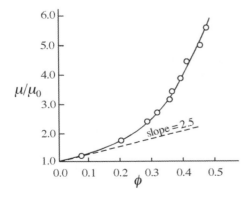

Fig. 8-8: Relative viscosity of a colloidal dispersion as a function of particle volume fraction. Data from [Roscoe, R., *Br. J. Appl. Phys.*, **3**, 267 (1952).]

The Peclet Number is in general the dimensionless ratio of the effectiveness of convective transport to diffusive transport, and in this case it refers to the transport of particles. It may be formulated as the ratio of the time required, t_B, for a particle to diffuse by Brownian motion a distance equal to its own diameter divided by the time it takes, t_C, to be convected that distance in a shear flow such as that of Fig. 8-1. It can be shown that

$$t_B = \frac{12\pi\mu a^3}{kT}, \text{ and} \tag{8.10}$$

$$t_C = \frac{2}{\dot{\gamma}}, \text{ so that} \tag{8.11}$$

$$Pe = \frac{t_B}{t_C} = \left[\frac{6\pi\mu_0 a^3 \dot{\gamma}}{kT}\right], \tag{8.12}$$

For $Pe \ll 1$, so that the flow only minimally distorts the equilibrium structure of the dispersion, Batchelor[7] derived $b = 6.2$, and the series of Eq. (8.9) truncated after this term applies up to $\phi \approx 0.15$. At greater concentrations, terms of $O(\phi^3)$ and higher are required, and one must resort to empirical data-fitting.

There are a number of terms used for describing the viscous behavior of Newtonian dispersions (or polymer solutions). Definitions of the most commonly used of them are summarized in Table 8-3, together with their values for an Einstein dispersion, *i.e.*, a dispersion obeying Eq. (8.6). The intrinsic viscosity yields valuable information on the nature of the dispersed phase particles. It has the value of 2.5 only for the case of monodisperse spheres, but will take on other values under other conditions.

Table 8-3: Viscosity terms used to describe Newtonian dispersions.

Term	Symbol (definition)	Value for Einstein dispersion
Viscosity of dispersion medium	μ_0	
Viscosity of dispersion	μ	$\mu_0(1 + 2.5\phi)$
Relative viscosity	$\mu_r = \mu/\mu_0$	$(1 + 2.5\phi)$
Specific viscosity	$\mu_{sp} = \mu_r - 1$	2.5ϕ
Reduced viscosity	$\mu_{red} = \mu_{sp}/\phi$	2.5
Intrinsic viscosity	$[\mu] = \lim_{\phi \to 0} \mu_{red}$	2.5

A useful equation for describing denser and polydisperse dispersions (of non-interacting hard spheres) and free of adjustable parameters was developed by Mooney,[8] who considered the effect of adding an increment $d\phi_i$ to the volume fraction of particles in the free liquid medium, so that the

[7] Batchelor, G. K., *J. Fluid Mech.*, **41**, 545 (1970);
  Batchelor, G. K., *Ann. Rev. Fluid Mech.*, **6**, 227 (1974);
  Batchelor, G. K., *J. Fluid Mech.*, **83**, 97 (1977).
[8] Mooney, M., *J. Colloid Sci.*, **6**, 162 (1951).

viscosity of the dispersion increased from μ to $\mu + d\mu$. The *increment* in viscosity was assumed to be given by the Einstein Law, since the *added* volume fraction $d\phi_i$ was small enough to conform to the required degree of dilution. In other words, the medium to which the new material was added had the viscosity μ that accounted for the solids already present. Thus: $d\mu = 2.5\mu d\phi_i$. The increment to the volume fraction in the particle-free medium, $d\phi_i$, is related to the increase in the overall volume fraction of particles by: $d\phi_i = d\phi/(1-\phi)$, so that

$$d\mu = 2.5\mu\frac{d\phi}{(1-\phi)},\qquad(8.13)$$

which, upon integration from $\phi = 0$ (where $\mu = \mu_0$) to $\phi = \phi$ yields the Mooney Equation:

$$\mu = \mu_0(1-\phi)^{-2.5}.\qquad(8.14)$$

Krieger[9] later modified and generalized the Mooney Equation in two ways. It was shown that a more appropriate expression for $d\phi_i$ was

$$d\phi_i = \frac{d\phi}{(1-\phi/\phi_{max})},\qquad(8.15)$$

where ϕ_{max} was the maximum achievable volume fraction of the particles for the given system (typically ≈ 0.64). This corrects for the tacit assumption in Eq. (8.12) that ϕ may approach unity. Also, the numerical coefficient of 2.5 in Mooney's Equation was replaced with the more general intrinsic viscosity, $[\mu]$. This leads to:

$$\mu = \mu_0\left[1 - \frac{\phi}{\phi_{max}}\right]^{-[\mu]\phi_{max}},\qquad(8.16)$$

known as the Krieger-Dougherty Equation. It has met with good success for spherical particles for both low and high shear rates, but often not for intermediate shear rates, as discussed below. For low shear rates, in a given study, putting $\phi_{max} = 0.632$ gave the best results, while for $Pe \rightarrow \infty$, the best-fitting value was 0.708.[10]

3. Dilute dispersions of non-spherical particles

Dispersions of non-spherical particles have a far more complex rheology than spherical particles, unless they are sufficiently small that Brownian motion effectively prevents them from aligning with the streamlines. For this case, Simha[11] derived expressions for the intrinsic

[9] Krieger, I. M., *Adv. Colloid Interface Sci.*, **3**, 111 (1972).

[10] De Kruif, C. G., van Iersel, E. M. F., Vrij, A., and Russel, W. B., *J. Chem. Phys.*, **83**, 4717 (1985).

[11] Simha, R., *J. Phys. Chem.*, **44**, 25 (1940);
Simha, R., *J. Chem. Phys.*, **13**, 188 (1945).

viscosity of both prolate and oblate ellipsoids of revolution, results of which are shown in Fig. 8-9. For prolate particles, Simha's result is

$$[\mu] = \frac{14}{15} + \frac{(a/b)^2}{15[\ln(2a/b) - \lambda]} + \frac{(a/b)^2}{5[\ln(2a/b) - \lambda + 1]},\qquad(8.17)$$

where (a/b) is the ratio of the major to the minor axis, and λ is 1.5 for ellipsoids and 1.8 for cylindrical rods. It is evident that for strongly acicular particles, the viscosity is considerably greater than that for spheres of similar size, and that viscosity measurements may be used to obtain information regarding particle shape.

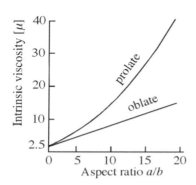

Fig. 8-9: Calculated results of Simha theory for dilute dispersions of ellipsoids of revolution.

D. Non-Newtonian rheology

1. *General viscous behavior of dispersions of non-interacting particulates*

The foregoing discussion has assumed that the shear rates involved are sufficiently low ($Pe \ll 1$) that the equilibrium microstructure of the dispersion remains unaltered. Experiments show that as the shear rate is increased above these low values, the effective viscosity of the dispersion becomes dependent on the shear rate (except for the trivial case of very low volume fractions, *i.e.*, $\phi \leq 0.02$, the Einstein limit). The dispersions are thus *non-Newtonian*. For dispersions of hard, non-interacting (*i.e.*, non-aggregating) spherical particles, it is found that the effective viscosity, η, varies in a systematic way with Pe, as shown in Fig. 8-10. Newtonian behavior is observed at low shear ($Pe \ll 1$) with a low- shear limit viscosity $\eta(0) = \mu$. The dependence of the low-shear viscosity on particle volume fraction is usually well described by the Mooney Equation or the Krieger-Dougherty Equation, as detailed in the previous section. At sufficiently high shear rates ($Pe \gg 1$), if they can be achieved, a high-shear limit Newtonian viscosity, $\eta(\infty) = \mu_\infty$, is observed. It is found to be lower (sometimes much lower) than the low shear rate limit, but the Krieger-Dougherty Equation describes its dependence on ϕ reasonably well, with different values of the

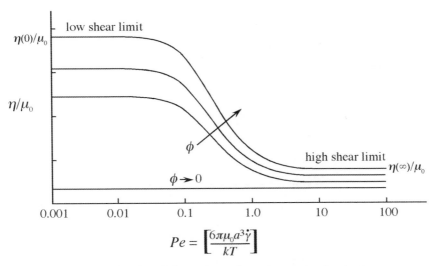

$$Pe = \left[\frac{6\pi\mu_0 a^3 \dot{\gamma}}{kT}\right]$$

Fig. 8-10: Viscous behavior of dispersions of non-interacting hard spheres as a function of Peclet Number (nondimensionalized shear rate).

parameters ϕ_{max} and $[\mu]$.[12] At intermediate shear rates, the viscosity of the dispersion is a function of shear rate, *i.e.*, $\eta = \eta(Pe)$. One of the semi-empirical representations of this intermediate behavior is given by[13]:

$$\frac{\eta(Pe) - \eta(\infty)}{\eta(0) - \eta(\infty)} = \frac{\sinh^{-1}(Pe)}{Pe}. \tag{8.18}$$

It is seen that for a given shear rate (*i.e.*, Pe) the viscosity function depends only on the particle volume fraction, ϕ, or conversely, for a given volume fraction, depends only on shear rate. It is independent of particle size. While the magnitude of the limiting Newtonian viscosities depends on ϕ, the Peclet Number range over which the change occurs remains essentially the same. Extensive data obtained with monodisperse polymer latices verify the behavior shown in Fig. 8-10,[14] which may be understood with the help of Fig. 8-11. In the low shear rate regime, Brownian diffusion maintains the same random equilibrium dispersion structure that exists in the absence of shearing. As the shear rate increases ($Pe \approx 1$) the particles begin to arrange themselves into layers parallel with the flow, so that shearing is eased, but some of the diffusion between the layers persists. In the high-

---

[12] The treatment of the low and high shear rate limits has also been extended to include particles which are ellipsoids of revolution rather than spheres: Hinch, E. J., and Leal, L. G., *J. Fluid Mech.*, **52**, 683 (1972).

[13] Maron, S. H., and Pierce, P. E., *J. Colloid Sci.*, **11**, 80 (1956).

[14] Woods, M. E., and Krieger, I. M., *J. Colloid Interface Sci.*, **34**, 91 (1970); Paper, Y. S., and Krieger, I. M., *J. Colloid Interface Sci.*, **34**, 126 (1970).

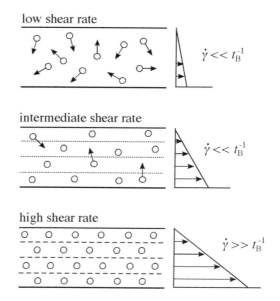

Fig. 8-11: Structural changes in dispersions as a function of shear rate.

shear limit, which cannot always be realized in practice, nearly all the particles are confined to layers moving past each other with negligible particle diffusion between them. It should be noted that at very high shear rates, these layers may become hydrodynamically unstable and buckle, leading to an increase in effective viscosity with shear rate.[15]

The relationship of Eq. (8.18) is more commonly expressed in terms of the Cross Equation:[16]

$$\frac{\eta(\dot{\gamma}) - \eta(\infty)}{\eta(0) - \eta(\infty)} = \frac{1}{1 + (C\dot{\gamma})^p}, \tag{8.19}$$

where C [=] s, is the "Cross constant," and p is the "Cross exponent." It is often used in the simpler form of Eq. (8.20), in which the high shear rate viscosity is regarded as negligible, either because it is very small or because it is not achievable for the system of interest.

$$\frac{\eta(\dot{\gamma})}{\eta(0)} = \frac{1}{1 + (C\dot{\gamma})^p}. \tag{8.20}$$

The behavior observed at intermediate shear rates in dispersions of hard, non-interacting spheres described above is the simplest category of non-Newtonian behavior, i.e., the viscosity function depends only on shear rate $\dot{\gamma}$. Over much of the intermediate shear rate range, Eq. (8.20) takes the even simpler functional form of

[15] Hoffman, R. L., J. Colloid Interface Sci., **46**, 491 (1974).
[16] Cross, M. M., J. Colloid Sci., **20**, 417 (1965).

$$\eta = \eta(\dot{\gamma}) = m|\dot{\gamma}|^{n-1}, \tag{8.21}$$

where the Cross constants have been re-written in terms of the parameters "m" and "n." Equation (8.21) defines a "power law" fluid, and the constitutive equation that generalizes Newton's Law of viscosity for such a fluid is

$$\tau = -\eta\dot{\gamma} = -m|\dot{\gamma}|^{n-1}\dot{\gamma}. \tag{8.22}$$

For n =1, the fluid is Newtonian, with m = μ. For n < 1, the shear stress and the viscosity function decrease with increasing shear rate, behavior termed *shear thinning*, or "pseudo-plastic." For n > 1, the shear stress and viscosity increase with shear rate, producing *shear thickening*, or "dilatant" behavior. The more general definitions of shear thinning and shear thickening behavior are not tied specifically to the power law model. Plots of the shear stress and the viscosity function for both are pictured in Fig. 8-12.

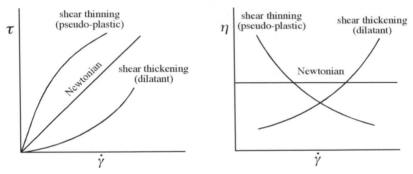

Fig. 8-12: Dependence of shear stress (left) and viscosity function (right) on shear rate for simple non-Newtonian fluids.

Shear thinning is very commonly observed in dispersions of moderate concentration of non-interacting, hard-sphere dispersions, but is also often associated with systems of anisotropic particles (which become favorably aligned at higher velocity gradients) and with systems that are weakly flocculated (in which the velocity gradients break down the flocks), as pictured schematically in Fig. 8-13.

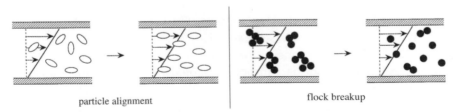

Fig. 8-13: Common origins of shear thinning behavior in colloidal dispersions.

Shear thickening is often observed in *very* dense colloids in which there is barely enough continuous phase to fill the voids between particles that are essentially in contact, as shown in Fig. 8-14. The transition to shear thickening behavior as particle loading is increased thus often coincides with the critical pigment volume concentration, *CPVC*, defined in Chap. 5.A.3. Shearing requires that this dense structure be broken down to permit particles to flow past one another, and the accompanying expansion leaves insufficient liquid to fill the voids. This is commonly observed when walking on wet beach sand. The sand becomes rigid and seems to become dry (water is flowing into the expanded voids) when stepped on. However, if one very slowly presses down on such sand, it will be quite fluid and yielding.

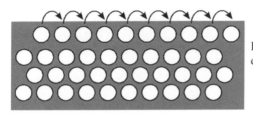

Fig. 8-14: Shear-thickening in dense dispersions.

It is clear that the capillary viscometer is not a good way to study non-Newtonian materials since it involves a parabolic rather than a linear velocity profile and therefore, at any axial location, a range of shear rates prevail. Nonetheless, high-pressure capillary viscometers are available commercially, but their use requires rather extensive calibration for each system. The parallel plate configuration also suffers this limitation, but to a lesser extent. The remaining two rotational viscometer arrangements (concentric cylinder and cone-and-plate) produce very nearly constant rates of shear and are therefore more versatile. They may be operated to perform a "shear sweep" covering a large range of shear rates, and this is usually done. Sometimes rather complex behavior is observed, in which over a certain range of shear rates a given set of power-law constants is observed, but over other regions, another set of constants is required.

2. Fluids with a yield stress

Another common observation is that of *plastic* behavior, which refers to the existence of a yield stress, τ_y, below which the material behaves as a solid. When the yield stress is exceeded, the liquid may behave as a Newtonian fluid (and is termed a *Bingham plastic*), or it may exhibit shear thinning behavior, and is termed *viscoplastic*. Both are illustrated schematically in Fig. 8-15.

The existence of a yield stress is important in the coatings industry since it is desirable that the coating dispersion flow only when being brushed

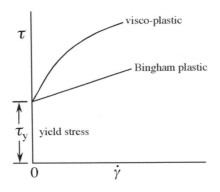

Fig. 8-15: Fluids exhibiting a
yield stress.

onto the surface (sheared at a given level) and immediately after deposition,
but not thereafter. One may recall the television advertisement for a
particular brand of spray paint promising "no drips, no runs, no errors."[17]
Such behavior is often observed in particulate systems that have
"percolated," i.e., a continuous contact between dispersed particles has been
established. This may result from gel formation ("sol gel") or completed
flocculation. The breakdown of the gels structure is pictured schematically
in Fig. 8-16. A percolated condition need not require a dense dispersion. In
this state, the system has solid-like properties, and a certain yield stress is
required to initially break down the structure.

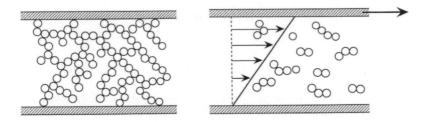

Fig. 8-16: Breakdown of a percolated structure upon shearing (Bingham
plastic or viscoplastic dispersion).

The yield stress of weakly aggregated dispersions depends on the
strength of the particle-particle contacts in the structure, F_s, times the
number of such contacts per unit area. The latter scales as ϕ^2/a^2, where a is
the particle radius, so that

$$\tau_y \propto \frac{\phi^2}{a^2} F_s, \tag{8.23}$$

where F_s is the force required to disjoin a particle-particle contact. It may be

---

[17] The ad featured the renowned baseball player, Johnny Bench.

be estimated from the particle interaction potential function as follows.[18] When a pair of particles is aggregated, at some separation distance $S_{0\text{-min}}$, the potential energy of the pair, Φ, is at a minimum. As one attempts to pull these particles apart, the structure experiences a resisting force, proportional to $d\Phi/dS_0$, and which increases from zero to a maximum before it again falls off to zero. This maximum is identified with F_s, and may be approximated by $-\Phi(S_{0\text{-min}})/S_{0\text{-min}}$, so that substituting into Eq. (8.23) gives

$$\tau_y \propto -\frac{\phi^2}{a^2}\frac{\Phi(S_{0\text{-min}})}{S_{0\text{-min}}} \tag{8.24}$$

The potential function for particles with surfaces of constant charge density is obtained by combining Eqs. (7.7) and (7.74), viz.

$$\Phi = -\frac{Aa}{12S_0} + 2\pi\varepsilon\varepsilon_0 a\psi_\delta^2\exp(-\kappa S_0), \tag{8.25}$$

which, upon substitution into Eq. (8.24) and replacing the Stern potential, ψ_δ, with the zeta potential, ζ, gives

$$\tau_y \propto \frac{\phi^2}{a}\left(\frac{A}{12S_{0\text{-min}}^2} - \frac{2\pi\varepsilon\varepsilon_0}{S_{0\text{-min}}}\zeta^2\ln\left[\frac{1}{1-\exp(-\kappa S_{0\text{-min}})}\right]\right). \tag{8.26}$$

The predicted dependence of the yield stress on $-\zeta^2$ has been observed in experiments with aqueous titania dispersions,[19] and the predicted dependences on ϕ^2 was roughly corroborated. Interestingly, in dealing with clay dispersions, where flock structure is dominated by edge-to-face hetero-aggregation, the yield stress was found to depend instead on $+\zeta^2$.[20]

From a practical point of view, one would like to accurately measure the yield stress of a percolated dispersion. This is often challenging as the material is often highly shear thinning in the vicinity of the yield point. An effective method is to seek the break point in the $\log\eta$ vs. $\log\tau$ plot,[21] as shown in Fig. 8-17.

3. Time-dependent rheology

Frequently percolated systems exhibit time-dependent rheological behavior, but such time-dependence may also be observed in systems not exhibiting a distinct yield stress. Three categories of such behavior have been identified:

---

[18] Larson, R. G., **The Structure and Rheology of Complex Fluids**, p. 351, Oxford Univ. Press, New York, 1999.

[19] Leong, Y. K., and Ong, B. C., *Powder Tech.*, **134**, 249 (2003).

[20] Laxton, P. B., and Berg, J. C., *J. Colloid Interface Sci.*, **296**, 749 (2006).

[21] Barnes, H. A., **A Handbook of Elementary Rheology**, p. 73, Univ. of Wales, Aberystwyth, 2000.

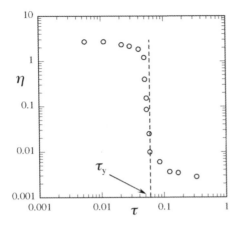

Fig. 8-17: Experimental
determination of the yield stress
of a percolated dispersion.

1) *thixotropic* behavior: the viscosity function of the system decreases not only with shear rate, but also with time at constant shear rate. In this case there is a time-dependence associated with the structure-breakdown process. The greater the constant shear rate, the faster is the breakdown of the structure and the loss of viscosity. The net structure loss is the result of breakdown occurring more rapidly than rebuilding, with the maximum rebuild rate occurring when the system is at rest.

2) *anti-thixotropic* behavior: the viscosity function of the system increases with time at constant shear rate. This behavior suggests that shearing assists in structure building.

3) *rheopectic* behavior: similar to thixotropic, but with rebuild rates occurring faster at higher shear rates. Thus the net breakdown over time is slower at faster shear rates.

Time-dependent rheology is often conveniently studied by performing upward-followed-by-downward shear sweeps ("loop tests"). A thixotropic system will show a sharp decrease in the viscosity function as shear rate is increased, followed by a slower recovery as the shear rate is decreased. The hysteresis will widen as the rate of the shear rate change is increased.

4. *Viscoelasticity*

Many colloidal systems of interest exhibit both liquid-like and solid-like characteristics, *i.e.*, they are *viscoelastic*. One of the ways their behavior may be directly exhibited through the use of creep-compliance curves, as shown in Fig. 8-18. In such studies, the *extent* of strain (for example shearing strain) is monitored as a function of time, while the applied (shear) stress is maintained constant. The *creep compliance*, shown as the ordinate in Fig. 8-18, is defined as the (amount of strain)/(constant applied stress). For the system shown, the creep compliance rises sharply at first and then after some time (called the relaxation time) characteristic of the system and the stress level, increases linearly. At some later time the stress is removed, and the system recovers while at rest. This also occurs over some

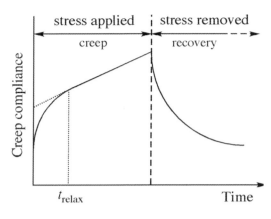

Fig. 8-18: Creep-compliance-time behavior of a colloidal dispersion exhibiting viscoelasticity.

characteristic relaxation time, t_{relax}. If the system's relaxation time is large relative to the time over which changes are imposed, t_{exptl}, the system will behave as a solid. This relationship is quantified in terms of the Deborah Number,[22] De:

$$De = \frac{t_{relax}}{t_{exptl}}, \tag{8.27}$$

from which it is evident that $De \gg 1$ implies solid-like behavior, while $De \ll 1$ implies liquid-like behavior. In the intermediate range, $De \approx 1$, viscoelasticity may produce some unusual effects (if one thinks only in terms of Newtonian behavior), such as the swelling of such liquids as they are extruded from a die ("die swell") or the climbing of such liquids around a stirring rod (the "Weissenberg effect"). More examples are tabulated in Bird *et al.* cited earlier. Experiments of the type suggested by Fig. 8-18 are rich in information concerning the colloid structure.

Another important way to study viscoelastic systems is to subject them to a stress (or strain) that varies periodically in time (*oscillatory rheometry*). This may be achieved, for example, with a rotational viscometer operated in an oscillatory mode. The amplitude of the strain variation is kept small to prevent plastic structure breakdown and thus to remain within the *linear viscoelastic* region of behavior. Both the stress and the strain are varied sinusoidally, but in general will be out of phase with one another, as shown in the bottom graph of Fig. 8-19. At any frequency, the data can be interpreted using either a generalized form of Hooke's Law:

$$\tau = G^* \gamma, \tag{8.28}$$

where γ is the strain level, and G^* is the *complex* shear modulus (ratio of stress/strain), or a generalized form of Newton's Law:

$$\tau = -\eta^* \dot{\gamma}, \tag{8.29}$$

---

[22] Named for the prophetess, Deborah, who, in the Bible: Judges 5:5, foretold that "the mountains melted from before the Lord."

where $\dot{\gamma}$ is the strain (shearing) rate, and η^* is the *complex* viscosity function.

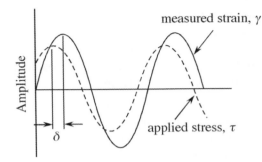

Fig. 8-19: Sinusoidal variation of stress and strain in oscillatory rheological studies.

Considering the generalized Hooke's Law representation, the complex shear modulus may broken down into real and imaginary parts:

$$G^* = G' + iG'', \qquad (8.30)$$

where G' is the *storage*, or real shear modulus (component of the stress in phase with the strain, divided by the strain) and G'' is the *loss* or imaginary shear modulus (component of the stress out of phase with the strain, divided by the strain). G' is associated with the storage of recoverable elastic energy during deformation (solid-like property), and G'' is associated with the dissipation or degradation of energy to heat during deformation (fluid-like property). The phase angle δ, shown in Fig. 8-19, is defined such that

$$\tan\delta = G''/G'. \qquad (8.31)$$

Alternatively, one may write out the complex viscosity as:

$$\eta^* = \eta' - i\eta'', \qquad (8.32)$$

where $\eta' = \eta$, the real (dynamic) viscosity, and η'' is associated with the solid-like behavior.

The first oscillatory measurement to be made in the study of a given dispersion is a *strain sweep*, in which the storage modulus is measured as the *amplitude* of the oscillation (in %) is varied at a constant frequency, usually 1 Hz, as shown in Fig. 8-20. Strain % is defined in different ways,

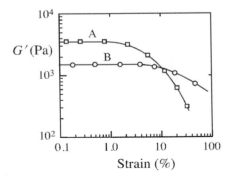

Fig. 8-20: Strain sweep for two different dispersions A and B.

depending on the instrument, but the angular displacement corresponding to 100% strain is usually less than 1°. The region of amplitude below which the storage modulus G' is constant defines the *linear viscoelastic* region. It is to be noted that for sample A, only strain amplitudes of 1% or less can be used without breaking down the structure of the dispersion, whereas dispersion B has a linear viscoelastic region up to almost 10% strain. This is considered a measure of the internal strength of the dispersion structure.

Within the linear viscoelastic region of a given dispersion, the frequency dependence (angular velocity = ω) of the moduli (G', G'') or (η', η'') gives information on the relaxation processes that are occurring. Figure 8-21 shows the dependence of G', G'' and the magnitude of the viscosity function $|\eta^*| = \sqrt{\eta'^2 + \eta''^2}$ as a function of frequency for a given dispersion. The various moduli are related to each other in a simple way: $G' = \eta''\omega$, and $G'' = \eta'\omega$, so that

$$|\eta^*| = \frac{1}{\omega}\sqrt{G'^2 + G''^2}.$$

(8.33)

The concentric-cylinder, parallel plate and cone-and-plate rheometers are easily used to perform oscillatory rheometry. What one commonly finds are plots of G', G'' and the magnitude of the complex viscosity, $|\eta^*|$ *vs.* the angular velocity, ω. The characteristic time, t_{exptl}, in such experiments is ω^{-1}. Thus at low enough frequencies, we may observe liquid-like behavior and at high enough frequencies, solid-like behavior. The results of Fig. 8-21 show a system that percolates at a particular frequency, termed a *gel point*.

Many other types of behavior can be observed. Systems of strongly aggregated particles may show G' in excess of G'' over the entire frequency range. In a weakly associated or non-associated system, the reverse would be true. It is also generally found that in associated systems, $|\eta^*|$ decreases with ω, while in non-associated systems (more nearly Newtonian), it remains nearly constant with frequency.

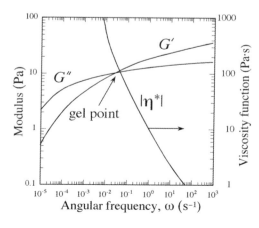

Fig. 8-21: An example of viscoelastic behavior of a latex dispersion as function of frequency in the linear viscoelastic region.

E. Electroviscous effects

Additional important rheological phenomena, particularly in aqueous media, are the *electroviscous* effects. They are treated briefly here, but in more detail in Hunter,[23] and are closely related to the electrokinetic effects discussed earlier. Electroviscous effects may be manifest in the flow of liquid through fine capillaries or porous media bounded at their inner surfaces with diffuse electrical double layers. When the flow caused by an imposed pressure drop sets up a streaming potential and the attendant electro-osmotic flow in the reverse direction, the "apparent" viscosity is larger than that determined when no electrokinetic effect is occurring (as when the double layer is collapsed). The chief concern here, however, is with the rheology of colloidal dispersions. The effects are most pronounced in dense colloids, but may also be important in rather dilute dispersions. Three electroviscous effects have been identified[24], as shown in Fig. 8-22. The *primary* electroviscous effect is the result of resistance to the deformation of the diffuse double layer in a shear field (the *relaxation* effect). The problem was first examined by Smoluchowski in 1916 for the case of thin double layers ($\kappa a > 100$) and low zeta potential ($\zeta \leq 25$ mV) in a 1-1 electrolyte. It was investigated in greater detail by others, and in particular by Booth[25] in 1950. His result, which was limited to the same conditions as Smoluchowski, but was for both large and small values of κa, is conveniently expressed in terms of intrinsic viscosity as:

$$[\mu] = 2.5\left\{1 + \frac{4\pi(\varepsilon\varepsilon_0\zeta)^2}{k\mu_0 a^2}\left[(\kappa a)^2(1 + \kappa a)^2 Z(\kappa a)\right]\right\}, \tag{8.34}$$

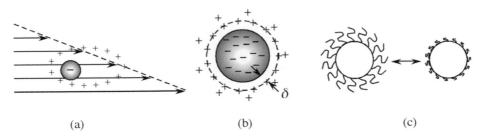

(a)	(b)	(c)

Fig. 8-22: The three electroviscous effects. (a) Distortion (relaxation) of the diffuse double layer in a shear field; (b) Increase in the effective hydrodynamic radius of a particle due to the presence of its double layer, and (c) change in adlayer configuration due to changes in electrolyte conditions.

---

[23] Hunter, R. J., **Zeta Potential in Colloid Science**, pp. 179-204, Academic Press, New York, 1981.

[24] Conway, B. E., and Dobry-Duclaux, A., in **Rheology Theory and Applications**, Vol. 3, F. Eirich (Ed.), p. 83ff, Academic Press, New York, 1960.

[25] Booth, F., *Proc. Roy. Soc.*, **203A**, 533 (1050).

where k is the specific conductivity of the continuous phase. The function $Z(\kappa a)$ was evaluated for large κa (> 100) as: $(3/2\pi)/(\kappa a)^4$ and for small κa (< 10) as: $(200\pi\ \kappa a)^{-1} + (11\kappa a/3200\pi)$. Booth's result was also limited to low Peclet Numbers (defined earlier), but was extended somewhat to larger Pe by Russel[26] as long as $Pe << \kappa a$, yielding:

$$[\mu] = 2.5\left\{1 + \frac{6\pi(\varepsilon\varepsilon_0\zeta)^2}{k\mu_0 a^2}\frac{1}{1+Pe^2}\right\}. \tag{8.35}$$

The *secondary* electroviscous effect refers to the increase in effective hydrodynamic radius due to the thickness of the diffuse double layer. This is usually of greater importance than the "primary" effect. It is an over-simplification to assume the increase in effective radius is just κ^{-1}. The effective thickness must consider the effects of Brownian motion and shearing. For low shear rates ($Pe << 1$), Russel[27] derived the following expression for the effective collision diameter of spherical particles, $2a_{eff}$, for arbitrary values of ζ and κa:

$$2a_{eff} = \kappa^{-1}\ln\left[\frac{\alpha}{\ln(\alpha/\ln\alpha)}\right], \tag{8.36}$$

where α is a dimensionless parameter representing the ratio of electrostatic forces to "Brownian forces," *viz.*

$$\alpha = \frac{4\pi\varepsilon\varepsilon_0\zeta^2 a^2\kappa}{kT}\exp(2\kappa a). \tag{8.37}$$

Russel's expression for the reduced viscosity of the dispersion was:

$$\mu_{red} = [\mu]_B\left(1+\frac{2}{5}\phi\right) + \frac{3}{40}\ln\left(\frac{\alpha}{\ln\alpha}\right)\left[\ln\frac{\alpha}{\ln(\alpha/\ln\alpha)}\right]^4\frac{\phi}{(\kappa a)^5}, \tag{8.38}$$

where $[\mu]_B$ is the intrinsic viscosity obtained by Booth for the primary electroviscous effect.

An interesting experiment highlighting the secondary electroviscous effect examines the effect of counterion molarity on an electrostatically stabilized dispersion, as shown for a dense ($\phi = 0.456$) dispersion of 0.139 μm diameter polystyrene/vinyl-toluene latices, as shown in Fig. 8-23. At first the viscosity decreases due to the shrinkage of the diffuse double layer, but eventually flocculation is so strongly enhanced that viscosity again rises.

---

[26] Russel, W. B., *J. Fluid Mech.*, **85**, 673 (1978).
[27] Russel, W. B., *J. Fluid Mech.*, **85**, 209 (1978).

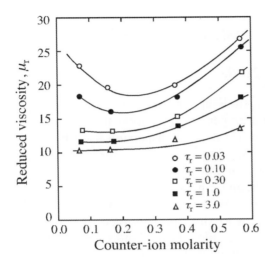

Fig. 8-23: Reduced viscosity of a dense ($\phi = 0.456$) dispersion of 0.139 μm diameter latex in water as a function of counterion molarity for different reduced shear rates,

$$\tau_r = \tau a^3 / kT$$

After [Woods, M.F., and Krieger, I.M., *J. Colloid Interface Sci.*, **34**, 91 (1970).]

Finally, the *tertiary* electroviscous effect is due to the expansion or contraction of the adlayers, often polyelectrolyte or ionizable polymers, upon changes in the ionic environment or *pH* of the medium. The effect can be very large in specific situations, and is an important factor in many coating formulations.

Some fun things to do:
Experiments and demonstrations for Chapter 8

1. *Dilatant behavior in a bottle*

When very dense dispersions are sheared, the particles are forced to move out their close-packed arrangement to support the shear deformation, as shown in Fig. 8-14. This means that the dispersion itself becomes stiffer as it is sheared, and the dispersing liquid flows into the expanded structure. This is observed as one walks on wet beach sand. The sand is quite rigid when you first take a step on it, and it noticeably whitens around the edge of your foot as water is drawn inward. If you stand on the sand long enough, however, you will slowly begin to sink in. This type of behavior can be observed with wet sand in a flexible plastic bottle. When the bottle is squeezed, the sand inside it is sheared and made to expand its volume, while the total volume inside the bottle is essentially constant. The contents of the bottle will become rigid, and water will be sucked into the interior of the sand.

Materials:

- Flexible plastic bottle (250 mL is a good size)
- Enough sand to fill the bottle
- ¼ inch glass tube, 6 inches in length

Procedure:

1) First wet the sand in a beaker and then fill the bottle to about 90% of its volume. Add enough water to reach a level about 2 mm above the level of the sand. First squeeze the bottle while looking at the sand surface through the opening, and notice that it becomes dry.

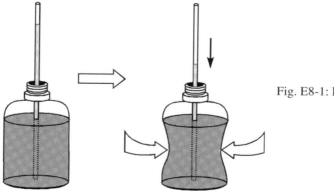

Fig. E8-1: Dilatancy bottle.

2) Inset the glass tube into the sand nearly to bottom of the vessel, and add water to the tube so that the meniscus stands 1-2 inches above the top of the bottle. Now squeeze the bottle and notice

the sudden drop in the water level, a counter-intuitive result, unless ones knows about dilatancy.

2. Yield stress of a colloidal gel

The colloidal gel formed by the percolation of Laponite clay described in Experiment 2 after Chapter 7 is an example of a viscoplastic material, *i.e.,* it possesses a yield stress, beyond which it is a shear-thinning liquid. The yield stress may be determined by measuring the shear stress required to shear a layer of the gel of given thickness between flat plates. This can be done by applying varying gravitational stress to a pair of tilted plates confining the gel layer, as shown in Fig. E8-2. When the angle of inclination is β, the shear stress on the gel layer is given by

$$\text{Shear stress} = \frac{\rho w g}{\delta}\sin\beta,$$

where w is the upper plate thickness, ρ its density, and δ is the gel layer thickness. A common problem that arises when one attempts to measure the rheological properties of a gel is that the surface of the tool in contact with the gel may slide on a thin film of liquid without engaging the gel itself. This may be addressed in many cases by roughening the surface of the tool.

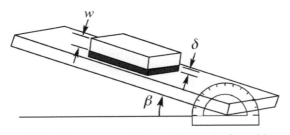

Fig. E8-2: Measurement of the yield stress of a gel layer.

Materials:
- Laponite® RD powder as described in Experiment 1, Chap. 7.
- 50-mL beaker and stirring rod
- Two aluminum plates, 2×6×1/4 inch, one with a smooth surface and the other roughed with coarse-grit sand blasting.
- Two aluminum plates, 1.5×1.5×0.25 inch, one with a smooth surface and the other roughed with coarse-grit sand blasting.
- Protractor
- Two thin metal plates (6×1/4×1/64 in.)
- Teflon tape or Parafilm
- Scalpel or razor blade

Procedure:

1) Prepare two frames for casting the gel samples as follows. Bend the thin metal strips around one of the 1.5×1.5×0.25-in. aluminum blocks and hold the frames together using Teflon tape or Parafilm®.

2) Place one of the frames on the center of the smooth surface of the 2×6×1/4-in. aluminum plate and the other at center of the large plate with the roughened surface.

3) Mix 1.4 g of Laponite® powder in 40 mL of water in a 50-mL beaker and immediately pour half of it into each of the frames, as the dispersion will gel within one minute.

4) Wait 15 minutes and then carefully cut the gel from the frames using a scalpel or razor blade and remove the frames.

5) Place the 1.5×1.5×0.25-in. aluminum blocks on top of the gel samples, smooth facing smooth and roughened facing roughened. Measure the thickness of the gel specimens.

6) For the system with smooth-surfaced plates, begin slowly tilting the lower plate, as shown in Fig. E8-2, until the top plate begins to slide. Use the protractor to measure the angle. Notice that the top plate slides over the top surface of the gel deforming the sample only very slightly.

7) Repeat the above step using the roughened plates, and note that the angle of inclination required for movement is greater, and that during the motion of the top plate the entire gel specimen is being deformed. Calculate the yield stress of the gel.

Chapter 9

EMULSIONS AND FOAMS

A. General consideration of emulsions

1. *Classification of emulsions*

Emulsions and foams are dispersions (often colloidal) of one fluid phase in another, stabilized by the presence of strong surfactant adsorption (or its equivalent) at the fluid interface. Emulsions refer to liquids dispersed in liquids, while foams refer to gases dispersed in liquids. The dependence of the boundary tension on the adsorption of the emulsifier or foaming agent, and its dependence, in turn, on bulk composition, temperature, *etc.*, plays a key role in the hydrodynamics of the coalescence phenomenon that determines the survivability of emulsions and foams.

In describing emulsions, one must first distinguish between *macro*emulsions and *micro*emulsions. Microemulsions have been discussed briefly in Chap. 3, and are best regarded as ultra-swollen micelles, with diameters of the order of 5 - 50 nm or more. They are thus thermo-dynamically stable solutions, and are discussed later in this chapter. Our immediate concern is with macroemulsions, which shall henceforth be referred to simply as emulsions. The droplets are usually at least an order of magnitude larger than in microemulsions (usually with a high degree of polydispersity) and constitute a separate phase within the dispersion medium. They are thus lyophobic colloids, and although such systems may have a high degree of metastability, they are not thermodynamically stable. Most emulsions consist either of some water-insoluble organic liquid (known generically as "oil") dispersed in water (O/W) or alternatively, an aqueous phase dispersed in an oil (W/O), as shown in Fig. 9-1. The volume

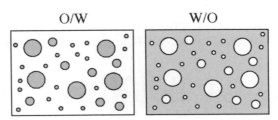

Fig. 9-1: Types of emulsions: Oil-in-water (O/W) or water-in-oil (W/O).

fraction of the dispersed phase can vary from a few ppm up to very high values. While the maximum packing density of equal-sized spheres is $\approx$

74%, polydispersity allows for greater values. In some cases, the liquid drops (owing to their deformability) form closely packed polyhedra with a volume fraction exceeding 95%. These are referred to as "biliquid foams."

Emulsions are potentially unstable with respect to all the processes described in Chap. 5 for lyophobic colloids: phase segregation (sedimentation or creaming due to gravity), flocculation (clustering together of emulsion droplets) and coalescence (the merging of droplets into larger droplets), and drop size disproportionation. The distinguishing process by which an emulsion "breaks," however, is that of coalescence, in which adjacent droplets merge to form larger droplets, as shown in Fig. 9-2, until a

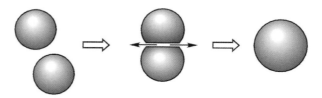

Fig, 9-2: The breakdown of emulsions through flocculation and coalescence.

system of two self-contiguous bulk liquid phases is formed. The coalescence is preceded by flocculation, in which the drops are drawn together by net long-range attraction, although they may be mechanically brought, by gravity (sedimentation or creaming), filtration, *etc.*, into sufficient proximity to coalesce. The liquid interface between adjacent droplets is flattened into opposing circles separated by a thinning film of the external phase liquid. The liquid film must drain to some "critical thickness," h_c, found to be of the order of 50 - 100 nm, following which it may remain for a finite period of time ("resting" time). It then suddenly ruptures. In the most common cases, the coalescence time may be equated to the drainage time, but in other cases, long resting times are achieved, as drainage is halted before the critical film thickness is reached.

2. Emulsifiers and emulsion stability

If an emulsion is to show any significant longevity, it must contain a third constituent (in addition to oil and water) known as an emulsifier or emulsifying agent. Three principal classes of such agents can be distinguished, as pictured in Fig. 9-3. The first important class of emulsifiers

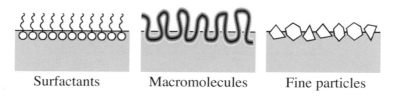

Surfactants   Macromolecules   Fine particles

Fig. 9-3: Types of emulsifiers.

is that of surfactants, which are strongly adsorbed and oriented at the interface. They play both mechanical and physicochemical roles in emulsion stabilization, as described below. Other emulsifiers are macromolecular substances. These might include proteins, gums, starches and derivatives from such substances (such as dextrin, methyl cellulose, lignosulfonates, *etc.*) as well as certain synthetic polymers or polyelectrolytes. These materials are also strongly adsorbed at the interface and confer stability primarily through "steric" and mechanical effects. Finally, finely divided solids may act as emulsifiers, yielding "Pickering emulsions," as discussed in Chap. 4.J.4. The particles cling to the interface because neither liquid will totally wet them out relative to the other, and they act to hold the approaching droplet surfaces apart. To the above, a fourth type of emulsifier is sometimes added, *viz.*, certain inorganic anions which adsorb to the interface in sufficient quantity to confer some electrostatic stabilization. Common among these are the thiocyanate ions: CNS^-.

The first two types of emulsifiers assist to some extent in the *formation* of emulsions in that they reduce interfacial tension, hence the thermodynamic work ($\sigma\Delta A$) required, as discussed further below, but this is usually a small fraction of the mechanical work (*e.g.*, jetting, mixing, sonication, *etc.*) actually required to form the emulsion. The principal mechanism by which these emulsifiers act to preserve the emulsion is by impeding the re-coalescence of the droplets once they have been produced by mechanical means. Surfactant films resist both shear and dilational distortion, as described in Chap. 3.K.2. The resistance to dilational distortion (Gibbs elasticity) makes the surfaces of the draining film hydrodynamically rigid, reducing the rate of drainage. The assumption of hydrodynamic rigidity is made in the derivation of the Reynolds drainage equation, Eq. (5.11). Equally important, the elastic-like properties stabilize the films against the effect of macroscopic mechanical disturbances otherwise leading to premature rupture, as pictured in Fig. 9-4. As the film seeks to thin itself,

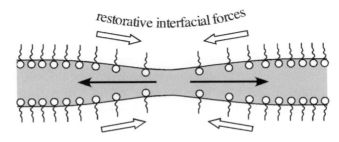

Fig. 9-4: The mechanism of Gibbs elasticity stabilizing the film between emulsion droplets.

it locally dilates, depleting surfactant from the region of dilation, producing σ-gradients that oppose the process (Gibbs elasticity), so that the thin spot re-thickens. A thermodynamic expression for the Gibbs elasticity of an interface, E_G may be given by:

$$E_G = \frac{d\sigma}{d\ln A_2} = -\frac{d\sigma}{d\ln \Gamma_{2,1}}, \tag{9.1}$$

where A_2 is the specific area of the surfactant, and $\Gamma_{2,1}$ its adsorption. Gibbs "elasticity" is thus obtained by differentiating the interfacial equation of state (the σ - A relationship), but an evaluation of the *effective* elasticity must also take into account the dynamics of the restoration of the surfactant to the dilated interface, which in the simplest case is controlled by diffusion from the bulk liquid to the film. Correlations between emulsion stability and interfacial shear viscosity exist as well.

The early stages of the drainage-coalescence event depend on three factors: 1) the magnitude of the energy of droplet impact (influenced by external stirring and temperature), 2) the droplet-droplet interaction potential function, Φ, or more appropriately the disjoining pressure isotherm $\Pi(h)$, and 3) the hydrodynamics of the drainage of the inter-droplet film. If $\Pi(h)$ is such that coalescence is prevented completely (as in the case of strong steric stabilization by polymers or macromolecules) or if the energy barrier in $\Pi(h)$ is so high that the flocculation rate even under the influence of vigorous stirring is insignificant, the emulsion may be considered "stable," or at least metastable, until either external physical changes occur, or the passage of time changes the system's chemistry.

In the study of coalescence, investigators for many years have made direct observations of isolated films in air and under another liquid and have also looked at the coalescence of macroscopic drops at phase interfaces. These all have the disadvantage of presenting much larger areas of contact than those corresponding to the contact surfaces between colloidal droplets, leading to the importance of macroscopic irregularities in the drainage process, factors that would not be important in the smaller systems. The time scale for the drainage is also grossly larger than that for the systems it seeks to model. The length of time of typical Brownian interaction is thought to be only $\approx 10^{-5}$ s. Nonetheless some important patterns of behavior have emerged from such studies. The drainage sequence usually follows one of the following courses:

1) The film drains uniformly in accord with expected hydrodynamics down to a reasonably reproducible critical rupture thickness, h_c, at which point it ruptures. Emulsions forming inter-particle films with such behavior are clearly unstable with respect to coalescence. The model proposed by Reynolds[1] in 1886 was based on lubrication theory, *i.e.*, the set of simplifying assumptions used in formulating the momentum equation for viscous flow in thin layers. The flow velocity component and all property gradients in the direction normal to the layer are neglected, together with the inertial terms (giving creeping flow). For steady radial flow in a horizontal layer, this leads to:

---

[1] Reynolds, O., *Phil. Trans. Roy. Soc.*, **177A**, 175 (1886).

$$0 = -\left(\frac{dp}{dr}\right) + \mu\left[\frac{d}{dr}\left(\frac{1}{r}(rv_r)\right)\right]. \tag{9.2}$$

The assumptions that the drainage surfaces are flat and immobile produces reasonable prediction of drainage rates in accord with:

$$t_{\text{drain}} = \frac{3\mu A}{4\pi p}\left[\frac{1}{h_c^2}\right], \tag{9.3}$$

where t_{drain} is the time required for the film to drain to a thickness of h_c under the influence of a pressure p pushing the drops together over an area of contact, A; μ is the viscosity of the liquid. In the context of the colloidal droplet problem, both the pressure, p, and the area of contact, A, can be computed in terms of the droplet size and the potential energy function, Φ. In the macroscopic cases studied, the origin of the pressure, p, is usually the buoyancy and the suction at the Plateau borders of the film, as discussed later in the section on foams, and possibly some external applied force. The area of contact is usually a constant both in the colloidal drop case and the macroscopic case, but the pressure, p, for the drainage between colloidal droplets is the negative of the disjoining pressure. Recall from Chap. 2.K.1 that the disjoining pressure, Π, is the net inter-droplet force per unit area acting to push the droplet surfaces apart,

$$\Pi = -\left(\frac{d\Phi}{dh}\right)A^{-1}, \tag{9.4}$$

which is clearly dependent on h. One might try to remedy this situation by differentiating to get:

$$-\frac{dh}{dt} = -\left(\frac{4\pi\Pi}{3\mu A}\right)h^3, \tag{9.5}$$

which, in a formal sense is not correct because one cannot in general recover the unsteady state hydrodynamic behavior of a system by time-differentiating a steady-state solution. Nonetheless, critical thicknesses have been determined experimentally to fall in the range between 10 and 50 nm for this type of system, and drainage rates have been found to vary as the cube of the film thickness. Coalescence by this mechanism might be termed "fast coalescence," and (except when the medium is highly viscous) emulsions coalescing by this pattern must be regarded as quite unstable.

2) Another pattern of coalescence behavior occurs when the film drains as described above, but reaches a certain "equilibrium" thickness at which drainage stops. These "equilibrium" thicknesses are reproducible for a given system and lie in the same range as the critical thicknesses described above. They correspond to situations where the driving force for drainage, $-\Pi$, goes to zero due to a balance between attractive (van der Waals) forces and electrostatic (and/or) steric repulsion, and the derivative of the disjoining

pressure isotherm, $d\Pi/dh$, is negative to assure stability. The "equilibrium" film, especially for droplets 1 μm in diameter or larger corresponds to thick thin films of Type IV disjoining pressure curves in Fig. 2-69. They also correspond to the secondary minimum in the Φ curve. These metastable films may exist for time varying from a few seconds to several hours, but they will eventually either rupture or decompose into another structure when a disturbance of sufficient amplitude is encountered. Several investigators have formulated hydrodynamic stability analyses of such films, but none has yet been able to achieve quantitative agreement with experiment. Films which rupture after their relatively brief resting time at "equilibrium" thickness must still be regarded as corresponding to relatively unstable emulsions.

3) A third coalescence pattern emerges when the breakdown of the equilibrium film (as the result of a small but finite disturbance) results not in rupture but rather in a patch of very thin but coherent film known as a "black film" or a "Perrin black film," more often in the context of foam films than in those between emulsion droplets. It is so named because it is so much thinner than the wavelength of ordinary light that it is invisible to direct observations. These films may actually coincide with the thinnest of the "equilibrium" films described above and correspond to the thin thin films of Type III disjoining pressure curves in Fig. 2-69. Called also "coalescence stable" films,[2] they are usually of the order of 60 nm in thickness, and often represent a balance between electrostatic and van der Waals forces. Their thickness is just slightly in excess of twice the thickness of the surfactant layers adsorbed on the surfaces. The formation and thickness of such a film is critically dependent upon surfactant type and content and upon electrolyte content.

Metastable structures leading to long-term stability are more often produced by macromolecular or polymeric stabilizers, which form a gel in the film, bringing drainage to a halt. A similar effect may be produced by appropriate finely divided solids. Their entrapment between particles may prevent the film from draining to a thinness that is capable of leading to rupture by ordinary disturbances.

The most direct way of examining the rate of emulsion coalescence is to obtain direct observations of the time rate of change of droplet concentration and size distribution. Such studies have revealed that in a floccule of droplets, the coalescence of one of the droplets has no effect on the other droplets (process is first-order) and that plots of the log of the total number of droplets in a given volume, N, becomes linear *vs.* time. The slopes of such log N *vs.* t plots reveal rate constants varying from 10^{-3} s^{-1} or higher (for "rapid coalescence") down to 10^{-7} s for more "stable" films.

---

[2] Sonntag, H., and Strenge, K., **Coagulation and Stability of Disperse Systems**, Halsted Press, New York, 1964.

The mechanism of particle size disproportionation for the breakdown of emulsions may also play a role in some cases. If the disperse phase, or a component of it, is sufficiently soluble in the medium, the small droplets will decrease in size at the expense of the large ones (by the Kelvin effect) due to the enhanced solubility of the material in the smaller drops. Higuchi and Misra[3] determined that the solubility of n-decane in water (50 μg/L) is sufficient to cause the disappearance of micron and submicron droplets (relative to larger ones) of decane in water over reasonable periods of time. It has more recently been shown that the addition of even small amounts of insoluble, larger-chain hydrocarbons to such drops markedly improves their stability by reducing the equilibrium fugacity of the shorter chain compound in water.[4]

3. Thermodynamics of emulsification/breakdown

The formation and breakdown of an O/W emulsion is pictured schematically in Fig. 9-5. The free energy of a system consisting of a certain volume of oil and a certain volume of water is seen to be higher in the emulsified state than in the state of two bulk liquids, but the amount of

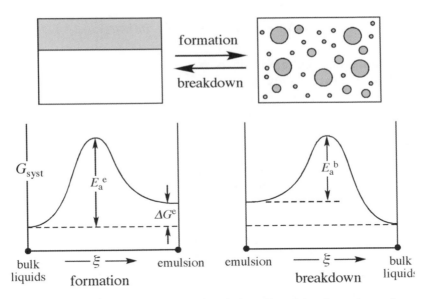

Fig. 9-5: Simplified thermodynamic description of emulsion formation and breakdown.

energy (in the form of an "activation energy") required to produce the emulsion is often many times greater than the indicated thermodynamic

[3] Higuchi, W. I., and Misra, J., *J. Pharm. Sci.*, **51**, 459 (1962).
[4] Ugelstad, J., *Makromol. Chem.*, **179**, 815 (1978);
   Ugelstad, J., Kaggerud, K. H., Hansen, F.K., and Berge, A., *Makromol. Chem.*, **180**, 737 (1979);
   Ugelstad, J., Mørk, P. C., Kaggerud, K. H., T. Ellingsen, T., and Berge, A., *Adv. Colloid Interface Sci.*, **13**, 101 (1980).

minimum, *i.e.*, $E_a^e \gg \Delta G^e$. Thus considerable mechanical energy, as described briefly below, is generally required to form the emulsion.

The *thermodynamic* requirement, ΔG^e, *i.e.*, the free energy change associated with the process whereby a mass of liquid (say oil) is broken up into N droplets and distributed throughout the volume of its emulsion in water is the sum of the free energy of increasing the interfacial area ($\sigma \Delta A$) and that associated with the increased *configurational* entropy of the oil droplets. The latter is given by:[5]

$$\Delta S^{\text{config}} = -Nk\left[\ln\varphi_0 + \left(\frac{1-\varphi_0}{\varphi_0}\right)\ln(1-\varphi_0)\right], \tag{9.6}$$

where k is Boltzmann's constant, and φ_0 is the volume fraction of the oil droplets in the emulsion. The entropy increase results from the accessibility of the oil to the entire volume of the system in the emulsified state but to only its own volume in the original un-emulsified state. Thus:

$$\Delta G^{\text{emul}} = \Delta G^{\text{area}} - T\Delta S^{\text{config}}$$

$$= \sigma \Delta A + NkT\left[\ln\varphi_0 + \left(\frac{1-\varphi_0}{\varphi_0}\right)\ln(1-\varphi_0)\right]. \tag{9.7}$$

The configurational entropy term is negative and thus subtracts from the thermodynamic free energy requirement for emulsification. For cases of moderate-to-high interfacial tension, the entropy term is usually negligible, but for sufficiently low σ, one may have

$$\Delta G^{\text{emul}} \leq 0, \tag{9.8}$$

leading to the possibility of "spontaneous emulsification." Putting $\Delta A = 4\pi a^2 N$, and considering the case of a given volume ratio of oil to water, and a given drop radius a, one gets

$$\sigma^{\text{crit}} = -\frac{kT}{4\pi a^2}\left[\ln\varphi_0 + \left(\frac{1-\varphi_0}{\varphi_0}\right)\ln(1-\varphi_0)\right]. \tag{9.9}$$

For $\varphi_0 = 0.5$, $T = 298$K and $a = 100$ nm, $\sigma^{\text{crit}} \approx 5 \times 10^{-5}$ mN/m. Such low interfacial tensions are sometimes achievable under equilibrium conditions by surfactants for which the critical packing parameter, CPP (Chap. 3.I.1), is near unity. This often leads to the formation of *microemulsions*, as described later in this chapter. Ultra-low interfacial tensions may also be obtained under transient conditions during the transfer of surfactants across the oil/water interface,[6] leading to possible spontaneous emulsification.

---

[5] Tadros, T. F., and Vincent, B., in **Encyclopedia of Emulsion Technology**, Vol. 1, P. Becher (Ed.), pp. 129-285, Marcel Dekker, New York, 1983.
[6] England, D. C., and Berg, J. C., *AIChE J.*, **17**, 313 (1970).

Figure 9-5 also suggests that while thermodynamics favors the spontaneous breakdown of emulsions to the parent bulk phases, the process is generally impeded by an effective "activation energy" barrier. The latter is symbolic of the time effects associated with the coalescence process. Breakdown as well as formation may be assisted through the addition of mechanical energy.

4. Preparation of emulsions

As indicated above, the preparation of an emulsion requires energy input, as suggested by Fig. 9-6. Sometimes simple continuous or intermittent shaking of the component liquids is sufficient, while in other cases, high energy input by way of a number one of a number of different devices is required. The process involves the breaking of the liquids into large drops at first through instabilities described in Chap. 5 and summarized in Fig. 9-6.

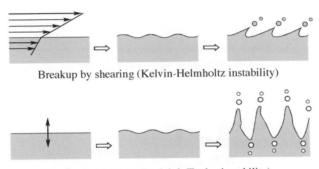

Breakup by shearing (Kelvin-Helmholtz instability)

Breakup by shaking (Rayleigh-Taylor instability)

Fig. 9-6: Hydrodynamic mechanisms of emulsification.

Additional energy input breaks these droplets down further by imparting shear to them, usually in the turbulent regime. The ultimate drop size distribution achieved depends in a complicated manner on many factors, such as the emulsifier type and concentration, the duration of energy input, the configuration of the device, *etc*. Nonetheless, for a given set of materials, hydrodynamic analysis, dimensional analysis and experimentation have led to useful empirical equations relating the maximum drop size to the power input per unit volume, the interfacial tension, and the density of the continuous phase. For simple mixers and ultrasonicators (which make use of cavitation), one finds relationships of the type[7]:

$$d_{max} \propto \tilde{\Pi}^{-2/5}\sigma^{3/5}\rho_c^{-1/5}, \tag{9.10}$$

where $\tilde{\Pi}$ is the power input/volume. The drop size is seen to decrease with the intensity of power input, and with decreasing interfacial tension. The dependence on continuous phase density is seen to be weak, and on

[7] Walstra, P., in **Encyclopedia of Emulsion Technology**, Vol. 1, P. Becher (Ed.), pp. 57-128, Marcel Dekker, New York, 1983.

dispersed phase density and on any system viscosity is negligible. A nearly identical expression is found for homogenizers and colloid mills.

B. O/W or W/O emulsions?

1. *Rules of thumb*

The formation and stabilization of emulsions is seen to be complex, and it would be unreasonable to expect to describe all the phenomena involved in terms of one or two simple "laws." One of the first problems is to determine which phase will be dispersed, as pictured in Fig. 9-7. Practitioners, faced with the problem of preparing and breaking or inverting emulsions of many different types, have developed some highly simplistic models and "rules of thumb" concerning emulsions. An apothecary's handbook dated approximately a century ago warned the chemist to hire only right-handed assistants for preparing emulsions, as the stirring of left-handed persons often led to poor results!

As suggested by Fig. 9-7, when sufficient energy is put into the system, emulsion droplets of *both* types are formed, but in the usual case, the

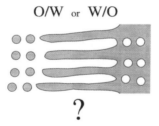

O/W or W/O

?

Fig. 9-7: Which type of emulsion forms depends on which type coalesces faster.

droplets forming one type of emulsion (O/W or W/O) will coalesce much faster than those of the opposite type. Which type is favored for survival is determined largely by the emulsifier used. For either type of emulsion to survive, the emulsifier must concentrate itself at the oil/water interface, but the emulsion type favored is the one in which the continuous phase has the stronger affinity for the emulsifier. Most of the "rules of thumb" for deciding which type of emulsion will be formed are variations of the above statement. One of the earliest of these was the *Bancroft Rule*.[8] In its simplest form, this rule states that the phase in which the emulsifying agent is the *more soluble* will become the continuous phase when the emulsion is formed. A rationale for the rule may be found by arguing, as pictured in Fig. 9-8, that under these conditions, the bulk of the emulsifier molecule at the interface must lie in the continuous phase to satisfy the curvature requirements for the emulsion droplet. A more direct application of this idea is found in the so-called *oriented wedge rule*. It requires the portion of the adsorbed emulsifier molecule that is larger (either the hydrophilic or the lipophilic part) to reside

[8] Bancroft, W. D., *J. Phys. Chem.*, **17**, 501 (1913);
  Bancroft, W. D., *J. Phys. Chem.*, **19**, 275 (1915).

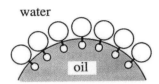
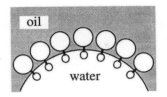

Fig. 9-8: The driving force for coalescence is related to curvature.

in the continuous phase. The classic example of this rule is the action of univalent and divalent metal soaps in stabilizing emulsions, as exemplified in Fig 9-9. A monovalent soap like sodium oleate will yield O/W emulsions because the larger-sized portion of the adsorbed molecule is the hydrophilic

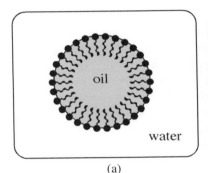
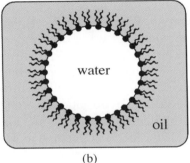

(a)   (b)

Fig. 9-9: Oriented wedge "theory." (a) Stabilization of an O/W emulsion by a monovalent metal soap and (b) stabilization of a W/O emulsion by a divalent metal soap.

part, the hydrated anionic group. Soaps of divalent metals like calcium or magnesium, however, form W/O emulsions since the divalent salt is essentially undissociated in water and the double hydrocarbon chain is larger than the head group. The oriented wedge idea also extends to explain which type of emulsion will be formed when they are stabilized by small, suspended particles (Pickering emulsions). The phase that preferentially wets the solid particles will be the external phase, as illustrated in Fig. 9-10.[9]

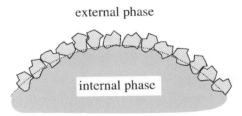

Fig. 9-10: A "Pickering emulsion:" oil droplets stabilized by a coating of particles preferentially wet by the external phase.

[9] Kitchner, J. A., and Musselwhite, P. R., in **Emulsion Science**, P. Sherman (Ed.), Academic Press, London, 1968.

Particles to be used as emulsifiers, or in combination with other emulsifiers, are often surface-treated by methods described in Chap. 4 to achieve the desired wettability preference.

Another way of stating the above rule uses the critical packing parameter (*CPP*), defined in Chap. 3 in connection with the spontaneous self-assembly of various nanostructures (micelles, vesicles, *etc.*) in surfactant solutions. Recall

$$CPP = \frac{v}{a_0 l_c},$$ (9.11)

where v and l_c are the molecular volume and the contour length of the hydrophobic chain (obtainable as the appropriate van der Waals dimensions), and a_0 is the effective cross-sectional area of the hydrophilic head group. When the *CPP* < 1, the oil will seek to be on the concave side of the interface, and when the *CPP* > 1, the water will seek to be the concave side. This criterion provides an explanation for the effects of changing salinity, *pH*, temperature, *etc.* on the type of emulsion formed. For example, an increase in ionic strength will reduce the effective area of an ionic head group, due to electrostatic screening of its charge that allows closer packing at the interface. For the case of a nonionic head group such as a polyethylene oxide chain, increasing temperature progressively reduces a_0 by dehydrating the group. The *pH* is important in many cases because the head group may ionize above or below a certain *pH* value, leading to a substantial increase in a_0. Progressive changes in any of these variables, as produced by titration, heating, *etc.*, may lead to an inversion of the original emulsion or more often to its breaking. Although a typical macro emulsion droplet size is far larger, relative to the emulsifier molecule size than suggested in Figs. 9-8 – 9-10, the interfacial curvature is evidently decisive.

Other factors being equal, the dispersed phase should tend to be the one present in the least volumetric amount. Also the manner and particularly the order of phase addition may be crucial in determining the emulsion type and properties. When one wants to form, for example, an O/W emulsion, one generally adds the oil *to* the water phase (not vice versa) in small increments with vigorous mixing between each addition. The emulsifier is generally present initially in the continuous phase or formed *in situ*.

2. *The hydrophile-lipophile balance (HLB) and related scales*

A useful parameter for quantifying the characteristics of emulsifiers is the *Hydrophile-Lipophile Balance* (*HLB*), developed by Griffen[10] and referring to the balance between the hydrophilic and lipophilic portions of the emulsifier molecule. An emulsifier that is hydrophilic in character (high water and low oil affinity) has a high *HLB*, whereas lipophilic compounds have low *HLB* values. For nonionic (polyethylene oxide) surfactants, the

---

[10] Griffen, W. C., *J. Soc. Cosmetic Chemists*, **1**, 311 (1949).

HLB is defined as the weight percent of the molecule that is hydrophilic (usually just the percent of the molecular weight made up of the polyethoxylate chain) divided by 5. The values should thus range from 0 to 20 with the midpoint (equal solubility) at 10. When two or more emulsifiers are blended, the *HLB* value for the mixture is taken to be the *mass* average of the individual *HLB*'s. The Spans® and Tweens® (surfactants based on cyclic ethers of mannitol, polyethoxylated to various degrees), whose structures are shown in Fig. 9-11, provided the original basis for establishing

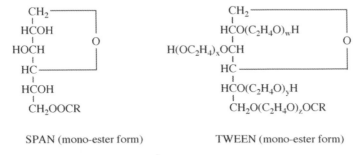

SPAN (mono-ester form)   TWEEN (mono-ester form)

Fig. 9-11: SPAN's and TWEEN's used to establish the *HLB* scale.

the scale, a more detailed picture of which is shown in Fig. 9-12. High values of *HLB* favor the formation of O/W emulsions, while low values favor W/O emulsions. Some *HLB*-values for various surfactants are shown in Table 9-1, and an empirical rule for estimating the *HLB* in terms of functional group contributions is shown in Table 9-2. Many of these values were obtained by studying the behavior of surfactants in admixture with the

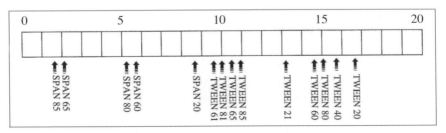

HLB Range	Use
4 - 6	W/O Emulsifiers
7 - 9	Wetting Agents
8 - 18	O/W Emulsifiers
13 - 15	Detergents
10 - 18	Solubilizers

Fig. 9-12: The *HLB* scale and some of its uses in guiding the selection of surfactants for given end use applications.

original SPAN-TWEEN scale. When ionic surfactants are used, their effectiveness is such that their indicated *HLB* numbers may reach as high as 40.

Table 9-1: *HLB* values for various surfactants and oils		
Surfactants		*HLB*
Oleic acid		1.0
Span 65 (Sorbitan tristearate)		2.1
Span 80 (Sorbitan mono-oleate)		4.3
Span 40 (Sorbitan monopalmitate)		6.7
Span 20 (Sorbitan monolaurate)		8.6
Tween 81 (PEO sorbitan mono-oleate)		10.0
Igepal Ca-630 (EO alkyl phenol)		12.8
Tween 80 (PEO sorbitan mono-oleate)		15.0
Sodium oleate		18.0
Oils	HLB (O/W)	HLB (W/O)
Cottonseed oil	7.5	-
Paraffin (household)	9.0	4.0
Mineral oil (Marcol GX)	10.0	4.0
Decanol	14.0	-
Benzene	15.0	-
Stearic acid	17.0	-

Table 9-2: *HLB* Group Numbers: $HLB = 7.0 + \Sigma(\text{Group Numbers})$	
Hydrophilic Group Numbers	
-SO_4Na	38.7
-COOK	21.1
-COONa	19.1
-N (tertiary amine)	9.4
Ester(sorbitan ring)	6.8
Ester (free)	2.4
-COOH	2.1
-OH(free)	1.9
-OH(free)	1.3
-O-	0.5
-OH (sorbitan ring)	38.7
Lipophilic Group Numbers	
-CH	-0.475
-CH_2-	-0.475
CH_3-	-0.475
CH-	-0.475
Derived Group Numbers	
-$(CH_2$-CH_2-O)-	0.33
-$(CH_2$-CH_2-CH_2-O)-	-0.15

The use of the *HLB* method in guiding the selection of emulsifiers is as follows. *HLB*'s are determined for a pair of different surfactants, and their ability, in different proportions, to emulsify a given oil in water is determined, as pictured in Fig. 9-13, by measuring the rate of droplet coalescence, $-d\ln N/dt$. *N* is the total number of emulsion droplets per unit volume viewed microscopically. The proportion yielding the most stable O/W emulsion (*i.e.*, the lowest coalescence rate) should give, in accord with the above mixing rule, an *HLB* value *for the oil*. Optimum *HLB*-values can be determined for the formation of both O/W and W/O emulsions with the given oil. Other surfactants can then be tested as emulsifiers for the same oil in a mixture with one or the other of the original pair. *HLB*'s for different oils could thus be obtained. Using extensive testing of this type over a period of years, Griffen and later others built up tables of *HLB*-values for many surfactants and for a number of different oils.

While the use of *HLB*'s to quantify emulsification is quite crude, it is useful as a starting point for more detailed testing. As a practical matter, it is believed that using two emulsifiers to produce a given *HLB* generally produces a more stable emulsion than a single emulsifier of the same *HLB*.

Combinations of oil-soluble and water-soluble agents, which are generally assumed to form interfacial complexes of the type shown in Fig. 9-14, have been shown to produce robust emulsions.

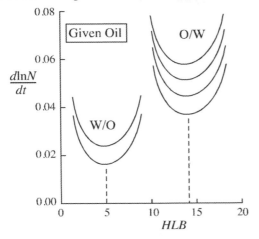

Fig. 9-13: Rate of coalescence *vs. HLB*, used for determination of the *HLB* values of a given oil.

As noted in Fig. 9-12, other surfactant properties (such as wetting, detergency, solubilization, *etc.*) also appear to correlate reasonably well with the *HLB*. There has thus been considerable motivation to put the concept on a more scientific basis and to develop methods relating *HLB* directly to surfactant structure. Thousands of emulsifiers are produced commercially, and they are catalogued in McCutcheon's handbook: **Emulsifiers and**

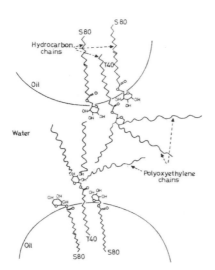

Fig. 9-14: Schematic representation of orientation of Tween 40 and Span 80 molecules in mixed films adsorbed at the oil-water interface. [Boyd, J., Parkinson, C., and Sherman, P. *J. Colloid Interface Sci.*, **41**, 359 (1972).]

Detergents, available annually in both North American and International editions, both in print and electronic forms. Trade names, formulae, *HLB*-numbers and applications are given for each material, as shown in a portion of a page from the Handbook, Fig. 9-15.

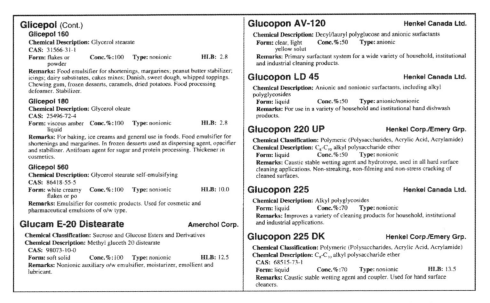

Glicepol (Cont.)

Glicepol 160
Chemical Description: Glycerol stearate
CAS: 31566-31-1
Form: flakes or   Conc.%:100   Type: nonionic   HLB: 2.8
   powder
Remarks: Food emulsifier for shortenings, margarines; peanut butter stabilizer; icings; dairy substitutes, cakes mixes; Danish, sweet dough, whipped toppings. Chewing gum, frozen desserts, caramels, dried potatoes. Food processing defoamer. Stabilizer.

Glicepol 180
Chemical Description: Glycerol oleate
CAS: 25496-72-4
Form: viscous amber   Conc.%:100   Type: nonionic   HLB: 2.8
   liquid
Remarks: For baking, ice creams and general use in foods. Food emulsifier for shortenings and margarines. In frozen desserts used as dispersing agent, opacifier and stabilizer. Antifoam agent for sugar and protein processing. Thickener in cosmetics.

Glicepol 560
Chemical Description: Glycerol stearate self-emulsifying
CAS: 86418-55-5
Form: white creamy   Conc.%:100   Type: nonionic   HLB: 10.0
   flakes or po
Remarks: Emulsifier for cosmetic products. Used for cosmetic and pharmaceutical emulsions of o/w type.

Glucam E-20 Distearate   Amerchol Corp.
Chemical Classification: Sucrose and Glucose Esters and Derivatives
Chemical Description: Methyl gluceth 20 distearate
CAS: 98073-10-0
Form: soft solid   Conc.%:100   Type: nonionic   HLB: 12.5
Remarks: Nonionic auxiliary o/w emulsifier, moisturizer, emollient and lubricant.

Glucopon AV-120   Henkel Canada Ltd.
Chemical Description: Decyl/lauryl polyglucose and anionic surfactants
Form: clear, light   Conc.%:50   Type: anionic
   yellow solut
Remarks: Primary surfactant system for a wide variety of household, institutional and industrial cleaning products.

Glucopon LD 45   Henkel Canada Ltd.
Chemical Description: Anionic and nonionic surfactants, including alkyl polyglycosides
Form: liquid   Conc.%:50   Type: anionic/nonionic
Remarks: For use in a variety of household and institutional hand dishwash products.

Glucopon 220 UP   Henkel Corp./Emery Grp.
Chemical Classification: Polymeric (Polysaccharides, Acrylic Acid, Acrylamide)
Chemical Description: C_8-C_{10} alkyl polysaccharide ether
Form: liquid   Conc.%:50   Type: nonionic
Remarks: Caustic stable wetting agent and hydrotrope, used in all hard surface cleaning applications. Non-streaking, non-filming and non-stress cracking of cleaned surfaces.

Glucopon 225   Henkel Canada Ltd.
Chemical Description: Alkyl polyglycosides
Form: liquid   Conc.%:70   Type: nonionic
Remarks: Improves a variety of cleaning products for household, institutional and industrial applications.

Glucopon 225 DK   Henkel Corp./Emery Grp.
Chemical Classification: Polymeric (Polysaccharides, Acrylic Acid, Acrylamide)
Chemical Description: C_8-C_{10} alkyl polysaccharide ether
CAS: 68515-73-1
Form: liquid   Conc.%:70   Type: nonionic   HLB: 13.5
Remarks: Caustic stable wetting agent and coupler. Used for hand surface cleaners.

Fig. 9-15: A portion of a page from McCutcheon's Handbook. From [McCutcheon Division; M C Publishing Co.; 175 Rock Road; Glen Rock, NJ 07452.]

Another useful method for classifying polyethoxylate emulsifiers was proposed by Shinoda and co-workers.[11] It attempted to address the major shortcomings of the *HLB* method, *viz.*, its inability to account for important chemical differences between different systems having the same nominal *HLB*-values and its inability to account for the very large changes in emulsifier behavior with temperature. The *HLB*-values in the literature are applicable only at room temperature (20-25°C). It was noted, for example, that an O/W emulsion prepared at room temperature may break or invert to a W/O emulsion as the temperature is raised, while a W/O emulsion prepared at a given temperature may invert to an O/W emulsion as temperature is lowered. The temperature at which such inversion occurs was defined as the *Phase Inversion Temperature (PIT)*. It was related to, but not identical with the cloud point of the surfactant as discussed in Chap. 3, *cf.* Fig. 3-25. The inversion was attributed to a decrease in the degree of hydration of the polyoxyethylene chains as temperature is increased, rendering the molecule less hydrophilic.

It was found that the *PIT* depended on more than just the particular emulsifier. For the surfactants studied, it depended also on the concentration of the surfactant, (particularly in the oil), the polarity of the oil, and the phase ratio of bulk phases. To standardize it, the *PIT* was defined as the inversion temperature observed for a particular system of chemicals when an emulsion was made with equal weights of oil and aqueous phases and 3-5% (by wt.) of surfactant. So defined, the *PIT* was an almost linear function of *HLB* for a given set of conditions, and in general

[11] Shinoda, K., and Arai, H., *J. Phys. Chem.*, **68**, 3485 (1964).

High *PIT* → high *HLB* → favors O/W emulsions
Low *PIT* → low *HLB* → favors W/O emulsions

For a surfactant with a given *HLB* value, the *PIT* increased with the polarity of the oil. Thus to keep the *PIT* the same, the *HLB* had to be decreased as the polarity of the oil decreased. This is qualitatively consistent with the notion that the *HLB* of an emulsifier system should be matched with the *HLB* for the desired dispersed phase. The *PIT* corresponding to a mixture of oils of different polarity is taken as the volume-averaged *PIT* of the components, *i.e.*,

$$PIT_{mix} = \phi_A PIT_A + \phi_B PIT_B .$$ (9.12)

It has been suggested by Shinoda and Saito[12] that O/W emulsions should be *prepared* at a temperature 2-4°C below the *PIT* (because an emulsion prepared very near the *PIT* has a very fine average particle size) and then cooled down to its storage temperature. Cooling it down increases the stability of the emulsion without increasing its average particle size. The criterion for selecting an O/W emulsifier is that it should have a *PIT* (for the system of interest) 20-60°C higher than the desired storage temperature. All of the foregoing criteria are reversed when dealing with W/O emulsions.

3. Double (or multiple) emulsions

Sometimes (and it is not particularly rare) one produces double emulsions, as pictured in Fig. 9-16: W/O/W or O/W/O.[13] These may arise in ordinary emulsification processes, as did the example shown, as the result of using a mixture of emulsifiers in which low *HLB* portions stabilize the W/O

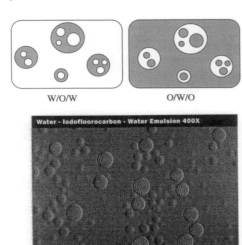

W/O/W   O/W/O

Fig. 9-16: Multiple emulsions. DIC micrograph shows W/O/W emulsions using an iodofluorocarbon oil, and a mixture of Pluronics® with an average *HLB* of 12 as the emulsifier system. The large droplets have an average diameter of ≈ 15 μm.

[12] Shinoda, K., and Saito, H., *J. Colloid Interface Sci.*, **30**, 258 (1969).
[13] Garti, N., and Bisperink, C., *Curr. Opinion Colloid Interface Sci.*, **3** (6), 657 (1998).

part and the high *HLB* components stabilize the O/W type. More often, they are made in a two-step process by first emulsifying the inner liquid in the middle liquid and subsequently emulsifying that emulsion in the outer fluid.[14] The result is usually a high degree of polydispersity with respect to both the inner and outer drop sizes. A much more controlled synthesis of multiple emulsions may be achieved using microfluidic devices to produce them a drop at a time. Weitz and coworkers,[15] for example, constructed a device in which the inner fluid was injected through a cylindrical nozzle into a stream of the middle fluid moving coaxially with it in a square cross-section glass duct, as shown in Fig. 9-17. In the most controlled mode of operation, the flow rate was such that coated drops were produced one a

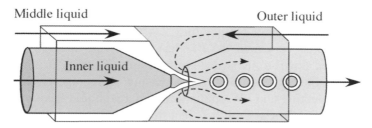

Fig. 9-17: Schematic of microfluidics device for production of W/O/W emulsions. After [Lee, D., and Weitz, D. A., *Adv. Mater.*, **20**, 3498 (2008).]

time at the tip of the nozzle. These were directed into a collection tube with the outer fluid around them. Highly monodisperse structures were obtained, and control over the size of both the inner and outer drops, and the number of inner droplets was maintained by controlling the relative flow rates of the liquids in the device. The single inner droplet configurations were later used as templates for the formation of colloidosomes by coating the inner and outer surfaces of the middle liquid with hydrophobic silica nano-particles, as shown in Fig. 9-18. A promising use of W/O/W emulsions is in the formulation of pharmaceutical slow release compositions.[16]

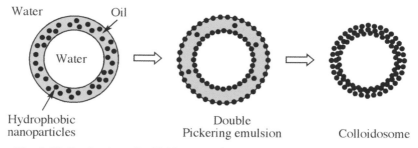

Fig. 9-18: Production of colloidosomes from multiple Pickering emulsions. After [Lee, D., and Weitz, D. A., *Adv. Mater.*, **20**, 3498 (2008).]

[14] Opawale, F. O., and Burgess, D. J., *J. Pharm. Pharmacol.*, **50**, 965 (1998).
[15] Utada, A. S., Lorenceau. E., Link, D. R., Kaplan, P. D., Stone, H. A., and Weitz, D. A., *Science*, **308**, 537 (2005).
[16] Cole, M. L., and Whateley, T. L., *J. Controlled Release*, **49**, 51 (1997).

C. Application of emulsions

1. *Formation/breaking in situ*

Many of the practical applications of emulsions are based on the delivery of the suspended material to some desired surface. Examples include topical application of cosmetics or medicines to the skin; application of waterproofers or anti-soils to textiles, carpets, *etc.*; application of herbicides or pesticides to plant or insect surfaces; application of wax to furniture or to floors, *etc.* In such situations, ultrahigh stability of emulsions is not always a desirable feature. Stable emulsions are required during storage (shelf-life), but when they are to be used, they must break and spread the contents of their dispersed phase onto the surface in question. This is sometimes accomplished by a post-treatment of the surface receiving the emulsion coating. For example, a fabric treated with an anionically stabilized emulsion (with a waterproofer) may be dipped in a bath with an aluminum salt (or other high valency cation) to destroy the O/W emulsifier.

Another device is to employ a transitory emulsifier, such as an ammonium salt of a fatty acid or a wood rosin. As the ammonia is lost by evaporation, the emulsifying agent becomes water-insoluble and the emulsion breaks. There are many variations on such procedures in current practice. One of the most important means of arranging for the application of an oil to a surface by means of an O/W emulsion is to stabilize the emulsion with an emulsifier which also strongly adsorbs onto the solid surface and in such a way that it renders that surface oleophilic. This simultaneously strips the droplets of the stabilizer and prepares the solid surface to be wet out by the oil, as shown in Fig. 9-19, illustrating emulsification of bitumen or tar by

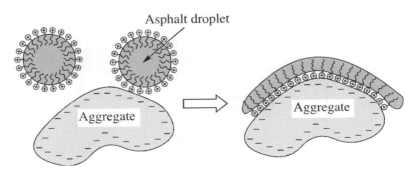

Fig. 9-19: Surface active agent linking the bitumen binder and aggregate.

cationic surfactants for use in treating sand and gravel for road construction. Various cationics are good emulsifiers for dispersing the tar in water. The sand and gravel carry a negative surface charge and so attract the cationic surfactants into "head-down" adsorption by ion pairing or ion exchange mechanisms. This exposes the hydrocarbon portions of the adsorbate to oil, which now spreads over them with ease.

Other situations exist wherein emulsions are entirely unwanted. They may arise in liquid extraction devices or as a consequence of other multiphase mixing operations, as pictured in Fig. 9-20, which shows a schematic of a standard mixer-settler unit. A fine dispersion is desired in the mixer in order to assist mass transfer, but in the settler, often a horizontal tank, the dispersion must break into contiguous streams of oil and water. An interesting illustration of the importance of the direction of mass transfer in a case of the above type is shown in Fig. 9-21. The data concern the rate of

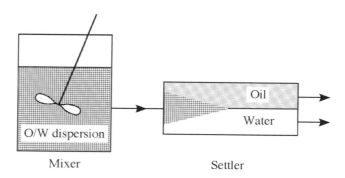

Fig. 9-20: Schematic of mixer-settler units used in liquid extraction.

coalescence of water droplets in benzene during which acetone is transferring either from the water drops into the external benzene phase, or

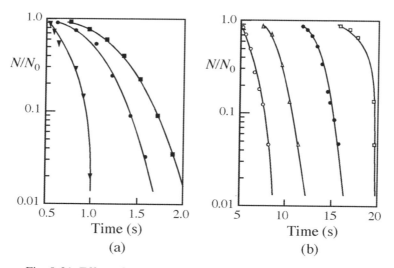

Fig. 9-21: Effect of mass transfer direction on coalescence time: water drops coalescing in benzene, with acetone transfer. (a) transfer of acetone out of water drops into benzene; (b) transfer of acetone out of benzene into water drops. After [Jeffreys, G. V., and Lawson, G. B., *Trans. Inst. Chem. Engrs.*, **43**, T294 (1965).]

vice versa. In the former case, the coalescence is rapid, but in the latter, more than ten times slower for the same driving force conditions. This result is explained in terms of the Marangoni effect, described in more detail in Chap. 10. The interfacial tension between the water and the benzene is reduced by the presence of acetone, as shown in Fig. 9-22. Thus when it is transferring outward, it saturates the thin film separating the water droplets, lowering the interfacial tension of the interface between the droplets relative to that of the rest of the drop. This produces surface forces that pull outward, promoting drainage and leading to rapid coalescence. During transfer in the opposite direction, the interfacial tension in the drainage zone is increased, producing an inward motion of the interfaces, retarding coalescence.

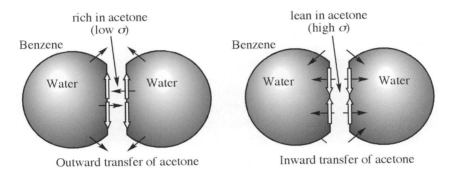

Fig. 9-22: Effect of mass transfer on coalescence.

2. *Demulsification*

Much of what has been stated concerning emulsion stabilization can be thought of "in reverse" to conceive of methods for breaking emulsions. For example, adsorptive displacement may be used whereby one adsorbate, that is a good emulsifier, may be displaced by another which is more surface active but not as good an emulsifier. Silicone liquids or fluorocarbon surfactants are often capable of such displacement. Electrostatically stabilized O/W emulsions may be swamped with electrolyte that may either destroy the double-layer repulsion or neutralize the adsorbent (which may invert the emulsion). Other ways of neutralizing, and therefore changing the emulsifying properties of, emulsifiers is to change the *pH* of the medium, add an ionic surfactant of opposite sign or to add some other reactant which will change the chemistry of the emulsifier. A variant of this is to use biological destruction by introducing micro-organisms which feed on the surfactant. For sterically stabilized emulsions, one can alter the temperature or add stabilizing-moiety non-solvent to achieve θ-conditions. Raising the temperature above or reducing it below the *PIT* of the emulsifier of the system may be effective, if the latter is accessible. Few emulsions can survive being frozen and re-thawed or boiled and condensed, but these are often unacceptably drastic means. Other means of breaking emulsions include diluting the continuous phase or using ultrafiltration, both with the idea of reducing the emulsifier concentration level below that which would correspond to stable emulsion conditions.

Emulsions of the W/O type can often be broken electrostatically, as pictured in Fig. 9-23(a). The water droplets, which generally bear a charge, are subject to electrophoretic motion. Coalescence can occur by collision of droplets as they migrate at different rates (due to differences in charge and

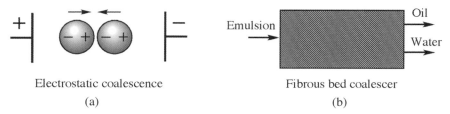

Electrostatic coalescence   Fibrous bed coalescer

(a)   (b)

Fig. 9-23: Methods for braking emulsions. (a) electrostatic coalescer for W/O emulsions; (b) fibrous bed coalescer.

size). Such collisions generally result in coalescence because the collisions are between the front of one drop and the back of the other, and due to drop polarization, these surfaces will be of opposite charge. Coalescence can also be achieved by inserting a semi-permeable barrier across the droplet migration path where the drops collect and coalesce. Such a device may be used in the absence of electrical migration by simply pumping an emulsion through a fibrous bed on which the dispersed phase collects, as shown in Fig. 9-23(b). The criterion for success of these units appears to be selective wetting of the fibrous media by the dispersed phase. This may be helped by electrostatics, with the fibers carrying a charge opposite to that of the droplets.

Other mechanical means for inducing coalescence include centrifugation and stirring at the right intensity to induce droplet contact but not droplet breakup.

D. Microemulsions

1. *Distinction between microemulsions and macro emulsions*

As stated earlier, microemulsions are best regarded as solutions of highly swollen micelles, either micelles in water swollen with oil, as pictured in Fig. 3-30, inverse micelles in oil swollen with an aqueous medium, or a so-called "middle phase" microemulsion, in which oil, water and surfactant exist in roughly equal proportions. As *solutions*, all three types are thermodynamically stable, single-phase entities. Their practical application parallels the use of micellar solutions in detergency and dry cleaning, and secondary valence catalysis, as well as that of ordinary emulsions in coatings and other products. They are also characterized by ultra-low interfacial tensions, as described below, which gives them application in tertiary oil recovery.

Microemulsions consist at a minimum of three components: water, oil and a surfactant, although other components, such as dissolved salt or a "co-surfactant" may also be present. Co-surfactants are usually medium-sized (3-8 carbons) aliphatic alcohols, amines or similar polar compounds that are unable to act as emulsifiers by themselves, but may be blended with traditional surfactants to alter the properties of the latter. The swollen micelles that make up microemulsions are generally in the size range of 5–50 nm radius so that they are optically clear. It is possible under some conditions to form ordinary emulsions having droplets nearly as small as the upper end of this range, referred to as "mini-emulsions," but these lyophobic systems should not be confused with microemulsions. Another distinction to be made is that while ordinary emulsions generally require significant energy input for their creation from bulk oil-water mixtures, microemulsion systems form spontaneously. The subject of microemulsions is complex, and it is treated only briefly here. More detail may be found elsewhere,[17] and the references therein.

Not all water-oil-surfactant combinations are capable of forming microemulsions. First of all, the surfactant must have sufficient monomer solubility in either the water or the oil to form micelles in either of these phases, and then the micelles formed must be structurally capable of imbibing a sufficient quantity of the opposite phase to reach the size associated with microemulsions. The range of states taking one from solutions of ordinary micelles with a small amount of solubilizate to a microemulsion is a continuum. This range may be traversed by variation of temperature or composition or both. Pressure, as in general for condensed phase media, is usually of secondary importance. It is the surfactant whose properties are decisive in a system's ability to form microemulsions. Surfactants that can imbibe a large amount of solubilizate are those whose monomer architecture produces critical packing parameters, $CPP = (v/l_c a_0)$ (as discussed in Chap. 3), near to unity. The CPP may be regarded, at least in a rough way, as the ratio of the effective size of the water-compatible moiety to that of the oil-compatible moiety or as a ratio of the water solubility of the molecule to its oil solubility. The effective CPP depends not only on the physical geometry of the surfactant molecule but also on its environment. For example, the effective area, a_0, of an ionic head group depends of the ionic strength of aqueous medium, becoming smaller as electrostatic screening is increased. Similarly, the area of a poly(ethylene oxide) head group of a nonionic surfactant is decreased by increasing temperature (due to progressive dehydration).

---

[17] Bourrel, M., and Schechter, R. S., **Microemulsions and Related Systems**, Surf. Sci. Ser. Vol 30, Marcel Dekker, New York, 1988;
Sottmann, T., and Strey, R., in **Fundamentals of Interface and Colloid Science**, Vol. V, J. Lyklema (Ed.), pp. 5.1-5.97, Elsevier, Amsterdam, 2005.

2. Phase behavior of microemulsion systems

The phase behavior of water-oil-surfactant systems is conveniently represented using a ternary diagram, the skeleton of which is shown in Fig. 9-24. The apices of the triangle represent the pure components, while the edges represent two-component mixtures. A point along one of the edges, such as Point **A** in Fig. 9-24, represents a binary mixture, in this case, one of

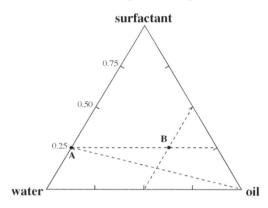

Fig. 9-24: Schematic of a triangular diagram for the representation of the composition of a three-component system.

water and surfactant, in the proportions given by the lever rule. The scale used may be mole fraction (or percent), volume fraction or mass fraction. Assuming mole fraction is used, Point **A** represents a mixture of 25 mole% surfactant and 75 mole% water. Any point in the interior of the triangle represents a three-component mixture, the relative proportions of which are given by the relative distances from the point to the respective apices. The dashed line drawn from the oil apex to Point **A** is the locus of all compositions with a constant surfactant-to-water ratio of 1 to 3, whereas the horizontal dashed line drawn from Point **A** to the oil-surfactant axis represents a constant surfactant mole fraction of 0.25. A similar line can be drawn from the oil-surfactant axis to the water-oil axis to fix the composition of Point **B** as $x_{surf} = 0.25$, $x_{oil} = 0.50$ and $x_{water} = 0.25$.

The phase behavior of the ternary system is described with the help of the Gibbs phase rule, Eq. (3.68). This gives the number of independent intensive variables required to fix the intensive state of such a system as $\mathsf{F} = 5 - \mathsf{p}$, where p is the number of phases. At constant temperature and pressure, a single-phase region thus requires the specification of two variables, *viz.*, two independent mole fractions that fix the system's composition. Figure 9-25(a) shows the phase behavior for one type of a microemulsion *system*, consisting of both a single and a two-phase region. The single-phase region is white, while the two-phase region is lightly shaded. The water and oil are seen to be almost totally immiscible in the absence of surfactant. Inside the two-phase region (at constant T and p), the system variance drops to one, *viz.*, the overall mole fraction of any component or its value in either of the phases. Point **A**, inside the two-phase region, consists of an aqueous phase (represented by Point **B)** and an oil phase (represented by Point **C**), joined by a *tie line*, *i.e.*, a line joining phase compositions in equilibrium. The relative proportions of the phases are given by the lever rule applied along the tie line. Most of the surfactant in this system is seen to be in the aqueous phase, Point **B**, which is presumed to be an O/W microemulsion. This is an example of a "Winsor Type I" microemulsion system, named after one of the earlier investigators in this

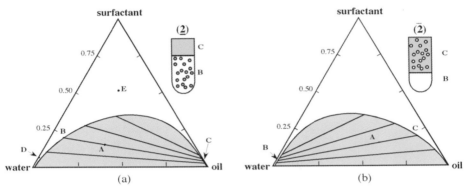

Fig. 9-25: Triangular diagrams showing the phase behavior of microemulsion systems. (a) System shows a water/oil phase split with the surfactant concentrated in the aqueous phase. Point **A** shows a Winsor I type system, denoted (2), in which an oil phase is in equilibrium with an O/W microemulsion. (b) System shows a water/oil phase split with the surfactant concentrated in the oil phase. Point **A** shows a Winsor II type system, denoted ($\overline{2}$), in which an oil phase is in equilibrium with a W/O microemulsion.

area.[18] Such systems are also denoted by "(2)" to suggest they are two-phase systems with an emulsion in the lower (aqueous) phase. Type I systems are formed using hydrophilic surfactants (*CPP* < 1; *HLB* > 10, *etc.*), and as pictured in the test tube of Fig. 9-23 (a), consist of an O/W microemulsion in equilibrium with an oil phase containing only a small amount of the monomeric surfactant. Point **D**, in the single-phase region of this ternary system, represents an aqueous solution of surfactant presumably above the *CMC*, with a small amount of oil solubilizate, while Point **E** in another part of the single-phase region may be a solution of other types of self-assembled surfactant structures, as described in Chap. 3.

If a *hydrophobic* surfactant had been used instead of a hydrophilic compound, one might obtain a phase diagram as shown in Fig. 9-25(b), and Point **A** in this figure would exemplify a "Winsor Type II, " ($\overline{2}$), system, *viz.* a W/O microemulsion in equilibrium with a water phase containing only a small amount of monomeric surfactant. The W/O microemulsion consists of inverse micelles, swollen with water.

[18] Winsor, P. A., *Trans. Far. Soc.*, **44**, 376 (1948);
    Winsor, P. A., **Solvent Properties of Amphiphilic Compounds**, Butterworth & Co., London, 1954.

Using a surfactant with equally balanced hydrophilic/hydrophobic character might produce phase behavior of the type represented in Fig. 9-26. At relatively low surfactant concentrations, as represented by Point **A**, the system produces water and oil phases, Points **B** and **C**, respectively, both containing monomeric surfactant or perhaps micelles in one or the other phase. As surfactant concentration is increased (at a constant water/oil ratio), one might reach a point such as Point **D**, inside a *three*-phase region. Here the variance has dropped to zero. *Any* point within the three-phase region is

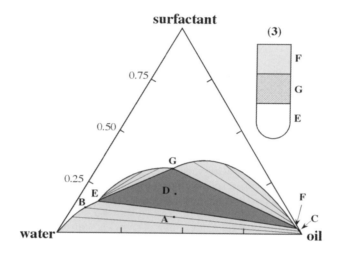

Fig. 9-26: Ternary diagram showing phase behavior of a Winsor III type system, consisting of water and oil phases in equilibrium with a middle phase microemulsion.

a mixture of phases of the compositions at Points **E**, **F** and **G**, in relative amounts given by the location of Point **D** within the *tie-triangle*. The phases at **E** and **F** are aqueous and oil solutions of monomeric surfactant, respectively, but the phase at **G** is something new. It is richest in surfactant, and contains oil and water in roughly equal proportions. It is termed a *middle-phase* microemulsion and believed to have a bicontinuous structure. Three-phase systems of this type are designated as Winsor III or "(3)," and the one corresponding to Point **D** is pictured schematically in Fig. 9-26.

It is possible to move through a sequence of system types from Winsor I through Winsor III to Winsor II with a given nonionic surfactant of the C_nE_x type, where C_n is an alkane chain, and E_x is a chain of ethylene oxide groups. The transformation is effected by raising the temperature, converting a hydrophilic surfactant at low temperature to a hydrophobic one at higher temperature. The changes can be represented with a stack of triangular phase diagrams, as indicated in Fig. 9-27. It is of interest to examine a section through this prism at a constant water/oil ratio, as shown. Such a section, as shown in Fig. 9-28, can reveal the phase states of the system as a function of temperature and overall surfactant mole fraction, z_s, and is known as an *isopleth*. (Other measures of surfactant content, such as overall mass or volume fraction, are also commonly used.) An isopleth is not a phase diagram, and thus does not contain tie lines. At very low surfactant concentrations, we have a two-phase system over the whole temperature range investigated, and as it is increased sufficiently, the aqueous phase will become an O/W microemulsion. Holding this concentration z_s fixed, as the temperature is raised, one may switch to a system in which the oil phase becomes a W/O microemulsion. The temperature at which this occurs has been identified earlier in the discussion of ordinary emulsions at the Phase

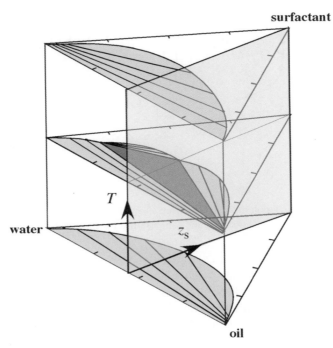

Fig. 9-27: Phase behavior of a microemulsion-forming system formed using a surfactant of the type C_nE_x as a function of temperature. Shaded rectangle is an isopleth, *i.e.*, a surface of constant oil/water ratio showing phase behavior as a function of surfactant concentration and temperature.

Inversion Temperature, *PIT*. Continuing to increase the surfactant concentration eventually leads to a point $z_s = z_{s0}$ where, at a given temperature, T^*, the three-phase region appears. This system is of the Winsor III type, consisting of both a water-rich and an oil-rich phase, both containing dissolved surfactant, and a middle-phase microemulsion. Further increases in surfactant concentration at T^* lead to greater proportions of the middle phase, until at $z_s = z_s^*$, the system consists entirely of that phase. This single phase persists as surfactant concentration is increased further, but its internal structure changes. At $z_s = z_s^*$, the system is a sponge-like bicontinuous structure, perhaps like that shown in Fig. 3-34(b), but at higher concentrations, it morphs into a lamellar "meso-phase" structure, *i.e.*, a stack of lamella as pictured in Fig. 3-35(b). The word "meso-phase" is used because it is not a distinct phase in the thermodynamic sense. Above and below T^* in this single-phase region, other meso-phases may appear, such as the hexagonal structure of Fig. 3-35(a), or "cubic" structures consisting of packed swollen micelles. Such micro-structural details are revealed using

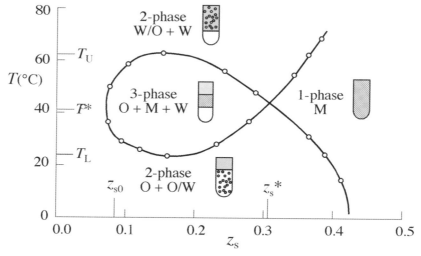

Fig. 9-28: Schematic isopleth "fish" diagram showing phase behavior as a function of surfactant concentration and temperature for the system of Fig. 9-27.

various light scattering techniques, particularly small-angle X-ray scattering (SAXS) and small-angle neutron scattering (SANS), as well as freeze-etch electron microscopy. The three-phase region also exists at temperatures above and below T^*, specifically between an upper limit of T_U and a lower limit of T_L as shown. The isopleth of Fig. 9-28 is known, for obvious reasons, as a *fish diagram*, and it has several characteristic temperatures, T^*, T_U and T_L, and surfactant concentrations, z_{s0} and z_s^*. T^* is generally the same temperature as that corresponding to z_s^*, and to good approximation, $T^* \approx \frac{1}{2}(T_U$ and $T_L)$. Above a surfactant concentration of z_{s0}, one does not observe a single *PIT* as temperature is increased, but instead a range of temperature between T_L and T_U over which the inversion is effected. Similar fish diagrams to that shown in Fig. 9-28 would be obtained at other water/oil ratios, but the gap between T_U and T_L would be reduced. Data for such a diagram, represented schematically by (o) in the figure, are obtained visually with jar tests.

The phase behavior shown in Figs. 9-25 – 9-28 is characteristic of that obtained with light hydrocarbon oils and small nonionic surfactants, C_nE_x, *i.e.*, with $4 < n < 8$ and $2 < x < 4$. For large n and small x, or small n and large x, microemulsion systems are not formed. Such systems might produce only W/O or O/W macro-emulsions. When the hydrophobic and hydrophilic moieties are more balanced, but both chains are made longer, microemulsion systems are possible, and two important changes emerge. First, the three-phase region appears at lower surfactant concentrations, and the critical concentration z_s^*, the lowest surfactant concentration producing a system entirely of middle phase, is reduced. The *efficiency* of the surfactant in producing a uniform oil-water-surfactant system is thus increased. A second difference is more subtle. Within the single-phase region to the right of z_s^* in the fish diagram, the system shows a stronger tendency to form lamellar meso-phases.

Entirely different types of surfactants may also be used to form microemulsion systems. Di-tail ionic surfactants, as exemplified by Aerosol OT® (Na di-2-ethylhexyl sulfosuccinate), can be made to traverse the Winsor I → Winsor III → Winsor II range by varying system salinity rather than temperature. The middle phase microemulsion in this system occurs for NaCl concentrations between approximately 0.05–0.07 M. Single-tail ionic surfactants can be made to produce microemulsion systems of the Winsor I type with the help of co-surfactants, as mentioned above in Section 1.

3. *Ultra-low interfacial tension*

An important characteristic of microemulsion systems is the ultra-low interfacial tensions that exist between the phases involved. While ordinary emulsions are characterized by interfacial tensions in the range of 1–10 mN/m, microemulsion systems produce values as low as 10^{-3} mN/m, or even lower. This fact is thought to enable the spontaneous formation of microemulsions as described by Eq. (9.9). It is also critical to the use of surfactant/polymer mixtures in tertiary oil recovery, in which the ultra-low interfacial tensions are required to reduce the flow resistance of the underground oil/water mixtures. Three different interfaces may be considered: the W–O/W interface in the Winsor I system; the O–W/O interface in the Winsor II system, and the O/W interface exposed in the Winsor III system when nearly all of the middle phase is removed. A typical result is shown in Fig. 9-29 for the case of $C_{10}E_4$ surfactant in the system of

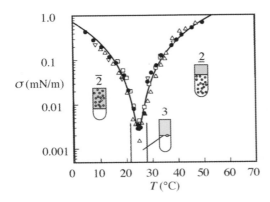

Fig. 9-29: Interfacial tension data for the water-*n*-octane-$C_{10}E_4$ system as a function of temperature obtained by various investigators. After [Sottmann, T., and Strey, R., *J. Chem. Phys.*, **106**, 8606 (1997).]

water and *n*-octane. Interfacial tension spikes downward over a narrow range of temperature. Similar results exist for the Aerosol OT® system at 25°C and a surfactant concentration of 0.05 M as the concentration of NaCl is varied from 0 to 0.10 M. Measurements of ultra-low interfacial tensions are obtained using either the pendant drop or the spinning drop techniques described in Chap. 2.

4. Interfacial film properties in microemulsion systems

Return finally to the surfactant criteria for the formation of microemulsion systems. It was stated earlier that the emulsifier under the conditions of application must have a close balance between hydrophilic and hydrophobic properties. In addition to the *CPP* (which must be near unity) or the *HLB* (which must be near 10), other useful criteria have been proposed. Winsor constructed a ratio *R*, for a surfactant layer (C) between a water (W) and an oil (O) phase, and defined as $R = A_{CO} / A_{CW}$, where A_{CO} is the cohesive interaction per unit area between the surfactant (regarded as a layer of finite thickness containing tail groups, head groups, oil and water) and the oil phase, and A_{CW} is the same parameter for the surfactant and the oil phase. Both could presumably be computed using molecular interaction parameters and assumed packing densities. The expression for *R* has been refined to account also for tail-tail, head group-head group, oil-oil, and water-water interactions,[19] but the idea is that if *R* > 1, the surfactant layer will seek to increase its area against the oil. Thus the interface will spontaneously bend to put oil on the convex side, *i.e.*, to favor the formation of W/O emulsions. *R* < 1 suggests interface bending to put water on the convex side, and *R* ≈ 1 favors the formation of microemulsions.

The tendency of an interface to bend in a given way has also been expressed in terms of the spontaneous curvature, κ_0, introduced in Chap. 2. The spontaneous curvature is that curvature adopted by the interfacial film with water on one side and oil on the other if no external forces, thermal fluctuations, or geometric or other constraints are imposed upon it. By convention, if $\kappa_0 < 0$, the interface bends to put oil on the convex side, and if $\kappa_0 > 0$, water will be on the convex side. The formation of interfacial structures that differ from κ_0 and/or have a Gaussian curvature different from zero is resisted by a free energy/area, F^σ given by the Helfrich Equation, Eq. (2.59). Its magnitude depends on the bending modulus and the saddle splay modulus of the film. Films that are more rigid tend to form larger, more rigid lamellar structures in middle phase microemulsions, a situation that leads, among other things, to a higher viscosity for these systems. Surfactants with longer alkane hydrophobes, C_n, produce more rigid interfacial layers, while those with side chains or unsaturation produce more flexible layers.

---

[19] Bourrel, M., and Schechter, R. S., **Microemulsions and Related Systems**, Surf. Sci. Ser. Vol 30, pp. 15ff , Marcel Dekker, New York, 1988.

E. General consideration of foams

1. *Nature and preparation of foams*

Foams are dispersions of gas in liquid. While the individual bubbles of gas are usually in the size range from about one mm to a few cm, foams can also be encountered with bubbles as small as 0.01 μm in diameter. Foams are similar to emulsion systems in that their breakdown generally depends upon the encounter of the dispersoid particles (bubbles) with one another followed by the drainage of the continuous phase liquid from the film between them and ultimately its rupture permitting the dispersoid particles to join (coalescence). The large interfacial area of both types of systems renders them inherently unstable with respect to such processes, which reduce this area. The systems are also similar in that foams generally require the aid of an additional component (foaming agent) to have any appreciable lifetime. Foaming agents like emulsifiers, are usually strong surface active agents or adsorbing macromolecular solutes. Also under the right conditions, finely divided solids may serve as foaming agents, creating perhaps the "armored bubbles" discussed in Chap. 4.

Foams also differ from emulsions in a number of important respects. The density difference between the bubbles and the dispersion medium is usually so great that they are highly unstable with respect to phase segregation by gravity. The bubbles quickly rise and are pressed together to form a polyhedral (honeycomb) structure, as shown in Fig. 9-30, similar to that of very dense emulsions. The "flocculation" process is thus nearly always irrelevant in foam systems. Separate bubbles are either brought together by impaction as they rise at different rates (due to size differences) or when they reach the top of the liquid phase. The bubbles in ordinary foams are much larger than the droplets in ordinary emulsions so that the draining contact films between them are much larger in area. This means that the drainage hydrodynamics may be more complex and irregular than that between emulsion droplets.

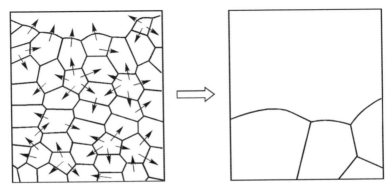

Fig. 9-30: Evolution of foam structure due to gas diffusion across foam lamellae and film breakage.

Foams also differ from emulsions in the importance of the rate of diffusive transport of dispersed phase molecules across the continuous phase membrane. The bubbles are often air or other small-molecule materials that diffuse rapidly across the thin liquid lamellae making bubble size disproportionation (big bubbles gaining at the expense of small ones) an important process. The larger Young-Laplace pressure inside the small bubbles provides the driving force, so that foams become courser over a period of time. Superimposed upon this disproportionation is a general collapse of the foam by diffusion of the gas out of the bubbles into the adjoining atmosphere (the equivalent of a bubble of infinite radius).

The relatively large size of lamellae in foam means that prototypes of them may be easily created as free films (open to the atmosphere on both sides) and examined in the laboratory, as the biconcave menicus shown in Fig. 2-64. Despite the often-inherent thermodynamic instability of these films, they may be made to exhibit extraordinary longevity. Metastable foam films correspond to disjoining pressure curves of Type III or IV, Fig. 2-69, as discussed in Chap. 2, and earlier in this chapter with respect to films between emulsion droplets. Dewar[20] maintained several soap films for over a year, and one for three years (!), by taking precautions against evaporation, mechanical and thermal shocks, and absorption of carbon dioxide.

Foams are prepared by one, or a combination, of methods, including:

1) spraying or sparging of gas into the interior of a liquid so that the gas leaves the orifice(s) as individual bubbles or as jets of gas that quickly break up into bubbles;

2) stirring or mixing liquids in contact with gas such that the stirring action creates waves and vortices at the liquid surface that entrain the pockets of gas;

3) plunging liquid masses (usually jets) from the gas into pools of liquids so that under the right conditions, enormous amounts of gas may become entrained , and

4) "precipitation" of dissolved gases from the liquid, often assisted by the presence of dispersed solid particles or irregularities on the liquid container walls that serve as heterogeneous nucleation sites for the gas bubbles. Precipitation is a means of obtaining micro- or nano-bubbles.

Foams are often created accidentally and are unwanted guests in many situations. Air entrained by leaking into process lines and especially pumps or by being sucked into mixing units or vapors generated in distillation units, boilers, evaporators, *etc.* may require shutdown of the entire operation. On the other hand, foams have many important uses. Their presence often has an important cleansing action (SC Johnson's "Scrubbing Bubbles®") when solids are being cleaned by liquids. They are particularly effective in

[20] Dewar, J., *Proc. Roy. Inst.*, **22**, 179, 359 (1917);
    Dewar, J., *Proc. Roy. Inst.*, **24**, 197 (1923).

capturing fine particulates from gases. Foams are sometimes important because the large liquid-gas area they provide facilitates gas-liquid mass transfer and/or chemical reactions. Liquid foams are also the precursors of solid foams (foamed plastics, like Dow's Styrofoam® thermal insulation and packing materials) that are important structural materials where there is a low weight/volume requirement. Foams are often very important in fire fighting where they may be used to create and hold a blanket of carbon dioxide over a fire or to disperse water over an oil fire (3M's "Light Water®" fire fighting foams). Foams are also the basis of several important separation processes including ordinary flotation, ion flotation, and foam fractionation.

2. Stages in foam lifetime

It is convenient to describe foams in terms of their "lifetime," the various stages they go through as they mature and eventually are destroyed, as shown in Fig. 9-31. Some, depending on their properties, may have their development arrested at some intermediate stage, in which state they may

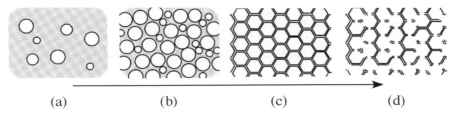

Fig. 9-31: Stages in a foam lifetime: (a) *Kugelschaum* period (independent gas bubbles), (b) gravity drainage period, (c) lamella thinning period, (d) film rupture.

exist for a long time (perhaps indefinitely) until they are destroyed without maturing further, analogous to the "coalescence-stable" or Perrin black films described earlier in the context of emulsions. Other foams may be "accident prone", *i.e.*, only weakly metastable, and be terminated at an early stage as a result of small disturbances. These are sometimes designated as "evanescent foams" or "froths." Still others may go through the full number of stages exhibited by foams, but the rate at which they undergo these processes may differ greatly from one system to another.

Foams are usually born as a dispersion of independent, spherical bubbles, as shown in Fig. 9-31(a). In this period of infancy, the system is termed a *Kugelschaum* ("spherical foam"). In liquids of ordinary viscosity, this stage is highly transitory, but if the medium is very viscous, slowing translational motion, the development may be arrested in this stage.

Next, the lamellae are in the thickness range from one mm to 10μm, as shown in Fig. 9-31(b) and during which the structure drains by gravity. This period generally lasts only a few seconds to minutes, and is followed by the longer lasting lamellae thinning period in which the bubbles are pressed together to form a polyhedral or cellular structure, as pictured in Fig. 9-31(c). The lamellae always meet at three-way intersections known as

"Plateau borders," and always at precisely 120° angles, in accord with "Neumann's triangle." More than three lamellae meeting at a common intersection is unstable.[21] Finally, when a critical thickness is reached, the lamellae burst, and the foam breaks down.

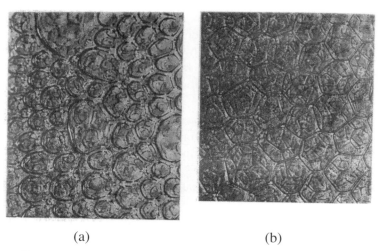

(a)   (b)

Fig. 9-32: Foams formed from 600-800 mg/L of Aerosol OT surfactant in water at 25°C. (a) gravity-draining froth; (b) lamellae thinning period. From [Hartland, S., and Barber, A., *Trans. Inst. Chem. Engrs.*, **52**, 43 (1974).]

3. *Stability mechanisms*

Foams could not exist at all if it weren't for the self-healing of their lamellae through Gibbs elasticity, as pictured in Fig. 9-4, so it is evident for their stability a surfactant is required. The lamellae are continually subject to minor stresses and strains tending to locally dilate or compress the surface. Unless the film can respond to such distortions by developing restoring forces, the lamellae will burst immediately. Because of its "elasticity," if the film suffers a local dilation (thinning), reducing the local amount adsorbed per unit area (even if temporarily), it generates surface tension forces that tend to drag liquid back to the thin spot, "healing" it. This motion is also resisted by the viscosity of the liquid in the film and sometimes by intrinsic surface rheology. The driving force for the restorative flow is the film elasticity E_f, twice (since there are two surfaces to the film) the Gibbs surface elasticity referred to in Chap. 3.K.2, and given by Eq. (9.1), *i.e.*

$$E_f = 2E_G = 2\frac{d\sigma}{d\ln A} = -2\frac{d\sigma}{d\ln\Gamma},\qquad(9.13)$$

---

[21] Plateau, J., **Statique expérimentale et théorique des liquides**, Vol. 2, Gauthier-Villars, Paris, 1873.

where subscripts on A, the interfacial area, and Γ, the relative adsorption have been dropped for brevity. Figure 9-33 shows qualitatively the relationship between the surface equation of state for the surfactant and the elasticity. The maximum elasticity is seen to occur close to, but below, the *CMC* or the collapse point of the surfactant. The effective elasticity realized in practice depends on more than this thermodynamic parameter. When the surface experiences a local dilation, it seeks to efface the resulting non-uniform distribution of surfactant by whatever processes are available to it. In addition to the restorative flow resulting from the surface tension gradient, surfactant will diffuse to the dilated region from the adjacent bulk phase, and in the case of an insoluble surfactant, will move to the dilated region by surface diffusion. The general problem is discussed in more detail in Chap. 10.D, while three scenarios are described below.

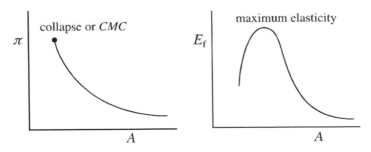

Fig. 9-33: Gibbs or Gibbs-Marangoni elasticity of foam film lamella.

The first scenario may be termed the "equilibrium Gibbs mechanism." If it is assumed that the patch of lamella that is dilated is of constant volume as it changes configuration, the increased area will consume a larger fraction of the total inventory of adsorbate, reducing the concentration in the bulk phase, and hence the position of adsorption equilibrium. For an area change dA (of a two-sided film), the inventory of solute in the bulk will change by

$$VdC = 2\Gamma dA, \tag{9.14}$$

but the volume of the patch of lamella is given by $V = Ah$. Thus

$$\frac{dA}{A} \equiv d\ln A = \frac{h}{2\Gamma}dC. \tag{9.15}$$

Then from the Gibbs adsorption equation, Eq. (3.91):

$$d\sigma = -\Gamma RT\frac{dC}{C}. \tag{9.16}$$

It has thereby been assumed that the concentration of solute is below its *CMC*. Substituting Eqs. (9.15) and (9.16) into Eq. (9.13) yields:

$$E_f = 4\frac{\Gamma^2 RT}{Ch}. \tag{9.17}$$

It is assumed that instantaneous adsorption equilibrium exists at all times between the surface and the bulk, and that there is no significant lateral transport of material in the bulk of the film during the dilation. Both assumptions are better the thinner the film, and this is also where E_f (by this mechanism) takes on its largest values. For typical values for surfactant solutions ($C = 10^{-3}$M; $\Gamma = 3$ μmole/m^2) and $T = 300$K, we may determine the value of h for which E_f will be 10 mN/m. The result is $h = 1$ μm. This mechanism may thus be significant for films thinner than this.

Another scenario is what is termed the Gibbs-Marangoni[22] mechanism. In it, film elasticity is developed assuming that there is no equilibrium of adsorption between surface and bulk following a dilation or compression. The times involved in foam production are of the order of 1-100 ms, and these are reasonable lifetimes for film dilations. During such a brief time, it may be unlikely that adsorption equilibrium can be restored by solute diffusing from the bulk to the surface. This would be particularly true for very dilute solutions of strong surfactants (large Γ, low C). Thus, there would be a strong temporary surface tension gradient opposing the dilation. A rough computation of this temporary elasticity is as follows. Following a dilation of 50%, i.e., $\Delta A/A = \Delta \ln A \approx 0.5$, so that to a first approximation, $\Delta \Gamma_0 = \Gamma_0/2$, the surface tension will change by

$$\Delta \sigma = \left(\frac{\partial \sigma}{\partial \Gamma}\right)\Delta \Gamma_0 = \left(\frac{\partial \sigma}{\partial \Gamma}\right)\frac{\Gamma_0}{2} = \left(\frac{\partial \sigma}{\partial C}\right)\frac{\Gamma_0 K}{2}, \tag{9.18}$$

where K is the adsorption equilibrium constant, Γ/C. Thus if no back diffusion occurs,

$$E_f = 2E_G \approx K\left(\partial \sigma/\partial C\right)\Gamma_0. \tag{9.19}$$

But if during the dilation time, t_d, a finite amount of back diffusion does occur, modifying the change in adsorption in accord with Eq. (3.150):

$$\Delta \Gamma_{diff} = \frac{2}{1000}\sqrt{\frac{Dt}{\pi}}C, \tag{9.20}$$

where Γ is in moles/cm^2 and C is in mole/liter. The net change in adsorption is thus:

$$\Delta \Gamma_{net} = \Delta \Gamma_0 - \Delta \Gamma_{diff} = \frac{\Gamma_0}{2} - \frac{2}{1000}\sqrt{\frac{Dt}{\pi}}C, \tag{9.21}$$

and the effective film elasticity would thus be

$$E_{Mar} = K\left(\frac{\partial \sigma}{\partial C}\right)\left[\Gamma_0 - \frac{4}{1000}\sqrt{\frac{Dt}{\pi}}C\right]. \tag{9.22}$$

---

[22] After C. Marangoni, an early (1870's) investigator of flows resulting from surface tension variations due to variations in surface temperature and composition.

Elasticity thus decreases as the bulk concentration C increases. The Marangoni mechanism applies to thick films as well as thin and is generally considered to be more important then the Gibbs mechanism. It more properly leads to an apparent dilational *viscosity* rather than an elasticity.

Finally, intrinsic surface rheology sometimes plays a role. Certain polymeric films (and possibly others) may possess real surface dilational viscosity and/or elasticity (aside from apparent rheology due to compositional rearrangement) that resists dilation and film breakup.

During the early stages, liquid is drained from the foam structure by gravity by way of the Plateau borders, in accord with[23]:

$$\frac{dh}{dt} = -\left(\frac{\mu H}{\rho g t^3}\right)^{1/2},$$ (9.23)

where h is the thickness of the thinning lamella, and H is the elevation of the foam. Following the brief gravity-drainage period, which has left the foam lamellae at a thickness of approximately 10 μm, a second type of drainage occurs. Liquid is drawn from the flat film portions by capillary suction toward the Plateau borders, as shown in Fig. 9-34. Consider the pressure

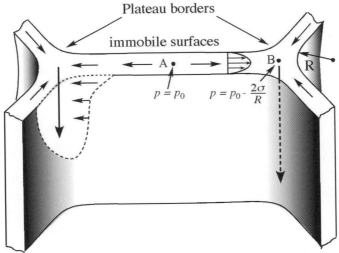

Fig. 9-34: Suction into and drainage down through Plateau borders.

differences developed between A and B. The pressure out in the gas phase is the same everywhere, p_0. Also, since the liquid film at A is ≈ flat, $p_A ≈ p_0$. The high curvature at the liquid surface at the Plateau border, however, requires that: $p_B < p_0$, and the foam lamellae drain into the Plateau borders, which themselves are drained primarily through gravity. The rate of film thinning by suction into the Plateau borders is given by:

[23] Hartland, S., and Barber, A., *Trans. Inst. Chem. Engrs.*, **52**, 43 (1974).

$$\frac{dh}{dt} = -\frac{4\pi(\Delta p)h^3}{3\mu A},$$

(9.24)

where Δp is the pressure difference between the liquid in the flat lamella and the Plateau border, and A is the area of the lamella. It may last several minutes to several hours.

This thinning generally proceeds until the films are in the range where disjoining pressure effects take over. If as in the usual case the disjoining pressure is negative, the film will eventually reach a hydrodynamically "critical thickness," h_c, of the order of a few tens of nm, where spontaneous rupture occurs. In other situations, which parallel the discussion of the drainage of films leading to droplet coalescence in emulsions, metastable structures may be achieved when the disjoining pressure goes to zero, as it is decreasing with h, at a thickness greater than h_c. Such disjoining pressure behavior results from electrostatic repulsion between the ions of surfactant stabilizing the films or steric effects, as shown in Fig. 9-35. Even more stable structures are formed if the bulk liquid of the lamellae contains macromolecular solutes capable of forming a gel. This is often the case when the foam is stabilized by natural rosins or proteins.

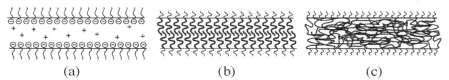

(a)   (b)   (c)

Fig. 9-35: Mechanisms of long-term foam stability: (a) electrostatic repulsion, under low salt conditions, (b) steric repulsion, and (c) gel formation.

Foams in which rupture does not occur immediately following drainage by capillary suction to a critical thickness are termed "persistent foams." Such systems almost always contain foaming agents that are ionic or macromolecular surfactants (or both). Sometimes finely divided solids may contribute to the formation of a persistent foam in a manner similar to the way they stabilize emulsions, as shown in Fig. 9-36(a). The requirement

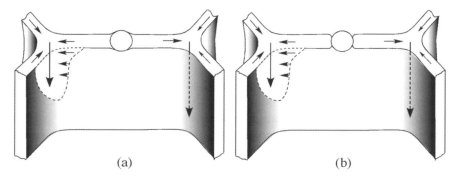

(a)   (b)

Fig. 9-36: Effect of particulates on foam stability. (a) foam stabilization by wetted particulates, (b) foam breaking by de-wetted particulates.

for a stabilizing effect in the case of foams is that the solid particles be wet by the liquid, *i.e.*, the contact angle be less than 90°. The particles are thus mainly in the liquid film and hold its surfaces apart. If the particles are wet out by the liquid ($\theta = 0°$), they will tend to accumulate or create a "log jam" in the Plateau borders reducing the drainage rate and increasing the effective liquid viscosity. This will also enlarge the Plateau borders, reducing their curvature and hence the capillary suction responsible for drainage. Highly concentrated ionic surfactant solutions (far beyond the *CMC*) may contain colloidal clumps of liquid crystal that are also capable of clogging the Plateau borders and influencing the effective flow cross-section and liquid viscosity. On the other hand, particulates which are de-wet by the foamate are capable of breaking foams by creating holes in them, as shown in Fig. 9-36(b) and discussed below in Section 5.

In addition to persistence, another aspect of foams may be termed "foamability." This refers to the ability of the solution to readily yield large volumes of foam when treated in one or more of the ways described in the earlier paragraphs on foam production. It is generally favored by low surface tension and low viscosity, whereas the persistence or stability of the foam that is produced is generally favored by high Gibbs elasticity and high viscosity. There are numerous examples of systems that readily produce very voluminous foams that nonetheless do not persist for any significant time. This is characteristic of aqueous systems with low molecular weight aliphatic alcohols, acids, amines, *etc.* used as foaming agents. On the other hand, there are many systems out of which it is quite difficult to form a foam, but once formed may have remarkable persistence. This is often the case for polymeric or macromolecular foaming agents.

4. Foam behavior and foaming agents

Practical tests of foaminess for sake of quantitative comparison should clearly distinguish between the above two aspects of the problem. Probably the most common test in industrial use is the Ross-Miles method [ASTM D 1173], shown in Fig. 9-37. In it, 200 mL of a solution of surfactant (usually 0.25 wt.%) contained in a pipette of specified dimensions with a 2.9 mm I.D. orifice are allowed to fall 90 cm onto 50 mL of the same solution contained in a cylindrical vessel maintained at a given temperature (often 60°C) by means of a water jacket. The height of the foam produced in the cylindrical vessel is read immediately after all the solution has run out of the pipette ("initial foam height") and then again after a given amount of time (generally 5 min). The Ross-Miles test thus provides a measure of both aspects of foaminess. Another common test is to pass air bubbles at a given rate through a given orifice into a standardized vessel of frothing liquid. The steady-state height of the column of foam so formed is balanced by the rate of foam production. The height under these conditions is thus a combined measure of both aspects of foaminess.

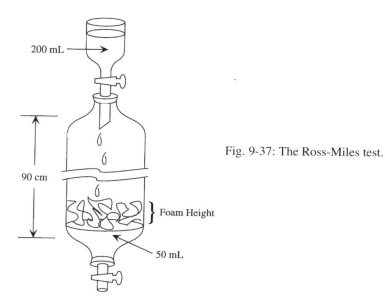

Fig. 9-37: The Ross-Miles test.

Another simple and useful test for determining foam stability is that of the single, standard-sized, bubble injected beneath the surface of a quiescent foamate solution, as shown in Fig. 9-38. The bubble rises to the surface, and its lifetime is measured.

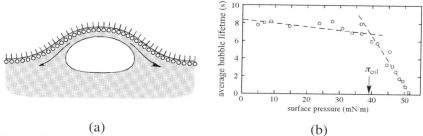

(a)   (b)

Fig. 9-38: (a) Single-bubble lifetime test. (b) Average bubble lifetime *vs.* surface pressure for octadecanol monolayer $T = 19.2°C$. Dashed lines are linear least-squares fits of data above and below the incipient collapse pressure of the film. Results suggest that bulk surfactant particles act as reservoirs whose presence destroys the film elasticity. After [Ternes, R. L., and Berg, J. C., *J. Colloid Interface Sci.*, **98**, 471 (1984).]

Tests of the above types have shown that among the best surfactants for producing foams are ionic *i.e.*, soaps and detergents such as the aliphatic sulfates and sulfonates. The more coherent the hydrophobic chains, the better, so that the straight-chain compounds are generally preferred over those with branched, aromatic, or unsaturated chains. Such surfactants produce high levels of surface tension reduction over narrow ranges of concentration change, therefore yielding high film elasticity. The optimum carbon chain length is about 12-14 at room temperature, and the optimum concentration is usually just below the *CMC*. Foam persistence is achieved in part by electrostatic repulsion between the opposing film surfaces as they approach critical thinness. The measured surface viscosity of a film of surfactant often correlates with foam stability, but this is most likely because both surface viscosity and foam stability correlate with film coherence. Surfactant film "brittleness" and collapse pressure are also important parameters. Films of surfactant may be brittle to the point of fracture when subjected to local compression when the hydrophobic chain length is too great, but even more dramatically when anionic surfactants contact polyvalent cations (like Ca^{++}) under *pH* conditions that yield an un-ionized salt. Film collapse usually leads to the formation of bulk phase fragments, which may destabilize the foam, as exemplified in the results shown in Fig. 9-38(b). It is thus critical for stable foam that the solubility limits of the surfactant *not* be exceeded. The excess surfactant in such cases is usually a powerful antifoaming agent.[24] When mixtures of surface active agents are used, often complex synergistic (or antagonistic) effects are observed.[25]

The influence of various system parameters on foam behavior may be summarized as follows.

1) Raising the temperature usually de-stabilizes foam to some extent, perhaps totally. It reduces the viscosity of the bulk film liquid, increasing drainage rate. More importantly, it may expand the hydrophobic chain layer and make it less than coherent. This destroys the no slip condition and may greatly increase drainage rates. As temperature is increased, the optimum carbon chain length goes up so that the increased van der Waals forces compensate for the thermal agitation. Thus at 60°C, the optimum alkyl sulfate surfactant has 16 carbon atoms (up from 12-14 at room *T*). At the boiling point of water, the C_{18} compounds are the best foamers.

2) High ionic strength, and/or the presence of polyvalent ions will collapse the diffuse double layer and destabilize a foam containing ionic surfactant. Thus calcium in hard water is a foam destabilizer for typical anionic soaps and detergents.

3) The *pH* is often important, and if it is adjusted so that an otherwise ionic surfactant cannot dissociate, foam will be destabilized. Thus many anionic surfactant foams are destabilized in acidic media, whereas cationic foams may be destabilized in basic media.

4) Moderate differences in pressure level do not strongly affect foam stability, but sudden changes in pressure (particularly decreases) may totally break down a foam. Such changes cause a rapid local expansion of the lamellae that often cannot be accommodated by the surfactant film.

[24] Ross, S., and Nishioka, G., *J. Phys. Chem.*, **79**, 1561 (1975).
[25] Frye, G. C., and Berg, J. C., *Chem. Eng. Sci.*, **43**, 1479 (1988).

5) The presence of organic solutes may have a very great effect, either to promote or inhibit foam persistence. Since maximum effects either way are often achievable at very low additive concentration levels, they present an important means of foam control or stabilization and have received much study. Additives that enhance foam stability may achieve their influence through at least three different effects, often working together. First, they reduce the rate at which the surface and bulk phase restore mutual equilibrium following a dilation or compression (Marangoni effect) by forming co-micelles with the parent surfactant and reducing the effective *CMC*. This increases the micelle population and decreases the monomer concentration in the bulk. The most effective additives are often long chain nonionic polar compounds of approximately the same chain length as the surfactant. Thus C_{12} alcohol is an effective stabilizing additive for surfactants like sodium dodecyl sulfate. The alcohols are solubilized in the palisade layer of the SDS micelles. Another important consequence of their presence is that they penetrate the surfactant film and exert a condensing effect on the hydrophobic chains, making them more coherent without becoming "brittle." The condensation occurs because the penetrant goes between the ionic surfactant molecules that are rather widely spaced due to lateral electrostatic repulsion. The film then becomes more slowly draining and has a higher Marangoni elasticity. A third consequence of stabilizing polar oils is the increased micellar population they foster which yields a greater capacity for solubilizing (and thus scavenging) apolar oils that generally have a negative effect on foam persistence. Apolar oils, if not fully solubilized, may be foam inhibitors. When a stable foam is desired, it is now routine to use an appropriate surfactant-organic additive combination rather than just a pure surfactant. For such foam-forming combinations, there is often a definite "transition temperature" above which the foam is fast-draining and less stable. This corresponds to the temperature above which significant additive penetration of the surfactant film does not occur. The transition temperature increases in a regular way with the relative amount of polar additive present. Typically, the amount of polar additive used is 1/20 to 1/50 of the surfactant.

5. Antifoam action

Antifoam action refers either to the inhibiting effect of some material in the solution which, except for its presence, would form stable foam, or to the ability of a material to kill a stable foam when added in small amount to the system. These are foam inhibiters and foam breakers, respectively. A number of mechanisms are at play,[26] but both of them generally act by getting into the surface and forming a non-coherent, fast-draining lamella. They also greatly reduce $d\sigma/dC$ so that Marangoni elasticity cannot manifest itself. Tributyl phosphate is a typical foam inhibitor that does both these

[26] Garrett, P. R., "The Mode of Action of Antifoams," in: **Defoaming: Theory and Industrial Applications**, P. R. Garrett, Ed., pp. 1-119, Marcel Dekker, New York, 1993.

things. Others include branched-chain alcohols, esters of long chain alcohols, low *HLB* nonionics, and compounds with multiple polar groups. Unsolubilized apolar oils also act in this way. Interestingly, if a stabilizing additive is added in excess of its solubility limit, the excess material may act as a powerful foam inhibitor. Its lenses in the interface are reservoirs of material to be spread out whenever there is a dilation, keeping σ constant and destroying the elasticity of the surface, as shown in Fig. 9-38(b). The same principles are true for foaming and its inhibition in non-aqueous media. As an example, the foaming properties of four lube oils were examined in the presence and absence of silicone oils.[27] In all instances, the silicone oils acted as pro-foamers up to the limit of their solubility, but at higher concentrations, they were foam inhibitors.

Foam breakers face the task of displacing an existing surfactant film from the surfaces of the lamellae. To do this they must have a positive spreading coefficient against the existing film, *i.e.*, either:

$$S = \sigma_A - \sigma_B > 0, \qquad\qquad (9.25)$$

where σ_A is the surface tension of the original film and σ_B the surface tension of the monolayer of film breaker which replaces it, or if spreading occurs from lenses of foam breaker:

$$S = \sigma_A - \sigma_B - \sigma_{A/B} > 0, \qquad\qquad (9.26)$$

where $\sigma_{A/B}$ is the interfacial tension between the foam killing oil and the bulk liquid of the lamellae. These requirements usually mean that the foam breaker has a lower surface tension than the surfactant-covered surface of the lamellae. An effective breaker will generally cause immediate collapse of foam as the bubble walls are thinned by the spreading action itself, and the film replacing the original is of the non-coherent, fast-draining type. The bubbles thus burst, and the properties of the foam breaker may retard additional foam for some time afterwards. The foam killer may be slowly solubilized, however, so that eventually foaming returns, and foam killer must be reapplied.

Some of the most powerful antifoams are hydrophobic particulates, usually hydrophobized silica (treated with silanes or silicones, as described in Chap. 4). These form a bridge across the lamella and are de-wet, causing rupture,[28] as pictured in Fig. 9-36(b). They are even more effective when used in combination with liquid antifoams of the type described above.[29] Foams may also be destroyed by salting them out, changing the *pH*, raising the temperature, dropping the pressure, or mechanically disturbing them. In addition, it is sometimes possible to break foams by forcing them through a porous bed or subjecting them to electric fields.

---

[27] Shearer, L. T., and Akers, W. W., *J. Phys. Chem.*, **62**, 1264, 1269 (1958).

[28] Frye, G. C., and Berg, J. C., *J. Colloid Interface Sci.*, **127**, 222 (1989).

[29] Frye, G. C., and Berg, J. C., *J. Colloid Interface Sci.*, **130**, 54 (1989).

6. *Froth flotation*

The large differences that exist or can be made to exist between the wettability of different solids by water or other liquids is the basis for an important separation process known as flotation, illustrated earlier in Fig. 4-61. Probably the most important application of this separation technique is the concentration of valuable mineral ores form gangue (ore flotation). It is also important in the de-inking of paper for reuse (flotation de-inking) and in the purification of a variety of pharmaceuticals and proteins from unwanted surface-active ions (ion flotation). Other applications are sure to be developed in the future. Ore flotation may be used to illustrate the mechanism of the separation process, and a good example is the selective flotation of lead, zinc and copper sulfide ores from siliceous gangue. Mineral sulfides are generally not very well wet by water; contact angles are typically $\approx 90°$. Silica, however, and other siliceous materials are generally wet out by water, *i.e.*, $\theta \approx 0°$. (The relative ability of these minerals to be wet by a nonpolar oil, however, is just the reverse.) If a mixture of sulfide ore and silica particles are thus suspended in water (by agitation) and air bubbles are introduced, the sulfide particles will attach themselves to the bubble surfaces (just the system's attempt to exhibit the appropriate contact angle), while the silica particles will shun them. The sulfide particles, thus attached, may be floated toward the surface. In practice, particles up to about 0.3 mm in diameter may be "floated" in this way.

There are two basic problems that must generally be overcome before the above-described process is an industrially feasible one. First, due to contact angle variability caused by roughness and contamination of the particle surfaces, the separation is often not good, *i.e.*, many sulfide particles are not attached to the bubbles ("collected") and significant numbers of the silica particles are attached. What is needed is the addition of a compound (called a "collector") which adsorbs preferentially to the sulfide surface and renders it even more hydrophobic, while either not adsorbing to the silica or adsorbing in a manner which does not render it hydrophobic. The second problem has to do with the bubbles. In pure water, air bubbles will immediately coalesce on contact and rapidly destroy surface area to which particles may attach. If they do manage to make it to the surface, they will burst there and discharge their cargo of particles to sink back into the water. What is needed is something to stabilize the bubbles (called a "frother"). Frothers are also surface active agents. These should be most attracted to the liquid-gas surface and not compete successfully with the collector for adsorption sites on the mineral ore to be collected, nor should they adsorb to the silica surface rendering it hydrophobic. The proper choice of a collector-frother combination is a delicate matter, and much "art" is involved. [30]

The separation of PbS (galena) from SiO_2 (silica), a mixture representative of what is found in a lead ore, is relatively simple, and is

---

[30] Leja, J., **Surface Chemistry of Froth Flotation**, Plenum Press, New York, 1982.

representative of other sulfide ores. A particularly effective family of collectors for sulfide minerals is the xanthates, alkyl dithiocarbonates:

$$R-O-C\overset{\displaystyle S}{\underset{\displaystyle S^-K^+}{\big\|}}$$

These show strong specific adsorption at sulfide surfaces and expose the hydrophobic R group (which need not be long chain) to the water. They adsorb minimally to oxide surfaces, and this can be depressed by increasing the *pH*. Many different "weak" surfactants may be used as frothing agents, but a common and inexpensive one is pine oil (terpineol). When the desired mineral is collected with adequate selectivity in a stable froth, the separation may be carried out in the simple flotation cell shown schematically in Fig. 4-61. The valuable minerals are merely skimmed off or allowed to flow from the surface in the froth. In addition to the collector and the frothing agents, most flotation recipes contain additional ingredients with particular objectives in mind. For example:

1) Electrolytes with potential-determining ions may be added to enhance the adsorption of the collector on the desired mineral. This technique is important when such adsorption is by the mechanism of ion pairing. For example, if the mineral to be collected has H+ and OH- as potential-determining ions, and the collector to be used is an anionic surfactant, adsorption may be enhanced by lowering the *pH*.

2) Certain substances, known as "activators" react with a specific mineral to improve the adsorption of the collector, *e.g.*, $CuSO_4$ for the adsorption of surfactants on ZnS.

3) Other substances, known as "depressants" are specifically adsorbed to block subsequent adsorption of the collector on particles which one does not want to collect in a given case.

7. *Foaming in non-aqueous media; general surface activity near a phase split*

The discussion of foams thus far has tacitly assumed the foaming medium was aqueous, and that foam stability was usually traceable to the presence of surface active agents. Nonetheless, significant foaming or frothing is often observed in such vapor-liquid contacting operations as distillation, gas absorption or gas stripping even when no water and no surfactants are present. The resulting liquid entrainment and back-mixing, referred to as *foam flooding*, can be a serious problem.[31] When it occurs, it is usually centered on only a few trays in a tray column or only in a certain region of a packed column, but not elsewhere. The origin of such behavior is

[31] Ross, S., in **Interfacial Phenomena in Apolar Media**, H.-F. Eicke and G. D. Parfitt (Eds.), Surf. Sci. Ser. 21, 1, Marcel Dekker, 1987.

now well understood and requires a more general view of surface activity than that associated with aqueous surfactants.

Many mixtures of two or more components are not miscible in all proportions over the temperature range of interest. Instead, at a given temperature, the solution may exhibit a split into phases of different composition. To keep things simple, consider a two-component system of liquids A and B. The components may be fully miscible at a given temperature, but undergo phase splitting as the temperature is lowered, producing a phase diagram of the type shown in Fig. 9-39 (a). The horizontal tie lines at any temperature below T_C, the critical solution temperature, join the compositions of the A-rich and B-rich phases. This system exhibits an *upper critical solution temperature* (UCST), but there are others that show phase splitting when the temperature is raised and therefore exhibit a *lower critical solution temperature* (LCST). There are a few systems that exhibit both a UCST and an LCST. In general, the two components have different

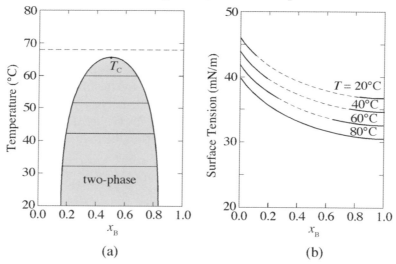

(a)   (b)

Fig. 9-39: (a) Phase diagram for the system of components A and B showing a phase split below the critical solution temperature, T_C, into an A-rich phase (left) and a B-rich phase (right). (b) Surface tension behavior of the binary system of A and B.

surface tensions, so that at a temperature just above T_C, (dashed line in Fig. 9-39(a)), the surface tension may vary with composition as shown in Fig. 9-39(b). In this example, B has the lower surface tension. In moving from pure A toward the critical point composition, one is approaching a point of incipient phase splitting. As this occurs, one would expect the relative adsorption of component B to increase, since adsorption is a precursor to an imminent phase separation. As adsorption of one of the components (the one of lower surface tension) increases, the range of phenomena associated with surface activity begins to emerge. Important for the present context is the ability of the adsorbed component to support an evanescent foam.

If one has a thermodynamic description of the solution, *i.e.*, the activity coefficients of the components, γ_A and γ_B as functions of composition, the relative adsorption may be derived using the Gibbs adsorption equation, Eq. (3.89), as a function of composition and temperature:

$$\Gamma_{B,A} = -\frac{x_B}{\left[1+(\partial\ln\gamma_B/\partial\ln x_B)_T\right]}\left(\frac{\partial\sigma}{\partial x_B}\right)_T. \qquad (9.27)$$

For a system exhibiting a symmetrical phase split of the type shown in Fig. 9-39(a), the non-ideality of the solution may be well represented by a regular solution model,[32] for which (when the molecular sizes of A and B are not too different):

$$\ln\gamma_A = \frac{\beta}{RT}x_B^2 \quad \text{and} \quad \ln\gamma_B = \frac{\beta}{RT}x_A^2 , \qquad (9.28)$$

where β is a constant. For such a case, the relative adsorption is given, using Eqs. (9.27) and (9.28), as

$$\Gamma_{B,A} = -\frac{x_B}{\left[1+(2\beta/RT)x_A x_B\right]}\left(\frac{\partial\sigma}{\partial x_B}\right)_T. \qquad (9.29)$$

Given the shape of the surface tension curves in Fig. 9-39(b), Eq. (9.29) produces a result for a given temperature (dashed lines in Fig. 9-40(a)) showing a maximum in $\Gamma_{B,A}$ slightly to the left of the critical point composition. This weighting of the adsorption toward a composition richer in the component of higher surface tension is due to the steeper slope of the $\sigma(x_B)$ curve in this region.

Ross, Nishioka and coworkers[33] examined a number of binary systems and produced results in the form of contours of constant relative adsorption, $\Gamma_{B,A}$, as exemplified in Fig. 9-40(a) for the system of diethylene glycol (A) and ethyl salicylate (B). The thesis of the above argument is that surface activity should correlate with adsorption. This was in fact found for frothing behavior measured in the same system as that of Fig. 9-40(a), and shown in Fig. 9-40(b). Froth lifetimes were measured as a function of temperature and composition producing a set of contours of constant froth lifetime, Σ, termed "isophroic" curves. Froth was formed by passing dry N_2 through 50 mL of a solution of desired composition atop a fritted disk in a 30 mm jacketed (thermostatted to $\pm0.1°C$) column. A steady state froth height, h, was reached for a given linear gas flow rate, V_{gas}, and the foam lifetime was defined as $\Sigma = h/V_{gas}$. At low gas flow rates, Σ was constant. The similarity

---

[32] See, *e.g.*, Prausnitz, J. M., Lichtenthaler, R. N., and Azevedo, E. G., **Molecular Thermodynamics of Fluid-Phase Equilibria**, 3rd Ed., pp. 313ff, Prentice-Hall, Upper Saddle River, NJ, 1999.

[33] Ross, S., and Nishioka, G., *J. Phys. Chem.*, **79** [15], 1561 (1975);
Ross, S., and Nishioka, G., *Colloid & Polym. Sci.*, **255**, 560 (1977);
Nishioka, G. M., Lacy, L. L., and Facemire, B. R., *J. Colloid Interface Sci.*, **80**, 197 (1981).

between the iso-Γ contours and iso-Σ curves in Fig. 9-39 is striking. One would not expect them to be identical because frothing depends on properties, notably viscosity, other than adsorption.

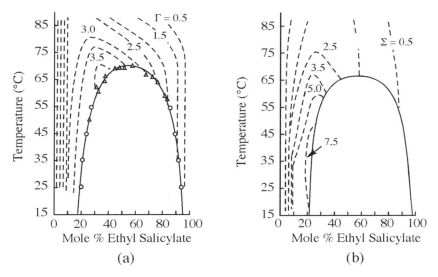

Fig. 9-40: (a) Phase diagram for the system of Diethylene glycol and Ethyl salicylate showing contours of constant adsorption, Γ, in units of μmole/m^2. (b) Lines of constant foam lifetime, isophroic curves, in seconds, for same system as that of (a). After [Ross, S., and Townsend, D. F., *Chem. Eng. Commun.*, **11**, 347 (1981).

Another observation concerning foaming in systems that are near to a phase split is that slight changes in conditions may lead to a sudden and complete disappearance of frothing. This may also be easily understood. Consider the system in Fig. 9-40(a) at a temperature of 50°C in the single-phase region just a little to the left of the phase boundary. If the temperature dropped slightly, or the mole fraction x_B were increased by a small amount, the system could move into the two-phase region. In this case, the small amount of B-rich conjugate phase that would be produced would have a lower surface tension than the A-rich phase it split from, and would spread spontaneously on its surface destroying foam in its wake, much like an anti-foam. The spreading coefficient would be:

$$S_{B/A} = \sigma_A - \sigma_B - \sigma_{A/B} > 0,$$ (9.30)

where A and B are shorthand for the A-rich and B-rich conjugate phases, respectively. This situation is to be contrasted with what would occur if one were to move into the two-phase region from the B-rich side of the phase diagram. In this case, the small amount of A-rich conjugate phase produced would *not* spread on the parent B-rich phase, and foaming would not be inhibited. It can be shown in general that if $S_{B/A} > 0$, then $S_{A/B} < 0$.

Some fun things to do:
Experiments and demonstrations for Chapter 9

1. *Making, testing and breaking macro emulsions*

Macro-emulsions (or simply, emulsions) are formed when oil and water are mixed with an appropriate emulsifier. These are commonly surfactants, with water-soluble compounds (high *HLB*) producing oil-in-water (O/W) dispersions or oil-soluble compounds (low *HLB*) producing water-in-oil (W/O) emulsions. One of the simplest ways to determine if a given emulsion is O/W or W/O is to attempt to dilute it with a dye. A water-soluble dye, such a food coloring, will readily dissolve into an O/W emulsion, but not a W/O emulsion. One of the easiest ways to break an emulsion is to change the nature of the emulsifier, either by adding an additional component or changing the temperature. For example, an O/W emulsion stabilized with a soap (salt of a fatty acid, such as Na oleate) will break when a sufficient amount of acid is added to convert the oleate anion to oleic acid, a very water-insoluble compound. Another method is to add a salt with a divalent cation, such as $CaCl_2$. This will produce the undissociated calcium di-tail compound, which is water insoluble. An O/W emulsion stabilized with a nonionic, polyethoxylate surfactant will break or invert upon heating as the ethylene oxide groups progressive dehydrate and lose their hydrophilicity. In these experiments, several different emulsions will be formed, examined and in some cases broken.

Materials:

- 150 mL light, white mineral oil
- Solution prepared by dissolving 2 g of sodium oleate in 100 mL of water
- 20 mL of Joy® dishwashing liquid
- 25 mL of 1N HCl
- 25 mL of 1M $CaCl_2$
- Some everyday emulsions, such as hand lotion, salad dressing, cream, melted butter, *etc*.
- 2–3 mL of red food coloring (a water-soluble dye)
- Four 120-mL jars with tight-fitting caps
- 6 disposable 1-mL pipettes
- 2 watch glasses

Procedure:

1) Prepare an O/W emulsion (Emulsion A) by adding 20 mL of white oil and 60 mL of the sodium oleate solution described above to a 100-mL jar, cap it and hand shake for about 30 s, let stand for about 30 s, repeating for about 5 minutes. The end result will be an O/W emulsion, but may actually form into two layers: a sparse emulsion of

small oil droplets in water on the bottom and a O/W biliquid foam layer on top. Also, the top space in the jar will fill with foam. Small variations in the recipe and the mixing procedure may lead to fairly large differences in the end results. The bottom-most layer may appear clear, but shining a laser through it (using a laser pointer, black background and darkened room) will reveal a distinct Tyndall cone verifying the presence of the dispersed phase.

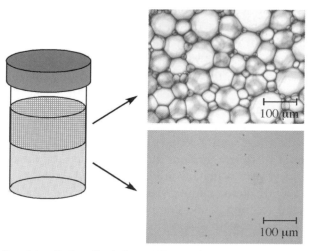

Fig. E9-1: Emulsion B described above. The top layer is found to be a W/O biliquid foam with cell sizes of the order of 10-100 μm, while the lower is an O/W emulsion with oil drops 5-10 μm.

2) Test either or both of the layers formed in Emulsion A by extracting a small portion of the layer, depositing it in a clean watch glass and adding a drop of red food coloring. It will be evident that the dye dissolves in the liquid of both layers, indicating that they are water-continuous.

3) Prepare a second emulsion (Emulsion B) by combining 5 mL of Joy® dishwashing liquid, 45 mL of de-ionized water and 50 mL of light white oil in a 120-mL jar, cap tightly and shake vigorously for about 10 s and let stand about 30 min. The system will split into two liquid layers, a slightly turbid layer on the bottom, a white layer on top, and a foam layer in the empty space above the liquid. The foam will gradually collapse.

4) Test both of the liquid layers formed in Emulsion B by extracting a small portion of the layer, depositing it in a clean watch glass and adding a drop of red food coloring. It will be evident that the dye dissolves in the lower liquid layer, indicating that it is water-continuous, but not in the top layer, indicating that it is oil-continuous. The commercial detergent used consists of a number of different compounds so that it is able to stabilize both water-continuous and

oil-continuous dispersions. If a microscope is available, the two layers will reveal structures as shown in Fig. E9-1.

5) Test a number of everyday emulsions using food coloring to determine which phase is continuous.

6) Break Emulsion A by adding one drop of 1N HCl to 10 mL of the emulsion, and by adding 1 drop of 1M CaCl$_2$ solution to 10 mL of the emulsion. Add one drop of the acid to 3 mL of Half&Half coffeee creamer and see what happens. (This emulsion will be found to invert from O/W to W/O.)

2. *Making a microemulsion*

A microemulsion may be thought of as a solution of micelles swollen to a very large size (5 – 50 nm) by oil solubilizate. They may also take on other forms, such as the bicontinuous structure shown in Fig. 3-34. They may be formed in a number of different ways by different materials, as described in Chap. 9, but one of the simplest recipes is that of water, oil and surfactant in *roughly* comparable proportions, together with a co-surfactant which is commonly a short-chain alcohol, such as butanol or pentanol. The co-surfactant intercalates itself into the surfactant palisade of the micelle allowing it to swell in size by more than an order of magnitude.

Materials:

•   In addition to materials listed above, 15 mL of *n*-butanol

Procedure:

Combine 15 mL of Joy® detergent, 15 mL of de-ionized water, 55 mL of light white oil and 15 ml of *n*-butanol in a 120 mL capped jar. Mix gently by rocking the jar a few times and let stand for 2-3 hrs. The system will split into three liquid layers, the central layer being a middle phase microemulsion. The lower phase is an O/W emulsion and the upper phase a W/O emulsion, both visibly turbid. The middle phase will appear absolutely clear, but shining a laser through it will reveal a Tyndall cone.

3. *Making and breaking foams*

Foams are readily made by shaking soap or detergent solutions in a closed container with ample gas space. With the right surfactant and composition the foam may show extraordinary longevity. Common ways for breaking them include adding salt, which collapses the electric double layer whose repulsive electrostatic interactions retard lamella thinning, and adding hydrophobic particles, which span the lamellae and are subsequently de-wet, causing film breakage.

Materials:

- Joy® dishwashing liquid
- Three 120-mL jars with tight-fitting caps
- About 5g sodium chloride
- About 5g hydrophobic silica, ≈ 1-5 μm diameter
- About 5g hydrophilic (ordinary) silica, ≈ 1 μm diameter

Procedure:

1) Prepare the foam by shaking 1 mL of the dishwashing liquid with 4 mL of water in the three 120-mL jars marked 1 - 3.

2) In Jar 1, sprinkle sodium chloride over the top of the foam. In Jar 2, sprinkle hydrophobic silica over the foam. In Jar 3, sprinkle hydrophilic silica over the top of the foam. In the first two cases, the foam will be seen to break, while in the last, the foam will remain intact.

Chapter 10

INTERFACIAL HYDRODYNAMICS

A. Unbalanced forces at fluid interfaces

"Interfacial hydrodynamics" refers to fluid flows that are generated or at least influenced by the existence boundary tension at fluid interfaces and its variations with temperature, composition, and surface charge density. These have already been encountered in this text in numerous contexts, including wetting, spreading and wicking, foam and emulsion stabilization, the behavior of systems with monomolecular surfactant films, electro-capillarity, *etc*. The fascinating manifestations of these phenomena have been reported since ancient times. The damping of waves through the use of olive oil that spontaneously spread on water was reported by Pliny the Elder and Plutarch in ancient Rome,[1] and has been illustrated in Fig. 1-4 of this text. Another phenomenon known to the ancients was the "camphor dance," mentioned briefly in Chap. 3.K.1, referring to the swimming and spinning of camphor particles when placed on water.[2] The nineteenth century saw capillary hydrodynamic phenomena occupy the attention of many of the most prominent scientists of the day, and some of their work is summarized in the now-famous (and still useful) entry on "Capillary Action" (by J. C. Maxwell and Lord Rayleigh) to the 11th (1911) Edition of the Encyclopedia Britannica.[3]

It was clear that motion, or the damping of motion, at fluid interfaces could be traced to unbalanced forces there. A patch of fluid interface may spring into motion in response to such forces, the motion being just a consequence of Newton's Second Law. To pursue this, it is convenient to resolve the forces on a patch of fluid interface into components normal to and tangential to the interface, as pictured in Fig. 10-1.

1. *Unbalanced normal forces*

Unbalanced normal forces arise whenever the pressure difference across the interface does not balance the capillary pressure required to support its curvature, *i.e.*, when the shape and/or location of the interface

---

[1] Gaines, Jr., G. L., **Insoluble Monolayers at Liquid-Gas Interfaces**, p. 1, Interscience Publ., New York, 1966.

[2] Tomlinson, C., *Phil. Mag.*, Ser. 4, **38**, 409 (1869).

[3] Maxwell, J. C., and Lord Rayleigh (J. W. Strutt), **Encyclopedia Britannica**, 11th Ed., Vol. 5, pp. 256-275, 1910. The original article, by Maxwell alone, appeared in the 9th Ed. (1876).

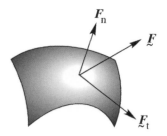

Fig. 10-1: Resolution of
surface forces into
components normal and
tangential to the surface.

fails to satisfy the Young-Laplace equation at each point. The local normal
stress at each point is thus given by:

$$\frac{F_n}{A} = \Delta p - \frac{2\sigma}{R_m},$$ (10.1)

where Δp is the local hydrostatic and/or hydrodynamic pressure difference
across the interface at the point of interest, and R_m is the local mean radius of
curvature. The hydrodynamic component of Δp may consist of an inertial
(dynamic) component: $\frac{1}{2}\rho v_z^2$, and a normal viscous component: $-2\mu(\partial v_z/\partial z)$,
where ρ is the fluid density and μ its viscosity, and z is the coordinate
normal to the surface at the point of interest.

Examples of flows due to imbalances of normal forces at the interface
include the oscillations of a non-spherical liquid drop (assuming the absence
of gravity). A second example is that of a cylinder of liquid (*i.e.*, a jet, as
pictured in Fig. 1-6 or Fig. 5-14) that finds itself unstable with respect to
axial corrugations along its surface. These grow in amplitude, eventually
pinching the jet off into drops. The classical problem of the breakup of
capillary jets is considered further below in Section B.1. The oscillation of a
liquid jet emerging from an elliptical orifice, pictured in Fig. 2-48, is another
example. The motion of and growth of capillary waves, including those
described in Chap. 5.C.2, as the Rayleigh-Taylor and Kelvin-Helmholtz
problems provide further examples of flows due to normal force imbalances.
Finally, it is these forces that are responsible for the rise of liquid inside a
capillary tube (the surface of which it wets) or of wicking into porous media,
described in Chap. 4. H.

2. Tangential force imbalances: the Marangoni effect

When surface or interfacial tension varies from point to point in the
interface, a tangential force equal to the gradient of the tension is developed,
as shown in Fig. 10-2. It is directed from the point of lower tension toward
the point of higher tension, and the magnitude of the resulting surface stress
is given by

$$\frac{F_t}{A} = \nabla_{\text{II}}\sigma,$$ (10.2)

where $\nabla_{II}\sigma$ is the interfacial gradient of the boundary tension, *i.e.*, the gradient taken with respect to orthogonal coordinates tangential to the

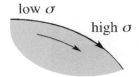

low σ

high σ

Fig. 10-2: Flow induced by unbalanced tangential forces at a fluid interface: the Marangoni effect.

interface. Variations in tension frequently arise during the transfer of heat or chemical species across the interface, owing to variations in interfacial temperature or composition as occur in many of the multiphase separation processes used by chemical engineers. Movement along the patch of interface is transmitted to the adjacent bulk phases via the "no slip" condition, and significant bulk convection may result. Historically, descriptions of these flows emerged in the 19th century and have come to be termed *Marangoni effects*,[4] defined as bulk flows generated by a spatial variation in boundary tension. They are named for the Italian physicist, Carlo G. M. Marangoni, who provided one of the early studies of the phenomenon in his PhD dissertation from the University of Pavia, Italy, in 1865, and in subsequent publications. Both observations and appropriate explanations, however, pre-date Marangoni. One of the earliest reports was that of the English microscope maker, Cornelius Varley, who in 1836 noticed with the aid of a microscope "motions of extremely curious and wonderful character in fluids undergoing evaporation."[5] Much of the early work has been summarized, with significant referencing, by Scriven and Sternling in 1960.[6]

One example of Marangoni convection occurs when there is mass transfer of a solute across a curved meniscus, as shown in Fig. 10-3. If the solute is transferring upward, it will become preferentially depleted from the region beneath corner of the meniscus due to the geometric asymmetry of the system. If the solute decreases interfacial tension, its depletion from the meniscus region will locally increase interfacial tension there, leading to a force directed toward the wall. This may generate convection eddies in either or both the bulk phases, as pictured in Fig. 10-3, which shows the schlieren image produced during the transfer of acetic acid across the interface between water and a benzene/chlorobenzene solution of just slightly greater density than water in a capillary cell. This particular Marangoni flow is macroscopic in nature, *i.e.*, it is generated and sustained by macroscopic asymmetry in the system, and may be designated as *macro* Marangoni convection.

The first quantitative example of macro Marangoni convection documented in the literature concerned the phenomenon of "tears of strong

[4] Marangoni, C., *Nuovo Cimento*, [2] **16**, 239 (1871); [3] **3**, 97 (1878).

[5] Varley, C., *Trans. Soc. Arts*, **50**, 190 (1836).

[6] Scriven, L. E., and Sternling, C. V., *Nature*, **187**, 186 (1960).

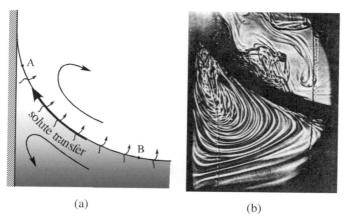

(a)   (b)

Fig. 10-3: (a) Macro-Marangoni convection generated during the transfer of a solute across a curved meniscus. (b) Example of convection generated during transfer of acetic acid across the interface between benzene/chlorobenzene (on bottom) and water. Dimensions of meniscus are approx. 3 cm x 3 cm. After [Kayser, W. V., and Berg, J. C., *Ind. & Eng. Chem. Fund.*, **10**, 526 (1971).]

wine," described by J. J. Thomson in his 1855 article entitled: "On certain curious Motions observable at the Surfaces of Wine and other Alcoholic Liquors."[7] A film of wine, an ethanol/water solution, forms a meniscus around the glass similar to that shown in Fig. 10-3. The alcohol preferentially evaporates from the meniscus because it is more volatile than water, leaving behind a water-rich film of relatively high surface tension. More liquid is then drawn up the side of the glass where it accumulates in a ring that eventually breaks into droplets ("tears") that flow back down into the wine. The effect can be enhanced by swirling the wine in the glass to produce an enlarged meniscus. A photograph of the phenomenon is shown in Fig. 10-4, and on the cover of this book.

Fig. 10-4: Tears of wine shown in the shadow of this glass of 20.0% Porta Roca ruby port wine.

[7] Thomson, J., *Phil. Mag.*, **19**, 330 (1855).

Another type of Marangoni convection may occur as a result of a system's inherent instability with respect to small disturbances. Consider, for example, desorption of acetone from a shallow aqueous pool infinite in lateral extent (therefore free of menisci), as pictured in Fig. 10-5. During the evaporation, the surface region becomes enriched in water and will therefore have a higher surface tension than that corresponding to the bulk solution beneath. Such a system, as all systems, is subject to small disturbances. One component of such a disturbance may be a local dilation, bringing an eddy of acetone-rich liquid up to the surface. The surface tension will thus be locally reduced, and a dilational force established. This will further dilate the surface, bringing still larger amounts of acetone-rich liquid to the surface, *etc.*, until the disturbance has amplified itself to macroscopic proportions, usually within a fraction of a second. Because this bulk flow results from the amplification of microscopic disturbances, it may be referred to as a *micro* Marangoni effect, as opposed to the macro Marangoni effect described earlier. It is perhaps evident that such convection would not occur if the transfer direction were reversed, *i.e.*, if the acetone vapor were being *ab*sorbed into the water. Under these circumstances, a dilational eddy would bring *water*-rich liquid to the surface, locally *in*creasing the tension and generating a converging flow that destroys the disturbance. When it *does* occur, the self-amplification of the disturbance is opposed by the eroding influence of diffusion and viscosity, and only when a certain critical concentration gradient exists will the disturbance grow as shown. A similar phenomenon occurs when a liquid pool is heated from below (or cooled

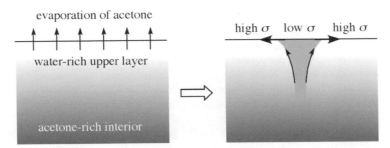

Fig. 10-5: Flow generated by self-amplification of small disturbances
(micro-Marangoni effect resulting from hydrodynamic instability).

from above, as by evaporation). Since surface tension decreases with temperature, an eddy bringing warmer liquid to the surface from the interior may be self-amplified. The instability criterion, as determined in an example given below (thermocapillary instability in a shallow pool) turns out to be quite easily satisfied. The result in shallow pools of liquid may be a highly regular pattern of hexagonal cells, as shown in Fig. 10-6.

This pattern is referred to as one of *Bénard cells* (after Henri Bénard,[8] who observed and photographed them as they occurred in shallow pools of molten wax). Figure 10-6 shows drawings of Bénard and a photograph of a

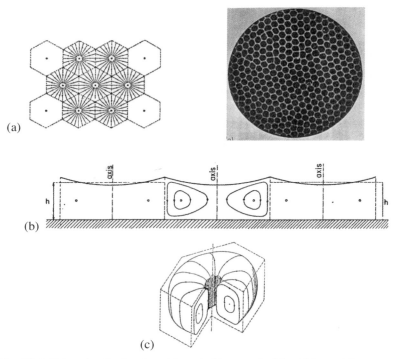

Fig. 10-6: Bénard cells developed in a shallow pool of liquid heated from below: (a) top, (b) side, and (c) perspective views. Photograph shows a 4-cm diameter pool, 3 mm deep, of molten wax heated from below. (b) shows that flows produce surface relief, with elevation differences of the order of 1 μm. [Avsec, D., "Tourbillons thermoconvectifs dans l'air – application a la métérorologie," Publ. Sci. Tech. Min. d L'air, N° 155, Paris (1939).]

shallow layer of molten spermaceti (cetyl palmitate) heated from below. The flow is one of regular hexagonal cells with liquid rising at the cell centers and descending around the edges. The size scale of the cells is in proportion to the depth of the layer. A surface relief pattern (with elevation differences of the order of one μm) is produced, and is the basis for many of Bénard's beautiful photographs. As the liquid layer becomes deeper, the cell structure becomes less regular, and eventually breaks down entirely into a pattern of random "tessellations," as described by Thomson.[9] As discussed below, such flows in deep layers are generally more the result of buoyancy-driven natural

[8] Bénard, H., *Rev. Gen. Sci. Pures Appl. Bull. Assoc. Franc. Avan. Sci.*, **11**, 1261, 1309 (1900);
  Bénard, H., *Ann. Chim. Phys.*, [7] **23**, 62 (1901);
  Bénard, H., *Compt. Rend.*, **154**, 260 (1912); **185**, 1109 (1927); **185**, 1257 (1927);
  Bénard, H., *Bull. Soc. Franc. Phys.*, **266**, 1125 (1928);
  Bénard, H., *Compt. Rend.*, **201**, 1328 (1935).
[9] Thomson, J. J., *Proc. Roy. Phil. Soc. Glasgow*, **13**, 464 (1882).

convection than a Marangoni effect. The formation of Bénard cells sometimes occurs during the drying of paint films (which is accompanied by the desorption of volatile solvents) and produces *pigment flotation, i.e.,* the convection of pigment to the borders of the cells. The result, upon solidification or curing, is known by paint-makers as "fish scale," pictured in Fig. 10-7, and is a highly undesirable consequence of the Marangoni effect.

Fig. 10-7: Pigment flotation in a drying paint film ("fish scale"). Diameter of figure ≈ 2 cm. From [Bell, S. H., *J. Oil Colour Chemists Asscoc.,* **35**, 373 (1952).]

Sometimes the micro Marangoni effect is manifest as small-scale chaotic flows referred to as "interfacial turbulence." It is not to be confused, however, with the turbulence associated with high Reynolds Number flows.

Usually a system unstable with respect to Marangoni convection for transfer of a solute in one direction is stable for transfer in the other direction. In the stable case, however, macro Marangoni convection may occur. This is pictured for the transfer of acetic acid across an interface between a drop of ethyl acetate at the tip of a capillary tube and ethylene glycol, as pictured in Fig. 10-8. The system is unstable with respect to micro

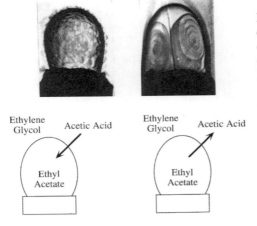

Fig. 10-8: Marangoni convection during mass transfer as a drop of ethyl acetate emerges from a 3 mm diameter nozzle into ethylene glycol. Micro Marangoni convection ("interfacial turbulence") occurs during inward transfer of acetic acid, but a single macro Marangoni vortex is observed during outward transfer. From [Bakker, C. A. P., van Buytenen, P. M., and Beek, W. J., *Chem. Eng. Sci.,* **21**, 1039 (1966).]

Marangoni convection, manifest in this case as "interfacial turbulence," when the solute is initially in the ethyl acetate phase, but not for the reverse transfer direction. In the latter case, a single toroidal eddy is formed in the drop, as the interfacial tension at the tip of the drop is lower than at the base.

3. Boundary conditions at a fluid interface

The mathematical origin of the phenomena described above is contained in the extensions that must be made to the boundary conditions to the equations of fluid mechanics, *i.e.*, the Navier-Stokes equations,[10] to account for "capillary effects." These are formulated below for the case of a quasi-flat interface normal to the *z*-axis and dividing phases (') and ("), as shown in Fig. 10-9, but may be formulated for other macroscopic interface geometries as needed. η is the elevation above the flat datum surface, which is taken as normal to the *g*-vector.

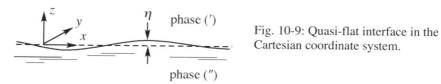

Fig. 10-9: Quasi-flat interface in the Cartesian coordinate system.

The first boundary condition is the *kinematic condition*, which express the continuity of velocity across the interface:

1. $$\underline{v}'' = \underline{v}' = \frac{\partial \eta}{\partial t}$$ (10.3)

It is the tangential component of Eq. (10.3) that is referred to as the "no slip" condition. Continuity of the normal component of velocity assumes that a rapid phase change is not occurring. In such a case, the normal component of Eq. (10.3) would need to be

$$\rho'' v_z'' - \rho' v_z' = \dot{m},$$ (10.4)

where $\dot{m}$ is the mass rate of phase change from phase (") to phase (') per unit area.

The second boundary condition, the so-called *dynamic condition*, expresses the continuity of forces at the interface. Following the decision to resolve these into normal and tangential components, we write first, for the balance of normal forces:

2. $$p'' - p' + (\rho' - \rho'')g\eta + \tau_{zz}'' - \tau_{zz}' + \sigma \nabla_{II}^2 \eta = 0.$$ (10.5)

The pressures p'' and p' may have dynamic (inertial) as well as hydrostatic components, which are referenced back to the datum plane. The next term is the hydrostatic adjustment to any elevation difference between the actual interface and the datum plane. The terms τ''_{zz} and τ'_{zz} are the normal viscous stresses, which for Newtonian incompressible fluids are given by:

$$\tau''_{zz} = -2\mu'' \left(\frac{\partial v''_z}{\partial z} \right), \; etc. \tag{10.6}$$

The final term on the left hand side of Eq. (10.5) is the Young-Laplace curvature term, in which the curvature, as given by Eq. (2.32), has been simplified for the case in which the surface deflection is small, so that $\kappa \approx \nabla_{\text{II}}^2 \eta$.

Had the surface in Fig. 10-9 been that of an axisymmetrically perturbed right circular cylinder of unperturbed radius a_0, the curvature term would take the form:

$$\kappa = \left(\frac{1}{R_1} + \frac{1}{R_2} \right) = \frac{1}{a_0 + \eta} - \frac{d^2\eta}{dz^2} \approx \frac{1}{a_0} - \frac{d^2\eta}{dz^2}, \tag{10.7}$$

where z is the cylindrical axial coordinate, and $\eta(z)$ is the surface perturbation.[11]

If rapid phase change is occurring at the interface, an additional term must be included in Eq. (10.5) to describe the resulting "vapor recoil," viz.

$$\dot{m}^2 \left[\frac{1}{\rho''} - \frac{1}{\rho'} \right]. \tag{10.8}$$

It has been shown that at pressures under approximately 100 kPa, the vapor recoil effect may produce violent convection and noticeable depressions in the surface of an evaporating liquid.[12]

The remaining dynamic condition (the Marangoni equation) expresses the balance of tangential stresses:

3. $$\underline{\tau}''_z - \underline{\tau}'_z + \nabla_{\text{II}}\sigma + \underline{\tau}^s = 0, \tag{10.9}$$

where $\underline{\tau}''_z$ and $\underline{\tau}'_z$ are the viscous tractions of the adjacent bulk phases, viz.

$$\underline{\tau}''_z = \underline{e}_x \tau''_{zx} + \underline{e}_y \tau''_{zy}, \; etc., \tag{10.10}$$

where $\underline{e}_x$ and $\underline{e}_y$ are unit vectors in the x and y directions, respectively, and assuming Newtonian fluids

---

[11] More general expressions for the curvature of perturbed cylindrical and spherical surfaces may be found in: Levich, V. G., **Physicochemical Hydrodynamics**, pp. 377-378, Prentice-Hall, Inc., Englewood Cliffs, NJ, 1962.

[12] Palmer, H. J., and Berg, J. C., *J. Fluid Mech.*, **75**, 487 (1976).

$$\tau_{zx}'' = \mu'' \left(\frac{\partial v_z''}{\partial x} + \frac{\partial v_x''}{\partial z} \right) \text{ and } \tau_{zy}'' = \mu'' \left(\frac{\partial v_z''}{\partial y} + \frac{\partial v_y''}{\partial z} \right). \tag{10.11}$$

$\underline{\tau}^s$ represents the contribution of any intrinsic interfacial rheology. Such effects are generally present only when there are surfactant films at the interface and are described more fully in Section D below.

Equation (10.9) is a vector equation yielding two scalar equations, for the x and y components, respectively. It is instead often more convenient to extract a pair of scalar equations by taking i) the surface divergence, and ii) the normal component of the surface curl of Eq. (10.11).[13] This results (for Newtonian incompressible fluids) in:

$$3(i) \quad -\mu'' \left[\left(\frac{\partial^2 v_z''}{\partial z^2} \right) + \nabla_{\rm II}^2 v_z'' \right] + \mu' \left[\left(\frac{\partial^2 v_z'}{\partial z^2} \right) + \nabla_{\rm II}^2 v_z' \right] + \nabla_{\rm II}^2 \sigma + \nabla_{\rm II} \cdot \underline{\tau}^s = 0 \tag{10.12}$$

and

$$3(ii). \quad -\mu'' \underline{e}_z \cdot \nabla_{\rm II} \times \left(\frac{\partial v_{\rm II}''}{\partial z} \right) + \mu' \underline{e}_z \cdot \nabla_{\rm II} \times \left(\frac{\partial v_{\rm II}'}{\partial z} \right) + \underline{e}_z \cdot \nabla_{\rm II} \times \underline{\tau}^s = 0. \tag{10.13}$$

The term in Eq. (10.9) describing the Marangoni effect is the one that describes the variation in the boundary tension caused by variations in interfacial temperature, composition and electrical potential, *i.e.*,

$$\nabla_{\rm II} \sigma = \left(\frac{\partial \sigma}{\partial T} \right)_0 \nabla_{\rm II} T + \left(\frac{\partial \sigma}{\partial C} \right)_0 \nabla_{\rm II} C + \left(\frac{\partial \sigma}{\partial E} \right)_0 \nabla_{\rm II} E, \tag{10.14}$$

where E is the local electrical potential. The last term represents electrocapillary effects, which are important in a number of situations, particularly in metallurgical operations,[14] but will not be considered further here. The terms on the right hand side are seen to be leading terms in a Taylor expansion. Equation (10.14) reveals that the Navier-Stokes equations are coupled, through the boundary conditions, to the thermal energy equation, and the convective diffusion equations. The boundary conditions for the thermal energy equation at the interface are:

$$T' = T'', \text{ and } q' + q'' + S_{\rm T} = 0, \tag{10.15}$$

where q' and q'' are heat fluxes from the adjacent bulk phases, generally given in terms of Fourier's Law:

$$q' = -k' \frac{\partial T'}{\partial z}, \text{ etc.}, \tag{10.16}$$

---

[13] Miller, C. A., in **Surface and Colloid Science**, Vol. 10, E. Matijevic (Ed.), pp. 227-293, Plenum Press, New York, 1978.

[14] Brimacombe, J. K., Graves, A. D., and Inman, D., *Chem. Eng. Sci.*, **25**, 1817 (1970); Brimacombe, J. K., and Richardson, F. D., *Trans. Inst. Min. Metall.*, **80**, C140 (1971); Brimacombe, J. K., and Weinberg, F., *Metall. Trans.*, **3**, 2298 (1972).

where k is the thermal conductivity. S_T represents the generation (or consumption) of heat (per unit area) arising from any chemical reaction occurring at the interface.

For "ordinary" solutes, *i.e.*, those showing no significant accumulation at the interface, the boundary conditions for the convective diffusion equation (for a species i) are

$$C_i' = mC_i'', \text{ and } j_i' + j_i'' + \mathcal{R}_i = 0, \tag{10.17}$$

where m is the distribution equilibrium constant for species i, and j_i' and j_i'' are fluxes of species i from the adjacent bulk phases, generally given in terms of Fick's Law:

$$j_i' = -D_i' \frac{\partial C_i'}{\partial z}, etc., \tag{10.18}$$

where D_i is the diffusivity of species i and C_i its concentration. $\mathcal{R}_i$ represents the generation (or consumption) of species i (per unit area) arising from any chemical reaction occurring *at the interface*. Equations (10.17), as well as a modified boundary condition to the convective diffusion equation, are developed further below to consider the special properties imparted to fluid interfaces by surface active agents.

B. Examples of Interfacial Hydrodynamic Flows

1. *The breakup of capillary jets*

As an example of flows governed by normal force imbalances at fluid interfaces, consider the classical problem of the breakup of capillary jets as pictured in Figs. 1-6 and 5-14. Whenever a liquid emerges from the end of a small-diameter round tube or circular orifice into an immiscible fluid at sufficient velocity, a capillary jet is formed. It inevitably disintegrates into droplets under the influence of surface tension forces. Under some conditions, other forces may also be involved. Prediction of the characteristics of the jet breakup in terms of fluid properties and system parameters is one of the classical problems in fluid mechanics. Of particular interest are the unbroken length of the jet and the size (or size distribution) of the resulting drops. Capillary jets are of considerable practical importance. Examples include fuel injection, spray coating, sieve-tray liquid extraction, fiber spinning, jet reactors, ultrasonic emulsification, ink jet printing, jet cutting, and others.

Plateau[15] first demonstrated that a cylinder of liquid becomes unstable with respect to disintegration into droplets when its length exceeds its circumference, a result that was proven mathematically at about the same

---

[15] Plateau, M. T., *Phil. Mag.* **12**, 286 (1856).

time by Beer.[16] The cylinder of liquid will become unstable when an axisymmetric, sinusoidal disturbance of the jet surface, as shown in Fig. 10-10, results in a reduction of surface area. This is because a reduction in surface area results in a reduction in system free energy, characteristic of a spontaneous process.

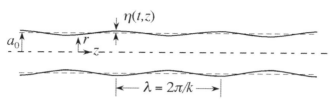

Fig. 10-10. Axisymmetric disturbance on a liquid cylinder.

The system is continually subject to small disturbances of all wavelengths (nozzle vibrations, motion in the surrounding fluid, *etc.*). For short-wavelength undulations, the area of the surface is greater than that of the undisturbed cylinder, but as the wavelength increases, this area augmentation decreases until finally a wavelength is reached for which there is an area *decrease*. This critical wavelength turns out to be essentially equal to the circumference of the liquid cylinder. While this result explains qualitatively why capillary jets are subject to breakup, it cannot be used to predict the jet length or drop size under various conditions. The first analysis leading to such predictions was that of Lord Rayleigh[17] in 1879. Rayleigh's analysis is discussed after summarizing some of the experimental observations concerning jet breakup.

When a capillary jet issues from a circular nozzle, the axisymmetric disturbances to which it is subject grow in amplitude until they reach the magnitude of the radius of the jet itself, pinching it off into droplets as shown in Fig. 5-14. These corrugations are generally too small and fast moving to be seen by the naked eye, and indeed the train of droplets often appears to be part of the unbroken jet. High-speed photographs, as Fig. 1-6, however, reveal a pattern remarkably close to that suggested by the drawing. The generally observed dependence of jet length L on velocity for a given liquid is shown in Fig. 10-11. Very low flow rates produce the "dripping regime," when no jet is formed. As velocity is increased, a laminar jet forms, and its length increases linearly with velocity over what is called the "Rayleigh region." Breakup in this range is highly regular and generally yields drops of uniform size and spacing, although in many cases, large drops (of radius comparable to that of the jet) alternate with tiny "satellite" drops. As velocity increases still further, a maximum in jet length is observed, beyond which jet length decreases. Breakup is then not as regular, and there may be noticeable side-to-side thrashing, so that the drops spread out over a fairly broad area.

[16] Beer, M., *Poggendorf's Annalen*, **96**, 1 (1855).
[17] Lord Rayleigh (J. W. Strutt), *Proc. Roy. Soc. (London)*, **29**, 71 (1879).

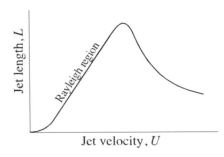

Fig. 10-11: Observed
dependence of jet length
on velocity.

Further increases in velocity lead to greater irregularity, until finally a
velocity is reached for which the jet seems to burst into a spray of droplets
(atomization). At this point, jet length becomes hard to define. It is assumed
that the liquid emerging from the orifice is now in turbulent flow. If it is, the
breakup will be highly irregular and will in general not show the above
dependence on velocity. The liquid may under some conditions actually
burst into a spray immediately upon leaving the orifice.

Rayleigh modeled the jet, as shown in Fig. 10-10, as an initially
quiescent cylinder of inviscid, incompressible liquid of surface tension σ and
unperturbed radius a_0 subject to axisymmetric perturbation, in the absence of
gravity, made up of components of all wavelengths. Because the disturbance
amplitudes are all very small, their description is effected with the linearized
equations of change, and the total disturbance is obtained by the simple
superposition of all the disturbance components. It thus suffices to describe a
single perturbation component of general wavelength λ. Rayleigh assumed
that out of all the disturbance components present, there would be one of
some wavelength (whose value would depend on system properties) that
would be amplified *much* more rapidly than all the others, so that it by itself
would determine the morphology of the breakup. The regularity of the
breakup as observed in high-speed photography validates Rayleigh's
assumption.

The perturbed jet radius, a, is given by

$$a = a_0 + \eta(t,z), \tag{10.19}$$

where the perturbation $\eta(t,z)$ is assumed to be of the form

$$\eta = \eta_0 \exp(\beta t + ikz) \tag{10.20}$$

where η_0 is an initial disturbance level, z is the axial coordinate moving with
the velocity of the jet, and r is the radial coordinate. (Recall that θ-variations
were not considered, so that the class of disturbances is limited to those that
are axisymmetric.) k is the wave number of the disturbance, related to the
wavelength by $k = 2\pi/\lambda$. The disturbance is imposed along the whole
length of the jet and allowed to grow *temporally*, when in fact what actually
happens is that a disturbance is imposed at one end of the jet and then grows
spatially. A spatial stability analysis of jet breakup paralleling the temporal

analysis of Rayleigh has been performed,[18] and the results coincide with Rayleigh's over the range of meaningful wavelengths. The imposition of disturbances in the jet profile produces corresponding disturbances in the other dependent variables in the system, *viz.*, the axial and radial velocity components and the pressure, all of which must obey the relevant equations of change (the Navier-Stokes equations) subject to the appropriate boundary conditions. The velocity components and pressure take the form:

$$v_r = R(r)\exp(\beta t + ikz),$$ (10.21)

$$v_z = Z(r)\exp(\beta t + ikz), \text{ and}$$ (10.22)

$$p = p_0 + \frac{\sigma}{a_0} + P(r)\exp(\beta t + ikz),$$ (10.23)

where p_0 is the pressure in the external phase. The final term in Eq. (10.23) is seen to be the pressure perturbation.

The linearized r and z-components of the equation of motion for an inviscid, incompressible fluid in the absence of gravity are

$$\frac{\partial v_r}{\partial t} = -\frac{1}{\rho}\frac{\partial p}{\partial r}, \text{ and}$$ (10.24)

$$\frac{\partial v_z}{\partial t} = -\frac{1}{\rho}\frac{\partial p}{\partial z},$$ (10.25)

and the continuity equation (in cylindrical coordinates) is

$$\frac{1}{r}\frac{\partial}{\partial r}(rv_r) + \frac{\partial v_z}{\partial z} = 0$$ (10.26)

Substitution of Eqs. (10.22) – (10.23) into Eqs. (10.24)–(10.26) gives:

$$\beta R = -\frac{1}{\rho}\frac{dP}{dr},$$ (10.27)

$$\beta Z = -\frac{ik}{\rho}P, \text{ and}$$ (10.28)

$$\frac{dR}{dr} + \frac{R}{r} + ikZ = 0.$$ (10.29)

Elimination of Z in Eq. (10.29) using Eqs. (10.27) and (10.28) gives

$$r^2\frac{d^2R}{dr^2} + r\frac{dR}{dr} - (1 + k^2r^2)R = 0,$$ (10.30)

a modified Bessel equation of order 1, whose general solution is given by

---

[18] Keller, J. B., Rubinow, S. I., and Tu, Y. O., *Phys. Fluids*, **16**, 2052 (1973).

$$R(r) = C_1 I_1(kr) + C_2 K_1(kr),$$ (10.31)

where $I_1(kr)$ and $K_1(kr)$ are modified Bessel functions of the first and second kind, respectively.

Next, the boundary conditions are applied. First, at the centerline, $r = 0$, the velocity components must remain finite. Since $K_1(kr)$ diverges as $r \rightarrow 0$, $C_2 = 0$, and

$$R(r) = C_1 I_1(kr).$$ (10.32)

The first boundary condition at the surface of the jet, $r \approx a_0$, is the kinematic condition

$$v_r = \frac{\partial \eta}{\partial t}.$$ (10.33)

Substitution of Eqs. (10.20), (10.21) and (10.32) into Eq. (10.33) gives $C_1 I_1(ka_0) = \eta_0 \beta$, so

$$R(r) = \eta_0 \beta \frac{I_1(kr)}{I_1(ka_0)}.$$ (10.34)

The remaining variables, v_z and p may now be obtained using the expression for $R(r)$. For example, using the above solution for $R(r)$ with Eq. (10.27) gives

$$P(r) = -\frac{\eta_0 \rho \beta^2}{k} \frac{I_0(kr)}{I_1(ka_0)},$$ (10.35)

in which use of the fact that $\int_0^r I_1(kr) = I_0(kr)$ has been made. Substitution of Eq. (10.35) into Eq. (10.23) then gives

$$p = p_0 + \frac{\sigma}{a_0} - \frac{\eta_0 \rho \beta^2}{k} \frac{I_0(kr)}{I_1(ka_0)} \exp(\beta t + ikz).$$ (10.36)

The second boundary condition at the surface of the jet is the balance of normal forces:

$$p = p_0 + \sigma \left(\frac{1}{R_1} + \frac{1}{R_2} \right),$$ (10.37)

where R_1 and R_2 are the principal radii of curvature. R_1 is the local radius of the jet, so that, making use of Eq. (10.20)

$$\frac{1}{R_1} = \frac{1}{a_0 + \eta} \approx \frac{1}{a_0} - \frac{\eta}{a_0^2} = \frac{1}{a_0} - \frac{\eta_0}{a_0^2} \exp(\beta t + ikz),$$ (10.38)

while, in accord with Eq. (10.9), and again making use of Eq. (10.20)

$$\frac{1}{R_2} = -\frac{d^2\eta}{dz^2} = \eta_0 k^2 \exp(\beta t + ikz). \tag{10.39}$$

Substitution of Eqs. (10.36), (10.38) and (10.39) into Eq. (10.37) yields what is termed the *dispersion equation* for the system, *i.e.*, a relationship between the growth rate constant and the wave number (hence wavelength) of the perturbation:

$$\beta^2 = \frac{\sigma k}{\rho a_0^2} \frac{I_1(ka_0)}{I_0(ka_0)} \left(1 - k^2 a_0^2\right). \tag{10.40}$$

It is evident that perturbations grow in amplitude (*i.e.*, $\beta > 0$) only if

$$ka_0 \equiv \frac{2\pi a_0}{\lambda} < 1, \tag{10.41}$$

that is, the jet is unstable only to perturbations whose wavelength exceeds the jet circumference. The wavelength producing the maximum growth rate is found by setting the derivative of β^2 with respect to k equal to zero and solving for k (hence λ), obtaining the preferred wavelength as

$$\lambda^* \approx 9.02 a_0, \tag{10.42}$$

predicting a drop volume of

$$V_{\text{drop}} = \pi a_0^2 \lambda^* \approx 28.34 a_0^3, \tag{10.43}$$

corresponding to a drop radius about 1.89 times the radius of the orifice. This is about what is observed in practice.

Substitution of Eq. (10.42) back into Eq. (10.40) gives the maximum growth rate constant as

$$\beta_{\text{max}} \approx \left(0.12 \frac{\sigma}{\rho a_0^3}\right)^{1/2}. \tag{10.44}$$

This allows computation of the predicted jet length. If it is presumed that the exponential growth of the disturbance amplitude persists all the way until it reaches the value of the jet radius, we may substitute $\eta = a_0$ into Eq. (10.20) and solve for the time of flight required to reach the end of the jet. This time is also L/U, the length of the jet divided by the jet velocity:

$$t_{\text{breakup}} = \frac{1}{\beta_{\text{max}}} \ln \frac{a_0}{\eta_0} = \frac{L}{U}. \tag{10.45}$$

Thus the jet length is given by:

$$L = \frac{U}{\beta_{\text{max}}} \ln \frac{a_0}{\eta_0} \propto U \left(\frac{\rho a_0^3}{\sigma}\right)^{1/2}. \tag{10.46}$$

The predicted dependence of the jet length on all of the properties and parameters of the system has been verified by experiment.

Three major limitations of the Rayleigh analysis were relaxed by Weber[19] in 1931. He permitted the jet liquid to have finite viscosity, the external fluid to have finite density (although it remained inviscid), and disturbances to have asymmetrical (θ-direction) components. Weber's dispersion equation is considerably more complicated than Rayleigh's, but to a good approximation for the case of a zero-density external fluid, yields the following results for the maximum growth-rate constant and the corresponding wave number, respectively:

$$\beta_{max} = \left[\left(\frac{8\rho a_0^3}{\sigma} \right)^{1/2} + \left(\frac{6\mu a_0}{\sigma} \right) \right]^{-1} \tag{10.47}$$

$$k* = \left[2a_0^2 + \left(\frac{\mu^2 a_0^3}{2\rho\sigma} \right)^{1/2} \right]^{-1/2}, \tag{10.48}$$

where μ is the jet liquid viscosity, and ρ is the density of the jet liquid. The effect of viscosity may thus be easily computed, and it can be shown that the jet may be regarded as effectively inviscid until μ reaches about 10 cP.

When external-phase density is included, the results for β_{max} and $k*$ cannot be expressed in simple closed form, but they do not show significant deviations from Eqs. (10.47) and (10.48) until the dimensionless group now known as the Weber Number (We) exceeds approximately 0.1, $i.e.$,

$$We = \frac{U^2 \rho_e 2a_0}{\sigma} > 0.1, \tag{10.49}$$

where ρ_e is the density of the *external medium*. Beyond $We = 0.1$, β_{max} begins to increase with We, so that the jet length begins to *decrease*. Thus Weber was able to predict the observed maximum in jet length as velocity is increased. The physical reason for the increased growth rate of the disturbances is that the relative motion of the external phase (of non-zero density) over the corrugated surface of the disturbed jet causes (via the Bernoulli effect) a longitudinal variation in pressure external to the jet. The pressure is lower over the crests of the disturbances (where local relative velocity is higher) and higher over the troughs, thus accelerating the necking-down process. This is also called the "aerodynamic effect."

The possibility of the spontaneous growth of non-axisymmetric disturbances is allowed for by writing the disturbance parameters in the more general form (*i.e.*, instead of Eq. (10.20)):

---

[19] Weber, C., *Zeit. Angew. Math. Mech.*, **11**, 136 (1931).

$$\eta = \eta_0 \exp(\beta t + ikz + ih\theta),\qquad\qquad (10.50)$$

where "h" is a "symmetry factor" which can assume values 0, 1, 2, ..., etc. h = 1 corresponds to a planar sinuosity of the type shown in Fig. 10-12 and leads to side-to-side thrashing of the jet if this disturbance component grows to sufficient amplitude before the jet is broken up by the growth of

Fig. 10-12: Asymmetric disturbances of planar sinuosity, h = 1.

axisymmetric disturbances. h = 2 corresponds to jets of non-circular cross-section, and higher values of h correspond to still more complex deformations. Weber considered only h = 1, and found that such disturbances should be capable of producing noticeable thrashing only when aerodynamic effects begin to be important, i.e., when $We > 0.1$. Indeed the growth of such disturbances depends entirely upon aerodynamic effects. All of Weber's predictions, when properly tested (primarily by the extensive experiments of Haenlein[20] in 1932), have been at least semi-quantitatively validated in the laboratory.

Tomotika (1935) examined the effect of external-phase viscosity on the wavelength of the most unstable disturbance in the absence oaf relative motion (and found its effect to be significant, with higher viscosity producing longer wavelengths). Others[21] have generalized this study to include relative motion so that the breakup of liquid jets in liquids could be described. The effects of mass (or heat) transfer, chemical reaction and a variety of other circumstances on the breakup characteristics of capillary jets have also been investigated.[22]

2. Steady thermocapillary flow

One of the simplest examples of a macro Marangoni effect is that of steady thermocapillary flow in a shallow pool, as pictured in Fig. 10-13. A liquid is contained in a shallow trough, whose lateral edges are maintained at different temperatures. The higher temperature at the left edge produces a lower surface tension there relative to the tension at the right, and this sets the surface into motion, pulling fluid to the right. This must be balanced in the steady state by an equal volumetric flow to the left beneath the surface.

[20] Haenlein, A., NACA Tech. Memo No. 659 (1932).
[21] Coyle, R. W., Berg, J. C., and Niwa, J. C., Chem. Eng. Sci., 36, 17 (1981).
[22] Burkholder, H. C. and Berg, J. C., AIChE J., 20, 863 (1974);
    Burkholder, H. C. and Berg, J. C., AIChE J., 20, 872 (1974);
    Tarr, L. E. and Berg, J. C., Chem. Eng. Sci., 35, 1467 (1980);
    Nelson, N. K., Jr. and Berg, J. C., Chem. Eng. Sci., 37, 1067 (1982);
    Coyle, R. W. and Berg, J. C., Physicochem. Hydrodynamics, 4, 11 (1983);
    Coyle, R. W., and Berg, J. C., Chem. Eng. Sci., 39, 168 (1983).

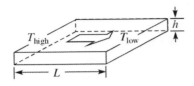

Fig. 10-13: Example of simple
thermocapillary flow in a
shallow pan supporting a
lateral temperature gradient.

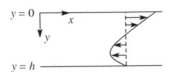

The profile generated is qualitatively as shown in Fig. 10-13. The Navier-Stokes equations reduce in this case, neglecting edge and end effects, to:

$$\mu \frac{d^2 v_x}{dy^2} = \left(\frac{dp}{dx}\right),$$ (10.51)

which is to be solved with the boundary conditions:

    i) (no-slip condition at bottom surface)

$$\text{at } y = h, \quad v_x = 0$$ (10.52)

    ii) (balance of tangential forces at upper surface)

$$\text{at } y = 0, \quad \mu \frac{dv_x}{dy} + \left(\frac{d\sigma}{dT}\right)\left(\frac{dT}{dx}\right) = 0$$ (10.53)

In writing the second boundary condition above, it has been assumed the temperature variation across the surface from 60°C to 20°C is linear. When the above equation is solved, the pressure gradient is eliminated by requiring that the total volumetric flow rate per unit width, Q/W, be zero:

$$Q/W = 0 = \int_0^h v_x \, dy.$$ (10.54)

The resulting velocity profile is given by:

$$v_x = \frac{1}{\mu}\left(\frac{d\sigma}{dT}\right)\left(\frac{\Delta T}{L}\right)(h - y) - \frac{3}{4h\mu}\left(\frac{d\sigma}{dT}\right)\left(\frac{\Delta T}{L}\right)(h^2 - y^2).$$ (10.55)

For the case of $L = 10$ cm, $h = 1$ mm, $\Delta T = -40°C$, $\mu = 5$ cP, and $(d\sigma/dT) = -0.1$ mN/m·°K, the resulting surface velocity is 2 mm/s.

    Similar results may be found for *diffusio*-capillary flow maintained by lateral solute concentration differences instead of temperature differences. Rather complex cross-effects may arise when gradients in both composition and temperature are present.[23]

---

[23] Azouni, M. A., Normand, C., and Pétré, G., *J. Colloid Interface Sci.*, **239**, 509 (2001).

3. *The motion of bubbles or drops in a temperature gradient*

Another interesting example of thermocapillary flow is that associated with bubbles or drops that find themselves immersed in a temperature (or by analogy, a concentration) gradient. Imagine, as shown schematically in Fig. 10-14, a small drop of liquid immersed in another liquid of equal density, but supporting a vertical temperature gradient. The interfacial tension will therefore be lower at the top, and its gradient along the surface of the drop will draw liquid around the surface from the top to the bottom, setting up an internal circulation and propelling the drop upward, *i.e.*, toward the region of higher temperature. It is well known that small drops or bubbles tend to swim up a temperature gradient in this manner, and the fact is made use of in a number of microfluidics devices.

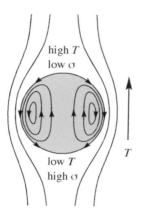

Fig. 10-14: The swimming of a drop "up a temperature gradient."

Perhaps the most well known example of such flow is that of bubbles in a vertical temperature gradient, as described by Young, Goldstein and Block.[24] They provided both experimental data and hydrodynamic analysis that delineated the conditions under which the rise of a small bubble in a vertical temperature gradient, increasing *downward*, would be arrested. They generalized the problem of the rise or fall of a bubble or drop in a viscous medium due to density differences first described by Hadamard[25] and Rybcznski,[26] which itself was a generalization of the Strokes' Law[27] for the steady rise or fall of a small sphere in a viscous medium.

The steady creeping flow form of the Navier-Stokes equations in spherical coordinates, centered at the center of the drop, are the same for both the inner and outer flows, and the boundary tension is assumed large enough that the bubble or drop remains spherical in shape. The Hadamard-Rybczynski solutions for the velocity profiles is obtained using the boundary conditions expressing the equality of the inner and outer fluid velocities at

[24] Young, N. O., Goldstein, J. S., and Block, M. J., *J. Fluid Mech.*, **6**, 350 (1959).]
[25] Hadamard, J., *C. R. Acad. Sci., Paris*, **152**, 1735 (1911).
[26] Rybcznski, D., *Bull. Acad. Sci. Cracovie*, **1**, 40 (1911).
[27] Stokes, G. G., *Trans. Cambridge Phil. Soc.*, **9**, 8 (1851).

the surface of the sphere as well as the balance between viscous tractions on both phases at the interface. Assuming there to be a density difference between the two phases, the drop or bubble will move at a steady velocity, v_∞. Use is also made of the fact that all variables remain finite at the center of the sphere, and that far from the sphere, the velocity in the outer phase approaches a uniform flow at v_∞, relative to a coordinate system centered on the drop. Once the velocity components are obtained, the viscous shear stress on the surface of the sphere is evaluated and integrated over the surface to give the viscous drag. Pressure is integrated over the surface to give the sum of the buoyancy force and the form drag on the sphere. Finally, the sum of the buoyancy and drag forces are equated to the gravitational force on the sphere to give an expression for the terminal velocity. Stokes' Law for the sedimentation (or creaming) of a solid sphere has been given earlier, Eq. (5.51):

$$v_\infty = \frac{2a^2(\rho' - \rho)g}{9\mu},$$

(5.51)

where v_∞ is the downward velocity, a is the sphere radius, and the prime (') refers to properties of the sphere . The Hadamard-Rybczybski result is:

$$v_\infty = \frac{2a^2(\rho' - \rho)g}{3\mu}\left(\frac{\mu + \mu'}{2\mu + 3\mu'}\right).$$

(10.56)

When $\mu' \to \infty$, i.e., the sphere is a solid, and Stokes' Law is recovered. When the sphere is a gas bubble, ρ' and $\mu' \approx 0$, and the above equation becomes

$$v_\infty = -\frac{a^2\rho g}{3\mu}.$$

(10.57)

which shows that a bubble will rise in a gravitational field at 3/2 times the rate given by Stokes' Law.

The boundary condition expressing the balance of viscous shear stresses in the Hadamard-Rybczybski problem is, at $r = a$:

$$\tau'_{r\theta} = \tau_{r\theta}, \quad \text{or} \quad \mu'\left(\frac{\partial v'_\theta}{\partial r} - \frac{v'_\theta}{r}\right) = \mu\left(\frac{\partial v_\theta}{\partial r} - \frac{v_\theta}{r}\right).$$

(10.58)

Young, Goldstein and Block modified this equation by including a term for the surface or interfacial tension gradient, resulting from the temperature gradient superimposed on the system, i.e.

$$\mu'\left[\left(\frac{\partial v'_\theta}{\partial r}\right)_{r=a} - \frac{v'_\theta}{a}\right] = \mu\left[\left(\frac{\partial v_\theta}{\partial r}\right)_{r=a} - \frac{v_\theta}{a}\right] + \frac{1}{a}\left(\frac{d\sigma}{dT}\right)\left(\frac{\partial T}{\partial \theta}\right)_{r=a},$$

(10.59)

where θ is the polar angle. The expression for $(\partial T / \partial \theta)$ at the surface is obtained from the simultaneous solution of the thermal energy equation, which for the case of creeping flow, reduces simply to the conduction equation:

$$\nabla^2 T = 0, \tag{10.60}$$

subject to the condition that far from the sphere, a linear temperature profile is recovered:

$$T = T(z) = T_0 + \left(\frac{dT}{dz}\right)(z - z_0), \tag{10.61}$$

where z is the upward vertical coordinate. Then applying the thermal boundary conditions of Eq. (10.13) at the interface of the sphere, one obtains:

$$\left(\frac{\partial T}{\partial \theta}\right)_{r=a} = -\left(\frac{dT}{dz}\right)\left(\frac{3}{2 + k'/k}\right)\sin\theta, \tag{10.62}$$

where k' and k are the thermal conductivities of the drop and the medium, respectively, for substitution into Eq. (10.59). Finally, subject to this modified boundary condition, Young, Goldstein and Block obtained (for velocity in the downward, *i.e.*, in this case -z, direction):

$$v_\infty = \frac{2\mu}{3(2\mu + 3\mu')}\left[\frac{3a\mu}{(2 + k'/k)}\left(\frac{d\sigma}{dT}\right)\left(\frac{dT}{dz}\right) + a^2(\rho' - \rho)g(\mu + \mu')\right]. \tag{10.63}$$

When the temperature is uniform, *i.e.*, $(dT/dz) = 0$, the result reduces to the Hadamard-Rybczybski case, Eq. (10.56).

With the inner phase a gas bubble, k' as well as ρ' and $\mu' \approx 0$, and setting $v_\infty = 0$ to recover the case in which the bubble is suspended, leads to

$$\left(\frac{dT}{dz}\right)_{v_\infty = 0} = \frac{2}{3}a\rho g \left(\frac{d\sigma}{dT}\right)^{-1}. \tag{10.64}$$

The temperature gradient required for bubble suspension is seen to be directly proportional to the bubble radius and independent of the liquid viscosity.

Young *et al.* obtained semi-quantitative verification of Eq. (10.64) with air bubbles ranging in radius from 10–100 μm suspended in three different silicone oils (Dow-Corning DC 200 oils of kinematic viscosity 20, 200 and 2000 cS). Bubbles were observed in a cylindrical liquid bridge held in the gap between the anvils of a machinist's micrometer, shown in Fig. 10-15. The temperatures of the top and bottom surfaces were determined using mercury thermometers mounted in copper blocks borne by the anvils. The temperature of the lower block was raised by increasing the current to a Nichrome wire wrapped around it. Bubble diameters were measured with a microscope used to track them near the centerline of the liquid bridge. As an example of the results, a 50-μm bubble required a temperature gradient of approximately 50°C/cm to be suspended.

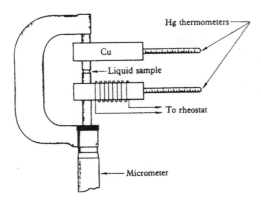

Fig. 10-15: Schematic of apparatus used by Young, Goldstein and Block to study the motion of gas bubbles in a liquid supporting a temperature gradient. From [Young, N. O., Goldstein, J. S., and Block, M. J., *J. Fluid Mech.*, **6**, 350 (1959).]

A phenomenon closely related to that of bubbles or drops swimming in a temperature or composition gradient is that associated with Marangoni driven circulation around bubbles in sub-cooled nucleate pool boiling. As the temperature of a solid surface is raised to the point where vapor bubbles are nucleated, the efficiency of heat transfer generally increases sharply before nucleate boiling begins. Although increased levels of natural convection may be part of the reason for this, it cannot be the total explanation because the phenomenon is observed even in zero-gravity experiments and even when the heated surface is facing downward. It is believed that the enhancement is traceable to the presence of bubbles formed by dissolved air and attached to the heated surface.[28] Because such bubbles have a higher temperature near their base, where they are attached to the solid surface, than at their apex, surface tension gradients develop, and the resulting surface motion induces convection in the adjacent liquid, as shown schematically in Fig. 10-16. Significant enhancements in heat transfer were obtained in comparison to the case in which the bubbles were not present.

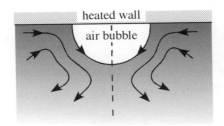

Fig. 10-16: Marangoni convection generated around an air bubble attached to a heated surface in subcooled nucleate boiling.

---

28 Petrovich, S., Robinson, T., and Judd, R. L., *Intl. J. Heat Mass Transf.*, **47** [23], 5155 (2004).

4. *Marangoni instability in a shallow liquid pool – Bénard cells*

Some of the most important examples of interfacial hydrodynamics
are formulated in terms of hydrodynamic stability analysis,[29-30] as the case of
laminar jet breakup treated earlier. Another important example is that of a
quiescent shallow pool of liquid supporting an "adverse" temperature or
concentration gradient during heat or mass transfer normal to the pool
surface. The gradients are considered "adverse" if, when sufficiently steep,
they lead to instability and spontaneous flow such as Bénard cells or other
structures. In this case, the gradient supports a layer of liquid at the surface
whose surface tension is higher than that corresponding to the liquid in the
interior.

The gradients may simultaneously produce either adverse or
stabilizing density stratification, which may (for the adverse case) lead to
what is termed simply "natural convection." Density-driven natural
convection in a liquid pool, is the result of what is termed "Rayleigh
instability,[31]" and often coexists with surface tension driven flow, although
the phenomena were initially investigated separately. It is Marangoni
instability that is outlined briefly below.

Surface-tension-caused instability in a shallow pool heated from
below was first considered by Pearson[32] in 1958, while the analogous case of
instability during the transfer of a solute across a liquid-liquid interface was
investigated by Sternling and Scriven in 1959.[33] The situation considered by
Pearson is pictured in Fig. 10-17 in which a pool of depth h and infinite

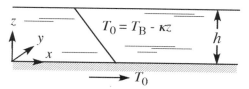

Fig. 10-17: Liquid pool
supporting an "adverse"
temperature profile.

lateral extent supports, in the unperturbed state, a linear temperature profile
$T_0(z)$:

$$T_0(z) = T_B - \kappa z, \tag{10.65}$$

where T_B is the temperature at the bottom of the pool, assumed constant, and
κ is the magnitude of the temperature gradient,

$$\kappa = \left| \frac{dT_0}{dz} \right|. \tag{10.66}$$

[29] Miller, C. A., in **Surface and Colloid Science**, Vol. 10, E. Matijevic (Ed.), pp. 227-293,
      Plenum Press, New York, 1978.
[30] Johns, L. E., and Narayanan, R., **Interfacial Instabilities**, Springer, New York, 2002.
[31] Lord Rayleigh (J. W. Strutt), *Phil. Mag.*, **32** [6], 529 (1916).
[32] Pearson, J. R. A., *J. Fluid Mech.*, **4**, 489 (1958).
[33] Sternling, C. V., and Scriven, L. E., *AIChE J.*, **5** [4], 514 (1959).

At a particular instant ($t = 0$), the pool is subjected to a perturbation of infinitesimal amplitude. The temperature is perturbed from its initial value: $T = T_0 + T^*(t,x,y,z)$, and a fluid velocity develops with components in all three directions. The perturbations must satisfy the relevant flow (Navier-Stokes) equations and the thermal energy equation. It is assumed that the fluid is Newtonian and incompressible, and that the temperature dependence of its density and viscosity are negligible. Thus the only property depending on temperature is the surface tension. In terms of dependent variables that are all first-order small, the equations are linear, becoming (in Cartesian scalar form):

Continuity:

$$\frac{\partial v_x}{\partial x} + \frac{\partial v_y}{\partial y} + \frac{\partial v_z}{\partial z} = 0 \tag{10.67}$$

x, y, and z – Momentum

$$\frac{\partial v_x}{\partial t} = -\frac{1}{\rho}\frac{\partial \mathcal{P}}{\partial x} + v\left[\frac{\partial^2 v_x}{\partial x^2} + \frac{\partial^2 v_x}{\partial y^2} + \frac{\partial^2 v_x}{\partial z^2}\right],$$

$$\frac{\partial v_y}{\partial t} = -\frac{1}{\rho}\frac{\partial \mathcal{P}}{\partial y} + v\left[\frac{\partial^2 v_y}{\partial x^2} + \frac{\partial^2 v_y}{\partial y^2} + \frac{\partial^2 v_y}{\partial z^2}\right], \tag{10.68}$$

$$\frac{\partial v_z}{\partial t} = -\frac{1}{\rho}\frac{\partial \mathcal{P}}{\partial z} + v\left[\frac{\partial^2 v_z}{\partial x^2} + \frac{\partial^2 v_z}{\partial y^2} + \frac{\partial^2 v_z}{\partial z^2}\right],$$

where $\mathcal{P}$ is the pressure-gravity term: $p + \rho g z$, z being the elevation above a datum plane, and v is the kinematic viscosity (μ/ρ). The thermal energy equation is

$$\frac{\partial T^*}{\partial t} - \kappa v_z = \alpha\left[\frac{\partial^2 T^*}{\partial x^2} + \frac{\partial^2 T^*}{\partial y^2} + \frac{\partial^2 T^*}{\partial z^2}\right], \tag{10.69}$$

where T^* is the temperature perturbation, κ is the undisturbed temperature gradient, and α is the thermal diffusivity ($c_p\mu/k$), with c_p the heat capacity and k the thermal conductivity. The independent variables in Eqs. (10.67)–(10.69) are nondimensionalized as

$$(\tilde{x}, \tilde{y}, \tilde{z}) = (x/h, y/h, z/h), \text{ and } \tilde{t} = t\alpha/h^2, \text{ so that we have}$$

$$\frac{\partial v_x}{\partial \tilde{x}} + \frac{\partial v_y}{\partial \tilde{y}} + \frac{\partial v_z}{\partial \tilde{z}} = 0, \tag{10.67'}$$

$$\frac{\partial v_x}{\partial \tilde{t}} = -\frac{h}{\rho \alpha} \frac{\partial \mathcal{P}}{\partial \tilde{x}} + Pr\left[\frac{\partial^2 v_x}{\partial \tilde{x}^2} + \frac{\partial^2 v_x}{\partial \tilde{y}^2} + \frac{\partial^2 v_x}{\partial \tilde{z}^2}\right],$$

$$\frac{\partial v_y}{\partial \tilde{t}} = -\frac{h}{\rho \alpha} \frac{\partial \mathcal{P}}{\partial \tilde{y}} + Pr\left[\frac{\partial^2 v_y}{\partial \tilde{x}^2} + \frac{\partial^2 v_y}{\partial \tilde{y}^2} + \frac{\partial^2 v_y}{\partial \tilde{z}^2}\right], \qquad (10.68')$$

$$\frac{\partial v_z}{\partial \tilde{t}} = -\frac{h}{\rho \alpha} \frac{\partial \mathcal{P}}{\partial \tilde{z}} + Pr\left[\frac{\partial^2 v_z}{\partial \tilde{x}^2} + \frac{\partial^2 v_z}{\partial \tilde{y}^2} + \frac{\partial^2 v_z}{\partial \tilde{z}^2}\right], \text{ and}$$

$$\frac{\partial T'}{\partial \tilde{t}} - \frac{h^2 \kappa}{\alpha} v_z = \left[\frac{\partial^2 T^*}{\partial \tilde{x}^2} + \frac{\partial^2 T^*}{\partial \tilde{y}^2} + \frac{\partial^2 T^*}{\partial \tilde{z}^2}\right], \qquad (10.69')$$

where Pr is the Prandtl Number (ν/α).

All of the dependent variables in Eqs. (10.67') – (10.68') are first-order small and expected to exhibit a periodic structure in the horizontal coordinates (x and y), and since time appears only as a derivative with respect to time, dependence on it must be exponential. Thus assuming separation of variables, one may express them as the sum of terms of the form

$$f^z(\tilde{z})\exp[i(\ell_j \tilde{x} + m_j \tilde{y}) + \beta(\ell_j, m_j)t], \qquad (10.70)$$

where the superscript z represents the z-component of the function. Each set of (ℓ_j, m_j) values represents a different "mode" of the perturbation, and they may be summed ("superposed") to give the total disturbance because they are sufficiently small. The relative values of ℓ and m give the planform structure of the disturbance mode, with $\ell = m$ corresponding to square cells, $\ell = 0$, and m finite corresponding to 2-d roll cells, *etc.* It is common to put these together by defining

$$a_j^2 = \ell_j^2 + m_j^2, \qquad (10.71)$$

with a_j being the dimensionless wave number of the disturbance mode j, *i.e.,*

$$a_j = 2\pi h / \lambda_j, \qquad (10.72)$$

for 2-d roll cells, where λ_j is the wavelength of the disturbance mode.

A system is deemed unstable if there is but one mode of the disturbance that allows itself to be amplified, *i.e.*, if there is but one mode for which the real part of the growth constant β is positive. Thus one seeks to identify the state(s) of *marginal stability*, in which the real part of β is zero. In many, but not all, cases it can be proved that if the real part of β is zero, the imaginary part is also zero (called the principle of "exchange of stabilities"[34]). Pearson assumed this to be true in the present case.

Substitution of expressions of the type of (10.70) for the velocity perturbations into Eqs. (10.68′)–(10.69′) yields

$$i\ell v_x^z + imv_y^z + Dv_z^z = 0 \tag{10.73}$$

$$\beta\ell v_x^z = -i\frac{\ell h}{\rho\alpha}\mathcal{P}^z + Pr(D^2 - a^2)v_x^z$$

$$\beta\ell v_y^z = -i\frac{mh}{\rho\alpha}\mathcal{P}^z + Pr(D^2 - a^2)v_y^z \tag{10.74}$$

$$\beta\ell v_y^z = -\frac{h}{\rho\alpha}D\mathcal{P}^z + Pr(D^2 - a^2)v_z^z,$$

where D is the operator $d/d\tilde{z}$. Equations (10.73)–(10.74) may be combined into single equation in terms of v_z^z alone, viz.,

$$(D^2 - a^2)\big[Pr(D^2 - a^2) - \beta\big]v_z^z = 0. \tag{10.75}$$

Substitution of the expression for the temperature perturbation into the thermal energy equation, (10.75), yields:

$$(D^2 - a^2 - \beta)T^{*z} = -\frac{\kappa h^2}{\alpha}v_z^z. \tag{10.76}$$

Finally, Eqs. (10.75) and (10.76) may be combined to give a single governing equation in terms of the temperature perturbation alone,[35] viz.

$$(D^2 - a^2)(D^2 - a^2 - \beta)\big[Pr(D^2 - a^2) - \beta\big]T^{*z} = 0. \tag{10.77}$$

To describe the condition of marginal stability, one sets $\beta = 0$ in Eq. (10.77), yielding:

$$(D^2 - a^2)^3 T^{*z} = 0. \tag{10.78}$$

One then seeks non-trivial solutions to Eq. (10.78) that satisfy the relevant boundary conditions, three each at the lower and upper surfaces. Different thermal boundary conditions may be considered, but it shall be assumed that the lower surface is conducting, so that

$$\text{(at } \tilde{z} = 0) \qquad T^{*z} = 0. \tag{10.79}$$

---

[34] See, e.g., Pellew, A., and Southwell, R. V., *Proc. Roy. Soc.*, **176A**, 312 (1940).
[35] It is not possible in this case to formulate a single governing equation in terms of v_z^*.

At the upper surface, essentially all possibilities are covered using a "Newton's Law of cooling" or "radiation" type of boundary condition, *viz.*

$$q = -k\frac{dT}{dz} = h_q(T - T_{\text{ref}}),$$ (10.80)

where q is the heat flux from the upper surface, h_q is the heat transfer coefficient in the vapor phase, and T_{ref} is the temperature in the vapor phase. When nondimensionalized, and upon substitution of the assumed form of the temperature perturbation, Eq. (10.71) becomes

(at $\tilde{z} = 1$) $(D - Bi)T^{*z} = 0,$ (10.81)

where Bi is the Biot Number, $(h_q h/k)$. Putting $Bi \to 0$ recovers the insulating (or constant-flux) condition, while $Bi \to \infty$ recovers the conducting condition.

Two hydrodynamic boundary conditions are required at each surface. They may be converted to their "temperature forms" using Eq. (10.23). The bottom surface is assumed rigid and supporting a no-slip condition. Thus

(at $\tilde{z} = 0$) $v_z^z = 0,$ and $Dv_z^z = 0.$ (10.82)

The temperature forms for Eqs. (10.82) are:

(at $\tilde{z} = 0$) $(D^2 - a^2)T^{*z} = 0,$ and $D(D^2 - a^2)T^{*z} = 0.$ (10.82′)

At the top surface, the surface tension is regarded as sufficiently large to render the surface non-deformable (*i.e.*, rigid), so that

(at $\tilde{z} = 1$) $v_z^z = 0,$ or (10.83)

(at $\tilde{z} = 1$) $(D^2 - a^2)T^{*z} = 0.$ (10.83′)

The remaining condition at the free surface expresses the balance between the surface tension gradients (resulting from the temperature variations accompanying the convection) against the viscous traction from beneath. The most convenient form of the required equation is the surface divergence of the force balance as given in Eq. (10.12). Neglecting the viscosity of the vapor phase and any intrinsic surface rheology, this becomes

$$-\mu\left[\left(\frac{\partial^2 v_z}{\partial z^2}\right) + \nabla_{II}^2 v_z\right] + \nabla_{II}^2\sigma = 0.$$ (10.84)

Assuming the surface is non-deformable, v_z is everywhere zero at the surface, so $\nabla_{II}^2 v_z = 0$, and we have

$$\nabla_{II}^{2}\sigma \equiv \frac{\partial^{2}\sigma}{\partial x^{2}} + \frac{\partial^{2}\sigma}{\partial y^{2}} = \mu\frac{\partial^{2}v_{z}}{\partial z^{2}}.$$ (10.84')

Surface tension is a function of temperature, so we may write the divergence of Eq. (10.14) as:

$$\nabla_{II}^{2}\sigma = \left(\frac{d\sigma}{dT}\right)_{0}\nabla_{II}^{2}T + \cdots,$$ (10.85)

where the right hand side is the leading term in a Taylor series. Combining Eqs. (10.85) and (10.84'), gives

$$\left(\frac{d\sigma}{dT}\right)_{0}\nabla_{II}^{2}T = \mu\frac{\partial^{2}v_{z}}{\partial z^{2}}.$$ (10.85')

Substituting expressions for v_z and T^* of the form of (10.70), with (10.71) into Eq. (10.85') gives, for the final boundary condition:

$$(\text{at } \tilde{z} = 1) \qquad a^{2}\left(\frac{d\sigma}{dT}\right)_{0}T^{*z} = \mu D^{2}v_{z}^{z}.$$ (10.86)

Using Eq. (10.76) puts this final boundary condition into terms of the temperature perturbation alone, i.e.,

$$D^{2}(D^{2}-a^{2})T^{*z} = a^{2}\left[\left(\frac{d\sigma}{dT}\right)_{0}\frac{\kappa h^{2}}{\mu\alpha}\right]T^{*z}.$$ (10.87)

The dimensionless group on the right hand side is now commonly called the Marangoni Number, Ma:

$$Ma = \left(\frac{d\sigma}{dT}\right)_{0}\frac{\kappa h^{2}}{\mu\alpha} \equiv \left(\frac{d\sigma}{dT}\right)_{0}\left|\frac{dT_{0}}{dz}\right|\frac{h^{2}}{\mu\alpha}.$$ (10.88)

It should be noted parenthetically that had the system been one supporting an "adverse" concentration gradient of a surface tension lowering solute rather than an adverse temperature gradient, the mathematics would have been identical, but the Marangoni Number would have been:

$$Ma = \left(\frac{d\sigma}{dC}\right)_{0}\left|\frac{dC_{0}}{dz}\right|\frac{h^{2}}{\mu D},$$ (10.89)

where D is the solute diffusivity.

The general solution to the governing equation, Eq. (10.78), can be written in terms of exponential functions, viz.,

$$T^{*z}(\tilde{z}) = \sum_{i=1}^{6}A_{i}e^{s_{i}\tilde{z}},$$ (10.90)

where the s_i are roots to the indicial equation

$$(s_i^2 - a^2)^3 = 0, \qquad\qquad (10.91)$$

and the coefficients A_i are determined by requiring that Eq. (10.90) satisfy the six boundary conditions. Setting all the $A_i = 0$ provides one solution, of course, but non-trivial solutions exist only when the 6 x 6 determinant of the coefficients of the equations obtained by substituting Eq. (10.81) into the six boundary conditions is equal to zero. This process leads to the *characteristic equation* of the system. Using a somewhat different but equivalent approach, Pearson[36] obtained the following characteristic equation for the case in which the bottom surface is conducting (constant temperature):

$$Ma = \frac{\left[\dfrac{2B_1}{a} - a + (B_1 - 1)\tanh a\right]}{\left(\dfrac{B_1 + 3}{4a}\right) - \left(\dfrac{1}{4} + \dfrac{B_1}{4a^2} + B_2\right)\tanh a}, \text{ where} \qquad (10.92)$$

$$B_1 = a\coth a - 1, \text{ and}$$

$$B_2 = \frac{a^2 \cosh^2 a + a\sinh a\cosh a + \sinh^2 a + Bi(a^2 + a\sinh a\cosh a + \sinh^2 a)}{4a^2 \sinh a(a\cosh a + Bi\sinh a)}.$$

Results are plotted in Fig. 10-18 for $Bi = 0$, 2 and 4, representing increasing values of the efficiency of heat transfer from the top surface.

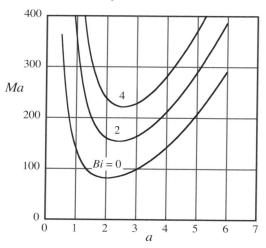

Fig. 10-18: Neutral stability curves for Marangoni convection in a pool of liquid supporting an adverse temperature gradient for the case of a conducting (*i.e.*, constant temperature) bottom surface, in which the critical value of the Marangoni Number, *Ma*, is plotted as a function of the dimensionless wave number, *a*, of the perturbation. Results are shown for different values of the Biot Number, *Bi*.

[36] Pearson, J. R. A., *J. Fluid Mech.*, **4**, 489 (1958), Eq. (27), with errors corrected.

Regions below any of the curves correspond to stability with respect to Marangoni convection, while those above are unstable, leading possibly to the formation of Bénard cells.

The minimum critical Marangoni Number is 80, corresponding to a dimensionless wave number of 2, suggesting, in accord with Eq. (10.70) a cell size (wavelength) of approximately three times the pool depth. Both observations are in at least qualitative accord with experimental observation.[37] Pearson's analysis of the case for which the bottom surface was insulating (constant flux) led to a critical Ma of 48, for $Bi = 0$, at a wave number of 0, corresponding to an infinite disturbance wavelength. Such a situation would not be expected to produce observable convection. Finite values of Bi led to critical Marangoni Numbers and wave numbers similar to those obtained for the conducting bottom case.

In their landmark paper in 1959, Sternling and Scriven[38] described the stability of a pair of liquid layers with respect to mass transfer of a solute species between them. More complex results than those of the Pearson analysis were obtained in part because the "exchange of stabilities" assumption was not made, and indeed oscillatory instabilities were found. The qualitative results of Sternling and Scriven, however, can be summarized as follows: Interfacial turbulence may be expected in either or both of the liquid phases in which the solute lowers the interfacial tension and its concentration is sufficiently steep, if the solute transfer is out of the phase of higher kinematic viscosity and lower solute diffusivity. The usual case is that either both or neither of the last two criteria are satisfied.

As mentioned earlier, liquid pools as shown in Fig. 10-16, supporting adverse temperature (or concentration) profiles are subject to buoyancy-driven as well as Marangoni instability. It was the publication of Bénard's results near the turn of the 20th century that motivated Rayleigh to perform his stability analysis of adverse density stratification. Rayleigh's result was expressed in terms of a dimensionless group now known as the Rayleigh Number, Ra:

$$Ra = \frac{g\xi\kappa h^4}{\alpha v},$$ (10.93)

where g is the gravitational acceleration, ξ is the coefficient of volumetric expansion, and as before, κ is the temperature gradient, h the depth of the pool, α the thermal diffusivity and v the kinematic viscosity. The critical value of the Ra for the boundary conditions relevant to Eq. (10.83), with $Bi = 0$, is 669.[39] It is possible to make a direct comparison between the "critical depth" of a pool of a given liquid with a given temperature drop as predicted

[37] Berg, J. C., Acrivos, A., and Boudart, M., "Evaporative Convection." in **Advances in Chemical Engineering**, Vol. 6, pp. 61-123, T. B. Drew, J. W. Hoopes, Jr., T. Vermuelen and G. R. Cokelet (Eds.), Academic Press, New York, 1966.

[38] Sternling, C. C., and Scriven, L. E., *AIChE J.*, **5**, 514 (1959).

[39] Sparrow, E. M., Goldstein, R. J., and Jonsson, V. K., *J. Fluid Mech.*, **4**, 743 (1964).

by the Marangoni *vs.* the Rayleigh mechanism. For most liquids, including high MW alkanes (such as the molten wax layers studied by Bénard), with a 1° temperature drop across the layer, one finds a critical depth for Rayleigh instability of the order of 3-10 mm, whereas for Marangoni instability, it is less than 1 mm. Thus the cellular convection observed by Bénard was clearly due to surface tension effects. Considering both Marangoni and Rayleigh convection simultaneously, Nield[40] found the processes to be tightly coupled so that a pool under conditions producing a *Ra* half its critical value would require a *Ma* about half *its* critical value in absence of density effects for instability to be predicted, *etc.*, as shown in Fig. 10-19. These predictions have been at least semi-quantitatively confirmed by experiment, as shown in the figure. A relatively more recent re-working of

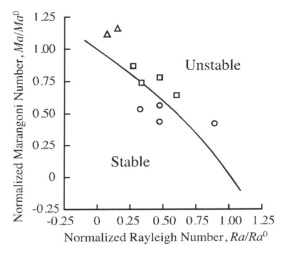

Fig. 10-19: Nield stability diagram (solid line) for pool heated from below with constant temperature lower surface and $Bi = 0$ at the top. Ma^0 and Ra^0 are the critical Marangoni Number in the absence of buoyancy and the critical Rayleigh Number in the absence of surface tension variations, respectively. Data shown (from Palmer and Berg) are for (Δ) 10 cS, ($\square$) 50 cS, and (o) 200 cS DC 200 silicone oils. Data from [Palmer, H. J., and Berg, J. C., *J. Fluid Mech.*, **47**, 779 (1971).]

the Pearson-Rayleigh problem, in which the Marangoni Number, Rayleigh Number, Biot Number and disturbance wave number are all allowed to vary simultaneously has produced results of the type shown in Fig. 10-20, which are in qualitative but not quantitative agreement with Pearson and with Nield. For example, the global minimum *Ma* is found to be approximately 223 (rather than 80) at $Bi = 1.54$, and a dimensionless wave number $a = 2.33$.

---

[40] Nield, D. A., *J. Fluid Mech.*, **19**, 341 (1964).

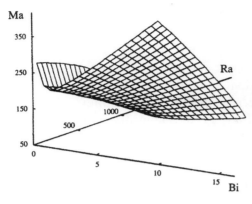

Fig. 10-20: Neutral stability surface in $Ma - Ra - Bi$ space for coupled thermocapillary-buoyancy convection. From [Rabin, L. M., *Phys. Rev. E*, **53** [3], 53 (1996).]

An interesting difference between the surface tension and buoyancy driven cases concerns the surface deformations they produce. The surface relief accompanying Bénard cells, as shown in Fig. 10-6, shows depressions above the upwelling liquid at the cell centers. Hershey[41] obtained an estimate of the elevation difference produced:

$$\Delta h = \frac{3(d\sigma / dT)_0 \Delta T_{\text{lat}}}{\rho g h},$$ (10.94)

where ΔT_{lat} is the maximum temperature difference existing along the surface. Equation (10.94) gives an elevation different of 1.58 μm for a 1.2 mm deep layer of spermaceti (the molten wax studied by Bénard) at 100°C and assuming $\Delta T_{\text{lat}} = 1°C$, in good agreement with Bénard's measurements and with an analysis of Scriven and Sternling[42] who relaxed the assumption of free surface non-deformability. In contrast, for buoyancy-driven convection, Jeffreys[43] has shown that the surface must be convex at the cell centers, as was later reported by Spangenberg and Rowland[44] for deep layers of evaporating water.

With respect to stability, the effect of surface deformability on the Pearson and Rayleigh-Pearson (Nield) analyses was examined by Scriven and Sternling[45] and Smith,[46] and it was found that for realistic wavelength disturbances, the original results were not significantly affected.

[41] Hershey, A. V., *Phys. Rev.*, **56**, 204 (1939).
[42] Scriven, L. E., and Sternling, C. V., *J. Fluid Mech.*, **19**, 321 (1964).
[43] Jeffreys, H., *Quart. J. Mech.*, **4** [3], 283 (1951).
[44] Spangenberg, W. B., and Rowland, W. R., *Phys. Fluids*, **4**, 743 (1961).
[45] Scriven. L. E., and Sternling, C. V., *J. Fluid Mech.*, **19**, 321 (1964).
[46] Smith, K. A., *J. Fluid Mech.*, **24**, 401 (1966).

C. Some Practical Implications of the Marangoni Effect

1. *Marangoni effects on mass transfer*

Mass transfer during the contact of fluid phases, such as occurs in gas absorption or stripping, distillation or liquid extraction, may be strongly influenced by the Marangoni effect.[47] The effects are manifest in two general ways. First, the presence of flow in the immediate vicinity of fluid interfaces, as either macro or micro Marangoni convection (the latter often termed "interfacial turbulence") may increase the overall mass transfer efficiency by factors up to ten or more. Second, in some cases such flows may result in large changes in the interfacial area available for transfer, either increasing or decreasing it from the case when boundary tension gradients are absent.

Many reports of mass transfer enhancement have been given, but a couple of examples of long standing suffice to document the phenomenon. Brian, Vivian and Mayr,[48] for example, determined liquid phase mass transfer coefficients for a propylene tracer during the desorption of acetone (in a situation analogous to that pictured in Fig. 10-4), as well as methyl chloride, ethyl ether, and triethylamine from water in a short wetted-wall column, as shown in Fig. 10-21. Mass transfer enhancement factors over penetration theory for the absorbing propylene depended on the driving force for desorption of the various solutes, all of which strongly reduced surface tension, and varied between 2.0 and 3.6 when the initial

liquid

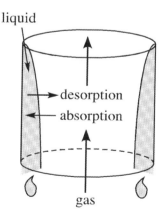

Fig. 10-21: Wetted-wall column for carrying out gas-liquid mass transfer experiments. During transfer of acetone and other surface tension lowering solutes between water and air, the mass transfer coefficient during *de*sorption from water was 2.0 – 3.6 times larger than for transfer in the opposite direction. This is explained by the existence of Marangoni convection during desorption.

concentration of the desorbing solute was set at 1 wt%. The measured stability criteria, as determined by the minimum concentration of desorber required to give mass transfer enhancement, were generally higher than predicted by Pearson theory, and in fact methyl chloride produced no enhancement whatever. Such observations could be explained by the

---

[47] Berg, J. C., in **Recent Developments in Separation Science**, Vol. 2, N. N. Li (Ed.), pp. 1-33, CRC Press, Cleveland, OH, 1972.

[48] Brian, P. L. T., Vivian, J. E., and Mayr, S. T., *Ind. Eng. Chem. Fund.*, **10**, 75 (1971).

presence of small amounts of "surfactant" contamination, as described below.

Interfacial turbulence effects have also been amply documented for liquid-liquid mass transfer. For example, Olander and Reddy[49] found enhancements up to a factor of 4.0 during the transfer of nitric acid from an isobutanol phase to water in a continuous-flow stirred tank contactor, in comparison with the case of transfer in the opposite direction. The Sternling-Scriven criteria for interfacial turbulence were satisfied for transfer from isobutanol to water, but not for the reverse case, *i.e.*, the kinematic viscosity of the isobutanol phase exceeded that of the water phase, and the diffusivity of nitric acid was less in the isobutanol phase than in the water. This system was investigated in a mass transfer cell of the type shown in Fig. 10-22(a).[50] Schlieren images show the convection accompanying mass transfer, producing results such as shown in Fig. 10-22(b) for the case of transfer of nitric acid from the upper (isobutanol) phase to the water phase beneath, while no convection was observed for transfer in the reverse direction. The Marangoni flows observed are accompanied by buoyancy-driven natural

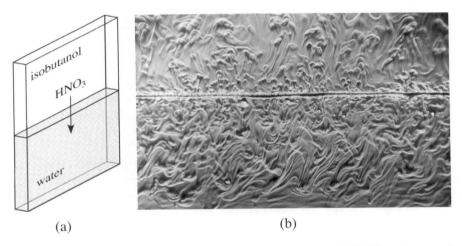

(a)   (b)

Fig. 10-22: (a) Mass transfer cell, dimensions: 4 × 5 × 1/4 in., (b) Schlieren image of convection accompanying the transfer of nitric acid from an isobutanol phase (upper layer) in which in initial concentration is 2.0 M, into a water phase below. Image is 15 cm across and was acquired approximately 30 s after phases were contacted. Both Marangoni and buoyancy-driven convection are present. The convection exhausts itself as the system equilibrates over approximately 45 minutes.

convection in the case shown, as the nitric acid increases the density of both the isobutanol and the water phases in proportion to its concentration. The role played by such effects has been investigated using a system in which the lower phase was a solution of benzene (lighter than water) and chlorobenzene (denser than water) and the upper phase was water. By

[49] Olander, D. R., and Reddy, L. B., *Chem. Eng. Sci.*, **19**, 67 (1964).
[50] Berg, J. C., and Morig, C. R., *Chem. Eng. Sci.*, **24**, 937 (1969).

varying the benzene/chlorobenzene ratio, the extent of the density
stratifiation accompanying mass transfer could be quantitatively controlled.[51]
Figure 10-23 shows the instantaneous mass transfer coefficients obtained
over time during the transfer of acetic acid from the lower
(benzene/chlorobenzene) phase to water under different buoyancy
conditions. Enhancements of nearly a factor of 40 are obtained at early
times, when the density difference is maximum.

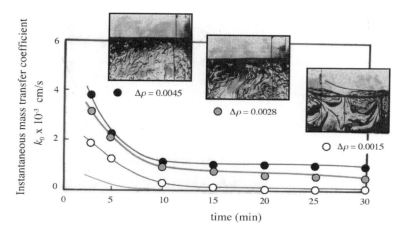

Fig. 10-23: Instantaneous mass transfer coefficients measured during the
transfer of acetic acid from a benzene/chlorobenzene phase (below) to
water under different density difference conditions. The lower curve
corresponds to the simple diffusion case. The schlieren images are 10 cm
in width, taken approximately 30 s after phase contact in each case.

Interestingly, in many of the reported studies of liquid-liquid mass
transfer, convection was observed even when the Sternling-Scriven criteria
were not satisfied, even for cases in which buoyancy-driven natural
convection was absent. This happenstance is generally explained in terms of
the existence of macro Marangoni convection generated at the menisci of the
interface against the confining walls of the vessel.

The other important way that the Marangoni effect influences mass
transfer rates in fluid contacting devices is through its influence on fluid
interfacial area. This is a manifestation of macro Marangoni effects in
systems that contain thin liquid films, which may be either "supported" (as
on a solid substrate) or "unsupported" (with a vapor phase on both sides).
Supported films are exemplified by liquid flowing over packing elements in
a packed tower, while unsupported films occur in froths or foams.
Zuiderweg and Harmens[52] identified two broad types of multicomponent
systems with respect to the relative surface tension of the components in
comparison to their relative volatility. Surface tension positive (σ^{pos})

[51] Berg, J. C., and Haselberger, G. S., *Chem. Eng. Sci.*, **26**, 481 (1971).
[52] Zuiderweg, F. J., and Harmens, A., *Chem. Eng. Sci.*, **9**, 89 (1958).

systems were defined as those for which the more volatile component has the lower surface tension, while surface tension negative (σ^{neg}) systems were those for which the more volatile component has the higher surface tension. Surface tension positive systems are more common, but surface tension negative systems represent perhaps 20-25% of the systems encountered in practice. If a liquid film of either type of system experiences a variation in its thickness, the thinner area will equilibrate more quickly than the surrounding area, and therefore be enriched in the less volatile component, as suggested by the shaded regions in Fig. 10-24. In a σ^{pos} system, this leads to higher surface tension in the thin area. Surface tension forces then develop opposing further thinning, leading to "self-healing" of the film. In contrast, σ^{neg} systems will exhibit a reduced surface tension in the thinner area as equilibration is approached, leading to an amplification of the thinning process, and possibly the formation of a bare patch or film rupture. These different behaviors are diagrammed in Fig. 10-24. A surface tension neutral system is identified as one for which either the driving force and/or the surface tension difference between the components is too small for sufficient surface tension differences to develop between thin and thick portions of the film. Differences of the order of 1–2 mN/m appear to be necessary.

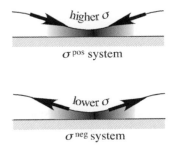

Fig. 10-24: Surface tension variation accompanying local film thinning in a surface tension positive system contrasted with that occurring in a surface tension negative system.

In supported area distillation equipment, σ^{pos} systems and surface tension neutral systems produce liquid films that wet the solid support more or less evenly, yielding relatively high interfacial area, whereas in σ^{neg} systems, the liquid film breaks up into rivulets and droplets that run rapidly through the column, reducing both interfacial area and holdup. In a packed distillation column, the effect may be as much as a 50% reduction in efficiency. This has been demonstrated using a system of water and formic acid, whose vapor-liquid equilibrium data at atmospheric pressure are shown in Fig. 10-25. The system is seen to exhibit an azeotrope at 43.5 mole% water. At water concentrations below that of the azeotrope, formic acid is the more volatile component, and the system is σ^{pos} (since the surface tension of formic acid is substantially lower than that of water), while above the azeotrope point, the system is σ^{neg}. Distillations carried out in a 4 cm

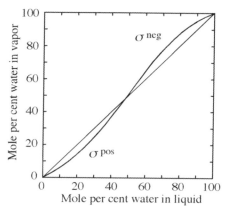

Fig. 10-25: Vapor-liquid
equilibrium data for the
water-formic acid system at
atmospheric pressure.

diameter, 76.2 cm long column packed with 4mm unglazed ceramic Berl saddles, yielded a height equivalent to a theoretical plate (HETP) of 9.14 cm over the σ^{pos} side of the azeotrope and 13.26 cm on the σ^{neg} side. For unsupported area distillation equipment, such as perforated plate columns, stable froths are formed in σ^{pos} systems, but not in σ^{neg} or neutral systems, resulting in substantially higher interfacial area in the former, and generally up to 100% greater efficiencies.

2. Marangoni drying

An important application of the Marangoni effect is *Marangoni drying*,[53] a process designed for the cleaning of semiconductor wafers, medical and dental instruments and other surfaces. Water is used as a rinsing medium, but upon its removal from hydrophilic surfaces by conventional means, a thin film of water remains, which upon evaporation leaves behind non-volatile residues (water spots). Even spin-drying does not produce quantitative drying unless the dynamic receding contact angle is greater than 0°, a condition which is rarely satisfied. A simple configuration for Marangoni drying is pictured in Fig. 10-26, in which a smooth plate slowly withdrawn from a water-rinsing bath. At distances far above the bath surface, the steady state thickness of the retained film is given by[54]

$$h_\infty = 0.93 \left(\frac{\mu v_\infty}{\rho g} \right)^{1/2} Ca^{1/6}, \tag{10.95}$$

where Ca is the Capillary number, $Ca = \mu v_\infty / \sigma$. For water at room temperature, even with a withdrawal rate as low as one cm/s, the thickness of retained water is about 7 μm. With a spin dryer operating at 5000 RPM, the thickness can be reduced to about 0.1 μm, but this is still not low

[53] Leenaars, A. F, M., Huethorst, J. A. M., and van Oekel, J. J., *Langmuir*, **6**, 1701 (1990); Marra, J., and Huethorst, J. A. M., *Langmuir*, **7**, 2748 (1991).
[54] Levich, V. G., **Physicochemical Hydrodynamics**, pp. 674-683, Prentice-Hall, Inc., Englewood Cliffs, NJ, 1962.

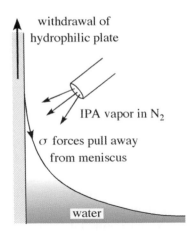

withdrawal of
hydrophilic plate

IPA vapor in N_2

σ forces pull away
from meniscus

water

Fig. 10-26: Schematic of Marangoni drying process. A water-soluble, low surface tension liquid is directed toward the receding water meniscus as the plate to be dried is withdrawn. Isopropyl alcohol (IPA) is commonly used. The lowered surface tension causes the meniscus liquid to pull back from the interline, leaving a dry surface behind.

enough to avoid levels of contamination that are presently unacceptable for semiconductor wafer surfaces. Even at equilibrium on a vertical surface, if the liquid fully wets the surface, there will be a nano-film of liquid often extending several mm above the nominal interline, as discussed in Chap, 2. K.4. In Marangoni drying, a current of vapor of a water-soluble, surface tension lowering liquid is directed toward the receding meniscus, as shown in Fig. 10-26. A commonly used material is isopropyl alcohol (IPA), whose surface tension is 21.3 mN/m at 20°C, delivered in a nitrogen atmosphere. The vapor, which must not condense on the solid surface leaving behind an evaporative film of its own, dissolves into the receding meniscus, reducing the surface tension, setting up Marangoni forces that pull the liquid away from the interline, leaving behind a completely dry surface, in a process which is the reverse of the wine tear phenomenon described earlier. Withdrawal rates must be kept to a few mm/s for the process as pictured, but when combined with spin-drying, the process may be speeded by orders of magnitude. The rate of delivery of IPA (or other drying vapor) may be increased by introducing it in the form of an aerosol.[55] The technology of Marangoni drying has been incorporated into an integrated process for continuous etching or wet-treating, rinsing and drying of large surfaces in a single consolidated operation.[56]

3. *Marangoni patterning*

The unwanted phenomenon of "pigment flotation" that may arise during the curing of paint films, and shown in Fig. 10-7, has recently been suggested[57] as a means for patterning films of colloid particles. When a shallow pool of a colloidal dispersion is subjected to an adverse vertical temperature or composition gradient, as by heating from below, cooling from above or through evaporation or the desorption of a surface tension

[55] http://www.globalmanufacture.net/home/communities/engineering/dry.cfm
[56] Britten, J. A., *Solid State Tech.*, **6** [October], 143 (1997).
[57] Harris, D. J., Hu, H., Conrad, J. C., and Lewis, J. A., *Phys. Rev. Lett.*, **98** [14], 148301 (2007); Harris, D. J., and Lewis, J. A., *Langmuir*, **24**, 3681 (2008).

lowering solute, a pattern of Bénard cells may result. The colloidal particles are convected about with the flow, and depending on the relative density of the particles to the medium, the particles may accumulate either around the cell edges or centers. Recent work has sought to control the pattern of convection by placing a mask with patterned holes just above the evaporating surface, leading to what the authors term "evaporative lithography." [58]

Another recent example of Marangoni patterning has led to the supra-assembly of fibrous surfactant crystallites at the water/air surface, producing what the authors[59] term "micro-pottery." In a study of the slow crystallization of sodium myristate from its aqueous solution, needle-like crystals first appeared at the water surface. As they grew, they were seen to bend into arcs and then into closed circles approximately one µm in diameter. Subsequently formed needles were seen to wrap themselves around these rings, eventually forming three-dimensional pottery-like structures. The mechanism is believed to be as pictured in Fig. 10-27. First, an incipient needle is bent through the influence of Brownian processes. As

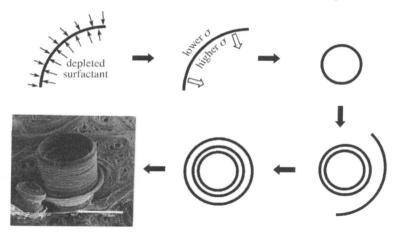

Fig. 10-27: Formation of "micro pottery" by Marangoni forces. The initial needle-like crystallite is bent into a ring by differential surface tension forces, and collects additional filaments until a three-dimensional pot-like structure is formed. Scale bar in inset is 10 µm. Inset from [Raut, J. S., Bhattad, P., Kulkarni, A. C., and Naik, V. M., *Langmuir*, **21**, 516 (2005).]

the bent needle continues to grow, it preferentially depletes dissolved surfactant from its concave side, leading to an increased surface tension on that side relative to that on the convex side. This difference enhances the bending, eventually forming the needle into a ring. The same type of depletion will draw additional crystallite needles to the ring, eventually forming a circular structure.

[58] Harris. D. J., and Lewis, J. A., *Langmuir*, **24**, 3681 (2008).
[59] Raut, J. S., Bhattad, P., Kulkarni, A. C., and Naik, V. M., *Langmuir*, **21**, 516 (2005).

D. The Effect of Surface Active Agents

1. *Gibbs elasticity*

The Marangoni effects described so far have led to the spontaneous generation of flows. What appears to be another type of Marangoni effect leads to just the opposite results, *viz.*, the *suppression* of flow. It is associated with the presence of surface active agents, and is pictured in Fig. 10-28 (and earlier in Fig. 3-39, for the case of an insoluble monolayer). Recall that a surfactant is a solute at the end of the spectrum of surface tension lowering solutes, one that produces large reductions in surface tension at low concentrations. When a disturbance eddy dilates the surface locally, the concentration of the surfactant is reduced there and cannot be immediately restored. The cleared area has higher (often much higher) surface tension than its surroundings, tending to pull the dilated area back to its original configuration, and to suppress the dilational flow beneath the surface. If on the other hand the bulk flow disturbance were one causing a local *compression* in the surface, rather than a dilation, the surfactant concentration at that point would be *increased*, leading to a *decrease* in surface tension, and to surface forces seeking to re-dilate the surface. The automatic redistribution of surfactant in response to any surface distortion, producing forces opposed to the distortion, gives surfactant-laden fluid interfaces an elastic character, referred to as *Gibbs elasticity*. As first described by Gibbs,[60] the behavior is not to be confused with intrinsic or structural interfacial elasticity. The key feature of Gibbs elasticity is the fact that the restoring forces are temporal, as the dilated patch will seek to heal itself by adsorption of surfactant from the subphase and, if it is insoluble in

restoring forces acting in opposition to the
dilational disturbance of the surfactant monolayer

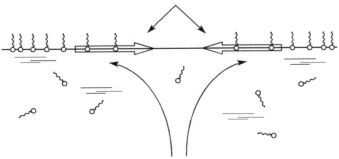

Fig. 10-28: The mechanism of Gibbs elasticity.

the subphase, by surface diffusion in the interface. What determines whether or not the developed surface forces oppose or have no effect on the initial disturbance is whether or not the surface concentration of the surfactant can

[60] Gibbs, J. W., *Trans. Conn. Acad.*, **3**, 343 (1878).

be restored over the time scale of the disturbance. The combination of a bulk liquid layer supporting an "adverse" concentration gradient of a surfactant (or just a surface tension lowering solute) together with rapid surface-subsurface equilibration in fact leads to instability and disturbance growth, so that Gibbs elasticity and Marangoni instability are seen to be just opposite ends of a continuous spectrum of behavior.

The consequences of Gibbs elasticity are often dramatic, particularly in aqueous systems. A teaspoon or so of olive oil (or other appropriate surfactant) is all that is needed to suppress capillary waves over large areas of lakes, *etc.*, as discussed briefly in Chap. 3 and shown in Fig. 1-4. Also, a tiny amount of olive oil will completely suppress "wine tears," and a drop of oleic acid will bring an abrupt end to the camphor dance. Surfactants can completely suppress the formation of Bénard cells, in aqueous and many nonaqueous systems. Figure 10-29 shows the suppression of pigment flotation in an oil base paint, with a drop of silicone oil, which acts as a surfactant in this system. Micro-bubbles or drops in aqueous systems invariably rise or sediment in accord with Stokes' Law, suggesting that their interfaces are stagnant, being totally covered with surfactant whose incipient redistribution generates boundary tension gradients sufficient to cancel the viscous traction of the bulk fluids. The "surfactant" may be just the small amount of adventitious contamination that is always present unless scrupulous means are taken to eliminate it. The suppression of convection with oleic acid during mass transfer of a solute across a water-oil interface in the schlieren cell of Fig. 10-22 is shown in Fig. 10-30.

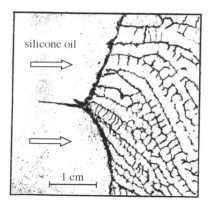

Fig. 10-29: Suppression of Bénard convection and pigment flotation in a curing oil base paint film. A drop of a silicone oil has been deposited at the right hand side and is spreading from left to right.

The behavior of vapor-liquid contacting processes such as distillation and gas absorption or stripping may also be greatly affected by the presence of surfactants. A small amount of the surfactant 1-decanol was shown to be capable of converting the σ^{neg} behavior observed in a packed distillation column treating the water-formic acid system shown in Fig. 10-25 effectively to a σ^{pos} system, resulting in a nearly 100% improvement in column efficiency.[61] The uniform wetting and irrigation produced was also

---

[61] Francis, R. C., and Berg, J. C., *Chem. Eng. Sci.*, **22**, 685 (1967).

clearly observable. In a study with the same system in a sieve tray column (unsupported area equipment), the use of the same surfactant produced modest improvements in efficiency on the σ^{neg} side of the azeotrope, and a significant increase in frothing was observed.[62]

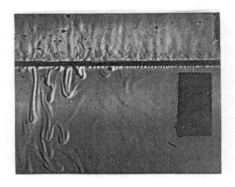

Fig. 10-30: Suppression of convection generated during transfer of nitric acid from isobutanol (above) to water (below) by the addition of a drop of oleic acid at the right side of the interface. [Berg, J. C., and Baldwin, D. C., *The Trend in Engineering*, University of Washington, pp. 13-17, October (1965).]

2. The boundary conditions describing the effects of surfactants

To describe the hydrodynamic effects of surfactants quantitatively, we require further elaboration of the tangential force balance at the interface, Eq. (10.7). This will require 1) development of the term τ^s, representing intrinsic interfacial rheology, generally thought to be important only when a concentrated surfactant monolayer is present, and 2) an additional term in Eq. (10-12) to account for the interfacial distribution of surfactant, and its consequences, *i.e.*, Gibbs elasticity. Accounting for the presence of a surfactant, Eq. (10.14) becomes:

$$\nabla_{\text{II}}\sigma = \left(\frac{\partial \sigma}{\partial T}\right)_0 \nabla_{\text{II}}T + \left(\frac{\partial \sigma}{\partial C}\right)_0 \nabla_{\text{II}}C + \left(\frac{\partial \sigma}{\partial \Gamma}\right)_0 \nabla_{\text{II}}\Gamma_i \ \text{......} \tag{10.96}$$

where Γ_i is the local surface adsorption of a surfactant "i". In many situations, the final term in Eq. (10.96) completely dominates the other two.

To obtain the surface gradient of the surfactant concentration requires solution to the convective diffusion equation for the surfactant species, subject to the appropriate boundary condition expressing its conservation. The latter requires generalization of Eqs. (10.17). The first of these may be supplemented with an adsorption isotherm expressing the distribution equilibrium between the interface and one of the subjacent bulk layers. With reasonable generality, this may be given by the Langmuir isotherm, *cf.* Eq. (3.104):

$$\Gamma_i = \frac{\Gamma_\infty C_i}{a + C_i}, \tag{10.97}$$

[62] Brumbaugh, K. H., and Berg, J. C., *AIChE J.*, **19**, 1078 (1973).

where Γ_∞ is the saturation adsorption, obtained for close-packed monolayers as C_i becomes large, and a is a constant. For dilute systems, the isotherm reduces to its linear form: $\Gamma_i = KC_i$, where $K = \Gamma_\infty/a$.

Generalization of the second of Eqs. (10.17) requires consideration of a number of processes that may affect the local concentration of surfactant in the interface, as pictured in the Fig. 10-31, which diagrams the material balance for it in a small patch of interface at the location of interest, delineated by a small circle drawn in the interface. In words, the balance expresses the fact that the rate of accumulation of the surfactant in the patch equals the net rate at which it is transported into the patch from both the adjacent bulk phases as well as adjacent areas of the interface, plus any generation (or consumption) of surfactant within the patch by chemical reaction ($\mathcal{R}_i$) or spreading from (or collapse into) bulk fragments of surfactant within the patch (j_{sp-col}). The flux of surfactant from the adjacent bulk phases has been expressed in Eq. (10.17) as j_i' and j_i'', as for any solute species. For a surfactant, the rate of the process may be dominated by either diffusion or by the kinetics of the adsorption or desorption processes. Thus we may have

$$j_i' = -D_i'\left(\frac{\partial C_i'}{\partial z'}\right)_{interf} \quad \text{or} \quad = k_1'C_i' - k_{-1}'\Gamma_i, \text{ etc.,} \qquad (10.98)$$

shape change: $\Gamma_i\left|v_n\nabla_{II}^2\eta\right|$

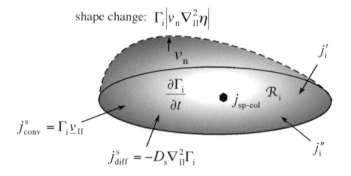

Fig. 10-31: Surfactant material balance on a patch of surface.

where k_1 and k_{-1} are adsorption and desorption rate constants, respectively, assuming the processes to be first order. If transport of surfactant to or from the interface is adsorption/desorption rate controlled, the adsorption equilibrium expression of Eq. (10.97) is replaced by equating the two expressions of Eq. (10.98).

The rate of convective interfacial transport of surfactant is given by $\Gamma_i\underline{v}_{II}$, where $\underline{v}_{II}$ is the local interfacial velocity. The net rate of efflux per unit area of the patch due to interfacial convection is the divergence of this flux vector: $\nabla_{II}\cdot\Gamma_i\underline{v}_{II}$. The rate of surface diffusion is given by a 2-d analog of Fick's Law, viz., $-D_s\nabla_{II}\Gamma_i$, where D_s is the surface diffusivity of the surfactant. The net rate of efflux per unit area of the patch by surface diffusion is given by the divergence of this flux: $-D_s\nabla_{\text{II}}^2\Gamma_i$. The amount of surfactant in the patch is given by Γ_i times the patch area, A_p, and the latter may be changing if the shape of the interface is changing. The rate of accumulation of the surfactant in the patch delineated by the fixed circle in the interface is thus given by:

$$\frac{\partial(\Gamma_i A_p)}{\partial t} = A_p \frac{\partial\Gamma_i}{\partial t} + \Gamma_i \frac{\partial A_p}{\partial t}, \tag{10.99}$$

or on a per unit patch area basis:

$$\frac{\partial\Gamma_i}{\partial t} + \Gamma_i \frac{\partial\ln A_p}{\partial t}. \tag{10.100}$$

The fractional patch area change, for small interface deformations, is:

$$\frac{\partial\ln A_p}{\partial t} = \left|v_n\nabla_{\text{II}}^2\eta\right|, \tag{10.101}$$

where v_n is the fluid velocity component normal to the interface. Putting all of these pieces together yields the boundary condition:

$$\frac{\partial\Gamma_i}{\partial t} + \nabla_{\text{II}}\cdot\left(\Gamma_i\underline{v}_{\text{II}} - D_s\nabla_{\text{II}}\Gamma_i\right) - \Gamma_i\left|v_n\nabla_{\text{II}}^2\eta\right| + j_i'' + j_i' + \mathcal{R}_i \pm j_{\text{sp-col}} = 0. \tag{10.102}$$

It would be rare in any given case that all, or even a significant number, of the terms in Eq. (10.102) would be in play, but there are important cases in which any of the various combinations of the terms might be important.

Next one must consider the effects of intrinsic interfacial rheology, codified thus far as $\underline{\underline{\tau}}^s$, and embodying the constitutive equation for the interface. The state of stress in the interface may be given in terms of the interfacial rheological stress tensor, $\underline{\underline{\tau}}^s$. Using the coordinate system of Fig. 10-9, and assuming a flat interface:

$$\underline{\underline{\tau}}^s = \begin{bmatrix} \tau_{xx}^s & \tau_{xy}^s \\ \tau_{yx}^s & \tau_{yy}^s \end{bmatrix}. \tag{10.103}$$

Phenomenologically, for a 2-d compressible Newtonian fluid, the interface resists expansion at a finite rate in accord with

$$\tau_{xx}^s = \tau_{yy}^s = \kappa_s \frac{1}{A}\frac{\partial A}{\partial t} = \kappa_s\nabla_{\text{II}}\cdot\underline{v}_{\text{II}}, \tag{10.104}$$

where κ_s is the dilitational surface viscosity. Secondly, the interface resists shearing in accord with:

$$\tau_{yx}^{s} = \tau_{yx}^{s} = -\mu_{s}\left[\frac{\partial v_{x}}{\partial y} + \frac{\partial v_{y}}{\partial x}\right], \tag{10.105}$$

where μ_{s} is the surface (or interfacial) shear viscosity. Both κ_{s} and μ_{s} have the units of 3-d viscosity divided by length. For example, a "surface poise" is 1 g/s. As discussed in Chap. 3.K.5, the surface shear viscosity has been measured for many systems, but the surface dilitational viscosity is difficult (if not impossible) to separate from the generally much greater effects of Gibbs elasticity.

The interfacial viscous force per unit area needed in Eq. (10.11) is obtained by taking the divergence of the interfacial viscous stress tensor: $\underline{\tau}^{s} = \nabla_{II}\underline{\underline{\tau}}^{s}$, which, after some manipulation takes the form:[63]:

$$\underline{\tau}^{s} = (\mu_{s} + \kappa_{s})\nabla_{II}^{2}\underline{v}_{II} + \mu_{s}\underline{e}_{z} \times \nabla_{II}(\underline{e}_{z} \cdot \nabla \times \underline{v}_{II}), \tag{10.106}$$

where $\underline{e}_{z}$ is the unit vector normal to the interface. In the usual case, one ignores κ_{s} and also assumes irrotational interfacial flow (so the curl of the interfacial velocity, $\nabla \times \underline{v}_{II} = 0$), reducing Eq. (10.106) to

$$\nabla \cdot \underline{\tau}^{s} = \mu_{s}\nabla \cdot \nabla_{II}^{2}\underline{v}_{II} = -\mu_{s}\nabla_{II}^{2}\left(\frac{\partial v_{z}}{\partial z}\right) \tag{10.106'}$$

Unfortunately, it appears that many of the surfactant monolayers of interest cannot be modeled as Newtonian, so rather complex models have been proposed for non-Newtonian behavior.[64] Also, for completeness, the bending moments and saddle-splay deformations discussed in Chap. 2.G.4, should be taken into account.[65]

3. The effect of surfactants on bubble or droplet circulation

The effects of surfactants in interfacial hydrodynamics is well illustrated by revisiting a couple of the problems considered earlier in this chapter, the first concerning the circulation occurring within droplets or bubble descending or ascending in a gravitational field, or "swimming" in a temperature or concentration gradient, as pictured in Fig. 10-14. Experimentally it had been possible to reproduce the suspension of air bubbles in a suitable temperature gradient only for a series of silicone liquids, whose surface tension was so low that the bubble surfaces were unlikely to be contaminated. Young, Goldstein and Block noted that even hexadecane was subject to sufficient "surfactant" contamination (presumably by traces of silicone), that bubble suspension was not possible. The classical results of Hadamard and Rybczinski for the terminal velocity

[63] Scriven, L. E., *Chem. Eng. Sci.*, **12**, 98 (1960).
[64] Wasan, D. T., Djabbarah, N. F., Vora, M. K., and Shah, S. T., in **Lecture Notes in Physics No. 105: Dynamics and Instability of Fluid Interfaces**, T. S. Sørensen (Ed.), pp. 205-228, Springer-Verlag, Berlin, 1978.
[65] Eliassen, J. D., **Interfacial Mechanics**, PhD Dissertation, University of Minnesota, 1963.

of drops or bubbles in a quiescent liquid were observed in the laboratory only for rather large drops or bubbles, usually diameters (in excess of 5 – 10 mm), while smaller bubbles or drops appeared to follow Stokes' Law more closely, especially if one of the fluids was water. The presence of surfactant "contamination" was clearly the explanation.

The key boundary condition used by Hadamard and Rybczinski in deriving the flow patterns inside and outside a sedimenting drop balanced the shear stress inside and outside the drop as:

$$\tau'_{r\theta} = \tau_{r\theta}, \quad \text{or} \quad \mu'\left(\frac{\partial v'_\theta}{\partial r} - \frac{v'_\theta}{r}\right)_{r=a} = \mu\left(\frac{\partial v_\theta}{\partial r} - \frac{v_\theta}{r}\right)_{r=a}, \tag{10.57}$$

where the primed quantities referred to the drop phase. Young, Goldstein and Block modified this boundary condition to account for the drop or bubble being in a temperature field:

$$\mu'\left(\frac{\partial v'_\theta}{\partial r} - \frac{v'_\theta}{r}\right)_{r=a} = \mu\left(\frac{\partial v_\theta}{\partial r} - \frac{v_\theta}{r}\right)_{r=a} + \frac{1}{a}\frac{\partial \sigma}{\partial \theta}, \tag{10.58}$$

in which the last term was expanded as: $\frac{1}{a}\frac{\partial \sigma}{\partial \theta} = \frac{1}{a}\left(\frac{d\sigma}{dT}\right)\left(\frac{\partial T}{\partial \theta}\right)_{r=a}$ to give Eq.

(10.59). Specifically, these authors considered bubbles immersed in a temperature field whose gradient was collinear with the gravitational field and directed such that the bubbles sought to swim downward, and conditions were found in which their rise could be suspended.

Levich[66] was perhaps first to address the problem of droplet sedimentation in the presence of surfactants. When a surfactant is present, Eq. (10.58) takes the form (with the subscript i suppressed):

$$\mu'\left(\frac{\partial v'_\theta}{\partial r} - \frac{v'_\theta}{r}\right)_{r=a} = \mu\left(\frac{\partial v_\theta}{\partial r} - \frac{v_\theta}{r}\right)_{r=a} + \frac{1}{a}\left(\frac{d\sigma}{d\Gamma}\right)\left(\frac{\partial \Gamma}{\partial \theta}\right). \tag{10.107}$$

Levich argued that as a result of the drop's motion, surfactant would be swept toward the rear stagnation region of the interface, as shown in Fig. 10-32 for the case of sedimentation. He considered that the distribution of surfactant about the sphere could be governed by a balance between interfacial convection and: 1) transport from one of the bulk phases controlled by adsorption/desorption kinetics, or 2) controlled by bulk diffusion, or 3) for an insoluble surfactant, controlled by surface diffusion. The appropriate expression for $(\partial \Gamma / \partial \theta)$ is sought for each case. In all cases,

[66] Levich, V. G., **Physicochemical Hydrodynamics**, pp. 395-429, Prentice-Hall, Englewood Cliffs, NJ, 1962.

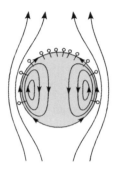

Fig. 10-32: Distribution of
surfactant around a circulating
drop, as envisioned by Levich.

it was assumed that the adsorption would be perturbed by only a small
amount from its equilibrium value, *i.e.*:

$$\Gamma = \Gamma_0 + \Gamma*,$$ (10.108)

where the perturbation is designated with an asterisk to avoid confusing it
with the phase identification superscript (').

For case (1), the material balance on surfactant, Eq. (10.90) becomes

$$\nabla_{\mathrm{II}} \cdot \left(\Gamma_i \underline{v}_{\mathrm{II}}\right) + j'_i = 0.$$ (10.109)

Substituting Eq. (10.108) and noting $\underline{v}_{\mathrm{II}} = \underline{e}_\theta v_\theta$, gives

$$\Gamma_0 \nabla_{\mathrm{II}} \cdot \underline{e}_\theta v_\theta + j'_i = 0,$$ (10.110)

wherein the second-order small quantity $\Gamma * v_\theta$ has been neglected. Putting
the divergence operator into spherical coordinates, and substituting
$v_\theta = v_0 \sin\theta$, where v_0 is the velocity at the equator, to be determined later,
the first term in Eq. (10.110) becomes

$$\Gamma_0 \nabla_{\mathrm{II}} \cdot \underline{e}_\theta v_\theta = \frac{\Gamma_0}{a \sin\theta} \frac{\partial}{\partial \theta}(v_\theta \sin\theta) = \frac{2\Gamma_0 v_0 \cos\theta}{a},$$ (10.111)

The next task is to determine j'_i, the net rate of transfer of surfactant from
the bulk liquid to the interface. Levich's Case 1 corresponds to Eq. (10.98).
The bulk solution is assumed to be at the uniform concentration C_0, so
rewriting (10.98) using Eq. (10.108) gives

$$j' = k_1 C_0 - k_{-1}\Gamma = k_1 C_0 - k_{-1}(\Gamma_0 + \Gamma*)$$

$$= k_1 C_0 - k_{-1}\Gamma_0 - k_{-1}\Gamma* = -k_{-1}\Gamma*$$ (10.112)

$k_1 C_0 - k_{-1}\Gamma_0 = 0$, because C_0 and Γ_0 are equilibrium values. Substituting Eqs.
(10.111) and (10.112) into (10.109) gives

$$\frac{2\Gamma_0 v_0 \cos\theta}{a} - k_{-1}\Gamma* = 0.$$ (10.113)

Using Eq. (10.113) we have

$$\frac{\partial \Gamma}{\partial \theta} = \frac{\partial \Gamma^*}{\partial \theta} = -\frac{2\Gamma_0 v_0 \sin\theta}{ak_{-1}} = -\frac{2\Gamma_0 v_\theta}{ak_{-1}} \qquad (10.114)$$

for substitution into Eq. (10.107). This provides the needed boundary condition on v_θ. Implementing it leads finally to the expression for the terminal velocity of the drop in the form

$$v_\infty = \frac{2a^2(\rho'-\rho)g}{3\mu}\left(\frac{\mu+\mu'+\gamma}{2\mu+3\mu'+3\gamma}\right), \qquad (10.115)$$

where

$$\gamma = \frac{2\Gamma_0}{3k_{-1}a}\left(\frac{\partial\sigma}{\partial\Gamma}\right). \qquad (10.116)$$

When γ is zero, Eq. (10.115) reverts to the Hadamard-Rybczynski equation, *i.e.* Eq. (10.56), but when finite, is multiplied by 3 in the denominator and unity in the numerator, so its presence is seen to have a retarding effect. When γ is sufficiently large, the terminal velocity is as predicted by *Stokes' Law*.

For Levich's Case 2, Eqs. (10.109) and (10.111) are still valid, but j_i' takes the approximate form

$$j_i' \approx -D\frac{\Delta C}{\bar{\delta}}, \qquad (10.117)$$

where D is the bulk diffusivity of the surfactant, ΔC is the concentration difference between the bulk and a layer immediately subjacent to the interface, and $\bar{\delta}$ is the average diffusion boundary layer thickness. ΔC is given by

$$\Delta C = \frac{\partial C}{\partial\Gamma}\Delta\Gamma = -\frac{\partial C}{\partial\Gamma}(\Gamma-\Gamma_0) = -\frac{\partial C}{\partial\Gamma}\Gamma^*, \qquad (10.118)$$

whence combining Eqs. (10.114) and (10.115)

$$\frac{\partial\Gamma}{\partial\theta} = \frac{\partial\Gamma^*}{\partial\theta} = \frac{2\bar{\delta}\Gamma_0}{aD}\left(\frac{\partial\Gamma}{\partial C}\right)v_\theta. \qquad (10.119)$$

This leads to an expression for the terminal velocity the same as that of Eq. (10.116), except that instead for γ we have

$$\gamma = \frac{2\Gamma_0\bar{\delta}}{3aD}\left(\frac{\partial\Gamma}{\partial C}\right)\left(\frac{\partial\sigma}{\partial\Gamma}\right), \qquad (10.120)$$

with essentially same type of result as Case 1.

For Case 3, the monolayer is insoluble, and Eq. (10.102) takes the form

$$\nabla_{II}\cdot\left(\Gamma_i \underline{v}_{II} - D_s\nabla_{II}\Gamma_i\right) = 0, \qquad (10.121)$$

which upon substitution of Eqs. (10.108) and (10.109) gives

$$2\Gamma_0 v_0 \cos\theta = \frac{D_s}{a\sin\theta}\frac{\partial}{\partial\theta}\left(\sin\theta\frac{\partial\Gamma^*}{\partial\theta}\right),\qquad(10.122)$$

and integrating:

$$\Gamma^* = -\frac{2a\Gamma_0}{D_s}v_0\cos\theta.\qquad(10.123)$$

Finally

$$\frac{\partial\Gamma}{\partial\theta}=\frac{\partial\Gamma^*}{\partial\theta}=\frac{2a\Gamma_0}{D_s}v_\theta,\qquad(10.124)$$

and the retardation coefficient becomes

$$\gamma=\frac{2\Gamma_0 a}{3D_s}\left|\frac{\partial\sigma}{\partial\Gamma}\right|.\qquad(10.125)$$

Levich noted that "certain limitations are imposed on this conclusion" in that the retardation term was predicted in this case to *increase* with drop size, contrary to both experiment and intuition.

While Levich's analysis was highly insightful and qualitatively in agreement with the observation that the presence of surfactant retarded the steady state rate of rise or fall of bubbles or droplets, it soon became clear that it was quantitatively incorrect in at least two respects. First, Levich's prediction was one of a *gradual* change in the terminal velocity with drop or bubble size or with the addition of increasing amounts of surfactant, a direct consequence of the assumption of Eq. (10.118). It had long been known that the change was not gradual but abrupt.[67] Levich's analysis also predicted a gradual diminution of the fluid velocity throughout the drop and its vicinity with increasing surfactant, whereas careful observations of Savic[68] documented much different behavior, as pictured in Fig. 10-33. The surfactant was swept toward the rear where it formed a stagnant cap beneath which there was no circulation. The cap, consisting of a condensed monolayer, gave the interface the hydrodynamic properties of a solid surface. In front of the cap, the interface was virtually free of surfactant. Davis and Acrivos[69] analyzed this situation and showed that the cap size, in any given case, was determined by the collapse pressure of the monolayer, assuming this condition was reached at the rear stagnation point. At steady state, the rate at which surfactant arrived and was swept backward was balanced by the rate the monolayer was shed at the rear stagnation point.

[67] Bond, W. N., and Newton, D. A., *Phil. Mag.*, **5**, 794 (1928).
[68] Savic, P., Report M22, Div. of Mechanical Engineering, National Research Council of Canada, 1953.
[69] Davis, R. E., and Acrivos, A., *Chem. Eng. Sci.*, **21**, 681 (1966).

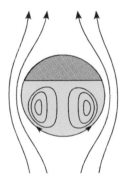

Fig. 10-33: Distribution of surfactant around a circulating drop, showing a stagnant cap in the rear stagnation region, with circulation confined to the region in front of the cap, as documented by Savic.

The result obtained above for droplets or bubbles is indicative of behavior observed in general for sparingly soluble surfactants at interfaces undergoing macroscopic flow, *i.e.*, the distribution of the surfactant along the interface is dominated by interfacial convection, and depending on circumstances, the interface will be roughly divided into regions nearly free of surfactant and regions covered by a close-packed, solid-like monolayer. The steady state distribution may be governed by the total amount of surfactant present relative to the interfacial area available, or by the collapse characteristics of the monolayer. The boundary condition, Eq. (10.102) may reduce effectively to

$$\nabla_{II} \cdot \left(\Gamma_i \underline{v}_{II} \right)_i - j_{col} = 0.$$ (10.126)

In situations involving moderate bulk concentrations of soluble surfactants and/or cases in which the interfacial velocities are very very small, more gradual or continuous variations in interfacial surfactant concentration are observed. Such is the case for linear stability analyses of systems with surfactants, as discussed below.

4. The effect of surfactants on the stability of a pool heated from below

Consider again the problem of the stability of a shallow pool heated from below (or, analogously, supporting an adverse concentration gradient) treated earlier, but with the presence of a surfactant.[70] The linear stability analysis remains the same in all respects except that the dynamic boundary condition at the upper surface, replacing Eq. (10.85'), becomes

$$-\mu \left(\frac{\partial^2 v_z}{\partial z^2} \right) + \nabla_{II}^2 \sigma + \nabla_{II} \cdot \underline{\tau}^s = 0$$ (10.127)

The second term is given by Eq. (10.96):

$$\nabla_{II}^2 \sigma = \left(\frac{d\sigma}{dT} \right)_0 \nabla_{II}^2 T + \left(\frac{d\sigma}{d\Gamma} \right)_0 \nabla_{II}^2 \Gamma,$$

[70] Berg, J. C., and Acrivos, A., *Chem. Eng. Sci.*, **20**, 737 (1965).

and with Eq. (10.106'):

$$\nabla \cdot \underline{\tau}^{s} = \mu_{s} \nabla \cdot \nabla_{II}^{2} \underline{v}_{II} = -\mu_{s} \nabla_{II}^{2} \left(\frac{\partial v_{z}}{\partial z} \right)$$

gives

$$\mu \left(\frac{\partial^{2} v_{z}}{\partial z^{2}} \right) + \left(\frac{d\sigma}{dT} \right)_{0} \nabla_{II}^{2} T + \left(\frac{d\sigma}{d\Gamma} \right)_{0} \nabla_{II}^{2} \Gamma - \mu_{s} \nabla_{II}^{2} \left(\frac{\partial v_{z}}{\partial z} \right) = 0. \tag{10.128}$$

The interfacial distribution of surfactant is given in general by Eq. (10.102), but for the case of an insoluble monolayer, it takes the form

$$\nabla_{II} \cdot \left(\Gamma \underline{v}_{II} - D_{s} \nabla_{II} \Gamma \right) = 0. \tag{10.129}$$

Assuming the temperature T and the adsorption Γ are equal to their undisturbed values plus a perturbation, i.e., $T = T_{0} + T*$ and $\Gamma = \Gamma_{0} + \Gamma*$, Eq. (10.127) becomes:

$$\mu \left(\frac{\partial^{2} v_{z}}{\partial z^{2}} \right) - \mu_{s} \nabla_{II}^{2} \left(\frac{\partial v_{z}}{\partial z} \right) - \left(\frac{d\sigma}{dT} \right)_{0} \nabla_{II}^{2} T* + \left(\frac{d\sigma}{d\Gamma} \right)_{0} \frac{\Gamma_{0}}{D_{s}} \left(\frac{\partial v_{z}}{\partial z} \right) = 0, \tag{10.130}$$

which is now used in place of Eq. (10.86). When the dependent variables are assumed to be of the form given by Eq. (10.70) and the equation is nondimensionalized, it takes the form

$$D^{2} v_{z}^{z} + (a^{2} Vs + Es) D v_{z}^{z} - a^{2} Ma T *^{z} = 0, \tag{10.131}$$

in which, as before, D is the operator $d / d\tilde{z}$. It is noted that in addition to the Marangoni Number, Ma, Eq. (10.131) has two additional dimensionless groups, characterizing the effect of the surfactant monolayer:

The Surface Viscosity number: $Vs = \dfrac{\mu_{s}}{\mu h}$ \hfill (10.132)

The Elasticity number: $Ei = \left| \dfrac{\partial \sigma}{\partial \Gamma_{0}} \right| \dfrac{\Gamma_{0} h}{D_{s}}$ \hfill (10.133)

Vs is seen to depend inversely on pool depth, while Ei (in which the "i" is a reminder that the surfactant has been assumed insoluble) depends directly upon it, so to obtain numerical values for them in typical cases, h shall be taken as 1 mm. Two cases will be considered: a close-packed monolayer and a "gaseous" monolayer, presumed to have a surface concentration 1/100 that of the close-packed film, both on a water substrate at room temperature. As reported by Berg and Acrivos, and references therein, values for the surface viscosity in the two cases may be estimated as 10 g/s and 10^{-3} g/s, respectively, giving values for Vs of about unity and 10^{3}, respectively. For Es, it is reasonable to put $\Gamma_{0} = 8.3 \times 10^{-12}$ mole/cm^2 (and 8.3×10^{-10} mole/cm^2 for the gaseous monolayer). Typical values of $|\partial \sigma / \partial \Gamma_{0}|$ for the two cases would be 1.34×10^{12} and 2.44×10^{10} mJ/m^2, respectively. Their estimate for the surface diffusivity of $\approx 10^{-5}$ cm^2/s was later validated by measurements.[71] These physical parameters put Ei at approximately 10^9 and 10^5 for condensed and gaseous monolayers, respectively.

The stability analysis, which proceeded in the same way as Pearson's analysis, led to new predictions for the critical Marangoni Number as a function of the Surface Viscosity number and the Elasticity number. For the numbers computed, Vs appeared to have only a small to negligible effect on the critical Ma, but Ei had an enormous effect. Even a gaseous monolayer, with a surface pressure of less than 1 mN/m, moved the critical Marangoni Number from 80 up to more than 100,000, while the condensed monolayer boosted it to over 10^9! Thus even small amounts of surfactant are capable of completely shutting down micro Marangoni convection. Owing to its inverse dependence on pool depth, it appears that Vs may begin to play a role for very thin layers ($h < 1$ μm).

The situation is not as clear-cut in the case of a soluble surfactant.[72] It is reasonable to assume no kinetic barrier to adsorption/desorption and negligible surface rheology. In this case, the boundary condition, Eq. (10.128) takes the form

$$\mu \left(\frac{\partial^2 v_z}{\partial z^2} \right) + \left(\frac{d\sigma}{dT} \right)_0 \nabla_{\text{II}}^2 T + \left(\frac{d\sigma}{d\Gamma} \right)_0 \nabla_{\text{II}}^2 \Gamma = 0, \tag{10.134}$$

with Eq. (10.102) for the surfactant distribution:

$$\nabla_{\text{II}} \cdot \left(\Gamma \underline{v}_{\text{II}} \right) - D \left(\frac{\partial C_0}{\partial z} \right) = 0. \tag{10.135}$$

This leads to another form of the Elasticity number, $viz.$

$$Es = \left| \frac{\partial \sigma}{\partial C_0} \right| \frac{\Gamma_0}{\mu D}, \tag{10.136}$$

in which the "s" is a reminder that the surfactant has been assumed soluble. While strong stabilization can again be the result of the presence of surfactant, intermediate levels of stabilization are also possible.

---

[71] Sakata, E. K., and Berg, J. C., *Ind. Eng. Chem. Fund.*, **8**, 570 (1969).
[72] Palmer, H. J., and Berg, J. C., *J. Fluid Mech.*, **51**, 385 (1972);
    Palmer, H. J., and Berg, J. C., **75**, 487 (1976);
    Palmer, H. J., and Berg, J. C., *AIChE J.*, **23**, 831 (1977).

Some fun things to do:
Experiments and demonstrations for Chapter 10

Some of the ideas in this section and others were inspired by the 1969 film: **Surface Tension in Fluid Mechanics** by Lloyd Trefethen at Tufts University.[73]

1. *Tears of strong wine*

When strong wine (> 15 proof) or brandy evaporates from a glass, a corona of droplets spontaneously forms above the meniscus, as first described by J. J. Thomson in 1855, and illustrated on the cover of this book. This is the result of preferential evaporation of the alcohol from the meniscus, leaving behind a thin film of water whose surface tension is higher than the alcohol solution beneath it. Surface tension forces draw the water upward into a ring of liquid that breaks up into droplets that run back down into the liquid pool.

Materials:

- Strong wine (Sauterne, Port, *etc.*) or brandy in a large open wine glass or brandy snifter.

Procedure:

Fill the wine glass to about 1/3 and first allow it to stand and observe droplets forming just above the meniscus. The effect is enhanced by gently swirling the wine in the glass to form a thin film reaching above the meniscus about one cm, setting the glass down and observing the tears form.

2. *Thermocapillary flow*

The surface tension of a liquid decreases with temperature so that if a portion of a liquid surface is warmer than the neighboring regions, surface tension forces develop which dilate the surface where it is warmer. As the surface expands, the liquid beneath it is also drawn outward by virtue of its viscosity. Conversely, if a liquid surface is locally cooled, the adjacent surface is drawn inward toward the cool region.

Materials:

- Small ring stand
- Thin aluminum plate or heavy foil ($\approx 4\times15\times0.065$ cm)
- Two C-clamps

---

[73] **Surface Tension in Fluid Mechanics**, produced by Education Development Center under the direction of the National Committee for Fluid Mechanics Films, with the support of the National Science Foundation; Encyclopedia Britannica Educational Corp., Chicago, IL.

- Soldering gun
- Ice cube
- 10 mL of 1-centistoke Dow Corning 200 Fluid (silicone oil)

Procedure:

1) Place the aluminum plate on top of the ring stand and secure it using the clamps. Smear a small puddle of the oil on the plate.

2) Heat up the soldering gun and place its tip against the metal plate beneath the center of the puddle. The heated spot will form a crater as the oil flows away from it.

3) Use the tip of the ice cube to cool a spot beneath the center of the puddle and note that a hump is formed as oil flows toward it.

3. *Bénard cells*

Marangoni flow in the form of Bénard cells, as shown in Fig. 10-6, occurs when a shallow liquid pool is heated from below or cooled from above. The phenomenon can be observed by placing suitable tracer particles in the liquid or by observing the surface of the liquid in oblique light at a shallow angle, which renders the accompanying surface relief visible.

Materials:

- 50 mL of 500-centistoke Dow Corning 200 Fluid (silicone oil)
- 10 g aluminum powder
- 100-mL beaker
- 15-cm diameter Petri dish
- Hot plate
- Sheet of white paper
- Overhead projector

Procedure:

1) Mix about 1 teaspoon ($\approx$ 5g) of aluminum powder into 30 mL of the silicone oil until the particles are uniformly distributed.

2) Turn on the hot plate and wait until it is heated. Then place a sheet of white paper (to help make the flow more visible) and the empty Petri dish on top of it, and pour the aluminum powder-in-oil dispersion into the dish. The dish will be filled to about half its depth. In about 2-3 min, the Bénard cells will be clearly visible, as shown in Fig. E10.1. Examine the surface of the oil at a very shallow angle to observe the surface relief.

3) Repeat Step 2 above using only $\approx$ 10 mL of the dispersion to form a very shallow pool. When the convection has developed, the dish may be carefully transferred to a pre-focused overhead projector for observation of the Bénard cells on the screen. Note that the cell size is smaller than that for the deeper pool.

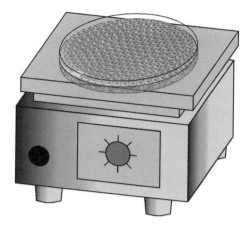

Fig. E10.1: Observation of Bénard cells by heating a pool of oil with suspended particles.

4. *Jet breakup*

Capillary liquid jets break up into a regular train of drops, as described by Lord Rayleigh and in Section B.1, and shown in Fig. 1-6. For larger-diameter and faster-moving jets, the break is more irregular. The structure of the breakup is not discernable by eye because it is too rapid, but can be readily captured by flash photography in a dark room.

Materials:

- Water faucet with a hose-connection nozzle
- White back plate large enough to encompass the disintegrating jet ($\approx$ 6×20 in)
- Digital camera with flash
- Ring stand or tripod for mounting camera
- 250-mL beaker
- Stopwatch

Procedure:

Place a white plate behind the faucet in the sink and turn on the valve to produce the slowest flow rate that gives a continuous stream of liquid. Mount and focus the camera on the jet. Then turn off the room lights and take a flash picture of the jet. The flash duration is only a few microseconds, fast enough to capture a sharp image of the breakup process. The flow rate can be measured by determining (with a stop watch) the amount of time required to fill the 250-mL beaker. The flow rate can be varied and photographs taken for each. Figure E10.2 shows the breakup of jets issuing from a 4 mm diameter nozzle. Smaller diameter jets can be produced by putting reducers on the nozzle.

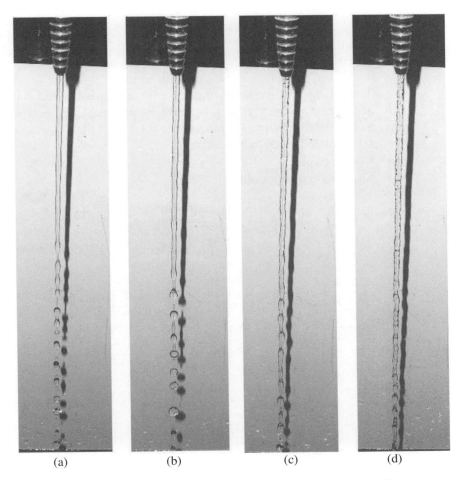

Fig. E10-2: Flash photographs of jets issuing from a 4 mm diameter orifice. (a) 14, (b) 18, (c) 25 and (d) 42 cm³/s.

4. *Marangoni effects in foam films and puddles*

A foam film suspended on a loop provides a convenient way to study Marangoni effects, as shown in Fig. E10.3. The detergent solution used has a surface tension of approximately 30 mN/m. When contacted with different liquids, the soap film may remain intact, rupture or exhibit some other changes. Both the surface tension of the contacting liquid and its solvency in the aqueous soap film are involved in the explanation of what happens. The film will also respond differently to being locally heated or cooled. Shallow puddles of water will also respond differently to the addition of single drops of different solvents, depending on their surface tension and solubility.

Materials:

- 250-mL beaker
- 100 mL of 50/50 Joy® dishwashing detergent in water
- 1 - 1.5-in. diameter plastic ring with handle
- Ten mL each of the following solvents in tightly capped vials: acetone, ethylene glycol, *n*-hexanol, isopropanol, and octane.
- Six 1-mL disposable PE pipettes
- Glass plate, about 4×4 in. area.
- Cotton swabs
- Straightened paper clip
- Matches
- Ice water

Fig. E10.3: Testing the integrity of soap films with respect to drops of various solvents.

Procedure:

1) Form a soap film by dipping the plastic frame into the soap solution, draining it for about 3 s and then resting the frame horizontally on the edge of the empty 250-mL beaker. Using one of the pipettes, drop a droplet of the solvent to be tested onto the film from a distance of a few mm above it, and note what happens. Repeat this at least four times for each solvent. Explain the results obtained in each case in terms of the Marangoni effect.

2) Heat the end of a straightened paper clip using a match (far away from the solvents!) and poke it into the soap film. Do the same thing with the end of wire chilled by being held in the ice batch. Explain any differences observed in terms of the Marangoni effect.

3) Using a cotton swab, smear a small puddle of pure water onto the glass plate. Deposit a drop of each of the above solvents at the center of such a puddle and observe the result. Try this also with a drop of the soap solution.

Appendix 1

EXERCISES

Chapter 1

1. Provide provisional answers to as many as you can of the twelve questions posed in Table 1-1.

2. The following prefixes are used to describe small things. If they are applied to *meter*, try to think of something that fits each size category.

$$
\begin{aligned}
\text{mille} &= \times 10^{-3} & \text{femto} &= \times 10^{-15} \\
\text{micro} &= \times 10^{-6} & \text{atto} &= \times 10^{-18} \\
\text{nano} &= \times 10^{-9} & \text{zepto} &= \times 10^{-21} \\
\text{pico} &= \times 10^{-12} &
\end{aligned}
$$

3. Name at least five different colloids that you have encountered in the past 24 hours.

Chapter 2

1. Estimate the surface tension of benzene and nitric acid at their respective normal boiling points using the following information.

Compound	MW	T_b	T_c	σ (at 20°C)	ρ (at 20°C)
Benzene	78.11	353.2K	562.2K	28.9 mN/m	0.879 g/mL
Nitric acid	63.02	356.2	520	41.2	1.502

2. The Lennard-Jones parameters for benzene are: $\delta = 5.270$ Å, and $\varepsilon/\kappa = 440(°K)$, where $\kappa =$ Boltzmann's constant. Evaluate the molecular separation distance corresponding to the minimum of the potential well. If the mean molecular separation in liquid benzene at a given temperature is just 1% greater than the separation at the potential minimum, calculate the attractive force between a pair of molecules at this distance.

3. The Hamaker constant A may be estimated from surface tension data using Eq. (2.14). Make this estimate using surface tension data at 20°C for dodecane (25.35 mN/m) and water (72.8 mN/m) and compare with the independently determined Hamaker constants of 5.04×10^{-20} J for dodecane and 3.70×10^{-20} J for water. It has been argued that a better estimate is obtained for self-associating liquids, like water, if one uses the dispersion component of the surface tension, σ^d, rather than the full value. Check this out for water, using data from Table 2-2.

4. Surface tension data for mercury at 20°C are ($\sigma = 485$, $\sigma^d = 200$ mN/m) and for water ($\sigma = 72.8$, $\sigma^d = 21.1$ mN/m). Use these data to calculate the interfacial tension between mercury and water using both the Antanow Law and the Fowkes equation, and compare with the experimental value of 415 mN/m.

5. Consider the ellipsoid of revolution obtained by rotating the ellipse whose equation in the x-y plane is

$$\frac{x^2}{a^2} + \frac{y^2}{b^2} = 1$$

about its major axis (the x-axis in this case). The major and minor half-axes of the ellipse are a and b, respectively. Evaluate the curvature of the surface of the ellipsoid, in terms of a and b, at its apex and at its equator.

6. Consider the meniscus of water against a vertical flat plate with which it makes a contact angle of 0° and calculate the distance away from the plate where the elevation of the meniscus is one micrometer above the undisturbed surface.

7. By integrating the appropriate form of the Young-Laplace equation, show that the height h of a meniscus made against a vertical wall is given by:

$$\left(\frac{h}{a}\right)^2 = 1 - \sin\theta,$$

where a is the capillary length and θ is the contact angle.

8. Determine the shape of the slightly deformed circular surfaces shown below. Case (a) is a soap film that is deformed by the application of a constant Δp applied uniformly across the surface of the film, and Case (b) is the meniscus of liquid of density ρ emerging from a circular horizontal orifice. In both cases, the surface tension is σ, the radius is R, and the deformations are sufficiently small that $y' \ll 1$.

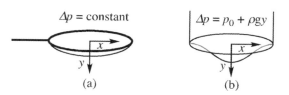

$\Delta p = $ constant $\Delta p = p_0 + \rho g y$

(a)   (b)

9. Use the Sugden tables to predict the capillary rise of methyl iodide ($\sigma = 51$ mN/m, $\rho = 2.279$ g/mL) in a capillary tube of i.d. $= 2.0$ mm. If the measured capillary rise is only 80% of this value, estimate the contact angle that is implied.

10. Calculate the temperature, at atmospheric pressure, required to "boil" water out of 8-nm diameter pores into which it has condensed.

11. If n-heptane at 20°C at a partial pressure equal to 98% of its vapor pressure (4.74 kPa) is found to produce a 18 nm thick film of condensate of on a horizontal quartz surface, evaluate the implied disjoining pressure and the effective Hamaker constant.

Chapter 3

1. It can be shown that if the position of a patch of interface of curvature κ is shifted outward along its normal, z, its area A changes in accord with

$$\frac{d\ln A}{dz} = \kappa$$

and the corresponding volume change on the concave side of the interface is $dV = Adz$. Using these facts, provide a "thermodynamic" derivation of the Young-Laplace equation by using Eq. (3.20) to describe the quasi-static isothermal expansion (for which $dF = 0$) of a patch of surface.

2. The surface tension of an aqueous solution of n-butanol at 20°C is given (up to a molality of $m_2 = 0.4$ mol/kg) to good accuracy by

$$\sigma\,(\text{mN/m}) = 72.2 - 15.93\ln(1+12.27m_2).$$

Determine the adsorption isotherm and the surface equation of state up to a molality of 0.4 mol/kg, assuming ideal dilute solution behavior.

Actually, the solution does exhibit some deviation from ideality, and the activity coefficient for the solute butanol (up to a molality of $m_2 = 0.25$ mol/kg) is given by

$$\ln\gamma_2 = -0.0014 - 0.8418m_2 + 3.1554m_2^2 - 5.4245m_2^3$$

Using these data, compute the relative adsorption of the butanol at $m_2 = 0.10$ mol/kg and compare it to the value obtained when ideality was assumed.

3. Surface tension data for a given C_{14} anionic surfactant in pure water at 25°C are fit by the equation:

$$\sigma(\text{mN/m}) = 72.0 - 10.8\ln\left[1 + \frac{C_2}{3.90\times10^{-5}}\right].$$

   i) Determine the "efficiency" of the surfactant, *i.e.*, its pC_{20} value.

   ii) If the surface tension of a solution for which C_2 is 0.10 M is found to be 31.0 mN/m, estimate the CMC of the surfactant.

iii) If the surfactant above were the C_{16} member of the same series, estimate how the surface tension equation and the pC_{20} value would change?

iv) If the substrate were a 0.1M solution of NaCl instead of pure water, estimate (qualitatively) how would expect the surface tension behavior of the system would change.

4. Consider an ideal dilute aqueous solution of two solutes (2) and (3), whose surface pressure is given by

$$\pi = RTB \ln\left(1 + \frac{C_2}{a} + \frac{C_3}{b}\right).$$

i) Derive general equations permitting one to evaluate $\Gamma_{2,1}$ and $\Gamma_{3,1}$ from surface tension data for the two-solute system

ii) For a particular case at $T = 298K$, $B = 3.0 \times 10^{-10}$ mol/cm²; $a = 10^{-3}M$ and $b = 2.8 \times 10^{-3}M$, evaluate $\Gamma_{3,1}$ when $C_2 = C_3 = 5 \times 10^{-3}M$.

5. Integrate the BET isotherm, Eq. (4.162), for the non-specific adsorption of a gas on a solid, in accord with the Gibbs adsorption equation, Eq. (3.93), to obtain the surface pressure equation. Integrating all the way to saturation provides the equilibrium spreading pressure.

6. The *CMC* for hexadecyl trimethylammonium bromide (HTAB) in water has been measured at 25 and 40°C to be 0.92 and 1.00 mM, respectively. Using Eq. (3.129), which relates the *CMC* to the free energy of micelle formation:

$$\Delta G_{mic}^{\Theta}\Big|_{ionic} = (1 + y)RT\ln(CMC),$$

with y, the fraction of anionic surfactant molecules to which cations are bound (assumed to be << 1), evaluate $\Delta G_{mic}^{\Theta}, \Delta H_{mic}^{\Theta}$ and ΔS_{mic}^{Θ}, referring to the formation of micelles from one mole of monomers.

7. Dipalmitoyl lecithin (lung surfactant) molecules, as shown in Fig. 2-6, are known to spontaneously form vesicles in water, suggesting a critical packing parameter ½ < *CPP* < 1. Using formulas such as those of Eq. (3.144), estimate the implied range of values for the effective area, a_0, of the phosphoryl choline head group.

Chapter 4

1. The Zisman critical surface tension of a given smooth specimen of polyethylene is 22.0 mN/m, and the contact angle made by α-bromonaphthalene against it is measured at 66°.

    i) Obtain an estimate of the solid-liquid interfacial energy, σ_{SL}.

    ii) Calculate the work of adhesion, W_A.

    iii) Determine the contact angle α-bromonaphthalene would make against polyethylene if it were roughened to a rugosity of 1.8.

2. If the contact angle made by water (saturated with benzene) against a given solid is $\theta_W = 70°$, and that by benzene against the same solid is $\theta_W = 30°$, calculate the contact angle of a water drop (saturated with benzene) on the solid surface immersed in benzene. The surface and interfacial tensions under the given conditions are: $\sigma_W = 62.2$ mN/m; $\sigma_B = 28.8$ mN/m, and $\sigma_{B/W} = 35.0$ mN/m.

3. Show that if a liquid A spreads on another immiscible liquid B, liquid B cannot spread on liquid A.

4. In a plunge tank experiment, as in Fig. 4-29(b), the dynamic wetting of a polymer film by a particular solvent (with $\sigma = 32$ mN/m, $\mu = 2.5$ cP) is being investigated. The static contact angle is found to be $\theta_0 = 20°$, and the dynamic advancing contact angle observed when the film is drawn into the liquid at a rate of 4.4 cm/s is 35°. Estimate the "rise canceling velocity," *i.e.*, the velocity for which the dynamic advancing angle is 90°.

5. A liquid resin is sprayed (in droplets of volume equal to 10^{-3} mm^3 each) onto a smooth solid surface on which they have a contact angle of 35°. They solidify into an array of bumps that provide an anti-stick coating. Calculate the height of the bumps.

6. It is desired to determine the advancing contact angle of water ($\sigma = 72.4$ mN/m, $\mu = 1.0$ cP) against a fine powder. For this purpose, the powder is packed into a tube, and water is permitted to wick into the tube. The time required for the water to wick a distance of 3.0 cm into the tube is 3.5 min. The time required for the reference liquid, a silicone oil (with $\sigma = 16$ mN/m, $\mu = 1.5$ cP) to wick the same distance is 4.7 min. The silicone is assumed to wet out the powder. Evaluate the contact angle of the water against the powder, and calculate the time required for the water to wick a distance of 5.0 cm.

7. Consider the immiscible displacement of one liquid by another in a horizontal tube, as shown in Fig. 4-69. Liquid A is an oil, and liquid B is water (both with identical bulk fluid properties, $\rho = 1.0$ g/mL, $\mu = 1.0$ cP), and the tube is initially filled with water. The oil preferentially wets the tube wall so that the meniscus is hemispherical, convex to the oil, and the oil/water interfacial tension is 50 mN/m. The tube diameter is 0.2 mm, its length is 0.20 m and its volume is negligible relative to that in the reservoirs. Calculate the time for the oil to completely displace the water from the tube. Is the format of the Washburn equation followed? Explain.

8. In a spray coating operation, droplets are formed on the surface in what is assumed to be a uniform array with an initial distance between their centers of 3.00 mm. When the drops first impact the surface they cover a negligibly small area, but it is desired that they spread out and merge to form an ultra-thin coherent film. The time required for this process is the "film-forming time," t_F. It is observed that the drops spread from their initial size to $r = 1.0$ mm in 5 s. Estimate t_F.

9. In a BET experiment with a particular activated charcoal, use of N_2 (with an adsorbed cross-sectional area of 16.2×10^{-20} m²/molecule) as the probe gas produces a specific area of $\Sigma = 940$ m²/g, while use of n-butane, whose cross-sectional area may be computed using Eq. (4.173), gives a value of $\Sigma = 520$ m²/g. If the charcoal surface is believed to have a fractal structure, calculate the indicated fractal dimension.

Chapter 5

1. For a particular virus ($\rho_p = 1.03$ g/mL) dispersed in water at 298K, ultracentrifuge measurements have yielded a sedimentation coefficient of 4.4×10^{-11} s, while dynamic light scattering measurements have given a diffusivity of 1.8×10^{-8} cm²/s.

    i) Determine the virus particle mass, m_p. (Do not assume spherical particles.)

    ii) Estimate the average linear distance a virus particle travels by diffusion in one hour.

    iii) Assuming the particles to be prolate ellipsoids of revolution, evaluate their aspect ratio $\beta = b/a$.

2. Output from a particular particle-sizing instrument is in the form of a histogram giving the cumulative distribution of the particle diameter based on particle mass. The results for a given system are:

Class boundaries	Cumulative percent, q_i	Class boundaries	Cumulative percent, q_i
0.8-1.0 μm	1	2.0-2.2 μm	74
1.0-1.2	9	2.2-2.4	83
1.2-1.4	21	2.4-2.6	90
1.4-1.6	36	2.6-2.8	96
1.6-1.8	50	2.8-3.0	99
1.8-2.0	63	3.0-3.2	100

    i) Obtain and plot the frequency distribution of the particle diameter.

    ii) Calculate the mean, standard deviation and skewness of the above distribution.

iii) Assuming the particles to be spheres, convert the above frequency distribution to one based on particle area and on particle number.

iv) Evaluate the Sauter mean diameter and the Z-average particle diameter.

3. A given polystyrene latex (ρ_p = 1.05 g/mL) consists of monodisperse spherical particles of diameter 1.2 μm in water at 20°C. Calculate the time required to centrifuge this colloid (*i.e.*, settle out all the particles) in a 10-cm long centrifuge tube spun at 1500 RPM about an axis 3.0 cm from the top of the tube.

4. Consider a colloidal dispersion of uniform-sized (diameter 100 nm) polystyrene latex particles (ρ_p = 1.05 g/mL) in water at 298K in a vessel (of uniform cross-section normal to g) of depth 5 cm. Initially, the particle concentration in the vessel is uniform at 10^{10} particles/cm^3. The dispersion is permitted to reach diffusion-sedimentation equilibrium. Compute the particle concentration at the top and at the bottom of the vessel.

5. For the colloid in Prob. 4 above, calculate the particle diffusivity and estimate the root mean square displacement to be expected in 30 min, 1 hr, and 24 hr. Re-calculate your results if the particles were of the same mass but prolate ellipsoids of revolution with an aspect ratio of 2.

6. A particular tobacco mosaic virus has the shape of perfect, thin cylindrical rods of diameter d = 1.2 nm and length L (with $L \gg d$). These virus particles, with refractive index n_{part} = 1.39, are dispersed in water (n_{water} = 1.33). When a light source of λ_0 = 633 nm is used, a Guinier plot has yielded a slope of -320.0 nm^2.

i) Evaluate L of the rods.

ii) Estimate the value of the form factor for the rods at a scattering angle of 20°.

7. A dispersion consists of equal numbers of spherical particles of radius 20 and 40 nm. Calculate the average particle radius that will be indicated by a classical light scattering measurement.

Chapter 6

1. Consider a glass surface against a 5×10^{-3}M aqueous solution of KNO$_3$ at 25°C, with a *pH* of 6.1. The point of zero charge (*PZC*) for the glass is a *pH* of 5.0.

i) Evaluate the surface potential, ψ_0. Is the Debye-Hückel approximation valid for this case?

ii) Evaluate the double layer thickness. Do the hydronium and hydroxyl ions contribute in any significant way to this?

iii) Determine the concentration of K^+ and NO_3^- ions a distance of 1.0 nm from the surface.

iv) Evaluate the surface charge density.

v) Evaluate the capacitance of the double layer

2. Consider a capillary tube of the same glass and containing the same electrolyte solution as that of Prob. 1 above.

i) Estimate the zeta potential, assuming it to be the potential that exists at a distance of 2Å from the surface.

ii) Evaluate the electro-osmotic velocity, V_e, in the tube when the electric field strength is $E = 20$ Volt/cm. The dielectric constant for water is $\varepsilon = 80$, and its viscosity 1.0 cP.

3. Derive Eq. (6.33) for the potential profile for the boundary condition of constant surface potential.

4. The electrophoretic mobility of 1-μm diameter silica particles in an aqueous solution of 10^{-4}M KNO_3 at 25°C has been measured at 3.80 (μm/s)/(V/cm). Evaluate the apparent zeta potential:

i) using the Helmholtz-Smoluchowski equation;

ii) using the Hückel equation, and

iii) using the O'Brien and Hunter analysis, Eq. (6.121). The limiting ionic molar conductances are: Λ_+^0 (for K^+) = 64.4, and Λ_-^0 (for NO_3^-) = 61.7 [ohm·cm²·mol]$^{-1}$.

5. An expression for the streaming potential developed when an aqueous electrolyte solution is made to flow through a capillary tube of radius R and length L by an imposed pressure drop Δp, as shown in Fig. 6-29, can be developed as follows:

i) Determine the electric current i (in the $-z$ direction) by integrating the convection of electricity over the cross-section of the tube

$$i = -2\pi \int_0^R \rho_e(r)v(r)r\,dr \,,$$

where $\rho_e(r)$ is the space charge density and $v(r)$ is the local velocity, both functions of the radial coordinate r. Since only the thin layer (relative to radius of the tube) enveloping the double layer contributes to the integral, it may be replaced by an integral over the thickness of a flat layer of horizontal dimensions $2\pi R \times L$, *i.e.*, with respect to the distance $y = R - r$ measured away from the wall.

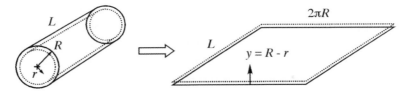

Show that the velocity profile for laminar flow in a capillary is given by the Hagen-Poiseiulle law:

$$v(r) = \frac{R^2 \Delta p}{4\mu L}\left[1 - \left(\frac{r}{R}\right)^2\right] \xrightarrow{y<<R} \frac{R\Delta p}{2\mu L} y$$

Next substitute for the space charge density from the Poisson-Boltzmann equation to obtain:

$$i = \frac{\varepsilon\varepsilon_0 \pi R^2 \Delta p}{\mu L} \int_0^\infty \frac{d^2\psi}{dy^2} y\, dy$$

ii) Integrate the above equation by parts, noting that at $y = 0$, $\psi = \zeta$; and as $y \to \infty$, $(d\psi/dy) = 0$, to obtain:

$$i = \frac{\varepsilon\varepsilon_0 \pi R^2 \Delta p}{\mu L}\zeta$$

iii) Finally, obtain the expression for the streaming potential, ΔE_S, by using Ohm's Law, noting that the resistance is given by

$$\text{Resistance} = \frac{L}{C\Lambda \pi R^2},$$

where C is the molar ion concentration, and Λ is their molar conductance, to obtain

$$\Delta E_S = \frac{\varepsilon\varepsilon_0 \Delta p}{\mu C\Lambda}.$$

The laminar volumetric flow rate in a round tube is given by

$$Q = \frac{\pi R^4 \Delta p}{8\mu L} \quad [\textit{cf.} \text{ Eq. (8.2)}]$$

so alternatively,

$$\Delta E_S = \frac{8\varepsilon\varepsilon_0 L}{\pi R^2}\frac{Q}{C\Lambda}\zeta.$$

6. Electrowetting may be defined as the process whereby the contact angle of a liquid against a solid may be reduced by the application of a

potential difference across the liquid surface or the liquid/solid interface or both. The basis for the phenomenon is the Lippmann equation, Eq. (6.69) or (6.71), which shows that for polarizable interfaces, the application of the potential reduces the interfacial energy. Consider the simple case where an electrode is in a drop of liquid (of finite conductivity) resting atop a thin layer of polymer coated over the surface of a metal plate that acts as the second electrode.

Schematic of electrowetting. When a potential E is applied across the solid-liquid interface, there is a reduction in σ_{SL} and consequently the contact angle.

i) By integrating Eq. (6.71) and noting that σ_{SL} is at its maximum value of $\sigma_{SL}{}^0$ when $E = 0$, show that:

$$\sigma_{SL} = \sigma_{SL}^0 - \frac{\overline{C}_{SL} E^2}{2},$$

where $\overline{C}_{SL}$ is the capacitance of the double layer against the solid-liquid interface.

ii) Show that the contact angle is changed in accord with:

$$\cos\theta = \frac{\sigma_S - \sigma_{SL}^0 + 1/2\overline{C}_{SL} E^2}{\sigma_L}$$

7. When the *pH* is measured in the interior of a particulate dispersion, it is often found to be different from the value measured in the supernatant, with which it is in equilibrium. This is an example of the "suspension effect" attributable to Donnan equilibrium. Consider a dispersion (in contact with clear supernatant) of 10^{16} particles/mL, each carrying a negative charge of -150e. The only dissolved electrolyte in the system is HCl. The *pH* measured in the supernatant is 3.00. Estimate the *pH* that will be registered by an electrode in the dispersion.

Chapter 7

1. Use the Lifshitz approach, in the form of Eq. (7.39) to compute the Hamaker constant for silica particles (2) dispersed in acetone (1) at 25°C. For these materials, $\varepsilon_1 = 21$, $\varepsilon_2 = 4.3$; $n_1 = 1.359$, $n_2 = 1.500$.

2. Assume that the geometry of the tip in an AFM system can be modeled as a paraboloid of revolution, as shown. Use the Derjaguin approximation to derive the expression for the electrostatic interaction force, F_R,

between the tip and a flat surface with identical double layers from the parallel plate interaction force given by Eq. (7.64).

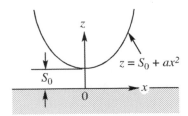

3. It is found for a particular aqueous colloid for which $\psi_0 = -135$ mV, the critical coagulation concentration (CCC) for NaCl is 85 mM. Estimate the CCC for the following electrolytes: $Ca(NO_3)_2$; Li_2SO_4; $Fe_2(SO_4)_3$.

4. Derive the expression for the critical coagulation concentration (CCC), for which Eqs. (7.86) and (7.87) are limiting cases, from the DLVO equation in the form of Eq. (7.80). Use the result, together with the appropriate Hamaker constant found in Table 7-1, to prepare and plot a curve of CCC vs. the effective surface potential, ψ_δ, for an aqueous colloid of 1-µm diameter silica particles at 25°C.

5. Determine the aggregate size distribution for a colloid aggregating in accord with Smoluchowski theory, Eqs. (7.124) and (7.125), at $t = t_{1/2}$ and $t = 2t_{1/2}$.

6. During the rapid aggregation of an electrocratic colloid (for $C > CCC$), aggregates of mass fractal dimension $d_f \approx 1.7$ are formed, while for slow aggregation ($C \ll CCC$) $d_f \approx 2.1$. Assuming aggregates of 100 primary spherical particles each are formed and sediment to the bottom of the vessel where hexagonal close packing exists, estimate the relative sediment heights achieved for a given total number of primary particles for the cases of no aggregation, slow aggregation and rapid aggregation. While the result should be qualitatively correct, the assumptions are unrealistic for a number of reasons. Discuss these.

Chapter 8

1. Consider a dispersion of non-interacting spheres at a volume fraction of $\phi = 0.10$ and predict its relative viscosity using (i) Eq. (8.9), with $b = 6.2$, as suggested by Batchelor, (ii) the Mooney equation, Eq. (8.12), and (iii) the Krieger-Dougherty equation, Eq. (8.14).

2. Some data have been obtained using a Couette viscometer for the viscosity of a given dispersion at different shear rates as follows: $(\dot{\gamma}, \eta)$ [=] (s⁻¹, cp) = (1, 14.1); (5, 11.0); (10, 9.9); (50, 7.8). Determine if

this dispersion can be characterized as a power law fluid over the range of shear rates investigated, and if so, evaluate the constants m and n.

3. *Estimate* the yield stress of a weakly percolated dispersion of silica particles (at $\phi = 0.2$) in an aqueous 0.05N KNO$_3$ solution at 25°C using Eq. (8.24). The zeta potential has been measured to be 10 mV, and the minimum particle separation may be taken as 0.2 nm. The Hamaker constant can be found in Table 7-1, and κ computed using Eq. (6.26).

4. Weak gels, as might be formed by percolated dispersions, can be used to suspend particles of greater density than the medium against sedimentation. This requires for spheres of diameter d that a dimensionless yield parameter, defined as

$$Y = \frac{\tau_y}{d(\Delta\rho)g},$$

where τ_y is the yield stress and $\Delta\rho$ is the density difference between the spheres and the medium, exceed a critical value of approximately 0.06 and that tanδ for the dispersion be ≤ 1.[1] Consider spheres of density 1.30 suspended in a medium of density 1.05, tanδ = 0.90 and the properties shown in Fig. 8-17 (where τ [=] kPa). Determine the maximum size of spheres that may be stabilized against sedimentation.

Chapter 9

1. In the production of a given emulsion using an ultrasonicator it is found that the maximum drop size in the dispersion was 20 μm. If it is desired to reduce this to 12 μm, estimate the increase in power input that would be required.

2. Consider a pair of 5-μm diameter emulsion droplets of dodecane in water at 25°C, as pictured in Fig. 9-2. Assume that the contact area has a radius of 1.5 μm, and the pressure drawing the drops together is 0.2 N/m². Estimate the drainage time to coalescence assuming a critical film thickness of 50 nm.

3. Aerosol OT, or Sodium bis(2-ethylhexyl) sulfosuccinate (C$_{20}$H$_{37}$NaO$_7$S) is an important surfactant with many applications. Its structure is shown below. Using the group contribution method, estimate the *HLB* for this compound. Assume the sulfate and sulfonate groups have the same contribution. What type of emulsions would you expect this emulsifier to produce?

[1] Laxton, P. B., and Berg, J. C., *J. Colloid Interface Sci.*, **285**, 152 (2005).

4. The isotherm at 17°C from the classic data of Adam[2] for the $\pi - A$ behavior of myristic acid monolayers on an acidified aqueous substrate are well fit in the liquid-expanded and intermediate regions by:

- liquid-expanded region: $(33.5 < A < 52)$ Å^2/molecule

  π(mN/m) $= kT/A = 411.5/A(\text{Å}^2)$ [=] mN/m , and

- intermediate region: $(22 < A < 33.5)$ Å^2/molecule

$$\pi = 11.9 + 0.917\left[\frac{33.5 - A}{A - 21.2}\right]$$

Evaluate the monolayer elasticity in the two regions, and comment on the effectiveness of this surfactant for stabilizing foam lamellae.

5. In detergency, the objective is the detachment of oil droplets from solid surfaces under water ("roll up") through the use of an appropriate surfactant (detergent). In other situations, it might be desirable to do just the opposite, i.e., to induce *spreading* of the oil drop on the solid surface under water. What type of surfactant would you recommend for this purpose, and how would it contrast with that used in detergency?

Chapter 10

1. Calculate the size of drops created by a jet emerging from an orifice of diameter 50 μm. What assumptions are being made?

2. Consider the movement of a 1-mm diameter droplet of a perfluorinated liquid ($\sigma' = 13$ mN/m; $\mu' = 0.80$ cP; $k' = 0.09$ W/m°C) immersed in an immiscible oil ($\sigma = 35$ mN/m; $\mu = 5.0$ cP; $k = 0.15$ W/m°C) of equal density. Estimate the rate of movement in a temperature field of 20°C/cm.

---

[2] Adam, N. K., **The Physics and Chemistry of Surfaces**, p. 59, Dover Publ., New York, 1968.

3. The Sternling-Scriven criteria, in their simplest form, for the occurrence of interfacial turbulence during the transfer of a solute across a liquid-liquid interface, suggest that instability should be expected when the transfer is out of the phase of higher kinematic viscosity and lower solute diffusivity (assuming the presence of solute at the interface lowers its interfacial tension). Check these criteria for the case of nitric acid transferring between isobutanol and water, the system pictured in Fig. 10-22.

4. Cellular convection in pools of liquid heated from below is dominated by surface tension (Marangoni) effects when the pools are shallow (≤ 1 mm) and by gravity (Rayleigh) effects when they are deeper (≥ 1 cm). In both cases, the warmer liquid wells up at the cell centers. For the Marangoni case, there is a surface depression at the cell center, as shown in Fig. 10-6(b), but for the gravity case, the surface is convex upward at the cell center. Explain this difference on phenomenological grounds. If in a given experiment with a deep pool heated from below it is found that the cell centers are elevated by 2 μm, estimate the velocity of the upwelling liquid. The cells may be considered circles of radius 0.5 cm, and the fluid properties are those of water at 25°C.

5. When a surfactant film is present, surface tension driven instability in a shallow pool heated from below is prevented, as the free surface takes on the hydrodynamic character of a solid wall. If the adverse temperature gradient is made steep enough, however, it should be possible to produce gravity-induced natural convection. The critical Rayleigh Number for the case of solid walls at top and bottom is approximately 1700. Estimate the critical temperature drop required to produce convection in a 1-mm deep pool of water covered with a strong surfactant.

Appendix 2

THE TOP TWELVE

Among the approximately 1200 equations in this book, twelve of them are singled out below as being of the greatest fundamental importance to the development of interfacial and colloid science.

1. The Young-Laplace Equation, Eq. (2.38):

$$\Delta p = \sigma \left(\frac{1}{R_1} + \frac{1}{R_2} \right)$$

Also known as the differential equation of capillary hydrostatics, its solution gives the shape and location of fluid interfaces in a gravitational field at rest or in rigid body motion. As such, it provides the driving force for such phenomena as the spontaneous wicking of liquids in porous media. It is generalized to include additional hydrodynamic effects in Eq. (10.5), the normal force balance boundary condition to the Navier-Stokes equations, and it may also be generalized to include the effect of disjoining pressure in thin fluid films, as in Eq. (2.124), or to include the effects of other external force fields acting on the system.

2. The Kelvin Equation, Eq. (2.92):

$$p_r^s = p_\infty^s \exp\left(\frac{2\sigma v^L}{rRT} \right)$$

This expresses the dependence of the vapor pressure of a liquid droplet on its radius, but more generally gives the dependence of the fugacity of a substance on its interfacial curvature. It is thus the basis for the instability of a polydisperse dispersion with respect to particle size disproportionation, for the spontaneous annealing of irregular or erose surfaces and for the description of phase change by the mechanism of nucleation and growth.

3. Young's Equation, Eq. (3.43):

$$\cos\theta = \frac{\sigma_{SG} - \sigma_{SL}}{\sigma_{LG}}$$

This expresses the dependence of the contact angle of a liquid against a solid surface in terms of the surface (free) energies of the solid-gas and solid-liquid interfaces and the liquid-gas surface tension, but more generally relates the contact angle of any fluid interface against a solid to the energies of the three interfaces involved. It may be generalized to account for adsorption to any of the interfaces, as in Eq. (4.3), or to include line tension, as in Eq. (3.47). It relates wetting, spreading, wicking and adhesion to the energies of the interfaces involved.

4. The Gibbs Adsorption Equation, Eq. (3.91):

$$\Gamma_{2,1} = -\frac{C_2}{RT}\left(\frac{\partial\sigma}{\partial C_2}\right)_T$$

This expresses the adsorption isotherm for a solute (2) at the interface of its solution (assumed ideal or ideal-dilute) against another fluid phase in terms of the dependence of the surface or interfacial tension on solution composition. More generally, as in Eq. (3.86), it is the analog of the Gibbs-Duhem equation for interfacial systems and may be applied to solutions that are non-ideal and consist of any number of components.

5. Stokes' Law, Eq. (5.50):

$$v_\infty = \frac{d^2(\rho_p - \rho)}{18\mu}$$

This gives the steady rate of movement (sedimentation or creaming) of a spherical particle, of density different than its medium, in a gravitational field. It may be generalized to non-spherical particles, as in Eq. (5.54), or to centrifugal fields, as in Eq. (5.70), or to any phoretic process, as for example, electrophoresis, in Eq. (6.102).

6. The Einstein-Smoluchowski Equation, Eq. (5.82)

$$D = \overline{\ell^2} / 2t$$

This relates the mean square displacement of objects moving under the influence of their intrinsic translational kinetic energy to their Fick's Law diffusivity.

7. The Einstein Diffusivity Equation, Eq. (5.83)

$$D = \frac{kT}{f}$$

This relates the Fick's Law diffusivity of particles to the hydrodynamic resistance to their motion. Equations 6 and 7 together are the basis for measuring and describing Brownian motion.

8. The Rayleigh Equation for light scattering by a small particle, Eq. (5.107):

$$\frac{I_\theta}{I_0} = \frac{8\pi^2 a^6}{r^2 \lambda^4} \left[\frac{m^2 - 1}{m^2 + 2} \right] (1 + \cos^2 \theta)$$

This expresses the intensity of scattered light from a single, non-absorbing spherical nano-particle ($a < \lambda/10$) in terms of its radius, a, wavelength of light, λ, refractive index contrast, m. It is the basis for more general scattering theory, for dilute dispersions of nano-particles, as in Eq. (5.108), to larger (and non-spherical) particles, as for example, in Eqs. (5.128) and (5.140), and for absorbing particles.

9. The Nernst Equation, Eq. (6.6):

$$\psi_0 = \frac{RT}{z_i F} \ln \frac{C_i}{(C_i)_{PZC}}$$

This relates the electrical potential at an interface against a solution to the chemical composition of the solution, in particular to the activities of species involved in a reversible redox reaction at the surface.

10. The Poisson-Boltzmann Equation, Eq. (6.20)

$$\frac{d^2\psi}{dx^2} = -\frac{1}{\varepsilon\varepsilon_0}\sum_i z_i e n_{i,\infty} \exp\left(\frac{-z_i e\psi}{kT}\right).$$

This equation describes the spatial variation (written here for the x-direction only) of potential resulting from the presence of a spatial array of charges. It is the basis for the description of the electrical double layer.

11. The DLVO equation, Eq. (7.79):

$$\Phi_{net} = \Phi_A + \Phi_R$$

This expresses the fact that the net potential energy of interaction between a pair of colloid particles at a given distance of separation is the sum of the attractive (Hamaker or Lifshitz) interaction, as given for example by Eq. (7.7), and the electrostatic repulsion between the particle surfaces, as given for example by Eq. (7.78).

12. The Marangoni equation, Eq. (10.9):

$$\underline{\tau}''_z - \underline{\tau}'_z + \nabla_{II}\sigma + \underline{\tau}^s = 0$$

This is the boundary condition for the Navier-Stokes equations expressing continuity of tangential stress at a fluid interface. It balances the boundary tension gradient $\nabla_{II}\sigma$ against the viscous traction of the bulk phases and any intrinsic interfacial rheological resistance to distortion. It is the basis for the many manifestations of the Marangoni effect, including spontaneous flows accompanying heat or mass transfer and Gibbs elasticity expressing the resistance of surfactant monolayers to interfacial deformation.

Appendix 3

OTHER SOURCES

A number of comprehensive textbooks dealing with interfaces and colloids are currently available and listed below. They differ significantly from one another in terms of the topics covered and emphasized and in their styles and perspectives. They are divided into Introductory and Intermediate-to-Advanced levels, although in some cases this was a difficult distinction to make. In addition to the textbooks listed below, a large number of handbooks, encyclopedias and more specialized monographs, compendia and conference proceedings are available.

Introductory

1. Barnes, G., and Gentle, I., **Interfacial Science, An Introduction**, Oxford Univ. Press, Oxford, UK, 2005.

2. Cosgrove, T. (Ed.), **Colloid Science**, Blackwell Publ., Oxford, UK, 2005.

3. Myers, D., **Surfaces, Interfaces, and Colloids: Principles and Applications**, 2nd Ed., Wiley VCH, New York, 1999.

4. Pashley, R. M., and Karaman, M. E., **Applied Colloid and Surface Chemistry**, Wiley, New York, 2004.

5. Shaw, D. J., **Introduction to Colloid and Surface Chemistry**, 4th Ed., Butterworth Heinemann, Oxford, UK 1992.

Intermediate-to-Advanced

1. Adamson, A. W., and Gast, A. P., **Physical Chemistry of Surfaces**, 6th Ed., Wiley-Interscience, New York, 1997.

2. Evans, D. F., and Wennerström, H., **The Colloidal Domain**, 2nd Ed., VCH, New York, 1999.

3. Goodwin, J., **Colloids and Interfaces with Surfactants and Polymers**, Wiley, New York, 2004.

4. Hiemenz, P. C., and Rajagopalan, R., **Principles of Colloid and Surface Chemistry**, 3rd Ed., Marcel Dekker, New York, 1997.

5. Hunter, R., **Foundations of Colloid Science**, Oxford Univ. Press, Oxford, UK, 2001.

6. Lyklema, J., **Fundamentals of Interface and Colloid Science**

    I. Fundamentals, Academic Press, London, 1991.

    II. Solid-Liquid Interfaces, Academic Press, London, 1995.

    III. Liquid-Fluid Interfaces, Academic Press, San Diego, 2000.

    IV. Particulate Colloids, Elsevier, Amsterdam, 2005.

    V. Soft Colloids, Elsevier, Amsterdam, 2005.

7. Miller, C. A., and Neogi, P., **Interfacial Phenomena**, 2nd Ed., Marcel Dekker, 2007.

8. Morrison, I. D., and Ross, S., **Colloidal Dispersions**, Wiley-Interscience, New York, 2002.

9. Russel, W. B., Saville, D. A., and Schowalter, W. R., **Colloidal Dispersions**, Cambridge Univ. Press, Cambridge, UK, 1992.

Index

Page entries where **definitions** of the listed items are found are in **bold face**.